T0140084

Advances in Intelligent Systems and Computing

Volume 1283

The series "Advances in Intelligent Systems and Computing" contains publications on theory, applications, and design methods of Intelligent Systems and Intelligent Computing. Virtually all disciplines such as engineering, natural sciences, computer and information science, ICT, economics, business, e-commerce, environment, healthcare, life science are covered. The list of topics spans all the areas of modern intelligent systems and computing such as: computational intelligence, soft computing including neural networks, fuzzy systems, evolutionary computing and the fusion of these paradigms, social intelligence, ambient intelligence, computational neuroscience, artificial life, virtual worlds and society, cognitive science and systems, Perception and Vision, DNA and immune based systems, self-organizing and adaptive systems, e-Learning and teaching, human-centered and human-centric computing, recommender systems, intelligent control, robotics and mechatronics including human-machine teaming, knowledge-based paradigms, learning paradigms, machine ethics, intelligent data analysis, knowledge management, intelligent agents, intelligent decision making and support, intelligent network security, trust management, interactive entertainment, Web intelligence and multimedia.

The publications within "Advances in Intelligent Systems and Computing" are primarily proceedings of important conferences, symposia and congresses. They cover significant recent developments in the field, both of a foundational and applicable character. An important characteristic feature of the series is the short publication time and world-wide distribution. This permits a rapid and broad dissemination of research results.

** Indexing: The books of this series are submitted to ISI Proceedings, EI-Compendex, DBLP, SCOPUS, Google Scholar and Springerlink **

More information about this series at http://www.springer.com/series/11156

John MacIntyre · Jinghua Zhao ·
Xiaomeng Ma
Editors

The 2020 International Conference on Machine Learning and Big Data Analytics for IoT Security and Privacy

SPIoT-2020, Volume 2

 Springer

Editors
John MacIntyre
David Goldman Informatics Centre
University of Sunderland
Sunderland, UK

Jinghua Zhao
University of Shanghai for Science
and Technology
Shanghai, China

Xiaomeng Ma
Shenzhen University
Shenzen, Guangdong, China

ISSN 2194-5357   ISSN 2194-5365 (electronic)
Advances in Intelligent Systems and Computing
ISBN 978-3-030-62745-4   ISBN 978-3-030-62746-1 (eBook)
https://doi.org/10.1007/978-3-030-62746-1

This Springer imprint is published by the registered company Springer Nature Switzerland AG
The registered company address is: Gewerbestrasse 11, 6330 Cham, Switzerland

Foreword

SPIoT 2020 is an international conference dedicated to promoting novel theoretical and applied research advances in the interdisciplinary agenda of Internet of Things. The "Internet of Things" heralds the connections of a nearly countless number of devices to the Internet thus promising accessibility, boundless scalability, amplified productivity, and a surplus of additional paybacks. The hype surrounding the IoT and its applications is already forcing companies to quickly upgrade their current processes, tools, and technology to accommodate massive data volumes and take advantage of insights. Since there is a vast amount of data generated by the IoT, a well-analyzed data is extremely valuable. However, the large-scale deployment of IoT will bring new challenges and IoT security is one of them.

The philosophy behind machine learning is to automate the creation of analytical models in order to enable algorithms to learn continuously with the help of available data. Continuously evolving models produce increasingly positive results, reducing the need for human interaction. These evolved models can be used to automatically produce reliable and repeatable decisions. Today's machine learning algorithms comb through data sets that no human could feasibly get through in a year or even a lifetime's worth of work. As the IoT continues to grow, more algorithms will be needed to keep up with the rising sums of data that accompany this growth.

One of the main challenges of the IoT security is the integration with communication, computing, control, and physical environment parameters to analyze, detect, and defend cyber-attacks in the distributed IoT systems. The IoT security includes: (i) the information security of the cyber space and (ii) the device and environmental security of the physical space. These challenges call for novel approaches to consider the parameters and elements from both spaces and get enough knowledge for ensuring the IoT's security. As the data has been collecting in the IoT and the data analytics has been becoming mature, it is possible to conquer this challenge with novel machine learning or deep learning methods to analyze the data which synthesize the information from both spaces.

We would like to express our thanks to Professor John Macintyre, University of Sunderland, Professor Junchi Yan, Shanghai Jiaotong University, for being the keynote speakers at the conference. We thank the General Chairs, Program Committee Chairs, Organizing Chairs, and Workshop Chairs for their hard work. The local organizers' and the students' help are also highly appreciated.

Our special thanks are due also to editors Dr. Thomas Ditzinger for publishing the proceedings in Advances in Intelligent Systems and Computing of Springer.

Organization

General Chairs

Bo Fei (President) Shanghai University of Medicine & Health Sciences, China

Program Chairs

John Macintyre (Pro Vice Chancellor) University of Sunderland, UK

Jinghua Zhao University of Shanghai for Science and Technology, China

Xiaomeng Ma Shenzhen University, China

Publication Chairs

Jun Ye Hainan University, China
Ranran Liu The University of Manchester

Publicity Chairs

Shunxiang Zhang Anhui University Science and Technology, China
Qingyuan Zhou Changzhou Institute of Mechatronic Technology, China

Local Organizing Chairs

Xiao Wei Shanghai University, China
Shaorong Sun University of Shanghai for Science and Technology, China

Program Committee Members

Paramjit Sehdev	Coppin State University, USA
Khusboo Pachauri	Dayanand Sagar University, Hyderabad, India
Khusboo Jain	Oriental University of Engineering and Technology, Indore, India
Akshi Kumar	Delhi Technological University, New Delhi, India
Sumit Kumar	Indian Institute of Technology (IIT), India
Anand Jee	Indian Institute of Technology (IIT), New Delhi, India
Arum Kumar Nachiappan	Sastra Deemed University, Chennai, India
Afshar Alam	Jamia Hamdard University, New Delhi, India
Adil Khan	Institute of Technology and Management, Gwalior, India
Amrita Srivastava	Amity University, Gwalior, India
Abhisekh Awasthi	Tshingua University, Beijing, China
Dhiraj Sangwan	CSIR-CEERI, Rajasthan, India
Jitendra Kumar Chaabra	National Institute of Technology, Kurkshetra, India
Muhammad Zain	University of Louisville, USA
Amrit Mukherjee	Jiangsu University, China
Nidhi Gupta	Institute of Automation, Chinese Academy of Sciences, Beijing, China
Neil Yen	University of Aizu, Japan
Guangli Zhu	Anhui Univ. of Sci. & Tech., China
Xiaobo Yin	Anhui Univ. of Sci. & Tech., China
Xiao Wei	Shanghai Univ., China
Huan Du	Shanghai Univ., China
Zhiguo Yan	Fudan University, China
Jianhui Li	Computer Network Information Center, Chinese Academy of Sciences, China
Yi Liu	Tsinghua University, China
Kuien Liu	Pivotal Inc, USA
Feng Lu	Institute of Geographic Science and Natural Resources Research, Chinese Academy of Sciences, China
Wei Xu	Renmin University of China, China
Ming Hu	Shanghai University, China

2020 International Conference on Machine Learning and Big Data Analytics for IoT Security and Privacy (SPIoT-2020)

Conference Program

November 6, 2020, Shanghai, China

Due to the COVID-19 outbreak problem, SPIoT-2020 conference will be held online by Tencent Meeting (https://meeting.tencent.com/).

Greeting Message

SPIoT 2020 is an international conference dedicated to promoting novel theoretical and applied research advances in the interdisciplinary agenda of Internet of Things. The "Internet of Things" heralds the connections of a nearly countless number of devices to the Internet thus promising accessibility, boundless scalability, amplified productivity, and a surplus of additional paybacks. The hype surrounding the IoT and its applications is already forcing companies to quickly upgrade their current processes, tools, and technology to accommodate massive data volumes and take advantage of insights. Since there is a vast amount of data generated by the IoT, a well-analyzed data is extremely valuable. However, the large-scale deployment of IoT will bring new challenges and IoT security is one of them.

The philosophy behind machine learning is to automate the creation of analytical models in order to enable algorithms to learn continuously with the help of available data. Continuously evolving models produce increasingly positive results, reducing the need for human interaction. These evolved models can be used to automatically produce reliable and repeatable decisions. Today's machine learning algorithms comb through data sets that no human could feasibly get through in a year or even a lifetime's worth of work. As the IoT continues to grow, more algorithms will be needed to keep up with the rising sums of data that accompany this growth.

One of the main challenges of the IoT security is the integration with communication, computing, control, and physical environment parameters to analyze, detect, and defend cyber-attacks in the distributed IoT systems. The IoT security includes: (i) the information security of the cyber space and (ii) the device and environmental security of the physical space. These challenges call for novel approaches to consider the parameters and elements from both spaces and get enough knowledge for ensuring the IoT's security. As the data has been collecting in the IoT, and the data analytics has been becoming mature, it is possible to conquer this challenge with novel machine learning or deep learning methods to analyze the data which synthesize the information from both spaces.

We would like to express our thanks to Professor John Macintyre, University of Sunderland and Professor Junchi Yan, Shanghai Jiaotong University, for being the keynote speakers at the conference. We thank the General Chairs, Program Committee Chairs, Organizing Chairs, and Workshop Chairs for their hard work. The local organizers' and the students' help are also highly appreciated.

Our special thanks are due also to editors Dr. Thomas Ditzinger for publishing the proceedings in Advances in Intelligent Systems and Computing of Springer.

Conference Program at a Glance

Friday, Nov. 6, 2020, Tencent Meeting		
9:50–10:00	Opening ceremony	Jinghua Zhao
10:00–10:40	Keynote 1: John Macintyre	
10:40–11:20	Keynote 2: Junchi Yan	
11:20–11:40	Best Paper Awards	Xiaomeng Ma
14:00–18:00	Session 1	Shunxiang Zhang
	Session 2	Xianchao Wang
	Session 3	Xiao Wei
	Session 4	Shaorong Sun
	Session 5	Ranran Liu
	Session 6	Jun Ye
	Session 7	Qingyuan Zhou
	Short papers poster	

SPIoT 2020 Keynotes

The Next Industrial Revolution: Industry 4.0 and the Role of Artificial Intelligence

John MacIntyre

University of Sunderland, UK

Abstract. The fourth industrial revolution is already approaching—often referred to as "Industry 4.0"—which is a paradigm shift which will challenge our current systems, thinking, and overall approach. Manufacturing industry will be transformed, with fundamental changes possible in how supply chains work, and how manufacturing becomes agile, sustainable, and mobile. At the heart of this paradigm shift is communication, connectivity, and intelligent manufacturing technologies. Artificial intelligence is one of the key technologies that will make Industry 4.0 a reality—and yet very few people really understand what AI is, and how important it will be in this new industrial revolution. AI is all around us already—even if we don't realize it!

Professor John MacIntyre has been working in AI for more than 25 years and is Editor-in-Chief of Neural Computing and Applications, a peer-reviewed scientific journal publishing academic work from around the world on applied AI. Professor MacIntyre will give a picture of how Industry 4.0 will present both challenges and opportunities, how artificial intelligence will play a fundamental role in the new industrial revolution, and provide insights into where AI may take us in future.

Unsupervised Learning of Optical Flow with Patch Consistency and Occlusion Estimation

Junchi Yan

Shanghai University, China

Junchi Yan is currently an Associate Professor in the Department of Computer Science and Engineering, Shanghai Jiao Tong University. Before that, he was a Senior Research Staff Member and Principal Scientist with IBM Research, China, where he started his career in April 2011. He obtained the Ph.D. in the Department of Electrical Engineering of Shanghai Jiao Tong University, China, in 2015. He received the ACM China Doctoral Dissertation Nomination Award and China Computer Federation Doctoral Dissertation Award. His research interests are mainly machine learning. He serves as an Associate Editor for IEEE ACCESS and a Managing Guest Editor for IEEE Transactions on Neural Networks and Learning Systems. He once served as Senior PC for CIKM 2019. He also serves as an Area Chair for ICPR 2020 and CVPR 2021. He is a Member of IEEE.

Oral Presentation Instruction

1. Timing: a maximum of 10 minutes total, including speaking time and discussion. Please make sure your presentation is well timed. Please keep in mind that the program is full and that the speaker after you would like their allocated time available to them.
2. You can use CD or USB flash drive (memory stick), make sure you scanned viruses in your own computer. Each speaker is required to meet her/his session chair in the corresponding session rooms 10 minutes before the session starts and copy the slide file(PPT or PDF) to the computer.
3. It is suggested that you email a copy of your presentation to your personal inbox as a backup. If for some reason the files can't be accessed from your flash drive, you will be able to download them to the computer from your email.
4. Please note that each session room will be equipped with a LCD projector, screen, point device, microphone, and a laptop with general presentation software such as Microsoft PowerPoint and Adobe Reader. Please make sure that your files are compatible and readable with our operation system by using commonly used fronts and symbols. If you plan to use your own computer, please try the connection and make sure it works before your presentation.
5. Movies: If your PowerPoint files contain movies please make sure that they are well formatted and connected to the main files.

Short Paper Presentation Instruction

1. Maximum poster size is 0.8 meter wide by 1 meter high.
2. Posters are required to be condensed and attractive. The characters should be large enough so that they are visible from 1 meter apart.
3. Please note that during your short paper session, the author should stay by your short paper to explain and discuss your paper with visiting delegates.

Registration

Since we use online meeting way, no registration fee is needed for SPIoT 2020.

Contents

Authentication and Access Control for Data Usage in IoT

Data-Driven Co-design
of Communication, Computing
and Control for IoT Security

Design of a Force Balance Geophone Utilizing Bandwidth Extension and Data Acquisition Interface

Xiaopeng Zhang$^{(\boxtimes)}$, Xin Li, Tongdong Wang, and Weiguo Xiao

Northwest Institute of Nuclear Technology, Xi'an 710024, China
zhangxiaopeng_nint@eiwhy.com

Abstract. This paper reports on the design, simulation and hardware implementation of a force balance geophone comprising a high order $\Sigma\Delta M$ with the proportional-integral-derivative (PID) feedback loop. The system is simulated and analyzed by Simulink modeling, and obtains a low-frequency and stable system. The prototype engages a conventional low cost 4.5 Hz SM-6 geophone cascaded by two distributed feedback (DF) electronic integrators for high-order noise shaping. Our resulting force balance geophone confirms an extended bandwidth to 0.16 Hz and a simulated NEA of 17 ng/$\sqrt{\text{Hz}}$. These performances shows its potential for highly effective and low-cost seismic exploration.

Keywords: Force balance geophone · Bandwidth extension · Data acquisition interface · Exploration

1   Introduction

Deep Earth seismic explorations require geophones to have good low-frequency bandwidth to measure waves with a good signal-to-noise (SNR) [1]. However, the widely used conventional geophones only have a better sensitivity above a few Hz since the open-loop resonant frequency is restricted by the mechanical system [2]. In addition, there are problems of massive weight, expense and insufficiency for the conventional exploration instruments because of the separated system structures of sense and data acquisition equipment [3]. According to these shortages in response limitation and hardware cost, it is with great insights to demonstrate a digital geophone with the good low-frequency response for the effective and economical exploration equipment.

In this paper, we designed a 4th-order sigma-delta modulator ($\Sigma\Delta M$) for a conventional low cost geophone. The force balance accelerometer (FBA) principle is carried out to a conventional geophone for bandwidth extension with the single electromagnetic coil served as both sensor and actuator [4]. The over sampling technique is applied to shape out the quantization noise and meanwhile sample the analog voltage [5]. Principle of the single-coil force balance geophone is first reviewed in Sect. 2. Section 3 discusses the Simulink modeling, performance analysis and simulated results. Section 4 puts forward the hardware implementation and experimental results. Section 5 draws the conclusions.

J. MacIntyre et al. (Eds.): SPIoT 2020, AISC 1283, pp. 3–9, 2021.
https://doi.org/10.1007/978-3-030-62746-1_1

2 Single-Coil Force Balance Geophone

The diagram of a force balance geophone with its intrinsic coil is explained in Fig. 1 [4]. A conventional geophone is connected between the negative input and the output terminals of Operational Amplifier B. The control voltage expressed as V_{ctrl} will modify the coil current as V_{ctrl}/R_1 and interact an electromagnetic force to proof mass. This force, expressed as GV_{ctrl}/R_1, would affect the mass motion relative to geophone housing by driving a frame acceleration $K_{dr}V_{ctrl}$, where $K_{dr} = G/(R_1 m)$ is the drive gain, and G is the coil transduction constant. In the circuit related to Operational Amplifier A, resistance R_2 is deployed with the same values of R_3, and R_1 is also arranged to equal to the resistance of the geophone coil. Therefore, the output voltage V_{out} can be derived to

$$V_{out} = GM\left(\ddot{Y} - \frac{F_L}{m}\right), \tag{1}$$

where

$$M = \frac{\dot{X}}{\ddot{Y}} = \frac{s}{s^2 + 2\zeta_0\omega_0 \cdot s + \omega_0^2} \tag{2}$$

is the Laplace form of the transfer function of a geophone. $\ddot{Y}$ is the ground motion which drives a relative $\dot{X}$ between proof mass and magnetic field. ζ_0 is the damping ratio, and ω_0 is the resonant frequency. As shown in Eq. (1), the close loop transfer function can be expressed as:

$$\frac{V_{out}}{\ddot{Y}} = \frac{GM}{1 + GMH_{ctrl}K_{dr}}. \tag{3}$$

According to the FBA principle, Eq. (3) goes to a constant sensitivity:

$$\frac{V_{out}}{\ddot{Y}} \approx \frac{1}{H_{ctrl}K_{dr}}, \tag{4}$$

when the electronic architectures results in a great system loop gain. Thus the close-loop behaviors of the force balance geophone are only decided by the feedback loop.

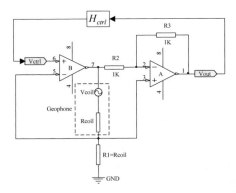

Fig. 1. Diagram of a single-coil velocity feedback geophone

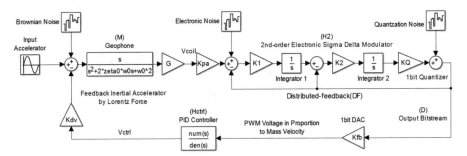

Fig. 2. Simulink model of the force balance digital geophone based on the M architecture

Table 1. Parameters of the force balance geophone

Parameter	Value	Parameter	Value
Geophone model	SM-6 from ION Inc.	Mechanical sensitivity: G	28.8 V/(m/s)
Coil resistance: R_{coil}	375 Ω	Frequency corner of H_2	2.7e^3 Hz
Resonance: $\square_0$	4.5 Hz	Damping constant of H_2	0.707
Mass: m	11.1 g	Quantization gain K_Q	10^5
Damping constant: $\square_0$	0.56	Feedback gain: K_{fb}	10^{-4}

3 System Modeling and Analysis

3.1 System Modeling

As shown in Fig. 2, the force balance digital geophone is modeled with Simulink. The conventional geophone converts the input accelerator into the relative velocity and served as a 2nd-order band pass filter. The output voltage of the geophone is picked off by Operational Amplifier A in Fig. 1. Another purely electronic 2nd-order low pass filter H_2 is connected to the pickoff circuit with a distributed-feedback (DF) topology [6].

K_1 and K_2 are the feed-forward gains of integrator 1 and 2. The 1bit quantizer has a sampling rate much higher than twice of the bandwidth of the input signal and delivers the over sampled digital bitstream D. The controller filters the pulse width modulated (PWM) voltage from a 1bit DAC for the control voltage to the Operational Amplifier B.

For analysis of the transform properties, the quantizer is considered as a gain of K_Q added by a white quantization noise N_Q [7, 8]. Using superposition principle and assuming the input signal sufficiently busy, the digital output can be expressed [9] as $D = STF(\ddot{Y}+B)+QNTF \cdot N_Q+ENTF \cdot N_E$, where K_{pa} and K_{fb} are the gains of pick-off amplifier and the DAC, respectively. B is the Brownian noise, N_E is the electronic noise [10], and N_Q is the quantization error. For this digital geophone, the transfer functions of the 4th-order $\Sigma\Delta M$ geophone, quantization error, and electrical noise, respectively, can be expressed as follows:

$$STF_4 = \frac{D}{\ddot{Y}} = \frac{GMK_{pa}H_2}{1 + GMK_{pa}H_2K_{fb}H_{ctrl}K_{dv}},\tag{5}$$

$$QNTF_4 = \frac{D}{N_Q} = \frac{1}{\left(1+\dfrac{GMK_{pa}K_1K_2K_QK_{fb}H_{ctrl}K_{dv}}{s^2}+\dfrac{K_1K_2K_Q}{s^2}+\dfrac{K_2K_Q}{s}\right)},\tag{6}$$

$$ENTF_4 = \frac{D}{N_E} = \frac{H_2}{1 + GMK_{pa}H_2K_{fb}H_{ctrl}K_{dv}}.\tag{7}$$

Table 1 shows the parameters of a conventional geophone model SM-6 from ION inc. and the readout interface of our digital geophone. The bode plots of Eqs. (6), (7) and (8) are predicted in Fig. 3(a). At the mechanical resonance 4.5 Hz of SM-6 geophone, it can be observed that an inflection occurs in $QNTF_4$, $QNTF_2$ and $ENTF_4$ share the same notches, respectively. Assuming a 100 nV/$\sqrt{Hz}$ electronic noise and a 1 micro-g input (60 Hz), Fig. 3(b) shows the simulated power spectral density (PSD) of the output, and the predicted noises form mechanical and electrical systems. The noise floor of 2nd order geophone is –80 dB and quantization noise dominates other noises. Since the integration of mechanical elements, it only gains a noise shaping of the first order. With regard to the same PSD of 4nd-order $\Sigma\Delta M$, the noise level decreases 50 dB and reaches to about –130 dB (equivalent to 17 ng/$\sqrt{Hz}$). It indicates that Brownian motion is the major noise in-band, and the $\Sigma\Delta M$ architecture we presenting has a quantization error shaping performance of order 3.

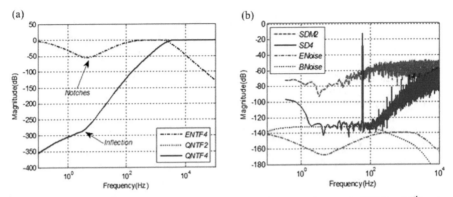

Fig. 3. (a): Predicted bode plots of the noise transfer functions $ENTF_4$ and $QNTF_4$ of 4th-order geophone, the reduced $QNTF_2$ of 2nd-order geophone is also predicted for comparison; (b): PSD of the output bitstream with a 1 mg, 60 Hz acceleration input signal; predicted electronic and Brownian noises observed in D

3.2 System Analysis

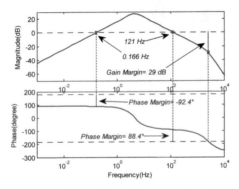

Fig. 4. Stability analysis of the gain and phase margins of the feedback loop

In practical, to isolate the long period seismic fluctuation, and to reduce the high frequencies noise, we deployed the PID controller in the band limited form:

$$H_{ctrl} = K_p + K_i \frac{1}{s} + K_d s. \tag{8}$$

With expressions of Eqs. (2) and (8), Eq. (5) would define the close-loop geophone as band-pass system with a bandwidth witch is given by Eq. (9).

$$Bw = \frac{2\zeta_0\omega_0 + K'K_p}{1 + K'K_d} \tag{9}$$

According to Eq. (4), the sensitivity of the force balance geophone becomes to

$$STF_{sensitivity} = \frac{D}{\ddot{Y}} = \frac{GK_{pa}H_2}{2\zeta_0\omega_0 + K'K_p}. \tag{10}$$

According to parameters in Table 1, the stability analysis of the gain and phase margins of the feedback loop are shown in Fig. 4. The feedback loop obtains phase margins of −92.4° and 88.4° at ether crossover frequency 0.166 and 121 Hz, which satisfies the stable requirement.

4 Prototype Implementation and Test Results

As shown in Fig. 5(a), the hardware of $\Sigma\Delta$M interface of the force balance geophone was implemented on analog and digital printed circuit boards (PCBs). The analog PCB detects the analog coil voltage of the geophone and gets the 1-bit digital signal to form a PWM voltage by operating the switches (ADI. ADG812). An analog PID controller shapes the PWM wave to drive the balanced force. The digital PCB accounts for a 16-bit ADC (ADI Inc. AD7606) that samples the coil voltage and an FPGA operates a discrete time filtering of H_2.

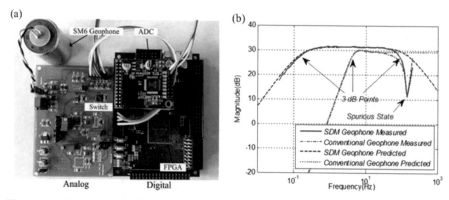

Fig. 5. (a): Hardware of the force balance geophone with a SM-6 geophone connected; (b): Bode plots of force balance geophone and conventional geophone

The measurements and predictions of the bode plots of the force balance geophone and conventional geophone are illustrated in Fig. 5(b). The in-band sensitivity illustrated by Eq. (10) is also shown in this figure. It can be observed that the lower corner frequency is extended to 0.16 Hz by the $\Sigma\Delta$M architecture, which proves the efficiency of our force balance method.

5 Conclusion

In this manuscript, we designed a force balance geophone utilizing the bandwidth extension and data acquisition interface which is made up by a PID controlled feedback loop in the $\Sigma\Delta$M architectures. The simulated noise floor of this close-loop geophone is about 17 ng/$\sqrt{\text{Hz}}$. And testing results show that this close-loop feedback and digital readout technique could extend the low-frequency response to 1/20 of the resonance of the cascaded conventional geophone. Analysis of the instrument noise model and experimental evaluations call for further studies in future work.

References

1. Mougenot, D.: Pushing toward the low frequencies. World Oil, pp. 25–28, September 2005
2. Laine, J., Mougenot, D.: Benefits of MEMS based seismic accelerometers for oil exploration. In: 2007 International Solid-State Sensors, Actuators and Microsystems Conference (TRANSDUCERS 2007), Lyon, France, June 2007, pp. 1473–1477 (2007)
3. Savazzi, S., Spagnolini, U., Goratti, L., Molteni, D., Latva-aho, M., Nicoli, M.: Ultra-wide band sensor networks in oil and gas explorations. IEEE Commun. Mag. **51**(4), 150–160 (2013)
4. Vanzandt, T.R., Manion, S.T.: Single-coil force balance velocity geophone. U.S. Patent 6 075 754, 13 June 2000
5. Lemkin, M.A., Ortiz, M.A., Wongkomet, N., Boser, B.E., Smith, J.H.: A 3 axis surface micromachined sigma delta accelerometer. In: IEEE Solid-State Circuits Conference, San Francisco, USA, pp. 202–203 (1997)
6. Dong, Y., Kraft, M., Redman-White, W.: Higher order noise-shaping filters for high-performance micromachined accelerometers. IEEE Trans. Instrum. Meas. **56**(5), 1666–1674 (2007)
7. Chen, F., Yuan, W., Chang, H., Yuan, G., Xie, J., Kraft, M.: Design and implementation of an optimized double closed-loop control system for MEMS vibratory gyroscope. IEEE Sens. J. **14**(1), 184–196 (2014)
8. Chen, F., Li, X., Kraft, M.: Electromechanical sigma-delta modulators ($\Sigma\Delta M$) force feedback interfaces for capacitive MEMS inertial sensors: a review. IEEE Sens. J. **16**(17), 6476–6495 (2016)
9. Almutairi, B., Alshehri, A., Kraft, M.: Design and implementation of a MASH2-0 electromechanical sigma-delta modulator for capacitive MEMS sensors using dual quantization method. J. Microelectromech. Syst. **24**(5), 1251–1263 (2015)
10. Barzilai, A., VanZandt, T., Kenny, T.: Technique for measurement of the noise of a sensor in the presence of large background signals. Rev. Sci. Instrum. **69**(7), 2767–2772 (1998)

Application of 3ds Max Technology in Archaeology

Qiwang Zhao[1](✉) and Shaochi Pan[2]

[1] School of History and Culture, Sichuan University, Chengdu, Sichuan, China
zachery.ok@eiwhy.com
[2] Chengdu Institute of Cultural Relics and Archaeology Research, Chengdu, Sichuan, China

Abstract. Combined with relevant examples, this paper explores the application of 3ds Max technology in archaeology from perspectives of topography of the site (environment), excavation unit, stratum overlying and artifact restoration in detail. The purpose of this paper is to present the sites and relics intuitively, which is of great significance to promote the development of public archaeology and the exhibition of archaeological museums.

Keywords: 3ds Max · Three-dimensional modeling · Public archaeology

1 Introduction

In recent years, it becomes increasingly common to apply new technologies, such as GIS and 3D laser scanning techniques, in archaeology. The combination of these new technologies with archaeology will undoubtedly promote the development of archaeology to a large extent. In this study, we applied the technology of 3ds Max to archaeology, aiming to explore the application of 3ds Max technology in archaeology. The combination of this technique and archaeology will contribute to the development of public archaeology and the research of many archaeological problems.

2 Introduction to 3ds Max

The full name of 3ds Max is 3D Studio MAX. It is a three-dimensional animation production and rendering software developed by the Autodesk Company. From the end of the last century to now, after more than 20 years of development, the technology of 3ds Max software has become quite mature.

The version of software used in this paper is 3ds Max 2012. In this version, the software has inserted a very advanced graphite modeling tool, which can easily realize 3D modeling of complex geometry. It is of great significance in the process of archaeological relics restoration.

At present, 3ds Max has been applied in many fields at home and abroad, and its achievements have attracted people's attention. It mainly includes eight aspects: "architectural visualization, film and television special effects, game design and development,

J. MacIntyre et al. (Eds.): SPIoT 2020, AISC 1283, pp. 10–16, 2021.
https://doi.org/10.1007/978-3-030-62746-1_2

film title and column packaging, film and television advertising, card cartoon, industrial design visualization, and multimedia content creation." [1].

From the above eight aspects, we can see that in the main application fields of 3ds Max technology, architectural visualization, industrial design visualization and multimedia content creation have great reference significance for us to explore the specific application of this technology in archaeology.

This time, we introduced the technology of 3ds Max into archaeology. The main reason is that 3 ds Max technology can realize virtual reality. "Virtual reality is a higher level of multimedia technology development; it is the integration and penetration of these technologies at a high level. It gives users more realistic experience. Because of various reasons, it is not convenient for people to directly observe the movement and change of things. This technique provides great convenience for people to explore the macro and micro world." [2].

Through the above introduction, we can draw a conclusion that the combination of 3ds Max technology and archaeology is very promising, and that exploring the specific aspects of its application is also a very meaningful process. Now, by citing some examples, we will combine the technology of 3ds Max with archaeology, and make some positive explorations. I am glad that other scholars can criticize and correct the shortcomings of this paper.

3   The Specific Application in Archaeology

In this paper, we will briefly introduce the application of 3ds Max technology in archaeology from four aspects: the topography of the site (the large environment), the excavation unit (stratum), stratum overlying and the restoration of artifacts.

3.1   Topography of the Site

With the powerful function of terrain making in the 3ds Max software, we can realize the three-dimensional representation of the specific environment of the site. Through the three-dimensional modeling of the topography of the site, we can see the environment of the site very intuitively. The specific methods are as follows. Firstly, we need to obtain the contour map of the site, and calculate the actual size of the scene through the scale of the map. Secondly, we can make a plane equal to its actual size in the 3ds Max software, and paste the contour map on the plane for display. Thirdly, through the line command in 3ds Max software, we need to copy each contour line in the contour map, and pulls the height of each contour line to the actual height marked in the map. Fourthly, selecting all contour lines, the three-dimensional planning can be realized through the terrain command under the compound object of the software. In the three-dimensional view, we use different colors to represent different features and mark them, and mark the specific location of the site in the three-dimensional map, so that we can clearly see the general environment of the site, as shown in Fig. 1.

We believe that the cost of this process is much lower than that of aerial photography. Therefore, it has advantages. As for its significance, now we think that it is mainly used as a display language. Through the presentation of the three-dimensional terrain of the

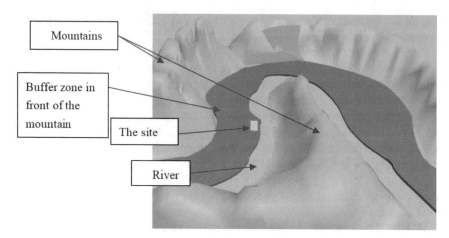

Fig. 1. Location of the site

site, we can see the environment of the site clearly, which is conducive to the display of archaeology in the museum and can promote the development of public archaeology.

Excavation Unit (Stratum)

The main application of 3ds Max technology in excavation unit is the three-dimensional modeling of strata. There have been some researches on the method of stratum modeling. "There are three main types of stratum modeling methods: facial model, volumetric model and mixed model." [3]. We mainly use the method of facial model in stratum modeling. First of all, we need to explain the facial model. "A facial model establishes the three-dimensional model of a solid by describing the geometric shapes of the outer surface of the spatial object. Its advantages include small data storage, fast modeling speed, and convenient data display and update. The disadvantage is that it is difficult to obtain the internal attributes of spatial entities, and it is not easy to conduct spatial analysis." [4]. Because our research focuses on the open exhibition of field archaeology, we use this method in modeling. Our exploration in this part of the application focuses on stratum restoration based on data left on the profile of the square diaphragm. In addition, this method is still of certain significance in some aspects of archaeological analysis. "From the three-dimensional view of the cross-section, we can find the macroscopic spatial characteristics of the archaeological strata in a certain area. It also clearly reflects the thickness change and rules of strata overlying, as well as the breaking phenomenon on the ground level." [4].

According to data obtained in field archaeology, 3D solid modeling has different methods to realize. Some of them have been studied by predecessors. "3D solid modeling is carried out according to archaeological data, and various modifiers and tools are widely used in the modeling process. For example, the river is made by squeezing and bending, and then the river is formed on the ground model by Boolean method. A combination method is used to combine several objects into one object, such as a complex house. In order to enhance the authenticity of the model and reduce the size of generated file, some details can be directly imitated with pictures without modeling, that is, using mapping and texture mapping technology." [5]. This time, we used two different methods to model the stratum.

First, we adopt a traditional method to carry out 3D modeling. Firstly, we import stratum section data obtained from the field into AutoCAD software to draw the corresponding plane map, and then import the plane map into the 3ds Max software for 3D modeling. The four walls of each excavation unit are respectively drawn and modeled, and finally the four walls are connected in sequence. Under the condition of real and effective data, the four walls of each excavation unit can realize seamless docking, as shown in Fig. 2. The advantage of this modeling method is that modeling data is very close to real data. It reduces the error to the greatest extent, and the results obtained are more effective, which greatly increases the credibility of using this model in later analysis. However, its disadvantages cannot be ignored. To adopt this method needs two programs, and the amount of data used in AutoCAD software is large. The workload is greatly increased, so the actual operation is difficult and the efficiency is relatively low. Therefore, this method is suitable for the excavation unit of small-scale sites [6]. In the early stage of exploration, we used this method. The effect was very good, but the progress was too slow. Therefore, this method is not suitable for the 3D modeling of strata of large-scale sites.

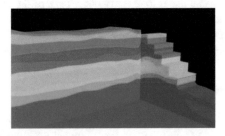

Fig. 2. Chart of the excavation unit obtained from the stratum plane made by CAD

Second, we adopted a method similar to the three-dimensional modeling method in paper, *Using 3D Virtual Modeling Technology to Reproduce the Cultural Scene of Archaeological Sites*. It "directly uses pictures to imitate, that is, using mapping and texture mapping technology". The specific method is as follows. When we excavate, the specification square of the excavation unit is $5 \times 5 \, \text{m}^2$. Excluding 1 m left in the east and north sides of the diaphragm, the actual excavation area of each unit is $4 \times 4 \, \text{m}^2$. We first import the photos of each excavation unit into photoshops software to intercept the stratum section. Since the length of each square wall is fixed, we can calculate the actual height of the side walls shown in the photos according to the length width ratio of the screenshot [7]. Then, a plane of the same size is created in the software of 3ds Max, and the intercepted figure is pasted on the created plane with the bitmap command by using the material editor provided by the software; stratum lines are copied according to the figure. In order to ensure that the stratum lines copied are more accurate, we enlarge the view of 3ds Max appropriately, so that we can see more details and get more accurate lines. Finally, the 3D modeling is realized through extruding these lines, as shown in Fig. 3.

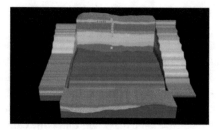

Fig. 3. Chart of the excavation unit obtained through copying stratum lines

The advantage of this method is that the amount of data required is relatively small. Although it still has two operation procedures, it has high mapping efficiency and faster speed compared with the first method; it is suitable for the exploration of large-scale excavation units [8]. Its disadvantage is that, photos obtained by camera are influenced by perspectives and other factors; the accuracy is slightly lower. Data in the later analysis is less reliable than the former. After the three-dimensional modeling of strata, we can directly display the two-dimensional or three-dimensional maps of these excavation units in the museum, or we can conduct some later analyses. For example, 3ds Max software has a layer manager. We can use it to classify each layer of the excavation unit. We can freely display or hide any layer in the software. Through this kind of processing, we can clearly see the distribution of each cultural layer in each location of the site, and use these data to make macro analysis on related problems of the stratum.

3.2 Stratum Overlying

Stratum overlying is a theory formed based on the need of showing to the public, which is of great significance in the exhibition of archaeological museums. The specific method is as follows. We build models on main cultural accumulation layers of the site in 3ds Max software, which can be combined with specific data of the site, or not combined with specific data. According to the time of each cultural accumulation layer, layers overlie each other from top to bottom. Finally the result is output in the form of video with simple dubbing introduction. Through this video, we can help the public to understand the general situation of the site, and achieve a demonstration purpose. In the development of public archaeology, it is of great significance to explore the theory of stratum overlying. We can assume that from top to bottom, the main cultural accumulation of a site includes, the topsoil layer, the Ming and Qing cultural accumulation, the Tang and Song cultural accumulation, the Northern and Southern Dynasties cultural accumulation, the Eastern Han cultural accumulation, the Eastern Zhou cultural accumulation, the Western Zhou cultural accumulation, and the Neolithic Age cultural accumulation. Neolithic cultural accumulation is the earliest layer in this site. First we put the raw soil layer at the very bottom, and then we superimpose them in the order; main relics contained in the cultural deposits of various periods are also displayed, as shown in Fig. 4. It should be pointed out that this experiment did not combine with specific data of the site. It only presents the general formation process of a site, in order to visually present the main overview of the site.

Fig. 4. Illustration of tomb complex of Eastern Zhou Dynasty

3.3 Restoration of Artifacts

The technology of 3ds Max is of great significance in the restoration of archaeological artifacts. In field excavation, we can find many incomplete artifacts. In order to better study cultural factors reflected in the site, we need to restore these damaged artifacts. Previous experience tells us that some potteries and other artifacts excavated in the field have relatively regular shapes, so they can be easily restored in the 3ds MAX software, as shown in Fig. 5.

Fig. 5. Restoration of artifacts

For the restoration of irregular objects, the advanced graphite modeling tool in 3ds Max software can easily realize the modeling of irregular geometry shapes and the restoration of objects with irregular shapes. What's more, the result generated is not only convenient for the storage of electronic data, but also convenient for scholars to extract in the future [9].

In addition, scholars at home and abroad have done a lot of research on the restoration of tombs, especially the restoration of large-scale tombs. The significance of restoring tombs is that it provides convenience for further research on their shapes. Since many achievements have been made in this respect, we will not further explore the issue here.

4 Conclusion

The application of 3ds Max technology in archaeology has considerable prospects. We believe that 3ds Max technology will play an important role in future exhibition of archaeological museums and promote the development of public archaeology. In view of the limitations of knowledge, we only explored the application of this technique in four aspects. As for the application of 3ds Max technology in other aspects of archaeology, we will continue to explore in the future. It is also hoped that other scholars can participate in the discussion, so that the 3ds Max technology can make more contributions to the development of archaeology.

References

1. Mars Age: 3ds Max 2011 Platinum Handbook I, pp. 3–7. People's Posts and Telecommunications Press, Beijing (2011)
2. Ye, X.S.: Three dimensional representation of archaeological sites. Wenbo **5**, 70–74 (2002)
3. Kan, Y.K., et al.: Three dimensional reconstruction of archaeological strata. Comput. Appl. Res. **24**(3), 302–305 (2007)
4. Zhu, L.D., et al.: Research on 3D visualization and application of archaeological strata: taking Jinsha site in Chengdu as an example. J. Chengdu Univ. Technol. (Nat. Sci. Ed.) **5**, 564–568 (2007)
5. Liu, Y.L., et al.: Reappearance of cultural scenes of archaeological sites using 3D virtual modeling technology. Wenbo **1**, 87–90 (2009)
6. Liu, J., Geng, G.H.: 3D realistic modeling and virtual display technology of cultural sites. Comput. Eng. **36**(020), 286–290 (2010)
7. Lin, B.X., Zhou, L.C., Sheng, Y.H., et al.: 3D modeling method of archaeological sites based on hand drawn maps. J. Earth Inf. Sci. **016**(003), 349–357 (2014)
8. Zhang, J.: Research and Application of 3D Reconstruction Technology of Site Scene Based on Laser Scanning Data, Northwestern University (2012)
9. Li, W.Y., Zhang, F.S., Yang, J.: Application of 3D scanning and rapid prototyping in cultural relic restoration. Wenbo **6**, 78–81 (2012)

The Application of Virtual Reality Technology in ESP Teaching

Xuehong Yuan$^{(\boxtimes)}$

Air Force Aviation University, Changchun, Jilin, China
cnafau@163.com

Abstract. Discuss how virtual reality (VR) technology can improve the learning efficiency of Chinese students' ESP (English for special purpose), especially if it is applied to early childhood education. VR mainly revolves around scenes, interactivity, fun, and immersive experiences. The research on the theory of information processing and learning proves that the information delivered to students by ESP teaching using virtual reality technology is more acceptable than the information delivered to students by ordinary ESP teaching. Accordingly, this article analyzes and studies the application of VR technology in ESP teaching.

Keywords: Virtual · Virtual reality · Virtual technology · ESP teaching

1 Introduction

English for Specific Purposes, also known as ESP, refers to English categories related to certain fixed industries or subjects. It is an important component of English language teaching and an outstanding branch of applied linguistics. It involves a very wide range, not only involving linguistic knowledge, pedagogy knowledge, but also involving professional knowledge [1]. In traditional English teaching, teachers always explain English as pure knowledge, spending a lot of time and resource in the classroom, allowing students to mechanically train and learn by rote. The application of VR technology has brought a series of changes to secondary vocational English teaching. It can stimulate students' various senses, activate the classroom atmosphere, attract students' attention, expand students' thinking space, and stimulate students' learning interest. It is great importance for improving classroom teaching. Therefore, teachers should realize the effective integration of VR technology and courses, so that multimedia can assist English classrooms more effectively [1].

2 VR Technology Overview

2.1 Definition of VR Technology

VR technology is an extremely important research branch of simulation. At present, one challenging frontier subject and research field is to combine simulation technology with computer graphics interface technology, multimedia technology, sensor technology and

J. MacIntyre et al. (Eds.): SPIoT 2020, AISC 1283, pp. 17–23, 2021.
https://doi.org/10.1007/978-3-030-62746-1_3

network technology, so as to achieve similar effects to the real world. The environment is a real-time 3D realistic image generated by a computer, while the VR technology uses these generated images to simulate nature, environment, and perception skills. Among them, the VR technology should also include the real human perception and way of thinking [1]. This not only refers to the usage of computer graphics technology to produce corresponding vision, but also requires the production of movement, touch, hearing and other perceptions, even including taste and smell, which we call multi-perceptual phenomena. With perception, we naturally think of making certain responses to the sensed signals that is, producing corresponding actions, which is called natural skills. The so-called natural skills refer to the movement of human body parts or other human actions. This requires the computer to record data and predict the possible physical movements of the participant in the next step, and transmit the data and other information to the user in real time [2].

2.2   Characteristics of VR

The quasi-reality technology integrates a variety of scientific technologies, such as simulation technology, graphics technology, sensor technology, display technology, etc. The application of these technologies realizes the creation of a virtual information environment in a multi-dimensional information space, so that users have an immersive experience, and the environment has perfect interaction and helps to stimulate ideas [2]. Mainly have the following characteristics, as shown in Fig. 1:

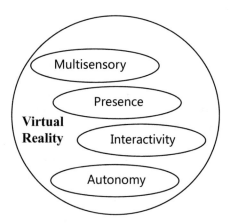

Fig. 1. Characteristics of VR

1) Multi-sensitivity. Refers to the interaction of multiple sensors with the virtual environment of multi-dimensional information. In addition to the computer visual perception, there are tactile perception, auditory perception, motion perception, and even smell, taste, perception, etc. [3]. The ideal VR should have all the perception functions of the human body. It is these characteristics that make VR technology

present in front of people in an intelligent form and enjoy a certain degree of security in exploring new knowledge.

2) Sense of presence. It refers to the counter-touch of the human body through a variety of sensors, making participants feel immersive in the virtual environment and feel the real existence in the simulated environment. The ideal simulation environment should make it difficult for users to distinguish true from false [3]. In the virtual world, users can get a different experience from reality, and perfectly integrate cognition and perception.

3) Interactivity. It means that while participants control and feel things in the virtual environment through a variety of sensors, the virtual environment uses sensors to counteract the participants, and each and every information exchange is mutual. Interactivity reflects the difference between VR and traditional 3D animation and models generated by traditional CAD systems. The construction of VR is not only to provide visual enjoyment similar to 3D movies, but also to construct dynamic, transformable and interactive effects. An obvious advantage of the VR system lies in its interactivity [4]. In this system, users can negotiate with objects in the virtual environment through interactive devices, and can effectively control the triggering of certain actions according to their personal thinking. It can also provide users with suggestions in time according to their needs. It reflects the user's degree of objects manipulation in virtual environment. The natural degree of feedback and the method of rearranging the virtual environment of the system [5].

4) Autonomy. Means that the virtual environment is designed and constructed by participants according to their own needs. It is in a dominant position in the operation of VR. It is designed and produced according to the physical objects in the real world and the laws of physical motion [4]. In VR, various the operation of things must conform to the objective laws of reality.

3   The Advantages of VR Technology in ESP Teaching

In the whole learning process of students, it is not a simple information input, storage, and extraction, but a process of interaction between old and new experiences, which involves creating scenarios, establishing hypotheses, collecting information, exploring hard, verifying hypotheses, and earnestly. Thinking and a series of links. Everyone can experience and feel for themselves, which is actually more convincing than empty and abstract preaching. There is a difference in nature between active interaction and passive indoctrination [5]. VR uses modern electronic information technology to establish various virtual laboratories, and has advantages that traditional laboratories cannot match:

3.1   Cost Saving

When talking about VR systems, people first think of expensive electronic equipment and high development costs, thinking that the cost is too high. However, considering the sharing of hardware and software, the saved costs of housing and construction, as well as the repeated use and saved material costs, the unit cost of the test is relatively low [6]. Due to hardware constraints such as equipment, venues, and funding, many experiments cannot be carried out. Using the VR system allows students to do various experiments without leaving home.

3.2   Avoid Risks

Many students are exposed to the experiments they have learned for the first time. There are often various dangers in real experiments or operations. If the operation is a little careless, they may agree with very serious consequences, and this pressure often makes It is difficult for students to receive the due ESP ESP teaching effect due to stress or failure [7]. Using VR technology to conduct virtual experiments not only allows students to observe the whole process of the experiment clearly and vividly, but also to experience the feelings of on-site operation. Students can safely do various dangerous experiments in the virtual experiment environment. Repeat the test until the goal of ESP ESP teaching is achieved [6].

3.3   Expanding Space

With the technology development, the natural science knowledge that students need to master is more extensive and subtle. Under realistic conditions, some experiments are difficult to achieve. However, the use of VR technology can break the limitations of space and time. Without leaving the classroom and in a virtual laboratory, students can go to the sky and the sea, feel the vastness of the universe and the depth of the seabed, and enter these objects to observe the movement of particles., And even observe the chemical reaction process [8].

4   The Application of VR Technology in ESP Teaching

4.1   Application of Information Processing

Learning is a gradual process. Generally, it is divided into several stages and iteratively to truly understand a piece of knowledge. The process is tortuous, and the accumulation of quantitative changes will cause a qualitative leap. At these stages, it is necessary to process the knowledge learned. In these processes, new things are often produced [9]. The so-called review of the past to learn the new, and this phenomenon can be collectively referred to as learning. Learning refers to the process of processing inside the human brain, and it is an important part of learning information processing theory. Therefore, in the process of teaching, we must take into account the ability of each student to study and process, and try to cultivate and improve the ability of students in this area. Therefore, the teacher's teaching phase and the student's learning phase are complementary. The teaching process must be carried out in accordance with the basic principles and basic strategies of learning [8]. After determining the learning outcomes, they must be arranged in an appropriate order according to the teacher's teaching goals.

4.2   The VR Technology Application in the Improvement of ESP Learning

Learning motivation is a dynamic mechanism of learning activities, which determines whether the learning activities can be realized smoothly, that will directly affect the learning effect of students. Ordinary ESP teaching, students only get audio-visual ESP stimulation from books and teachers. This kind of stimulation is monotonous and boring for students. Students can only connect with this kind of stimulation through basic cognition of previous things. Obviously, this connection is weak and indirect [9]. The use of VR technology can greatly improve the connection between this stimulus and cognition. Due to its immersion, VR technology can allow students to be fully immersed in an environment full of ESP atmosphere, and its interactivity can be Provide students with more sources of stimulation, so that students can not only get stimulation from books and teachers, because of its conception, students can not only be exposed to the stimulation of known things, but also can get stimulation from unknown things [10]. In addition, the VR technology application in ESP teaching is becoming more and more widespread. It is estimated that the market will reach 40 million US dollars in 2020, and the compound annual growth rate will remain double digits, as shown in Fig. 2.

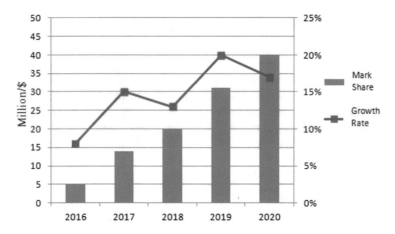

Fig. 2.   The application status of VR in ESP teaching

4.3   Store and Manage the Daily Behavior Data of College Students

4.4   VR Technology Realizes Operating Conditional Learning Theory

The main thing is to figure out what is a "trigger reaction" and what is a "spontaneous reaction", and these two reactions correspond to two behaviors: reaction behavior and operation behavior. Generally speaking, reaction behavior is more passive and controlled by stimuli; operational behavior means that the body can actively adapt to the environment and is controlled by behavioral results. Therefore, two learning forms have emerged: one is used to shape the reaction behavior of the organism, which is called classical conditioning; the other is used to shape the operation behavior of the organism, which we call operational conditions [10].

4.5 The Enhanced Stimulation of Operational Conditioning by VR Technology

A very important principle is mentioned in the behavioral learning theory. The result-following behavior is more effective than the result delay. Instant feedback plays an important role in VR technology: first, it more intuitively reflects the connection between behavior and results, and second, it makes the feedback information more valuable. The interactive nature of VR makes it easy to achieve instant feedback [11]. Similarly, in traditional classroom teaching, the teacher preached and solved the puzzles by karma, while the students listened in a muddled manner. Students are passive listeners and basically have no chance to respond in general, and they often respond in a positive way. Traditional textbooks cannot avoid being boring, and even make students feel disgusted, which hinders the learning of the important knowledge points covered in the book [11].

5 Conclusion

In summary, due to the immersion, interactivity, and visualization of VR technology, comparing students with ordinary ESP learning can effectively stimulate students' interest in learning and thus improve their enthusiasm for learning, and can achieve effective feedback and timely feedback. Give students appropriate positive reinforcement and negative reinforcement (positive reinforcement), which can ultimately greatly improve students' ESP learning ability.

References

1. Cui, L.: Overview of virtual reality technology and its applications. Fujian Comput. **11**(25), 11–14 (2018). (in Chinese)
2. Liu, W.: Analysis of the application of virtual technology in English teaching. Road Success **8**(34), 59–60 (2019). (in Chinese)
3. Cong, R.: The design strategy of "virtual technology" in junior high school English classroom teaching. Curric. Educ. Res. **12**(14), 129–130 (2019). (in Chinese)
4. Tang, C.: Design and application of junior middle school English "micro course". English Teach. **18**(16), 97–99 (2018). (in Chinese)
5. Weng, J.: Analyze the application of micro-classes in English grammar teaching. English Teach. **9**(19), 91–92 (2015). (in Chinese)
6. Hu, T.: The new trend of the development of regional education information resources in virtual reality technology. Audio-Visual Educ. Res. **13**(10), 61–65 (2017). (in Chinese)
7. Lei, Y.: The role and application of virtual reality technology in high school English classroom teaching. Master Online **22**(6), 91–92 (2019). (in Chinese)

8. Xia, M.: Research on the design and application of virtual reality technology in high school English grammar teaching. Sci. Consultation Educ. Res. **8**(12), 137–139 (2018). (in Chinese)
9. Li, Q.: Research on the application of virtual reality technology in high school English grammar teaching. English Teach. **11**(12), 110–116 (2018). (in Chinese)
10. Wang, F.: Research on high school English teaching practice based on virtual technology. Overseas English **5**(12), 20–21 (2018). (in Chinese)
11. Li, L.: Virtual reality technology short video teaching high school English grammar and language points flexibility and timeliness exploration. Mod. Commun. **33**(2), 192–193 (2018). (in Chinese)

Application of Simulation Method Based on Computer Bionic Design

Bin Wang[⊠]

School of Digital Arts and Design, Dalian Neusoft University of Information, Dalian, China
wangbinar@163.com

Abstract. With the development of science and technology and the advancement of computer technology, computer bionic design has become a new direction for exploring concepts and form of design. Through the integration of computer algorithms and impact factors, design innovation based on the shape of biomorph has become one of the emerging design directions. With the development and change of design concepts, many new design methods and theories have emerged. The simulation method is one of the most significant design methods. This article explains the influence and application of the simulation method on design form in modern design through the application of the simulation method in different design fields under computer bionic design.

Keywords: Computer bionic design · Modern design · Simulation method · Architecture design · Product design

1 Introduction

Computer has become an essential tool in the field of modern design. Due to the increase in the performance and speed of computer, the efficiency of design can be greatly improved to fulfill the requirements that manpower cannot meet.

In terms of bionic design, as an interdisciplinary subject, it relies on the results obtained by bionics, sets the algorithm through computer technology, and incorporates multifaceted influence factors for simulation calculations to form design results with bionic properties.

This paper explains the application of simulation method in design practice from the perspective of computer bionic design to apply and explore the bionic design more intelligently [1].

2 Computer Bionic Design

2.1 Basics of Bionic Design

When it comes to computer bionic design, we must first understand two concepts: bionics and bionic design.

J. MacIntyre et al. (Eds.): SPIoT 2020, AISC 1283, pp. 24–29, 2021.
https://doi.org/10.1007/978-3-030-62746-1_4

Bionics refers to a method of studying various characteristics of the biological world as objects of imitation for improving or innovating modern science and technology. Although science and technology have developed to a very high level, there are a large number of biological phenomena in nature, which are incomparable with the existing technology of human beings. Observation and analysis of biological characteristics can effectively promote the development and progress of human society.

Bionic design refers to a design activity that simulates the characteristics of biological information and laws in nature and is also a design activity carried out by imitating or receiving inspiration from living things. As an interdisciplinary combination of bionics and artistic design, bionic design is a new direction for the development of human science in the early 21st century.

2.2   Principles of Computer Bionic Design

Computer bionic design is one of the forms using simulation method. In computer bionic design, algorithmic formulas and software is used to combine specific design environments or influence factors of design requirements, and establishes a bionic system and structural model through differentiated repetitive calculations and morphological evolution. Computer bionic design is non-repetitive. Even with the same environment and impact factors, because of the difference of time and design concepts, despite the repeated iterations of the algorithm, the final design model will be significantly different.

Computer bionic design involves two important phases: model construction and model analysis. The scientificity, accuracy, completeness, and systematicness of the model construction have a fundamental role in the later model analysis and problem optimization. Model analysis refers to testing and perfecting the constructed model from various aspects such as morphology, structure, function, and environment. At the same time, it looks for the optimal design solution by observing changes in the model.

2.3   Principles of Computer Bionics

The integration and analysis of bionic models through computer is the basis of simulation design. By building models, many problems that were solved using complex, micro, unreasonable but conventional methods can be settled. During the model building phase, it should be observed from a dynamic perspective, including multiple factors such as structure, function, and form. The model can be a mathematical model, a physical model, a dynamic model, a discrete model, etc.

Functional principle: Different biological forms contain different characteristics, and different samples determine different characteristics of the model structure. Therefore, at the beginning of sample selection and model construction, functional requirements should be determined to select bionic samples in order to meet the requirements of accurately extracting functional features.

The principle of simplicity: In the process of establishing the bionic model, the main characteristics of the sample should be extracted to reduce the influence of interference factors. The main factors are extracted, the secondary factors are weakened, the primary and secondary relationships are clear, and the bionic design is optimized.

The principle of simplicity: In the process of establishing the bionic model, the main characteristics of the sample should be extracted to reduce the influence of interference factors. Extract the main factors and weaken the secondary factors to optimize the bionic design.

The principle of accuracy: The design of bionic models by computers is the basis for subsequent analysis, observation, and improvement. The accuracy of the model is the guarantee for the smooth progress of subsequent research. So accuracy determines whether the bionic model can succeed.

The principle of identifiability. In the process of model construction, bionic samples are the main source of data, so the model constructed should have obvious sample characteristics, which is the source of identifiability. At the same time, the function, morphology, texture, and characteristics of the organism are directly related to the shape, therefore the bionic model should contain characteristics of identifiability.

3 Application of the Simulation Method in Product Design

Conceptually, the simulation method in modern design refers to the reproduction of the appearance, nature, pattern, and feature of a prototype or pattern by means of an object or process, a scientific analogy method that uses the similarities and correlations of different objects in the design behavior. The image reproduced by the simulation method is created by simulating the system of diverse types and materials. Therefore, the simulation design method is an important scientific method to research and create new objects and products by correspondence theory. The prototype is created through creative thinking.

3.1 The Essence of Product Design

In product design, the goal product, but the ultimate purpose is user, that is, the ultimate purpose is to meet the needs of people in manipulation and life regardless of the style of the product. The purpose of design is people, and the ultimate purpose is to meet people's increasing material and spiritual needs. Human runs through the entire process of product design. The success of product design depends on two aspects, which are "Whether it works" and "Whether it works well". A good product should be able to meet user's needs to the greatest extent. On the one hand, the material needs, on the other hand, the spiritual needs.

3.2 The Simulation Method in Product Design

The appearance and modeling design of a product is one of the cores of product design. Nowadays, product design not only meets user's needs on function but also pays more attention to spiritual communication with people. The simulation method in architecture design takes the internal organization and external form of organisms as research objects to explore the scientific and reasonable construction laws contained in it, and applies the results to architecture design, in order to enrich architecture design methods and improve functions of buildings, and achieve people's needs [2–4].

4   Application of the Simulation Method in Architecture Design

4.1   Application of the Simulation Method in Construction Outward Appearance Design of Architecture

Applying the simulation method to the appearance of a building is innovative. Studying the appearance of various organisms in the natural to explore the inherent scientific rules, and to future explore the possibility of applying those rules to architecture design. This can not only integrate the functions and forms of architecture but also create innovate and varied architectural forms. The Ronchamp Chapel (Fig. 1) designed by Le Corbusier is a typical representative of the application of the simulation method in architecture design. The structure of the Ronchamp Chapel is dynamic. It looks like a bird with wings flying high in the air. Kennedy Airport (Fig. 2), designed by Eero Saarinen, is a classic modernist design, and the simulation method is also adopted when designing the appearance. John F Kennedy International Airport is like a soaring eagle with its wings spread, which exactly matches the function of the airport. In addition, the Bird's Nest in Beijing and the National Centre for the Performing Arts of China are also an outstanding representative of the simulation method in the construction outward appearance design of architecture.

Fig. 1. La Chapelle de Ronchamp   **Fig. 2.** John F. Kennedy International Airport

4.2   Application of the Simulation Method in the Function Design of Architecture

It is common to use the simulation method in the function design of architecture, which can combine good practicality and functionality, as well as flexible expression method. Nature is a magical treasure, which designers can draw infinite inspiration and apply to design, to create innovative space and functionality of a building. Through studying the structure of the ant nest, people found that there is a complete temperature regulation system inside the ant nest, and the temperature and humidity in the nest can be adjusted by the transformation of the anthill and the internal channel. This construction is also used in architecture design. The design of the new energy-saving building constructed in the capital of Zimbabwe is imitating the structure of an ant nest, which not only effectively adjusts the indoor temperature but also saves energy.

4.3   Application of the Simulation Method in Structure Design of Architecture

Along with the society progress and development, the style of architecture design has also changed a lot. Large-span, high-strength, and distinctive shape are becoming more and more popular in architecture design. The structural the simulation method is based on organisms in nature, which is inspired by theoretical research and applied in the practice of the structure design of buildings. The eggshell is something we are all familiar with. A whole egg shell is difficult to crumb though the shell is thin. This is because its shape is curved. When it is subjected to external pressure, the force will be evenly distributed along the curved surface. It is because of the inspiration of egg shells that many innovative architectures with shell structures are designed. For example, the National Centre for the Performing Arts of China (Fig. 3) is a typical example of shell structure. The long span and sleek appearance make the architecture both beautiful and functional. The Sydney Opera House (Fig. 4) was designed and built by Jon Woo Chung. It is a typical shell structure. Its appearance is like a standing white shell facing the sea, as well as a white sail sailing in the wind. Today, Sydney Opera House is not only the temple of art and culture in Sydney, but also the soul and logo of Sydney [5–7].

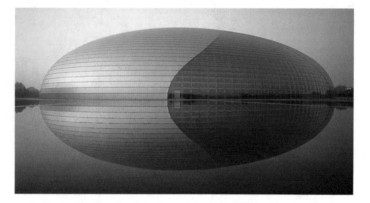

Fig. 3. The National Theatre of Beijing

Fig. 4. Sydney Opera House

5   Conclusion

This article interprets the biological morphology language from the perspective of computer bionics and simulation. It explains the basic methods and principles of bionic design. Bionic is a specific form of simulation, the simulation method is a significant design method through the reproduction and creation of a prototype. Every creature in nature could be the inspiration of design and expressed through simulation. Varying expressions and colorful colors are the characteristics of the simulation method which are different from other design methods. The continuous deepening of the application of simulation in design will certainly lead modern design to a more colorful world.

References

1. Liu, G., Shen, J.: Products Basic Design, pp. 124–132. China Light Industry Press, Beijing (2001). (in Chinese)
2. Yu, F., Chen, Y.: Bionic Plastic Design, pp. 211–237. Huazhong University of Science and Technology Press, Wuhan (2005). (in Chinese)
3. Gao, J., Liu, X.: Preliminary Study on Bionic Building Structures, pp. 221–234. China Architecture & Building Press, Beijing (2001). (in Chinese)
4. Yao, X., Shen, M.: Bionic Architecture Design, pp. 105–141. Shanxi Architecture, Shanxi (2006). (in Chinese)
5. Yao, X., Shen, M.: Bionic Architecture Design, pp. 104–112. Shanxi Architecture Press, Shanxi (2006). (in Chinese)
6. Cai, J., Wang, J.: Bionics Design Study, pp. 168–201. China Architecture & Building Press, Beijing (2013). (in Chinese)
7. Ma, S., Zhu, X.: Artificial Intelligence, pp. 94–113. Tsinghua University Press, Beijing (2004). (in Chinese)

The Implementation and Application of Computer Simulation Technology in PE Teaching

Junhua Lu[✉]

Liaoning Institute of Science and Engineering Physical Education College,
Jinzhou 121000, Liaoning, China
qtds_7608@163.com

Abstract. Implementation method of computer simulation technology in physical education as a subject to study. First of all, the paper discusses the basic concepts of computer simulation technology, the application scope and were briefly reviewed in this paper, the main process. Secondly, the obstacle factors of computer simulation technology in physical education teaching are summarized and analyzed from three perspectives. Thirdly, the application method of computer simulation technology in the field of sports science is introduced from two aspects. Finally, combining with the teaching practice of physical education, this paper comprehensively demonstrates the implementation method of computer simulation technology in the practice of physical education from three dimensions. The research results can provide positive and beneficial inspiration and help for related experts and researchers.

Keywords: Computer simulation technology · Physical education · Implementation and application

1 Introduction

Since the beginning of the 21st century, computer technology has developed rapidly. Computer simulation technology has played its unique advantages in early warning and prediction, evaluation and evaluation, practical experiment, system analysis, and principle design. Based on computer simulation system and the basis of the research and development of all kinds of models, not only reflects the elements in the process of modeling, associated with each system object, the internal relation between each pixel unit and the objective law, and with the maturity of the computer simulation technology and its related fields of large data process function, the interaction between the virtual reality, graphics geometry more and more to promote the development of college physical education teaching, as a result, the current research on virtual human body and its movement became one of the hot topics. With the reform of physical education and teaching in colleges and universities in our country in full swing, computer simulation

J. MacIntyre et al. (Eds.): SPIoT 2020, AISC 1283, pp. 30–38, 2021.
https://doi.org/10.1007/978-3-030-62746-1_5

technology is bound to get new development opportunities centering on the general principle of "total balance, component disassembly and momentum control" in the process of physical education [1]. Therefore, combined with the overall planning of the 14th five-year plan, it is not difficult to predict that the advantages and disadvantages of the application and implementation methods of computer simulation technology in physical education will directly affect the effectiveness of sports training, and even gradually improve the importance of sports teaching or competition training projects.

2 Overview of Computer Simulation Technology

2.1 Basic Concepts of Computer Simulation Technology

Computer Simulation is a realistic and dynamic imitation of the structure, function and behavior of the system and the thinking process and behavior of the people participating in the control of the system by using an electronic Computer.

Computer simulation technology has its own uniqueness, it is not only a very effective quantitative analysis method and tool, but also a highly intuitive descriptive tool. In simple terms, it is to describe a specific system or a complex process vividly by means of modeling. With the help of experimental working methods, its internal characteristics are described in detail and presented in the form of data, so as to provide timely and effective decision-making Suggestions for the decision-making and management levels.

2.2 Application of Computer Simulation Technology

The application of computer simulation technology is more and more extensive, and it plays a positive and important role in people's production and life planning, natural environment maintenance, military prediction and analysis, human physiological simulation and many other fields. Quantitative analysis was carried out on the set model, trial and error and draw final conclusion.

2.3 The Main Flow of Computer Simulation Technology

Computer simulation technology firstly completes mathematical modeling through continuous exploration, then carries out computer simulation experiments according to the simulation field, and finally obtains conclusions through the comprehensive analysis of experimental results in three main stages to realize the entire process of computer simulation. From the perspective of process, computer simulation technology is not only an experimental analysis of a single individual or a single model, but a comprehensive review of the three stages. Generally speaking, the process steps of computer simulation technology include the following steps:

2.3.1 Introduction of the Question

Through a preliminary description and analysis of the problems, to clarify the basic characteristics of the respondents, the computer and then determine the final purpose of test, and gradually perfect all kinds of demand factors, then, according to the previous set of requirements, a reasonable plan and the subsequent restriction conditions about the design of computer simulation technology as well as the structure and size of the system itself.

2.3.2 The Establishment of Mathematical Model

The existence characteristics of computer simulation technology depend on the accuracy of mathematical model establishment, which will directly affect the deviation degree and reliability of simulation results. Therefore, this step is also called the core step.

2.3.3 Quadratic Mathematical Modeling

According to the computer's own operation law, simulation mode and precision account-ing, on the basis of the original first mathematical modeling, it is further transformed into a new model that can be designed according to the computer program. Through the transformation, the simulation system really has the theoretical basis of mathematical modeling, and at the same time, the purpose of secondary modeling is better realized.

2.3.4 Programme

As a computer simulation technology in the process of physical education, programming is more important, in the process of the integration of the two technologies should be used in different ways, more effective methods, mathematical model programming, and continue to carry out professional debugging, finally achieve the ideal effect.

2.3.5 Promotion of Simulation Experiment

According to the requirements of the regulations, the simulation system is deduced on the spot, and the experimental data are collected and sorted out.

2.3.6 Comprehensive Analysis and Model Validation

The application of computer simulation technology in the field of physical education is still in its infancy, and each technology fusion needs continuous verification to find the best technical scheme. It can make the accuracy of the system model more reasonable and finally fully analyze the effectiveness of simulation technology.

3 Analysis on the Obstacle Factors of Computer Simulation Technology in Physical Education Teaching

Computer simulation technology is a brand new teaching concept, college physical education teachers need a period of adaptation to the new food, to be able to popularize and develop the need for everyone to emancipate the mind, serious study. Actively overcome adverse factors. Analysis of the reasons is mainly limited to the following aspects:

3.1 The Practical Limits of Scientific and Technological Power

Sports discipline subject classification is more, so, the teaching content is relatively complex, social institution or university itself, generally lack of perfect computer for sports discipline echelon personnel, owing to lack of emphasis, inadequate funding, caused in the field of sports skill in exclusive computer simulation software research and development ability are relatively weak, already can't adapt to, different month demand new sports teaching reform and teaching practice. At the same time, there are relatively few technical service companies aiming at college physical education in the society [2]. Either the resources distribution is not uniform, or the technical companies are not professional enough. In addition, the lack of close cooperation between universities and enterprises has exacerbated the shortage of scientific research and technology.

3.2 Interdisciplinary Communication is Difficult

Computer simulation technology is a high degree of specialization, the complex routing is a relatively new concept, in the whole simulation technology into the teaching process, we need to provide necessary advice and guidance of physical education teachers in colleges and universities, also need more powerful sports scientific research team in colleges and universities, puts forward higher requirements to the computer simulation technology is insufficient [3]. The reality is that some universities lack research teams in sports institutes, some cannot find the right balance between the two, and some fail to connect the two. Therefore, how to seamlessly link the professional knowledge of physical education with the thinking of computer scientists is an urgent problem to be solved.

3.3 The Defect of Educational Equipment

Although said that the computer simulation technology is a software development, the application, the promotion of the work, but also needs some necessary hardware equipment facilities to do the support. At present, there are many levels of colleges and universities in China, and the gap between colleges and universities is also obvious. Some remote areas, low-level colleges and non-professional colleges are generally unable to guarantee the basic hardware ratio.

It is highlighted in the following aspects:

1) the related computer equipment and facilities are aging,
2) the lack of necessary teaching and exhibition places,
3) the lack of the necessary interactive communication platform, and
4) the lack of some basic teaching tools.

Therefore, the lack of educational equipment will also restrict the application of computer simulation technology to some extent.

4 The Application of Computer Simulation Technology in the Field of Sports Science

4.1 The Application of Computer Simulation Technology in Physical Education Teaching

As a new multimedia teaching method, computer simulation technology can change the traditional teaching method. This technology can show the difficult points and key points of the knowledge in different ways, and can show the teaching process more vividly, so that students can participate in the learning in a more interesting way [4]. Differentiated group teaching can be achieved before, during and after teaching, so as to ensure the maximum overall learning effect, and finally achieve the synchronous improvement of classroom teaching and training teaching.

The intervention of simulation technology in teaching, through the introduction of three-dimensional teaching, on the one hand, visual simulation technology can be used to achieve vivid teaching, so that students have a sense of immersive experience in the learning process; On the other hand, the advantages of the software module of the simulation system can also be used to enhance students' cognitive level. Differentiated group teaching can be achieved before, during and after teaching, so as to ensure the maximum overall learning effect, and finally achieve the synchronous improvement of classroom teaching and training teaching.

4.2 The Application of Computer Simulation Technology in Sports Training

The discipline of computer graphics, which is very mature and widely used, captures real system data by simulating a certain phenomenon or by simulating a certain stage to optimize the quality of process control. In sports training, college physical education subjects pay attention to actual combat, pursue the data accuracy of sports variables, and stress the practicability, applicability and usability. From the mathematical model to the simulation model, and then upgraded to the model data that can be realized by the computer program, the repeated analysis of the data in this process will help us to achieve the purpose of computer simulation.

5 The Implementation and Application of Computer Simulation Technology in Physical Education Teaching Practice

5.1 Increase the Application of Computer Simulation Technology in Classroom Teaching

With the continuous improvement of science and technology and the continuous development of information technology, the traditional teaching mode should be reformed in the teaching of physical education in colleges and universities. As a new teaching form, computer simulation technology should be boldly tried. Specific methods are as follows:

5.1.1 Preview Before Class

Teachers should consciously intercept the important knowledge points of key links to do a preview, guide and properly ask students to develop the awareness and habit of pre-class learning. In order to be able to accurately guide the student to the good habits, physical education teachers should do their homework, such as early rushed out the next phase of the teaching emphasis and difficulties, clarify the cross between knowledge and be relations, at the same time, the physical education teachers to find a way to stimulate students' thirst for knowledge and curiosity, using a herd mentality and ambience, strengthen the pre-course reading effect.

5.1.2 During the Lecture

Teachers should reasonably arrange the curriculum design, mainly reflected in the following aspects: first, increase the overall proportion of computer simulation technology in time allocation. Physical education teachers can consider to spend no less than 35% of their time to do computer simulation demonstration, and then carry out about 35% of the traditional teaching, according to the simulation and the traditional teaching model grasp about 30% of the students to summarize and understand. Through this design, the proportion of computer simulation technology in classroom teaching can be strengthened, and the original teaching mode can be completely improved. Second, if we want to further guarantee the application quality of computer simulation technology, we should consider the redistribution of this period of time on the basis of 30% time allocation.

5.1.3 Review and Consolidate After Class

It is necessary to use the quantitative assessment method of "time" to comprehensively supervise students' after-class learning results, timely find out problems and provide follow-up Suggestions for improvement. To be specific, on the one hand, the traditional after-class examination method should be reformed, and descriptive assessment of technical situation should be introduced. On the other hand, students' effective learning time on the information platform should be assessed by virtue of the advantages of information platform.

5.2 The Application of Computer Simulation Technology in Sports Training is Introduced

The important content of college physical education teaching work is sports training. Considering the outstanding characteristics of sports training, such as practical operation and landing, it should be "fragmented" to insert the auxiliary means of rectifying the deviation of computer simulation technology. Specifically, it should be implemented from the following aspects: [5].

5.2.1 Group Mode

So that the trainees can establish a sense of competition, not only pay attention to their own team learning effect, but also observe the learning effect of other groups, make full use of the advantages of computer simulation technology, form an intuitive contrast. Through horizontal comparison, timely fill their own defects, and actively learn the advantages of other groups.

5.2.2   Adopt the Mode of "One-to-One" Support

Excellent first-tier students were selected from the parallel group, while the relatively weak members of the last tier were also selected. By playing the big data effect of computer simulation technology, the data differences between them were refined to make targeted rectification and improvement plans.

5.2.3   Take a "Growth Record" Approach

Establish a growth profile for key students, and provide scientific advice that can be consulted, classified and guided by data records and comparisons, so as to narrow the gap among team members and effectively manage the gap improvement.

5.2.4   Take Video Recording Method

After collecting videos of the technical movements of the typical representatives of the student groups, they are synchronized with the standard video movements in the same picture. The teacher makes a comparison and explanation on this basis, and the teaching effect is also very good.

5.3   The Application of Computer Simulation Technology is Introduced in the Connection of Teaching Plates

In the traditional teaching process, different learning stages should be based on the students' grasp of the situation, to teach students to consolidate and improve the content. To solve this problem, computer simulation technology can provide more alternative solutions by teaching students in accordance with their aptitude, which is a good solution. We mainly carry out the following sections:

5.3.1   Similar Teaching Plate

The "splicing" process of similar teaching plates can enable computer simulation technology to play a special role. For example, training data, drill pictures, routine movements, teaching concepts and learning modes can be connected by means of editing. The splicing and series method can not only make use of the series mode of each knowledge point to complete the docking, but also complete the effective coherent connection between modules and technical actions.

5.3.2   Large Differences in the Teaching Plate

The application of computer simulation technology can greatly reduce the "incompatibility" of knowledge points between teaching plates with large differences, so as to reduce the difficulty of students in the learning process. In other words, through the application of computer simulation technology, the seemingly unrelated sports knowledge points are connected into a line, breaking the thinking and learning mode under the traditional perspective and traditional thinking framework.

5.3.3 Teaching Modules for Different Learning Groups

In order to avoid a wide range of personality differences when learning the same knowledge structure in different learning groups, computer simulation technology can be used to avoid the phenomenon of lagging or leading in teaching progress. In view of the "lagging phenomenon", physical education teachers can give full play to their unique advantages of "passing, helping and leading", and use computer simulation technology to form a one-to-one private teaching model to reinforce individuals. In view of the "leading phenomenon", physical education teachers can use the method of "collapse, pressure, correction, setback" to form a local control situation by using the computer simulation technology, which not only meets the individual's desire for progress, but also controls the adverse influence brought by the individual to the group.

6 Conclusion

In the era of information technology, computer simulation technology has been more and more fields, in order to further improve the computer technology is widely used in the field of sports, must want to have more support, is also inseparable from the technology platform constantly improve, human consciousness constantly improve and the industry as a whole application atmosphere formed a series of factors such as support [6]. It is gratifying to see that with the continuous improvement of management consciousness, the continuous enrichment of management measures, the continuous expansion of management ideas, the continuous innovation of management mode, the continuous clarity of management logic, the continuous maturity of management culture, the increasing importance of the educational and teaching reform of physical education. With the improvement of the degree of attention, the biggest change is that the progress of the informatization reform and improvement of physical education is gradually accelerated. Can really improve the teaching quality and efficiency, can effectively reduce the time cost, risk components, as long as the thinking is clear, the method is appropriate, the application of lasting, will be able to achieve and exceed the expected effect [7].

References

1. Wang, L.L.: The research on computer simulation technology in physical education and scientific. Global Hum. Geogr. **18**(5), 70–73 (2012). (in Chinese)
2. Yan, G.D., Sun, J.H.: Application of computer simulation technology in PE teaching. Sport Sci. Res. **29**(5), 68–70 (2008). (in Chinese)
3. Li, H.C., Li, Y.N.: The application of computer simulation technology in teaching. J. Jiamusi Vocat. Inst. **182**(1), 396–398 (2018). (in Chinese)
4. Han, Z.Y.: Research on the effect of multimedia teaching on the technique teaching effect of table tennis players in college. J. Tianjin Univ. Sport (2015). (in Chinese)

5. Sun, Y.: The influence of educational technology application level on the teaching satisfaction of higher vocational education. Hebei University of Technology (2015). (in Chinese)
6. Zhang, F.: Computer simulation technology and its applications. Comput. Knowl. Technol. (Acad. Exch.) **41**(10), 233–238 (2007). (in Chinese)
7. Sun, Z.: A study on the educational use of statistical package for the social sciences. Int. J. Front. Eng. Technol. **1**(1), 20–29 (2019)

Construction of College Communities in the New Media Based on Network Environment

Min Qiu[✉]

Jilin Engineering Normal University, Changchun, Jilin, China
qiumin1376@sohu.com

Abstract. The new media represented by Internet and mobile Internet are changing people's life style, which has brought great changes to all fields of the world. The exploration and practice of information construction in colleges and universities has been carried out at home and abroad by using a variety of new media integration presently. However, the theoretical research on the construction of college communities by using new media based on network is obviously lagging behind, which cannot reflect the laws that should be followed in the construction and development of college communities under the actual conditions. On the basis of summarizing the development rules and models of college communities, this paper comprehensively uses a variety of research methods to propose related strategies for college communities construction based on the network new media environment. Based on the results of comparative research and questionnaire analysis, it summarizes the influence degree of the new media based on network during the construction of college communities to improve the competitiveness of college student associations and guide practice.

Keywords: New media based on network · College communities · Construction

1 Introduction

Although the operation logic of the network media platform overlaps with that of the traditional mass media, it is obviously different [1]. Which leads to different ways of producing content, distributing information and using media [2]. New media has been used in various places. Schwartz [3] shares a knowledge method of using new media online in the classroom. In the classroom space, new media can be used to properly develop the design produced by learners. Bose [4] investigated the impact of mobile devices on teaching practice, student learning and course learning outcomes. Gan [5] gained experience in the context of a mobile learning project supporting the iPad during the knowledge management course to support the University's technology support learning vision. Niemi [6] provides a set of conceptual mediators for student driven knowledge creation, collaboration, network and digital literacy to empirically study students' participation in learning skills in the 21st century.

The successful participation of college communities in promoting social and economic development is closely related to the degree of prediction of participation in

J. MacIntyre et al. (Eds.): SPIoT 2020, AISC 1283, pp. 39–44, 2021.
https://doi.org/10.1007/978-3-030-62746-1_6

practice in university core policies and practices [7]. Universities can provide design, planning, economic development and other strategies, concepts and policies through interdisciplinary outreach and participation in and use of service learning [8]. Brand [9] provides a theoretical basis for college communities' connection mechanism to explain why extracurricular participation is such an important part of university experience. Xypaki [10] discusses how a bottom-up, top-down incentive and capacity-building support structure of innovative student led extracurricular activity sustainability project can achieve sustainability results and inspire other UK higher education institutions and decision makers. The partnership between university and communities can play an important role in curriculum development, but there are few reports on the role of communities' institutions in curriculum design [11].

It is the trend of future development to manage and display all kinds of activity resources of colleges and universities comprehensively by using new media based on network, so as to realize intellectualization and digitalization on the network platform [12]. This paper intends to provide some application strategies for the construction of college communities through a variety of network new media, so as to realize the cross disciplinary resource sharing among various associations in colleges and universities, and finally improve the students' innovative practice ability.

2 Method

Colleges and universities are a starting point for college students to enter the society. Various college communities are an important part of campus culture. While cultivating students' interests and hobbies, the communities can let the members of the associations get relevant exercises, so that college students can benefit from the experience of the associations and grow up in the future when facing employment, which is an important core of college communities' construction. As a practical base for college students to cultivate team spirit and good values, college community is an effective platform for self-education and self-management of college students. In the process of building college community culture, the main problems faced by colleges and universities are as follows:

1) unreasonable organization structure distribution of the communities
   At present, the theoretical research-oriented or stylistic activities of college communities are relatively independent. Because most of the activities held by theoretical research are reading salons participated by internal members, the effect of the activities is not good, and the attraction to college students is low. However, the structure and activities of the members of the literary and sports associations are more diversified and more attractive to college students, but they cannot be integrated with the theoretical research associations together.

2) lack of understanding and control of the guiding concept of community development
   The purpose and concept of the development of college communities are the basis for distinguishing from other associations. Only around the center of the purpose and concept of the development of associations, can we better carry out the activities of college communities and all the subsequent development of communities. If every member of the communities does not fully understand and grasp the development

purpose and concept of the communities, it will affect the later development process of the communities work.

3) weak innovation and creativity

In the era of rapid development of Internet technology, most communities s fail to make full use of emerging media in the way of publicity. The traditional posters cannot fully express the activities of the community and the students, and the attraction to the students is not enough to achieve the expected publicity goals. At present, there are few communities with exclusive websites, some of which have slow update speed, insufficient maintenance and poor publicity effect.

3 Research and Design of College Communities Construction Based on the Network New Media Environment

3.1 Research Methods

This study intends to take the rational use of new network media as the starting point, to improve the construction of college communities culture as the goal; through the field survey, comparative research, inductive summary and other research methods to explore the construction of a resource sharing platform for college students' communities, and according to the results of comparative study and questionnaire analysis, this paper summarizes the impact of new media on the construction of university communities.

3.2 Questionnaire Design

500 college students who have participated in the college students' communities are taken as the survey objects in this paper. In order to better ensure the validity of the questionnaire, random samples are used to fill in the questionnaire, and the interviewees are allowed to enter the online questionnaire page to fill in. The main part of the questionnaire is about the perception, attitude and participation of the students to the existing college communities analysis, and draws relevant research data and conclusions. A total of 495 online questionnaires were received and 483 were valid.

4 Investigation and Analysis of the Effect of College Communities in the New Media Environment

4.1 Analysis of Application Effect by Popularity and Perceived Importance

According to the relevant data, this paper analyses the impact of the new media environment based on the network on the construction of college students' communities through the in-depth mining of the network survey data. This paper uses popularity and perceived importance to investigate the importance of college students in the construction of college students' communities constructed by different network new media. The results are shown in Fig. 1.

It can be found from Fig. 1 that college students think that the campus website is the most important among the four different network new medias, 100% of students think

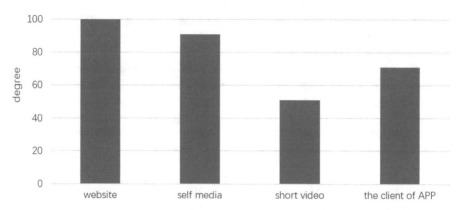

The importance of different network new media to the construction of college students club

Fig. 1. the importance of the construction of college communities in different network new media

that the campus website is very important for the construction of community culture, while the "small video" is the least important digital new media, but more than 50% of students recognize its importance in the construction of College communities.

The results of calculating perceived importance show that most of the sub items in the survey are more important than their popularity in this paper. The differences of the sub items of digital video among them are the largest, while the three items of campus website, microblog and wechat are not significantly different. The data of perceived importance proves the importance of these sub items to the construction of college communities, and measures must be taken to improve the use of these sub items.

4.2   Research Result

It is found through the above investigation and research that the application strategies of the construction of college communities in the new media environment based on network are as follows:

1) using the network new media platform to improve students' perception of college communities construction

The 483 students who have participated in the survey all think that it is necessary to construct the college communities by using the new media based on network. 97.91% of the students think that they can more vividly show the development process and objectives of the community culture through the new media on the Internet, and improve the atmosphere of the community culture construction in a more comprehensive way.

2) provide corresponding innovation mechanism of community with the help of new media based on network technology

College communities should be clear about the needs of the organizations, constantly adjust and innovate on the basis of the development of different periods, and improve the quality of the development of associations. 93.37% of the students think that they can use the power of new media to tap the comprehensive ability of the members of the community and carry out the assessment criteria, so as to provide guarantee for the cultivation and promotion of the backbones in the later period.

5 Conclusion

The construction of college communities is a complex and huge systematic project. Detailed planning and reasonable distribution are essential to ensure the sustainable and healthy development of community construction. The purpose of this study is to integrate all kinds of new media communication tools and media on the network platform, to make them play a collaborative role in the construction of college students' community culture, to spread consistent and coherent college students' community information to college students, to establish a stable two-way communication relationship with them, and to use new media on the network The effect of community construction is evaluated dynamically and circularly, so as to reach the goal of Influencing College Students' values and learning behaviors.

Acknowledgements. The Funds Program: The 13th Five- year Plan of Jilin Province Science of Education the key Research topics of the new ways of Ideological and political education based on the activities of College Students' Associations ZD19093.

References

1. Hurrah, N.N., Parah, S.A., Loan, N.A., Sheikh, J.A., Elhoseny, M., Muhammad, K.: Dual watermarking framework for privacy protection and content authentication of multimedia. Future Gener. Comput. Syst. **94**, 654–667 (2019)
2. Klinger, U., Svensson, J.: The emergence of network media logic in political communication: a theoretical approach. New Media Soc. **17**(8), 1241–1257 (2015)
3. Schwartz, L.H.: A funds of knowledge approach to the appropriation of new media in a high school writing classroom. Interact. Learn. Environ. **23**(5), 595–612 (2015)
4. Bose, D., Lowenthal, P.R.: Integrating mobile devices into the classroom: a qualitative case study of a faculty learning community. Int. J. Soc. Media Interact. Learn. Environ. **4**(4), 319–332 (2016)
5. Gan, B., Menkhoff, T., Smith, R.: Enhancing students' learning process through interactive digital media: new opportunities for collaborative learning. Comput. Hum. Behav. **51**, 652–663 (2015)
6. Niemi, H., Multisilta, J.: Digital storytelling promoting twenty-first century skills and student engagement. Technol. Pedagogy Educ. **25**(4), 451–468 (2016)
7. Mtawa, N.N., Fongwa, S.N., Wangenge-Ouma, G.: The scholarship of university-community engagement: interrogating Boyer's model. Int. J. Educ. Dev. **49**, 126–133 (2016)
8. Laninga, T., Austin, G., McClure, W.: University-community partnerships in small-town Idaho: addressing diverse community needs through interdisciplinary outreach and engagement. J. Community Engagem. Scholarsh. **4**(2), 2 (2019)

9. Branand, B., Mashek, D., Wray-Lake, L., Coffey, J.K.: Inclusion of college community in the self: a longitudinal study of the role of self-expansion in students' satisfaction. J. Coll. Student Dev. **56**(8), 829–844 (2015)
10. Xypaki, M.: A practical example of integrating sustainable development into higher education: green dragons, city university London students' union. Local Econ. **30**(3), 316–329 (2015)
11. Lewis, L.A., Kusmaul, N., Elze, D., Butler, L.: The role of field education in a university–community partnership aimed at curriculum transformation. J. Soc. Work Educ. **52**(2), 186–197 (2016)
12. Thakur, S., Singh, A.K., Ghrera, S.P., Elhoseny, M.: Multi-layer security of medical data through watermarking and chaotic encryption for tele-health applications. Multimedia Tools Appl. **78**, 3457–3470 (2019)

Political and Ideological Personnel Management Mode Based on Computer Network

Siqi Pu[✉]

Scientific Bureau, China West Normal University, Nanchong 637009, Sichuan, China
714920371@qq.com

Abstract. Computer network as a representative of a new economic form, under the impact of its wave of reform, the prevalence of emerging technologies of mobile Internet represented by big data, cloud computing, and the Internet of Things will definitely trigger new round of technological revolution and Industrial change, at the same time, these industrial developments that are transformed and transformed by computer networks are inseparable. Therefore, research on the development status of the integration of vocational education, education and education in the context of computer networks, with the help of computers Advantages and characteristics of the network It is of practical significance to improve the problems in talent-training model of production and education, thus improving the quality of political and ideological personnel training and enhancing the ability of talents to serve the society. This paper reforms the management mode of political and ideological talents from the aspects of educational concept, educational content structure, teaching methods, evaluation system and teaching staff, promotes the connotative development of political and ideological talent management mode, and realizes the improvement of teaching quality and quality of personnel training. The cause of socialism with Chinese characteristics fosters high-quality and innovative talents with both ability and political integrity and comprehensive development, providing sufficient human support and intellectual security to improve independent innovation capabilities and build innovative countries.

Keywords: Computer network · Talent training · Political and ideological · Talent management

1 Introduction

With the development of the economy and the rapid advancement of science and technology, competition among countries in politics, culture, military and other fields has intensified, and these competitions are ultimately the competition of talents [1–4]. The current innovative talent training as a major measure to improve China's overall national strength is getting more and more attention, and college students as the main source of talents and an important force to promote future social development can not be ignored, so the study of political and ideological talent management model has become The era has given important historical tasks to higher education [5–7]. As an important base for

J. MacIntyre et al. (Eds.): SPIoT 2020, AISC 1283, pp. 45–52, 2021.
https://doi.org/10.1007/978-3-030-62746-1_7

the cultivation of high-quality talents, universities should adhere to the cultivation of innovative talents and new requirements in the new era, focus on the innovation of the management mode of students' political and ideological talents, enrich the connotation of higher education, promote the modernization of higher education, and provide more for the long-term development of society [7]. Solid talent protection [8–11]. At the same time, the management of political and ideological talents of college students is also a measure of the cultivation of innovative talents in the new era, which helps to provide a solid guarantee for the cultivation of innovative talents and provides more long-term motivation support [12]. As an important educational force, political and ideological talent management must adapt to the development of the times and constantly make self-adjustment in order to achieve long-term development. The task of cultivating talents is not only the mission of the management of political and ideological talents, but also the innovation of political and ideological education. The important driving force and main content of development [13, 14]. Through the practical exploration of talent management and the experience of advanced talent management at home and abroad, this paper discusses the new challenges and characteristics in the talent management practice during the period of innovation and development, and discusses the characteristics, effects and safeguards of the new talent management model. In the analysis, we put forward targeted suggestions on talent management methods, key links, and implementation.

2 Analysis of the Reasons Affecting the Management Model of Political and Ideological Talents of College Students

(1) Outdated Ideas of political and ideological Talents Management

Educational concept is the theoretical guidance for carrying out educational work. To achieve good results in education, scientific educational concept is needed. The management of political and ideological talents is an important part of ideological work. It shapes the world outlook, outlook on life and values of College students, and needs more guidance from scientific concepts [15]. However, most of the political and ideological personnel management workers in Colleges and universities have outdated educational concepts, mistakenly regard the political and ideological personnel management of college students as campus stability, so as to ensure that college students do not have accidents. This concept makes some political and ideological personnel management workers focus on the management of students. In addition, some educators only regard the management of political and ideological personnel of college students as a means of livelihood, and have not really studied how to do a good job of education. In addition, a considerable number of political and ideological personnel management workers are not specialized in education, have not received systematic education training, lack of understanding of students, nor have they carefully analyzed the characteristics of students, and are carrying out political and ideological personnel management work. At the same time, neglecting students' personality and adopting the "one-size-fits-all" approach arouse students' disgust and greatly reduce the educational effect.

(2) The content of political and ideological personnel management lacks attraction.

At present, the content of political and ideological talent management in Colleges and universities mainly includes the curriculum of political and ideological theory course. Although the curriculum is compiled by authoritative experts organized by the Ministry of Education, because the curriculum itself is an introduction to the basic theory of Marxism, and because the political nature of the curriculum is strong, the attraction of the textbook itself is not enough; some teachers of political and ideological theory course do not prepare lessons carefully when they teach. Without analyzing the needs of students, they only copy the contents of textbooks, ignoring the knowledge that students must absorb in their development, which makes the content of education dull and lacks the attraction for students. Some managers of political and ideological personnel blindly preach to the students, express their words over and over again, without new ideas, which arouses students'disgust. It can not only fail to achieve the effect of political and ideological personnel management, but also cause students' rebellious psychology to some extent, weakening the effectiveness of political and ideological personnel management.

(3) The single management mode of political and ideological talents

Rich education and teaching methods are very important in the process of teaching and help to improve the educational effect. Therefore, in order to improve the effectiveness of political and ideological talent management, we should try our best to adopt a variety of educational methods. At present, there are serious deficiencies in the management of political and ideological talents. Political and ideological theory course is the main channel for the management of College Students' political and ideological talents. It should play an important role in the management of College Students' political and ideological talents. However, the current political and ideological theory classroom has the problems of single mode of education and lack of innovation in education and teaching. Campus activities are obsolete in theme, immobilized in form and lack of novelty. Students are required to participate in the activities in the form of organization and apportionment. It seems that the activities are carried out well, but in fact they lack effective educational effect. The singularity of education mode makes students lose interest in participating in learning, which directly affects the effectiveness of education.

3 Bayesian Network

In order to better study the attribute relationship of ideological and political talent management and establish research models, this paper studies the management of ideological and political talents with the advantage of Bayesian network. Bayesian network node relationships can describe conditional independent assumptions, The Bayesian network specifies that each node V_i condition in the graph is independent of any subset of nodes given a non-descendant node by the parent node of the node, That is, if we use $A(V_i)$ to represent any subset of nodes formed by not descendant node of V_i, and $F(V_i)$ to represent the parent node of V_i, then the conditional probability of V_i node can be expressed as formula 1.

$$p(V_i|A(V_i), F(V_i)) = p(V_i|F(V_i)) \tag{1}$$

1) A conditional probabilities table associated with each node. The conditional probability $p(V_i|F(V_i))$ represents the relationship between the node V_i and the parent node. A node without any parent node is called a priori probability, so a joint probability can be expressed in terms of a conditional probability link, and the expression can be described as formula 2:

$$p(V_1, V_2, \ldots, V_k) = \mathop{F}_{i=1}^{n} P(V_i|V_{i-1}, \ldots V_1) \tag{2}$$

4 New Trends in Talent Management Brought About by Changes in the Competitive Landscape of the Industry

China's macro-economy has entered a new normal state. By building a multi-level capital market and responding to the call of "mass entrepreneurship and innovation", meeting the needs of entrepreneurial, innovative and growth-oriented enterprises in the capital market is a new historical mission facing various industries. At the same time, it is also a new transformation and development opportunity for the industry. In the capital and talent-intensive industries, the rapid development of the industry has brought about the hunger for talents, and the environment for talent development has been increasingly optimized. Both the national and regulatory levels of the institutional and cultural environment have made great progress, and they are the industry's talents. Create a vibrant, innovative and orderly environment. The industry has become more attractive to talents, and the number and quality of the talent team have continued to show a good trend. However, there is still a certain gap between the overall level of talent development in the industry and the actual demand for economic and social development and the increasingly international requirements for capital market development. The change in the market competition pattern has brought about a new trend in industry talent management:

(1) Diversified talent competition. With the increase of competition subjects, the competition for high-end talents in the industry will become more intense, and it will become the new normal of talent competition. How to attract and retain key talents becomes the key to talent management.
(2) Integration of talent capabilities. With the diversification of business and the need for innovation and development, the direction of talent cultivation is a comprehensive talent with comprehensive ability. The integration of talents and abilities can be seen from the thoughts and achievements, as shown in the Fig. 1.
(3) Diversified talents. Due to the limitations of the system and mechanism, the industry is currently focusing on short-term incentives, lack of long-term incentive tools such as equity, and the loss of excellent talents and the frequent cross-cutting of the

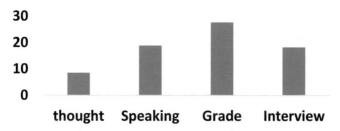

Fig. 1. Comparison of talent comprehensive ability

industry have affected the healthy development of the industry. With the intensified competition within the industry and the external impact of various new formats, enterprises have begun to attach importance to long-term incentives such as research and use of equity. In the context of the domestic market-oriented reforms and the deepening of state-owned enterprise reforms, enterprises are in the incentive mechanism. It will also be in line with international standards.

(4) Systematization of talent management. First, the key to increasing organizational support is the maximization of employee positions and a supportive work environment. While strengthening employee engagement and loyalty, it is becoming more and more important to improve organizational support from the aspects of building corporate culture and improving talent management system. new trend. Second, with the 80s and 90s, it has become a new force that cannot be ignored in the workplace. They are uniquely challenging the management of the workplace. Whether it is the view of career choice, the reason for leaving the job, the acquisition of professional accomplishment, or even the setback situation, the thinking scope after 80 and 90 has exceeded the experience gained by traditional talent management, and has become a challenge that needs to face in talent management. In the eyes of the new generation, attention to personal growth has replaced salary and become the most important gravitation of a job. The third is the informatization of management tools. After the introduction of ERP, CRM and other information management tools in talent management, with the promotion and application of information technology, contemporary Internet technologies represented by big data and cloud computing provide a new perspective for in-depth study of talent management [16] means. The comparison is shown in the Fig. 2.

Fig. 2. Contrastive analysis of management

5   Effective Ways of Political and Ideological Talents Management Based on Computer Network

(1) Adhere to the guiding position of Marxist theory and arm the minds of College Students

As the backbone of building socialism in the future, innovative talents must have correct political standpoint and attitude to ensure the correct direction of innovative activities and to choose the right innovative targets, so that their innovative activities and innovative achievements can promote social development and progress. The teaching of Marxist Zhuangben Principle is enough to guide the correct political direction and road. It is necessary to arm the minds of college students with the latest theoretical achievements of the Sinicization of Marxism, guide them to consciously use Mao Zedong Thought and the theoretical system of socialism with Chinese characteristics to guide their innovative practice, and keep the innovative practice in the right direction all the time, with advanced, revolutionary and revolutionary nature. People's character will eventually merge into the historical torrent of the construction of socialism with Chinese characteristics rolling forward.

(2) Enhancing College Students' Sense of Social Responsibility with Patriotism as the Core of National Spirit

With patriotism as the core, the national spirit inspires college students with the glorious course of the Chinese nation's self-improvement and perseverance, and the sense of mission and responsibility of realizing national rejuvenation. It provides them with a stable and lasting source of power in the process of transforming into innovative talents, as well as a driving force for their continuous innovation. This sense of responsibility makes them regard serving national progress, national development and human happiness as the starting point and ultimate destination of innovation activities. With this moral responsibility and humanistic care, we can truly feel the call of practice, the pulse of the times, and find problems, raise problems and solve them in rich and colorful practice, so as to create new achievements [17].

(3) Ideas of political and ideological Education should be renewed

In order to renew the teaching idea, efforts should be made to renew the educational thinking. Traditional political and ideological education of college students generally uses ready-made thinking mode, and under the guidance of this thinking mode, political and ideological education in Colleges and universities presents the characteristics of abstract isolation and solidification of education. To enhance the effectiveness of political and ideological education, we need to update the educational concept, change educational thinking, and guide political and ideological education with generative thinking. The generative way of thinking emphasizes that we should pay attention to human development and follow the law of human growth and development. political and ideological education of college students should pay attention to students themselves, regard students as the main body of political and ideological education, pay close attention to students'ideological trends at all times, adjust education content and teaching methods according to changes, make students the center of attention of education at all times, and then do a better job of political and ideological education and improve the effectiveness of political and ideological education.

6 Conclusion

Political and ideological education should play the role of guidance, development, construction and regulation in the training of talents under the background of computer network. However, it is affected and restricted by such factors as classroom teaching, evaluation system, teachers and external environment, and the effect of its function is unsatisfactory. The management mode of political and ideological talents for college students should be a systematic project that the whole society makes joint efforts. The factors and variables involved in the construction and improvement of the training system are numerous and complex. This paper only analyses the factors that restrict its function in the cultivation of innovative talents for college students from the perspective of internal factors and external environment of political and ideological affairs, and puts forward some ways and suggestions accordingly. The management mode, mechanism and system of political and ideological talents for college students have not been discussed in depth. Of course, the approaches and suggestions put forward in this paper are only for reference.

References

1. Baowei, J.: Research on improving the political and ideological qualities under the background of network information technology. In: Seventh International Conference on Measuring Technology & Mechatronics Automation, pp. 2–6. IEEE (2015)
2. Li, F.: Research method innovation of college students' political and ideological education based on cognitive neuroscience. NeuroQuantology **16**(5) (2018)
3. Qiu, J.: A kind of political and ideological education management platform design based on web technology. In: IEEE 2017 International Conference on Robots & Intelligent Systems (ICRIS) – Huai'an, 15–16 October 2017, pp. 1–4 (2017)
4. Kljun, N., Sprenger, M., Schär, C.: Frontal modification and lee cyclogenesis in the Alps: a case study using the ALPEX reanalysis data set. J. Hubei Correspondence Univ. **78**(1–2), 89–105 (2016)
5. Westerdale, J.: Walter Ruttmann and the cinema of multiplicity: avant-garde – advertising – modernity by Michael Cowan (review). German Stud. Rev. **40**(2), 435–437 (2016)
6. Ren, Z.-Y., Li, X.-X.: Reflecting on the wisdom of great minds to promote the, double first-rate construction. Univ. Educ. Sci. **123** (2017)
7. Ewees, A.A., El Aziz, M.A., Elhoseny, M.: Social-spider optimization algorithm for improving ANFIS to predict biochar yield. In: 8th International Conference on Computing, Communications and Networking Technologies, ICCCNT 2017 (2017). 8203950
8. Kaminski, M.M., Nalepa, M.: Suffer a scratch to avoid a blow? Why post-communist parties in Eastern Europe introduce lustration. Center for the Study of Democracy (2017)
9. He, L., Ding, X., Zhang, Z.: Research on extracurricular training and contributing factors of innovative talents-based on the investigation of University-Enterprise Clubs. Adv. Sci. Lett. **22**(8), 2057–2061 (2016)

10. Li, G., Fang, H., Zheng, L., Yang, B.: Research on optimal design of the injection mold parting direction based on preference relation. Int. J. Adv. Manuf. Technol. **79**(5–8), 1027–1034 (2015)
11. Huang, S., Wang, L., Liu, L., et al.: Nanotechnology in agriculture, livestock, and aquaculture in China a review. Agron. Sustain. Dev. **35**(2), 369–400 (2014)
12. Ai-Ju, L., Hua, T.: Social entrepreneurship education: new development in, college entrepreneurship education. Univ. Educ. Sci. **125** (2017)
13. Salys, R.: The pattersons: expatriate and native son. Russ. Rev. **75**(3), 434–456 (2016)
14. Kaplis-Hohwald, L.: Building the temple in verse: a reading of Juan de Jáuregui's Octavas in praise of Philip IV. Hispanófila **173**, 117–129 (2015)
15. Shankar, K., Elhoseny, M.: Trust based cluster head election of secure message transmission in MANET using multi secure protocol with TDES. J. Univers. Comput. Sci. **25**(10), 1221–1239 (2019)
16. El-Hasnony, I.M., Barakat, S., Elhoseny, M., Mostafa, R.R.: Improved feature selection model for big data analytics. IEEE Access **8**(1), 66989–67004 (2020)
17. Krishnaraj, N., Elhoseny, M., Lydia, E.L., Shankar, K., ALDabbas, O.: An efficient radix trie-based semantic visual indexing model for large-scale image retrieval in cloud environment. Softw. Pract. Experience (2020, in Press)

Analysis of Mapping Knowledge Domain on Health and Wellness Tourism in the Perspective of Cite Space

Cheng Wang[1(✉)], Huifan Luo[2], and Ronghong Wang[3]

[1] School of Tourism and Economy Management, Lijiang Teachers College, Lijiang 674199, Yunnan, China
995083294@qq.com
[2] School of International Culture Exchange, Lijiang Teachers College, Lijiang 674199, Yunnan, China
[3] Tourism Research Center, Lijiang 674199, Yunnan, China

Abstract. This research is aimed to reveal the situations of studies on health and wellness tourism, which will also provide theoretical guide for those who has been involved in or will be involved in this field. Taking Cite space as the analysis tool this research proceeds by means of online visual analysis of database, which has been collected through mapping knowledge domains including keyword burstiness, keyword cluster and sigma. It has been discovered that the research frontier of China's health and wellness tourism has been developing and diversifying including wellness tourism, towns of health and wellness tourism, destinations of health and wellness tourism, health and wellness tourism productions, health and wellness tourism industries, rural revitalization, ecological tourism and development paths etc. Conclusions have been arrived that in recent years, studies on China's health and wellness tourism have been started, at a stage of exploring, whose theoretical study lags behind practice. It is necessary to improve the quality of theoretical study of China's health and wellness tourism to guide its green development, ecological development and healthy development with unlimited potentials.

Keywords: Health and wellness tourism · Mapping knowledge domain · Research frontier · Cite space

1 Introduction

Outline of China's 2030 Health Plan noted that: "It is vital to promote the relations between wellness and retirement, tourism, internet, entertainment, fitness and food, in order to foster new industry, new forms of industry and new modes of health." [1] China National Tourism Administration issued Standards on Demonstrative Bases of National Health and Wellness Tourism (LB/T051—2016) (abbr. Standards) and authorized first 5 National Demonstrative Bases [2]. Furthermore, in March 2018, Ministry of Culture

© The Editor(s) (if applicable) and The Author(s), under exclusive license
to Springer Nature Switzerland AG 2021
J. MacIntyre et al. (Eds.): SPIoT 2020, AISC 1283, pp. 53–59, 2021.
https://doi.org/10.1007/978-3-030-62746-1_8

and Tourism of PRC authorized first 73 companies which launched national health and wellness tourism of Chinese Traditional Medicine. It is thriving a new mode of tourism industry: "tourism + health and wellness", which arouses people's concern on tourism and health theoretically and practically [2]. Through the bibliographic analysis, it explores that western countries have a thorough study on health and wellness tourism, which practically emphasizes more on how medical tourism policies influence travelers of medical tourism and residents of tourism destinations [3–9]. However, compared with the splendid history of people's wellness, China's health and wellness tourism has just launched with a quite short history and there's a long way to go to compare with western health and wellness tourism [3]. Taking "health and wellness tourism" as the keywords and retrieving in CNKI, this study finds no results reported on bibliometrics based on Cite space concerning health and wellness tourism.

Hence, this research is to show China's situations, trends and developments of studies on health and wellness tourism, in order to offer theoretical guides for those who has been involved in or will be involved in this field. To conclude, visual analysis online of health and wellness tourism is of great academic importance and immediate significance.

2   Research Methods and Data Sources

2.1   Research Methods [10–12]

Cite space, an online visual software, was created by Prof. Chaomei Chen of Drexel University. Taking Cite space (5.5.R2 edition) as the tool, this study conducted a case study on bibliometrics of China's health and wellness tourism in 2009, which analyzed the data collected through mapping knowledge domains including keyword burstiness, keyword cluster and sigma.

The meaning of 3 mapping knowledge domains is as follows: first, keyword burstiness is to identify the variation tendency of subjects according to the time sequence and then to find the research frontier in this field. Second, keyword cluster is to assemble key words of close relations which are to be clustered. Each key word is given a number and the key word with largest number is to be elected as the representative with a tag. The smaller the numbers are, the more the keywords are. Third is for sigma, formula is as follows:

$$\Sigma = (centrality + 1)^{burstiness} \tag{1}$$

2.2   Data Sources

This research sets CNKI as the retrieval source, health and wellness tourism as the retrieval subject, 2019 as the time of literatures, all journals as the literature resource, through which 275 Chinese articles are retrieved and 262 are chosen to be samples, in which 13 passages are canceled because there's re no authors or the subjects are not appropriate.

3   Online Visual Analysis Based on Citespace V [10–12]

3.1   Analysis of Keyword Burstiness

When Cite space (5.5.R2 edition) was set as the tool in order to calculate the literature database of health and wellness tourism, the parameters are set as the following: ① Configure the detection model:

$$f(x) = ae^{-ax}\left(a_1 \big/ a_0 = 2,\ a_i \big/ a_{i-1} = 2\right) \tag{2}$$

② The number of states = 1. ③ $\Upsilon = 0.4$, Υ ranged [0, 1]. ④ Minimum Duration = 1.

The calculation is in Table 1. The result indicates that among the strongest citation keyword burstiness, research frontiers of health and wellness tourism are sequenced

Table 1. Top 20 keywords with the strongest citation bursts

Keywords	Year	Strength	Begin	End	Timespan: 2009–2019
H & W Tourism	2009	8.6999	2015	2017	□□□□□□■■■□□
H&W Industry	2009	4.097	2015	2017	□□□□□□■■■□□
Forests H&W	2009	3.2412	2016	2017	□□□□□□□■■□□
Panzhihua City	2009	1.9623	2015	2015	□□□□□□■□□□□
H&W	2009	1.8788	2016	2017	□□□□□□□■■□□
Wellness Tourism	2009	1.5647	2017	2019	□□□□□□□□■■■
Distinctive Towns	2009	1.5131	2018	2019	□□□□□□□□□■■
Holiday Resort of H&W Tourism	2009	1.3343	2015	2015	□□□□□□■□□□□
Travelers of H&W Tourism	2009	1.1609	2017	2017	□□□□□□□□■□□
PRC	2009	1.1609	2017	2017	□□□□□□□□■□□
Sichuan	2009	1.1609	2017	2017	□□□□□□□□■□□
Hot Spring Tourism	2009	1.1609	2017	2017	□□□□□□□□■□□
Leisure Travel	2009	1.1609	2017	2017	□□□□□□□□■□□
H&W Tourism Destination	2009	1.1302	2018	2019	□□□□□□□□□■■
H&W Tourism Product	2009	1.1302	2018	2019	□□□□□□□□□■■
H&W Tourism Industry	2009	1.1302	2018	2019	□□□□□□□□□■■
Rural revitalization	2009	1.1302	2018	2019	□□□□□□□□□■■
Ecological Tourism	2009	1.1302	2018	2019	□□□□□□□□□■■
Products Development of H&W Tourism	2009	1.1302	2018	2019	□□□□□□□□□■■
Development Paths of H&W Tourism	2009	1.035	2017	2019	□□□□□□□□■■■

Notes: ① the form of key words is slightly changed according to the analysis. ② ■signified Keywords with the Strongest Citation Bursts. ③ Health and Wellness (abbr. H&W).

in descending order as the following: wellness tourism (1.5647), distinctive towns (1.5131), tourism destination (1.1302), H&W Tourism Products (1.1302), tourism industry (1.1302), rural revitalization (1.1302), ecological tourism (1.1302), tourism productions development (1.1302), development paths (1.035).

3.2 Evolutionary Path of Research Subjects

Citespace.5.5.R2 was set as a tool, to assemble and cluster literature data of China's health and wellness tourism. There are two standards to judge the results of cluster: ① Modularity is the Q, indicating a remarkable cluster structure when Q is higher than 0.3. ② Mean Silhouette is the S, indicating a reasonable remarkable cluster structure when Q is higher than 0.5. The bibliometrics of China's health and wellness tourism shows: ① Q = 0.6796, revealing a remarkable cluster structure ② S = 0.451, revealing a reasonable cluster structure. ③ There exists 7 clusters of China's bibliometrics. From the greatest to the least, they are development paths (#0), influence factors (#1), construction of health and wellness tourism demonstrative bases (#2), health tourism (#3), wellness tourism (#4), forests health and wellness tourism (#5), health and wellness industry (#6).

On the basis of cluster analysis, this research takes Citespace.5.5.R2 as a tool to calculate, whose calculation is two indexes: Centrality and Burstiness. Centrality and burstiness are compounded into a Sigma, which is to identify creative literatures and subjects (in Table 2) and analyze results of cluster in order to study the evolutionary paths and evolutionary situations of China's health and wellness tourism.

The main evolutionary laws are as follows: ① From the perspective of research subject, research on China's health and wellness started as the publication of the article the International Tourism Island: Health and Wellness Tourism in Hainan [13]. There are approximately two developing stages revealed in the research results: stagnating in 2009–2014 and rising in 2015–2019. The evolutionary trend is becoming diversified, in which some traditional research subjects are continuously substituted by new ones.

To be specific, firstly, parts of the research subjects last for a short time, for instance, some key words only existed in one year including Panzhihua City, Travelers of Health and Wellness, Health and Wellness Resort Area, China, Sichuan, Hot Spring Tourism, Leisure Tourism. In addition, little results have been reported in recent years, concerning health and wellness tourism, health and wellness tourism industry, and forests health and wellness tourism etc. Secondly, newly- developing subjects are diversifying into different branches as wellness tourism, distinctive towns, tourism destinations, tourism products, tourism industry, rural revitalization, ecological tourism, tourism products development, development paths etc. After studying the literatures of China's health and wellness tourism, a main problem has aroused that the research subjects are distracting, with a deficiency of profound and systematic studies. Chinese scholar Ren Xuanyu maintained in 2016 that there was no profound academic study on health and wellness tourism at home and abroad [14].

Table 2. Summary table sorted by sigma (CC ≥ 3)

Citation Counts (CC)	Burstiness	Centrality	Sigma	Keyword
82		0.57	1.00	H&W Tourism
25		0.46	1.00	Forests H&W
19	8.70	0.25	7.19	Tourism
17		0.47	1.00	All-for-one Tourism
10		0.11	1.00	H&W
8		0.00	1.00	Forests H&W Tourism
8	4.10	0.10	1.47	Industry
8		0.31	1.00	Development paths
7		0.00	1.00	Wellness Industry
6		0.10	1.00	Wellness Tourism
6	3.24	0.15	1.56	Forests
5		0.20	1.00	Rural Tourism
5		0.05	1.00	Panzhihua City
5		0.45	1.00	Medical &Wellness
5		0.06	1.00	SWOT
5		0.03	1.00	AHP
5		0.00	1.00	H&W Tourism Industry
4		0.13	1.00	Distinctive Towns
4		0.08	1.00	Development
3		0.01	1.00	Tourism Products
3		0.10	1.00	Space Structure
3		0.01	1.00	Health Tourism
3		0.02	1.00	Shandong Province
3		0.00	1.00	Rural Revitalization
3		0.00	1.00	Influence factors
3		0.00	1.00	Tourism Destination
3		0.00	1.00	Tourism Industry
3		0.05	1.00	Ecological Tourism
3		0.00	1.00	Industry Integration
3		0.05	1.00	Tourism Production Development

4 Conclusion

Conclusions have been come as the following: ① China's research on health and wellness tourism are diversifying, whose research frontiers consist of wellness tourism, distinctive towns, destination, products, industry of health and wellness tourism, as well as rural revitalization, ecological tourism, production development and development paths of health and wellness. ② There are 7 clusters of China's bibliometrics, including development paths, influence factors, construction of health and wellness tourism demonstrative bases, health tourism, wellness tourism, forests health and wellness tourism and health and wellness industry. ③ Studies on China's health and wellness tourism have been started and is proceeding, at a stage of exploring, whose theoretical study lags behind practice. It is necessary to improve the quality of theoretical study of China's health and wellness tourism in order to guide its green development, ecological development and healthy development with unlimited potentials.

Acknowledgments. This work was supported by Tourism Management, Key Major of Advanced Vocational Education of Yunnan (project number: YN 2018083).

References

1. Yang, X.-c., et al.: The spatial distribution characteristics and influence factors of health and wellness tourism resources in Fujian province. J. Fujian Normal Univ. (Nat. Sci. Ed.) **35**(5), 106–116 (2019). (in Chinese)
2. Liu, A.-l., et al.: Investigation and evaluation of health and wellness tourism resources in Liupanshui. J. Liupanshui Normal Univ. **30**(6), 18–23 (2018). (in Chinese)
3. Xie, X., et al.: The research of China's healthy tourism pattern in the regional characteristic towns. Ecol. Econ. **34**(9), 150–154 (2018). (in Chinese)
4. Snyder, J., Crooks, V.A., Johnston, R., et al.: Medical tourism's impacts on health worker migration in the Caribbean: five examples and their implications for global justice. Global Health Action **8**(S2), 1–9 (2015)
5. Crooks, V.A., Labonté, R., Ceron, A., et al.: "Medical tourism will…obligate physicians to elevate their level so that they can compete": a qualitative exploration of the anticipated impacts of inbound medical tourism on health human resources in Guatemala. Hum. Resour. Health **17**(1), 1–11 (2019)
6. Jain, V., Ajmera, P.: Quantifying the variables affecting Indian medical tourism sector by graph theory and matrix approach. Manag. Sci. Lett. **8**(4), 225–240 (2018)
7. Honda, N.H., Aoki, K., Kamisasanuki, T., et al.: Isolation of three distinct carbapenemase-producing Gram-negative bacteria from a Vietnamese medical tourist. J. Infect. Chemother. **25**(10), 811–815 (2019)
8. Mahmoudifar, Y., Tabibi, S.J., Nasiripour, A.A., et al.: Effective factors on the development of medical tourism industry in the West Azerbaijan Province, Iran: pattern presentation. Int. J. Med. Res. Health Sci. **5**(7S), 620–630 (2016)
9. Carly, J., Jeremy, S., Crooks, V.A., et al.: Exploring isolation, self-directed care and extensive follow-up: factors heightening the health and safety risks of bariatric surgery abroad among Canadian medical tourists. Int. J. Qual. Stud. Health Well-Being **14**(1), 161–3874 (2019)
10. Chen, C.: Information visualization. Wiley Interdiscip. Rev. Comput. Stat. **2**(4), 387–403 (2010)

11. Chen, C., Ibekwe-SanJuan, F., Hou, J.: The structure and dynamics of co-citation clusters: a multiple-perspective co-citation analysis. J. Am. Soc. Inf. Sci. Technol. **61**(7), 1386–1409 (2010)
12. Chen, C.: Science mapping: a systematic review of the literature. J. Data Inf. Sci. **2**(2), 1–40 (2017). Expert review
13. Wang, Z.: The international tourism Island: health and wellness tourism in Hainan. Hainan Today **12**(12), 12 (2009). (in Chinese)
14. Ren, X.: Health and wellness tourism: connotation analysis and development paths. Tour. Tribune **31**(11), 1–4 (2016). (in Chinese)

Application of Smart Retail Mode in Suning.Com

Jinzhi Zhai[(⊠)]

Liaoning Vocational Technical College of Modern Service, Shenyang, Liaoning, China
jinzhouzjz@126.com

Abstract. The core connotation of "smart retail" is to use Internet and Internet of Things technologies to feel consumption habits, predict consumption trends, guide production and manufacturing, and provide diversified and personalized products and services for consumers. Smart retail has realized the deep integration of real industries and new technologies, which breaks the situation of unilateral development online and offline. According to the actual needs of consumers, it has injected new vitality into the development of traditional retail through the mode of online and offline integration. Taking Suning.com as the research object, this paper analyzed the main problems faced by the application of smart retail mode, and put forward feasible development paths to promote the long-term development of China's traditional retail industry.

Keywords: Suning.com · Smart retail mode · Application

1 Introduction

With the change of customers' shopping preferences and the vigorous rise of online retail, the original marketing methods of retail enterprises are no longer adapted to the new changes. Copying the traditional retail development mode leads to many disadvantages such as high cost and lagging response. The "smart retail mode" is the key attempt to solve this major problem. The "smart retail mode" mode, which is still in the exploratory stage, needs benchmarking enterprises that dare to set up flags to do something. Therefore, this paper deeply studied the reform and innovation measures taken by Suning.com (hereinafter referred to as Suning), a traditional retail enterprise.

2 Importance of Traditional Retail Enterprises Opening Smart Retail Mode

The traditional retail industry is experiencing sluggish growth, sharp decline in profits and serious loss of customers. Reports of "closing tide" are constantly appearing in the media. In 2018, China's total retail sales of social consumer goods reached 38.1 trillion yuan, up 9% year-on-year, a slight decrease from the same period last year. Under the

J. MacIntyre et al. (Eds.): SPIoT 2020, AISC 1283, pp. 60–67, 2021.
https://doi.org/10.1007/978-3-030-62746-1_9

"stagnancy" of slowing down the overall growth rate of the retail industry, Suning's rapid expansion and outbreak against the trend in recent years have precisely grasped the opportunity of the retail industry reform. The scene that Suning vigorously promotes, the Internet, has become the consensus of the future development of the retail industry.

Facing the grim situation, traditional retail enterprises urgently need to innovate their business modes. At the same time, with the popularization of mobile Internet, E-commerce has been further developed. Online shopping has also been accepted by more and more consumers and has gradually become the most mature and popular way of shopping. It is also clear to us that, after years of development, with the improvement of income level and the change of consumption habits, the advantages of low-cost, low-price and convenient online shopping have begun to weaken, the growth rate of e-commerce has begun to slow down, the flow dividend of e-commerce enterprises has begun to shrink, the defects of e-commerce itself have begun to emerge, and have become the shackles restricting the development of e-commerce enterprises [1]. In this process, consumers pay more and more attention to the shopping experience of consumer terminals, which is also the key reason why e-commerce is not as good as traditional retail enterprises. Therefore, neither the traditional offline retail operation mode nor the single e-commerce mode is suitable for the future development of retail industry, and a new retail mode combining online and offline arises at the historic moment. In the new retail mode, retail enterprises no longer play a single role of earning intermediate profits, but expand the extension of goods and services, using new technologies, new logistics and new finance to create brand-new online and offline shopping scenes to provide customers with a more comprehensive shopping experience.

3 Current Situation of the Application of Smart Retail Mode in Suning

3.1 Brief Introduction of Suning

Established in 1990, Suning is the first home appliance chain enterprise listed on IPO in China. It was renamed "Suning Yunshang Group Co., Ltd." in 2013 and renamed "Suning.com Co" again in 2018 in line with the online APP strategy. Suning continues to promote its smart retail strategy, which insists on full category management, full channel operation and global expansion to realize a ubiquitous one-stop service experience [2]. Suning has pushed forward the construction of smart marketing projects around "more accurate service" and created a new marketing mechanism that is shared, open, "online, automated and smart". Successfully completed the research and development of basic functions of the project, built a basic framework for smart marketing, and supported the design, deployment, overall planning, implementation and evaluation of marketing strategies matching the four elements of "customer-product-event-contact". At the same time, the introduction of streaming data technology has greatly improved the response speed of marketing, and the change of production funds triggers the marketing function. By the end of 2018, Suning had more than 10,000 self-operated innovative Internet stores and outlets. Suning has entered the top three in China's B2C market through self-operated, open and cross-platform operations, leading the growth rate among mainstream e-commerce [3].

3.2   Current Situation of Application of Smart Retail Mode in Suning

(1) Smart retail strategy and selection background

The overall development of Suning has been at the leading level in the home appliance retail industry since its establishment in 1990 until Suning.com went online in 2009, but its growth has plummeted (Table 1 and Fig. 1).

Table 1. Suning's business background

	2019-09	2019-06	2019-03	2018-12	2018-09	2018-06
Profit margin on total assets (%)	5.01	0.95	0.05	6.34	2.95	3.17
Profit margin of main business (%)	14.34	13.65	15.61	14.63	14.34	14.07
Net profit margin of total assets (%)	5.40	1.00	0.05	7.09	3.23	3.38
Cost and expense profit rate (%)	8.47	1.48	0.10	5.77	3.52	5.43

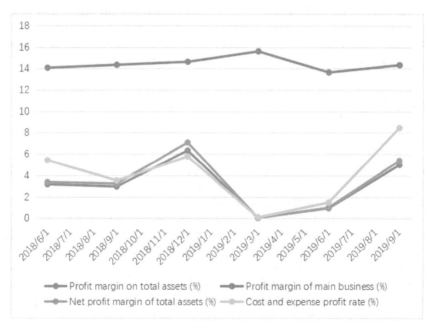

Fig. 1. Suning's business background from 2018 to 2019

From Suning's financial indicators, it can be seen that Suning's total asset profit rate, total asset net profit rate and cost profit rate decreased from February to March 2019 and increased after June. In order to adapt to the background of the Internet era and keep up with the changing consumption patterns of customers, Suning started the

O2O strategy in 2013, emphasizing that Internet retail is the main body to open online e-commerce platforms and offline physical stores to provide consumers with a full range of shopping experiences. In 2017, Suning further upgraded its six-year O2O strategy to an smart retail mode, fully integrate online and offline resources, realize the integration of information technology and physical stores, and enter a new era of smart retail mode. Smart retail mode refers to perceiving customers' consumption habits through Internet of Things technology and providing personalized products or services for customers [4]. The smart retail mode is another innovation of Suning's ten-year development strategy of "technological transformation and smart service". The wisdom of smart retail is mainly reflected in: smart retail provides services to consumers by sensing their consumption habits, realizes retail big data traction, and smart retail uses socialized customer service to realize personalized service and accurate marketing.

(2) Smart retail system

On the basis of O2O operation strategy, Suning's smart retail system can be summarized as "Two Major, One Small and Multi-specialization". "Two Major" refers to the entrance that uses Suning Square and Suning.com Square to create consumption scenes and carry retail technology applications and user experiences. "One Small" refers to connecting Suning's small shop to complete the last kilometer distance between the goods and the consumers' hands, and implanting Suning's products into the offline markets around the consumers. "Multi-specialization" refers to Suning's franchise stores. Suning makes the consumption scene complete and dense through the smart retail layout of "Two Major, One Small and Multi-specialization", thus meeting the different consumption needs of consumers at any time and space. Suning uses SMRAT smart marketing strategy to lay out and implement smart retail system. SMART marketing strategy represents the marketing strategy of S (scene), M (media), A (data), R (user) and T (industry chain) chain. Suning, with the help of O2O operation strategy, integrates all scenes online and offline, and relies on all media forms to obtain all data of consumers, thus covering the whole population, opening up the industrial chain, and achieving the effect of maximum integration of resources, as shown in Fig. 2.

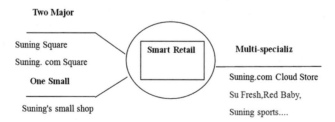

Fig. 2. Suning smart retail system

4   Problems in the Application of Smart Retail Mode in Suning

4.1   The Level of Using Network Marketing Tools is not High

The level of Suning's online marketing needs to be improved in two aspects: First, Suning does not attach importance to consumer satisfaction management. Suning carries out community integration and dissemination through online marketing methods such as fan interaction, prize-winning sign-in and promotion tools, but the actual operation result is that Suning integral mall BBS has more complaints about after-sales service, Suning has not responded to consumer complaints, and users are prohibited from giving business reviews on the review system. Second, Suning's use of online marketing tools to improve the level of customer traffic needs to be improved. However, in fact PPTV did not achieve the expected results. Suning did not convert users on PPTV into product sales by means of drainage and product trial, thus divesting this loss-making business.

4.2   Cost Pressure is Difficult to Relieve

Suning is facing the problem of high operating costs in the process of applying smart retail mode. In recent years, the state has issued a series of tax and fee reform policies, which have greatly reduced the operating costs of the retail industry [5, 6]. However, at the same time, with the rise of prices, the labor costs, logistics costs, financing costs and rental costs in Suning's operations are continuously rising. According to statistics, in 2016, the labor cost of retail enterprises above designated size in our country increased by 4%, while the management cost and financial cost increased by 2%. The retail industry was struggling and could not afford enough funds to carry out the transformation and upgrading of new retail.

4.3   Online Platforms and Offline Physical Stores Form a Zero-Sum Game

Suning's starting point for operating O2O strategy is to integrate e-commerce platform and physical stores, thus improving profitability. From the actual operation, it can be seen that the actual operation result is that e-commerce channels obviously divert customers from physical stores, and the proportion of online business in the total operating performance rose from 31.87% in the same period to 45.02%. Due to Suning's active competitive strategy, online gross profit margin remains at a relatively low level. The rapid growth of online sales led to a year-on-year decline in overall gross profit margin.

5   Improve the Application Strategy of Smart Retail Mode in Suning

5.1   The Application of Digital Technology has Provided Consumers with a Smart Retail Experience

Smart retail requires high technology. Smart technologies such as automation, intelligence and digitalization have created a brand-new Suning full-scene retail. In addition to the integration of online and offline development, Suning is also continuously committed to transforming and reshaping consumption scenes with smart technologies to improve consumers' shopping experience. As smart retail is a subversive change to traditional retail, Suning has realized the digitalization of retail scenes and the intellectualization of retail operations with the help of various digital technologies, such as big data and AI. Take Suning Store as an example. Its 24-h store not only has its own self-service cash register system, passenger flow analysis system, but also smart customer service system. In addition, the smart mechanical arm of Suning Store can sell cold drinks, coffee and packed meals in the form of self-service [7]. Suning Store can also realize the face swiping payment method of unmanned stores. Today, Suning Smart Retail perfectly adapts to the retail demand of CBD, large communities, office buildings and other large passenger flows, providing consumers with more high-quality and smart smart retail experiences. Suning focuses on the development of smart retail through online and offline channels, based on the application of more than 80 business engines and leading technologies such as cloud technology and big data. Suning Smart Home System integrates more than 100 brands and more than 5,000 smart home products. Consumers can also control through Suning Smart APP and Smart Speaker to form a set of interconnected whole-house smart solutions to create personalized smart families. In smart families, consumers can use home products according to their own preferences, making life full of personal colors and satisfying consumers' smart retail experience.

5.2 Reconstruct the Formation Mechanism of Online and Offline Prices

Starting from the viewpoint of market competition, the new price theory holds that the online price of the same product should be lower than its offline price in order to compete with competitors. In order to prevent the online channels from distributing the performance of offline stores due to price differences, on the one hand, it is possible to provide products specially earmarked for online funds, so that there are differences in product categories between online platforms and offline sales, which have different product modes but similar product functions; On the other hand, the biggest advantage of offline physical stores is to provide perfect customer experience service. Suning should improve the function of physical stores through good customer service from the perspective of meeting customers' shopping needs, and attract customers to physical stores for product experience and consumption [8].

5.3 Integration of Online and Offline to Promote the Development of Smart Retail

The integration of online and offline development is not only a response to the call of national policies, but also a realistic need for enterprise development. Online and offline consumption experiences are quite different and complementary. As consumers pay more and more attention to shopping experiences, Suning attaches great importance to the integration and development of online and offline retail formats. The integration of online e-commerce and offline entities is no longer a new concept. Breaking through the concept of online and offline is the trend to push forward the development of retail industry. The main connotation of online and offline dual-channel operation in the retail industry under the "new retail" is to be responsible for guiding traffic and placing orders online and for experience and consumption offline [9].

With the continuous maturity and development of information technology, there will be a large number of artificial intelligence to replace the original human resources, which can not only save costs to a great extent, but also improve work efficiency. It is also an inevitable path for the development and progress of e-commerce platforms to continuously promote the form of smart retail and gradually replace the existing retail form with smart retail and unmanned retail. Diversified retail forms will become the common mode of future retail. This new retail form can effectively improve the retail efficiency of e-commerce platforms.

No matter Suning's "Two Major, One Small and Multi-specialization" smart retail format group, it continuously creates a consumption environment and space across online and offline scenes. The smart retail mode of "One Body, Two Wings, Three Clouds and Four Ends" is also to better integrate online and offline with the aim of achieving full coverage of sales channels. Therefore, in order to develop smart retail, it is very important to integrate online and offline development, and it is also a link that retail enterprises should pay most attention to.

6   Conclusion

In the new retail era, traditional retail enterprises need to adjust their business philosophy. Only by combining online and offline can they gain a firm foothold in the highly competitive market. Suning accurately grasped the new situation of e-commerce development in its business process. A full-scene intelligent retail ecosystem has been constructed through taking consumers as the core, accurately positioning target consumers, accessing traffic to enhance competitiveness, and opening up social and community retail. It realizes full coverage from online to offline and from cities to villages and towns, and sets up a smart retail scene that can be seen and touched at any time for users. It meets the needs at any time, any place and any service, provides consumers with a better shopping experience, and further promotes the steady development of traditional retail and e-commerce.

References

1. Pantano, E., Priporas, C.-V.: Managing consumers dynamics within the emerging smart retail settings: Introduction to the special issue. Technol. Forecast. Soc. Change **124**, 225–227 (2017)
2. Hammond, R.: Smart retail: how to turn your store into a sales phenomenon. Diplomasi Revolusi (05), 78–86 (2013)
3. Vazquez, D., Dennis, C.: Understanding the effect of smart retail brand – consumer communications via mobile instant messaging. Comput. Hum. Behav. **77**, 425–436 (2017)
4. Wu, X.: Practice and reflection on the comprehensive practice course of "entity operation". Henan Educ. (Vocat. Adult Educ.) (12), 20–29 (2019)
5. Chen, Y.: Chairman of BOE (BOE): harmony and symbiosis, empowering scenes. Tsinghua Manag. Rev. (12): 23–27 (2019)
6. Li, Z.: Opening and enabling, Suning retail cloud brings new consumption vitality to sinking market. Electr. Appliance (12), 18–20 (2019)

7. Sun, W.: Suning's full scene smart retail will continue to grow in six directions. Chin. Foreign Manag. (12), 62–63 (2019)
8. Qiu, Q.: 2019 Smart Retail TOP100. Internet Wkly. (22), 32–35 (2019)
9. Zhang, X., Han, Y.: Research on the energy distribution and tendency of blockchain in the E-commerce field. Int. J. Front. Eng. Technol. **1**(1), 88–97 (2019)

Construction and Development of High-Tech Smart City

Quan Liu[✉]

Guangzhou College of Commerce, Guangzhou, Guangdong, China
quanliu82info@163.com

Abstract. Information technology is becoming more and more mature, and the construction of smart cities can achieve accurate economic and social development. Modern electronic information technology, represented by Internet of things technology, is the key force and core technology for building smart cities. This paper firstly studies the characteristics and connotation of smart cities, and then draws the main characteristics of smart city development. Then it focuses on the IoT from the aspects of architecture, transportation, tourism, and medical care briefly analyze the technology in building smart cities.

Keywords: Construction · Development · High-tech smart city

1 Introduction

The development of intelligent city construction is in full swing in our country at present stage, especially in the rapid growth of social economy, there will be more and more people crowded into cities, thus strengthen the represented by the IoT application and innovation of modern electronic information technology, to become the top priority of construction wisdom city [1]. Unlike agricultural and industrial societies, human beings mainly produce smart products in a smart society. Human physical and mental power will be liberated greatly, innovative thinking and emotions will become the main aspect of human values, and human beings are moving towards the realm of "freedom". Because cities are the main homes of modern human society. Therefore, smart cities have become the main carrier of smart civilization and have far-reaching strategic significance [2].

Smart city is a further extension of the digital city [3]. Use big data, cloud computing, IoT, optical network and other technology to restructure and optimize the core system of urban operation. With the continuous acceleration of the global informatization process, governments and related organizations around the world have issued grand plans [4]. Today, more than 100 cities in the world are vigorously promoting the practice and exploration of smart cities. The Chinese government has also kept up with the pace and actively promulgated relevant policies [5].

This paper firstly studies the characteristics and connotation of smart cities, and then draws the main characteristics of smart city development. Then it focuses on the IoT from the aspects of architecture, transportation, tourism, and medical care briefly analyze the technology in building smart cities.

J. MacIntyre et al. (Eds.): SPIoT 2020, AISC 1283, pp. 68–73, 2021.
https://doi.org/10.1007/978-3-030-62746-1_10

2 Construct the Framework and Characteristics of Smart City

2.1 Build the Architecture of Smart Cities

Smart city refers to the application of modern information and communication technology and other scientific and technological means in urban construction to enable the city to operate, construct, innovate and make decisions more intelligently, to maximize urban operation benefits and achieve sustainable development. Smart city is to change the interaction mode of various resources in the city through the application of information technology, such as IoT, cloud computing and big data, and make rapid and intelligent responses to various demands of urban traffic, people's livelihood, environmental protection and security, so as to improve the efficiency of urban operation, as shown in Fig. 1. It is field of future urban form and the development of various countries [6].

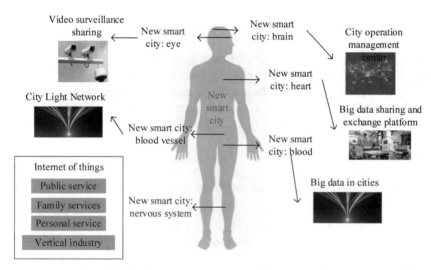

Fig. 1. Operation concept map of new smart city

Network information technology is the supporting foundation of smart cities. On this basis, various types of application software for smart cities have been developed, mainly including three major areas: industry, people's livelihood and management. As a leading development area, network information technology has a direct impact on the smart city construction process. The smart infrastructure and public services related to the people's life and consumption in a smart city are collectively called smart people's livelihood. Smart shopping, smart transportation and other fields are directly close to the mass consumer market. Therefore, development is rapid.

2.2 Main Features of Smart City Construction

IBM believes that smart cities have three main characteristics, namely perception, interconnection and intelligence [7]. It can also be summarized as fully interconnected, thorough perception, extensive integration, collaborative operation, incentive innovation and intelligent services. However, the construction of a smart city mainly includes the following four characteristics [8]:

(1) Broadband economy

Broadband economy is the foundation of smart city construction. Broadband networks provide opportunities for small, remote cities to move to economic centers.

(2) Knowledge labor

The knowledge economy's rapid development can promote the progress of constructing smart city, and knowledge labor is the core factor to promote the knowledge economy's rapid development.

(3) Broad integration

Smart cities are closely related to frequent human activities, constructing a sound and perfect institutional system for people's lives and production. In addition, make full use of the physical digital infrastructure to achieve the deep integration and high utilization of urban social, economic, living, cultural and other social information and environment, resources, water quality, transportation and other physical information.

(4) Cooperative operation

Collaborative operation closely connects and cooperates with each other and runs in an orderly and relaxed manner.

2.3   Overview of Smart City Construction

Since the beginning of this century, especially in the past 10 years, a new technology has been continuously developed, laying the hardware foundation for smart cities [9]. Many countries develop network information technology and take the lead in developing smart cities firstly. Many leading companies using network information technology have been formed in Silicon Valley, such as Microsoft, Apple, and Google. United States and Japan have become examples of smart cities.

China plays a great importance role in the construction of network information technology in smart cities, and network technology companies have emerged in Beijing, Hangzhou, Shenzhen and other places, such as BAT, Huawei, ZTE, Inspur, etc. [10]. The Central City Working Conference was held during the Central Economic Working Conference at the end of 2015, which further clarified the need to build a new city. Because green cities can be achieved through the path of smart cities, human cities are the crystallization of human wisdom and must be demonstrated.

After years of practice, great progress has been made in building smart cities in some parts of China, with pilot cities such as Nanjing, Hangzhou, Yinchuan, and Guiyang leading the way. In general, the construction of smart cities in China is still in its infancy, and the development level is uneven [11].

3 Application of Iot Technology in Smart City

(1) Smart building

When we construct smart cities, the IoT is the key technologies. To put it simply, the Internet has created the IoT under the combination of communication networks, so its application has also become an extension and extension of the Internet. Figure 2 is a conceptual diagram of the IoT technology. Devices with sensing capabilities can be responsible for sensing and identifying various types of information in the physical world, and through the interactive transmission function with the Internet, can make information and docking between people and things smoothly, to complete the connection [12]. Under the IoT technology, they can store and calculate massive amounts of data information in a more intelligent way, so that whether it is production technology, cultural life, or other aspects of the city, it can be more intelligent, thereby promoting sustainable development of the city.

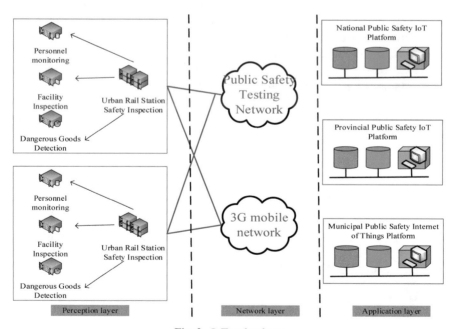

Fig. 2. IoT technology

(2) Smart transportation

Smart transportation refers to the use of the IoT technology to integrate various vehicles into a unified network, so that people and vehicles can establish perception and interaction in a unified transportation network, thereby effectively improving transportation efficiency, ensuring service levels, and achieving maximize resource benefits. Take smart parking lot as an example. By installing monitoring probes in the parking lot, and using sensing technology and mobile Internet technology to connect it to the main control room, a comprehensive monitoring of the parking lot is achieved. Monitoring equipment in parking will transmit and exchange data with the toll collection system and property management system to achieve effective resource sharing and improve the effectiveness of transportation management.

(3) Smart Medical

With the IoT to establish personal health records including all changes in the body from birth to death, and further expand the scope of sharing medical and health information resources. Therefore, when the patient is admitted to hospital, the doctor can quickly understand the patient's condition in a short time, and improve the efficiency and accuracy of diagnosis.

Remoting rescue, installing monitoring equipment in the ambulance by using IoT can help emergency personnel collect various vital signs information of patients. After that, the information is directly transmitted to the emergency command center or hospital emergency center through the CDMA network, to win precious treatment time for patients and improve their survival rate.

IoT can also be used to establish public health systems including medical devices and drug information in smart cities. By optimizing and integrating medical information resources and actively sharing them, the government can make relevant decisions at the fastest speed based on various medical emergencies, such as infectious diseases, and save precious time, as shown in Fig. 3.

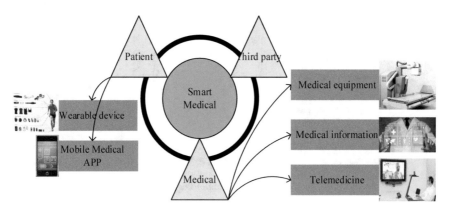

Fig. 3. Smart healthcare

4 Conclusion

With the advent of the information age, the importance of combining urban construction with IoT technology becomes more and more obvious. Facts have proved that smart buildings and smart cities can indeed provide convenience in many aspects by using IoT. When it comes to promoting smart cities, it is also necessary to implement a perfect combination with IoT technology, Internet technology, and other aspects in medical care, education, and housing. Further, promote the wisdom and intelligence of urban construction, strive to create a harmonious and comfortable living environment for urban residents, and truly achieve sustainable urban development.

References

1. Jiang, D.: The construction of smart city information system based on the Internet of Things and cloud computing. Comput. Commun. **150**, 158–166 (2020)
2. Ivanyi, T., Biro-Szigeti, S.: SMART CITY: an overview of the functions of city marketing mobile applications. Perspect. Innov. Econ. Bus. **18**(1), 44–57 (2018)
3. Naranjo, P.G.V., Pooranian, Z., Shojafar, M., et al.: FOCAN: a fog-supported smart city network architecture for management of applications in the Internet of Everything environments. J. Parallel Distrib. Comput. **132**, 274–283 (2019)
4. Komninos, N., Bratsas, C., Kakderi, C., et al.: Smart city ontologies: improving the effectiveness of smart city applications. J. Smart Cities **1**(1), 31–46 (2019)
5. Li, D., Deng, L., Gupta, B.B., et al.: A novel CNN based security guaranteed image watermarking generation scenario for smart city applications. Inf. Sci. **479**, 432–447 (2019)
6. Dai, C., Liu, X., Lai, J., et al.: Human behavior deep recognition architecture for smart city applications in the 5G environment. IEEE Netw. **33**(5), 206–211 (2019)
7. De Falco, S., Angelidou, M., Addie, J.P.D.: From the "smart city" to the "smart metropolis"? Building resilience in the urban periphery. Eur. Urban Reg. Stud. **26**(2), 205–223 (2019)
8. Dameri, R.P., Benevolo, C., Veglianti, E., et al.: Understanding smart cities as a glocal strategy: a comparison between Italy and China. Technol. Forecast. Soc. Change **142**, 26–41 (2019)
9. Ismagilova, E., Hughes, L., Dwivedi, Y.K., et al.: Smart cities: advances in research—an information systems perspective. Int. J. Inf. Manage. **47**, 88–100 (2019)
10. Chen, M., Gong, Y., Lu, D., et al.: Build a people-oriented urbanization: China's new-type urbanization dream and Anhui model. Land Use Policy **80**, 1–9 (2019)
11. Yigitcanlar, T., Foth, M., Kamruzzaman, M.: Towards post-anthropocentric cities: reconceptualizing smart cities to evade urban ecocide. J. Urban Technol. **26**(2), 147–152 (2019)
12. Liu, Y., Yang, C., Jiang, L., et al.: Intelligent edge computing for IoT-based energy management in smart cities. IEEE Netw. **33**(2), 111–117 (2019)

Design and Implementation of Self-service Tourism Management Information System Based on B/S Architecture

Yue Meng, Jing Pu, and Wenkuan Chen[✉]

Sichuan Agricultural University, Chengdu, China
wenkuan_chen@163.com

Abstract. Tourism informationization is a trend in today's society. Tourists are no longer satisfied with traveling with tourist groups, but hope to experience their own journey in the form of self-help. Aiming at this demand, this paper develops a self-service tourism management information system based on the B/S structure, using object-oriented analysis and design. Through the self-service tourism management information system designed in this article, it is convenient for tourists who want to travel on their own, so that they can determine all related matters of tourism through the system, and improve the management level of tourism.

Keywords: B/S architecture · Self-help tourism · Information system

1 Introduction

With the increase of per capita income, tourism consumption has become a popular leisure and entertainment method. With the maturity of computer science, most tourism companies still adopt the traditional manual data management method, and the disadvantages of its management model are increasingly apparent [1, 2]. First of all, the tourism industry is a multi-industry industry involving "food, food, housing, travel, tourism, shopping, and entertainment" [3]. Due to the unreasonable scheduling of tourism resources by traditional tourism management methods, a lot of costs are spent on labor. The second is insufficient protection of customer personal data, and a large amount of information leakage has occurred [4]. Thirdly, it is impossible to deal with the needs of tourists in a timely manner, and the lack of a unified evaluation standard for the quality of tourism services is not conducive to the improvement of the tourism industry [5]. Fourth, due to the asymmetry of travel information, many travel agencies are unfamiliar with modern travel management information systems and are unable to promote the company's business through the Internet, resulting in the loss of a large number of customers [6].

Travel agencies want to make great progress in the fierce tourism market, they must use modern technology to improve their service levels. Compared with traditional methods such as the use of electronic documents and forms such as WORLD and EXCEL

J. MacIntyre et al. (Eds.): SPIoT 2020, AISC 1283, pp. 74–81, 2021.
https://doi.org/10.1007/978-3-030-62746-1_11

for tourism information management, the application of a tourism information management system enables tourism managers to achieve economic benefits. In addition, tourism activities are affected by a variety of natural, economic, political, and cultural factors. Therefore, how to make tourists understand relevant information about tourism destinations in a short period of time is also the key to the success of travel agencies.

2 Feasibility Analysis

2.1 Demand Analysis

As more and more people travel, the number of travel agencies is also increasing year by year [7]. These travel agencies arrange various tourism services for tourists and solve a series of problems during the journey of tourists. Travel agencies often contact industries such as accommodation, catering, and transportation to serve tourists together. However, this one-stop service does not satisfy all tourists. Some people prefer to travel freely and enjoy themselves. They generally learn about some travel information through friends' introduction and recommendations, or through the Internet, and then rely on their own travel experience to travel together. However, whether it is a travel agency or individual tourism, the acquisition of tourism information is crucial. The success of a tourism is directly related to the acquisition of tourism information. The amount of tourist information includes information on tourist attractions, historical and cultural backgrounds of tourist attractions, traffic conditions around tourist attractions, accommodations around tourist attractions, eating conditions around tourist attractions, local customs and conditions around tourist attractions, and precautions for tourist attractions. Travel agencies often have most of the travel information, but individuals have very few. This has led to unsatisfactory travel experiences brought by personal travel. To solve this problem, many tourist attractions have developed their own online platforms for tourists to browse. In order to facilitate the search of tourists, a tourist information system has been created, which integrates information of multiple tourist attractions and is a comprehensive tourist information management platform.

2.2 Technical Analysis

The self-service tourism management information system is designed as a management information system based on the B/S structure. Users can log in to the management information system on any host that has a browser and is connected to the Internet. All the operations of the system are completed by the server, and the client is used to display the results calculated by the server. Such a structure facilitates the operation and use of the user, and saves the cost of the client, so that the client only needs a low-configured machine to complete the system's functions, and this approach eliminates the possibility of the client to modify the data, which improves the system security. Since all data is calculated on the server side, the performance requirements of the server are relatively high. Only the server has enough computing power to be able to process all the clients' requests at once, and when the server fails, all clients of the system cannot continue to use it.

The B/S framework also facilitates system maintenance. When a system fails, the administrator only needs to resolve the server's failure to enable the system to run again. At present, there are many ways to alleviate the problem of excessive computing pressure on the server. Establishing a distributed system is one of the solutions: the server is not just a host, but multiple hosts cooperate to divide the work, and finally summarize their respective work. Process to get the final result. This way, when one host crashes, the other hosts can continue to work, which improves the robustness of the system to a certain extent and makes the system more fault-tolerant. But everything has its shortcomings [8]. This distributed method must solve the problem of cooperative work between different hosts, and must determine the sequence of data processing to ensure the accuracy of the calculation results. For the self-service travel management information system, since the number of customers operating simultaneously is not large, and the current hardware level server can fully cope with the simultaneous access of more than one thousand customers, the topology structure of the self-service travel management information system with B/S structure is shown in Fig. 1 [9].

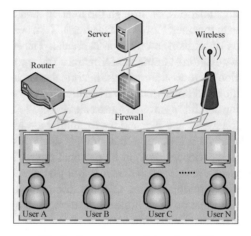

Fig. 1. System network topology [9]

2.3   Operational Analysis

In the Internet era, most network users generally have a certain network operation knowledge and can operate the functions independently. The management personnel of this system need a certain professional knowledge background, which shows that this tourism management information system has certain operability.

3   System Design

3.1   System Module Design

What is more important in designing a management information system is its logical structure. According to the previous needs analysis of the system, the logical structure framework of the self-service tourism management information system can be obtained [10]. The self-service travel management information system is designed in seven logical blocks, as shown in Fig. 2.

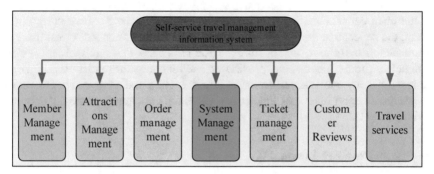

Fig. 2. System module diagram

Member management module: The member management module mainly completes the related functions of system user management. Member management mainly includes functions such as registration of member users, maintenance of member information, and recharge of member balances. The maintenance of member users is established on the premise of registration of member information: after the user completes registration, relevant information of the user is stored in the system database, and the user can modify this part of the information at any time. This information affects the functions users' use of the system.

Attractions management module: The attractions management module mainly implements the functions of attractions in the system, including the addition and modification of attractions, the maintenance of attractions information, the maintenance of tourist routes, and so on. In addition to the introduction of the attraction, the information of the attraction also includes the comment of the attraction and the tourist route. After viewing the attraction information, the user can see the introduction of the attraction, user reviews of the attraction, and recommended travel routes.

Ticket management module: The ticket management module mainly implements ticket-related functions. Including the addition and deletion of tickets, purchase of tickets and so on. Adding and removing tickets is a function that can only be used by administrator users, while purchasing tickets is a function for tourist users. After the administrator has set the tickets for the scenic area, the tourist users can purchase tickets for the scenic area.

Order management module: The order management module mainly implements order-related functions. The order management module includes the generation of orders, the maintenance of orders, the deletion of orders, and a refund function. The user can view the details of the order by querying the order information. When the order status is paid, the user can choose to refund the ticket.

3.2   System Main Class Design

In object-oriented methodology, everything is represented by objects, and classes are abstractions of objects. The design of an information system is inseparable from the design of a class. The class of the system directly reflects the main design of the system and the main connections between objects. According to the database design of the system, the conceptual design of the system can be further obtained. Guest users and system administrator users inherit from the user base class. Questions, orders, reviews, travel notes, photos, albums, etc. provide system functions for tourist users. Charging standards, tickets, question responses, customer service and other categories provide system administrator functions. Enterprises, attractions, and tourist routes all depend on scenic spots [11].

3.3   System Database Design

The database design of the system is the core part of the system design. The self-service tourism management information system based on B/S architecture is mainly the processing of data, so the design of the database will be the key content of the system. Database development on the platform not only facilitates the centralized management of system data, improves data reusability and consistency, but also facilitates the development and maintenance of system programs. The database table of the entire system includes the following parts: administrator information table, user information table, business information table, attractions information table, route information table, hotel information table, travel information table, order information table, user evaluation information table, and so on. The E-R of the main information table of the database is shown in Fig. 3.

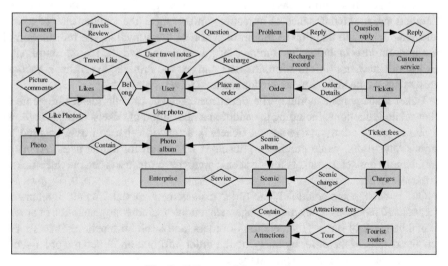

Fig. 3. E-R map of the attraction

4   System Function Implementation

4.1   Front-End Function Implementation

Take the main interface as an example to introduce some functions of the self-service travel management information system. There are mainly member management, attractions management, ticket management, order management, user reviews, travel services and system management.

Member management module is a module for managing all users in the system, including user registration and login, user information query and so on. User registration and login are the foundation of a management information system. To use the system, the user must first log in to the system. The management of user information is a major part of system management. The user information involves various functions and services of the system, so the maintenance of user information is very important. Figure 4 shows the user login and registration interface.

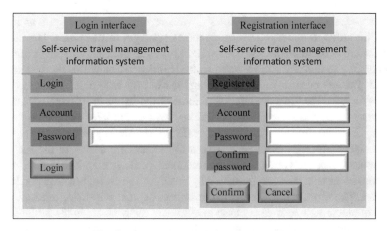

Fig. 4. User login and registration interface

Attractions management is the main function of the self-service tourism management information system. The attractions management module manages all attractions in a unified manner, making it easy for users to query and purchase tickets. In the management of attractions, the system categorizes them according to the provinces where the attractions are located. Among the specific attractions, there will be introductions to the attractions, tourist route planning, and related services near the attractions.

Ticket management mainly solves the problem of ticket purchase by users. Ticket management is mainly the interaction of data in the system database. After selecting a scenic spot, the user purchases tickets for the scenic spot through a ticket purchase operation. Depending on the user's level, the system gives users different discounts on ticket purchases.

Order management is a module for systematically managing all user orders. Order management mainly relies on the storage of database tables. The system reads the database table to personally display the order information that the user wants to know. The order management module depends on other modules in the system, and its function is only to manage all user orders. In order to complete their respective functions, other modules may call the functions of the order management module. Associated with the order management module are a member management module and a ticket management module. In member management, the member's personal information includes the member's order information.

4.2 Back-End Function Implementation

After logging in, the administrator enters the system management. Taking travel logs as an example, the administrator can add, delete, modify, view, authorize, and freeze the posts posted by members.

4.3 System Test

This process is an important link in the software development process and an important link that the software must go through in the early stages of its launch. The purpose of system testing is to make the software system run with high stability and reliability. In the process, try to find and solve system vulnerabilities to ensure that the system is more complete. The key purpose of implementing the test is to ensure that when the obtained system is delivered to the user, many problems of the user can be handled. During the implementation of the system design, the key task in the later stage is to be able to effectively test and process the obtained system. Considering that testers are involved in many aspects during the implementation of the specific design of the system, it is cumbersome to ask each other about the relationship. It is difficult to obtain functional effects consistent with the actual situation by means of subjective investigation by testers themselves. Therefore, it is necessary to perform a test process by using a specific process after the implementation of each phase to eliminate many errors in the initial situation. The specific purpose of the test implementation is as follows: first, the test processing for the error performance in the program; second, the specific test arrangement for the errors that are not shown in the program for the time being; and the third is for the program Test arrangements performed within error conditions not found by testers.

5 Conclusion

The tourism management information system based on the B/S architecture combines tourism management ideas with computer technology according to the current development situation and future development trends of the tourism industry. The systematic management of tourist management attractions information, hotel information, and route information makes the management of the tourism industry enter the ranks of information management, but the system still has many areas to be improved, such as the front-end interface module can be improved Some, the background data is more complete. In short, the B/S-based tourism management information system provides a simple and effective management information system to provide a reference for the development of the tourism service industry. It is believed that the tourism industry will receive more and more attention in the near future. The impact of people's lives will become more important.

References

1. Xiang, Y.: Set self-service sales and online customization in one of the product network marketing system construction and management research. Cluster Comput. **22**(4), 8803–8809 (2019)
2. Yi, R.: Research on development status and trend of unmanned wisdom hotel system based on artificial intelligence service—take Bangwei Company as an example. Financ. Market **4**(1), 36–43 (2019)
3. Lee, L.Y.S.: Hospitality industry web-based self-service technology adoption model: a cross-cultural perspective. J. Hosp. Tour. Res. **40**(2), 162–197 (2016)
4. Chiang, C.F., Chen, W.Y., Hsu, C.Y.: Classifying technological innovation attributes for hotels: an application of the Kano model. J. Travel Tour. Market. **36**(7), 796–807 (2019)
5. Taillon, B.J., Huhmann, B.A.: Strategic consequences of self-service technology evaluations. J. Strateg. Market. **27**(3), 268–279 (2019)
6. Considine, E., Cormican, K.: Self-service technology adoption: an analysis of customer to technology interactions. Procedia Comput. Sci. **100**(100), 103–109 (2016)
7. Dwivedi, Y.K., Rana, N.P., Jeyaraj, A., et al.: Re-examining the unified theory of acceptance and use of technology (UTAUT): towards a revised theoretical model. Inf. Syst. Front. **21**(3), 719–734 (2019)
8. Oh, H., Jeong, M., Lee, S., et al.: Attitudinal and situational determinants of self-service technology use. J. Hosp. Tour. Res. **40**(2), 236–265 (2016)
9. Considine, E., Cormican, K.: The rise of the prosumer: an analysis of self-service technology adoption in a corporate context. Int. J. Inf. Syst. Project Manag. **5**(2), 25–39 (2017)
10. Kairu, M.M., Rugami, M.M.: Effect of ICT deployment on the operational performance of Kenya Revenue Authority. J. Strateg. Manag. **2**(1), 19–35 (2017)
11. Ko, C.H.: Exploring how hotel guests choose self-service technologies over service staff. Int. J. Organ. Innov. **9**(3), 16–27 (2017)

Chinese Culture Penetration in Teaching Chinese as a Foreign Language in the Era of Mobile Internet

Ying Liu[✉]

Shandong Vocational College of Light Industry, Zibo, Shandong, China
LiuYing1983@haoxueshu.com

Abstract. The rise of the global Chinese fever has brought opportunities for the development of teaching Chinese, and it has also challenged the teaching models and methods of teaching Chinese. At present, Internet technology has achieved certain results in the change of teaching modes, the construction of online teaching platforms, and the development of APPs on mobile devices. The flexible use of Internet technology in teaching Chinese as a second language is conducive to teachers' improvement of teaching methods and the leading role of teachers; learners improve acquisition efficiency and use learning resources more reasonably and efficiently; and increase the reliability and validity of management and teaching evaluation; realize the modern teaching concept of taking students as the main body, teachers as the lead, and technology as the auxiliary.

Keywords: Teaching Chinese · Internet technology · Utility · Penetration of Chinese culture

1 Introduction

China's foreign exchanges continue to deepen, and its international status is increasing, the world is eager to learn about China, the trend of learning Chinese has spread, and the number of students studying in China has continued to increase. How to further improve international Chinese teaching and promotion when the "Internet +" era is coming is a problem that domestic researchers and experts have been working hard to explore. At the same time, the complexity and breadth of Chinese culture has determined the need for local universities to carry out corresponding Chinese cultural infiltration in teaching Chinese. Therefore, in the context of the "Internet +" era, how to innovate the teaching model of Chinese in combination with local characteristic resources, and to infiltrate regional culture into an efficient and practical teaching system that keeps pace with the times has become a profound research topic two. The important role of Chinese cultural penetration in teaching Chinese and culture are interdependent and affect each other. Language is an important part of culture, and the two are inseparable, especially in teaching Chinese as a foreign language. Today, the combination of language teaching and cultural teaching is the new situation, new task and new task faced by the flourishing

J. MacIntyre et al. (Eds.): SPIoT 2020, AISC 1283, pp. 82–89, 2021.
https://doi.org/10.1007/978-3-030-62746-1_12

development of the teaching of Chinese and the in-depth development of disciplines as China's national strength increases [1]. Professor Liu Yan, a famous scholar in teaching Chinese, has repeatedly emphasized: "Language teaching is inseparable from cultural teaching, and language teaching itself should include the cultural content necessary for successful communication using the target language." [2]. However, teaching Chinese is relatively for other language teaching, it is inherently special. The complexity and breadth of Chinese culture determines that local universities need to implement corresponding Chinese cultural infiltration in teaching Chinese. As far as the Chinese language is concerned, it has been passed down for thousands of years, and has a rich cultural heritage and complicated cultural content. From the perspective of the teaching object, learners in teaching Chinese are a special group. Before they came into contact with Chinese, they had formed a completely different cultural behavior and thinking mode from Chinese culture in their native language environment [3]. Therefore, it is necessary to enrich and improve the teaching content of Chinese through the penetration of Chinese culture. Cheng Shuqiu believes that the degree of adaptation of the Chinese cultural environment plays a decisive role in the successful acquisition of Chinese by foreign students [4]. While studying basic skills such as reading, writing, and translating, foreign students find entry points from Chinese culture, and truly feel the charm of Chinese culture through local dialects, diet, wedding and funeral customs, folk customs, folk beliefs, and dwellings. And the depth of Chinese, make Chinese teaching more lifelike, situational and communicative.

2 Application Status of Internet Technology in Teaching Chinese as a Second Language

2.1 Combine Online and Offline Teaching to Change the Original Teaching Mode

The application of Internet technology has changed the teaching of Chinese. In the past, the teaching method of "blackboard" and the teaching of "Yitangtang" by teachers has innovated the teaching mode to a certain extent. MOOC is a large-scale Internet-based open online course. "School Online" is currently the largest MOOC platform in China [5]. It offers elementary, intermediate and ancient Chinese language business courses. Flipping the classroom means that learners use the Internet to watch videos, listen to podcasts, and read e-books before the lesson to complete the learning of knowledge by referring to relevant material information through the network. Teachers can use precious classroom time to guide students and interact with students. Interact and communicate to promote personalized learning. At present, there are education platforms in China such as micro-classroom online, flip classroom, and easy-to-know flip classroom. O2O (Online To Offline) mode combines offline Chinese teaching with the Internet, making the Internet a platform for online teaching courses for offline trading, combined with the WeChat public platform to share online resources and interacting with offline classrooms Advantages, the introduction of multiple mixed, multiple interactive modes into Chinese teaching [6].

2.2 Network Teaching Platform to Broaden the Chinese Learning Path

Use the Internet to establish a network teaching platform for teaching Chinese. Throw the olive branch of Chinese to all parts of the world. Let learners experience "real" Chinese culture in the "virtual" of the Internet [7]. Online Beiyu is a professional website that can provide online Chinese distance education and non-degree education worldwide. Learners can obtain credits and degrees through platform learning and assessment. In addition, "Easy to Learn Chinese", "Internet Confucius Institute" and "Great Wall Chinese" are comprehensive platforms for Chinese teaching and education. Internet technologies are used to create interactive Chinese learning environments for Chinese learners in all regions of the world, providing rich and personalized learning Chinese [8]. In addition to the above-mentioned online platforms built by domestic universities or relevant national departments, there are also some Chinese language education websites operated by enterprises, such as eChineseLearning, which are currently the world's largest online Chinese language teaching service providers. Hansheng Chinese, Huayi Chinese Online School and other institutions have launched their own online Chinese language training courses. These Chinese language learning online platforms run by individuals pay more attention to communication with each learner, and the courses and corresponding services are more targeted, providing Chinese learners with more diversified choices and broad development prospects.

2.3 Chinese Learning APP Improves Chinese Learning Efficiency

In addition to the PC terminal, the development of connected mobile devices such as mobile phones and PADs is also very rapid. In recent years, various Chinese learning APPs have sprung up. Whether it is a comprehensive "Chinese skill", "Hello Daily", or "Hello HSK" for HSK exams, and special training on Chinese knowledge, etc., the app enables learners to start Chinese learning anytime, anywhere, with flexibility Highly-targeted and fragmented learning methods are conducive to improving the learning efficiency of learners, as shown in Table 1.

3 The Practical Value of Internet Technology in Teaching Chinese as a Foreign Language

With its advantages, Internet technology can expand the breadth and depth of "teaching", increase the initiative and efficiency of "learning", and increase the reliability and validity of the teaching achievement test. Internet-based computer multimedia technology has provided tremendous technical support for teaching Chinese, as shown in Fig. 1.

3.1 Combining Audio-Visual and Mobilizing the Senses

According to the research on memory of cognitive psychology, if information can be combined with certain situations through visual, auditory and other sensory multi-channel inputs, the rate of forgetting will be greatly reduced. The development of the Internet has made it possible to create a realistic language environment across China.

Table 1. Chinese learning APP statistics

Name	Category	Main content	Date of initial release
Chinese Skill	Comprehensive	Vocabulary and Grammar	2014/8/3
Hello Daily	Comprehensive	Vocabulary, Reading and Listening	2015/3/27
Learn Chinese by Talking Learn	Comprehensive	Vocabulary, Reading and Writing	2015
Phoenix Chinese	Comprehensive	Vocabulary, Dialogue, Speech, Writing	2016/8/23
Hello HSK	HSK	Vocabulary, Reading, Listening, Grammar	2015/1/27
Art of Chinese	Special	Chinese Characters	2012/5/5
Upright Miles	Special	Voice (Tone, Pinyin)	2000/5/1
Skritter Chinese	Special	Chinese Character	2012/6/1

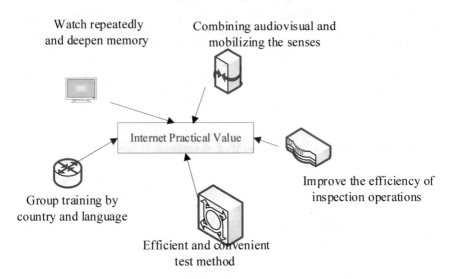

Fig. 1. Practical value of Internet technology in teaching Chinese

Through Flash animations, micro-teaching videos and even movies, the text and sound of pictures can be fully mobilized to fully mobilize the learner's sensory organs, so that learners can maximize accept and process the Chinese information learned [8]. The rich and colorful human-computer interaction method treats people as an information processing system that encodes, stores, and extracts sensory input, which is conducive to motivating students' learning motivation and obtaining better learning results.

3.2   Repeatedly Watch and Deepen the Memory

Learners can make full use of the videos and audios of teaching Chinese on the Internet for preview and review, and can click and watch repeatedly to repeat the study of the important and difficult points in each Chinese class until it is completely completed. understanding. To learn a language point, the learner needs to go through a process of repetitive stimulation and response to achieve long-term memory processing [9]. This iterative process can also delay the psychological anxiety caused by language barriers to learners, and help learners to adapt and overcome the discomfort caused by cultural conflicts.

3.3   Achieve Group Training by Country and Language Department

Whether it is group study in a Chinese university class or classroom group training, learners tend to choose partners with the same or similar mother tongue background for cooperation. Of course, this can reduce the barriers to communication at the initial stage, but when the learner's Chinese level reaches a certain level, communicating with learners with different mother tongue backgrounds is easier to point out the other party's errors due to negative mother tongue transfer, which is not conducive to intermediate and advanced stages [10]. Chinese learners in overseas classrooms have no choice of cooperative learning partners. Internet technology can provide learners with a wider range of choices for practice objects. Learners can interact online through various online interaction platforms, video and audio software, and other professional forums to achieve online interactions in different countries and languages. The practice of grouping is helpful for learners to develop the habit of thinking and thinking in Chinese and to speed up the progress of Chinese learning.

3.4   Improve the Efficiency of Inspection Operations

The forms and contents of operations arranged through Internet technology are more diverse, novel, and closer to reality. The way teachers use internet technology to check assignments is more flexible and convenient. Teachers can rely on free online social software such as Skype, QQ, Wechat to let learners upload audio and video for submitting dialogue exercises, increase learners 'interest in homework, and really practice elementary Chinese speaking in conjunction with students' actual lives, and when learners record their homework, they will also promote the Chinese element to the surrounding people to a certain extent, and expand the influence of Chinese.

3.5   Efficient and Convenient Testing Methods

Using the Internet technology, learners' learning can be detected through video, time-limited electronic test papers, and answer questions. For example, for beginners, especially zero-based learners, a computer-based test of correct pronunciation is the easiest, most direct and effective way to learn phonetics. Learners can answer the exam questions after carefully watching online instructional videos and completing simulation exercises. The test is submitted and the computer judges the test results of correct pronunciation and tone level. After you answer all the questions correctly, you can proceed to the next step. This kind of test is designed to test the learner's mastery of a specific concept. The test results are not only objective, but also fully motivate the learner's subjective initiative.

4 Innovative Research on O2O Chinese Culture Infiltration Model

The O2O culture teaching model is characterized by large amount of information, fast and real-time, which can effectively ensure that the content and form of teaching Chinese keeps pace with the times and is full of timeliness [6]. Among them, the empirical that best reflects its advancement with the times is the learning of new vocabulary on the Internet by Chinese learners [7]. Based on this, the O2O Chinese culture infiltration model designed by this project integrates offline traditional classroom teaching and extra-curricular practice activities with online teaching resources, fully utilizing the interactive advantages of offline classrooms and the advantages of sharing online resources. Meet the international students' mobile access to Chinese language and culture knowledge, and finally form an ideal teaching mode that is fragmented, personalized, and autonomous, as shown in Fig. 2.

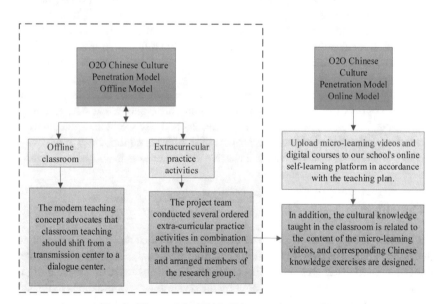

Fig. 2. The model of O2O Chinese culture teaching

4.1 Online Model of O2O Chinese Culture Infiltration Model for Teaching Chinese

The online quality courses are mainly based on this project's own micro-courses. At the same time, the latest "Culture Tour: China" supplemented by the online digital

culture course Unipus (the largest domestic university foreign language teaching platform in China) as the online Chinese culture learning content Blueprint [11]. Upload
micro-learning videos and digital courses to our school's online self-learning platform
in accordance with the teaching plan. International students can log in to computers,
mobile terminals and mobile smart devices anytime and anywhere for mobile access
and fragmented learning. In addition, the cultural knowledge taught in the classroom is
related to the content of the micro-learning videos, and corresponding Chinese knowledge exercises are designed. Through online self-learning, learn about relevant cultural
teaching content and complete online related assignments. Encourage international students to record problems encountered during self-study, communicate with teachers
and classmates online, and solve problems, or ask teachers in offline classes. Another
advantage of online learning is that students can use mobile smart devices to play back
teaching videos, consolidate assessments of key knowledge points, and conduct instant
evaluations of classroom effects.

4.2 Off-Line Model of O2O Chinese Culture Penetration Model for Teaching Chinese

The traditional teaching concept of offline classrooms believes that the classroom is the
process by which teachers impart knowledge to students. In teaching Chinese, teachers
instill a large amount of knowledge of Chinese into the classroom. The modern teaching concept advocates that classroom teaching should shift from a transmission center
to a dialogue center. Teaching is not a process of one-way information transmission
from teachers to students, but a process of information complementation and knowledge
generation through the dialogue between teachers and students, students and students.
Learning is not only a process of passively receiving information and strengthening and
consolidating, it is also a process in which learners actively select and process external
information, and actively construct internalization of information. Only in this way can
the effect of learning and utilization be achieved, otherwise it will only fall into the
dilemma of "dumb foreign language" [12].

5 Conclusion

In the context of the global "Chinese fever", research on the spread of Chinese language
is in the ascendant. The overseas teaching of Chinese language has an important role that
cannot be ignored in the spread of Chinese culture. Based on the era of "Internet +", meet
the challenges and opportunities brought by the development of the times to the spread
of Chinese language and culture, promote the smooth implementation of the "Internet
+" action plan, and highlight innovation, democracy, openness, freedom, interaction,
sharing, and common development. Innovation, co-governance, and win-win. Infiltrate
Internet thinking into the training of Chinese teachers, and improve the level of teaching practice; deeply integrate the innovations of the Internet into the field of teaching
Chinese, make full use of the Internet platform and information and communication technology to upgrade the original teaching of Chinese, and improve Chinese as a foreign
language Innovation in teaching. Absorb and utilize existing achievements, make continuous innovations and breakthroughs, promote the development of teaching Chinese,

effectively disseminate Chinese language and culture, and increase Chinese cultural and economic output. In short, the continuous updating of new ideas and new technologies in the "Internet +" era promotes the development of teaching model research. The O2O regional culture infiltration mode uses online excellent courses O2O mode and WeChat public platform O2O mode to integrate online teaching resources with offline traditional classroom teaching and extracurricular practice activities, giving full play to the advantages of sharing online resources and offline classrooms. Interactive advantage. This research model has played an important role in quickly stimulating learners' interest in learning Chinese, and has helped them to form a fragmented, personalized, and autonomous teaching model.

References

1. Ho, W.Y.J.: Mobility and language learning: a case study on the use of an online platform to learn Chinese as a foreign language. London Rev. Educ. **16**(2), 239–249 (2018)
2. Yang, Y., Qi, T.: Research on the influence of internet on extracurricular learning and life of English major college students. Theory Pract. Lang. Stud. **7**(8), 695–700 (2017)
3. Ai-qin, Q., Ping, F.: An analysis of the reasons for the prevalence of diurnal vocabulary in Chinese. J. Jiamusi Vocat. Inst. **1**, 180–181 (2018)
4. Liu, C.: Exploration on Italian language teaching by "we-media" in the mobile internet era: taking "Himalaya FM" app as an example. Stud. Lit. Lang. **16**(3), 68–73 (2018)
5. Le, A.S.H.: The studies of Chinese diasporas in colonial Southeast Asia: theories, concepts, and histories. China Asia **1**(2), 225–263 (2019)
6. Wei, W.: The integration of mobile micro-learning with college English classroom teaching in the internet plus era. J. Jiamusi Vocat. Inst. **2**(8), 236–237 (2017)
7. Ning, Z., Ling, Z., Zhenyu, T.: Academic libraries' subject services in the mobile internet era. Libr. Work Coll. Univ. **2**(2), 14–15 (2016)
8. Ye, G.A.N.: Study on college English mobile teaching based on constructivism. J. Zhejiang Univ. Sci. Technol. **1**(2), 14–15 (2019)
9. Zhang, S.J., Yu, G.H.: Mobile learning model and process optimization in the era of fragmentation. Eurasia J. Math. Sci. Technol. Educ. **13**(4), 3641–3652 (2017)
10. Meng, F.: New paradigm for China's college education reform in the era of the fourth industrial revolution. J. Educ. Res. **16**(3), 33–53 (2018)
11. Qiu, J.: A survey and analysis of mobile learning of local undergraduate college students. J. Lang. Teach. Res. **10**(6), 1245–1250 (2019)
12. Wang, X.: Cultivating online English learner autonomy in internet plus era: a DST perspective. Stud. Lit. Lang. **13**(5), 14–19 (2016)

Application and Outlook of Digital Media Technology in Smart Tourism

Jing Pu[⊠]

Sichuan Agricultural University, Chengdu, China
pujing_614@163.com

Abstract. With the development of digital media technology, the concepts of a smart earth and a smart city have been gradually recognized. In recent years, smart tourism has gained much more attention. Smart tourism is the product of the continuous improvement of the tourism standard and different type of information technology. This article analyzes the information art design situation from the connotation of information art design, and establishes the research context from the connecting between information application design, media, and user experience. Moreover, the media regulates the form and content of information design. This article takes the development trend of media integration as the starting point, and determines the research ideas for cross-media integrated design. Based on the design examples, the types and components of the information art design of the digital service platform of the scenic spot are summarized and demonstrated.

Keywords: Information technology · Digital media · Smart tourism

1 Introduction

The tourism industry is highly integrated and highly dependent on information. The characteristics of diversified subjects, complex environments, strong mobility, and large amounts of information determine that the development of all levels of tourism activities requires information technology as a support [1]. The world-famous futurist John Naisbitt predicted in "megatrends": in the recent decades, telecommunications in the tourism industry will lead the economic development effectively, but the organic combination among the tourism factors will grow into a more powerful strength. This technology will not only provide a broad stage for the development of the telecommunications and information technology industries, but also give the tourism industry tremendous development potential [2]. The advancement of tourism informatization, and experienced the initial stage of internal information management and simple release [3]. At this stage, Internet applications have not yet become widespread, and the construction of information sites is mainly used for internal information management [4]. After 10 years, the tourism information network began to cover and gradually popularize Internet-based application services. It can provide tourists with some basic single-event online services, such as ticket booking, booking and other services, and provide electronic payment [5].

J. MacIntyre et al. (Eds.): SPIoT 2020, AISC 1283, pp. 90–96, 2021.
https://doi.org/10.1007/978-3-030-62746-1_13

However, the practical scale of system construction Small, single function, lack of inter-activity. In the recent decades, the platform of tourism development integration has begun. During this period, smart terminals have become mainstream, and online travel booking, online payment, online marketing, and online services have enjoyed unprecedented growth. The original single-function service gradually became more diverse. Functional integration services change.

As an advanced stage of the development of tourism informatization, smart tourism will be the general trend of the broad enhancing of tourism in the future [6]. In a word, the increasing of tourism informatization has roughly gone through four stages, namely tourism informatization, tourism digitization, tourism intelligence, and tourism intelligence. The four stages are both continuous and advancing with the times. The development stages of tourism informatization, digitization, and intelligence are mainly based on the processing and dissemination of tourism information [7]. The starting point is to enhance the operation of tourism enterprises and facilitate the management of tourism agencies. The biggest feature of smart tourism is that tourists are the center, which reflects the "people-oriented" concept, through the acquisition and effective integration of information resources, through the setting of mobile terminals, to provide tourists with timely, comprehensive, rich and convenient travel information, is a continuation of the broad enhancing of tourism technology, but also development of tourism industry advanced stage.

Although the concept of tourism informatization has developed to a certain extent, and experts and scholars have made some progress in the study of smart tourism, on the whole, smart tourism is still in the exploration stage of its initial development internationally [8]. Experts and scholars are still at a shallow level in terms of theoretical research on smart tourism, elaboration of connotations, research on system architecture, and research on technical aspects of applied research [9, 10]. There is still little research on operation and management issues after the construction of smart tourism. The greatest theoretical significance of this research is to fill the research gaps in the field of operation of "smart tourism" after completion, and further enrich and develop the theory of "smart tourism".

2 Definition of Smart Tourism: "Smart Tourism" and "Smart City"

The concept of "smart tourism" is derived from the wisdom concept provided by "smart earth" and "smart city". IBM CEO Sam Palmisano first proposed the concept of a "smart earth" in 2008. "Smart Earth" is a general term for modern high-tech with "Internet of Things" and "Internet" as the main operating carriers, and it is an active solution to many major problems facing the world today [11]. The smart earth is a highly integrated and universal methods, and this part can help us to realize organic combination of entities. Later, IBM introduced the concept of "smart city". Simply put, smart cities are an advanced form of urban informatization, the next generation of innovation based on a knowledge society, and the full use of new generations of information technology in all walks of life in cities. Generally speaking, smart cities include public services such as smart transportation, smart education, smart healthcare, and smart power, as well as smart tourism industries, agriculture, and finance. Among them, "smart tourism" has become an important content, which is regarded as a part of the construction of "smart cities", which can be seen in Fig. 1.

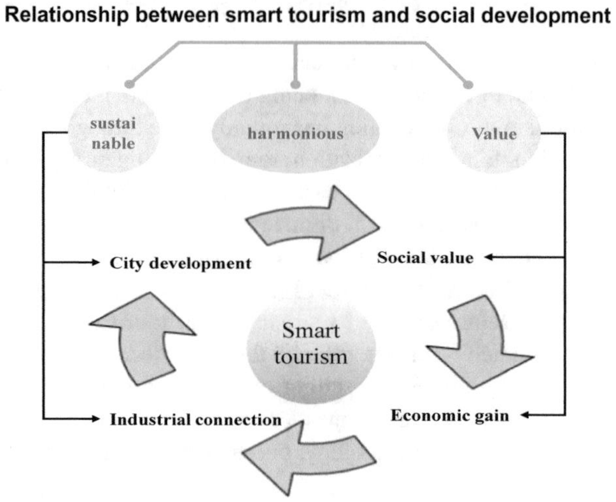

Fig. 1. Relationship between smart tourism and social development

In this respective, there are several steps to help building a much smarter earth and also some tourism applications. First, as a new concept of urban development, smart cities are driving the conceptual change, prompting people to break through the original single, mechanical development model, and highlighting the construction of a more intelligent tourism model; second, The practical exploration of smart city construction provides corresponding practical experience for the development path, development model, and infrastructure of smart tourism. In addition, as the tourism industry itself has a strong correlation, components of smart cities-smart transportation, smart communities, smart banks, smart cultural creative industries, etc.-can build a support platform for smart tourism. Regarding the concept of smart tourism, there is currently no unified expression. Based on the differences between smart tourism and traditional tourism, and the conclusions of our experts and scholars on the concept of smart tourism, we can draw the relevant concepts and connotations of smart tourism. On the whole, smart tourism can be understood in two levels: smart tourism in a broad sense and smart tourism in a narrow sense. In a broad sense, the so-called smart tourism refers to the ever-changing tourism needs of tourists.

In all aspects and stages of the development of the tourism industry, by exerting intelligent thinking, condensing the strength of a smart team, and using smart means to continuously meet the needs of tourists, so that Achieve low-cost, high-efficiency, personalized development results. This level of smart tourism emphasizes the full development and utilization of smart thinking to help tourists achieve smart tourism goals. Smart tourism in a narrow sense is explained from a technical perspective, that is, smart tourism is to use cloud computing, the Internet of Things, big data and other technologies to publish tourism resources in real time through electronic devices such as portable smart terminals through the Internet, mobile Internet, and communication networks Information on tourism, tourism economy, tourism activities, etc., as well as tourist-related information, to help information seekers to realize proactive perception and ease of use, to adjust and arrange work and travel plans, which is important to develop tourists interact with the network in real time, and allow tourism to enter the "touch era". In other words, smart tourism, as the result of the continuous development, is sustainable product of the tourism industry and information technology. Through the highly systematic integration of tourism resources, it can better provide services to tourists, destination residents, tourism enterprises and tourism governments. Smart tourism is not only a brand-new tourism concept, but also a brand-new tourism operation mode. In essence, smart tourism is to create a smart effect by introducing new concepts, applying new technologies, and operating new models to achieve the entire process of tourism activities, the entire process of tourism operations, and the entire chain of tourism services to create smart values.

3 Digital Media Technology in Smart Tourism

3.1 Cloud Computing and Smart Tourism

Cloud computing is a virtualized resource that can be accessed anytime, anywhere on demand through the Internet, and is dynamic and easily extensible. The "cloud" in cloud computing refers to the network that provides resources. Cloud computing has the characteristics of great flexibility, agility, scalability, and low cost. According to the service content, cloud computing mainly includes three major service models: software as a service (SaaS), platform as a service (PaaS), and infrastructure as a service (IaaS). According to specific deployment methods, cloud computing can be divided into three types: public, private, and hybrid cloud. The application of cloud computing in smart tourism is mainly manifested in six aspects: tourism cloud data centers, tourism industry information clouds, tourism software application clouds, tourism e-commerce clouds, cloud computing call centers, and cloud computing translation centers. The functions that tourism cloud computing can achieve are mainly divided into three levels: the tourism infrastructure service layer (T-IaaS), the tourism platform service (T-PaaS) layer, and the tourism software system service (T-SaaS) layer. The tourism infrastructure service layer (T-IaaS) can help the government and related enterprises build IT systems. In addition, it provides communication infrastructure services to facilitate related enterprises to develop tourism services. At the level of tourism software system services (T-SaaS), cloud computing provides system software services for tourists, governments and tourism enterprises. Among them, providing community-based software services for tourists, while providing a platform for mutual communication, communication, feedback, evaluation and complaints, it also includes some mobile smart terminal applications. Software service systems for governments and enterprises mainly include various internal management applications, tourism portal systems and evaluation systems, as shown in Fig. 2.

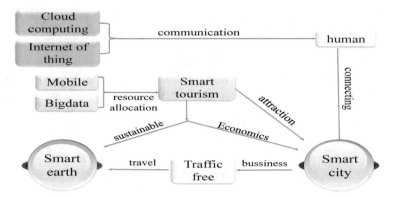

Fig. 2. Schematic diagram for the role of digital media technology played in smart tourism

3.2 IoT and Smart Tourism

The Internet of Things (IoT) is a kind of newly network technology. It mainly connects any items and items in the world, realizing the exchange, connection and communication of information, and can intelligently identify, locate, and track items and regulation. The Internet of Things uses a variety of sensing devices, currently mainly including global positioning systems, radio frequency identification devices (RFID), laser scanning and infrared sensors. At the same time, China's network industry will achieve a scale of more than 500 billion yuan, with an average annual growth rate of more than 10%. The Internet of Things technology can provide tourists with "full-process" tourist information services. Before traveling, the Internet of Things can provide tourists with comprehensive information related to eating, living, traveling, travelling, shopping, and entertainment. During the tour, tourists can enjoy intelligent navigation and guide services through the positioning function of the Internet of Things. Enhance the experience. After the tour, tourists can use the web2.0 technology provided by the Internet of Things to express opinions and share feelings through the network platform, and they can also feedback or complain about related issues. For tourism enterprises, IoT technology can form a virtual resource library to help tourism enterprises realize informatization construction, and it can also analyze the behavior of tourists to help enterprises achieve accurate Internet marketing. For tourist attractions, the Internet of Things can sense the space-time location and status of tourist resources in real time, and track, locate, monitor and manage the density of people in the tourist attractions, facilitating managers to guide passenger flow and rationally control passenger flow. In a word, IoT technology can effectively integrate tourism information resources, which could help realize the resource use, and also help tourism industry management department, maintaining a good tourism order, and deal with problems in the tourism process in a timely manner.

3.3 Mobile Internet and Smart Tourism

Mobile Internet, as a new type of network activity, mainly uses smart mobile terminals to obtain relevant transactions and services through mobile wireless communications. A complete mobile Internet includes three parts: terminal, software and application.

Among them, the terminal mainly includes smart phones, tablet computers, mobile Internet terminals MID (Mobile Internet Devices) and so on. From 2006 to 2012, the growth rate of China's mobile Internet market exceeded 80%. The 31st Statistical Report on Internet Development in China shows that by the end of 2013, the number of mobile Internet users in China had exceeded 618 million. After the year-on-year increase of 37.6% at the end of 2012, it increased by about 10%. China's smart terminal market share is also rapidly increasing, and China's mobile Internet market has a lot of room for development. The development of smart tourism has benefited from the promotion and popularization of mobile Internet and smart terminals, and "smart tourism" has changed from idea to reality. The combination of mobile Internet and smart tourism will innovate tourism applications, provide tourists with more convenient and efficient products and services, and provide a more personalized experience. Among them, Location-Based Service (LBS) is one of the breakthrough applications. LBS provides users with a value-added service. The application platform is GIS (Geographic Information System), which is a geographic information system. The way to obtain location information is through the wireless communication network of a telecommunications operator or other external positioning methods such as GPS (Global Positioning System), which is a global positioning system. Wait. LBS has strong applicability to the tourism industry and has great advantages in terms of tourist location information and location services. LBS-based technologies and related concepts have spawned many valuable intelligent terminal applications in the tourism industry. LBS provides an effective promotion method for merchants. Merchants can sell their products to tourists through information push function, send coupons, etc. to attract tourists. For tourist attractions, the terminal information display can be used to provide tourists with timely and accurate tourist information displays, and electronic tickets for scenic spots can be issued with barcodes and QR codes printed. Tourists can also download the APP application software through the mobile client to implement scenic area navigation, preview the real scene, and enjoy a comprehensive and multi-angle tour of the scenic area.

3.4 Big Data Technology and Smart Tourism

Big data is also named as large amount of data, which means that the number of data covered by internet has reached a significant level in accordance with the current mainstream software and tools. The information can't help the business operation and decision-making. Big data technology is featured as the followed points, that is, volume, speed, diversity, and low value. Big data technology can't be easily handled according to used data, but a professional processing of meaningful massive data, and find what is really valuable for decision-making. Big data is another subversive technological change in the information technology industry after cloud computing and the Internet of Things. The tourism industry generates hundreds of millions of tourists and resources, which are not only derived from external data of tourists, tourism companies and tourism industry management departments, but also from a lot of gaining tourists and places gotten from internal management of hotels and scenic spots. The data, complicated types, and low value density are in line for the improvement of the resources, which are important reasons for applying a smart tourist places. In terms of smart tourism services, through the establishment of data centers, the creation of user-end and server-side platforms, the use of advanced analysis techniques and methods to analyze and integrate the relevant data of the tourism industry, provide scientific and effective tourism consulting for tourists, and assist tourism Decision-making can also continuously improve tourism public information services and improve tourist satisfaction.

4   Conclusion

The development of smart tourism meets the needs of the development of urban informatization. On the basis of urban modernization, the functions of the city have been further extended and sublimated through new-generation information technology. The development of smart tourism involves the comprehensive operation of multiple systems. While applying smart technology to tourist attractions and related services, it should expand the scope of application in a timely manner, adaptively change the way of economic development, and promote urban public services and social management. The informatization development of the city enables cities to focus on intelligently discovering and solving problems, and has a stronger ability to innovate and develop, improve the soft power of the city, change the image of the city, and improve the quality of life of people.

References

1. Li, Y., Hu, C., Huang, C., et al.: The concept of smart tourism in the context of tourism information services. Tour. Manag. **58**, 293–300 (2017)
2. Jovicic, D.Z.: From the traditional understanding of tourism destination to the smart tourism destination. Curr. Issues Tour. **22**(3), 276–282 (2019)
3. Huang, C.D., Goo, J., Nam, K., et al.: Smart tourism technologies in travel planning: the role of exploration and exploitation. Inf. Manag. **54**(6), 757–770 (2017)
4. Del Vecchio, P., Mele, G., Ndou, V., et al.: Creating value from social big data: implications for smart tourism destinations. Inf. Process. Manag. **54**(5), 847–860 (2018)
5. Koo, C., Shin, S., Gretzel, U., et al.: Conceptualization of smart tourism destination competitiveness. Asia Pac. J. Inf. Syst. **26**(4), 561–576 (2016)
6. Brandt, T., Bendler, J., Neumann, D.: Social media analytics and value creation in urban smart tourism ecosystems. Inf. Manag. **54**(6), 703–713 (2017)
7. Park, J.H., Lee, C., Yoo, C., et al.: An analysis of the utilization of Facebook by local Korean governments for tourism development and the network of smart tourism ecosystem. Int. J. Inf. Manag. **36**(6), 1320–1327 (2016)
8. Huang, A.: The First International Conference on Smart Tourism, Smart Cities, and Enabling Technologies (the Smart Conference), Anatolia, vol. 30, no. 3, pp. 431–433 (2019)
9. Romao, J., Neuts, B.: Territorial capital, smart tourism specialization and sustainable regional development: experiences from Europe. Habitat Int. **68**, 64–74 (2017)
10. Chung, N., Tyan, I., Han, H.: Enhancing the smart tourism experience through geotag. Inf. Syst. Front. **19**(4), 731–742 (2017)
11. Koo, C., Mendes Filho, L., Buhalis, D.: Smart tourism and competitive advantage for stakeholders. Tour. Rev. **74**(1), 1–128 (2019)

Accounting Informationization in Computer Network Environment

Shirong Qin[⊠]

School of Accounting, Harbin University of Commerce, Harbin, China
sdqsr@163.com

Abstract. With the advent of the information age, people's lives have changed greatly, and computer networks have emerged. Although China is continuously developing and computers have been widely used, there are more and more problems, which seriously threaten Computer network security. In the computer network environment, accounting informationization is also facing new bottlenecks, so in this situation, the path of accounting informationization is constantly explored.

Keywords: Computer network · Accounting information · E-Commerce

The gradual improvement of science and technology has led to the rapid development of internet technology. China has entered an ever-changing era of e-commerce, and there is nothing wrong with it that has had an important impact on traditional accounting in the past. Under the computer network environment, e-commerce has developed rapidly. This situation the characteristics of accounting informatization have gradually emerged. Only by building an enterprise-based information system on the basis of the Internet, readjusting the more important business processes in the enterprise, realizing the integration of information systems in the true sense, and effectively controlling the operation of the enterprise, only in this way can we get the top spot in the increasingly fierce market competition.

1 Characteristics of Accounting Informationization in Computer Network Environment

1.1 Paperless

With the advent of the e-commerce era, the most basic feature of accounting information is paperlessness. The development of e-commerce has further promoted the networking of enterprise information systems, and the accounting information system has gradually developed in an open state. Accounting systems and other The business systems are effectively connected [1–3]. In the manual accounting environment, the most obvious feature of paper data is that the relevant business data of the company in the paper accounting

J. MacIntyre et al. (Eds.): SPIoT 2020, AISC 1283, pp. 97–102, 2021.
https://doi.org/10.1007/978-3-030-62746-1_14

account is easily left after being modified. However, e-commerce in the computer network environment is a kind of Paperless transactions, the paper-based accounting data in the past is properly saved using USB disks and hard disks as media. In addition, after the electronic data is modified, there will be no marks left, making it possible for criminals to take advantage of it. Data Security.

1.2 Openness

With the continuous development of the Internet and e-commerce, accounting informatization uses advanced science and technology to realize automatic data processing and sharing of accounting information and data resources. It transfers accounting information to internal systems of enterprises through network channels, such as production departments, business departments, etc. And external systems, such as banks, prompt internal and external related systems to collect corporate information directly. In addition, other departments within the enterprise and external agencies can collect relevant information directly through the network on the basis of authorization. This also shows that under the computer network environment the accounting information is open to development [4–6].

1.3 Diversification

Faced with the increasingly diversified information needs of users, it has led to the diversified development of accounting information. Accounting information is the display of the company's operating results, and the company's operating results and quotas belong to accounting information users. As the primary accountant of a company the task is to provide information to users to meet their needs. Although there are great differences in user information needs at different levels, there are also many similarities. Given the different interest relationships between users who use accounting information, the need for accounting information there is also a difference, which is also the main reason for the difference. Users are diversified due to different information need, which also creates the characteristics of diversified accounting information.

1.4 Intelligent

The characteristics of intelligent accounting information are mainly reflected in the automated processing of corporate accounting operations and the sharing of information and data resources, which can collect accounting information in real time and report in a timely manner, readjust the accounting model with advanced information technology, and modernize management models to the traditional Reorganize the accounting organization model and business, improve it, and use modern organization methods and management models for online office and virtual office.

2 The Impact of Computer Network Environment on Traditional Accounting

2.1 Impact on Accounting Assumptions

(1) Impact on the going concern assumption

Here, continuous experience usually refers to the assumption of accounting experience so that the company can develop in a certain period of time. It can effectively avoid corporate liquidation and bankruptcy. Based on the continuous development of e-commerce, accounting entities will also develop at high speed. Can't keep up with the times and will be eliminated, it is clear that companies are no longer suitable for use. Therefore, in the computer network environment, companies use liquidation accounting methods or optimize and create accounting systems and methods need to be clear [7–9].

(2) Impact on accounting entity assumptions

Accounting entities pretend to standardize corporate accounting work and implement clear regulations on the space. Under the premise of clear assets, liabilities, and the rights and interests of all personnel, they can be attributed to accounting factors above the space. However, e-commerce is temporary and not reasonable Plan the scope of the space, so in the computer network environment, the accounting entity should be redefined, and the accounting entity assumptions can also be modified.

(3) Impact on currency measurement assumptions

In the computer network environment, in the process of e-commerce transactions, both virtual companies and online banks no longer use banknotes for transactions, and banknotes are gradually replaced by electronic money, becoming the main form of online payment. Due to the generation of electronic money, In the daily registration of books, it indicates that the uniqueness of the currency of the measurement unit is gradually weakened, improving the operating efficiency and utilization rate of corporate and bank funds, and prompting them to complete capital decisions in a short period of time. Intangible increase in currency risk.

2.2 Impact on the Form of Accounting

The so-called form of accounting is that an enterprise processes various accounts and analyzes and compares them with accounting vouchers and books, and then verifies whether the statements are in line with the actual situation, and accurately analyzes the financial situation of the enterprise. In the accounting work, manual accounting is often used, and the work is not effective. However, in the computer network environment, the accounting efficiency has been greatly improved, and the accounting form has also changed significantly. Online banking has greatly improved the settlement of funds. Efficiency, while sharing information and data resources within the enterprise, enhancing the transparency of accounting work and improving the level of financial management, in addition to reducing the workload of accounting staff [10].

2.3 Impact on Payment and Settlement Methods

In the computer network environment, the settlement method has the characteristics of paperlessness, informatization and digitization, and the calculation method is more convenient and flexible. The Internet foundation, the establishment of online banking, compared with the traditional settlement methods in the past, online banking is more efficient and it is convenient and easy to operate. After completing the transaction, you can still issue the corresponding electronic voucher, which will help the accounting staff to organize the account information and improve the settlement effectiveness. Moreover, online banking is consistent with people's needs at this stage, but if you want online banking In order to give full play to its role, it should also ensure its security and confidentiality. In addition, electronic money used in online transactions or load-bearing often has shortcomings such as insufficient funds or repeated use. Therefore, it is necessary to ensure the safety of online banking and the electronic money for normal use, a sound network system security management system and a sound economic and financial management system should be formed in order to promote the continuous optimization and update of settlement methods.

3 The Construction Strategy of Accounting Information in the Computer Network Environment

3.1 Strengthening Awareness of Cyber Security Prevention

In the computer network environment, first of all, it is necessary to strengthen the safety awareness of financial personnel, strengthen the financial management of enterprises, and make relevant financial personnel aware of the importance of accounting information data and the serious consequences of leakage of accounting information. Second, the introduction of advanced science and technology Prevent potential accounting information risks within the enterprise, set up professional login accounts for financial staff's work content, and set access to select all, effectively avoiding the use of accounting information systems by staff in other departments at will, preventing the occurrence or damage of accounting information Finally, a firewall is installed on the computer that analyzes and processes accounting information, and external systems are strictly prohibited from accessing internal corporate systems. In addition, backups of more important accounting information data are effective to avoid data loss.

3.2 Improving the Level of Accounting Operations and Computer Operation Capabilities

Regardless of the nature of the enterprise, talents provide a source of power to promote its development. As far as the current situation is concerned, China's accounting professionals are relatively scarce, and the existing accounting personnel have a low level of business, and their accounting capabilities and management capabilities are low. In addition, the computer operation ability is also poor, and the above-mentioned reasons have a certain impact on the accuracy of the accounting work, and can not provide a reliable basis for enterprise managers to make decisions. First, improve the access standards of accountants, and at the same time Existing accountants have increased their training efforts, urged them to understand and master the latest accounting information, strengthened their computer operation capabilities, repositioned the accounting work, and cultivated the sense of ownership of accountants. Second, better create good quality for accountants. Social and school environment, and set up specialized training organizations to effectively guide the business level of accounting staff. Third, closely integrate computer networks with accounting to improve the effectiveness of accounting informationization, maximize the use of the advantages of computer networks and give full play to come out, enhance its level of accounting information processing, and rationally allocate accounting information resources.

3.3   The Establishment of Electronic Information Integration System

Increase the digital management of basic information of enterprise accounting informatization, improve the knowledge level and professional skills of staff, integrate core business processes in the enterprise, and integrate modern management concepts and technologies in the management process to achieve the current e-commerce era. The demand for accounting information. As an information system where the enterprise should be centralized and decentralized from the source, it is necessary to create a central database to make full use of the transaction execution status of each subsystem in the enterprise's accounting information system to improve and update the information in the central database in a timely manner.

4   Conclusion

Computer network is the guarantee condition for enterprises to realize accounting informatization. At the same time, the reasonable allocation of accounting information resources promotes the proper coordination of various information resources within the enterprise and promotes the development of accounting informatization. Under the computer network environment, accounting informatization is important for e-commerce. Realization has played an important role, and at the same time, the financial management level of the enterprise should be improved. The financial personnel's safety awareness can be continuously improved to improve the level of accounting personnel and operational capabilities. Establish an integrated electronic information integration system and create a central database In order to help enterprises gradually realize the informatization of accounting, improve and improve the third-party payment system, and continuously increase the market share of enterprises, so as to better promote sustainable development.

References

1. Guo, X.: Accounting informationization in computer network environment. Mod. Mark. (Inf. Edn.) (08), 60 (2019)
2. Li, Y.: Basic architecture and function realization of accounting information system based on computer network environment. Microcomput. Appl. **34**(03), 77–79 (2018)

3. Dai, X.: Research on internal control of accounting computerization in computer network environment. Educ. Teach. Forum (10), 197–198 (2017)
4. Zhang, Q., Zhang, X.: Design and implementation of network accounting information system for large and medium-sized enterprises. Manag. Inf. Syst. (06) (1997)
5. Fan, X.: Discussion on the optimization of accounting information system of group companies——taking Haike Group as an example. Finan. Account. Newslett. (34) (2018)
6. Tu, W.: Problems and countermeasures of modern accounting information system management. Econ. Res. Guide (03) (2019)
7. Wang, Y.: Research on the problems and countermeasures of accounting informationization in the internet + background. Mod. Market. (Bus. Edn.) (04) (2019)
8. Cai, J.: Implementation method and process of accounting information system. Heilongjiang Sci. (07) (2019)
9. Lin, M.: Probe into the security problems of accounting information system. Account. Chin. Townsh. Enterp. (05) (2019)
10. Zhang, J.: Problems in current accounting information system and improvement measures. Chin. Foreign Entrep. (18) (2019)

Mobile Phone GPS and Sensor Technology in College Students' Extracurricular Exercises

Zhikai Cao[✉]

Jianghan University, Wuhan 43000, Hubei, China
307473627@qq.com

Abstract. Fainting and sudden death often occurred in college students' Physical Fitness Test, which has worried all people in college and the whole society. The college students' physical education and the credit system for extracurricular exercise performed practically no function. Combination with mobile Internet, mobile phone GPS and sensor technology was adopted. There were altogether 3642 students as the research objects in this experiment. Results for the students' extracurricular exercise and physical fitness test without using mobile phone GPS and sensor technology were taken as experimental control groups. The experiment was divided into two stages. It was found that students' enthusiasm for extracurricular exercises could be enhanced by the application of mobile phone GPS and sensor technology; mobile phone GPS and sensor technology were helpful for analyzing and optimizing the frequency and effect of students' extracurricular exercises, and could monitor their exercise levels and scientifically improve their physical fitness.

Keywords: GPS and sensor technology · Extracurricular exercise · Exercise levels monitoring

1 Introduction

1.1 Research Background

At present, some colleges and universities in China have adopted measures such as morning exercises, daily running, and checking in at morning and evening, all in group. According to the results of physical fitness test, these measures have improved the physical health of college students to a certain extent However, at the meantime, there are a lot of problems: inflexible timetable, it cannot be adjusted according to personal timetable or schedule; potential safety hazard, group activities lead to greater safety hazards in various aspects such as transportation and walking; increasing the extra workload of teachers, there are tedious tasks such as organizing, supervising and performance managing, which affects Teachers' routine teaching work; at the same time, the exercise status of students cannot be actually monitored, and students' exercise cannot be effectively supervised [1–4].

J. MacIntyre et al. (Eds.): SPIoT 2020, AISC 1283, pp. 103–108, 2021.
https://doi.org/10.1007/978-3-030-62746-1_15

Based on the current situation mentioned above, in consideration of differences of campus and site specificity, in order to achieve comprehensive supervision of students' extracurricular exercise, and allow students to do their own extracurricular exercises at will, this study proposes to combine mobile Internet technology with mobile phone GPS and sensor technology, etc. In the end, students' extracurricular exercise habits can be changed so as to achieve the ultimate goal of improving their physical fitness.

1.2   Mobile Phone GPS and Sensor Technology

Almost all smart phones on the market have GPS and sensor modules. Mobile phone GPS is mainly used as positioning and navigation. A sensor is a detection device that can be used to sense the information, transform the information into electrical signals or other required forms of information in accordance with certain rules to meet the requirements for transmitting, processing, storing, displaying, recording and controlling. Based on the above two characteristics, developing a mobile program to apply it to students' extracurricular exercises can achieve a real-time monitoring of students' extracurricular exercises [5–8].

With the help of Internet technology, mobile phone application software can be designed to meet the expectations of the experiment. It can collect all student data after they registered and uploaded their personal information. When students participate in extracurricular exercises, they take their mobile phones, they open the APP and GPS before the exercise starts. Then, GPS will record student's positioning and sensor will record student's movement and determine the exercise status. After the exercise, the data will be automatically uploaded to the management platform, which is convenient for the college's supervision and management for student's exercise situation. Based on these data, the college displays student's exercise situation and the score of the physical education curriculum, to provide feedback and reminders to the students who failed to reach the exercise goal. The mobile phone GPS and sensor technology can be used to preliminarily supervise students' extracurricular exercises, display students' exercise scores, and provide exercise data. This technology can be fully applied to college students' extracurricular exercises through experimental comparison.

2   Experimental Methods and Steps

2.1   Experimental Methods

Comparative experiment was used here: extracurricular exercise and the physical fitness test results of those who did not have access to mobile phone GPS and sensor technology were taken as the control group, while the results of those who do have access to these technologies were taken as the experimental group. The experiment was divided into two phases. At phase one, the initial trial period lasted one semester, starting from September 2018 to January 2019, during which students completed registration and achieved their extracurricular exercise goals on schedule, without other requirements. This phase was mainly aimed at testing Students' initiative and practicability of mobile phone GPS and sensor technology. Phase two was for comprehensive testing, lasted one semester from

March 2019 to July 2019, during which, students must complete registration and achieve their extracurricular exercise goals. This phase mainly meant to solve the problems of the first phase and fully applied the mobile phone GPS and sensor technology. The experiment targeted students enrolled in 2017 and 2018, with a total of 3,642 students, 51.3% are boys while 48.7% are girls, a little difference in the male-to-female ratio.

2.2 Experimental Step

At the first phase, a set of simple mobile app for students and a simple back-stage management were adopted. Before exercising, the college uniformly set the running distance, speed requirements and time range for the one-time exercise. The students need to trigger the collection of positioning information while running. They turned on the timer to do the timekeeping. The system calculated the distance and the running speed based on the positioning and the time range. After exercising, the system automatically determined whether the running distance and speed uploaded by the students were within the required goals. It would be recorded as a verified exercise when the date met the requirements [9, 10]. Each student must achieve at least 10 verified exercises in a semester. After students continuously exercised in this way for a month, they could independently use this app and uploaded exercise data. Every day, many students ran with mobile phones, no need for being organized by teachers. Students could also decide by themselves when or where to exercise, as long as the requirements were met. This phase lasted one semester, from February 2019 to July 2019.

At the second phase, the mobile app and back-stage management were further optimized. Students must register and use the required mobile phone software. The sensor electronic fence was used so that students were ensured to exercise within the safe range. Orienteering mode was adopted to prevent cheating because students' routes were different. At the same time, sensor data analysis was applied to judge the difference between the simulator and the real person. In terms of the number and intensity of exercise, the logic of limitation of one or two verified exercises per day proved to be reasonable in order to prevent a series of operations such as overexercising and cheating that might appear at the first phase. This phase lasted one semester, starting from September 2019 to November 2019.

3 Experimental Results

3.1 Helped Improve the Participation Rate of Students' Extracurricular Exercises

The results of the experiment at the first phase were good according to experimental data. Students actively participated in extracurricular exercises. Some of them gradually developed habits of physical exercise so that it achieved the goal of extracurricular exercises. At the same time, new problems were discovered through actual investigations: in order to quickly complete the goal, some GPS deception tools were used to simulate exercising, students would exercise multiple times a day, some just went for imposters, some chose to run outside the campus, which was not safe, and so on.

At the second phase, extracurricular exercise gradually became part of students' lives, so that their physical qualities was really improved. Experimental data indicated that during this phase, an increasing number of students participated in running, and they became more enthusiastic about extracurricular exercises. Except for the extreme weather, there were always a lot of students exercising after class every day. Moreover, some data and information of great research value were found. For example, in terms of time, usually students run in the morning or evening. Among them, most ran at evening, accounting for more than 65%; in terms of gender, girls exercised relatively less than boys, generally 1 to 2 km, with a stable frequency of fixed times per week, while boys exercised heavier, ran quicker than girls, generally 1.5 to 3 km, lack of frequency stability. Some boys might have no verified exercise record in a week, but multiple records in the next week; some students were unwilling to exercise, they did it only for the task. However, after exercising for a period of time as required, they started to accept it, with frequency and amount of exercise obviously increasing.

By comparing the results of the first and second phases of the experiment with the students' extracurricular exercise when mobile phone GPS and sensor technology were not applied, it was found that problems of cheating and insecurity were solved with the help of the mobile Internet, mobile phone GPS and sensor technology. The experiment was important to both students and teachers, good for teachers' supervision, testing and management of extracurricular exercises. Students exercised by themselves so that teachers had more time to be well prepared for the good lessons. Overall, this method obviously improved students' motivation in extracurricular exercises (Fig. 1).

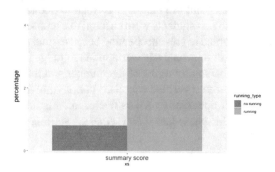

Fig. 1. Running participation of students with and without the technology

Students should be encouraged to do extracurricular exercises, which is an important way to enhance students' physical fitness. With the help of mobile phone GPS and sensor technology, the number of students exercising tripled.

3.2 Effectively Monitored Students' Exercise Levels and Instructed Students to Exercise Scientifically

The figure above showed that the relationship between them is like an inverted "U" shape curve. When a student exercised very less or very more a week, his test scores were not

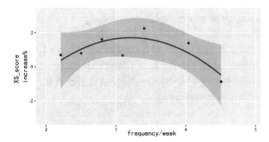

Fig. 2. Relationship between the weekly running frequency and students' physical fitness test results growth

likely to be impacted. The results of student's physical fitness test clearly showed how their physical health could be. In terms of the frequency of extracurricular exercises, it could be found that an average of 3 to 4 effective exercises per week had the most positive effect on enhancing students' physical health. When the frequency increased or decreased, there was less growth in students' physical health (Fig. 2).

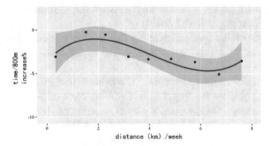

Fig. 3. Relationship between the weekly running distance and students' running performance growth

It was concluded from the figure above that the relationship between them is like a "N" shape curve. When a student's weekly running distance was less, his endurance increased slowly. When a student's weekly running distance was around 6 km, his endurance increased the most. After 6 km, the performance tended to be stable. Therefore, college should arrange extracurricular exercises based on this conclusion: make weekly running distance a reasonable number so that students get better endurance running performance and enhanced physical qualities (Fig. 3).

4   Conclusion

Judging from the experimental results, it is very practicable and reliable to apply mobile Internet, mobile phone GPS and sensor technology to extracurricular exercises. It provides a very convenient way for the actual data of students' exercises to be collected, for objective data to be quantitatively analyzed.

The data can be used to accurately understand and grasp students' exercises situation, and to establish various models based on big data analysis, which are the important theoretical basis for relevant decisions and measures. Except basic running, other types of exercises can be reflected in the form of data through sensors and smart Internet of Things in the future, which will be the foundational research for the further development of students' physical fitness monitoring.

References

1. Wang, Q.: Discussion on the approaches and significance of improving physical education efficiency under the big data background. Bull. Sport Sci. Technol. **26**(1), 91–93 (2018)
2. Zhang, B., Zi, X., Zhu, X., Ting, X.: Research on the construction of university sports information platform based on smart campus. J. Zhejiang Bus. Technol. Inst. **16**(2), 84–88 (2017)
3. Yang, R., Zheng, Z., Zhang, C.: Research on the effect of college students' independent extracurricular exercise based on running app. China School Phys. Educ. **5**(11), 29–32 (2018)
4. Ding, Y.: Development and implementation of physical education curriculum management under smart platform. J. Shandong Agric. Eng. Univ. **36**(10), 147–148 (2019)
5. Xiao, Y.: New thoughts on college students' physical fitness test in the era of media. Sports World (Scholarly) **01**(11), 166–167 (2019)
6. Wang, H.: Research on the application of big data in the integrated management and service of higher vocational sports under the background of informatization. Sports Vis. **26**(09), 215–216 (2019)
7. Li, G., Shi, L.: Research on the application of UWB-based positioning and tracking test system in middle and long distance running. China Sport Sci. Technol. **54**(6), 103–108 (2018)
8. Chen, J., Qiu, K., Zhang, Y.: Construction plan of smart sports platform under the background of "Internet+". J. Nanjing Sports Inst. **16**(03), 152–130 (2017)
9. Pan, Y.: Research on data visualization of sports apps in the era of big data. Phys. Educ. Rev. **37**(03), 24–39 (2018)
10. Chen, L., She, Z.: The impact of smart phone exercise software on college students' running performance, exercise attitude and behavioral habits——based on the 18-month use of "joy running circle". J. Hanshan Normal Univ. **40**(03), 66–74 (2019)

Design of Networking Network Model Based on Network Function Virtualization Technology

Jin Bao[✉]

Zhaotong College, Zhaotong 657000, China
ztbaojin1989@sina.com

Abstract. With the development of society, the living standards of our people have been continuously improved. More and more people are changing from walking to driving. China's auto sales are also increasing year by year. The purpose of this article is to design a networking model for the Internet of Vehicles based on network function virtualization technology. In terms of method, this paper mainly designs the preliminary structure of the model, and proposes to improve it under the PON technology, and uses the PON system, which is mainly composed of optical line terminal (OLT), optical distribution network (ODN), and optical network unit (ONU) composition. This paper proposes the design of FTTH networking mode based on EPON technology. Since FTTH requires a large number of splitters and PON ports, the splitter method is used in computer rooms or optical intersections. In terms of experiments, the validity of the FTTH networking model proposed in this paper is mainly verified. By comparing the pass rate of the four subjects of the driver's license with those of the driver's license, it is found that the pass rate of the vehicle-connected FTTH networking mode is higher than that of the driver's license, and exceeds 96%.

Keywords: Internet of Things mode · Internet of vehicles technology · Virtualization technology · PON technology

1 Introduction

Car + network is an application that belongs to the network. With the development of society, more and more cars enter our lives. A huge market base and people's needs have emerged. According to relevant data, the total number of vehicles worldwide has reached 1.4 billion in recent years. According to incomplete data statistics, more than 86% of them are light vehicles, and they are shown through annual data analysis in 2050. With the emergence of more than 200 million cars, the development of the times is the improvement of people's living standards, which is the basis for building a huge automobile market. With the increase of cars, the development of the Internet has become more and more common. In a way, it has many facilitation effects on world traffic order, and it is mainly reflected in applications such as on-board electronic sensor devices. This can effectively promote the modernization of information exchange processes between cars,

J. MacIntyre et al. (Eds.): SPIoT 2020, AISC 1283, pp. 109–114, 2021.
https://doi.org/10.1007/978-3-030-62746-1_16

roads, cities and people. Further trends in the automotive and information technology industries will evolve towards networks and intelligence. He is fundamentally different from traditional transportation systems. Reasonable application of the car's Internet connection provides a smarter and safer driving environment for better users.

In-vehicle social networks are the development direction of automotive networks, and with the understanding of modern cars, many cars have an entertainment environment. In this way, cars and owners become more interactive. You can also install a satellite navigation system on the display, play videos and view vehicle safety monitoring data. By applying it to in-car social networks, you can drive more safely. By connecting an on-board computer, you can also intelligently control the vehicle [1]. Solve problems such as forgetting to lock car pillars, shutting down the engine, and prompting smart users to deal with them in a timely manner. Although the diversification of in-vehicle intelligent applications provides endless prospects for the development of the automotive Internet, the lack of network scalability and security of the traditional mobile Internet has also contributed to the development of automotive Internet. Brings challenges [2]. Today's car dialing equipment is mainly connected directly to human networks through 3G/4G network cards. Each car can only get one dynamic dial IP. The network information of the second and third layers of the vehicle terminal cannot be effectively transmitted to the intelligent vehicle application platform in the cloud computing node. Increase the difficulty of identifying each device on the platform [3]. With the rapid development of the Internet, its security has become increasingly prominent. There are potential security risks in using the mobile Internet to implement automotive networks, including transmitting personal and vehicle data, attacking smart car application platforms, and failing to ignore security issues.

In this article, we designed the initial architecture of the model, developed it using PON technology, and used a PON system. The PON system is mainly composed of an optical line terminal (OLT), an optical distribution network (ODN), and an optical network. This paper proposes a FTTH multi-network mode design based on wireless optical network EPON technology. Due to the high demand for FTTH and spectrometers and PON ports, please use spectrometers in the computer room or optical intersection.

2   Method

The vehicle network-based network model is an innovative service model of cloud network fusion. Its main design is as follows.

(1) Through a dedicated network, assign a fixed dedicated network IP to each 4G dial-up card, establish a fixed line channel, and completely improve the security of vehicles and personal data [4, 5].

(2) Use tunnel technology or large-scale two-layer technology to implement a fixed IP strategy for vehicle terminals to meet the requirements of rapid identification of terminals, avoid unnecessary application platform development; achieve loose coupling and improve network flexibility.

(3) Combining cloud computing and network function virtualization technology to achieve the flexible arrangements required for network functions, avoiding excessive one-time investment costs, effective cost savings and control have been achieved [6].

The PON system is mainly composed of three parts. Optical Line Terminal (OLT), Optical Distribution Network (ODN), Optical Network Unit (ONU). Here, the optical line terminal (OLT) carries various service signals, sends the access part to the end user, and classifies the signals from the end user according to various services, which are different from each other [7, 8]. Must be sent to the service network. Optical distribution network (ODN) provides optical transmission equipment between OLT and ONU, and has four different configurations, namely star, wood, bathroom and ring [9]. In this configuration, the optical network unit (ONU) is responsible for exchanging information with the optical line terminal (OLT), and can provide user access rights by integrating or removing user network interface equipment.

The PON system interface includes data downlink and data uplink; in the case of data downlink, the OLT can send data to any ONU at any time, and in the uplink direction, the OLT can simultaneously send an ONU [10].

3 Experiment

3.1 Experiment Purpose

The validity of the FTTH networking model proposed in this paper is verified.

3.2 Subjects

Connected vehicle FTTH networking mode.

3.3 Experimental Design

We researched and compared the feasibility of our network model in driving test. Regarding the intelligent driving test supervision platform network: through telecommunications, the Internet APN network dedicated to intelligent driving vehicles is realized, connected to telecommunications, and the fixed IP vehicle network LNS platform is supported. Intelligent information services are deployed on cloud nodes. The car dial device is designed in GRE mode and only supports GRE. In the telecommunication information service cloud node, we opened a set of total virtual FW and total ROS, and opened a set of driving school FW and ROS for the driving license school according to the actual online plan and the actual online plan. These all correspond to test vehicles under the same training. The intelligent test and monitoring platform VDC opens a GRE tunnel between the car's on-board dialing equipment and the ROS of the driving license school, drives the fixed IP of the school's on-board equipment, and directly accesses the cloud nodes that deploy comprehensive information services. Guaranteed you can do it. The communication intelligent information service cloud node will open the MPLSVPN business through the large channel on the B plane of the IP Castle LAN, and contact government agencies, driving schools and VPN groups, and supervise the intelligent driving test and platform, government management and central supervision of the agency The centre and the test centre of the driving school are connected seamlessly.

4 Discussion

4.1 Comparison of Different Access Mechanisms

The current research results based on the above access methods and their corresponding performance pairs are shown in Table 1. As can be seen from the table, compared with other mechanisms, the CSMA/CA mechanism has the characteristics of high throughput, low delay, and low packet loss rate. Therefore, 802.11p is the most widely used communication protocol in the Internet of Vehicles. It is at the MAC layer. EDCA channel access mechanism formed by adding differentiated service priorities based on CSMA/CA is used.

Table 1. Comparison of channel access mechanisms

Access form	Existing access mechanism	Throughput	Performance	Packet loss rate	Advantage	Disadvantage
Centralized access	CDMA	Medium	Delay	Low	Large system capacity and good confidentiality	Multiple access interference, high communication cost
	TDMA	High	Low	Low	Low energy consumption and strong anti-interference ability	Large time synchronization overhead and complex frequency allocation
Distributed access	CSMA/CA	High	Medium	Low	Flexible and fair	Conflict

Although the EDCA protocol divides the service priority, the division method is fixed, and after the AC queue is assigned, the channel conflict, information collision, and transmission failure in the message sending process, the calculation parameters of the competition window are also fixed, so Can not meet the needs for the efficient transmission of a large number of messages in the Internet of Vehicles. First, redefine the priority division method, then set the network status threshold, and dynamically calculate the backoff window value according to the threshold to achieve adaptive and dynamic information access and improve communication efficiency.

4.2 Validation of Networking Model

In this paper, we experiment with a driving school on the channel access mechanism and use centralized access to CDMA. We compare the effectiveness with the real-life vehicle networking model by comparing it with real people.

From the data in Fig. 1, it can be seen that the passing rate of the four subjects of the driving license test under the driving of the network model is higher than the passing rate of the four subjects of the driving license test by those who already have a driving license. In many vehicles, vehicle problems may also occur, resulting in reduced data. On the network, there may also be local network instability, resulting in data decline. Among them, in Course 1, the passing rate under network mode driving was 96.8%, and the passing rate under driving with a driver's license was 92.7%. Among them, in the second subject, the passing rate of driving in the networking mode was 98.5%, and the passing rate of driving with a driver's license was 91.6%. Among them, in Subject 3, the passing rate under the network mode driving rate was 97.1%, and the passing rate under the driving conditions of a driver with a license was 92%. Among them, in Subject 4, the pass rate under the network mode driving rate was 99.2%, and the pass rate under the driver's license was 93.4%. In contrast, under the networking model proposed by this article, the passing rate of the driving test in four subjects is above 96%, which proves that it is feasible. But in actual application, how much improvement is needed.

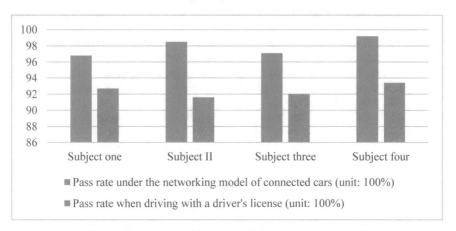

Fig. 1. Comparison with normal driving in a network model

5   Conclusion

In terms of method, this paper mainly designs the preliminary structure of the model, and proposes to improve it under the PON technology, and uses the PON system, which is mainly composed of optical line terminal (OLT), optical distribution network (ODN), and optical network (ONU). composition. This paper proposes the design of FTTH networking mode based on EPON technology. Since FTTH requires a large number of splitters and PON ports, the splitter method is used in computer rooms or optical inter-sections. In terms of experiments, the validity of the FTTH networking model proposed in this paper is mainly verified. By comparing the pass rate of the four subjects of the driver's license with those of the driver's license, it is found that the pass rate of the vehicle-connected FTTH networking mode is higher than that of the driver's license, and exceeds 96%.

References

1. Tang, H., Yuan, Q., Lu, G.: A model for virtualized network function deployment based on node-splitting in vEPC. J. Electron. Inf. Technol. **39**(3), 546–553 (2017)
2. Cotroneo, D., Natella, R., Rosiello, S.: NFV-Throttle: an overload control framework for network function virtualization. IEEE Trans. Netw. Serv. Manage. **14**(4), 949–963 (2017)
3. Sun, C., Bi, J.: HYPER: a hybrid high-performance framework for network function virtualization. IEEE J. Sel. Areas Commun. **35**(11), 2490–2500 (2017)
4. Li, H., Mao, X., Wu, C.: Design and analysis of a general data evaluation system based on social networks. EURASIP J. Wirel. Commun. Netw. **2018**(1), 109 (2018)
5. Pan, C., Zhang, R., Chen, L.: Research on variable current regenerative braking control strategy based on radial basis function neural network tuning PID control. J. Comput. Theor. Nanosci. **14**(1), 468–476 (2017)
6. Li, G., Zhao, G., Zhou, C.: Stochastic elastic properties of composite matrix material with random voids based on radial basis function network. Int. J. Comput. Methods **15**(1), 1750082 (2017)
7. Long, P.M., Sedghi, H.: On the effect of the activation function on the distribution of hidden nodes in a deep network. Neural Comput. **31**(12), 1–19 (2019)
8. Wu, W., Liu, M., Yu, S.: Current model analysis of South China sea based on empirical orthogonal function (EOF) decomposition and prototype monitoring data. J. Ocean Univ. China **18**(2), 305–316 (2019)
9. Liu, L., Cheng, X., Lai, J.: Segmentation method for cotton canopy image based on improved fully convolutional network model. Trans. Chin. Soc. Agric. Eng. **34**(12), 193–201 (2018)
10. Liu, J., Duan, S., Li, T.: Design of RBF neural network control system based on spintronic memristor for robotic manipulator. Yi Qi Yi Biao Xue Bao/Chin. J. Sci. Instr. **39**(8), 212–219 (2018)

Intelligent Media Technology Empowered Brand Communication of Chinese Intangible Cultural Heritage

Lei Cui[(✉)] and Xiaofen Shao

School of Marxism, Tongling University, Tongling 244000, Anhui, China
cuilei8208@163.com

Abstract. Intelligent media technology has the characteristics of immersive experience in all aspects, blended user experience and two-way interactive experience, creating a new pseudo-environment between users and information, capturing the attention of fast-paced information consumer audiences, and achieving accurate push. Expanded the intangible cultural heritage (ICH) communication groups and consumer subjects, and built a brand-new modern communication network for the communication of ICH, which is of great significance for building the national cultural soft power and improving the national image. Establish Chinese ICH brand communication system by improving the accuracy of branded communication, refining core brand symbols and values, strengthening the development and communication of brand creativity, and protecting the intellectual property rights of communication brands, but must also follow the rules of modern information communication to strengthen community protection of ICH, achieve an orderly and harmonious development.

Keywords: Intelligent media technology · Intangible cultural heritage · Brand communication · Pseudo-environment · Holographic image

1 Introduction

We begin this paper with a consensus: The intelligent media technologies represented by 5G and artificial intelligence integrate video and text, games and learning, touch control, etc, subverting the original social public opinion and discourse pattern, and creating a new situation between users and information that "moves with the heart". In a pseudo-environment, information can be transmitted at a high data rate, and the content is presented in a more vivid and concrete manner. It has established a high-level modern communication network for the communication of intangible cultural heritage (ICH), which is in line with the intelligent trend of the entire information production process.

ICH carries the cultural precipitation and essence of the country and the nation, and is the core cultural resource for the nation and the nation to become culturally confident. "Operational Directives for the Implementation of the Convention for the Safeguarding of the ICH" issued by UNESCO in 2018 states that "communications and media" can

J. MacIntyre et al. (Eds.): SPIoT 2020, AISC 1283, pp. 115–121, 2021.
https://doi.org/10.1007/978-3-030-62746-1_17

raise awareness of the importance of the public, foster social cohesion, show and express awareness of diversity, enhancing knowledge of local languages and culture, sharing of information and interactive exchange of information [1]. The development of intelligent media can achieve effective protection of ICH, and promote the communication of ICH, which is of great significance for building the national cultural soft power and improving the national image.

2 Background

The development of intelligent media technology has moved from concept to practice, which has brought about a huge revolution in the change of the media, and the medium for obtaining information has changed from smart phones to intelligent media. According to the 44th "Statistical Report on the Development of the Internet in China" issued by the China Internet Network Information Center (CCNIC) in June 2019, the scale of Internet users in China reached 854 million, the Internet penetration rate reached 61.2%, and the number of mobile Internet users reached 847 million, 99.1% of Internet users used mobile phones to access the Internet [2]. Smartphones have occupied the main channels for users to obtain information, and "terminals go with people, and information surrounds people" has become a new trend of information communication. The smart phone already constitutes the largest mobile communication network system. According to reports, China is expected to become the world's largest 5G market in 2025, users will reach 430 million. In the future, all information can be turned into data, which can be obtained with a single smartphone.

The short video of 2019 has become the "new social language of the data age." The survey report shows that since Tik Tok was officially launched in September 2016, the number of daily active users in China has exceeded 250 million and the monthly active users have exceeded 500 million. There are about 65 million short videos of various traditional cultures on the line, more than 4.4 billion likes, and more than 16.4 billion views. On April 2019, 1,372 national-level ICHs representative projects, 1,214 had disseminated relevant content on the Tik Tok platform, and the coverage rate exceeded 88%. These 1,214 items of national-level ICHs produced more than 24 million videos and more than 106.5 billion plays [3]. Tik Tok has actually become the largest platform for disseminating ICH. The feature documentary, as "Masters in the Forbidden City" and "The great Shokunin", defines the recording object as an ICH craftsman, faithfully records and shows the production process of unknown craftsmanship and exquisite artifacts, It brings a huge contrast experience to young audiences, and reminds the public that there is a heavy cultural accumulation behind each of the techniques that are on the verge of being lost, but it also reveals the helplessness of the successors.

Intelligent media technology has seized the public's desire to learn about ICH, especially targeted at the reality of the mainstream user group of young audiences, seized the attention of fast-paced information consumer audiences, achieved accurate push, and expanded ICH communication group has made the approval of young people. Diversified communication channels and display forms have greatly enriched the artistic content and communication forms of ICH, and built a brand-new communication system that meets the growing cultural needs of the contemporary public. Internal communication promotes the traditional culture of China to enhances the sense of national identity; external communication promotes communication and exchanges, enhances the sense of honor and pride of overseas Chinese, and provides a foundation for ICH communication and the expansion of consumer subjects.

3   Materials and Methods

Intelligent media technology has fundamentally changed the audience experience and the way of obtaining information. From the past text reading experience to the pursuit of immersive sensory experience, it presents the following characteristics in terms of audience information perception needs and information consumption experience:

3.1   All-Round Immersion Experience

Intelligent media technology can make users feel like they are on the scene through infinitely close to realistic image simulation, allowing users to have an immersive experience in all directions. Holographic communication technologies represented by Virtual Reality (VR) and Augmented Reality (AR) make users feel "immersive" when using the technology. Holographic images such as the "simulated world" created by VR equipment and "reality + virtual" created by AR equipment bring a new perception and cognitive experience through the superposition of reality and virtual scenes, triggering the audience to develop a rich imagination and reproduce the intangible culture The real historical geographic information and cultural image space of the heritage [4], activate knowledge experience and emotional experience. The application of intelligent media technology to the protection and communication of ICH will effectively improve the quality and efficiency of interaction, which will not only bring a new communication experience for the ICH to the public, but also increase the public's cognition.

3.2   Blended User Experience

Intelligent media technology emphasizes the user's psychological feeling and presence experience in the scene, produces a strong sense of presence, and improves the effectiveness of information communication. The holographic images in intelligent media technology have been applied in virtual exhibition halls or digital museums to manage information communication of folk arts and crafts in a digital manner and then present them in a living manner. For example, the "Global Kun Opera Digital Museum" and "Qingming shanghe Tu 3.0" interactive art exhibitions have digitally processed the documentary image data and combined with various immersive performances to reproduce the real scenes and create a 360-degree scene of teahouses, theaters, markets, neighborhoods, gardens, etc. [5]. It realized the reorganization of historical scenes and the fusion of spatial images, and improved user participation in space-time traversal and virtual worlds.

3.3 Two-Way Interactive Experience

Intelligent media technology, in comparison, pays more attention to the user's feelings and perceptions when they influence the interactive experience, and the communication and feedback in the two-way interactive experience between users and holographic images, users and virtual environments. Take the interactive experience of The Silk Road by VR to users, this sense of interaction can be analyzed from both the internal and external aspects of the video. The scene scheduling within the image triggers indirect feedback from the user, which is mainly achieved by creating an "environment", and then the user interacts with the "environment". The specific methods include adjusting the lighting shape to simulate the real light feeling, and adjusting the camera movement to simulate The Silk Road in people's perspective. The scene scheduling outside the image triggers the direct feedback from the user, which is mainly caused by the user's use of the haptic device to "sense" and then generate interaction, using gestures, virtual buttons, etc. [6].

4   Results

4.1 Improve the Accuracy of Branded Communication

The number of ICH under the fourth grade protection systems is huge. Four published batches listing 3,140 national ICH projects. There have great differences between different categories (Fig. 1). The characteristics, methods, and channels of technical communication of ICH should be selected and targeted to locate individualized brand communication channels and methods.

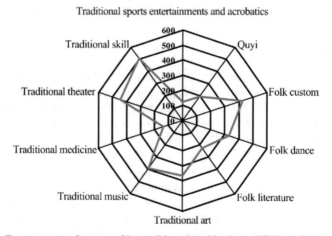

Fig. 1. Type-structure features of intangible cultural heritage (ICH) projects in China

ICH products have high recognition, simple production technology, broad market prospects and industrial standard production. Such as Huizhou stone carving, The craft of Shu-xi, Xiaoxian peasantry painting and other well-known folk arts, mass-produced by mechanized production lines according to market requirements, and enhance the brand's influence.

For labor-intensive products that rely mainly on manual production and high cultural creativity, they should protect and respect the individual's creative creation and production, appropriately introduce modern craftsmanship, position high-quality and sophisticated brands, and protect non-materials by exerting market functions. The core skills of cultural heritage products increase cultural added value to promote their products and services.

Projects with low market acceptance and unsuitable for industrial development are gradually shrinking due to the impact of modern social consumption methods and diversified leisure and entertainment channels. They should be positioned to create high-quality products, sort out brands, and use modern media technology to widely publicize and inherit. Encourage the combination of ICHs and cultural resources with local tourism resources, geographical landscapes and ecological resources, to promote the integrated development of the cultural industry, tourism and entertainment.

4.2 Refining Core Brand Symbols and Values

The brand iceberg theory points out that brand logos and symbols only account for 15%, and 85% are hidden values, wisdom and culture. These core values have the characteristics of exclusivity, appeal and compatibility, and are the key elements and core of brand communication basis.

Focus on inheriting core skills and cultural connotations. ICH contains profound humanistic connotation and cultural value. The key to its branding operation is to extract core brand symbols and values. These very new brand symbols and values are the connotation, essence or core skills condensed in the development process of ICH. It's the key factor for the public to identify them, allowing them to appreciate the superb traditional skills, the heavy historical and cultural connotations, and the simple practical value.

Focus on absorbing modern intelligent media technologies. In the traditional era, the inheritance of ICH is mostly dictated by personal experience, which limits the radiation of the inheritance of skills, resulting in the severance of some ICH skills. At this time, the proper introduction of modern technology can not only meet market demand, but also preserve and save traditional skills. For example, the development and revival of the three carvings in southern of Anhui, the production of rice paper, and the embroidery of the Miao nationality are the best proof. While adhering to manual techniques and maintaining traditional methods, they introduced modern new technologies and developed new products with modern flavor [7]. However, it should be noted that the protection of ICH cannot blindly deny the entry of modern technology, but cannot rely on new technology, and must maintain its intrinsic core value.

4.3 Strengthen the Development and Communication of Brand Ideas

Exploiting the potential cultural and economic value of ICH, enhancing the influence of ICH transmission, and transforming cultural resources into cultural productivity are the best ways to protect ICHs. The capital age is gone, and the creative age is here. Many traditional cultural resources in China have been developed abroad to develop their cultural industries, such as cartoons "Kung Fu Panda" "Mulan" "Dragon Ball" and "Three Kingdoms". The development and communication of ICH brand ideas has brought unexpected and great success. The "God's Enclosure" in the fantasy animation movie "Big Fish and Begonia" is based on the ICH of Hakka earth building. A series of successful creative development and communication at home and abroad shows that cultural creativity is a transformation bridge between traditional cultural resources and modern industrial resources. It will contain cultural genes and cultures created through historical accumulation and modern creation in ICH cultural products and cultural activities. The elements are reorganized and spread to build a new ICH brand. At present, the key to the creative development of ICH brand lies in innovation-driven, creative inspiration and shaping.

Excavate and develop the core cultural symbols and cultural values contained in the ICHs, and construct new cultural ideas. Originality and exclusivity are the key to ICH products to show cultural identity, gain social identity, and enhance market competitiveness. Under the premise of combing and integration, we must extract the core cultural symbols and cultural values of ICH, and build new ones. To expand the radiant power, attractiveness, influence and vitality of ICHs.

Choose the most bearable and expressive carrier to create a brand new ICH in the creative era. In response to consumers' aesthetic tastes and consumer desires, with the help of intelligent media technology, cultural resources can be transformed into cultural productivity, cultural creativity can be transformed into cultural products, and the brand's agglomeration, derivative, diffusion, amplification, drive, linkage and other communication effects can be created to new benefits, nurture cultural heritage, form a virtuous cycle of development, and truly protect, inherit and innovate the core values of ICH.

4.4   Protecting the Intellectual Property of Communication Brands

Because the identification of the ICH right holder is difficult to determine, whether traditional techniques constitute commercial secrets is still controversial, and rights protection, legal gaps, and lack of protection are the reasons. Intellectual property protection of the value of ICH brands must be:

At present, the relevant current laws on the protection of ICH require that the right holder of the ICH be clearly identified, which is representative, unique and subjective. The lack of knowledge accumulation and the poor orientation of economic interests of the grass-roots inheritors in the current protection have increased the difficulty of confirming the right holder's subjective status [8]. At the same time, the legal protection of intellectual property rights is pursuing "intra-generational equity", monopoly protection is limited to rights holders, not "intergenerational equity" between teachers and apprentices, fathers and sons.

The group inheritance of ICHs makes it difficult to determine the sole subject right holder. If it is forcibly confirmed by administrative means but a good benefit sharing mechanism is not established, it is likely to cause the isolation of the sole right holder. Traditional communities are the original domain of ICH inheritance. If the two are far away, it will cause alienation of ICH inheritance. Therefore, it is necessary to clarify the independent legal subjects of traditional communities and establish the intellectual property rights that are collectively shared by communities.

The rapid development of modern information technology has enriched the inheritance of ICH communication, and has also produced digital heritage. These digital heritages are embedded with the core cultural elements of ICHs identification [9], which enhances the comprehensive competitiveness of ICH brands and brings the necessity and urgency of intellectual property protection.

5 Discussion

The communication of ICH brands under the empowerment of intelligent media technology must follow the rules of modern information communication, innovative communication ideas and methods, excavate the core values of ICHs, and carry out communication activities with the concepts of original ecology, innovation and fashion.

The communication of ICH brand must be based on the protection of ICH. The community protection in the area where the ICH is inherited should be strengthened, and the protection of deep traditional skills of products should be strengthened. This is the protection of ICH. The essence is.

The essence of communicate ICH brands is to enable the traditional skills to be actively and orderly spread, to cultivate a cultural ecology, living environment, and inheritance methods suitable for it, and to develop harmoniously in tradition and modernity, material and spiritual, and national and world.

Acknowledgements. This work was financially supported by the Philosophy and Social Science Planning Project of Anhui Province (AHSKQ2016D86, AHSKYG2017D176); The Social Science Innovation and Development Research Project of Anhui Province (2019CX131).

References

1. Information on https://ich.unesco.org/en/directives
2. Information on http://www.cac.gov.cn/2019-02/28/c_1124175686.html
3. Li, Y.: When the intangible cultural heritage meet the live broadcast: new media empowered the communication of the intangible cultural heritage. China Culture Daily, 9 September 2017
4. Peng, D.: Research on future imaging in the context of intelligent media. J. People's Forum **24**, 40–49 (2018)
5. Information on http://world.people.com.cn/n1/2017/0826/c1002-29496035-2.html
6. McLuhan, E., Zingrone, F.: The Essence of McLuhan. Nanjing University Press, Nanjing (2000). pp. 152
7. Scoble, R., Isler, S.: The Coming Age of Scenes. Beijing United Publishing Co., Ltd., Beijing (2014). pp. 212
8. Huang, Y.L., Yu, H.: Application of intelligent media technology in the communication of Chinese intangible cultural heritage. J. Central China Normal Univ. (Hum. Soc. Sci.) **58**, 122–129 (2019)
9. Cheng, M., Zhan, L.Q.: The impact on digital survival and art perception in scenes of intelligent media. J. Commun. Univ. China **40**, 92–97 (2018)

Construction Strategy of Smart English Teaching Platform from the Perspective of "Internet + Education"

Zhenhua Wei[✉]

Xi'an Siyuan University, No. 28 Shui'an Road, Xi'an 710038, People's Republic of China
`james@xasyu.info`

Abstract. The advent of the "Internet + Education" era has brought smart education into a higher stage of development. Teachers' intelligent learning is not only a technological change, but more importantly, integrating network thinking into the teacher's learning. The concept of intelligent teaching allows students to learn more scientifically. English education is an important part of school education, which lays a good foundation for improving the overall quality of students. With the development of Internet technology, it provides a new platform for school English teaching. It provides favorable conditions for teacher-student interaction and improves the educational effect of school English and student learning. The purpose of this article is to study the construction strategy of smart English education platform from the perspective of "Internet + Education". In terms of methods, this paper proposes the coexistence of multiple learning modes to improve the original learning mode, build a personalized, experiential teaching ecosystem, and realize two-way interaction between online and offline to better improve learning efficiency. And establish a teacher-student interaction communication platform, teachers can adjust their own teaching progress, content, methods, etc. based on the feedback of students in the discussion group. Install assisted composition correction software to reduce the burden on teachers, thereby spending more time and energy to educate students. In terms of experiments, through the establishment of experimental classes and control classes, teaching is carried out separately, and the examinations are conducted before and after teaching. Send a questionnaire at the end of the period to see its satisfaction. It is found that the teaching platform of Internet + Education proposed in this paper is superior to traditional teaching in terms of learning effect and students' satisfaction with learning.

Keywords: Wisdom education · English teaching · Learning resources · Teaching effect

1 Introduction

In today's era, the computerization of the Internet is becoming more and more popular, and more and more people are using the Internet for learning and interaction. English is a global language and it is important to master English. The internet education platform

J. MacIntyre et al. (Eds.): SPIoT 2020, AISC 1283, pp. 122–128, 2021.
https://doi.org/10.1007/978-3-030-62746-1_18

combines students' hearing and vision, which maximizes the learning efficiency of students and improves their English level. Earlier, IBM introduced the idea of "intellectual education". It is derived from the "smart earth" inference put forward by IBM, as well as the philosophy of smart cities, smart communities and smart transportation. In essence, the next generation of information technology makes the Internet possible, makes everything on the planet smart, and is included in scientific and technological development plans by many developed countries. The Chinese government has made it clear that "Internet +" has been incorporated into the national strategy, and "Internet +" has been deeply integrated with various industries, and each industry is promoting optimization, growth and innovation. The proposal of the "Internet +" strategy has brought a new wave of wisdom, bringing smart education to a higher stage of development through technologies such as big data, artificial intelligence and mobile communications. There is no doubt that teacher training can also adapt to the development needs of the "Internet +" era, establish teachers 'wisdom and learning environment, and continuously improve teachers' learning level, awareness and communication ability.

In recent years, China's higher education has developed rapidly, students' English proficiency has been greatly improved, and English education has achieved great results. However, behind the achievement, there are still some problems in English teaching, especially in oral teaching [1]. On the student side, the vast majority of students lacked interest and motivation in spoken English; lacked self-confidence and did not dare to speak "English" boldly; they were too rigid in their knowledge of English, and did not know how to use it correctly, when and where to use it; And I don't know much about the culture of English-speaking countries, and there are some errors in the form of expression, way of thinking, internal structure, etc. [2]. Coupled with limited class time, the number of college English classes is about 50, students can not fully practice, and many schools do not include spoken language in the final examination, and the evaluation system is not scientific enough. As a result, students have not paid enough attention to spoken English, have no enthusiasm, talk about the "speaking" discoloration, and Chinese expressions [3]. In terms of teachers, college English listening and speaking courses are set together, and most teachers emphasize "listening" but not "speaking", and pay less attention to oral teaching.

In terms of methods, this paper proposes the coexistence of multiple learning modes to improve the original learning mode, build a personalized, experiential teaching ecosystem, and realize two-way interaction between online and offline to better improve learning efficiency. And establish a teacher-student interaction communication platform, teachers can adjust their own teaching progress, content, methods, etc. based on the feedback of students in the discussion group. Install assisted composition correction software to reduce the burden on teachers, thereby spending more time and energy to educate students.

2   Method

2.1   Achieving "Hybrid" Coexistence of Multiple Learning Modes

School education in the "Internet +" era has an inherent cultural gene. In classroom education, there is no good education content and education system [4, 5]. Transform

Internet + education into mobile social networks, transition classroom design, large-scale online courses, micro-courses, etc., to make these teaching carriers restrict and balance each other [6].

2.2 Build a Personalized and Experiential Teaching Ecosystem

Mobile Internet technology has been popularized in school education, creating multiple forms of education, including unique services [7]. In order to establish a more unique and more experiential education ecosystem in this type of education, we will promote personalized education, carry out various forms of education activities according to the learning needs of students, experience new models and establish a rich education environment [8]. Actively change education concepts, integrate education content and resources, and establish offline symbiosis ecosystems online.

2.3 Establish an Interactive Platform for Teachers and Students

English internet education platforms for schools can be set up to release the time and space restrictions on teachers and students, and help teachers provide online guidance for students [9, 10]. Create chat rooms and seminars to provide an effective way for dialogue between teachers and students. For example, students can discuss their questions at a seminar and discuss each other for valid answers. English learning experiences can also be shared at seminars. For teachers, the information feedback from students to the seminar can adjust their course progress, content and methods, and teach some questions by themselves, so as to really improve the effectiveness of the English classroom [11]. In addition, teachers answer questions on the Internet at fixed times each week and answer student questions. Courses on specific topics and specific English learning methods can be offered regularly. Improvements in learning skills such as reading comprehension and listening are not limited to overnight, nor is a single lesson that can produce significant effects [12]. However, by establishing a teacher-student communication platform, students will definitely improve their English level through long-term learning and communication. The self-study system can automatically track students' study time, effect, etc.

2.4 Install Auxiliary Composition Correction Software

Writing is an important part of English, and related research shows that many students have not specifically improved their English writing skills. Improving students 'English writing skills takes time and process, but installing an auxiliary writing editor can improve students' English writing skills to a certain extent. After the student has completed the manuscript of the dissertation using the assistant composition editor, place it on the education platform, and then click the "Revision" button to modify the text created by the student [13]. Corrected text has specific scores and corresponding comments. This method can help students understand writing level, provide guidance for understanding writing problems and improve writing level. It has a certain accuracy for errors in the overall structure. In a way, it can help improve English writing skills. In addition, it saves the teacher time to correct the composition. It helps teachers make the most of the time they save for teaching research.

3   Experiment

3.1   Experiment Purpose

Test the effectiveness of the combination of online and normal classes in the Internet + Education "teaching platform" proposed in this article.

3.2   Subjects

Two classes in a college of English at a university.

3.3   Experimental Design

Through the relevant experiments on two classes of a college of English in a university, the use of free supplementary lessons as benefits during the summer vacation, traditional education and online and normal classes in the "Internet + Education" teaching platform proposed in this article respectively Way of education. The traditional class is the control class, and the "Internet + education" teaching platform education is the experimental class. There are 50 students in each class, the average grades are similar, and the teachers are invited from outside, and the level is similar. Classes are usually simulated in accordance with the school's schedule. Classes are held five days a week, four classes in the morning, and two classes in the afternoon. The dictation and literacy classes are cross-cut, and the lessons in both classes are consistent. After returning home from the experimental class, the teachers will assign homework and check it next time. In the control class, homework is assigned by the teacher on the teaching platform after returning home from class, and is corrected using the auxiliary composition correction software. Usually on weekends, teachers will send some related learning materials. Finally, I maintained the teaching for 8 weeks, and conducted a bottom examination before and after the teaching to check the changes in the learning effect. After the end, a questionnaire survey was conducted on each student to check their satisfaction.

4   Discussion

4.1   School English Teaching Concept

At the same time that the mode of the Internet affecting various fields has changed, the school's English teaching philosophy has also changed and innovated, and information technology based on the Internet and big data platforms has opened a second English class. In order to meet the needs of personalized English learning, we have established a scientific and advanced English online education information system, and set up the collection, analysis, processing and application of English education data in the new school English education system. Using advanced digitization and information technology to achieve data processing in school English teaching (Table 1).

It can be seen from the data in the table above that before the summer vacation, the English levels of the two classes were similar. The average score in the experimental class was 0.3 points higher than the average in the control class, and the number of passing

Table 1. Mapping of experimental and control classes

	Average grade	Average grade	Passing number of experimental classes	Number of passers in the control class
Before studying	78.6	78.3	43	42
After studying	83.6	85.1	47	49

students was higher than the control class One. And the pass rate of both classes is above 80%, and their levels are high. However, there is a certain gap between the improvement of the two classes after the summer vacation. The average score in the experimental class increased by 5 points, and the average score in the control class increased by 6.8 points, 1.8 points more than the experimental class. 1.5 points more. In terms of passing numbers, there were 4 more experimental classes, 7 more control classes, and 3 more experimental classes. In general, the control class is better than the experimental class.

4.2 "Internet + Education" Massive Resources and Students' Personalized English Learning

Under the "Internet + Education" platform, school teachers can use the "Internet + Education" resources to focus on students' self-learning according to their learning needs. Collect and analyze relevant information materials and set conditions for students to learn English personally. The huge resource of "Internet + Education" requires teachers to prepare topics and materials for English courses. Students need to use resources on the Internet to find and read relevant information. Teachers organize and analyze these reading data to provide personalized English education for each student.

4.3 Satisfaction of Experimental and Control Classes

From the data in Fig. 1 below, it can be seen that 22 people in the experimental class were satisfied with the 8 weeks of summer vacation, and 38 people in the control class were satisfied with the 8 weeks of summer vacation. Among them, 7 were dissatisfied with the summer study in the experimental class, and only 2 were dissatisfied with the summer study in the control class. The platform proposed in this article supports online testing and online learning for students. During classroom learning and self-assessment, students are graded by the system and then a chart is created. Teachers use the data to understand the situation of students and better evaluate and reflect on each stage of education.

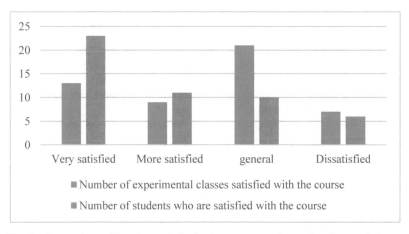

Fig. 1. Comparison of learning satisfaction between experimental and control classes

5 Conclusion

In terms of methods, this paper proposes the coexistence of multiple learning modes to improve the original learning mode, build a personalized, experiential teaching ecosystem, and realize two-way interaction between online and offline to better improve learning efficiency. And establish a teacher-student interaction communication platform, teachers can adjust their own teaching progress, content, methods, etc. based on the feedback of students in the discussion group. Install assisted composition correction software to reduce the burden on teachers, thereby spending more time and energy to educate students. In terms of experiments, through the establishment of experimental classes and control classes, teaching is carried out separately, and the examinations are conducted before and after teaching. Send a questionnaire at the end of the period to see its satisfaction. It is found that the teaching platform of Internet + Education proposed in this paper is superior to traditional teaching in terms of learning effect and students' satisfaction with learning.

References

1. Jan, S.U., Hussain, A., Ibrahim, M.: Use of internet by the teaching faculty of Peshawar Medical College, Peshawar, Khyber Pakhtunkhwa, Pakistan. J. Pak. Med. Assoc. **68**(3), 459 (2018)
2. Svantesson, D.: "Lagom Jurisdiction" – what viking drinking etiquette can teach us about internet jurisdiction and Google France. Masaryk Univ. J. Law Technol. **12**(1), 29–48 (2018)
3. Ge, H.: Research on the Chinese foreign English teaching quality assessment with intuitionistic fuzzy information. J. Comput. Theor. Nanosci. **15**(1), 278–281 (2018)
4. Yuan, B.: On the pragmatic functions of english rhetoric in public speech: a case study of Emma Watson's HeForShe. Engl. Lang. Teach. **11**(3), 113 (2018)
5. Wu, Q., Zhang, P.: E-learning user acceptance model in business schools based on UTAUT in the background of internet plus. J. Shanghai Jiaotong Univ. **52**(2), 233–241 (2018)

6. Kelly, C.: Learning to talk back to texts: multimedia models for students (and teachers). Pedagogy **18**(3), 271–286 (2018)
7. Yousuf, O., Mir, R.N.: A survey on the Internet of Things security: state-of-art, architecture, issues and countermeasures. Inf. Comput. Secur. **27**(2), 292–323 (2019)
8. Rahman, M.M., Pandian, A.: A critical investigation of English language teaching in Bangladesh: unfulfilled expectations after two decades of communicative language teaching. Eng. Today **34**(3), 43–49 (2018)
9. Granados-Beltrán, C.: Revisiting the need for critical research in undergraduate Colombian English language teaching. HOW **25**(1), 174–193 (2018)
10. Baa, S.: Lecturer perceptions toward the teaching of mathematics using English as a medium of instruction at the international class program (ICP) of mathematics department of the state. J. Phys. Conf. **1028**(1), 012131 (2018)
11. Skidmore, D., Murakami, K.: Dialogic pedagogy: the importance of dialogue in teaching and learning. Engl. Educ. **51**(2), 1–3 (2018)
12. Ruthven, K.: Characteristics and impact of the further mathematics knowledge networks: analysis of an English professional development initiative on the teaching of advanced mathematics. Teach. Math. Appl. **33**(3), 137–149 (2018)
13. Li, M.: Investigation on application of association rule algorithm in English teaching logistics information. Cluster Comput. **4**, 1–7 (2018)

Online Writing Effectiveness Under the Blended Teaching Mode of Moscotech APP

Juan Qian[✉]

School of Foreign Studies, Anhui Xinhua University, Hefei, Anhui, People's Republic of China
qianjuan0307@126.com

Abstract. The rapid development of modern information technology has brought great changes to People's Daily work and life, and the extensive application of network information technology in modern English teaching will bring about a series of innovations to the traditional English teaching mode. Based on the mixed English teaching mode promoted by the network environment, this paper constructs the mixed learning mode of Moscotech APP and applies it to English writing teaching. The results show that blended learning is more effective than traditional teaching. There is a significant positive correlation between the learning experience value and the final assessment result. This comparative study on the effectiveness of the large-scale online automatic correction software will be helpful to improve the teaching of college English writing.

Keywords: Blended teaching mode · College English writing teaching · Moscotech APP

1 Introduction

With the rapid development of the Internet and the deepening reform of modern education system, people pay more attention to the application of teaching methods and learning strategies. *China's Modernization of Education 2035* advoates that "Make full use of modern information technology to enrich and innovate curriculum forms" and "promote heuristic, exploratory, participatory and cooperative teaching methods" [1]. *Education Informatization 2.0 Action Plan* proposes that "Education informatization is the basic connotation and prominent feature of education modernization" and "adhere to the core concept of in-depth integration of information technology and education and teaching" [2]. These important programmatic documents provide ideas for the reform of curriculum construction and learning mode. The traditional teaching mode of classroom is characterized by teacher-centered features, which cannot resonate with the development trend of educational informatization, nor can it meet the learning needs of students, let alone improve the learning mode of students. In view of this situation, educators should think deeply about how to keep the upright and innovative learning mode and how to promote the deep integration of modern information technology and teaching so as to improve the learning effect.

J. MacIntyre et al. (Eds.): SPIoT 2020, AISC 1283, pp. 129–135, 2021.
https://doi.org/10.1007/978-3-030-62746-1_19

1.1 Definition of Blended Teaching Model

At the beginning of the 21st century, American scholars Smith, J and Alert Masier jointly put forward a new concept of learning mode, namely "Blended-Learning", which is an organic integration of traditional Learning and e-learning to improve the deficiencies of network teaching and form the blended Learning model. Subsequently, a large number of scholars and teaching workers at home and abroad have conducted in-depth research on hybrid learning and teaching, Among which, Russel T. Osguthorpe and Charles R. Graham put forward in the article that "blended learning combines face-to-face mechanism with distance mechanism… People who use blending learning environment are always trying to maximize the benefits of face-to-face learning and online learning." [3]. Margaret Driscoll and others believe that blended learning mode is an effective way to combine multiple network technologies, multiple teaching methods, and multiple forms of teaching techniques with face-to-face teacher guidance to complete teaching techniques and practical tasks [4]. Many domestic scholars and experts have studied the Blended teaching mode and written books. Professor Ho Kekang (2004) pointed out that "the blended teaching mode is a combination of various learning methods in traditional classroom teaching—The combination of audio-visual media teaching methods with traditional classroom teaching methods; The combination of CAI and traditional teacher's single instruction; the combination of students' independent learning and cooperative learning" [5, 6]. After long-term research, Professor Huang Ronghuai proposed that "blended learning is a strategy to implement teaching by integrating different learning theories, techniques and means as well as different application methods" [4].

1.2 Moscotech APP

Moscotech APP is a mobile teaching assistant APP. Teachers use mobile devices or computers to manage classes, send notifications, share resources, assign homework, organize class sign-ups, conduct discussions and answering questions, therefore Moscotech APP are an SPOC. Both the classroom and the after-school are learning sites. Teachers can carry out interactive teaching activities such as voting questionnaire, brainstorming, work sharing, timed quiz and unit test at any time, and students can get immediate feedback and comments. Each time students participate in learning activities, learning management platform and teachers can give students learning experience value. Moscotech APP dynamically records students' learning performance through learning experience value, which not only provides longitudinal and horizontal learning comparison, but also provides basis for formative evaluation.

2 Blended Learning Model Construction Based on Moscotech APP

As we all know, English learning is inseparable from the five skills of listening, speaking, reading, writing and translation. However, "writing can not only help consolidate the language materials input through reading and listening, promote the internalization of language knowledge, and improve the accuracy of language usage, but also lay a solid foundation for substantive oral competence" (Dong Yafen, 2003:5) [7]. According to the composition scores reflected by various automated Scoring, although Chinese college students have certain English grammar knowledge and vocabulary accumulation, many students' English writings are full of mistakes and the words fail to form sentences.

It's self-evident, it is necessary to set up the blended teaching mode based on Moscotech APP. As the helper and consultant for students to construct meaning, teachers use blended learning to guide students to acquire knowledge. The teaching process is divided into three stages, namely, before, after and after class. Before class, teachers upload learning resources of multi-modal writing to the learning platform to create learning situations and environments to stimulate students' interest and motivation. The learning resources include teaching videos, courseware and network links of knowledge points. Students carry out autonomous learning activities such as reading textbooks, watching teaching videos, browsing courseware and taking notes. Then carry out cooperative learning online, such as participating in question-answering activities, completing group tasks, etc., share the results of collective thinking, and preliminarily complete the meaning construction of the knowledge. When participating in learning activities, the learning platform can give learning experience values to students according to their participation records. In class, teachers implement inquiry learning, that is, problem-solving activities to help students construct knowledge, such as discussion and answering questions, brainstorming, etc. Teaching is no longer to transfer certain ready-made knowledge, but to activate the connection between students' original knowledge and new knowledge, to eliminate the blind spot of knowledge, promote the accumulation of knowledge and experience, improve the ability to solve problems, and improve the construction of meaning. In classroom learning, teachers grant learning experience value to students according to their classroom performance. After class, students make comprehensive use of the significance of the internalization of online and offline learning activities to construct the knowledge they have learned, including completing offline writing exercises, self-examination and self-reflection, and finally mastering the knowledge points and the significance of the internalization (Fig. 1).

3   Teaching Research Design – Take English Writing as an Example

Moscotech APP has an essay correction item. Teachers can set the task details, introducing the key points and requirements of the essay, the scoring method, the scoring points and their respective weights.

3.1   The Research Question

The purpose of this study is to investigate the effects of different online automatic composition correction software on college students' English writing effectiveness.

The specific research questions are as follows:

(1)   Is there any difference in learning effect between blended learning and traditional teaching mode?
(2)   How to further improve the learning effect of blended writing according to the comments on online literature?

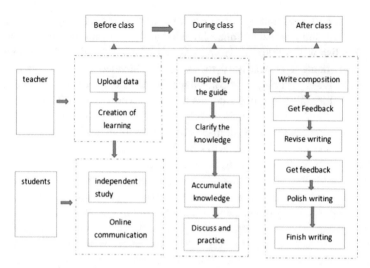

Fig. 1. The blended learning model of Moscotech APP

3.2 The Research Object

The subjects are students of two parallel classes in the third year of English major in the School of Foreign Languages. Each class has 40 students. The same teacher teaches two classes in English writing. Class A is A controlled class, implementing the teacher-centered traditional teaching method. Class B is an experimental class, implementing the learning-centered blended learning method.

3.3 The Research Tools

In this study, online writing review platform—Moscotech APP were used, which can automatically judge students' submitted essays online and give them overall comments, as well as feedback on grammar, vocabulary collocation and non-standard expressions. Students can review, revise and submit new compositions for many times, and the platform can give the scores submitted for each time and keep all text records. The basic data of this study comes from all kinds of data saved by Moscotech APP. The writing task of the students involved in this experiment is an English argumentative essay with no less than 180 words, the title of which was "How can you deal with stress?" Students must within two weeks of the assignment submit their essays directly online or revise their compositions according to the system feedback until the final essay is submitted.

3.4 Practice Effect and Reflection

(1) Comparison of the results of mixed learning method and traditional teaching method
① Pretest results
  At the beginning of the semester, the teacher collects the TEM-4 scores of the two classes to measure whether there are differences between the two classes in basic language skills. SPSS 19.0 was used to obtain the scores of two classes and independent sample T-test results, as shown in Table 1.

Table 1. Independent sample T test comparison of writing scores in pretest

	Average number	N	Standard deviation	Variance	Significance
Class A	79.352	40	−0.575	77	0.875
Class B	80.287	40			

The average grade of class A and Class B is 79.352 and 80.287. In the independent sample T-test, the significance level of English writing score was 0.875 > 0.05. There is no significant difference in English proficiency between the two classes. The pretest results show that the premise of this teaching experiment is valid.

② post-test experiments

After A semester of English writing in both classes, Class A teaches traditional English writing, while class B teaches mixed online writing, again using the final exam scores as A post-test. The grades of the two classes and the independent sample T-test results were collected again. The average scores of class A and Class B were 71.790 and 79.127 respectively (Table 2).

Table 2. Independent sample T test comparison of writing scores in post-test

	Average number	N	Standard deviation	Variance	Significance
Class A	71.790	40	−2.135	78	0.037
Class B	79.127	40			

The independent sample T-test results show that P = 0.037 < 0.05, there is a significant difference in the final assessment scores of students using the two teaching methods, that is, the learning effect produced by the mixed learning mode is better than the traditional teaching mode.

3.5 Reflection on the Effect of Blended Learning

Blended learning is essentially foreign language teaching in the context of "Internet +". "The in-depth integration of 'Internet +' makes foreign language teaching more ecological, independent and personalized, which can promote foreign language teaching and learning more effectively" [8]. The effectiveness of English writing results from the participation of various learning activities, so the improvement of blended learning needs to consider the factors involved in the learning process from both teaching and learning levels.

From the perspective of teachers, according to the characteristics of blended learning model, teaching strategies are related to learning resources, learning activities, learning monitoring, learning evaluation and other factors. Teachers should provide students with blended learning resources. Constructivism emphasizes the use of various information resources to support learning. In order to assist students to actively explore and meaning construction, teachers should set up the "one-stop" learning platform, to provide quality online courses, courseware, text, such as links to the knowledge network learning resources, all-round meet the learning needs of students, make its cast off the yoke of the single teaching materials, absorbing knowledge needed in multi-modal learning resources.

From the perspective of students, constructivism emphasizes that the ultimate goal of the learning process is to complete the construction of meaning, and students are the active constructors of meaning. Learning activities should be conducive to the completion and deepening of the meaning of the knowledge construction. Learning activities in the context of the Internet are more likely to attract students' attention. Students obtain learning content and resources from the "cloud" and choose learning content and activities at their own pace, regardless of time and place. On the learning platform, students can complete learning activities independently or cooperate in groups. Write your own opinion; Manual input of information; According to the platform revision Suggestions, study and polish again and again until the satisfactory work is finished.

For example, students benefit from feedback on text structure, vocabulary use, grammar, spelling and other details, and modify and polish them according to the feedback.

Example 1 "so just eliminate stress, and only in this way, can we have a smooth life".

Students soon received feedback from the correcting platform about this sentence, adding the preposition to between "just" and "eliminate", thus correcting the incorrect grammatical structure.

Example 2 "so the other will not has much complain to you".

Similarly, the online grading system shows that modal verbs and verbs are mismatched, suggesting have, and so on.

In addition, Moscotech APP pays more attention to sentence writing, frequently pointing out the "suspected Chinglish" expressions frequently seen in The English writing of Chinese college students.

Example 3 "my life pressure", "the high developing speed", and "I very like jogging to relax myself".

As Chinese and English belong to different language families, their grammatical systems are very different. Due to the fossilization of language learning, Chinese students often create similar Chinese English expressions or sentences in writing. Thanks to online writing correction system, they can disclose these errors in time, and correct them immediately.

In order to obtain the maximum learning benefit, students should often reflect on what they have mastered, what learning experience is worth promoting, what shortcomings exist, and what methods or strategies should be adopted to improve the learning effect. "The best education is self-education – enabling students to gain self-awareness and self-transcendence" [9]. Only through active reflection, can we find the weakness and improve the measures, and effectively construct the meaning.

4 Conclusion

With the rapid development of Internet technology, college English teaching should make full use of modern teaching means such as network technology, develop and utilize digital teaching resources, and construct a new teaching mode suitable for students' personalized learning and independent learning. In view of this, the author uses the "cloud class" learning platform to implement blended learning in English writing teaching, and the results show that blended learning achieves better results than the traditional teaching method. Through the data analysis and research of "Cloud class", teachers and students can understand the problems in English writing more intuitively and conveniently through these online correcting software, and make corrections according to the Suggestions provided by the online platform. This study provides some enlightenment and reference for the theoretical research, practical application of blended learning, stimulate the learning interest of students, and improve the writing efficiency of Chinese college students' English writing. However, in the writing teaching, the automated correction system plays the role of auxiliary teaching [10] and the teachers should rationally supervise the process of writing and make appropriate manual correction.

Acknowledgments. This research was supported by 2018 Provincial Quality Engineering Project of Anhui Higher Institute Massive Open Online Courses (MOOCs) demonstration projects (project number: 2018mooc429).

References

1. http://www.xinhuanet.com//mrdx/2019-02/24/c_137845629.htm, 24 February 2019
2. http://www.moe.gov.cn/srcsite/A16/s3342/201804/t20180425_334188.html, 18 April 2018
3. Qsguthorpe, R.T., Grahan, C.R.: Blended learning environments: definition and directions. Q. Rev. Distance Educ. **3**, 227–233 (2003)
4. Driscoll, M.: Blended Learning: Let's Get Beyond the Hype. MB Global Service, no. 11 (2011)
5. Ho, H.: Resistance from blending learning looking at the development of the theory of education technology research. J. Chin. Audio-Visual Educ. **3**, 1–5 (2004). (in Chinese)
6. Huang, R., Zhou, Y., Wang, Y.: Theory and Practice of Blended Learning. Higher Education Press, Beijing (2006). (in Chinese)
7. Guo, C., Qin, X.: Research on written feedback of foreign second language learners. J. PLA Foreign Lang. Instit. (5) (2006). (in Chinese)
8. Zhang, Q.: Psychological Principles of Foreign Language Learning and Teaching. Foreign Language Teaching and Research Publisher, Beijing (2011). (in Chinese)
9. Chen, J., Wang, J.: Normal changes and development in the informationization of foreign language Education: the foundation Visualization in Educational Informatization. Foreign Lang. Audio-Visual Teach. **2016**(2), 3–9 (2016). (in Chinese)
10. Lu, L., Wu, X., Wang, Y.: Research on the application of Juku Correct Network in English writing teaching. J. He Zhou Univ. **4**, 117–121 (2017). (in Chinese)

A Narrative Environment Model for the Sustainability of Intangible Cultural Heritage Under the 5G Era

Chen Qu, Ting Zhao, and Wei Ding[✉]

Southampton International College, Dalian Polytechnic University, Dalian, Liaoning, China
qc113@163.com

Abstract. With the arrival of the fifth-generation mobile networks (5G) era, it has been widely acknowledged that innovative technology has been involved in many aspects, such as social, culture and economy. In this context, the culture creative industries (CCI) are becoming increasingly fused with technologies. As intangible cultural heritage is recognized as a branch of CCI, to what extent and how technology serves as a tool in safeguarding the intangible cultural heritage (ICH), and what media can ICH interact with the audience should be considered. This study contributes to fill this gap and solve the issues by aiming at the Chinese ICH, thereby exploring a narrative environment model for the sustainability of ICH at World Heritage Sites (WHSs) based on the narratology and narrative environment theory under the 5G Era. The practical possibilities drew on the literature review and a develop model based on the narratology theory thereby contributing to the overall aim of this research. It was discovered that 5G technology applied to an interactive narrative space could probably promote the inheritance and protection for ICH and enhance the immersive audience experience in museums.

Keywords: Narrative environment · Intangible Heritage Culture (IHC) · Digital museum · 5G technology

1 Introduction

5G is the fifth generation of wireless technology. With the advent of the 5G era, the entire digital culture industry will undergo dramatic changes, which mainly refers to the combination of digital industry with all kinds of industries, making the industry itself produce many new formats, new applications and different application scenarios. Based on the advantages of accessing data more quickly and having a smoother experience, 5G technology can not only make user experience across scenes, but also realize the Internet of everything. For this trend, 5G and more and more new technologies will also change social ecology and promote the development of CCI, from exhibition, museums, the individual artists, display medium, and creation medium, to the latest changes in the cultural market… As the 5G is increasingly closer to daily life, the development of human society is increasingly accelerating and efficiency. The creators, viewers, operators and

J. MacIntyre et al. (Eds.): SPIoT 2020, AISC 1283, pp. 136–142, 2021.
https://doi.org/10.1007/978-3-030-62746-1_20

traders are facing the unprecedented challenges in how to adapt to the impact and changes brought by high technology.

The topic of the sustainability of ICH has become an international concern primarily through the work of United Nations Educational, Scientific and Cultural Organization (UNESCO). Recently, sorts of researchers state the digital reconstruction of preserving ICH and discusses how the innovative technologies play roles in inheriting ICH in China, Korea, even in Europe. However, little research has been done on the role of new technologies from an interactive narrative environment respective to explore the space-person relationships. Furthermore, few have discussed that how technologies can be utilized to exhibitions on ICH in relation to audiences' interactions.

Due to this gap, the article examines time and space, and 5G technologies in an interactive narrative way providing access to ICH and identifies gaps and constraints. It also explores the possibility of digital technologies being applied to exhibition environment on ICH in order to promote its preservation and sustainability. It draws on three dimensions of the literature review based on the theories of narratology and post-narratology, developed an interactive narrative environment model for ICH exhibition, and specified how specific 5G technologies can contribute to the display of ICH, thereby contributing to the sustainability of ICH.

2   Literature Review

2.1 ICH in China

CCI is new cross-sectoral and inter-departmental industry, combining with arts, culture, economy and science & technology [1]. As a branch of CCI, ICH can be considered to be a source of creativity, while creativity can be a medium to revitalize the culture [2].

It is important that ICH is a factor in maintaining cultural diversity in the face of growing globalization [3]. "An understanding of the intangible cultural heritage of different communities helps with intercultural dialogue, and encourages mutual respect for other ways of life." [3]. According to the "UNESCO" in 2003, the "intangible cultural heritage" is "traditional, contemporary and living at the same time", "inclusive", "representative", and "Community-based", which is constantly recreated by communities and groups in response to their environment, their interaction with nature and their history [4]. The "UNESCO" divided ICH into five categories, which are: (a) oral traditions and expressions, including language as a vehicle of the intangible cultural heritage; (b) performing arts; (c) social practices, rituals and festive events; (d) knowledge and practices concerning nature and the universe; (e) traditional craftsmanship [4]. It highlights the most important character of ICH is "authentic", and "dynamic and active" which plays a key role in maintaining cultural diversity. Social cohesion and individuals are enhanced and developed a sense of identity and belonging within a culture [4, 5]. The categories of national ICH and representative list examples of each type are shown in Table 1 below. Table 1 shows five categories classified by UNESCO used to classify the types of Chinese ICH on the UNESCO "Convention's Lists".

Table 1. National intangible cultural heritage in China [3, 9, 10]

	Category	Total	Representative list (Example)
a	Oral traditions and expressions	9	Hezhen Yimakan storytelling 2011; Hua'er, 2009
b	Performing arts	8	Pekng opera, 2010; Yueju Opera, 2009
c	Social practices, rituals and festive events	9	Qiang New Year festival, 2009
d	Knowledge and practices concerning nature and the universe	4	The Twenty-Four solar terms 2016 Chinese Zhusuan, 2013
e	Traditional craftsmanship	10	Wooden movable-type printing of China, 2010; Meshrep, 2010

2.2 Digital Technology for Preserving ICH

Utilizing digital technologies for preserving ICH can be widely acknowledged recently, such as computer-based visualization, cultural heritage 3D visualisation projects, IT platforms, etc. [7], which are newly devised operational methods to support the growth of cultural diversity and increase international networking for promotion of safeguarding and sustainable practices of ICH. They normally reconstruct or recreates historical buildings, artifacts and environments as audiences believe they were in the past [8]. The key thing is that it should "always be possible to distinguish what is real, genuine or authentic from what is not. In this sense, authenticity must be a permanent operational concept in any virtual archaeology project [8]. An application named "civilisations AR" produced by BBC has been applied to mobile phones, in which users could immerse in the museum space to feel the authenticity (see Fig. 1). This AR (Augmented Reality) technology applied to museums can draw public attention to enrich visitors' experience and broaden their horizons through the visual impact and abundant information provided by digital media.

Fig. 1. Application of "Civilisations AR" by BBC

2.3 Narrative Environment

Recently, individuals are provided a place where people can learn about history, culture, and other subjects through exhibitions, galleries or museums with a wealth of information in diverse fields [6]. On the basis of the audience-centered perspective, the space are encouraged to offer information in a narrative way that benefits audiences, enhancing the interactions from audiences and ICH. Therefore, to understand the relationships between space, time, and audiences, it is necessary to understand what is narrative environment, derived from the theory of narratology.

"Narrative theory, or narratology, is the study of narrative as a genre." [8]. The term narratology has been defined by some theorists in different way, but generally, these theorists explain that narratology is the study of the form and functioning of narrative. However, Amerian and Jofi (2015) argue that "narratology should make its borders larger" and need to "interact with other disciplines" reflected in the era of post-classical narratology. As Fig. 2 shows the components and elements of narrative text, which narratives are constructed and these components are related. The hierarchy identifies the two higher-levels are story and discourse, which lies at the heart of all structuralist approaches to narrative [8]. From Chatman's classification in Fig. 2, events and existents are other vital elements of narratives, and along with existents, events are the fundamental constituents of the story [8]. He also highlights the concept of "phenomenological aesthetics" can be used to explain the distinction between story, discourse, and manifestation.

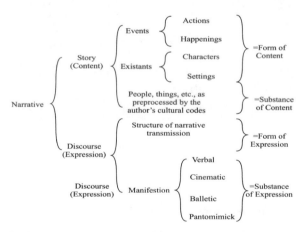

Fig. 2. Chatman's diagram of narrative, cited from Chatman [8]

3   Narrative Environment Model for ICH Exhibition

3.1 Narrative Environment Model

Figure 3 provides the narrative environment model for ICH exhibition, derived from the previous diagram of narrative from Chatman. According to the Chatman's diagram, the narrative text includes the main attributors, which are story and discourse [8]. There-fore, in terms of ICH, the story could be considered as ICH's content, and the ICH expression would be the discourse code. It means the form of expression can be recog-nized as the display of ICH. Because the vital factors of ICH is revise its "authentic" or "alive", from the audience perspectives, the form of expression should be considered from three attributes of ICH of, which are ideology, behaviour, and substance, to inter-active with audiences through an interactive narrative space. As the previous literature stated, utilizing technology must take consideration into the authenticity; therefore the "existent code" (ICH form) must have the scenario of people and context as a permanent operational concept in this model.

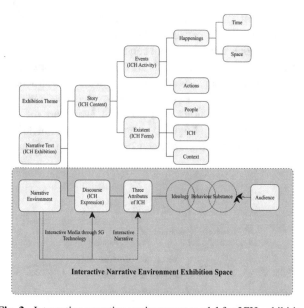

Fig. 3. Interactive narrative environment model for ICH exhibition

3.2 5G+Practical Contributions

Table 2 shows the "5G+Virual Reality (VR), Augmented Reality (AR), Mixed Reality (MR), Holographic" technologies on the exhibition of ICH in China. In this table, 5G is supported on VR, AR, MR and holographic projection technology to restore the occurrence, development and evolution of ICH works in a narrative way, and to enhance the interaction between artworks and audiences and the entertainment of exhibitions. 5G+VR will enhance the process of virtual technology to comprehensively represent ICH skills. 5G+AR augmented reality technology enables ICH artworks to present a brand-new visual effect with the combination of reality and reality, so as to restore their active state and show the cultural connotation behind them. 5G+MR mixed virtual reality technology enables viewers to see the ICH through their eyes to measure the scale and orientation of objects in real life, enhancing the interaction between the virtual world and the real world. 5G+ holographic projection technology enables the three-dimensional holographic projection portraits of intangible cultural heritage practitioners to be clearly presented, just like standing on the scene to show intangible cultural heritage technologies to the audience, the holographic projection realized by 5G technology at low delay and high rate will bring a new experience. The three technologies are integrated to break the limitation of time and space in the immersive space by means of linkage image, radar motion capture, and immersive digital sound effect, so as to realize the full-scene immersive experience ecosystem technology.

Table 2. 5G+New technology practical possibility

New technology content	New technology speciality	Possibility of new technology
5G+VR holographic 3D immerse VR system	Enhancing the process of virtual technology to comprehensively represent ICH skills. The 360-degree panoramic virtual simulation is a four-dimensional space for immersive viewing	Virtual wearing experience of ICH, visitors immersing themselves in ICH scenes
5G+AR material object scanner and 3D scanner	Augmented reality technology, which superimposes real environment and virtual objects on the same screen or space in real time	ICH display and physical scan, to achieve the perfect integration of traditional culture and technology
5G+MR hybrid display	A hybrid of augmented reality and augmented reality. Using MR hybrid technology to reproduce the advantages of reality to achieve the immersive effect and interactive effect of hybrid reality technology	Viewers can see the reproduction of ICH content and conduct real-time interaction around virtual cultural relics
5G+Holographic projection	3D holographic projection can be viewed directly without auxiliary equipment. 360, 270, 180 multi - latitude representation of the object's real 3D image	ICH documentary and theater projection. The suspended image shows the virtual ICH documentary and dramatic dance performance image through the light and shadow effect of the holographic image

4   Conclusion

This article tries to develop an interactive narrative environment model utilizing new technologies with 5G technology. The theoretical innovation of ICH exhibition model of interactive narrative environment, the theoretical innovation of science and technology and narratology, and the practical innovation of science and technology and culture have analytical identified in this research. First, it overviewed Chinese ICH at present, and subsequently it based on narrative theories but beyond the traditional narrative structure of exhibition, employing the narrative model of "post-classical narrative", and achieves the fusion of viewer and space, creating a new model of interactive narrative environment model for ICH exhibition through 5G technology. Lastly, the practical possibilities of 5G with interactive media (VR, AR, MR, Holographic) has been conducted from "text" to "graphics". These contributions and practical possibilities may serve as a guild line for the inheritors of ICH, researchers, artists, curators, and different stakeholders in sustaining the status of WHS.

Acknowledgements. This research is financially supported by the Educational Department of Liaoning Province. Project: 《The Theoretical and Practical Research of Intangible Cultural Heritage in Liaoning Province Based on the "Narrative Environment" Theory under the 5G Era》 (Project number: No. J2020079).

References

1. Wang, Y., Zheng, L.: Development research of cultural creative industry based on the intangible cultural heritage in Hangzhou. In: 2008 9th International Conference on Computer-Aided Industrial Design and Conceptual Design, Kunming, pp. 1293–1297 (2008). https://doi.org/10.1109/caidcd.2008.4730802
2. Tan, S.-K., Lim, H.-H., Tan, S.-H., Kok, Y.-S.: A cultural creativity framework for the sustainability of intangible cultural heritage. J. Hosp. Tour. Res. **44**(3), 439–471 (2020)
3. UNESCO: China and the 2003 Convention. https://ich.unesco.org/en/state/china-CN. Accessed 15 June 2020
4. Hahm, H., Lee, J., Jeong, S., Oh, S., Park, C.S.: A digital solution and challenges in the safeguarding practices of intangible cultural heritage: a case of 'ichngo.net' platform. In: Proceedings of the 2020 2nd Asia Pacific Information Technology Conference, January 2020, pp. 94–99 (2020)
5. Kim, S., Im, D., Lee, J., Choi, H.: Utility of digital technologies for the sustainability of intangible cultural heritage (ICH) in Korea. Sustainability **11** (2019). https://doi.org/10.3390/su11216117
6. Park, S.C.: ICHPEDIA, a case study in community engagement in the safeguarding of ICH. Int. J. Intang. Heritage **9**, 69–82 (2014)
7. Ioannides, M., Magnenat-Thalmann, N., Papagiannakis, G.: Mixed Reality and Gamification for Cultural Heritage. Spinger, Cham (2017)
8. Amerian, M., Jofi, L.: Key concepts and basic notes on narratology and narrative. Sci. J. Rev. **4**, 182–192 (2015). https://doi.org/10.14196/sjr.v4i10.1927
9. UNESCO: What is Intangible Cultural Heritage? https://ich.unesco.org/en/what-is-intangible-heritage-00003. Accessed 15 June 2020
10. Alivizatou-Barakou, M., Kitsikidis, A., Tsalakanidou, F.: Intangible Cultural Heritage and New Technologies: Challenges and Opportunities for Cultural Preservation and Development **6**, 59–67(2017)

Application Study of VPN on the Network of Hydropower Plant

Wenju Gao[(✉)]

Army Academy of Armored Forces, Changchun, Jilin, China
171257390@qq.com

Abstract. In this paper, the concept, as well as the advantages and the application areas of VPN (Virtue Private Network) are introduced; the security technology of VPN are discussed; a scheme about information exchange with safety among the stations of hydropower plant by means of VPN is put forward; and last, the process of establishing VPN by Windows is introduced in detail.

Keywords: VPN · The information network of hydropower plant · Tunneling technique · L2PT · Network security

1 Introduction

With the continuous deepening of information construction in China, especially the development of computer network in large state-owned enterprises, more and more information needs to be transmitted by means of network. Although the enterprise internal comprehensive information network of Northeast Power Grid Company is a wide area network system physically isolated from the Internet, which largely meets the needs of enterprises, many subordinate units still need to keep secret. However, it is hoped that while connecting to the integrated information network, some important internal information like notice, plan, etc. which can only be seen by the staff of the unit can be transmitted on the network [1].

The private network constructed by using public network is called virtual private network. The public network used to build VPN includes Internet, frame relay, ATM, etc. VPN built on the public network provides security, reliability and manageability just like the existing private network. The concept of "virtual" is relative to the construction of traditional private network. For WAN connection, the traditional networking is realized by remote dial-up connection, while VPN is realized by using the public network provided by the service provider.

2 Characteristics of VPN

2.1 Wide Area of Intranet

VPN can interconnect the main power plant with the branches and business points scattered around, for example, Baishan power plant is a hydropower plant with three

J. MacIntyre et al. (Eds.): SPIoT 2020, AISC 1283, pp. 143–148, 2021.
https://doi.org/10.1007/978-3-030-62746-1_21

power stations. Specifically, there are two hydropower stations in Baishan town and one hydropower station in Hongshi town in Huadian city [2]. The general plant is located in Huadian city. Baishan power plant can build the internal network of the unit in this way, so that the personnel scattered in different regions and outside the unit can use the internal network as safely and conveniently as in their own offices.

2.2   Reduce Operation and Management Costs

With the emergence of VPN, the operation cost of Wan has been reduced. With the wide area of Intranet, its importance is more prominent.

2.3   Enhanced Security

VPN is verified by using point-to-point protocol (PPP) user level authentication method, which is enforced by VPN server. For sensitive data, VPN connection can be used to separate highly sensitive data servers through VPN server. Users who have proper permissions on the Intranet can set up VPN connection with VPN server through remote access, and can access resources in sensitive department network.

2.4   Good Scalability

VPN supports the most widely used network protocols. Clients in IP, IPX and NetBEUI can use VPN without any difficulty. Therefore, applications that rely on special network protocols can be run remotely through VPN connection; in addition, VPN can support multiple types of transmission media, which can meet the requirements of high-quality transmission and bandwidth increase for new applications such as voice, image and data transmission at the same time.

3   Classification and Use of VPN

According to different needs, different types of VPN can be constructed. VPN is usually divided into three types: remote access virtual network (access VPN), internal virtual network (intranet VPN) and extended virtual network (extranet VPN).

3.1   Remote Access Virtual Network

Personnel may need to visit the intranet of the unit in time when they are on business trip.

3.2   Internal Virtual Network

Intranet VPN is a network that connects the LAN of each branch of an organization through public network. This type of LAN to LAN connection brings the least risk, because it is generally considered that each branch is reliable. Intranet is the extension or alternative form of traditional private line network or other enterprise network. This article applies this type of VPN [3].

The essence of using IP network to build VPN is to build VPN Security tunnel between routers through public network to transmit users' private network data. The tunnel technologies used to build this VPN connection include IPSec, GRE, etc. Combined with the QoS mechanism provided by the service provider, the network resources can be used effectively and reliably to ensure the network quality. VPN based on virtual circuit technology of ATM or frame relay can also achieve reliable network quality, but its disadvantage is that the interconnection area has greater limitations. On the other hand, building VPN based on Internet is the most economical way, but the quality of service is difficult to guarantee. When planning VPN construction, enterprises should weigh the above public network schemes according to their own needs.

3.3 Extended Virtual Network

Extranet VPN means to use VPN to extend the internal network to other units. Such extranet construction and maintenance is very expensive, because of the wide distribution of partners and customers.

4 Realization of Virtual Private Network

4.1 Security Technology of Virtual Private Network

As the private information is transmitted, VPN users are more concerned about the security and confidentiality of data. At present, VPN uses four technologies to ensure its security, which are tunneling technology, encryption & decryption technology, key management technology, user and device authentication technology.

Tunnel technology is the basic technology of VPN, just like point-to-point connection technology. It establishes a data channel (tunnel) in the public network (such as the enterprise internal integrated information network of Northeast Power Grid Company) to let the data packets transmit through this tunnel. Tunnel is generated by tunnel protocol, which is divided into two or three layers [4]. The second layer is to encapsulate all kinds of network protocols into PPP, and then package the whole data into tunnel protocol. The data packets formed by this two-layer encapsulation method are transmitted by the second layer protocol. The second layer of tunnel protocol includes L2F, PPTP, L2TP, etc. L2TP protocol is the current IETF standard, which is formed by IETF integrating PPTP and L2F. The third layer tunnel protocol is to install all kinds of network protocols directly into the tunnel protocol, and the data packets formed depend on the third layer protocol for transmission. The third layer of tunneling protocols includes VTP, IPSec, etc. IPSec (IP Security) consists of a group of RFC documents. It defines a system to provide services such as security protocol selection, security algorithm, and key used by services, so as to guarantee security at the IP layer. VPN is connected through dial-up, by sharing the tunnel of the enterprise internal comprehensive information network of Northeast Power Grid Company. It sends by one router on LAN, and then to another router on LAN, so that PPP packet flow can be transmitted [5]. Its essence is to replace the real dedicated lines with tunnels.

Encryption and decryption technology is a more mature technology in data communication. VPN can directly use the existing technology.

The key management technology is mainly about in how to transfer the key safely on the public network without being stolen. The current key management technology includes SKIP and ISAKMP/Oakley. SKIP transmits keys on the network by using Diffie-Hellman's algorithm; in ISAKMP, both parties have two keys, which are used for public and private purposes respectively.

The most commonly used authentication technology is user name and password or card authentication.

4.2   Implementation of Virtual Private Network

At present, the internal comprehensive information network of Northeast Power Grid Company has completed the full coverage of the basic units, providing a better information interaction platform for the basic units, but because many units and subordinate units are widely distributed in the region [6]. For example, Baishan power plant headquarters is located in Huadian City, and the three power stations are located in Baishan town and Hongshi Town, respectively. The distance between the plant headquarters and the subordinate power stations is far, which leads to the slow promotion of online office and paperless office. If we use the above-mentioned internal virtual network technology, we can easily connect several small LANs which are distributed in different regions and belong to the same unit into a large LAN, and realize the interconnection of all branches in the enterprise [7]. Although the information is transmitted through the internal public network of Northeast Power Grid Company, it can guarantee the security and confidentiality of information transmission. The system is shown in Fig. 1.

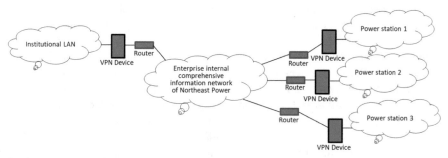

Fig. 1. Application of VPN in the enterprise internal integrated information network of Northeast Power Grid Company

The specific process of establishing VPN in Windows 2007 server is: first, install PPTP on Windows 2007 and input the number of VPNs supported by the server, then add VPN device as the port of RAS, configure encryption and verification methods, and then configure TCP/IP protocol for VPN tunnel. Finally, RAS routing is configured so that PPTP packets pass through the server to the network [7].

VPN device encryption adopts the special core password algorithm, and password management adopts the corresponding password management system. The composition structure and IP address planning of the system are as follows: LAN IP of central contact of the factory Department is 100.100.0.1, the subnet mask is 255.255 255.0, i.e. 100.100.0.0 network segment, addresses of each power station network segment are 100.100.1.0 to 100.100.3.0, IP addresses of VPN devices of each power station are 100.100.1.1 to 100.100.3.1, and masks are 255.255.255.0.

Each VPN device is composed of Windows 2007 system, and the configuration process is as follows:

a. Install the PPTP protocol on the windows 2007 server, and set the number of VPN private network channels to 2. These two channels are used to connect to the central point server.
b. Install VPN routing and RAS management software. When the installation is completed, the routing and RAS admin items will appear in NT program items.
c. Add VPN dial-up interface in dial-up network, and configure user name, password, running protocol and security authentication.
d. According to the requirements of the password system, configure the encryption relationship table between subnets.
e. For the configuration of routing options, we select static routing for configuration, where the destination IP address of VPN routing is 100.100.0.0 (mask is 255.255.0.0), that is, the route to the central contact. Here, the VPN IP address pool of the central point equipment is 100.100.100.0 to 100.100.100.255. The gateway of VPN routing in each power station is the first IP address of the central point VPN PPTP pool address pool, namely 100.100.100.1. Interface is the added VPN dial-up interface [8]. The route to the integrated information network is 0.0.0.0 (mask is 0.0.0.0), the gateway is the Ethernet port address of each power station, and the interface is the network card of the machine. Create a user for dialing in the device NT domain.

Configuration process of central contact VPN device:
The configuration process is basically the same as that of VPN devices in each power station, which means, the number of VPN dedicated network channels added is the sum added by each power station. To create VPN interfaces to each power station, and configure static routing for each interface. Also create each VPN user on the domain user manager of Windows 2007.

When the configuration of VPN equipment in the central point and each power station is completed, VPN connection can be carried out. In addition, it should be noted that the IP address of the client machine on each power station network segment should be configured as the planned address of the power station VPN, rather than the integrated information network address or others [9]. Only in this way can the client machine connect to the central contact through the local VPN device for resource access.

Through the above process, VPN can be easily established, and has good scalability and flexibility. Due to the establishment of VPN, hydropower plant enterprises can benefit from the following aspects, which are embodied in the realization of network security, simplification of network design, reduction of cost, easy expansion and full control of Initiative (for example, enterprises can give dial-up access to ISPs, and be responsible for the inspection and access rights of users). The establishment of VPN also supports the emerging applications (such as IP fax) [10].

5   Conclusion

VPN combines the advantages of private network and public network, allowing the unit with multiple sites to have a hypothetical completely proprietary network, while using the public network as the communication line between their sites. It ensures its security through reliable encryption technology. It is a low-cost and efficient scheme to provide reliable services in unreliable networks. With the development of IP technology and mobile communication technology, various technologies have emerged to realize VPN services, and constantly meet the requirements of users' security, confidentiality, efficiency and flexibility. Except for the enterprise internal comprehensive information network of Northeast Power Grid Corporation, most of the other network systems in our country are based on the TCP/IP protocol. VPN system can build a special network system for each business department to meet their own needs. Therefore, it has a broad application prospect.

References

1. Davis, C.R.: IPSec; VPN Security Implementation, pp. 14–15. Tsinghua University Press, Beijing (2019). Translated by Zhou Yongbin, Feng Dengguo, et al. (in Chinese)
2. Li, M.: MPLS VPN networking simulation and application design. Luoyang Electro Optic Control **6**, 45–48 (2009). (in Chinese)
3. Yang, G.: Baishan power plant's generating capacity reaches a record high [OL] (2015). http://finance.sina.com.cn/. (in Chinese)
4. Yu, C.: Computer Network and Information Security Technology. China Machine Press, Beijing (2018). (in Chinese)
5. Sun, W.: VPN tunnel technology. Comput. Appl. Res. **17**(8), 55–58 (2010). (in Chinese)
6. Brown, S.: Building a Virtual Private Network. People's Post and Telecommunications Press, Beijing (2009). Translated by Dong Xiaoyu, Wei Hong, Ma Jie, et al.
7. Chen, Y.: Application of VPN technology in hospital network expansion. Comput. Appl. Res. **28**(4), 35–36 (2017). (in Chinese)
8. Wan, L.: Research on firewall technology - implementation of virtual private network. University of Electronic Science and technology, Chengdu (2003). (in Chinese)
9. Gleeson, B., Lin, A., et al.: A Framework for IP Based Virtual Private Networks. IETF RFC 2764, February 2000
10. Johnson, D., Perkins, C., Arkko, J.: Mobility Support in IPv6. IETF RFC3775, June 2004

Prediction of Technology Trend of Educational Robot Industry Based on Patent Map Analysis

Yue Wu$^{(\boxtimes)}$ and Jian Fang

Jilin Engineering Normal University, Changchun, Jilin, China
28841898@qq.com

Abstract. This paper starts with the research of the education robot industry, conducts a detailed investigation, analysis and summary from the theory, analysis method and application of education robot, patent map research at home and abroad and other aspects. By collecting relevant patents of educational robots at home and abroad, the research status of educational robots is understood. Apply the research of patent map to the education robot technology, modify and supplement the existing patent map appropriately, dig the related patent literature of education robot deeply, draw the detailed patent map of education robot at home and abroad, and analyze it. This paper summarizes the research on hot technology of educational robot, explores the key technology of educational robot, and predicts the life cycle of key technology, so as to predict the key technology and hot technology of educational robot.

Keywords: Educational robot · Patent map · Technical trend prediction

1 Introduction

According to a variety of policies published in recent years around the world, the use of artificial intelligence technology to promote the reform and innovation of the education system has attracted the attention of countries around the world. In the future, under the background of the rapid development of education technology innovation and other aspects, education robots will develop rapidly. Had the state of education innovation, gradually pay attention to scientific and technological innovation, robot education has begun to gradually into the primary and secondary schools, let children early exposure to advanced interesting robot, by developing education function of the robot, the excavation of the intelligent robot education value, can effectively promote the innovation in education, promote the education reform and modernization of education, innovation and practical ability to improve primary and middle school students. In this paper, the application for a patent for nearly a decade of education robots at home and abroad to make objective statistical analysis, and then make a comparison on the development of future education robot reasonable life cycle prediction about the future of education robot production technology trends do detailed analysis, to make up for the market of education robot of blank in this field, provide the guide for the development of education robot in the future.

© The Editor(s) (if applicable) and The Author(s), under exclusive license
to Springer Nature Switzerland AG 2021
J. MacIntyre et al. (Eds.): SPIoT 2020, AISC 1283, pp. 149–155, 2021.
https://doi.org/10.1007/978-3-030-62746-1_22

2   The Development Status of Educational Robots

At first, educational robots were promoted and developed mainly through competitions. Later, they began to focus on teaching. However, competition robots still occupy a large share of the educational robot market. Although education robot in China started relatively late compared with abroad, but it has a very good development trend, our country has attached great importance to education in China since ancient times, Beijing, on average, a family for children education the teaching cost has been beyond the one-third of household spending, so good education product there will be a huge market demand in our country.

According to the China Academy of Commerce, the global market for educational robots totaled about $100 million in 2015 and is growing at a rate of about 20 percent annually. In 2017, the scale of global education machine market reached 850 million US dollars, with a year-on-year growth of 14.4%, and the overall market showed a steady rise. It is expected that the global education robot market will reach 970 million US dollars in 2018, with a year-on-year growth of 13.5%, and reach 240 million US dollars in 2020 [1].

According to the statistical data of Prospective Industry Research Institute, Since 2009, the number of patent applications for educational robots in China has been on the rise steadily. The number of patents filed in 2016, 2017 and 2018 was the highest, reaching 586 in 2017. It can be seen that China's educational robots have a good development trend with unlimited potential.

3   Comparative Analysis of Key Technology of Education Robot Patent

3.1   Application of Patent Analysis Method in Technology Foresight

Foresight (Technology Foresight) is proposed by the United States to combine science and technology and economic information, and to optimize the information integration of strategic management method, has been widely used in many countries today. There are also many methods for technology prediction. Taking the patent literature research method as an example, the patent analysis method using patent literature for technology foresight is also diversified. Firstly, the management level and technology level can be analyzed; secondly, the hot and cutting-edge technologies can be found through the statistics of patent reference profit frequency and the patent clustering network map; finally, the hot technologies and the forefront of technology can be found through the clustering layout of patent topics, and then the technology can be predicted [2].

At present, technology foresight has been widely used in the world, which provides a decisive basis for national strategic decision-making. With the continuous improvement of market demand for technology, technology is also in constant innovation. Therefore, enterprises need to do market research in a period of time, and make reasonable adjustments to the research and development of technology, in order to meet the market demand. First of all, enterprises can analyze the patent of relevant technology to understand the current development situation of this technology in the market, its related hot technologies, and what aspects this technology is applied to, so as to improve the market competitiveness. Secondly, the history and current development of this technology should be investigated in order to find the opportunity for technological innovation. Finally, find and solve the existing technical problems, so as to seize market opportunities. It can be seen that the analysis of patent literature is particularly important and has become one of the important methods of technology foresight [3].

3.2 Comparative Analysis of Patent Technology Map Between China and World Education Robot Industry

Robot in China started in the early 1970s, and slow development in recent years, the Chinese government attaches great importance to the development of robotics, and launched a series of policies to promote the development of robot industry, on the basis of the development of industrial robots, vigorously develop education robot, the current Chinese education robot with the world of education robot patent technology in the application field also has a lot of. Domestic educational robots such as Lego, Alpha, Chigo and BDS are typical representatives, while foreign educational robots such as NAO and MINIROB0T are typical representatives. Another important part mainly focuses on the education system. The leader in this aspect is Wonder Workshop. In this aspect, although their hardware technology is not as perfect as the previous part, it can be clearly seen that they combine the hardware system well with educational learning software [4]. This is more conducive to the user's learning, and can also better increase the user experience.

China is basically similar to the world in terms of applied technology. However, the similar part is generally the basic function of the robot, involving B25J17, B25J15 and B25J9, and the IPC classification number in this aspect refers to the basic technologies such as mechanical arm, mechanical wrist, mechanical finger and mechanical clip. Moreover, the proportion of these technologies is still quite large, accounting for 1/3. Therefore, it is not difficult to find that the development of educational robots pays more attention to institutional inventions such as the appearance of institutions [5]. In the future, the patent of educational robot will further strengthen the new creation, and the research on educational robot will lay more emphasis on artificial intelligence, which is also a new hot spot in the global scope.

4   Technology Trend Forecast of Educational Robot Industry

4.1 Theoretical Model

S type curve is a kind of long-term projections for growth curve model, and forecast data changes over time in accordance with the law of growth curve, namely the things happen through four stages, development, maturity, saturation, according to the principle of growth curve can change the process for a technology to use appropriate regression model analysis method, to obtain its growth curve, prediction technology in phase, the S curve mainly includes two kinds: one kind is symmetrical S curve, according to the Logistic curve; An asymmetric type of S curve is called the Gomper-TZ curve [6].

Logistic curve is more suitable for the study of educational robot technology in this paper. The curve equation is as follows:

$$y = \frac{k}{1 + a * e^{-bx}} \tag{1}$$

In the formula, Y is the number of patents accumulated; A is the slope of the S curve, which is the growth rate of the S curve; B is the time point of midpoint in the growth curve; K represents the saturation level of growth, namely saturation, which is defined as [K × 10%, K × 90%] and is also the time length t required in the growth and maturity stages.

4.2 Education Robot Industry Patent Technology Life Cycle Forecast

Education robot patent technology life cycle prediction data are based on the number of patent applications every year education robot industry, calculate every year in turn education robot application for a patent for the total amount, with an annual cumulants as dependent variable, an application for a patent for the year as the independent variable, Matlab software will each year accumulative total amount as a Logistic regression model, finally it is concluded that the three unknown parameters and S type curve equation, so that the prediction research of patent technology [7]. The prediction types in this paper are divided into two categories. Firstly, the patent technology cycle of educational robot industry in the world is predicted, and then the patent technology cycle of educational robot industry in China is predicted.

According to China Patent Network, the annual patent application volume of Education robot in China obtained by soopat patent network search engine is shown in Fig. 1.

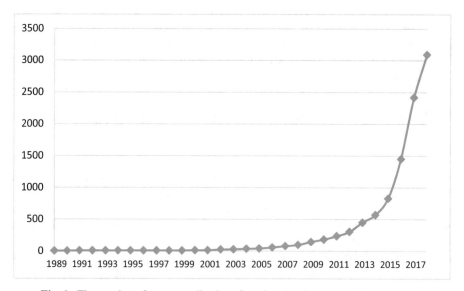

Fig. 1. The number of patent applications for educational robots in China per year

According to China Patent Network, the annual patent application volume of Education robot in China obtained by soopat patent network search engine. Based on Logistic Growth, taking the number of patents accumulated as the vertical axis and the year as the horizontal axis (time unit, year minus 1988), and use matlab software to each year accumulative total amount as a Logistic regression model, it is concluded that the three unknown parameters (see Table 1) and the scatter of existing data regression analysis get fitting curve of the Fig. 2, on the basis of this paint education robot S graph (see Fig. 3) [8].

Table 1. Value of regression model parameters

Parameter	The values
K	11536
a	2821400
b	0.4626

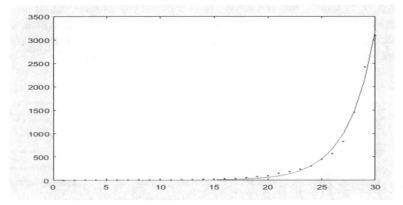

Fig. 2. Logistic regression fitting curve

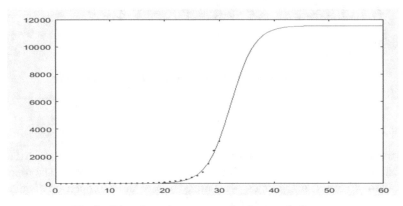

Fig. 3. Education robot patent technology cycle S curve

The four cycles of educational robot patent technology can be inferred from the graph, as shown in Table 2.

Table 2. Description of the growth stage of educational robotics Technology

Birth stage of technology	Technical growth stage	Technology maturity stage	Stage of technological decline
1989–2014	2015–2023	2024–2034	2035–

5   Key Technologies for the Future of the Education Robot Industry

Using China hownet analysis of the patent document can be related to education robot patent literature search and the center frequency and degree of high frequency keywords to search, in measuring visual analysis, to analyze all the retrieval, can according to the literature on patent issuing year distribution, main keyword distribution, subject distribution, distribution of the major categories, and can be part of literature pie chart and histogram for patent analysis [9]. The results are shown in Fig. 4.

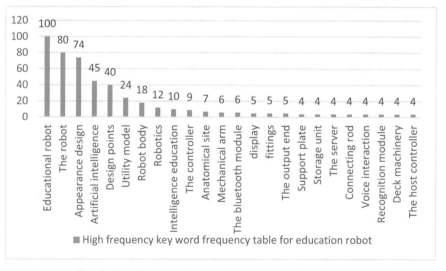

Fig. 4. High frequency keyword frequency of education robot

In the field of educational robot, key words: robot frequency of 100 times is the first. About 20% of the key words in the whole field of educational robotics. The robot was designed for 80 times and 74 times, accounting for 30% of the proportion of education robots. Artificial intelligence for 45 times, design essentials for 40 times, practical types for 24 times, robot body for 18 times, robot technology for 12 times, intelligent education for 10 times, these 6 aspects are the technologies of educational robot technology, which account for 30% of the proportion of educational robot technology. Technologies related to educational robots include controllers, anatomical sites, robotic arms, regulators, Bluetooth modules, display screens, connectors, output terminals, support plates, storage units, servers, connecting rods, voice interaction, recognition modules, deck machinery, master controllers, etc., which account for 20% of educational robotics. According to the analysis of high-frequency keywords in the field of international educational robotics, the research focuses on educational robotics, robotics, appearance design and artificial intelligence [10].

6 Conclusions

Based on the detailed comparison of the relevant educational robot industry patents in China and the world, and the prediction of the future key technologies of educational robots, this paper comprehensively presents the patent map of educational robots in China. In order to promote the healthy development of China's education robot industry, the education robot related industry is developing rapidly. Both at the world level and in China, educational robot patent applications in recent years are in the stage of rapid growth, and the trend of continuous growth. Although China has occupied a place in the world's total patent applications, but with the development of the past few years, the total number of Chinese patent applications in the next few years will still increase substantially, will certainly maintain its place in the world.

Acknowledgement. Jilin Science and Technology Development Project 《Patent analysis and strategic research of educational robot》 (Project number: 20190802025ZG).

References

1. Zhao, S., Zhang, Y.: Exploration and practice of robot application in the field of education. Robot Technol. Appl. (01), 39–43 (2016)
2. Huang, R., Liu, J.: Development status and trend of educational robots. Mod. Educ. Technol. **27**(01), 13–20 (2017)
3. Ge, Y., Li, W.: Design and application of Internet of Things education robot based on experimental teaching. Exp. Sci. Technol. (06), 18–22 (2015)
4. Hou, J.: Current situation and thinking of robot education in schools. Educ. Technol. **15**(12), 85–87 (2016)
5. Li, Z., Cao, Y.: Development status and future trend of speech recognition patent technology. Chin. Invent. Pat. **14**(S1), 55–59 (2017)
6. Yu, Y.: History and prospect of machine learning based on artificial intelligence. Electron. Technol. Softw. Eng. (4), 129 (2017)
7. Wu, M., Zhou, Y.: Introduction to language recognition technology. Mod. Comput. (19), 36–40 (2019)
8. Zhang, Z., Zhu, T.: Patent situation analysis in robot field. Chin. Invent. Pat. (04), 40–42 (2008)
9. Wu, Y.: Current situation, practice, reflection and prospect of robot education from the perspective of machine intelligence. J. Distance Educ. **36**(04), 79–87 (2012)

Coal Handling System of Power Plant Based on PLC

Zhijie Zhang[✉]

School of Electrical Engineering, University of Jinan, Jinan,
Shandong, People's Republic of China
2861515249@qq.com

Abstract. The main task of this design is to process coal from coal source into pulverized coal and supply it to the original coal bunker for combustion and power generation. In the design, Siemens CPU315-2DP is used as the main control unit to control all the components of the coal transportation system of the power plant. The PROFIBUS-DP communication method is used to implement the distributed control system through the hardware configuration of the ET200M distributed I/O site, and the distributed control area is scattered. Various equipment. While realizing the system's reverse coal flow sequence start, downstream coal flow sequence stop, fault interlock stop, severe fault emergency stop, automatic coal blending and other functions, it solves the serious problems of this type of system, such as the coal falling pipe is easy to block Problem; the problem that the conveyor belt is easy to tear and collapse; the problem of the deviation of the conveyor belt. After debugging, all the above functions have been realized.

Keywords: Siemens PLC · PROFIBUS-DP communication · Distributed I/O

1 Introduction

All along, the development level of the domestic power industry lags behind that of developed countries in the world to a great extent. The utilization rate of automation technology in the entire industry is low, the power generation capacity of a single generator is small, and the output of electrical energy needs to consume more coal. Poor reliability and stability, and the grim situation facing environmental sustainable development are all urgent issues that need to be resolved. Among them, the problems of automation and low electrification, small stand-alone power generation capacity, and high coal consumption for power supply are the bottlenecks in the development of the power industry, which severely restrict and restrict the progress of my country's power industry [1].

This requires us to study and apply new high-efficiency power grid technology, improve the reliability of power grid operation, reduce power generation energy consumption, increase power output, and make the greatest efforts to reduce coal consumption [2].

J. MacIntyre et al. (Eds.): SPIoT 2020, AISC 1283, pp. 156–162, 2021.
https://doi.org/10.1007/978-3-030-62746-1_23

2   Overall Design of Coal Transportation System

2.1   Design Task Analysis

The design requires the completion of the overall hardware design and software design of the power plant's coal transportation system to produce a system that can solve the actual problems such as low automation, tearing and collapsing of the conveyor belt, and clogging of the coal drop pipe. The design uses Siemens CPU315-2DP as the main control site, ET200M distributed remote I/O to achieve distributed control, and centralized management of the equipment in various places. In the design, the sensor detection technology is used to detect the running state of the conveyor belt and the coal bunker level, so as to monitor and control through the PLC in real time to ensure the stable operation of the system. The optimized design of software and hardware makes the whole system more reliable, and realizes the functions of sequential start and stop, fault interlock stop, and emergency stop of severe faults. In the entire design process, electrical wiring diagrams need to be drawn, hardware selection and software writing, debugging and integration of the entire system, and ultimately system requirements [3, 4].

2.2   Design of the Overall System Plan

Hardware part: Use Siemens PLC as the main control station, ET200M cooperates to realize remote distributed control, use sensors such as photoelectric switch, deviation switch, resistive coal level sensor to detect potential failure points, PLC reacts to all detection information, And execute the control plan to make the system run strictly in accordance with the requirements of PLC.

Software part: write high-quality control programs to improve the stability of the software, so that the system can achieve various functional requirements stably and efficiently, quickly analyze the fault information, and make accurate responses. The photoelectric switch sensor is used to detect the running position of the conveyor belt to make it work within the normal range [5]; the deviation switch is used to detect the deviation state to avoid the occurrence of heavy equipment failure; the resistance coal level sensor is used to detect the coal level, and the coal level is started according to the coal level [6]. Stop the program to realize automatic coal blending. Finally, a good control effect is achieved.

The upper computer part: configure the power control and PLC, control the resources of the PLC, realize the functions of automatic start and stop, fault interlock stop, emergency stop of severe faults, etc., and record the data, make an alarm record of the fault, and output Access database, information can be queried by accessing the database through the network [7, 8]. The overall control structure of the system is shown in Fig. 1.

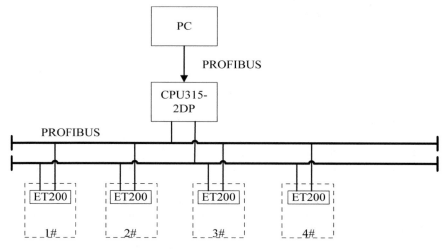

Fig. 1. Overall control structure of the system

3 Hardware Circuit Design

System hardware equipment is the basis on which the system depends, and strict hardware selection is required. It is essential to design reliable electrical circuits. This chapter analyzes and studies the hardware design, and finally designs a complete set of good hardware configuration [9].

3.1 PLC Selection

As the control unit of the system, PLC operates and monitors the whole process of the system control. Whether the PLC main controller is good or not is closely related to the reliability, safety, production efficiency, production cost and other issues of the overall system operation, so it is very important to choose a suitable PLC main controller.

Siemens CPU315-2DP integrates DP master-slave interface and MPI interface, has outstanding communication capabilities, easy to implement distributed structure, user-friendly interface, flexible operation mode, has a huge upgrade space, without adding any hardware equipment Under the premise of completing more complex tasks. PROFIBUS and MPI communication make communication networking simple and easy. Modular programming makes programming more flexible and can solve many complex tasks.

3.2 Distributed I/O Selection

This design uses distributed I/O to directly control the scattered devices in various places, and uses PROFIBUS-DP communication to communicate with the host CPU for information transmission. In all distributed I/O, ET200M can configure all modules of S7-300 PLC, which is easy to use, high efficiency, and has better scalability for more complex tasks [10].

3.3 Selection of Anti-tear Sensor

In the design and research of the entire coal transportation system, the belt conveyor of coal should be regarded as the key design and implementation process. During the movement of the belt conveyor, problems such as tearing or collapsing of the belt occur from time to time. In this design, sensor detection is used to determine the real-time running state of the conveyor belt, and the PLC reacts.

3.4 Selection of Anti-deviation Sensor

During the operation of the system, the conveyor belt will slip and run abnormally, which will cause serious problems such as belt deviation. This design uses a deviation switch to detect deviation information. When the deviation reaches a certain angle, the output level triggers an emergency stop signal to avoid serious deviation.

4 Software Design

The coal transportation system of the power plant in this design is different from the traditional coal transportation system. The control aspect is to complete the automatic control of the equipment in the system by writing a PLC program to complete the coal supply and supply task of coal-fired power generation. The order of system automatic start and stop, manual start and stop, automatic coal blending, fault interlock stop, and monitoring control requirements must comply with the principle: after the manual or automatic start signal is issued, the equipment in the PLC program control system extends in the direction of reverse coal flow Start in time sequence; after the automatic or manual stop signal is issued, all the equipment in the PLC program control system will stop according to the coal flow direction delay sequence; the system will automatically distribute coal according to the upper and lower limits of coal level detection, when the upper limit is reached, the system will be shut down, and when the lower limit is reached, Start the system; when one or more equipment failures occur in the system, immediately stop the failed equipment and all the equipment upstream of the failed equipment through interlock control; When a major accident such as an emergency stop, severe deviation, belt tear or collapse occurs, Forcibly stop all equipment in the system. The overall structure of the software is shown in Fig. 2.

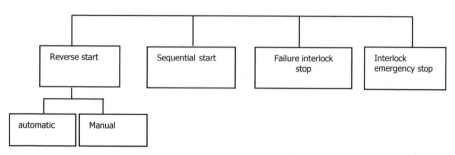

Fig. 2. Overall structure of the software

In the process of programming, it is necessary to set symbols for the I/O points used, and give simple and clear symbol names such as function blocks, storage units, timers, etc., so as to facilitate the staff to consult the program. The I/O points used in this design are 70 points, each point is defined with a simple and easy to understand symbol name, and an explanation is given.

5   Host Computer Design

During the operation of the system, due to the occurrence of various problems such as wear, aging and emergency conditions of the equipment, the system will inevitably be damaged or malfunction. This requires the staff to be able to find and solve the system problems in a timely manner and understand the operation status of all equipment in real time. The best way to prevent emergencies. In this design, the system monitoring adopts the force control networking monitoring mode, which is monitored through force control and PLC configuration. It monitors, starts and stops each remote I/O site, runs the conveyor belt conveyor, and each auxiliary conveyor equipment to monitor the operating status. Record data. When a fault occurs, an alarm message is issued to warn technicians to find and deal with it in time. And can record data and alarm information, output Access database, information can be checked through the network.

6   System Debugging

The first is hardware debugging, first verify the detection accuracy of the sensor. When detecting the photoelectric pair tube, use an oscilloscope to observe the change of the output level and the corresponding output voltage value when there is no object in front. The detection result photoelectric switch can reliably output a high level of about 23 V and a low level of about 1 V, and can accurately and reliably determine whether an object exists. The same detection method is used when detecting the deviation switch and the resistance type coal level detection sensor. The detection result of the deviation switch can stably output a high level of about 21 V and a low level of about 0.8 V, which can accurately and reliably determine the conveyor belt Whether to go off course. The detection result of the resistance coal level sensor is that it can stably output a high level of about 22 V and a low level of about 0.8 V, which can stably and accurately detect whether the coal level reaches the limit. The start-stop debugging of the equipment adopts a motor simulation experiment, which controls the contactor coil through the digital output point, and the coil controls the power supply of the motor. The experimental result is that the motor can be accurately started and stopped, that is, the equipment is started and stopped. The hardware debugging diagram is shown in Fig. 3.

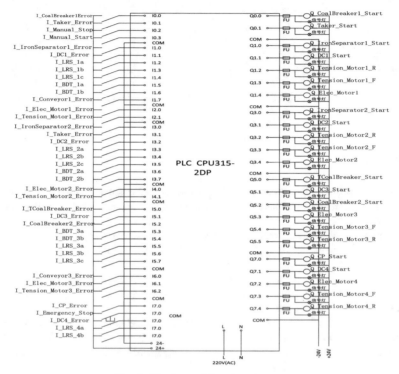

Fig. 3. Hardware debugging diagram

In terms of software debugging, online debugging and simulation of the program, automatic start and stop, fault interlocking, emergency stop for severe faults, automatic coal blending and other functions have been realized. The software simulation is shown in Fig. 4.

The functions realized by the power control host computer are basically the same as the system program. In addition, it also has an alarm function and a data record storage function

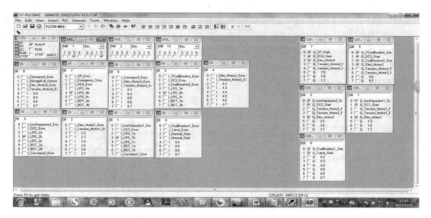

Fig. 4. Software simulation diagram

References

1. Escobar-Alvarez, H.D., Ohradzansky, M.: Research on integration and development of intelligent transportation and smart city construction. Digit. Commun. World **166**(10), 119–210 (2018)
2. Guerrero, J., Torres, J., Creuze, L.L., Peng, D.: Research and application of smart city transportation network system. Comput. Technol. Dev. **35**(1), 150–160 (2019)
3. Abolvafaei, M., Ganjefar, S.: Overview of key technologies of intelligent transportation system in construction of smart city. Renew. Energy **139**(2), 1437–1446 (2019)
4. Su, S.: On the development trend of intelligent transportation. Heilongjiang Transp. Sci. Technol. **9**(2), 280–281 (2017)
5. Amit, K.K.C., Jacques, L.: Discriminative and efficient label propagation on complementary graphs for multi-object tracking. IEEE Trans. Pattern Anal. Mach. Intell. **39**(1), 61–68 (2017)
6. Conesa, A., Madrigal, P., Tarazona, S., et al.: A survey of best practices for RNA-seq data analysis. Genome Biol. **17**(1), 181–185 (2018)
7. Mareda, T., Gaudard, L., Romerio, F.: A parametric genetic algorithm approach to assess complementary options of large scale wind-solar coupling. IEEE/CAA J. Autom. Sinica **4**(2), 260–272 (2017)
8. Planet, P.J., Parker, D., Cohen, T.S., et al.: Lambda interferon restructures the nasal microbiome and increases susceptibility to *Staphylococcus aureus* superinfection. Comput. Knowl. Technol. **7**(1), 111–115 (2018)
9. Sartoris, B., Biviano, A., Fedeli, C., et al.: Next generation cosmology: constraints from the Euclid galaxy cluster survey. Mon. Not. R. Astron. Soc. **459**(7), 968–974 (2019)
10. Ma, J.: Research on strategies to solve traffic congestion in smart cities under the background of big data. Comput. Knowl. Technol. **6**(6), 4262–4264 (2018)

Discussion on the Construction of Wireless Campus Network Based on SDN Architecture

Dongqing Hou[✉]

Information Department, West Yunnan University, Linxiang District,
Lincang 677000, Yunnan, China
17294202@qq.com

Abstract. Starting from the wireless network architecture, this article expounds the concept and principle of VXLAN technology, and proposes a SDN wireless campus network construction scheme that uses distributed Vxlan gateway networking and a director controller. This solution realizes the decoupling of the control plane and the user plane. According to the user attributes, the label is automatically and uniformly issued and the strategy is issued. Users and terminals in the entire network can be connected to the wireless network at any time and at any location without any location restrictions. Network resources and network permissions are always consistent.

Keywords: SDN · Wireless campus network · CAPWAP · VXLAN

1 Preface

Information technology has changed the way people work and learn, and has given the education industry new connotations and requirements. In addition to the traditional teaching environment such as classrooms, laboratories and training rooms on campus, teaching activities of higher education also take place in cyberspace based on information technology. However, the traditional wired network can only provide fixed location services, and its application is subject to greater restrictions and constraints, making it difficult to meet the needs of multiple terminals, multiple scenarios, and multiple services [1]. In this context, the spring of the development of wireless campus networks in colleges and universities has ushered in the increasing application of mobile office, mobile teaching and mobile scientific research. With the popularity of intelligent mobile terminals, IoE (Internet of Everything) has brought an explosive growth in the number of terminals, the workload of network management and configuration has increased dramatically, and the security policies have become more complicated.

The traditional wireless network architecture adopts a vertical layered model, coupling control plane and user plane data to network devices. Network devices must not only complete data forwarding [2], but also have control functions for generating forwarding and managing data forwarding rules [3]. The existing wireless network architecture is rigid and cumbersome under the new trend. In view of this, this article takes the

J. MacIntyre et al. (Eds.): SPIoT 2020, AISC 1283, pp. 163–170, 2021.
https://doi.org/10.1007/978-3-030-62746-1_24

West Yunnan Normal University of Science and Technology as an example, using the ADCampus technology proposed by H3C to build an intelligent SDN wireless campus network, providing network automation on-line, business end-to-end automated deployment, access user name and address binding, wired and wireless integration Management and other capabilities.

2 Overview of Key Technologies

2.1 CAPWAP

CAPWAP (Control And Provisioning of Wireless Access Points) wireless access point control and configuration protocol [4], established by the CAPWAP working group established by the IETF, is used for wireless terminal access points (AP) and wireless network controllers (AC) Communication interaction between the AC and the AC to achieve centralized management and control of its associated AP. Therefore, CAPWAP can be said to be one of the most important technologies in the thin AP solution.

CAPWAP is a UDP tunneling protocol that works at the application layer. UDP port 5246 is used to transmit control packets, and UDP port 5247 is used to transmit data packets. CAPWAP establishment needs to go through the following seven processes: AP discovers AC; AP obtains relevant information through DHCP; AP requests to join AC; AP automatically upgrades; AP configuration release; AP configuration confirmation; forwards data through CAPWAP tunnel [5].

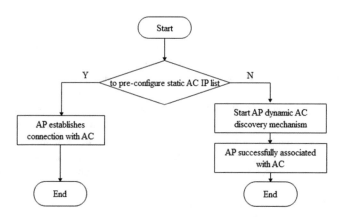

Fig. 1. CAPWAP establishment process diagram

The establishment process of CAPWAP is shown in Fig. 1. After the AP is powered on, if there is a static AC IP list manually pre-configured by the administrator, the AP starts the pre-configured static discovery process to establish a connection with the specified AC; if the AC IP is not configured in advance List, start the AP dynamic discovery mechanism, obtain the IP address, DNS server, domain name and other information through DHCP, and broadcast the discovery request to contact the AC, the AC that received the discovery request packet checks whether the AP has access authority, and if there is access The authority will reply to the confirmation message and establish a CAPWAP tunnel to forward user data.

2.2 VXLAN

VXLAN (Virtual eXtensible Local Area Network) is one of the NVO3 (Network Virtualization over Layer 3) standard technologies defined by IETF and is essentially a tunneling technology [6]. NVO3 is a general term for the technology of building a virtual network based on a three-layer IP overlay network. VXLAN is just one of NVO3 technologies. The original data frame is encapsulated in a UDP packet by adding a VXLAN header, and the UDP packet is forwarded in the transmission mode of a traditional IP network. After the packet reaches the destination endpoint, the outer encapsulated part is removed, and the original data frame is delivered to the target terminal [7].

VXLAN is to build a virtual layer 2 network on a layer 3 network to achieve layer 2 communication between hosts. As shown in Fig. 2, the endpoints of the VXLAN tunnel are called VTEP (VXLAN Tunnel Endpoints) and are responsible for the encapsulation and decapsulation of VXLAN packets. A pair of VTEP corresponds to a VXLAN tunnel. The source VTEP encapsulates the packet and sends it to the destination VTEP through the tunnel. The destination VTEP decapsulates the received packet.

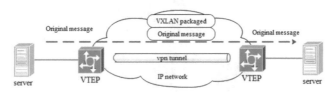

Fig. 2. VXLAN network model

Under the typical "Spine-Leaf" VXLAN networking structure, according to the deployment location of Layer 3 gateways, VXLAN Layer 3 gateways can be divided into centralized gateways and distributed gateways. A centralized gateway refers to the centralized deployment of a Layer 3 gateway on a Spine device, and all traffic across subnets is forwarded through the Layer 3 gateway to achieve centralized management of the traffic. A distributed gateway uses Leaf nodes as VXLAN tunnel endpoints VTEP. Each Leaf node can serve as a VXLAN Layer 3 gateway. Spine nodes are not aware of VXLAN tunnels and only serve as VXLAN packet forwarding nodes.

3   Architecture Analysis

According to the actual situation of West Yunnan Normal University of Science and Technology, a distributed Vxlan gateway networking method is adopted. The network is composed of access, aggregation, and core three-layer equipment. The external is equipped with a director controller. The aggregation layer is VTEP and user gateway, AC, Director, and DHCP. The server is deployed centrally, and the architecture is shown in Fig. 3. The architecture of the solution has the following features: the access layer uses vlan for communication, the aggregation layer is isolated by Vxlan, the aggregation and core run Vxlan to build an overlay network, and a logical large layer 2 network, while using distributed gateways to effectively suppress broadcasts Storm; Adopt business-oriented grouping mode strategy, divide security groups according to business scope, and formulate access strategies based on security groups; 5W1H-based authentication mechanism develops multiple dimensions to cover access scenarios; supports user terminals throughout the life cycle of mac and IP Strong binding to meet the strong security needs of loose campus network users. With Director-guided configuration, administrators can complete the automatic generation of configuration files without entering any command line. After the device is powered on, it can automatically load configuration files of corresponding roles without manual intervention.

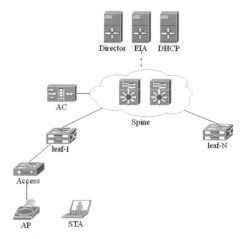

Fig. 3. Network architecture diagram

4   Implementation Process

The Director uses the VLAN4094/VXLAN 4094 interface address as the management IP to manage Spine, Leaf, Access, and wireless AC equipment; the wireless AC equipment establishes a CAPWAP tunnel through VLAN4093/VXLAN 4093 to manage the AP. The port where the ACCESS switch connects to the AP enables PoE, and is configured as pvid 4093 and trunk vlan all.

The AC controller is hung on the spine, communicates with the AP through vlan4093/vxlan4093, communicates with the director through vlan4094 and is included in the management of the ADCampus platform, and communicates with the authentication server for authentication. The core device configuration is as follows:

4.1 Wireless Controller

(1) Configure the Layer 3 interface VLAN4094 on the AC for the Director's management AC; configure the Layer 3 interface VLAN4093 as the communication interface with the AP, and the AC establishes a CAPWAP tunnel with the AP through the interface address.

```
#
interface Vlan-interface4094
  ip address 10.200.4.4 255.255.252.0
  interface Vlan-interface4093
  ip address 10.201.255.253 255.255.240.0
#
```

(2) Configure a local user, and the Director uses this user to connect and manage the AP.

```
#
local-user lcnc class manage
  password simple lcnc@12****
  service-type ftp
  service-type ssh telnet terminal http https
authorization-attribute user-role network-admin
#
```

(3) To use AP local forwarding mode, a configuration file needs to be generated locally and downloaded to the wireless AC. It is used to automatically configure all VLANs of the uplink trunk after the AP goes online, and the authorization acl required for MAC portal authentication.

```
#
interface interface Ten-GigabitEthernet1/0/19
  port link-type trunk
  port trunk permit vlan all
  port link-aggregation group 6
#
interface interface Ten-GigabitEthernet1/0/20
  port link-type trunk
  port trunk permit vlan all
  port link-aggregation group 6
#
acl advanced 3001
  rule 5 permit ip destination 10.200.4.2 0      #web portal server address
  rule 6 permit ip source 10.200.4.2 0           #web portal server address
  rule 11 permit ip destination 10.200.200.5 0   #AAA server address
  rule 12 permit ip source 10.200.200.5 0        #AAA server address
  rule 13 permit ip destination 10.200.200.7 0   #DNS server address
  rule 14 permit ip source 10.200.200.7 0        #DNS server address
  rule 100 deny ip
#
```

(4) Use map-configuration to specify the AP configuration file in the default ap group. When the AP goes online, the configuration file is automatically downloaded and effective.

```
#
wlan ap-group default-group
  map-configuration cfa0:/ad.txt
#
```

4.2 Sping Equipment

Configure the interface connected to the wireless controller as a trunk port and allow all VLANs to pass. And configure a server instance on the interface, bind vlan4094 and vsi4094 to the instance, which is used to realize the connection between the management channel and the control channel.

```
#
interface Bridge-Aggregation6
    description TO_WX5560H_BAGG6
    port link-type trunk
    undo port trunk permit vlan 1
    port trunk permit vlan 4093 to 4094
#
    service-instance 4093
        encapsulation s-vid 4093
        xconnect vsi vsi4093
#
    service-instance 4094
        encapsulation s-vid 4094
        xconnect vsi vxlan4094
#
```

4.3 Director Configuration

The core of the SDN network architecture is the Director controller, which is responsible for the generation and distribution of all network device configuration policies and effective management of Internet users. Deploy multiple servers in a cluster to install the Director controller platform and DHCP server, and connect the cluster server directly to the Spine core switch. After completing the above deployment, you can directly log in to the Director platform through HTTP protocol to complete the automatic launch of the device, management and policy settings.

(1) Campus planning

The "Campus Planning" module on the Director controller can plan the network management address segment, realize automatic deployment, and configure automatic deployment parameters and related strategies. According to the previous plan, divide the network into three layers: Spine, leaf, and Access. Set the deployment strategy template for the three-layer equipment and configure the network management address segment to 10.200.4.0/22. When the device goes online automatically, it can automatically obtain the management address and issue the policy template of the corresponding device, so that the device can go online automatically.

(2) Wireless management network configuration

Create a Layer 2 network domain and security group for wireless management through the "business" module on the Director controller. The configuration is as follows: the IP address segment is set to 10.201.240.0/22; vlan/vxlan are both configured to 4093; turn on arp proxy; configuration AC IP is the IP address of the wireless AC; add the corresponding security group, select the dhcp option to create an address pool on the DHCP server.

After the Director wireless network core function configuration is completed, the system will automatically send and receive vxlan and vsi interface configuration on Access, leaf, and spine to complete the device automatic online. Drictor can monitor the running status of events, fault information, health status and performance data of on-line equipment and related subsystems, and provide northbound API interface, with strong programmability to meet the integration requirements of automation platforms or cloud computing platforms [8].

5 Conclusion

The wireless campus network solution based on the SDN architecture proposed in this paper is based on VXLAN technology and uses Director as the core controller to provide network automation on-line, business end-to-end automated deployment, access user name and address binding, wired and wireless integrated management, etc. ability. Realize the decoupling of the control plane and the user plane, according to the user attributes, automatically and uniformly mark and issue the strategy, users and terminals in the entire network can be free of location restrictions, access to the wireless network at any time, any location, and obtain network resources Consistent with the network authority, to achieve the effect of the network with people [9, 10]. However, this solution is still not a complete SDN architecture, and not all equipment and functions can provide software definitions according to actual needs.

Acknowledgements. This work was supported by Scientific research fund project of Yunnan Education Department 2017. Project Name: Research On Campus Network Transformation Scheme Based On IPv6 In Frontier Universities, No.: 2017ZZX113.

References

1. Lv, S., Zhang, W.: Present situation and optimization strategy of campus wireless network construction in colleges and universities. Electron. World **9**(5), 13–14 (2020). (in Chinese)
2. Kim, H., Framster, N.: Improving network management with software defined networking. IEEE Commun. Mag. **51**(2), 114–119 (2013)
3. Hong, B., Wei, D.: SDN-based distributed wireless network architecture and controller configuration strategy. Firepower Command Control **44**(12), 158–162 (2019). (in Chinese)
4. Wang, Q.: Application of wireless controller AC and thin AP in campus network across multiple router environments. New Course **2**, 173 (2015). (in Chinese)
5. Yue, C.: Research on networking technology based on thin AP+AC mode. Sci. Technol. Innov. **20**, 69–71 (2018). (in Chinese)
6. Zhang, Z., Cui, Y., et al.: Research progress of software-defined networking (SDN). Journal of Software **26**(1), 63–80 (2015). (in Chinese)
7. Lin, W.: Research on the flat architecture design of campus network based on SDN and VXLAN. Comput. Knowl. Technol. **15**(26), 53–54 (2019). (in Chinese)
8. Zhang, C., Yang, F.: Research on the planning, construction and deployment of SDN wireless campus network. Electron. World **24**, 102–103 (2019). (in Chinese)
9. Bertaux, L., Medjiah, S., Berthou, P., et al.: Software de-fined networking and virtualization for broadband satellite networks. IEEE Commun. Mag. **53**(3), 54–60 (2015)
10. Erran, L., Morley, M.Z., Rexford, J.: Toward software-defined cellular networks. In: 2012 European Workshop on Software Defined Networking, vol. 5, pp. 7–12 (2012). (in Chinese)

Applicational Status Analysis of Artificial Intelligence Technology in Middle School Education and Teaching

Zhi Cheng[✉]

Hainan Tropical Ocean University, Sanya 572022, Hainan, China
Lijingtg109@sina.com

Abstract. With the development of the times, the progress of society, the continuous improvement of human science and technology and cultural level, the traditional middle school education and teaching has been unable to meet the learning needs of middle school students in the new era. In order to better make secondary education and teaching conform to the current trend of development, this paper puts forward the application of artificial intelligence technology in middle school education and teaching, in order to improve the teaching level of middle school education. Intelligent teaching in the field of artificial intelligence technology is a new product under the development of the current era, and has a high status in middle school education and teaching. Therefore, this paper analyzes the necessity of carrying out intelligent teaching in the field of artificial intelligence technology in middle school education and teaching, and systematically expounds the shortcomings in current middle school education and teaching, and makes a deep analysis on the reform needed in the future middle school education and teaching. Through the research and analysis, it is found that the artificial intelligence technology proposed in this paper has a great effect on the innovation and reform of middle school education and teaching.

Keywords: Science and technology culture · Middle school education · Artificial intelligence · Innovation and reform

1 Introduction

Artificial intelligence (AI) is an important branch of modern computer science [1, 2]. At present, in the field of base education in China, the research and implementation of artificial intelligence include two main aspects: one is to set up the course of artificial intelligence information technology. The main content includes the knowledge of artificial intelligence in the compulsory course of information technology in middle school education and the optional course of artificial intelligence information technology in primary school. On the other hand, the application of artificial intelligence in subject teaching, namely artificial intelligence [3–5], provides rich educational resources and scientific education evaluation methods for school education and teaching.

J. MacIntyre et al. (Eds.): SPIoT 2020, AISC 1283, pp. 171–178, 2021.
https://doi.org/10.1007/978-3-030-62746-1_25

From the current situation, the importance of carrying out intelligent education in secondary schools [6–8] is reflected in the following aspects: one feature is that intelligent education meets the application and development requirements of artificial intelligence. In recent years, with the development of Internet and artificial intelligence, artificial intelligence technology has gradually developed and penetrated into all fields of modern people's work and daily life. A new form of education and application of artificial intelligence has been formed. At present, artificial intelligence education is in the stage of rapid development. If the artificial intelligence education in a middle school is not fully applied and developed, then the ability of intelligent education learning activities and the efficiency and level of intelligent education learning of this middle school student may also be different and different with other artificial intelligence schools which can fully apply and develop artificial intelligence education. Second, give full play to the supporting role and value of middle school students. In addition to school education, high-tech education with artificial intelligence as the core is also one of the focuses of education. Middle school students are the pillar of the future development of artificial intelligence. The quality of intelligent education determines the role of middle school students in the field of artificial intelligence. The above situation puts forward higher requirements for the development of wisdom education [9, 10] in primary and secondary schools.

This paper mainly studies the innovation and reform of middle school education and teaching in the new era. In order to make contemporary middle school students better conform to the development trend of the times, keep pace with the development of the times and enrich their learning methods and learning contents, this paper puts forward the method of applying artificial intelligence technology to the innovation and reform of middle school education and teaching, and systematically expounds the current situation This paper analyzes the shortcomings in the teaching of learning education, and makes a deep analysis of the reform needed in the future middle school education and teaching. Through the research and analysis, it is found that the artificial intelligence technology proposed in this paper has a great effect on the innovation and reform of middle school education and teaching.

2 Application of Artificial Intelligence Technology in Middle School Education and Teaching

2.1 Artificial Intelligence Technology

Artificial intelligence is a science that studies how to use computers to simulate human intelligence. With the development of modern computer technology and the rapid development of social informatization, people pay more and more attention to the practical application of artificial intelligence and how to use artificial intelligence to change the production and life style of society. Education is the cornerstone of social progress, and the application of artificial intelligence is reasonable. First of all, it should be extended to the field of secondary education and teaching. Second, language is the most proud wisdom of mankind. Whether it is Turing test, natural language processing, machine translation, speech recognition technology, the application of artificial intelligence technology in language learning is natural.

2.2   Innovation and Development of Middle School Education and Teaching

In order to better develop middle school education and teaching, we can combine the characteristics of middle school intelligent education, select the appropriate classic artificial intelligence technology as auxiliary tools, and jointly improve the quality of intelligent education. For example, VR technology, as a classic artificial intelligence technology, has obvious advantages in the creation of scenarios and the construction of interactive environment. In the process of carrying out the reform of middle school intelligent education, the technology can be combined with intelligent education to create a virtual learning space for middle school students. Galileo ideal experiment in middle school physics as an example, in the teaching process, virtual reality technology can be used to create a smooth track and small ball composition to meet the needs of students to observe the trajectory and speed characteristics of the ball smooth track.

3   Experimental Correlation Analysis

3.1   Experimental Background

This paper investigates and studies the research and application of artificial intelligence information technology in the practice of artificial intelligence teaching in secondary vocational education in China, and makes a quantitative and qualitative analysis on the research of journals and literatures related to artificial intelligence information technology. Firstly, according to the statistical data of artificial intelligence published in journals, the mathematical intelligence analysis model is established, and the artificial intelligence index of the research object is analyzed and calculated by using the relevant mathematical intelligence model. Then, according to the numerical indicators of each object in the literature, combined with the actual situation of the first-line classroom and the latest development in the field of artificial intelligence, the nature, characteristics, development and change law of the object are judged and analyzed.

3.2   Experimental Design

From the number of journals published, artificial intelligence has become an important research field and development direction in the field of educational technology. The application of artificial intelligence in middle school education and teaching has also been widely concerned by experts and scholars. The following table shows the statistical results of artificial intelligence related literatures in recent five years. There are 79 literatures related to middle school education and teaching, accounting for 39.3% of the total length. The details are shown in Table 1:

Table 1. Statistics of the number of journal publications

Title	The number of articles directly related to middle school	Total number of articles
E-education research	15	50
China educational technology	23	66
Modern educational technology	41	82
Medical education technology in China	0	3
Total	79	201

4 Discussion

4.1 Analysis of the Current Situation of Middle School Education and Teaching Based on Artificial Intelligence

Under the background of the popularization of artificial intelligence in middle schools, basic intelligence education has been carried out. However, from the actual situation of wisdom education in middle schools, the wisdom teaching mode in different schools is generally lack of unity and standardization. The main reasons for the above situation are: the development time of artificial intelligence in China is relatively short, the experience of carrying out intelligent education is insufficient, and the related intelligent education system is not perfect, which eventually leads to the differences of intelligent teaching mode in this respect. Based on this, in order to better carry out the innovation reform

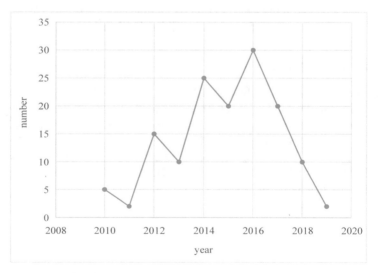

Fig. 1. Time series chart of the number of published articles

method of secondary education based on artificial intelligence, this paper investigates and analyzes the publication time series of artificial intelligence literature in China.

As shown in Fig. 1, according to the publication time series of journal literature, the number of literature presents a "convex" distribution. When artificial intelligence became the first simulated examination in the field of education, people were curious and expected that this new technology could imitate the human brain. Academic attention to this issue is increasing year by year. However, due to the complex discipline background of brain neuroscience, computer science, pedagogy, psychology and other complex disciplines, many experts and scholars have gradually weakened the research on artificial intelligence, so its development speed is slow, and the attention has been declining after 2016.

According to the development trend of research literature, this paper divides the research content into five categories. According to the research category, the research literature was statistically analyzed. The results are shown in Fig. 2. Among them, I refer to the research on the theory of artificial intelligence, II refers to the application research of artificial intelligence in education, III represents the research and development of artificial intelligence system, IV represents the teaching research of artificial intelligence, and V refers to others.

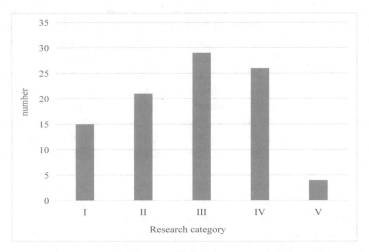

Fig. 2. Statistics of journal literature research categories

As can be seen from Fig. 2, the research contents are mainly focused on the theoretical research and system development of artificial intelligence, the application of artificial intelligence in education and the research in course teaching, and the distribution is relatively uniform. In the in-depth study, it is found that "robot teaching" is the main content in the teaching research of artificial intelligence course. The application of artificial intelligence education is mainly concentrated in Colleges and universities, and the application in middle school is relatively simple, such as the application of Z+Z intelligent education software in mathematics teaching.

4.2 Suggestions on the Development of Middle School Education and Teaching Based on Artificial Intelligence

With the development of artificial intelligence technology, intelligent education has gradually become one of the focuses of middle school education system. As there are still many deficiencies in the current middle school wisdom education, the role of wisdom education has not been fully played. How to improve the quality of wisdom education has become the main problem of wisdom education in middle schools. As the core of intelligent education, it is necessary to analyze the application of artificial intelligence in middle school.

In China, as a leading information technology, artificial intelligence is becoming more and more active in theoretical research and teaching practice in the field of education, which has laid a rich theoretical and practical foundation for digital education and educational management informatization in China. In particular, scholars in the field of educational technology in China are very keen on the application of artificial intelligence in the field of education, and gradually shift the focus of research from theoretical research to technological development research. Therefore, the research results in recent years are very significant, and the representative achievements are: the "intelligent teaching system" teaching system developed by Zhang Jingzhong, academician of Chinese Academy of Sciences, which is mainly aimed at the auxiliary teaching of mathematics. Its goal is to use the related technology of artificial intelligence to solve problems intelligently, automatically infer, draw dynamic diagram, and even interact with students. Through the use of various information technology and tools, the geometric thinking and spatial sense of teachers and students are fully cultivated, so as to effectively improve the teaching effect of mathematics. However, compared with the current foreign mathematics related research, it is relatively late in our country to expand the application of advanced artificial intelligence technology to the field of education and teaching research, and there is a certain gap with the research level of foreign countries. The scope of mathematics research in China is not wide enough and the degree is not deep enough.

Artificial intelligence technology, as a cutting-edge discipline that has crossed many disciplines, is changing the way of expression of modern people's daily thinking and people's traditional concepts, and improving the comprehensive application level of scientific knowledge and the theoretical and practical level of modern people's higher education. In the history of the development of higher education in China, the technology of bringing forth the new from the old often provides a strong support and impetus for the reform and innovation of modern education, which makes the development of teaching and scientific research more convenient and efficient, and makes education more and more fair, universal and popular. The intelligent education software can realize the vision, hearing, speaking and even understanding and feedback of the user's emotion or emotion, so that the user can communicate with various computer electronic devices naturally and fluently through language, text, gesture, expression and other ways, so as to realize the human-computer interaction. In a word, the reform brought about by the development of technology is to deepen the application of modern education and teaching. As the scientific and technological achievements of the three 20th century and new technologies such as space technology, atomic energy technology, artificial intelligence, etc., will be widely used in the field of modern education, and will have a profound impact on the concept of modern education, teaching process and teaching management in the future.

Science and technology management is undoubtedly one of the important components of the middle school education system. By analyzing the current situation of science and technology management in middle schools, the main obstacles and problems of science and technology management include: the management of scientific research fund is complex, the approval process is complex, time-consuming, and the evaluation of scientific research workers is very difficult. The integrated technology management and intelligent education, independent scientific research management functions can directly open the intelligent education platform, and the traditional artificial management mode can replace the scientific research management function and big data technology based on artificial intelligence technology in science and technology management. For example, in the aspect of scientific research fund management, instead of the traditional manual approval mode, researchers can directly log in their own accounts in the intelligent education platform of the scientific research management subsystem, enter the scientific research management background, account information of the association, and quickly input the relevant information platform of new scientific research projects. It can meet the funding needs of scientific research workers and provide certain funds for scientific research workers efficiently and conveniently.

5   Conclusions

In order to make contemporary middle school students keep up with the pace of the development of the times and enrich their knowledge and cultural life in the present era, this paper puts forward the method of applying artificial intelligence in middle school education and teaching to carry out innovation reform in middle school education and teaching Education and teaching in the situation, the development is not very full, there are many problems need to be taken seriously. Therefore, this paper systematically expounds the shortcomings of current middle school education and teaching, and makes an in-depth analysis of the reform needed in the future middle school education and teaching. Through the research and analysis, it is found that the artificial intelligence technology proposed in this paper has a great effect on the innovation and reform of middle school education and teaching.

Acknowledgements. Fundation: Educational Informatization Promote the Research of Educational Precise Poverty Alleviation in Hainan Minority Region (Serial number: RHDXB201703).

References

1. Roach, S., Sahami, M.: CS2013: computer science curricula 2013. Computer **48**(3), 114–116 (2015)
2. Al-Fedaghi, S.: Int. J. Adv. Comput. Sci. Appl. **7**(11), 1–6 (2016)
3. Jeavons, A.: What is artificial intelligence? Res. World **2017**(65), 75 (2017)
4. Lu, H., Li, Y., Chen, M., et al.: Brain intelligence: go beyond artificial intelligence. Mob. Netw. Appl. **23**(2), 368–375 (2017)

5. Crawford, E.D., Batuello, J.T., Snow, P., et al.: The use of artificial intelligence technology to predict lymph node spread in men with clinically localized prostate carcinoma. Cancer **88**(9), 2105–2109 (2015)
6. Ladd, H.F., Sorensen, L.C.: Returns to teacher experience: student achievement and motivation in middle school. Educ. Finance Policy **12**(2), 1–70 (2017)
7. Forgan, J.W., Vaughn, S.: Adolescents with and without LD make the transition to middle school. J. Learn. Disabil. **33**(1), 33–43 (2016)
8. Corey, C.G., Ambrose, B.K., Apelberg, B.J., et al.: Flavored tobacco product use among middle and high school students - United States, 2014. MMWR Morb. Mortal. Wkly. Rep. **64**(38), 1066–1070 (2015)
9. Martinez, M.: The pursuit of wisdom and happiness in education. Retour Au Numéro **39**(10), 745–774 (2015)
10. Hidayati, N.A., Waluyo, H.J., Winarni, R., et al.: Exploring the implementation of local wisdom-based character education among indonesian higher education students. Int. J. Instr. **13**(2), 179–198 (2020)

Virtual Enterprise Partner Selection by Improved Analytic Hierarchy Process with Entropy Weight and Range Method

Junfeng Zhao[1](✉) and Xinyi Huang[2]

[1] School of Mechanical and Electrical Engineering, Guangdong Polytechnic
of Industry and Commerce, 510510 Guangdong, China
junfengzhao_cn@163.com
[2] School of Mathematics, South China University of Technology, 510640 Guangdong, China

Abstract. Virtual enterprise is a new enterprise model born at the end of the 20th century, and it is a hot topic how to deal with virtual enterprises. Firstly, basic steps of traditional analytic hierarchy process (AHP) are introduced, on the basis of which, the range matrix and the entropy weight are applied to improve the judgment matrix between indicators. Secondly, entropy weight is presented to calculate the judgment matrix between the enterprise and the index. The weight of the matrix is recorded as objective weight. Then the two weights are coupled by the optimal weighting method. Furthermore, the coupling optimization weight is calculated by Lagrange multiplier method instead of the weight in the original analytic hierarchy process. Finally, compared with the standard AHP, the simulation results of the example by improved AHP are more reliable.

Keywords: Partner selection · Analytic hierarchy process (AHP) · Entropy method · Range method · Lagrange multiplier method

1 Introduction

As the further development of economy and society, the virtual enterprise has become an important part of the decision, especially when enterprise making decision to choose partner. In order to survive in the fierce market competition, the decision made by the enterprise of choosing partner must be effectively and time-saving. While in the analytic hierarchy process method to solve the partner selection, it has the biggest problem that the human factors of data have a great influence so that it can't give a correct advice to the enterprise manager.

Gui et al. [1] did some work about on in virtual enterprises partner selection base on the NSGA-II. Li [2] et al. introduced the multi-objective decision making in partner selection of virtual enterprises. The interval neutrosophic preference relations were presented by Meng and Wang [3]. Risk assessment approach for a virtual enterprise was given by Mahmood et al. [4] Liao et al. [5] committed to the benefit analysis of enterprise server virtualization and cloud computing. Huang et al. [6] coped with the multi-criterion

© The Editor(s) (if applicable) and The Author(s), under exclusive license
to Springer Nature Switzerland AG 2021
J. MacIntyre et al. (Eds.): SPIoT 2020, AISC 1283, pp. 179–186, 2021.
https://doi.org/10.1007/978-3-030-62746-1_26

partner selection problem. Form another perspective, Son et al. [7] addressed on the optimization research of partner selection and collaborative transportation scheduling in virtual enterprises by using GA. Shahrzad et al. [8] dealt with the design of a customer type for partner selection problem in practical life. Kohnke [9] dis some work about the risk and rewards of enterprise use of reality. The problem about partner selection under uncertain information about candidates was solved by Huang and Gao [10].

2   Establishment of Virtual Enterprise

From the perspective of information network, it refers to the enterprise organization without formal organizational structure connected by information network. The establishment of virtual enterprise is divided into four stages:

Step 1: Determine the objectives and goals to be achieved according to the market opportunities and their own resources.
Step 2: Evaluate the existing resources among enterprises, establish Virtual Enterprise model and select the compatible partners.
Step 3: Through the evaluation of enterprises, the organizational form is designed to distribute the benefits and risks of each enterprise at the same time.
Step 4: Integrate the resource implementation plan of the enterprise organization.

3   Standard Analytic Hierarchy Process (SAHP)

AHP is a decision-making method presented by Saaty of the United States in the 1970s, the concrete steps:

Step 1: Construct hierarchical structure model.
Step 2: Construct judgment matrix.
Step 3: Calculate weight vector by square root method (geometric average method).
Step 4: Consistency test. We introduce a consistency indicator CI. The smaller indicator CI, the greater consistency. When n is large, the average random consistency index RI is introduced to measure CI.

$$CI = \frac{\lambda - n}{n - 1}, RI = \frac{CI_1 + CI_2 + \cdots + CI_n}{n}. \tag{1}$$

Compare CI with RI to get the test coefficient CR, $CR = CI/RI$.
Step 5: Hierarchical total sequencing and consistency test.

4   Improved Analytic Hierarchy Process (IAHP)

One of the disadvantages of the standard analytic hierarchy process is that there are more qualitative data, which leads to a greater impact of subjective factors. Therefore, one of the ways to improve the method is to reduce the subjective factors and increase the credibility of the results, which is convincing. In reference [10], range method and entropy weight method are used to improve the standard analytic hierarchy process.

4.1 New Judgment Matrix Construction by Range Method

For judgment matrix $A = (a_{ij})_{m \times n}$, the ranking index is calculated: $r_i = \sum\limits_{j=1}^{n} a_{ij}$. Then a new judgment matrix C is constructed by range method:

$$C_{ij} = C_b^{r_i - r_j / R}, R = r_{\max} - r_{\min}, i = 1, 2, \ldots, m. \tag{2}$$

In the above formula, C_b is generally taken as 9, and R is the range.

4.2 Weight Calculation Steps by Entropy Weight Method

Step 1: Constructing m samples and n indexes judgment matrix $R = (r_{ij})_{m \times n}$;

Step 2: Normalize the judgment matrix to get the normalized judgment matrix $P = (p_{ij})_{m \times n}$, $p_{ij} = r_{ij} \left/ \sum\limits_{i=1}^{m} r_{ij}\right.$;

Step 3: Obtain the entropy value of the index $e_j = -\sum\limits_{i=1}^{m} p_{ij} \ln p_{ij}, j = 1, 2, \ldots, n$;

Step 4: Calculate the indicator weight $w_j = \left(1 - e_j\right) \left/ \sum\limits_{j=1}^{n} \left(1 - e_j\right)\right.$.

4.3 Weight Optimization by Lagrange Multiplier Method (LMM)

In reference [10], the subjective and objective weights (w_1 and w_2) are coupled to get the optimized weight, and the optimal weight is used for coupling optimization:

$$\begin{cases} \min D = \sum\limits_{j=1}^{n} \left(w_j \ln \frac{w_j}{w_j'} \right) + \sum\limits_{j=1}^{n} \left(w_j \ln \frac{w_j}{w_j''} \right) \\ \text{s.t.} \sum\limits_{j=1}^{n} w_j = 1, \ w_j > 0. \end{cases} \tag{3}$$

The LMM is used to solve the problem. The steps are:

Step 1: Introducing functions: $L(w_1, w_2, \ldots, w_n, \lambda)$, we can obtain

$$L(w_1, w_2, \ldots, w_n, \lambda) = f(w_1, w_2, \ldots, w_n) - \lambda g(w_1, w_2, \ldots, w_n), \tag{4}$$

$$f(w_1, w_2, \ldots, w_n) = \sum\limits_{j=1}^{n} \left(w_j \ln \frac{w_j}{w_{1j}} \right) + \sum\limits_{j=1}^{n} \left(w_j \ln \frac{w_j}{w_{2j}} \right), \tag{5}$$

$$g(w_1, w_2, \ldots, w_n) = \sum\limits_{j=1}^{n} w_j - 1. \tag{6}$$

Step 2: Let the partial differential be zero, so that we can obtain solutions.

Note that $w = (w_1, w_2, \ldots, w_n)^T$ is the coupling optimization weight.

5 Numerical Examples of Virtual Enterprise Partner Selection

Now an example is verified the effectiveness of IAHP. The data comes from reference [2]. An enterprise wants to establish a virtual enterprise, and the target criteria given are: quality a_1, reputation a_2, technical personnel content a_3, production capacity a_4, punishment a_5, price a_6 and completion time a_7. The first five are gain criteria and the last two are profit and loss criteria. The judgment matrix is as follows:

Table 1. Judgment matrix between objective criteria

Index	a_1	a_2	a_3	a_4	a_5	a_6	a_7
a_1	1	5	5	9	3	3	2
a_2	1/5	1	1/2	3	1	1/3	1/4
a_3	1/5	2	1	3	1/2	1/3	1/4
a_4	1/9	1/3	1/3	1	1/5	1/7	1/8
a_5	1/3	1	2	5	1	2	2
a_6	1/3	3	3	7	1/2	1	2
a_7	1/2	4	4	8	1/2	1/2	1

First, changing the judgment matrix of Table 1 to five scales, obtain

$$a_{ij} = \begin{cases} 1, & 1, \ 1/9 \le a_{ij} \le 1/7, \\ 2, & 1/5 \le a_{ij} \le 1/3, \\ 3, & 1/3 \le a_{ij} \le 3, \\ 4, & 3 \le a_{ij} \le 5, \\ 5, & 7 \le a_{ij} \le 9. \end{cases} \tag{7}$$

After the change, there is a five-scale judgment matrix, as shown in Table 2:

Table 2. Judgment matrix in five scales

Index	a_1	a_2	a_3	a_4	a_5	a_6	a_7
a_1	3	4	4	5	4	4	3
a_2	2	3	3	4	3	2	2
a_3	2	3	3	4	3	2	2
a_4	1	2	2	3	2	1	1
a_5	2	3	3	4	3	3	3
a_6	2	4	4	5	3	3	3
a_7	3	4	4	5	3	3	3

Then modify the data in Table 2 with the formula (2) in the range method, and the result is as shown in the table below:

Table 3. The modified judgment matrix by range method

Index	a_1	a_2	a_3	a_4	a_5	a_6	a_7
a_1	1.0000	3.2280	3.2280	9.0000	2.4082	1.5518	1.3404
a_2	0.3098	1.0000	1.0000	2.7781	0.7460	0.4087	0.4152
a_3	0.3098	1.0000	1.0000	2.7781	0.7460	0.4087	0.4152
a_4	0.1111	0.3587	0.3587	1.0000	0.2676	0.1724	0.1489
a_5	0.4152	1.3404	1.3404	3.7372	1.0000	0.6444	0.5566
a_6	0.6444	2.0801	2.0801	5.7995	1.5518	1.0000	0.8637
a_7	0.7460	2.4082	2.4082	6.7144	1.7967	1.1578	1.0000

Then use the square root method (geometric average method) to calculate the weight of the judgment matrix, which is the subjective weight, and the result is as follow:

$$w_1 = (0.2828, 0.0876, 0.0876, 0.0314, 0.1174, 0.1822, 0.2110)^T. \qquad (8)$$

The consistency test is carried out for the subjective weight w_1. Because $n = 7$, so $RI = 1.36$, test coefficient $CR = 0 < 0.10$, we can accept the result. From the results, the order of importance of the seven indicators is: quality, completion time, price, punishment, reputation and technical personnel content, production capacity. This correspond to the reality. Next, the objective weight is calculated, and the required data is shown in the table below:

Table 4. Value of candidate enterprises about each target criterion

Candidate enterprises	a_1	a_2	a_3	a_4	a_5	a_6	a_7
E1	5	9	30	9	0.86	2.87	150
E2	6	7	25	7	0.90	3.00	120
E3	7	8	50	6	0.89	2.96	135
E4	4	5	30	5	0.84	2.79	180
E5	8	6	28	8	0.91	3.02	115
E6	7	7	40	9	0.84	2.78	140
E7	5	8	35	4	0.96	3.21	125
E8	6	9	29	6	0.91	3.05	130

Table 5. Normalization rules for the data of Table 4

Index	1	2	3	4	5
a_1	4	5	6	7	8
a_2	5	6	7	8	9
a_3	< 30	$30 \leq a_{ij} < 35$	$35 \leq a_{ij} < 40$	$40 \leq a_{ij} < 45$	$45 \leq a_{ij} \leq 50$
a_4	≤ 5	6	7	8	9
a_5	≤ 0.86	$0.86 < a_{ij} \leq 0.88$	$0.88 < a_{ij} \leq 0.90$	$0.90 < a_{ij} \leq 0.92$	> 0.92
a_6	≤ 2.86	$2.86 < a_{ij} \leq 2.95$	$2.95 < a_{ij} \leq 3.04$	$3.04 < a_{ij} \leq 3.13$	> 3.13
a_7	≤ 128	$128 < a_{ij} \leq 141$	$141 < a_{ij} \leq 154$	$154 < a_{ij} \leq 167$	$167 < a_{ij} \leq 180$

The data in Table 4 after the specification becomes as follows:

Table 6. The enterprise value about target criterion after standardization

Candidate enterprises	a_1	a_2	a_3	a_4	a_5	a_6	a_7
E1	2	5	2	5	1	2	3
E2	3	3	1	3	3	3	1
E3	4	4	5	2	2	3	2
E4	1	1	2	1	1	1	5
E5	5	2	1	4	3	3	1
E6	4	3	4	5	1	1	2
E7	2	4	3	1	5	5	1
E8	3	5	1	2	3	4	2

The data in Table 6 is regarded as a judgment matrix of samples and indicators, and the objective weight w_2 is calculated by entropy weight method:

$$w_2 = (0.1493, 0.1498, 0.1371, 0.1403, 0.1398, 0.14480.1389)^T. \tag{9}$$

The subjective weight w_1 is coupled with the objective weight w_2, and then the LMM is used to calculate the coupling optimization weight.

$$w = (0.2145, 0.1196, 0.1144, 0.0693, 0.1388, 0.1696, 0.1788)^T. \tag{10}$$

The weight of seven objective criteria calculated by three methods is shown:

The weight w of coupling optimization is used to replace the weight of judgment matrix in reference [1]. The calculation formula is as follows:

$$s_{mk} = \frac{v_{km} - \min_{1 \leq m \leq M}(v_{km})}{\max_{1 \leq m \leq M}(v_{km}) - \min_{1 \leq m \leq M}(v_{km})}, \quad s_{mh} = \frac{\max_{1 \leq m \leq M}(v_{hm}) - v_{hm}}{\max_{1 \leq m \leq M}(v_{hm}) - \min_{1 \leq m \leq M}(v_{hm})}. \tag{11}$$

In the above formula, v represents the value of the enterprise on the target criterion, s_{mk} is the priority score under the gain criterion, and s_{mh} is the priority score under the profit and loss criterion. The total priority score S_m is calculated:

$$K_m = \sum_{k=1}^{K} w_k s_{mk}, \ K_h = \sum_{k=1}^{K} w_k s_{mh}, \ S_m = K_m + H_m. \tag{12}$$

The judgment matrix in Table 1 is calculated by the standard analytic hierarchy process, and the weight w_3 obtained by the square root method is as follows:

$$w_3 = (0.3547, 0.0639, 0.0706, 0.0258, 0.1568, 0.1673, 0.1609)^T. \tag{13}$$

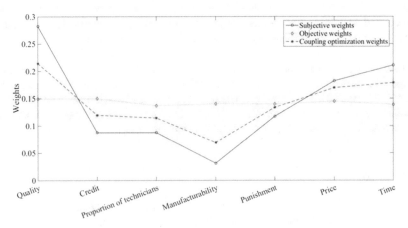

Fig. 1. Calculate weights by subjective, objective and couple optimization

According to formulas (11–12), calculate the total priority score of the enterprise:

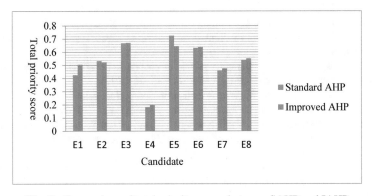

Fig. 2. Comparison of total priority scores between SAHP and IAHP

It can be seen from Table 6 and Fig. 2 that the priority of candidate enterprise selection obtained by the improved AHP is E3, E5, E6, E8, E2, E1, E7 and E4, and the priority of candidate enterprise selection obtained by the standard AHP is E5, E3, E6, E8, E2, E7, E1 and E4. There is no significant difference between the two, but the improved AHP reduces too many subjective factors, so it is more reliable.

6 Conclusion

The standard analytic hierarchy process and virtual enterprises are introduced. Then the standard AHP is modified by changing the judgment matrix. We apply the range method to modify the judgment matrix. The corresponding weight of the matrix is subjective weight. Then, the entropy method is adopted to calculate the judgment matrix between enterprises and indicators. Lastly, we apply Lagrange multiplier method to calculate the coupling optimization weight instead of the weight in the standard AHP. Furthermore, the modified AHP is applied to simulate the decision-making problem. The computation results show that our improved AHP is more effective than standard AHP.

Acknowledgements. This research was supported by the "Natural Science Foundation of Guangdong Province, No. 2019A1515011038", "Soft Science of Guangdong Province, and No. 2018A070712002, 2019A101002118", "Guangdong Province Characteristic Innovation Project of Colleges and Universities, No. 2019GKTSCX023". The authors are highly grateful to the referees and editor in-chief for their very helpful comments.

References

1. Gui, H.X., Zhao, B.L., Wang, X.Q, Li, H.Z.: Research on partner selection in virtual enterprises based on NSGA-II. In: ISICA2019, vol. 1205, pp. 307–319 (2020)
2. Li, L., Que, J.S., Zhu, Y.L.: Multi-objective decision making in partner selection of virtual enterprises. Comput. Integr. Manuf. Syst. CIMS **8**(2), 91–94 (2002). (In Chinese)
3. Meng, F.Y., Wang, N., Xu, Y.W.: Interval neutrosophic preference relations and their application in virtual enterprise partner selection. J. Ambient Intell. Humanized Comput. **10**(12), 5007–5036 (2019)
4. Mahmood, K., Shevtshenko, E., Karaulova, T., Otto, J.: Risk assessment approach for a virtual enterprise of small and medium-sized enterprises. Proc. Estonian Acad. Sci. **67**(1), 17–27 (2018)
5. Liao, H., Liu, S., Chiou, J.: Benefit analysis of enterprise server virtualization and cloud computing. Paper Aisa, v COMPENDIUM **1**(7), 35–41 (2018)
6. Huang, B., Bai, L., Roy, A., Ma, N.: A multi-criterion partner selection problem for virtual manufacturing enterprises under uncertainty. Int. J. Prod. Econ. **196**, 68–81 (2018)
7. Son, D.D., Kazem, A., Romeo, M.: Optimisation of partner selection and collaborative transportation scheduling in virtual enterprises using GA. Expert Syst. Appl. **41**(15), 6701–6717 (2014)
8. Shahrzad, N., Ahmet, M.O., Hakki, O.U., Sadik, E.K.: Design of a customer's type based algorithm for partner selection problem of virtual enterprise. Procedia Comput. Sci. **95**, 467–474 (2016)
9. Kohnke, A.: The risk and rewards of enterprise use of augmented reality and virtual reality. ISACA J. **1**, 16–23 (2020)
10. Huang, B., Gao, C.H.: Partner selection in a virtual enterprise under uncertain information about candidates. Expert Syst. Appl. **38**(9), 11305–11310 (2011)

Research and Implementation of Intelligent Tourism Guide System Based on Cloud Computing Platform

Xu Yan and Zhao Juan[✉]

Shandong Institute of Commerce and Technology, Jinan 250103, China
546718705@qq.com, zcr@126.com

Abstract. The rapid development of tourism industry will be accompanied by a large number of tourism data and information, and the development of tourism information system in different regions is unbalanced, and tourism information is basically not shared, so it is inevitable to adopt cloud computing technology to support the tourism system. This paper collects and analyzes the cases of cloud computing technology used in tourism industry at home and abroad, combined with its advantages and disadvantages, uses the idea of SOA to model services and data, and then puts forward the architecture of tourism cloud services. According to the characteristics of tourism business, the system construction of tourism cloud platform is divided into three parts: Tourism private cloud, tourism exclusive cloud and tourism public cloud. And the tourism private cloud platform for specific planning and design. In view of the rapid growth of tourism data, this paper also proposes a recommendation algorithm based on the information of tourists and scenic spots, which combines the collaborative filtering algorithm based on users and the information recommendation algorithm based on geographical location, so as to push more accurate information to tourists. Finally, due to the increasing number of free travelers, this paper develops an intelligent navigation system based on Android Application on the basis of cloud platform. The system mainly includes three service modules: scenic spot information service, tourists can query the basic information of scenic spots, tour group information, etc.; tour guide service, tourists can query the route, GPS Location, nearby search, weather forecast, etc.; tour strategy service, tourists can query the strategy of scenic spots, upload and release the strategy, and view and download offline maps. Through demand analysis and design, the Intelligent Tourism guidance system based on cloud computing platform is realized, which not only reduces the pressure of processing a large number of data and the state of insufficient information sharing, but also meets the needs of free travelers.

Keywords: Tourism cloud · Mobile navigation · GPS positioning

1 Introduction

With the progress of social economy and the rapid development of information technology, the profits of China's tourism industry have been growing by leaps and bounds.

J. MacIntyre et al. (Eds.): SPIoT 2020, AISC 1283, pp. 187–192, 2021.
https://doi.org/10.1007/978-3-030-62746-1_27

According to the calculation, in 2014, the tourism income of the mainland, Hong Kong, Macao and Taiwan can reach 2.9 trillion yuan, the number of tourists in the country has exceeded 3.25 billion people, the number of outbound tourists has also increased day by day, breaking through 97300000 person times, and the foreign exchange income of inbound tourism has broken through With us $47.8 billion, the number of new direct tourism employment has reached more than 500000 [1]. While promoting the development of tourism in the world, the focus of global tourism also flows to Asia, especially China. According to World Tourism According to the tourism outlook information released by the organization, by 2020, Europe will provide more than 700 million outbound tourists, which is still the region with the largest number of outbound tourists in the world. About 6% of the tourists will visit East Asia Pacific and South Asia; in 2020, America will also provide more than 200 million outbound tourists, about 9% of them will visit East Asia Pacific and South Asia; in 2020, East Asia Pacific More than 400 million outbound tourists will be born in the region, and up to 80% of the tourists will visit nearby countries, and the number may exceed the expectation. From 2000 to 2020, the number of inbound tourists in China will grow at an annual rate of 8%. It is expected that by 2020, China will face 137 million foreign tourists and become the first tourist attraction [2].

2 Cloud Computing Overview

Cloud computing is considered to be the third IT revolution after microcomputers and the Internet, and it is the inevitable trend of the development of information technology. The key to the continuous development and evolution of cloud computing as a technical support system lies in its openness, which enables the cloud to serve users in a standard and universal way. The lower the threshold of cloud computing capability, the more convenient it will be to use, and it can be more and more widely used. Open cloud computing capability not only brings convenience to users, but also enables the third party of enterprises to use open interfaces and services to provide more cloud computing services.

It is an inevitable choice to adopt cloud computing technology as the underlying support for information fusion. Through the cloud, users can gather, understand users' thoughts and behaviors through the Internet, gather all kinds of content on the Internet, produce and publish based on the Internet, and realize user centered content production and consumption. Through the cloud, we can ensure the fast online business, realize the natural growth of business, and survive the fittest. The infrastructure of cloud computing is based on the idea of SOA, which provides software and hardware resources as services to end users. Service developers provide more services by using the open interface provided by cloud computing platform, and mobile phone users can also customize services on cloud computing platform [3].

End user of tourism cloud: the user can obtain the information resources he wants through the application of computer, mobile phone, pad and other network devices, without having to master the cloud platform itself. There is no significant difference between the application services on the tourist platform and the previous tourism system. However, some services are not stored locally, but in the cloud. Moreover, these services need to be purchased. A service level agreement (i.e. purchase contract) is signed with the cloud platform operator. When the service purchased by the user reaches the service life, or upgrades the service, or requires a new service. You need to purchase from the cloud service platform.

3 Crisis Design of Recommendation Algorithm Based on Tourists and Scenic Area Information

The specific design of the algorithm is as follows: (1) in the collaborative filtering algorithm based on users, firstly, the interest score matrix rsmxn of tourists and scenic spots is established. The matrix is the specific scores of different tourists on each scenic spot project. There may be a value of 0 in the matrix because the tourist has not scored the project or the tourist is a new registered user. The scoring matrix is as follows:

$$
RSmxn = \begin{bmatrix} rs_{uls1} & rs_{uls2} & \cdots & rs_{ulsN} \\ rs_{u2s1} & rs_{u2s2} & \cdots & rs_{u2sN} \\ \cdots & \cdots & \cdots & \cdots \\ rs_{uMs1} & rs_{uMs1} & \cdots & rs_{uMsN} \end{bmatrix} \tag{1}
$$

The person similarity algorithm is used to calculate the similarity between the tourist u_i and the tourist u_j, such as Formula 1, $rs_{ui \cdot sk}$ represents the tourist u_i interest value of the scenic spot project s_k, and RSUI is the average value of the tourist u_i score on all scenic spots in the matrix.

$$
sim(u_i, u_j) = \frac{\sum sk \in US_{ui,uj}(rs_{ui,sk} - rs_{ui})(rs_{uj,sk} - rs_{uj})}{\sqrt{\sum sk \in US_{ui,uj}(rs_{ui,sk} - rs_{ui})^2}} \tag{2}
$$

Among them, the value of u_i and u_j is very important. The higher the value of $US_{ui,uj}$ the higher the similarity is.

According to the algorithm in Lars, we calculate the cost of scenic spot project location to tourists, that is, travel penalty (U, I) represents the average distance between scenic spot project I and historical behavior record of tourists U. Then, the interest value of tourist u to item I is recalculated by formula 3.

$$
RecSore(u, i) = P(u, i) - TravelPenalty(u, i) \tag{3}
$$

$RecSore(u, i)$ is the final interest value. According to the value, the recommended list is filtered twice to get a more accurate information list and recommend it to users.

Tourism cloud platform has been systematically planned and constructed, which can be divided into three forms: Tourism private cloud, tourism exclusive cloud and tourism public cloud. Then, it makes a specific research and Analysis on Tourism private cloud. In order to improve the efficiency of information recommendation, the recommendation algorithm based on the information of tourists and scenic spots is studied.

4 Demand Analysis of Tourism Guide System

With the improvement of human economic level and the rapid development of the Internet, the number and quality requirements of human outbound tourism are higher and higher. According to the statistics of the World Federation of travel agencies, people's travel mode is gradually changing, and the proportion of free travel is increasing. And with the increasing negative news from travel companies and tour guides, tourists will choose more free travel. Tourists can get the tour information they need through the Internet, and then make their own tour itinerary. Tourism guide system can not only provide information of all directions of tourist attractions for tourists to search before going out; at the same time, tourists can also get the peripheral information they want at any time in the process of playing, so that tourists can make the most suitable play plan; so that tourists can publish their personal experience on the platform after the trip.

4.1 The Design of Tour Guide System

The design of tourism guide system is divided into four parts: mobile application layer, tourism service layer, tourism data layer and infrastructure layer. The mobile application layer is only the function interface and control of the system terminal, which is responsible for displaying the scenic spot information, navigation information and strategy information to the user in the form of text, picture and video; when the user requests, it is transmitted to the tourism service layer through the encrypted and secure channel, so as to carry out the corresponding business logic processing; then, by accessing the tourism data layer, the data is transferred from the result of processing is obtained in the library and finally presented to the end user; the infrastructure layer is the tourism cloud platform environment.

The design of information management service can be from the perspective of the users of the service. The users of the service system include the users of the system and the managers of the system. The users of the system mainly include tourists, tourism companies and peripheral manufacturers; the system administrator is the system administrator. In this service, visitors can query the basic information of the scenic spot by registering their account on the client, including: scenic spot introduction, tour group information, peripheral manufacturer information, etc. System administrators, tourism companies, and surrounding manufacturers release tourism information through the open platform interface. At the same time, the system administrator also needs to manage the tourism information of scenic spots in a unified way to help scenic spots solve the problems of information sharing, security and expansion.

The following is the specific implementation of each component class of the service.

(1) Tourist information management the basic information of tourists is composed of address and contact information. Different users can have multiple addresses or contact information. The service includes functions of adding, deleting, modifying and checking, and these functions are applicable to all registered users of the system. As shown in the Fig. 1 below:

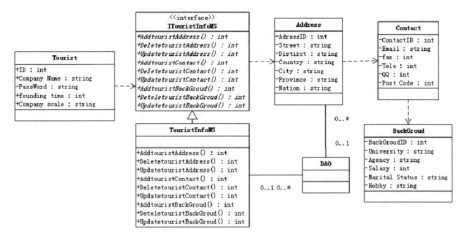

Fig. 1. Tourist information management class diagram

This paper introduces the design of tour guide system, including overall design, detailed design, database design, open interface design and so on. In the overall design, the level of the system is divided, and a detailed overview is given; in the detailed design of the system, the detailed design of each functional module is proposed; in the database design, not only the design overview of the key data table in the system, but also a detailed description of the authority control; at the same time, the system also uses rest The interface design of architecture makes the system have better maintainability and reusability.

5 Conclusions

According to the current development situation of tourism industry, this paper proposes and designs the solution of smart tourism cloud, and realizes the tourism intelligent navigation system based on Android platform on the basis of tourism cloud. Firstly, the paper studies the system planning of tourism cloud platform, and analyzes the specific architecture of tourism private cloud platform. Then, according to the development criteria of software engineering, a complete intelligent tourism system is designed and implemented, which integrates information management service, tourism guide service, tourism strategy service and information statistical analysis service.

Acknowledgements. A study on the civilization literacy of Chinese outbound tourists from multiple perspectives.

References

1. Yuliang, S., Li, P.: Research on tourism informatization based on cloud computing platform. Inf. Technol. Inf. **12**, 87–89 (2014)

2. Feng Ling, L., Ning, L.: Development environment and trend of tourism in the 12th Five Year Plan. Econ. Issues Explor. **9**, 24–28 (2011)
3. Liu, J., Fan, Y.: Composition, value and development trend of smart tourism. Chongqing Soc. Sci. **16**(10), 121–124 (2011)

Analysis of Financial Needs of New Agricultural Operators Based on K-Means Clustering Algorithm

Xiaohui Li[✉]

Department of Accounting and Statistics, Weifang Engineering Vocational College, Weifang 262500, Shandong, China
wfgcxylxh@163.com

Abstract. There are more levels and varieties of financial needs for the new agricultural operation entities. This paper introduces the current situation of the rural financial supply and the development of the new-type agricultural operation subject, analyzes the financial demand of the new-type agricultural operation subject, and puts forward some policy suggestions to meet the financial demand of the new-type rural operation subject. Clustering analysis is an important research topic in data mining. The goal of clustering is to aggregate data into different classes (or clusters) according to the similarity of data without any prior knowledge, so that the elements in the same class are as similar as possible.

Keywords: New agriculture · Business subject · Financial demand · Cluster analysis

1 Introduction

The formulation of "new type of agricultural operation subject" appeared earlier in some theoretical research and policy research articles, but it was officially proposed and received high attention after the 18th National Congress of the Communist Party of China. Agricultural operation subject refers to any individual or organization directly or indirectly engaged in the production, processing, sale and service of agricultural products. Although 2012 Since, the "new agricultural operation subject" has continued to appear in official documents, such as the opinions on speeding up the construction of policy system to cultivate new agricultural operation subject issued by the general office of the CPC Central Committee and the general office of the State Council, which is also a high-frequency vocabulary for the research of "agriculture, rural areas and farmers", but at present, there is no authoritative concept of "new agricultural operation subject", which mainly refers to family farms, large farmers, professional cooperatives and agricultural enterprises as the main development types of new agricultural operation [1]. In fact, there is no essential difference between family farms and large-scale farmers, so this study does not strictly distinguish family farms and large-scale farmers. The new type

J. MacIntyre et al. (Eds.): SPIoT 2020, AISC 1283, pp. 193–198, 2021.
https://doi.org/10.1007/978-3-030-62746-1_28

of agricultural operation subject has become the forerunner of agricultural supply side structural reform, with significant income effect and employment effect, but it still faces many development bottlenecks and problems. In terms of financial support, it is mainly reflected in financing difficulties and other issues. Scholars have done a lot of research on the financial support of the new agricultural operation subject, including whether the financial support can promote the cultivation of the new agricultural operation subject, the analysis of the financial supply and demand of the new agricultural operation subject, the analysis of the financing dilemma of the new agricultural operation subject, and the innovative financial support.

2   Whether Financial Support Promotes the Cultivation of New Agricultural Operation Subjects

In order to cultivate a new type of diversified agricultural operation subject, we need to innovate the agricultural operation system and mechanism, and increase the support of production factors such as rural finance. In terms of the mechanism of financial support for the new type of agricultural operation subject, sun Yongzhi and others believed that financial support is conducive to increase agricultural investment, improve the level of rural human capital, improve the level of agricultural technology, promote the transformation of agricultural production mode, promote the transfer of agricultural labor force, etc., so as to promote the development of rural economy. In terms of the scope of different financial types, Wang Yantao and others used the survey data of farmers in Jiaozhou City, Shandong Province, to demonstrate the impact of Rural Finance on the cultivation of new agricultural business entities [2]. It is found that the general finance provided by the government, the special finance provided by the financial institutions and the self owned funds of the new agricultural operation entities have different influences on the cultivation of different types of new agricultural operation entities. Among them, general financial support has a positive and promoting effect on the cultivation of family farms, farmer cooperatives and large farmers, but has no significant impact on agricultural enterprises; special financial support has a significant positive impact on the cultivation of agricultural enterprises, family farms and farmer cooperatives, but has no significant impact on the cultivation of large farmers; Self owned funds have a positive and significant impact on the cultivation of several major new agricultural business entities. Jiang Xianli, et al. Investigated the theoretical correlation and practical situation of the cultivation of new agricultural operation subjects and their financial services in China, and evaluated the financial service performance of the cultivation of new agricultural operation subjects by factor analysis. The results show that in the process of cultivating new agricultural operators, the overall performance level of fiscal and financial services to support agriculture has improved significantly, which ensures that China's grain output increases for many years in a row under the condition of a significant reduction of agricultural labor force, and makes a positive contribution to the national food security. The study confirmed the positive role of financial finance in the cultivation of new-type agricultural operators, but it is still questionable to regard financial finance as an important influencing factor of grain production under the condition of labor force reduction in China. Hong Zheng compared and analyzed the supervision efficiency of all kinds of new rural financial institutions and its impact on rural financing. The results showed that mutual supervision and contract interconnection can be effectively implemented by mutual fund cooperatives compared with commercial banks and small loan companies, and the joint development with professional cooperatives or leading companies can significantly improve rural financing.

3 Clustering

Clustering is a basic cognitive activity of human beings. Clustering is widely used. Including pattern recognition, data analysis, image analysis, and market research. Clustering is to divide data objects into several classes or clusters. The principle of clustering is that objects in the same cluster have higher similarity, while objects in different clusters are larger.

3.1 K Average Algorithm

K-means algorithm, also known as k-means, is the most widely used clustering algorithm. K-means algorithm takes K as a parameter, and divides an object into K clusters, so that the similarity within the clusters is high, while the similarity between clusters is low [3]. Firstly, the algorithm randomly selects K objects, each of which initially represents the average or center of a class. For each remaining object, it is assigned to the nearest cluster according to its distance from the center of each class. The average of each cluster is then recalculated. This process is repeated until the criterion function converges. The criterion functions are as follows:

$$E = \sum_{i=1}^{K} \sum_{x \in C_i} \left| x - \bar{x_i} \right|^2$$

Here e is the sum of the square errors of all objects in the database, $\bar{x_i}$ is the point in the space, representing the given data object. One is the average value of class C_j. This criterion tries to make the generated result classes as compact and independent as possible.

Input: number of clusters k and database containing ten objects
Output: k clusters, minimum square error criterion
(1) Assign initial value for means; // select any k objects as the initial cluster center
(2) REPEAT
(3) For j = 1 to n do assign each x_j to the cluster which has the closest mean
(4) FOR i = 1 to k DO $\bar{x_i} = \sum_{x \in C_i} x_i / |C_i|$; //average of updated clusters
(5) Compute

$$E = \sum_{i=1}^{K} \sum_{x \in C_i} x / \left| x - \bar{x_i} \right|^2; \text{ // calculation criterion function e}$$

(6) UNTIL E no longer changed significantly;
See Fig. 1 for the specific process.

Fig. 1. Clustering process of a group of objects in kedingk average method

4   Analysis of Financial Supply and Demand of New Agricultural Operation Subject

The main financial needs of the new agricultural operation entities are credit demand, insurance demand and Futures (b), which can be divided into productive financial demand and living financial demand according to their purposes. Ma Yanni et al. Studied 245 major apple loan fund uses in 11 counties, 36 townships and 82 villages in Shandong, Shaanxi and Shanxi Province, and found that major apple loan uses are relatively scattered and diversified, among which formal loan uses are mainly productive loan uses, informal, loan production and living uses are equally important. There are also other scholars

The productive lending of households is studied. Yu Bo and Wang Jia research show that the main purpose of family farm financing is productive expenditure. But for the agricultural listed company, which is a new type of agricultural operation subject, the situation is special. After studying the investment direction of the raised funds of the listed agricultural companies in China, Xing Suyuan found that the phenomenon of backfarming investment is very common. The funds are mainly used in non-agricultural industries, such as real estate, biopharmaceutical and other high-risk and high-yield industries. At the same time, there are a considerable number of companies that idle the funds raised for agricultural projects and have the characteristics of mass distribution.

The financial needs of the new type of agricultural operators are diversified. The new-type agricultural business entities applying for loans generally have credit needs, but those unexpressed or hidden credit needs will be ignored only based on this standard, such as the credit needs suppressed due to the mortgage guarantee requirements of financial institutions, the high cost of credit transactions, and the unwillingness to bear the potential risk of loss of collateral or failure to repay loans. Mushinski divides the credit demand of farmers into three categories: nominal credit demand, effective credit demand and potential credit demand, and puts forward the mechanism to identify and classify them. Zhou Yang used this analysis framework for reference to analyze the current situation and characteristics of credit and demand in the development of new agricultural business entities in Liaoning Province. Taking Sichuan Province as the research area, Wang Qiang et al. Analyzed the financial demand characteristics of the new agricultural operation subject, mainly including: strong willingness to borrow, relatively single borrowing channel, centralized use of borrowing, borrowing cycle in line with the growth cycle of agricultural products, high actual borrowing rate, large demand for loan amount, etc. Based on the analysis of the characteristics of the new type of agricultural operation subject, sun Ligang and others proposed that the new type of agricultural operation subject has new characteristics in financing demand, such as diversification of financing demand, coexistence of long-term and seasonal financing demand, and no financing cycle. Huazhongyu et al. Found through the investigation of 6 poverty-stricken counties in Gansu, Guizhou, Anhui provinces that the rural credit cooperatives are the main channel of financing for the new agricultural operation subject in poverty-stricken areas, and the new agricultural operation subject has a strong demand for loan for mortgage guarantee, and has the characteristics of large-scale financing amount and diversified use. Different types of new agricultural operators are also subject to different credit constraints. Using Heckman's selection model to study the credit demand, formal credit constraint and capital gap of new agricultural operators in Yangling Demonstration Area of Shaanxi Province, she found that different operators have different credit demand, credit constraint and capital gap in terms of resource endowment, operation ability and social capital. Compared with ordinary farmers, large growers and family farms, agricultural cooperatives and agricultural enterprises have more extensive collaterals and social capital, so they are subject to less formal credit constraints, resulting in a relatively small capital gap.

In terms of specific types of new-type agricultural operation entities, Liu Mingxuan, etc. made use of the sampling survey data of 10 provinces in China to analyze the demand for financial services and its influencing factors from two aspects of demand degree and demand type of farm type farmers, households, ordinary farmers with agriculture as the main part, and ordinary farmers with agriculture as the auxiliary part. It is found that the current demand for financial services of Chinese farmers is mainly concentrated in two aspects: loan service and insurance service. From the perspective of the demand for financial services, farm farmers are more and stronger than the latter, and the demand for loan service is the strongest in the demand type. The study points out that the type of farmers and regional variables are important factors affecting the demand of financial services.

5   Conclusions

Throughout the current research on the financial support of new agricultural operators, most of the studies analyzed the overall financial support of new agricultural operators, but most of them were case studies. On different types of new agricultural management subjects.

Research on financial support is still lacking. In the future, we should strengthen the research on the financial support of new agricultural operation subjects from the following aspects: ① financial needs and influencing factors of different types of new agricultural operation subjects; ② financial risks and risk management; ③ financial product innovation under the background of "Internet +"; ④ research on the construction of mortgage guarantee system; ⑤ research on the construction of credit evaluation system.

Acknowledgements. Research project of Weifang Engineering Vocational College in 2019 (Research on financing difficulties and countermeasures of New agricultural operation subject under the background of Rural revitalization – Taking Weifang city as an example) supporting achievements (20190712).

References

1. Yizhen, Z.: The current situation and development trend of the main body of agricultural management in China. Xinjiang Agric. Reclam. Econ. (5), 7–9 (1998)
2. There is no essential difference between family farm and major professional household. finance China net [EB/OL]
3. Zhaoyang, L.: Research on the welfare effect of the development of new-type agricultural operation subjects. Res. Quant. Econ. Technol. Econ. **6**, 41–58 (2016)
4. Zuhui, H., Ning, Y.: New type of agricultural operators: current situation, constraints and development ideas – an analysis of Zhejiang Province as an example. China's Rural Econ, **10**, 16–26 (2010)
5. Kong, X.: The role of cooperatives in the new type of agricultural operation. China Farmers' Cooperatives **11**, 29 (2013)

Research on the Application of Virtual Network Technology in Computer Network Security

Zhou Yun[(✉)] and Lu Rui

Wuhan Institute of Shipbuilding Technology, Wuhan 430050, Hubei, China
ruezhou@163.com, 606rui@163.com

Abstract. With the continuous progress and development of society, China has gradually entered the era of information society. Nowadays, the popularity of Internet and computer is no longer the luxury in the past. The sharing degree of information resources has greatly exceeded people's imagination. People have adapted to the convenience brought by the high-tech information age. And the development of computer network has entered a white hot stage of development, but there will be some corresponding problems in the process of development, which makes many experts puzzled. The virtual network technology in the computer network security has been the central topic that experts have been discussing, so this paper will discuss the virtual network technology in the computer network security. Mainly from the computer network security in the current development situation of virtual network technology, characteristics of specific content, type classification, factors affecting its development, future development prospects and its main specific forms of application to do a detailed explanation, hoping to help more scholars analyze the relevant computer network security in the virtual network technology application content.

Keywords: Computer network security · Virtual network technology · Application research

1 Introduction

Computer has become an irreplaceable important thing in today's society, its own role has been gradually magnified by the relevant experts. Today's computer can be used in many aspects of the content, and not only has prominent auxiliary features in work, but also can effectively improve people's quality of life in life, so the computer has been widely used It is not only a tool, it has been gradually intelligent, and is expected to move forward in a new direction. However, in the process of development, there will always be a lot of troubles and obstacles. The Internet can play the role of computer to the extreme, and at the same time, it can directly reveal the various drawbacks of the computer. Computer network security is another new challenge brought by the Internet to computer development. The virtual network technology in computer network security is one of the hot topics that experts study nowadays [1].

J. MacIntyre et al. (Eds.): SPIoT 2020, AISC 1283, pp. 199–204, 2021.
https://doi.org/10.1007/978-3-030-62746-1_29

2   Current Development Situation of Virtual Network Technology in Computer Network Security

Compared with other network technology means, virtual network technology in computer network security has a certain positive significance in the process of practical application, but for some technical fields, it is still only the content form at the basic level. If we continue to deepen the development, there will be some restrictive problems. The computer network security problem is not only the general security content work. It is a factor directly related to the vital interests of users. If we fail to recognize this point, we will make major mistakes in this aspect in the future. Nowadays, the computer network is very powerful, it can almost make people find any information and data they want to find in the computer network. However, if the resources are too strong, there will be some prominent problems. The network security problems have not been effectively controlled, which is why. Therefore, we need to apply the virtual network technology in computer network security to further improve the problem of negligence in this respect. The process of handling affairs needs to be standardized. It is not allowed to be too arbitrary or not to be carried out according to the rules and regulations.

But now the application of virtual network technology in computer network security is not very optimistic, there are problems in the mainstream direction of deviation and application of not hard. Examples will be relatively easy to understand: for example, a person's information is exposed on the network, but due to the special identity of this person, so far it has a serious adverse impact on society. Everyone's identity information is the most important information. Once used by lawless elements, it will cause great harm to the victims and even cause huge loss of property. Moreover, once it is rendered by public opinion, the ability of relevant units to handle affairs will be questioned, and the social situation will become worse The network technology needs to be further improved and standardized [2].

3   Influence Factors of Virtual Network Technology in Computer Network Security

3.1   Influencing Factors of Unauthorized Access

Illegal authorized access refers to the illegal and criminal behavior of visitors who, without the permission and authorization of users, gain personal or enterprise computer access rights by writing programs and changing programs without authorization, enter computer internal network programs, view and plagiarize important confidential documents such as data and resources. Its main purpose is to change the program to obtain access rights and other special rights, and some illegal visitors will damage the entire system of the user's computer, resulting in the paralysis of the system, resulting in the user's loss of huge property.

3.2   Natural Factors of Influencing Factors

There are many kinds of natural factors, such as common natural disasters, aging of equipment after a certain period of service, electromagnetic radiation and other natural factors. This kind of force majeure factors often cause direct damage and damage to the computer, but there are also some factors that only play indirect damage. The computer itself is a mechanical object afraid of water and fire, so natural factors are often the content requirements that need to be more careful.

3.3 Computer Virus Influencing Factors

When it comes to computer viruses, we will not feel strange, and we have more contact in our life. Computer virus is mainly through the computer itself to destroy and change the degree, especially to the computer internal important data and information for damage, resulting in the computer can not be normal use and operation, as well as instructions on the unrecognized and other unconventional error code. Chi virus once made people lose their courage for a period of time. It can directly destroy all the system files on the disk, and further infect all the logical drives on the disk. This virus starts from the main boot area of the hard disk and gradually infects the hard disk with garbage virus until all the data in the disk are damaged. Its destructive power is far more than that. It will also destroy the BIOS in the motherboard and return the CMOS parameters to the factory settings. If the user's CMOS is flashrom, the motherboard can no longer be used. Therefore, this virus is extremely destructive. We should pay attention to it. Computer viruses not only exist Chi virus, but also Yai virus, Christmas virus, funlove virus and so on [3].

3.4 Trojan Horse Program and Backdoor

Trojan horse program and back door is a kind of hacker intrusion technology, which is the earliest computer hacker technology found by people. At that time, hackers often used the back door technology to carry out many illegal infringement activities on the computer. The main damage program of the hacker was to control the system administrator first, let him remove the blocking authority, and make himself invisible. It could not be easily found by the system administrator, which reduced the time for illegal intruders to steal information. Later in the development, people also called this trojan horse as Trojan horse. This kind of remote control technology is one of the backdoor technology. Hackers can use this technology to control computer programs directly, and it has certain illegal authorization and concealment. Trojan horse generally includes basic control program and server program.

4 Network Model of Virtual Network Mapping

Physical topology is usually represented by weighted undirected graph $G_S = (N_S, E_S)$: each physical node ns has attribute set a (N_S), including CPU resource C (n_S) and geographical location loc (N_S), each physical link e_S has attribute set A (e_s), including bandwidth resource B (e_s); virtual network topology is also represented by weighted undirected graph $G_v = (N_v, E_v)$, and also has its own attribute set. In order to reduce the complexity of virtual network mapping, most algorithms divide virtual network mapping into two stages: node mapping and link mapping. For virtual node mapping, each virtual node should be able to map to a physical node. For virtual link mapping, it depends on whether the flow is separable. The flow is separable. Virtual link can adopt multiple physical path mapping, flow is not separable, and virtual link can adopt single path mapping. Since the main resource of a node is CPU, the main resource of a link is bandwidth. Therefore, when introducing the mathematical model of node mapping and link mapping, only CPU is considered for node constraints and bandwidth is considered for link constraints. Virtual nodes in different virtual networks can be mapped to the same physical node, but virtual nodes in the same virtual network must be mapped to different physical nodes. Each virtual node is mapped to a different physical node, which is represented as

$$F_N^{Ri} : N_V^{Ri} \rightarrow N_S^{Ri} \subseteq N_S : N_S^{Ri} = \left\{ n_s \in N_S \middle| R_{CPU}(n_S) \min_{n_V \in N_V^{Ri}} C(n_V) \right\}$$

N_S^{Ri} indicates that there is at least one physical node in the set, and the remaining CPU resources of the physical node can meet the CPU mapping requirements of one virtual node in the virtual network request RI. The remaining CPU resource ($R_{cpu}(n_S)$) of a physical node is defined as the total CPU of the physical node minus the CPU that has been mapped and used by the node:

$$R_{CPU}(n_S) = C(n_S) - \sum_{\forall n_V \uparrow n_S} C(n_V)$$

Each virtual link is mapped to a different physical path:

$$F_N^{Ri} : E_V^{Ri} \rightarrow P_S^{Ri} \subseteq P_S : P_S^{Ri} = \left\{ P \in P_S \middle| B(P) \geq \min_{e_V \in E_V^{Ri}} B(e_V) \right\}$$

Among them, P_S^{Ri} indicates that there is at least one physical path in the set, and the remaining bandwidth of the physical path can at least meet the bandwidth mapping requirements of a virtual link in the virtual network request Ri. The set of physical paths is represented by P_S, in which the source node is n and the destination node is n_s, which can be expressed as P_S (n_S, n_t). The available bandwidth B (P) of the physical path P is equal to the available bandwidth of the link with the least remaining bandwidth on the path

$$B(P) = \min_{e_s \in P} R_{BW}(e_s)$$

The remaining bandwidth resource $R_{BW}(e_s)$ of a physical link is defined as the total bandwidth of the physical link minus the bandwidth used by the link:

$$R_{BW}(e_s) = B(e_s) - \sum_{\forall e_V \uparrow e_s} B(e_V)$$

5 Application of Virtual Network Technology in Computer Network Security

5.1 Application of Virtual Network Technology in Computer Network Security Between Customers and Enterprises

The application scope of virtual network technology in computer network security is relatively wide. It involves many fields, which makes the computer network security become an important bridge between enterprises and customers. How to do a good job in computer network security is the key to the current primary problem, which will bring about cooperation between enterprises and customers Huge economic security. The application performance of virtual network technology in computer network security between enterprises and customers is reflected in the following aspects: first, in the cooperation between enterprises and customers, there is often the sharing of resources, data and files. By using the virtual network technology in computer network security, we can effectively ensure that these important information contents are not easy to flow out or be illegally divided Children get and use them. Second, if we can effectively apply the virtual network technology in computer network security, we can further strengthen the function of firewall setting. Try to perfect some protective settings, such as information verification and password update. These can not only reduce the communication obstacles, but also ensure to find the source of the problem in time, so as to solve the problem immediately and reduce the loss and trouble. Third, this virtual network technology in computer network security can temporarily set up client login to facilitate data search and utilization during access, promote cooperation and communication between both sides, improve work efficiency, maximize economic benefits, and avoid waste and loss of resources.

5.2 Application of Virtual Network Technology in Computer Network Security Between Enterprise Departments and Remote Departments

The virtual network technology in computer network security has been widely supported by many people. In today's computer network security development process, although the virtual network technology has been used, there are still some problems. Therefore, in the future development, we should pay attention to the application of this technology between enterprise departments and remote departments, so as to not only coordinate the specific work content between the two sides, but also effectively control the overall development of the whole. Once hackers and viruses attack, emergency treatment and rescue measures can be implemented in time to avoid the deterioration of the problem. In order to achieve the best resource sharing of enterprise information, enterprises should apply the virtual network technology in computer network security to effectively connect various enterprise departments and remote branches. This cross regional management mode can help enterprises to achieve strict control. The virtual network technology in hardware computer network security not only has strong encryption, but also is not easy to be destroyed, and has high efficiency.

6   Conclusions

The popularization of computer application is well-known, it can not only bring convenience to people's life and work, but also create favorable conditions for the development of the Internet. The sharing development mode of information resources has gradually moved towards the conventional development process. But the network security problem also along with produces, illegally obtains the access authority, the computer virus invasion and some network hacker's "rampant" has become the core hot topic of the network security question today, also brought the new challenge chemical work content to the expert in this aspect. This part of the problem is often difficult to root out, has been causing great trouble to customers, and the virtual network technology in computer network security is an important measure to solve these problems. In the computer internet security, it can effectively improve the efficiency of resource sharing among various users, reduce the waste of cost, and improve the computer network security problems. Therefore, only by fully studying the virtual network technology in computer network security can we really speed up the future development process of computer network security.

References

1. Qian, Z., Wei, W.: Analysis of the effect of virtual network technology in computer network security. Value Eng. **2014**(35), 196–197 (2014)
2. Chuanyang, L.: Application of virtual network technology in computer network security. Inf. Comput. (Theor. Ed.) **2015**(12), 114–115 (2015)
3. Li, M., Hongjie, L.: Application of firewall technology in computer network security. Comput. Knowl. Technol. **2014**(16), 3743–3745 (2014)

Application of Bionics in Underwater Acoustic Covert Communication

Kaiwei Lian, Jiaqi Shen, Xiangdang Huang[(✉)], and Qiuling Yang

School of Computer and Cyberspace Security, Hainan University, Haikou 570228, China
990623@hainanu.edu.cn

Abstract. In order to overcome the problems brought by traditional covert communication methods, based on bionics, this paper adopts real or simulated dolphin calls as communication signals to achieve the effect of covert communication and completely solving the safety problem of underwater covert communication. The main work of this paper is as follows: Step 1: Analyze the sound characteristics of cetaceans and classify them according to their sound characteristics. Create and store the database of these calls. Step 2: Analyze the actual application scenario, refer to our research and application, and select bionic signals from the database. Step 3: After the sound source is generated, select the appropriate modulation scheme. Step 4: Refer to the modulation scheme and channel characteristics, adopt appropriate channel estimation and equalization. Finally, the signal is extracted by any suitable receiver.

Keywords: Underwater acoustic communication · Bionics · Covert communication · Dolphin whistle · Dolphin tick-tock

1 Introduction

1.1 Application Background

With the development of modern detection technology, higher requirements are put forward for underwater combat platforms, especially for concealed underwater acoustic communication. The poor concealment of underwater acoustic communication is easy to expose the launching platform and lose the concealment advantage of underwater warfare. Therefore, it is very necessary to study the concealment of underwater acoustic communication.

The probability of detection and interception is proportional to the SNR. The high signal level exposes the communication platform. Therefore, the traditional hidden underwater acoustic communication is usually carried out under the condition of low SNR. However, with the decrease of SNR, the reliability of the system will be greatly reduced.

J. MacIntyre et al. (Eds.): SPIoT 2020, AISC 1283, pp. 205–210, 2021.
https://doi.org/10.1007/978-3-030-62746-1_30

2 Existing Challenges and Suggested Solutions

Traditional covert underwater acoustic communication is mostly realized by low SNR. The combination of low SNR covert communication and multi-carrier modulation technology is a research direction of low SNR underwater acoustic communication. GeertLeus et al. proposed a low SNR covert communication technology combined with OFDM modulation technology. In this communication system, the communication signal bandwidth is 3.6 kHz and is divided into 16 sub-bands. Experimental results show that when the SNR is reduced to -8 dB, the communication distance can reach 52 km and the communication rate is 78 bit/s. However, when the signal-to-noise ratio is too low, the communication system will collapse due to incorrect decoding [1, 2]. Document [3] also combines multi-carrier modulation to realize covert communication, and adopts turbo coding, frequency diversity and adaptive multi-band equalizer joint equalization technologies to further improve receiver performance. Although many scholars have made extensive research on low SNR covert communication, in the actual detection process, passive sonar can still detect such communication signals by long-time integration, especially in the position close to the sound source. Different from low SNR covert underwater acoustic communication, biomimetic covert communication emphasizes the method of biomimetic camouflage to induce the enemy to exclude the received signal and achieve the covert effect. Based on this idea, domestic and foreign scholars have carried out research on biomimetic camouflage concealed underwater acoustic communication. The advantage of this communication is that the enemy can detect the signal, but in the process of identification or classification, it will be excluded because it is the same as the original ocean noise.

2.1 Structure of the Paper

This paper focuses on the underwater acoustic communication of dolphin-like calls. The first chapter mainly introduces the research background of the thesis. The research status of low SNR covert underwater acoustic communication and bionic camouflage covert underwater acoustic communication at home and abroad are introduced respectively. The location and recognition function and communication function of dolphin sonar system are studied. The dolphin call signals are classified and introduced, and the signature whistle hypothesis is explained. The second chapter discusses the related work, analyzes and reviews the existing literature on the application of hidden underwater acoustic communication, and summarizes the shortcomings of the current work. The third chapter puts forward the suggested solutions: the first step, noise reduction and extraction of dolphin calls are proposed. In the second step, a dolphin-like whistle underwater acoustic communication is proposed. In the third step, underwater acoustic communication imitating dolphin ticking is proposed.

2.2 Related Work

Traditional covert underwater acoustic communication methods mostly start from the angle of low signal-to-noise ratio, hiding signals in ocean background noise to achieve the effect of LPI/LPD (Low Probability of Interception/Detection). Document [1, 2] proposes a multi-band OFDM modulation technique applied to low signal-to-noise ratio. Frequency diversity technique is used to improve the communication effect under low signal-to-noise ratio. The bandwidth in this system is 3.6 kHz and is divided into 16 sub-bands. When the communication distance reaches 52 km, the communication rate is 78 bit/s, the signal-to-noise ratio can be reduced to −8 dB, and under the condition of low communication rate, the signal-to-noise ratio can reach −16 dB. However, if the signal-to-noise ratio is too low, the communication system will collapse because the communication signal cannot be detected. Document [4] proposes an energy detector that is insensitive to phase jump based on spread spectrum technology. Using sea trial data, the correct demodulation of the data can be realized under the condition of the lowest signal-to-noise ratio of −10 dB. Document [5] estimates the time-varying channel impulse response using matched filter technology in the demodulation of spread spectrum signals, and the bit error rate is less than 10-2 under the condition of signal-to-noise ratio of −12 dB. Document [6, 7] uses incoherent underwater acoustic communication technology, and proposes biorthogonal modulation and biorthogonal differential phase shift keying, both of which use DSSS technology and RAKE receiver.

However, although the traditional low SNR covert underwater acoustic communication method reduces the average power of the transmitted signal, compared with the detection sonar, on the one hand, long-time integration can still detect the existence of the communication signal. On the other hand, reducing the transmission power of the communication signal to realize covert communication limits the communication distance. Based on this problem, from the perspective of bionic camouflage [8, 9], this paper proposes a method that can realize hidden underwater acoustic communication in full range. Using pulse modulation technology [10, 11] for reference, the information symbol form is improved.

3   Proposed Solutions

3.1   Noise Reduction and Extraction of Dolphin Calls

The noise reduction process is carried out according to the following steps: the first step is to use short-time Fourier transform to obtain the time-frequency spectrum image of dolphin whistle signal; the second step is to carry out smooth filtering to eliminate the high-frequency components in the image; the third step is to carry out direction selection filtering through a specific template to weaken the non target acoustic signal; finally, to suppress the harmonic of whistle signal to obtain a clear time-frequency spectrum of whistle signal fundamental wave.

3.2   Underwater Acoustic Communication with Dolphin Whistle

Composition of Whistle Signals. Dolphin whistle signal is an AM and FM harmonic signal, which can be expressed by a set of weighted sum FM sinusoidal signals as follows:

$$s[n] = \sum_{r=1}^{R} a_r[n] \sin(2\pi \varphi_r[n]).$$

Assuming a range of data, the signal is stationary. Using a window length of L, we can obtain the M th data block energy of the R th harmonic, Short-time Fourier transform. By using the interpolation method, the value of each sampling point in the block can be obtained, and finally the amplitude value of the Rd harmonic at each sampling point ar[n] can be obtained. The first harmonic energy of the whistler signal is larger than the second harmonic energy, but in a few sampling points, the second harmonic energy is larger than the first harmonic energy. Because the instantaneous frequency of a continuous-time signal can be expressed as a derivative of the phase, accordingly, the estimation of the phase of each sampling point can be expressed as an integral of the instantaneous frequency.

Underwater Acoustic Communication Technology Based on Simulated Whistle Signal. In the traditional underwater acoustic communication method, LFM signal is used as frame synchronization signal. But in the bionic communication system, in order to achieve the effect of bionic communication, the traditional synchronous signal can not be used. In the bionic communication, the original Dolphin whistle signal is used as the synchronization signal, and the zero sequence is inserted between the original whistle signal and the bionic modulation signal as the protection interval, prevent Crosstalk between synchronization signal and data symbol caused by multipath.

Imitation Whistle MSK Communication. MSK Signal is a special binary orthogonal 2FSK signal, in the process of signal change, can always maintain a constant envelope, phase continuous [12]. The simulated whistler MSK signal is composed of the time-frequency profile of the whistler signal and the MSK signal, and the communication information is loaded by the MSK signal.

3.3 Dolphin Click Underwater Communication

M-Ary Dolphin Tick-Tock Covert Communication. According to the characteristics of the tick-tock sound, a certain number of real tick-tock sound signals are selected for time-delay difference coding. In the delay difference coding system, a series of signals are continuously transmitted at the transmitter to modulate the delay length between adjacent signals to represent the communication information to be transmitted [13]. On the basis of this, the m-ary spread spectrum tick-tock delay difference code [14] uses multiple tick-tock sound signals as transmitting signals and modulates the information by changing the delay difference of adjacent tick-tock sound signals within a certain range, to achieve bionic covert communication. In order to synchronize the signal at the receiving end, different whistling signals are added at the beginning and end of the modulated M-ary tick-tock signal, in which a protection interval is inserted between the synchronous signal and the data symbol, used to resist crosstalk between synchronous whistles and modulated signals due to the multipath effect of underwater acoustic channels, where the protection interval is usually represented by a zero series.

Bionic Communication for UWB Signal. Ultra-wideband (UWB) signal is a narrow pulse impulse signal of nanosecond duration [15]. Ultra-wideband (UWB) communication, which uses UWB signal as information carrier, is a new radio communication technology with good concealment, high processing gain, high multipath resolution, large system capacity, low power consumption and low cost. In order to overcome the detection of low-frequency Detection Sonar, the high-frequency and narrow-pulse wideband signal is used to imitate the Dolphin's high-frequency tick-tock signal as the information carrier. Due to the characteristics of high frequency, wide band and narrow pulse, uwb signal has unique advantages in signal concealment and multipath resolution, so it is suitable for underwater acoustic communication. In time domain, the ticking signal of dolphins is a series of high frequency and narrow pulse signals with different time intervals. In order to simulate Dolphin tick sequences using ultra-wideband (UWB) signals, TH-PPM is used for information modulation.

4 Conclusions

Based on the principle of bionics, from the perspective of bionic camouflage, this paper proposes to use the inherent biological call as the symbol to realize the covert underwater acoustic communication. First of all, the underwater acoustic communication technology of dolphin-like whistle signal is studied. Modeling and Synthesis of osazone Acoustic Signal based on sinusoidal signal model. Based on the SWH hypothesis, the real dolphin whistle signal is modulated by time-frequency profile, and the bionic communication is realized by time-frequency profile shift modulation, stretch modulation and MSK modulation. The modulation and demodulation method of communication signal is given. Secondly, the underwater acoustic communication technology of dolphin tick signal is studied. In this paper, m-ary spread spectrum tick-tock delay difference coding is proposed, which uses Real tick-tock signal to modulate delay difference, and uses m-ary spread spectrum coding to improve communication rate. The ultra-wideband (UWB) signal is used to simulate the high-frequency dolphin tick-tock signal, and the time-hopping spread spectrum (TH-SS) technique is used to modulate multiple single pulses within each set of pulses according to the information bit transmitted, and each group of pulse as a whole PPM modulation.

Acknowledgment. This work was supported by the following projects: the National Natural Science Foundation of China (61862020); the key research and development project of Hainan Province (ZDYF2018006); Hainan University-Tianjin University Collaborative Innovation Foundation Project (HDTDU202005).

References

1. Leus, G., Van Walree, P.A.: Multiband OFDM for covert acoustic communications. IEEE J. Sel. Areas Commun. **26**(9), 1662–1673 (2008)
2. Leus, G., Van Walree, P., Boschma, J., Fanciullacci, C., Gerritsen, H., Tusoni, P.: Covert underwater communications with multiband OFDM. In: Oceans 2008, Quebec City, pp. 391–398 (2008)
3. Van Walree, P.A., Leus, G.: Robust underwater telemetry with adaptive turbo multiband equalization. IEEE J. Oceanic Eng. **34**(4), 645–655P (2009)

4. Yang, T.C., Yang, W.B.: Low probability of detection underwater acoustic communications using direct-sequence spread spectrum. J. Acoust. Soc. Am. **124**, 3632–3647 (2008)

5. Yang, T.C., Yang, W.B.: Performance analysis of direct-sequence spread-spectrum underwater acoustic communications with low signal-to-noise-ratio input signals. J. Acoust. Soc. Am. **123**, 842–855 (2008)

6. Jun, L.I.N.G., Hao, H.E., Jia, L.I.: Covert underwater acoustic communications. J. Acoust. Soc. Am. **128**, 2898–2909 (2010)

7. Ling, J., He, H, Li, J.: Covert underwater acoustic communications: transceiver structures, waveform designs and associated performances. In: Oceans 2010 MTS/IEEE Conferenc. Seattle, pp. 20–23 (2010)

8. Liu, S., Qiao, G., Ismail, A.: Covert underwater acoustic communication using dolphin sounds. J. Acoust. Soc. Am. **133**(4), EL300–EL306 (2013)

9. Liu, S., Qiao, G., Ismail, A., et al.: Covert underwater acoustic communication using whale noise masking on DSSS signal. In: OCEANS 2013 MTS/IEEE Conference, Bergen, Norway, pp. 1–6 (2013)

10. Yin, J.: Study on pattern delay coded underwater acoustic communication in multipath channels. Harbin Engineering University, Harbin, pp. 18–39 (2007)

11. Yin, J.: A study of pattern time delay coding communication underwater acoustic multipath channel. Harbin Engineering University, Harbin, pp. 18–39 2007

12. Changxin, F., Fuyi, Z., et al.: Communication Principles, pp. 34–35. National Defense Industry Press, Beijing (2001)

13. Liu, S., Liu, B., Yin, Y., Qiao, G.M.: Dolphin-like barking concealed underwater acoustic communication method. J. Harbin Univ. Eng. **35**(1), 119–125 (2014)

14. Shaozuo, L., Gang, Q., Yanling, Y.: A biomimetic underwater acoustic communication method using dolphin calls. J. Phys. **61**(14), 291–298 (2013)

15. Changyuan: Research on signal generation and modulation technology of UWB wireless communication system. Master's thesis, Harbin University of Engineering, p. 64 (2006)

Energy-Saving and Efficient Underwater Wireless Sensor Network Security Data Aggregation Model

Shijie Sun[1,2], Dandan Chen[2], Na Liu[2], Xiangdang Huang[2,3($\boxtimes$)], and Qiuling Yang[2,3]

[1] Information and Communication Engineering of Hainan University, Haikou 570228, China
[2] Computer Science and Cyberspace Security of Hainan University, Haikou 570228, China
990623@hainanu.edu.cn
[3] State Key Laboratory of Marine Resource Utilization in South China Sea, Haikou 570228, China

Abstract. Due to the inherent characteristics of resource constrained sensors, communication overhead has always been a major problem in underwater wireless sensor networks. Data aggregation is an important technology to reduce communication overhead and prolong network lifetime. In addition, due to the low bandwidth and high bit error rate transmission characteristics of underwater wireless media, underwater sensor nodes are more vulnerable to malicious attacks. Therefore, for such applications, data aggregation protocol must be energy-saving and efficient, and can prevent adversaries from stealing private data held by each sensor node. To solve this problem, this paper proposes an improved Energy-efficient Slice-Mix-Aggregate (ESMART) algorithm, which dynamically adjusts the number of data slices according to the amount of data sensed by the sensor nodes, which makes the total data transmission volume in the network lower than that of SMART (Slice-Mix-Aggregate) algorithm, that is, it does not introduce significant overhead on the sensor with limited energy, while maintaining the data security.

Keywords: Underwater wireless sensor networks · SMART algorithm · Data transmission

1 Introduction

Underwater wireless sensor network (UWSN) is one of the emerging wireless technology fields which has been widely developed in recent years. It establishes an underwater network, which monitors and communicates through fixed or mobile sensors. However, compared with the land sensor nodes, the underwater network nodes have the characteristics of energy limitation. Therefore, the design of Underwater Wireless Sensor Network (UWSN) wireless communication protocol must consider the energy consumption performance [1]. On the other hand, security is a crucial issue in UWSN. Compared with

J. MacIntyre et al. (Eds.): SPIoT 2020, AISC 1283, pp. 211–216, 2021.
https://doi.org/10.1007/978-3-030-62746-1_31

WSN, UWSN is more vulnerable because of its low bandwidth, large propagation delay, and high bit error rate.

For this reason, [2] proposed a data privacy protection scheme which can recover the original mining data and verify the integrity of the data, but the data fusion delay of this scheme is high, which reduces the efficiency of network operation. In [3], a secure data fusion scheme using private seeds to protect private data is proposed. The scheme has low communication overhead and computing load, but the security intensity is not high, and there is a large security risk.

In [4], a security data aggregation scheme of slicing, mixing and fusion is proposed. The data privacy protection performance of this scheme is better, but the communication overhead is huge.

In conclusion, this paper proposes ESMART algorithm. The algorithm dynamically selects the number of slices according to the amount of data sensed by the sub nodes, which will greatly reduce the communication load and prolong the network life cycle. The uncertainty of raw data segmentation also increases the difficulty of the attacker to restore the original data, effectively ensuring the confidentiality and integrity of the data.

2   Related Work

A lot of work has been done on data aggregation schemes in sensor networks, including [5–7]. All these works assume that all sensors are reliable and all communication is safe. However, in reality, sensor networks are likely to be deployed in an untrusted environment, such as underwater, so the link is easy to be eavesdropped, and the intruder may destroy the encryption key and manipulate the data.

In order to ensure data security and reduce the amount of information transmission, an Energy-efficient and high-accuracy secure data aggregation (EEHA) is proposed in [8]. Compared with SMART, EEHA can significantly reduce communication overhead and improve the accuracy of data fusion. Besides, a High energy-efficient and privacy-preserving secure data aggregation (HEEPP) is proposed in [9]. In slicing and hybrid technology, random distribution is introduced to determine the number of sliced data, to reduce the energy consumption of data collection and improve the security of data collection.

In [10], Energy- efficient adaptive slice-based secure data aggregation (ASSDA) scheme is proposed. In this scheme, a node's same data slice can only be sent once in the same time slot. ASSDA improves the efficiency of data slicing and reduces the energy consumption of nodes. By using the method of data decomposition, the number of slices is reduced and the communication cost is greatly saved.

Based on the above analysis, this paper proposes the ESMART algorithm, which not only retains the good privacy protection performance of the SMART algorithm, but also reduces the traffic of the whole network.

3   System Model

3.1   Network Model

In a 10000 * 10000 m underwater area, N sensor nodes are randomly distributed, and the following assumptions are made: 1) Node distribution density is random. 2) All nodes are assigned an ID to uniquely identify their identity in the network, and all nodes have the same initial energy. 3) Nodes can get their location data.

3.2 Clustering

A common technology of data aggregation is to establish aggregation tree and cluster aggregation. It is a directed tree formed by the union of all paths from sensor node to base station. These paths can be chosen arbitrarily, not necessarily the shortest path. The optimization of cluster structure is not in the scope of this paper. According to different application requirements, there are various methods of clustering. The clustering method in this article is described in [11].

3.3 Data Slicing

1) In order to save the energy consumption of cluster head, the cluster head node does not conduct data sensing and only performs data fusion operations;
2) The maximum number of slices preset by the system is recorded as N;
3) The amount of data collected by the sub-nodes in an induction cycle is different, before data segmentation, the amount of data collected by the sub-nodes is recorded as θ;
4) Record 2/3 of the maximum amount of data that a single node can collect as ρ, if $\theta > \rho$, the data will be divided into N pieces, otherwise the data will be divided into N−1 pieces;
5) The leaf node randomly keeps a data slice, encrypts the rest slices and sends them to its neighbor nodes.

Set the preset maximum number of slices N to 3. Take node 3 in the figure as an example, node 4 divides the data into two slices, keeps one slice, with the slice number of d_{33}, and sends the remaining slice to node 0, with the slice number of d_{30}. Data segmentation and slicing of other nodes follow the same rules (Fig. 1).

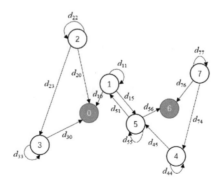

Fig. 1. Data slicing

3.4 Data Mixing

The node decrypts the received data slice and performs hybrid operation with the remaining original slices stored locally to generate a new data packet. If M_5 is the new data packet generated after hybrid operation of node 5, then can be obtained by $M_5 = d_{55}+d_{15}+d_{45}$. Similarly, mixed data of other nodes can be obtained (Fig. 2).

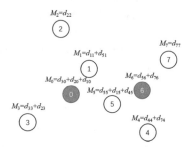

Fig. 2. Data mixing

3.5 Data Aggregation

The child node first encrypts the mixed result with the shared secret key, and then sends the encrypted data back to its corresponding cluster head nodes 0 and 6 respectively. The cluster head node will wait for a period of time so that all the leaf nodes can pass back the encrypted data. After receiving the encrypted data, the cluster head node decrypts it, merges it with the data slice it receives, and then encrypts the fusion result and continues to send it back to the base station (Fig. 3).

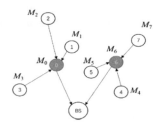

Fig. 3. Data aggregation

4 Model Analysis

4.1 Privacy Protection

The ESMART scheme proposed in this paper limits the data sensing behavior of cluster head nodes. At the same time, it uses the segmentation technology based on the amount of data collected by nodes to dynamically adjust the number of data slices, so that the total data transmission in the network is far less than the SMART scheme. At the same time, we know that the lower the amount of data communication, the lower the probability of privacy data exposure.

Secondly, the number of data slices in any sub-node of ESMART scheme is dynamically adjusted by the amount of data perceived by the sub node. Even if the attacker successfully attacks a certain point, it is still unable to determine whether he or she has heard all the data slices of the modified node, so it is impossible to determine whether the privacy data obtained by using the captured data is correct, which greatly increases the difficulty of the attacker to restore the original data.

Therefore, the ESMART scheme proposed in this paper has better performance in privacy protection.

4.2 Communication Cost

In SMART algorithm, the collected data is divided into N pieces before transmission, and N-1 pieces are distributed to neighboring nodes to generate N-1 packets, and the remaining one piece is reserved by itself. After the node data is mixed, the mixed data is transmitted to the upper node to generate one packet, so each node will generate N packets. Assuming that there are L nodes in the network, the total amount of data communication in the network is:

$$K = N \cdot L \tag{1}$$

The improved algorithm proposed in this paper is that the cluster head node does not carry out data sensing and data segmentation, the cluster head node only generates one packet of traffic, the sub node's segmentation of data is based on the uncertainty of the amount of data collected by the node, the proportion of the fusion node in the network is set as β, the probability of the leaf node dividing the data into N pieces is set as p, so the total amount of data communication in the network is:

$$K = \beta \cdot L + (1 - \beta) \cdot L \cdot N \cdot P + (1 - \beta) \cdot L \cdot (1 - P) \cdot (N - 1) \tag{2}$$

(1) and (2) can be obtained by subtraction:

$$\Delta = \beta \cdot L \cdot (N - 1) + L(1 - \beta) \cdot (1 - P) \tag{3}$$

Obviously, $\Delta > 0$. Therefore, the data traffic of the whole network is greatly reduced, that is, the communication cost is greatly reduced.

5   Conclusion

In this paper, we propose an energy-saving and efficient security data aggregation model for underwater wireless sensor networks. The model dynamically adjusts the number of data partitions according to the amount of data collected by the sub nodes, which ensures the data security without introducing significant additional overhead. Through the analysis of the model, we can find that the model provides good privacy protection for the original data with low cost.

Acknowledgment. This work was supported by the following projects: the National Natural Science Foundation of China (61862020); the key research and development project of Hainan Province (ZDYF2018006); Hainan University-Tianjin University Collaborative Innovation Foundation Project (HDTDU202005).

References

1. Su, C., Liu, X., Shang, F.: Vector-based low-delay forwarding protocol for underwater wireless sensor networks. In: 2010 International Conference on Multimedia Information Networking and Security, Nanjing, Jiangsu, pp. 178–181 (2010). https://doi.org/10.1109/mines.2010.46

2. Chen, C., Lin, Y., Lin, Y., Sun, H.: RCDA: recoverable concealed data aggregation for data integrity in wireless sensor networks. IEEE Trans. Parallel Distrib. Syst. **23**(4), 727–734 (2012). https://doi.org/10.1109/TPDS.2011.219

3. Yoon, M., Jang, M., Kim, H.I., et al.: A signature-based data security technique for energy-efficient data aggregation in wireless sensor networks. Int. J. Distrib. Sens. Netw. **2014**, 1–10 (2014)

4. Wenbo, H., Xue, L., Hoang, N., Nahrstedt, K., Abdelzaher, T.T.: PDA: privacy-preserving data aggregation in wireless sensor networks. In: 26th IEEE International Conference on Computer Communications, pp. 2045–2053 (2007)

5. Sattarian, M., Rezazadeh, J., Farahbakhsh, R., Bagheri, A.: Indoor navigation systems based on data mining techniques in internet of things: a survey. Wireless Netw. **25**(3), 1385–1402 (2019)

6. Brown, S.: An analysis of loss-free data aggregation for high data reliability in wireless sensor networks. In: 2017 28th Irish Signals and Systems Conference (ISSC), Killarney, pp. 1–6 (2017). https://doi.org/10.1109/ISSC.2017.7983622

7. Liu, C., Liu, Y., Zhang, Z.J.: Improved reliable trust-based and energy-efficient data aggregation for wireless sensor networks. Int. J. Distrib. Sens. Netw. **2013**(652495), 1–11 (2013)

8. Juan, L.H., Kai, L., Qiu, L.K.: Energy-efficient and high-accuracy secure data aggregation in wireless sensor networks. Compiter Commmun. **34**, 591–597 (2011)

9. Xun, L.C., Yun, L., Jiang, Z.Z., Yao, C.Z.: High energy-efficient and privacy-preserving secure data aggregation for wireless sensor networks. Int. J. Commun. Syst **26**, 380–394 (2013)

10. Pengwei, H., Xiaowu, L., Jiguo, Y., Na, D., Xiaowei, Z.: Energy- efficient adaptive slice-based secure data aggregation scheme in WSN. Procedia Comput. Sci. **129**, 188–193 (2018)

11. Madden, S., Franklin, M.J., Hellerstein, J.M.: TAG: a tiny aggregation service for ad-hoc sensor networks. OSDI (2002)

False Data Filtering in Underwater Wireless Sensor Networks

Na Liu, Haijie Huang, Xiangdang Huang, and Qiuling Yang[✉]

School of Computer Science and Cyberspace Security, Hainan University, Haikou 570228, China
990709@hainanu.edu.cn

Abstract. Nowadays, people pay more and more attention to the safety of Marine Territory and the development and utilization of marine resources. Underwater wireless sensor networks (UWSNs) [1] technology has become one of the hot research topics in the contemporary era. Underwater sensor nodes have the characteristics of poor defense ability, limited energy and easy to be breached. Once underwater sensor nodes are compromised by adversaries, the adversaries can use the sensitive information saved by the node to send false data injection attacks on legitimate reports. To address such attacks, this paper proposes a false data filtering strategy based on the basis nodes to mitigate the threat of these attacks. The strategy proposed not only improves the efficiency of network filtering and prolongs the network lifetime, but also reduces the energy consumption and resource waste.

Keywords: Underwater wireless sensor networks · Network security · False data injection attacks

1 Introduction

Underwater wireless sensor networks is a self-organized wireless communication network formed by randomly deploying a large number of micro sensor nodes to the monitoring water area and wireless communication between the nodes and autonomous underwater vehicle (AUV) [2, 3], the information of the sensing objects in the monitoring water area is sensed, collected and processed by cooperation between nodes, and then sent to the surface base station processing center (Fig. 1) [4].

The design of underwater wireless sensor networks should not only ensure the high efficiency and reliability of the network, but also consider the security of the network. Once underwater sensor nodes are compromised by adversaries, the adversaries can use the sensitive information saved by the node to send false data injection attacks on legitimate reports. At the same time, the energy saved by sensor nodes is limited, and malicious nodes constantly sending false data to the network will cause their own energy consumption, which will result in savings Point of rapid death. So it is very important to filter the false data in the underwater wireless sensor networks.

In order to resist the false data injection attacks, Ye et al. [5] proposed a Statistical En-route Filtering mechanism (SEF). In the data collection stage, the Center-of-Stimulus

J. MacIntyre et al. (Eds.): SPIoT 2020, AISC 1283, pp. 217–222, 2021.
https://doi.org/10.1007/978-3-030-62746-1_32

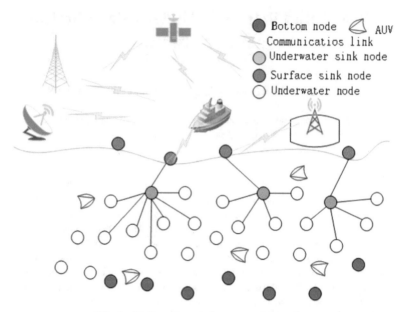

Fig. 1. Underwater wireless sensor networks.

(CoS) collects and summarizes the detection results by all detecting nodes, and produces a synthesized report on behalf of the group. Then, The cos then transfers the data to the sink node. This does not guarantee the authenticity and security of the data. Zhang Shuguang [6] proposed a false data filtering strategy based on neighbor node monitoring. In the phase of data packet generation, the cluster head node needs to communicate with the nodes in the cluster for many times to ensure the security of the data, and then the cluster head sends the data packets to the sink node in the form of single hop or multi hop. Due to the multiple communication between nodes, not only the energy consumption of nodes is large, but also the network life will be reduced.

None of the above strategies involve underwater wireless sensor networks. It not only causes the energy consumption of nodes, but also shortens the life cycle of the network, which can not guarantee the security of data. In view of the above problems, we propose a false data filtering strategy based on the basis of the nodes. In the packet generation phase, Within the overlapping area S of each sensing node in the cluster, a node is selected as the base Si. The received data is filtered falsely based on Si. In the transmission process, the false data is filtered again. The sink node as the last line of defense for filtering.

This paper is organized as follows. We survey related work on false data filtering in Sect. 2, We then describe our ideas and design in Sect. 3,We summarize this article in Sect. 4.

2   Related Work

According to the different data encryption technologies used, the existing false data filtering schemes for sensor networks can be divided into five categories: (1) False data filtering schemes based on symmetric keys; (2) False data filtering schemes based on public keys; (3) False data filtering schemes based on group key technology; (4) False data filtering schemes based on digital watermarking technology; (5) Filtering scheme based on time synchronization technology and time-varying parameters [7].

Symmetric key-based filtering is favored by sensor networks with limited energy because it not only has low computational complexity and energy consumption, but also is simple and practical to implement. Ye et al. [5] took the lead in proposing a Statistical En-route Filtering mechanism (SEF) by using symmetric key technology. During the data collection phase, the collected data is encrypted and transmitted directly to the CoS, and then forwarded by CoS to the sink node, which does not guarantee the security of the collected data. False reports may be forwarded in multiple hops before being filtered out, which wastes energy on nodes.

Ting Yuan [8] presented a false data filtering for wireless sensor networks based on key chain. During the report generation phase, In order to ensure the data security, the nodes in the cluster and the cluster head need to transmit multiple times.

Kim et al. [9] presented a fuzzy logic-based false report detection method (FRD). When an event occurs, the nodes together generate a report and attach their Mac to the report. The source node sends the report to sink. Even if some intermediate nodes find some wrong MAC reports, they can still be routed to sink until the threshold is reached. The fuzzy rule-based system on the sink is exploited to determine a threshold value by considering the number of false reports, the average energy of the forwarding path, and the number of MACs in the report.

Zhang Shuguang [6] proposed a false data filtering strategy based on neighbor node monitoring. The receiving node filters the false data by listening to the upstream node. The false data packet will be filtered out within one hop. In the phase of packet generation, this method needs to pass through multiple nodes in the cluster to ensure the security and authenticity of the data.

Liu et al. [10] proposed a geographical information based false data filtering scheme (GFFS). In the report generation stage, When an event happens, the CoS gets the location of the stimulus (Le) through the following way: First, it calculates the overlapping area (Do) of all the detecting nodes' sensing areas, then it randomly picks one point S from Do as the approximate value of Le. The CoS then broadcasts its reading e to all the detecting nodes. Upon receiving e, a detecting node S checks whether e is consistent with its own sensing value. Then CoS generates a report R, and finally sends the packet R to sink node through single hop or multi hop.

To sum up, in the existing false data filtering methods, nodes need to transmit data multiple times between the nodes in the cluster and the cluster head, or the cluster head encrypts the collected data and directly transmits it to the cluster head node, and then the cluster head transmits the data to the sink node. Data security and node energy consumption can not be both. At the same time, the above methods are not suitable for underwater wireless sensor networks.

3   False Data Filtering Strategy Based on the Basis Nodes

3.1   Network Models and Related Assumptions

Assuming that the sensing radius of the underwater wireless sensor networks nodes remains unchanged, the hierarchical clustering network is adopted in the network topology. In the cluster network, the cluster head knows the ID of the node in the cluster. Each node in the cluster transmits the collected data to the cluster head, and the cluster head generates the data packet, and transmits the data packet to the sink node or base station in one hop or multi hop mode, and the sink node or base station transmits the data to the water surface data center (Fig. 2).

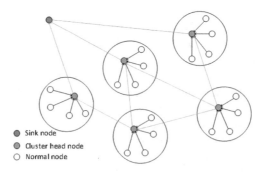

Fig. 2.   Network model diagram.

Assume that the network deployment and initialization phases are secure and not attacked. It is assumed that the location of nodes will not be changed in a large range after deployment, at the same time, it is ensured that nodes will not selectively discard data during transmission. It is assumed that sink node has strong security and it will not be captured.

3.2   False Data Filtering Strategy Based on the Basis Nodes

Key Distribution.   Similar to the SEF allocation strategy, sink maintains a global key pool composed of N keys. These keys are divided into M non overlapping partitions, each of which has Q keys. Each node selects t random keys from m partitions and stores them in the key space of the node together with the relevant key index. After the sensor node is deployed, each node recognizes any events it observes by generating a MAC on the report using its key.

Packet Generation.   The method consists of four steps: node selection, cluster-head data filtering, cluster head data comparison, and transmission filtering.

First, in the overlapping area S of each sensing node in the cluster, select a node with high reputation value as the basis S_i. In the overlapped area or the nearest node to the overlapped area is S_i, U5 and U6 (Fig. 3). If the reputation value of S_i node is the largest among the three nodes, then S_i is selected as the basis node. When the event occurs, each node in the cluster (including Si node and cluster head node) collects data and sends the data collected by each node in the cluster to the cluster head (including the base node Si).

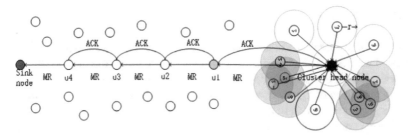

Fig. 3. Abstract model of false data filtering schemes.

Second, after the cluster head node receives the data, it looks up its neighbor node list to determine whether the node number (ID) is its neighbor node, otherwise it discards the data collected by the ID node. Then, the cluster head node checks whether the number of nodes of the collected data is greater than T, if not, the data is collected again in the cluster.

Third, Cluster heads compare the data collected by each node in the cluster based on Si. The comparison results retain the data within the specified error range. The cluster head checks whether the number of filtered nodes is greater than T, and if not, the data will be collected again in the cluster.

Fourth, the cluster head node packs the filtered data into MR, and transmits MR to sink node in the form of single hop or multi hop, and filters the false data in the transmission process. As the last line of defense, sink node filters MR.

En-Route Filtering. First, the node will judge whether to filter the packets according to the trust value of the upstream node of the forwarded packet MR. Secondly, the node will check whether the node forwarding MR is its own neighbor node, and verify the validity of one-way chain key in the packet. Then, the node checks whether the number of key index, ID and MAC in packet MR are equal to T. Then, the node checks whether the T key indexes come from T different key partitions. Then, the node checks whether there is the same key index as in MR. if there is the same key index, the node uses the key to verify its corresponding MAC. Then, the node checks the trust value of the upstream node forwarding MR, and determines whether to authenticate and then forward or forward and then authenticate MR. Then, within the specified time, the node monitors whether the last hop node sends ACK packets. Finally, the ID and one-way chain key of the sending node in MR are replaced by their own ID and new one-way chain key, and then they are forwarded to the downstream node, and the ACK packet is replied to the upstream node.

Sink Verification. After the sink node receives the packet MR, the sink checks whether the number of keys, ID and MAC in the packet MR is equal to T. The sink node then determines whether there are two key indexes from the same key partition. The sink node has all the information of node ID and corresponding key, and finally filters out the missing false data.

4 Conclusions

Aiming at the problem of excessive energy consumption in packet generation stage of the false data filtering strategy based on neighbor node monitoring proposed by Zhang Shuguang (2014), this paper proposes a false data filtering strategy based on the basis node. By analyzing our method, we can filter the false data significantly before the data packet is generated, which not only reduces the energy consumption and prolongs the network life cycle, but also ensures the data security.

Acknowledgment. This work was supported by the following projects: the National Natural Science Foundation of China (61862020); the key research and development project of Hainan Province (ZDYF2018006); Hainan University-Tianjin University Collaborative Innovation Foundation Project (HDTDU202005).

References

1. Wan, L., Zhou, H., Xu, X., Huang, Y., Zhou, S., Shi, Z., Cui, J.H.: Field tests of adaptive modulation and coding for underwater acoustic OFDM. IEEE J. Ocean. Eng. **40**, 327–336 (2015)
2. Wang, J., Chen, J.F., Zhang, L.J., et al.: Underwater wireless sensor networks. Acoust. Technol. **01**, 91–97 (2009)
3. Luo, Q.: Research on deployment of underwater wireless sensor networks. National University of Defense Technology, Changsha (2011)
4. Liu, Z.H., Zheng, J.J.: Ocean monitoring system based on underwater wireless sensor networks, 11–16 (2017). http://kns.cnki.net/kcms/detail/61.1450
5. Ye, F., Luo, H., Lu, S., et al.: Statistical en-route filtering of injected false data in sensor networks. IEEE J. Sel. Areas Commun. **23**(4), 839–850 (2005)
6. Zhang, S.G.: Research on false data filtering and malicious node location in wireless sensor networks. University of Science and Technology of China (2014)
7. Liu, Z.X.: False data filtering in wireless sensor networks. Central South University
8. Yuan, T.: Research on key management and false data filtering mechanism in wireless sensor networks. Fudan University (2009)
9. Kim, Lee, H.Y., Cho, T.H.: A fuzzy logic-based false report detection method in wireless sensor networks. Proc. Korea Simul. Soc. **17**(3), 27–34 (2008)
10. Wang, J., Liu, Z., Zhang, S., et al.: Defending collaborative false data injection attacks in wireless sensor networks. Inf. Sci. **254**, 39–53 (2014)

Research on Underwater Bionic Covert Communication

Jiaqi Shen, Kaiwei Lian, and Qiuling Yang[(✉)]

School of Computer Science and Cyberspace Security, Hainan University, Haikou 570228, China
13518034300@hainanu.edu.cn

Abstract. Due to the particularity of the underwater environment, underwater acoustic communication is an inevitable choice for underwater wireless sensor networks (UWSNs) communication. In recent years, with the development of underwater acoustic communication technology, the communication rate and communication quality of underwater wireless sensor networks have been greatly improved. At the same time, it has proposed the protection of the concealment of underwater equipment and the safety of underwater acoustic communication. New requirements. This paper reviews and summarizes the development status of underwater bionic covert communication. The analysis shows that the underwater bionic covert communication method using marine mammal vocalization as an information carrier has great advantages, which points out the direction for the research of underwater sensor network security communication.

Keywords: Underwater acoustic communication · Covert communication · Underwater wireless sensor networks · Bionic communication · Information hiding

1 Introduction

More than 70% of the earth's surface area is the ocean, which contains many untapped resources and is a natural treasure trove of resources. However, the current development and utilization of the ocean is still in its infancy, and there is an urgent need for new technological means for ocean development. Underwater acoustic communication technology is a necessary means of information exchange in the process of ocean development and research. Underwater wireless sensor networks (UWSNs) are powerful tools for detecting and developing marine resources, and their development is of national interest.

In recent years, our country has proposed to build a strong maritime country, continuously increase efforts in the development and utilization of marine resources, and propose the goal of building a digital ocean. In many emerging marine science and technology fields, underwater wireless sensor networks have attracted much attention because of their wide application in marine environment monitoring, marine life protection, seabed resource detection and development, and military fields. It is an inevitable choice for

J. MacIntyre et al. (Eds.): SPIoT 2020, AISC 1283, pp. 223–228, 2021.
https://doi.org/10.1007/978-3-030-62746-1_33

real-time transmission of marine data [1–3]. Underwater wireless sensor networks generally include a variety of sensor nodes, AUVs, buoys, etc. deployed underwater, and the data exchange between different devices mainly through acoustic signals. In the past, research in the field of underwater acoustic communication mainly focused on solving the problem of underwater acoustic communication from scratch, aiming to improve the communication rate and communication quality of underwater wireless sensor networks, ignoring the protection of information security during transmission [4]. With the development of underwater detection technology, underwater sensors can detect more sensitive information of the hydrological environment, which is related to national security. How to ensure the safety of underwater acoustic communication is gradually becoming the focus of research.

At the same time, excessive human underwater activities have also had a great impact on marine life. Studies have shown that the harmful effects of human activities on marine animals cannot be ignored [5]. In recent years, with the deployment and use of underwater sonar systems and underwater wireless sensor networks, some ecological events have occurred frequently. Therefore, research on high-speed, long-distance, secure, and environmentally friendly communication methods is the trend of a new generation of underwater acoustic communication technology.

2   Underwater Covert Communication

2.1   Overview of Underwater Covert Communication

Covert underwater acoustic communication is the concept of stealth communication in an underwater acoustic environment. There are two main methods for realizing subtle underwater acoustic communication, including low signal-to-noise ratio (SNR) hidden communication technology and bionic hidden communication technology [6].

Low signal-to-noise ratio concealed communication technology is to spread the communication signal spectrum or reduce the transmission power of the communication signal to send the communication signal in the form of a low signal-to-noise ratio, thereby reducing the probability of the enemy finding the communication signal and achieving the purpose of covert communication. Low signal-to-noise ratio covert communication technology is a common means of underwater covert communication, and its common implementation methods include OFDM [7], DSSS [8] modulation technology and direct sequence spread spectrum technology [9]. The low signal-to-noise ratio of covert communication technology means that the probability of the enemy detecting the communication signal is very small [9]. Low signal-to-noise ratio covert communication technology greatly limits the signal transmission rate and transmission distance. At the same time, the performance of low signal-to-noise ratio communication technology in cooperative communication networks is much lower than that of point-to-point communication And is not suitable for underwater sensor networks.

Bionic covert communication technology is to imitate the sounds inherent in the marine environment, including the sounds of marine mammals, the sounds of nature and human activities, or produce artificially similar sound signals similar to them as communication signals. These communication signals can be detected by the enemy, but in the signal processing process, the enemy treats these signals as marine environmental noise and excludes it from the recognition process, so as to achieve the purpose of covert communication [6]. The biggest feature of concealed bionic communication technology is that it can ensure the safety of communication and the concealment of launch platforms, and is bio-friendly.

2.2 Research Status of Underwater Bionic Covert Communication

In the early 21st century, the annual report of the US Naval Research Office first proposed the use of the sound of marine life or the sound of the ocean itself to conduct subtle underwater acoustic communication. Since then, researchers at home and abroad have proposed and implemented a variety of methods for concealed underwater acoustic communication based on bionic ideas.

In 2008, Dol, H.S and Quesson, B.A.J and others used dolphins' calls for underwater bionic communication for the first time and conducted underwater experiments, which verified the feasibility of underwater bionic communication for the first time [10]. Because the author uses multiple receivers to receive signals, but does not use channel equalization technology, the bit error rate is very high. A. EIMoslimany and others used signal modeling for the first time to construct FM signals to mimic the sounds of dolphins and whales for subtle underwater acoustic communication [11]. The experimenter first performs channel estimation through a matching tracking algorithm, then uses the maximum likelihood estimator to estimate the parameters of each received signal, and then decodes these parameters according to the mapping rules at the transmitter.

In 2012, Professor Liu Songzuo of Harbin Engineering University successfully experimented for the first time a method of covert communication using dolphin click sound as a communication signal carrier [12]. In this method, the dolphin's whistle sound is used as the synchronization signal, and the click sound is used as the information carrier to modulate the information on the time interval of each sound signal, thereby realizing covert communication. This method uses a matching tracking algorithm for channel estimation and a rake receiver for channel equalization [13]. In 2013, Liu and others proposed a new method of covert communication that mimics the high-frequency click sound of dolphins [14]. This method uses Rayleigh pulses that imitate the dolphin high-frequency click signal as the signal carrier. Each bit pulse is modulated by a time-hopping (TH) code, and the entire frame is modulated by pulse position modulation (PPM). In the same year, Liu and others also proposed a method of remote covert communication that mimics the sound of whales [15]. This method uses low-frequency, high-sound-level whale call signals as a covert carrier, hides the direct-sequence spread spectrum signal carrying information in the whale call signal range close to its frequency range, in order to achieve camouflage covert communication. Han Xiao studied the bionic communication method based on the dolphin whistle signal [16]. In this method, The confidentiality of the pattern delay difference coding is combined with the bionic effect of the dolphin whistle signal to increase the concealment effect of the communication. In addition, Jia Yichao and others from Nanjing University of Science and Technology also used the sound of sea lions to implement bionic underwater acoustic communication [17]. In this method, the sea lion call is used as the communication signal, and the bi-orthogonal modulation method is used to modulate the information on the sea lion call. In 2015, Liu proposed a bionic communication technology based on minimum-shift keying (MSK) modulation [18]. This method uses traditional MSK to modulate the information, and hides the modulated MSK signal in the spectral profile of the original dolphin whistle signal. In 2017, from the perspective of modeling, Liu and others used multiple chirp signals to imitate the real dolphin whistle to realize the subtle transmission of underwater information [19]. In 2017, Qiao Gang and his team designed a bionic portable modem for covert communication at short distances underwater [20]. The device stores different types of cetacean sound signals in the SD card of the modem for disguised communication effects in different scenarios. In 2018, Liu Zuozuo and others pioneered the selection of artificial noise such as marine piling noise as the overall signal carrier. Based on the characteristics that piling noise is arranged at approximately equal intervals, the secret information is modulated, and the synchronization signal is designed according to the time-hopping pulse position modulation method [21]. In 2018, Jiang Jiajia and others proposed a disguised sonar signal waveform design approach with its camouflage application strategy for underwater sensor platforms [22]. In 2019, Qiao Gang and others proposed a unique CUAC technology for simulating humpback whale songs [23]. Use Mary and PPM technology to modulate data in real humpback whale songs. In 2019, Muhammad Bilal and others proposed a new method of bionic Morse code that imitates humpback whale songs for underwater covert communication. Based on information entropy, this method compiles complex humpback whale songs into bionic Morse code, and develops standard imitation Morse code for English characters. In 2020, Jiang Jiajia and others designed a bio-inspired camouflage communication framework (BBICCF) based on the killer whale sound, which is used for safe underwater communication between military underwater platforms [24]. The research can choose the type of communication signal according to the actual application needs and it has practical significance for communication applications between underwater platforms.

3   Discussion

In summary, as a new research direction, underwater bionic covert communication is still in the initial development stage, and there are still many unresolved problems.

There are many types of ocean background sounds as information carriers, and it is very important to choose a suitable sound carrier. The sound of marine mammals is an ideal information carrier because of its proper frequency bandwidth and high sound source quality. Using the sound of marine mammals as an information carrier can control the power and frequency of the sounding unit very well, so it can reduce the impact of underwater wireless sensor networks on marine life to a certain extent, and also enhance the concealment of underwater equipment. Ensure the safety of the underwater acoustic communication network. However, the current research on marine organisms is not comprehensive enough. We are unable to determine the purpose of different vocal behaviors of marine organisms, and our understanding of the geographical distribution of marine organisms is also very limited. These issues are of great significance for the selection of information carriers.

Transmission rate, bit error rate, and concealment effect are the key indicators of underwater concealed communication. How to improve the transmission rate and concealment effect while reducing the bit error rate is a key issue. In this regard, there is an urgent need for more effective communication encoding and decoding methods. At the same time, the current evaluation of concealment effects lacks a unified index, and it is necessary to study and propose more complete standards to evaluate the disguise effect of concealed communications.

The current covert underwater acoustic communication emphasizes the covertness of the signal level rather than the covertness of the information level. Future research can introduce encryption at the information level, establish a hierarchy of information hiding, and choose different encryption levels for different information, reduce the resource consumption of underwater wireless sensor networks, improve work efficiency, and further improve the safety of underwater networks safety.

4 Conclusion

This paper introduces the development status of underwater bionic covert communication technology. After years of development, the underwater concealed communication technology has further matured, providing a good theoretical basis for the research of underwater communication equipment applications. Because of its unique low detection mechanism, underwater bionic covert communication can meet the needs of underwater wireless sensor networks for concealment, safety and transmission power, and has little impact on the ecological environment, which has the effect of improving the performance of underwater wireless sensor networks. Certain help. The current research in the field of underwater bionic covert communication has achieved some preliminary results, but there is still a lot of unresearched space that deserves our further research.

Acknowledgment. This work was supported by the following projects: the National Natural Science Foundation of China (61862020); the key research and development project of Hainan Province (ZDYF2018006); Hainan University-Tianjin University Collaborative Innovation Foundation Project (HDTDU202005).

References

1. Akyildiz, I.F., Pompili, D., Melodia, T.: Underwater acoustic sensor networks: research challenges. Ad Hoc Netw. **3**(3), 257–279 (2005)
2. Partan, J., Kurose, J., Levine, B.N.: A survey of practical issues in underwater networks. ACM SIGMOBILE Mob. Comput. Commun. Rev. **11**(4), 23–33 (2007)
3. Heidemann, J., Stojanovic, M., Zorzi, M.: Underwater sensor networks: applications, advances and challenges. Philos. Trans. R. Soc. A **2012**(370), 158–175 (1958)
4. Chitre, M., Shahabudeen, S., Freitag, L., et al.: Recent advances in underwater acoustic communications & networking. In: OCEANS 2008. IEEE (2009)
5. Li, S., et al.: Mid- to high-frequency noise from high-speed boats and its potential impacts on humpback dolphins. J. Acoust. Soc. Am. **138**(2), 942–952 (2015)
6. Gang, Q., Muhammad, B., Songzuo, L., et al.: Biologically inspired covert underwater acoustic communication—a review. Phys. Commun. (2018). S1874490717305608
7. Leus, G., Van Walree, P.: Multiband OFDM for covert acoustic communications. IEEE J. Sel. Areas Commun. **26**(9), 1662–1673 (2008)

8. Yang, T.C., Yang, W.-B.: Low signal-to-noise-ratio underwater acoustic communications using direct-sequence spread-spectrum signals. In: Oceans. IEEE (2007)
9. Yang, T.C., Yang, W.B.: Low probability of detection underwater acoustic communications using direct-sequence spread spectrum. J. Acoust. Soc. Am. **124**(6), 3632–3647 (2008)
10. Dol, H.S., Quesson, B.A.J., Benders, F.P.A.: Covert underwater communication with marine mammal sounds. Covert Underwater Communication with Marine Mammal Sounds Tno Repository (2008)
11. Elmoslimany, A., Zhou, M., Duman, T.M.: An underwater acoustic communication scheme exploiting biological sounds. Wirel. Commun. Mob. Comput. **16**, 2194–2211 (2016)
12. Liu, S., Qiao, G., Yin, Y.: A bionic underwater acoustic communication method using dolphin calls. Acta Phys. Sin. **14**, 291–298 (2013)
13. Liu, S., Qiao, G., Ismail, A.: Covert underwater acoustic communication using dolphin sounds. J. Acoust. Soc. Am. **133**(4), EL300 (2013)
14. Liu, S., Qiao, G., Zhang, L.: Biologically inspired covert underwater acoustic communication using high frequency dolphin clicks. In: OCEANS 2013 IEEE Conference, San Diego, vol. 58, no. 2, pp. 1–5 (2013)
15. Liu, S., Qiao, G., Ismail, A., et al.: Covert underwater acoustic communication using whale noise masking on DSSS signal. In: 2013 OCEANS – Bergen. MTS/IEEE. IEEE (2013)
16. Xiao, H., Jingwei, Y., Longxiang, G., et al.: Research on bionic underwater acoustic communication technology based on differential Pattern time delay shift coding and dolphin whistles. Acta Phys. Sin. **62**(22), 224301
17. Jia, Y., Liu, G., Zhang, L.: Bionic camouflage underwater acoustic communication based on sea lion sounds. In: International Conference on Control. IEEE (2015)
18. Liu, S., Ma, T., Qiao, G.: Bionic communication by dolphin whistle with continuous-phase based on MSK modulation. In: IEEE International Conference on Signal Processing. IEEE (2016)
19. Liu, S., Ma, T., Qiao, G., et al.: Biologically inspired covert underwater acoustic communication by mimicking dolphin whistles. Appl. Acoust. **120**, 120–128 (2017)
20. Qiao, G., Zhao, Y., Liu, S., Bilal, M.: Dolphin sounds-inspired covert underwater acoustic communication and micro-modem. Sensors **17**(11), 2447 (2017)
21. Liu, S., Wang, M., Ma, T., et al.: Covert underwater communication by camouflaging sea piling sounds. Appl. Acoust. **142**(DEC.), 29–35 (2018)
22. Jiang, J., Sun, Z., Duan, F., et al.: Disguised bionic sonar signal waveform design with its possible camouflage application strategy for underwater sensor platforms. IEEE Sens. J. **18**, 8436–8449 (2018)
23. Bilal, M., Gang, Q., Liu, S.: Covert mimicry communication using humpback whale song. In: IBCAST 2019 (2019)
24. Jiajia, J., Xianquan, W., et al.: A basic bio-inspired camouflage communication frame design and applications for secure underwater communication among military underwater platforms. IEEE Access **8**, 24927–24940 (2020)

Authentication and Access Control for Data Usage in IoT

The Application of Virtual Reality Technology in Architectural Design

Lili Peng[✉], Yihui Du, and Zijing Zhang

Chongqing College of Architecture and Technology, Chongqing, China
haoxueinfo@eiwhy.com

Abstract. Under the background of the rapid development of information technology and Internet globalization, the buildings and cities that people depend on have undergone earth-shaking changes, and architectural design tools and design models are constantly updated. Architecture is unique. The method continues to expand its development space. Construction engineering is a complex and large-scale dynamic system, which involves not only many but also more complicated links. Because there are complex structures inside general buildings. This article focuses on the application of virtual reality technology in architectural engineering design from the basic concepts and characteristics of virtual reality.

Keywords: Virtual technology · Virtual reality · Architecture · Design

1 Introduction

In today's big social background, with the changes of the times and the rapid development of science and technology, China's construction level has been continuously improved, the design plan has a stronger implementation feasibility, and the design concept of the building is therefore more diverse. At the same time, with the development of my country's economy and society, consumers' aesthetic concepts have undergone great changes, and architectural design styles and design concepts have also undergone great changes. Due to the acceleration of the pace of life, people's requirements for architectural design have changed from light luxury to simplicity. People have begun to pursue the beauty of simplicity and the inner beauty of things. Therefore, the transformation of architects' design style is to keep up with the pace of the times. It is also the embodiment of social aesthetics. This change actually reduces the difficulty of building construction, but on the other hand, it also imposes new requirements on architectural design. How to create a unique charm of architecture in a simple design style and lead the sustainable development of the industry. This will be an important subject of architectural design work [1].

2 Overview Virtual Reality Technology

VR (Virtual reality) was put forward by Jaron Lanier, an American entrepreneur. It means the integrated utilization of computer graphics system, multimedia technology, sensor

J. MacIntyre et al. (Eds.): SPIoT 2020, AISC 1283, pp. 231–236, 2021.
https://doi.org/10.1007/978-3-030-62746-1_34

technology, three-dimensional simulation technology and various devices to display and control interfaces; the computer can generate the interactive three-dimensional environment, which offers immersive techniques [2]. The core essence of VR is to visualize computer data, so as to achieve the interaction of information between human subjects and virtual digital objects [2].

The virtual environment system includes following disciplines and technologies: computer graphics, image processing, pattern recognition, voice processing, Internet techniques and so on. The generated data is processed and high-performance computing, and it is a new type of highly comprehensive high-tech information technology to model a large-scale integrated environment composed of sub-systems of considerable scale with different functions and levels [3]. Figure 1 is the embodiment of VR technology in architectural design.

Fig. 1. The embodiment of VR technology in architectural design

3   Characteristics of VR Technology

3.1   Immersion

VR technology is based on the physiological and psychological characteristics of human vision, hearing, and touch. The computer generates real visual three-dimensional images. The user wears interactive equipment such as helmet-mounted displays and data gloves, and even puts himself in a virtual environment In, become a member of the virtual environment. The interaction between the experiencer in the virtual world and various objects or scenes in the virtual environment will produce mutual influence, just like the interaction between humans and nature in the real world, all feelings are so realistic [4]. It's even a real, immersive experience.

3.2   Interactivity

Human-computer interaction in a VR system is that people use devices such as keyboards, mice, analog monitors, touch screens, etc. to enter the computer system. The computer system uses high-speed calculations and analysis to make a kind of human-like behavior, which is the initial embodiment of artificial intelligence.. Experiencers can inspect, manipulate, research, and calculate objects in the virtual environment through their own language, body language, or touch the keyboard, interface and other natural actions, and they can also generate big data records and archives [5].

3.3 Multi-sensitivity

The VR system is equipped with supporting devices for visual, audio, tactile and kinesthetic sensing and reaction, so that the experiencer can experience the realm of "ears and hearing is true, seeing is believing" in the virtual environment, and gain vision senses, hearing, touch, kinesthetics, etc., make the experiencer unable to distinguish between virtual and reality [6].

4 Application of VR Technology in Architectural Design

For architectural design, the charm of VR in the eyes of architects is not only to truly reflect the beauty of the building and the three-dimensional performance of human-computer interaction, but also to provide new information that other traditional expression methods cannot match [7]. Communication and editing interface: while experiencing the three-dimensional space in person, through real-time editing of architectural elements, real-time adjustment of three-dimensional scenes, system information integration and other technologies, it provides a powerful way for design comparison and design thinking, design features and related information display, highlighted in the following aspects:

4.1 Scientific and Accurate Analysis of Design Information

VR's powerful application program interface can import relevant complex system models for analysis, so it can provide architects with a global and objective understanding. It provides advanced optical systems, atmospheric analysis systems, geographic information systems (GIS), etc., which can make creative information analysis more scientific, real, and accurate [8].

4.2 VR Realizes Dynamic Spatial Design

VR technology can bring architects' abstract thinking performance closer to reality, free architects from the constraints of two-dimensional performance and the uncertainty of abstract thinking, and enable their ideas to be fully developed in the real-time generated three-dimensional world [8]. The conception of the created architectural space can be sculpturally free. We can think that the development of technology restricts the form, the requirements of environment and function affect the form, and the tools and media of design expression may also promote the emergence of new architectural forms, such as Fig. 2.

Fig. 2. VR realizes dynamic spatial design

4.3 VR Implements the Whole-Process Auxiliary Design

The application of the VR system in architectural design can run through the entire process, from the feasibility study at the initial stage, site condition analysis, to conceptual design, spatial shape combination at the development stage, architectural details, integration with the environment... until the technology at the construction stage Both simulation and VR systems can play their role [9].

4.4 VR Realizes Information System Design

In the design, through the "data interconnection" environment, the building model in the VR system is interconnected with other digital media information such as text, sound, screen, plan, three-dimensional model and 3D animation, so that a specific screen in VR can be Contains key information, such as: plan view, section view, related structure description, design content worthy of further scrutiny [10].

4.5 VR Realizes the Presence of Architectural Expression

The immersive nature of VR technology allows architects to be on the scene. Through the flight mode, they can take a bird's-eye view of the building and observe the relationship between the building and the surrounding environment; observe and experience the layout of the building space and landscape design through a full-scale perspective. Real-time roaming inside and outside the building can help you observe the details at close range, experience the building space, the appearance of the building, the size of the building components, the reinforcement of the components, etc., and you can even wear data gloves to touch the materials on the building surface, so that the architect can design [11]. The buildings will be more rational, scientific and beautiful.

4.6 Interactive Application of VR in Architectural Experience

VR technology realizes the interaction of architects with virtual environment objects through the use of virtual interactive interface equipment, and the space of the place is expressed as much as possible [7]. In VR, the viewing angle can be determined from the overall situation or from the details; the architectural elements can be edited; at important nodes, relevant data can be linked as needed to obtain more information, giving the architect more autonomy [11].

4.7 VR Realizes Creativity in Perception Activities

The VR system can provide architects with a free multi-angle and multi-dimensional human-computer interaction experience program. People can choose to observe the building carefully in a static state, or dynamically experience the architectural space in a variety of motions. At the same time, they can compare different plans in real time, compare multiple plans, and make judgments and choices. At the same time, the system can also simulate sunlight, related equipment and facilities, etc., making the overall expression of the building more comprehensive, more real, more scientific and more convincing [12], as shown in Fig. 3.

Fig. 3. VR in architectural design to view design effects from multiple angles

5 Conclusion

VR technology saves time and resources for the construction process by simulating and restoring the construction site, and at the same time adds a guarantee to the design of its buildings. Modern civil architecture design urgently needs the support of science and technology, and the emergence of VR technology is a booster for its rapid development using scientific and technological means. It is also an inevitable process of architectural design digitalization. The accuracy and scientificity of virtual technology will be the foundation of architectural design digitalization.

References

1. Zhao, N.: Research on the application of virtual simulation technology in architectural design. Changsha Univ. Sci. Technol. **5**, 23–26 (2019). (in Chinese)
2. Li, S.: Research on the application of virtual reality technology in architecture and urban planning. Changsha Univ. Sci. Technol. **12**(03), 112–115 (2018). (in Chinese)
3. Zhang, L.: Research on the application of virtual reality technology in architectural heritage protection. Jiangnan Univ. **11**(03), 34–37 (2019). (in Chinese)
4. Chen, Z., Li, X., Yan, W.: Application of virtual reality technology in the field of architecture. Eng. Constr. Des. **8**, 88–90 (2015). (in Chinese)
5. Wang, D., Zhang, J.: Building structure design based on virtual reality technology. Struct. Eng. **11**(5), 61–63 (2014). (in Chinese)
6. Zhu, X., Peng, S.: Application of virtual reality technology for construction safety. Anhui Sci. Technol. **3**(11), 44–46 (2014). (in Chinese)
7. Yang, Q., Lin, D.: Application and promotion strategy of VR technology in architectural design. Constr. Technol. **4**(08), 733–735 (2016). (in Chinese)
8. Bai, S., Zhang, Y.: Application value analysis of VR technology in prefabricated buildings. Constr. Econ. **6**(11), 106–109 (2015). (in Chinese)
9. Yuan, S., Wei, J., Zhai, M.: Building construction curriculum reform based on BIM virtual reality. J. Hunan Ind. Vocat. Tech. Coll. **6**(4), 97–99 (2017). (in Chinese)
10. Liu, J.: Talk about the application of VR technology in construction engineering. Archit. Eng. Technol. Des. **6**(08), 251–253 (2016). (in Chinese)
11. Gu, Z.: Analysis of application practice based on virtual construction technology. Constr. Eng. Technol. Des. **2**(08), 130–132 (2015). (in Chinese)
12. Liu, X., Wu, F.: The innovative application of BIM-based virtual reality technology (VR) in construction engineering. Chin. Foreign Archit. **4**, 134–136 (2018). (in Chinese)

Computer-Assisted Teaching and Cultivate Students' Innovative Thinking Ability

Feng Gao[(✉)]

South China Business College, Guangdong University of Foreign Studies, Guangzhou, China
caonan_lixin@eiwhy.com

Abstract. With the rapid development of computer-assisted teaching technology, it can not only improve the efficiency of classroom teaching and enhance the interaction between teachers and students, but also effectively activate the classroom atmosphere and increase students' interest in learning. As an important development direction of computer technology in the future, the application prospect in classroom teaching is very broad. By enabling knowledge to be presented in front of students more three-dimensionally, creating a more realistic classroom teaching atmosphere for students, achieving visual and auditory dual stimulation, so that students' learning initiative and learning effects are comprehensively enhanced. This article analyzes the application in computer-assisted teaching, and then discusses computer-assisted teaching and cultivating students' innovative thinking ability.

Keywords: Computer · Assisted teaching · Students · Innovative thinking

1 Introduction

Combining computer technology, information technology, etc., using computer simulation to create a three-dimensional space, providing users with sensory simulations of vision, hearing, touch, etc., so that users can have an immersive feeling. Virtual reality does not have time and space limitations, but transmits accurate 3D world images to users as users continue to move and explore. Through the application in classroom teaching, the content of the teaching can be displayed to students more realistically and vividly, allowing students to feel immersive, improving students' visual and auditory stimulation, and enabling students to be more engaged in teaching in the event [1].

2 The Application of Computer-Assisted Teaching

Computer-assisted teaching is a teaching method that assists students and teachers to complete discussion, lecture, review and other teaching links in the education process through computers. Its essence is to disseminate teaching information to students with more abundant means and efficiently [2], as shown in Fig. 1. In terms of its extension,

J. MacIntyre et al. (Eds.): SPIoT 2020, AISC 1283, pp. 237–242, 2021.
https://doi.org/10.1007/978-3-030-62746-1_35

computer-assisted teaching includes five modes. One is the application in student practice. It mainly refers to students completing various tasks assigned by the teacher through a computer system. The computer system provides students with the best results through intelligent analysis. The method of disintegration, and give reasonable supplementary exercises for each student's own weaknesses, and on this basis, provide each teacher with a learning plan for each student's characteristics in future teaching [3].

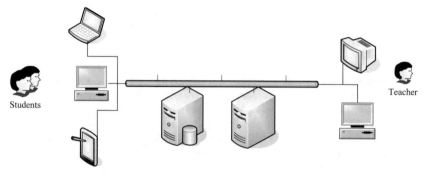

Fig. 1. Computer-aided teaching model

2.1   Computer-Aided Guidance to Students

In the traditional way of education, due to time and energy constraints, the guidance time for each teacher to students is always limited. Due to the different level of understanding of the knowledge taught by the teacher, each student has different problems with the content taught by the teacher, and the teacher cannot be comprehensive [4]. However, through computer-assisted teaching, the computer itself can give different guidance according to each student's weakness, which largely compensates for the teacher's inconsistent teaching due to time and energy constraints.

2.2   The Application of Computer-Assisted Communication Between Students

Teaching is not only a rigid answering and solving process, but also a process of forming a healthy personality for students through the education of emotion, will and reason. The answer to the exercises seems to be a rigid training process, but it also includes the life process of students asking questions, overcoming difficulties, and solving problems. In this process, students will inevitably have various emotional fears [3]. Computers can communicate well with students through artificial intelligence, and in this process, promote students to form a good attitude towards life and correct study habits.

2.3   Computer-Assisted Application in Teacher Teaching

In computer-assisted teaching, teachers can use computer-designed games to allow students to actively accept various knowledge points that the textbook wants to convey in the fun learning. At the same time, various complex formulas in physics, chemistry and mathematics can also be simulated by computer in the form of images and sounds, so that students can understand the inner meaning of these formulas more vividly, and they can more directly understand the meaning of these formulas [3].

2.4 Computer-Assisted Teaching Is Used in Problem Solving Process

Traditional teaching methods often only pay attention to the importance of answering questions. As long as one answering method can get the correct answer, even if this answering method is boring, it is good. However, through computer-assisted teaching, you can show the pros and cons of multiple solving methods, so that students can choose the most suitable method for solving problems during the learning process [5].

3 Advantages of Computer-Assisted Teaching

The essence of computer-assisted teaching is to disseminate teaching information to students efficiently with richer means. From this point of view, computer-assisted teaching has the following three advantages (shown in Fig. 2):

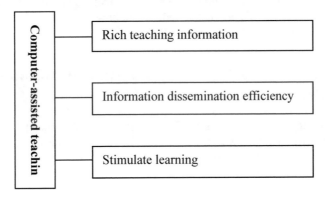

Fig. 2. The advantages of computer-assisted teaching

3.1 The Richness of Educational Information Dissemination

Traditional education's information dissemination method is often based on blackboard writing and exercises. This information dissemination method is mainly a way of disseminating abstract information from concept to concept. In this teaching method, teachers mainly focus on the explanation of concepts, and the dissemination of knowledge is often realized through enumeration and noun interpretation. In this teaching method, it is often difficult for students to understand what some difficult concepts are saying. Because this method of knowledge dissemination wants students to realize the most essential connotation of knowledge at the beginning of teaching. However, the process of human cognition is a process that slowly deepens from the surface [5]. Contrary to traditional teaching methods, computer-assisted teaching uses pictures, animations, videos, and sounds to allow students to understand knowledge from appearances, so that students can obtain all the graphical expressions of knowledge from the beginning [6].

3.2 The Efficiency of Educational Information Dissemination

Due to the limitation of the teacher's time, energy and educational resources, traditional education methods often fail to enable every student to effectively understand the meaning of knowledge itself [6]. However, through computer-assisted teaching, the system can automatically identify the difficulties and perplexities of each student in the education process. And give the most effective solution to these difficulties and puzzles. In this way, the efficiency of education and teaching is greatly improved.

3.3 Computer-Assisted Teaching Can Stimulate Students' Enthusiasm for Learning

The traditional teaching method is boring because it is from concept to concept. Computer-assisted teaching includes pictures, sounds, animations and other information dissemination methods, and provides different learning plans for each student. The concealed teaching methods are undoubtedly more targeted and more interesting, that make students better enter the classroom [7].

4 Computer-Assisted Teaching Trains Students' Innovative Thinking

The construction of a computer-assisted teaching training student innovation system should establish a computer-assisted teaching quality evaluation mechanism, a computer-assisted teaching quality monitoring mechanism, and a computer-assisted teaching quality incentive mechanism.

4.1 Establish a Computer-Assisted Teaching Quality Evaluation Mechanism

An effective way of quality management through quantitative methods. The establishment of a computer-assisted teaching quality evaluation mechanism must first establish an evaluation index system, secondly, clarify the evaluation object and the evaluation subject, and collect the original data according to the specific situation of the evaluation object, and finally according to the quantification [8]. The evaluation model calculates the specific evaluation value of each evaluation object, thereby laying the foundation for establishing an incentive system based on the quality evaluation results [8].

4.2 Establish a Computer-Aided Teaching Quality Monitoring Mechanism

A computer-assisted teaching quality and evaluation office should be established to uniformly track, evaluate and manage discipline construction and training quality. At the same time, a computer-assisted teaching supervision system was established to supervise and guide the education work of the unit. The members of the supervision team are composed of computer-assisted teaching teachers, who should be selected from among experienced and retired teachers with senior professional titles [9]. The main work of the supervision team is to integrate point-to-point and aspect-to-point, and to achieve the quality of computer-assisted teaching from the perspective of the entire life cycle of education Monitoring work.

4.3 Establish a Computer-Aided Teaching Quality Incentive Mechanism

Funding protection is the key to talent training. For example, Shenzhen Huawei Technology Company clearly stipulates in the "Huawei Basic Law" that the company shall use 4% of its total sales revenue for young teacher training expenses; P&G (China) shall recruit every year a large part of the new young teachers in China are sent to the United States to train and exercise for one year, and the company will bear all the expenses [9]. Therefore, in promoting the cultivation of talents in computer-assisted education, it is necessary to actively establish a talent training funding guarantee mechanism, form a capital investment system in which government finances and social entities participate together, and establish a teacher incentive mechanism based on the results of quality evaluation [10].

4.4 Establish an Intercommunication System for Computer-Assisted Education Institutions

The academic education level of computer-assisted education should be actively expanded to the postgraduate education level on the basis of junior college, undergraduate and other academic education levels, so as to meet the needs of modern computer-aided education objects for academic education on the basis of knowledge and skills education, and Realize advanced computer-assisted education [7]. Therefore, it is necessary to establish an intercommunication system between computer-assisted education institutions and ordinary universities, such as promoting the accumulation and conversion of learning achievements in ordinary universities, and initiating the certification of learning achievements and the accumulation of credits, so as to establish the advanced and long-term academic education based on computer-assisted education [10].

5 Conclusion

Computer technology is a rapidly developing subject. It is the requirement of the times to lay the necessary foundation for adapting to the study, work and life of the information society. It is also an indispensable ability for students to develop themselves. In the practice of computer-assisted teaching, teachers should create opportunities for students to actively explore, encourage and inspire students to think actively, be good at thinking, learn to think independently, and highlight the cultivation of innovative thinking. At the same time, teaching should emphasize the improvement of innovation ability, give full play to independent initiative, give full play to students' learning potential, and gradually improve practical ability and innovative spirit.

Acknowledgments. Project Fund: This paper is the outcome of the study, On the Construction of a Cooperative Teaching Mode for Teachers from Multiple Disciplines Based on the Development of Students' Innovative Thinking, which is supported by the Foundation for Provincial-level Projects of Young Innovative Talents in Colleges and Universities of Guangdong Province. The project number is 2019 WQNCW165.

References

1. Yi, L.: The concept and practice of computer-assisted teaching. China Contin. Med. Educ. **4**, 84–86 (2017). (in Chinese)
2. Baojiao, L.: A new idea of computer-assisted teaching in school classrooms. Educ. Teach. Forum **2**(03), 55–57 (2019). (in Chinese)
3. Ning, Z., Yingfang, F.: Application research of digital virtual technology in clinical teaching of hepatobiliary surgery. China Contin. Med. Educ. **10**(33), 16–19 (2018). (in Chinese)
4. Hao, C., Meiping, L.: Talking about the application of VR in the teaching of "Architectural Construction". Jiangxi Build. Mater. **6**(13), 54–55 (2018). (in Chinese)
5. Zhao Xi, H., Wenhua, Z.X.: Development and practical application of a certain radar virtual maintenance training system. Educ. Teach. Forum **8**(49), 76–78 (2018). (in Chinese)
6. Hang, Z., Canheng, Z.: Research on teaching experimental methods of civil aviation security inspection technology courses based on VR technology. Educ. Teach. Forum **12**(49), 273–274 (2018). (in Chinese)
7. Quan, S., Zhou, Yu.: Curriculum experimental teaching of combination of virtuality and reality and its application in civil aviation talent training. Educ. Teach. Forum **7**(9), 175–178 (2019). (in Chinese)
8. Hongxia, Z.: Some thoughts on the practice of computer multimedia assisted teaching in colleges and universities. Inf. Comput. Theory Ed. **11**(14), 255–256 (2018). (in Chinese)
9. Shuhui, L.: Computer-assisted teaching in colleges and universities. Sci. Technol. Econ. Trib. **22**, 181–184 (2017). (in Chinese)
10. Tingting, C., Zhengmao, L.: Exploration and research on teaching methods of computer basic multimedia courses. Shandong Ind. Technol. **3**(20), 184–187 (2019). (in Chinese)

The Reform Progress and Practical Difficulties of State-Owned Hospitals Under Information Age—Case Analysis Based on the Reform in a Medical Institution of a Group in China

Yilong Wang[1,2], Hongjie Bao[2(✉)], Xuemei Zhang[3], and Aihong He[4]

[1] School of Public Administration, Sichuan University, Chengdu 610041, China
wyl19881129@163.com
[2] School of Management, Northwest Minzu University, Lanzhou 730030, China
baohongjie88@126.com
[3] The Second Affiliated Hospital of Suzhou University, Suzhou, China
[4] School of Economics, Northwest Minzu University, Lanzhou 730030, China

Abstract. The reform of state-own hospital under information age is approaching. All state-owned hospitals will be separated from their parent system through reorganization, restructuring, transferring and centralized management under information age, but there are numerous practical difficulties in this reform under information age. This paper followed up the whole reform process in a medical institution of A Group, summarized practical difficulties in the reform and provided preliminary suggestions for reasonable development of the reform. Practical difficulties in the reform of state-owned hospitals include unclear ownership, backlash among employees and compensation dispute under information age. Integrating internal resources, seizing market opportunities, seeking differentiate development, filling the gap in technology, keeping pace with medical reform and playing a positive role in transformation are reasonable approaches that can improve the reform of state-owned hospitals and provide referential significance for the reform of health system in China under information age.

Keywords: Stated-Owned hospital · Reform · Difficulty · Hierarchical medical system · Information age

1 Introduction to the Reform of State-Owned Hospital

1.1 Enterprise Hospital

Enterprise hospital refers to all kinds of hospital established by different industries and other departments [1]. As the product of planned economy, it bears a unique mark of the time, and a typical example is the stated-owned hospital. Enterprise health resources, which is an important part of health resources in our country, exerts positive impacts on covering the shortage of local health resources, boosting productivity and improving

J. MacIntyre et al. (Eds.): SPIoT 2020, AISC 1283, pp. 243–250, 2021.
https://doi.org/10.1007/978-3-030-62746-1_36

people's health, especially the health of employees and their family members. However, common problems in most of the enterprise hospitals are that there are too many small-sized enterprise hospitals scattering in different places; the medical equipment is outdated; businesses are decreasing; resource efficiency is low and brain drain is severe [2]. In 2015, there are 985,181 health institutions and 26,314 hospitals in China, of which 3,100 hospitals are enterprise hospitals, accounting for 12% of the total hospitals [3]. Affected by advancement of reform of state-owned enterprise, separating administrative functions from business operation and solving historical problems attract attentions from the governments [4, 5].

1.2 Difficulties in Development

With the further implementation of new medical reform policy, enterprise hospitals are facing such problems as unsatisfactory compensation system, exclusion from health region planning, no allowance for community health service institution and unfair social security insurance. In particular, hierarchical medical service, cancellation of medicine markups and reform of payment in medical insurance greatly affect the development of enterprise hospitals. As different places are implementing the reform of hierarchical medical system and establishing medical partnerships, enterprise hospitals are being further marginalized. Besides, after canceling medicine markups, enterprise hospitals cannot make up its losses through increasing fees in diagnosis, treatment, surgery and medical care like public hospitals because of their limited technology, costs in examination, test and medicine account for a large proportion [6]. Under such circumstance, most hospitals have to adjust their performance distribution system; medical techniques become worsen; the number of nursing staff are decreasing. Therefore, it is necessary to put into practice the reform.

1.3 History of the Reform

With the gradual establishment and development of socialist market economy, the favorable environment for state-owned enterprise to establish hospital is disappearing, so it will be an irresistible trend to separate hospitals from the enterprises [7]. From 1990s, there were three large-scale reform and restructuring in enterprise hospitals. In 1995, government introduced a new policy to free enterprises from performing social functions, that means enterprise hospitals would be separated from their "parents" (the enterprises). The reform of separating the main businesses from auxiliary services were carried out in 100 state-owned hospitals in 18 pilot cities. The goal of this reform was to alleviate burdens on enterprise hospitals, but owning to increasingly large fiscal investment in medical health, better overall operation performance of state-owned enterprises and failure of reform in some enterprise hospitals, a large number of enterprise hospitals strongly insisted on staying in the system to clarify their public hospital identity, resulting to unsatisfactory effect in the reform [8]. In 2002, the state government published an announcement requiring hospitals established by medium- and large-scale enterprises to separate from the enterprises within 2–5 years. In this reform, three were about 1,800 local enterprise hospitals and 200 central enterprise hospitals separating from enterprises and became local self-supporting public institutions. In 2015, separating state-owned enterprises from medical services re-aroused public attention, and as medical health industry were set as strategic core industry, various investors attempted to seize the investment opportunities in the market of enterprise hospital reform. In March 2018, governmental announcement requested that offering allowance or subsidy to medical institutions set up by independent large-scale industrial and mining enterprises is not allowed, so separating hospital from parent enterprises is the only choice.

1.4 Reform Mode

In August 2017, Guiding Opinions on Deepening the Reform of Education and Medical Institutions Run by State-owned Enterprises published by State-owned Assets Supervision and Administration Commission (SASAC), State Commission Office for Public Sector Reform and other four commissions proposed 4 reform modes for the reform of medical institutions established by enterprises: handing over to local government, closing down, integrating resources and reorganizing.

(1) Encouraging handing over to local governments
    The central government encourage enterprises to reach agreement with the local governments. And local government are encouraged to take over management of non-profit medical institutions established by the enterprises and provide management service according to regulations of governmental medical institutions. If medical institutions established by the enterprises are not taken over by local governments, they can decide to close down or adopt other reform modes on their own.

(2) Orderly implementation of closing down
    For those medical institutions that have difficulties in operation and no competitive edge, it is allowed to close them down and employees' re-employment should be properly arranged.

(3) Actively integrate resources
    Support will be given to the operation of state-own enterprises focusing on health industry with government capital. Resource integration in medical institutions established by state-owed enterprises will be carried out through asset transfer, free transfer and trusteeship to realize specialized operation and centralize management, update and innovate medical health service and boost the development of heath care, healthy tourism and other industries.

(4) Promoting restructuring with aligned standards
    Specialized and powerful social capital should be introduced; restructuring reform for medical institutions established by state-owned enterprises should be taken orderly according to marketization principle; priority should be given to non-profit medical institutions in the reform. After the reorganization of enterprise operated medical institutions, it is supposed that state-owned enterprises that do not focus on health industry will not take part in reorganization, and enterprises intend to join the reorganization do not have reorganization responsibilities any more.

In the four reform modes, handing over to local government are favored by many hospitals established by state-owned enterprises, but as it is restricted by regional health resources planning, most local governments are unwilling to take over them. Therefore, infusing social capital is the most common practice in hospital restructuring. Starting from 2016, substantial acquisition of state-owned hospitals and reorganization of hospitals have never stopped. Most of the hospitals decided to transform into profit-making social medical institutions through resource integration or reorganization. As there are more detailed rules and regulations on centralized and specialized management being published, overall allocation will become the main reform mode.

2   Reform of Medical Institution of a Group

2.1   General Introduction

As an enterprise under SASAC, A Group has 24 medical institutions and 4,000 beds, and achieved an annual income of over RMB3 billion. In its 24 medical institutions, there are 17 hospitals and 7 clinics; 2 tertiary general hospitals, 9 level II general hospitals, 6 level I general hospitals and 7 no-grade medical institutions; 8 hospitals having independent legal entity and 16 having no independent legal entity. Among 24 institutions, there are 5 large-scale hospitals and they are public institutions. In 2015, A Group set up hospital management center to have centralized management on the 5 hospitals and take charge of the reform for medical institution reform under A Group.

2.2   Overveiw of the Reform

Hospital management center under A Group established committee of experts to conduct field investigation on medial business, hospital management, medical device and financial management in all hospitals under A Group. After several rounds of investigation, analysis and discussion, 9 medical institutions were adopted in the management system of the center while other 10 institutions were rejected (5 institutions were adopted in the management system of the center in 2015). Hospitals admitted to the center adopted different approaches in their admission based on different situations and natures, and different hospitals had different positioning and plans according to their current development, peripheral medical market resources and local medical health planning. For hospitals that did not get the admission to the center, they were advised to close down as they have such common problems as shortage in talents, old medical equipment or even no medical device and functions being replaced by community health center.

2.3   Reform Pattern

2.3.1   Gratuitous Transfer of Asset Between Enterprises

All budgeted staff will be accepted, asset will be gratuitously transferred as a whole, inheritance of medical qualification will be revised, and overall transfer of organizational system will be adopted in some medical institutions.

2.3.2 Gratuitous Transfer of Equity Between Enterprises

For hospitals that have changed to private non-enterprise hospitals (legal entity), originally affiliated enterprises should be responsible for their second restructure, and gratuitously transfer to hospital management centre after reorganizing to wholly owned subsidiary of the originally affiliated enterprises.

2.3.3 Regrading Medical Institutions Which Have Budgeted Employees But No Independent Legal Entity Under Public Institutions

(1) Gratuitous transfer of asset between public institutions: if A Group's public institutional hospitals has extensive management ability to accept medical intuitions around them, assets employed (including land and building) will be gratuitously transferred by originally affiliated enterprises; assets will be included in or centralized management will be carried out by another hospital near the original hospital; succession strategies will be developed for new hospital areas or hospital qualification will be created; employee status will be kept in the original hospitals.

(2) Gratuitous transfer of asset between public substitution and enterprise: if A Group's public institutional hospitals has no extensive management ability to accept medical intuitions around them, separation of staff and asset in public institutions will be adopted; overall restructure will be taken after separation. For those hospitals that cannot be restructured, originally affiliated enterprises will gratuitously transferred the assets to local hospital management company established by hospital management center and budgeted employees will be arranged with the policy that employees retired before the policy published will be granted basic pension according to the original regulations.

2.4 Practical Difficulties

2.4.1 Issues of Property Rights

Parent enterprises are closely linked with hospitals with unclear asset rights and liabilities in land and building, so transfer can be achieved only after further division.

2.4.2 Remaining Problems

In the first reorganization of some hospitals, they were changed to private non-public enterprises with capital from natural person, but the share capital repurchase and evaluation of the investors remains unsolved. Before transfer, these institutions must withdraw private non-public legal entity, reorganize the system, rearrange employees and solve the long-term housing allocation.

2.4.3 Multiple Interests

The reform of enterprise hospitals involves the interests of parent enterprise, local government, health department, employee, hospital, creditor and debtor in employee allocation, debt treatment, treatment for state-owned land, equity ratio and design of management structure in the hospital. As it is difficult to keep a good balance for the requirements from different parities, it leads to the bottleneck in the relationship between different parties and no progress in the reform.

2.4.4 Financial Problems

Hospitals and their affiliated enterprises have close service relationships and other medical institutions often get financial supports from hospitals through connected transactions, so there are problems in creditor's right and debt in historical connected transactions. In particular, medical service charges in the statements of some hospital remains unsettled.

2.4.5 Compensation

Most of the medical intuitions of A Group have outdated facilities and equipment, and maintain their operation hardly with support from the enterprises. After transfer, parent enterprise will not provide financial support to the hospitals. As an enterprise, the final goal of the hospital management center is to earn profit. At the beginning of the transfer, there are huge risks in making large investment, thus it takes a long time to negotiate the compensation.

2.4.6 Problems in Operation

Most of the hospitals have such common problems as unsatisfactory operation, inefficient profitability and short-term solvency, low income from main business, high medical cost, shortage of investment, high employee turnover rate, huge differences between enterprise hospital and public hospital and low medical insurance for enterprise hospitals, leading to the increase in accounts receivable in the hospital's financial account.

3 Reasonable Pathway for the Reform of State-Owned Hospitals

3.1 Integrate Internal Resource and Seize Market Opportunities

As our country has published policy to make great effort to boost the development of medical health industry, state-owned enterprises followed the guidance of "group operation, specialized management" while taking into account their own development, implement centralized management in the hospital and accept all medical institutions into hospital management center. By doing so, it can not only boost the development of the enterprises, but also avoid uncertainties brought by transferring or selling asset to local hospitals, which is of benefit for overall development of medical market in our country.

3.2 Seek Differential Development, and Fill the Gaps in Technology

Future development plan for medical institutions should be formulated in advance; dislocation competition with large-scale hospitals should be realized on the basis of combining unique characteristics with local requirements; great effort should be put into fields that large-scale hospitals have not gotten involved. Moreover, hospitals are encouraged to applied for hospital for special diseases or nationwide diagnosis and treatment center to expand their influences throughout China. Hospitals should also invest more capital in technology researches and bring their strengths into full play to become a specialized hospital.

3.3 Keep Pace with Medical Reform, Take Actions to Realize Transformation

Despite of the fact that enterprise hospitals confront enormous challenges under the background of new medical reform, they can also seize the opportunities of this reform to achieve transformation. For instance, as it is required in the medical reform that scale of public medical insinuation should be taken under control and wanton expansion is not allowed, after transferring to the local government, state-owned hospitals are encouraged to cooperate with public hospital to establish branch hospital, specialist sanatorium, elderly care institution and other medical institutions that are in urgent need.

Acknowledgements. 1) This work was supported by the Fundamental Research Funds for the Central Universities of Northwest Minzu University (Grant No. 31920180101);

2) This work was supported by the Fundamental Research Funds for the Central Universities of Northwest of Northwest Minzu University (Grant No. 31920190117);

3) This work was supported by the Fundamental Research Funds for the Central University of Northwest of Northwest Minzu University (Grant No. 31920190032);

4) This work was supported by National Social Science Fund Youth Project (Grant No. 19CGL061).

5) This work was supported by National Social Science Project (Grant No. 19BSH068)

6) This work was supported by The Ministry of Education of Humanities and Social Science Fund Youth Project in 2020 Year (Project Name: Research on the poverty reduction effectiveness of farmer cooperative economic organization in the Northwest Minority Areas from the Perspective of Rural Revitalization Strategy) (NO. 20YJC850001).

References

1. Abdel-Basset, M., Elhoseny, M., Gamal, A., Smarandache, F.: A novel model for evaluation hospital medical care systems based on plithogenic sets, artificial intelligence in medicine. 31 August 2019 (in Press)
2. Lu, L., Shaowei, W., Li, Z., Pengqian, F.: The historical roles and current development of enterprise hospitals. Chin. J. Soc. Med. (4) (2009)
3. National Health and Family Planning Commission of PRC: China Health Statistics Yearbook [EB/OL] (2015). http://www.moh.gov.cn/zwgkzt/ptjnj/list.shtml,2016.7.25
4. http://www.gov.cn/xinwen/2017-08/26/content_5220497.htm. Accessed 26 July 2016
5. The State Council of the People's Republic of China: Circular of the State Council on Printing and Distributing the Work Plan to Speed up the Divestment of State-owned enterprises from Their Social Functions and to Solve Historical Problems (Guofa[2016]No.19) [EB/OL]. http://www.gov.cn/xinwen/2017-08/26/content_5220497.htm. Accessed 26 July 2016

6. State-owned Asset Reform Commission. Guiding Opinions on Deepening the Reform of Education and Medical Institutions Run by State-owned Enterprises, No. 134 [EB/OL] (2017). http://www.gov.cn/xinwen/2017-08/26/content_5220497.htm. 26 Aug 2017
7. Haifeng, X.: Study on the development and reform trends of enterprise hospitals [EB/OL]. https://www.cn-healthcare.com/articlewm/20171123/content-1018895.html. 23 Nov 2017
8. CN-Health Care: Enterprise Hospital will become a history and Its Replacements Continue to Mushroom [EB/OL]. http://www.zyzpes.com/toutiao/3207/20180701A1BL9U00.html. Accessed 1 July 2018

Financing Efficiency of SMEs in New Third Board Market in the Information Times

Sheng Li[1] and Aihua Zhang[2(✉)]

[1] Management School, Wuhan University of Science and Technology, Wuhan, Hubei, China
[2] Management School, Hubei University of Education, Wuhan, Hubei, China
aiwa0716@163.com

Abstract. In the information times, the speed and method of information transmission have changed, and the level of enterprise informatization will have an uninterrupted influence on the expansion and competitiveness of enterprises. Many SMES face the problem of low financing efficiency. This paper focuses on the research of financing adeptness of SMEs in the New Third Board market. We collect monetary statistics of 50 SMEs enumerated; use SPSS to scrutinize the dynamics that affect the funding competence in the first year after SMEs are listed. Through the analysis, we can see that profitability has an encouraging sway on sponsoring productivity. The growth capacity of creativities has a constructive impression on financing efficiency; capital structure has a destructive sway on financing efficiency. Business cycle and liquidity ratio have no significant impact on financing efficiency. We can conclude that the financing efficiency of the current SMEs is not high, and the composition of the capital structure, solvency and operating capacity need to be improved.

Keywords: Information times · Financing efficiency · The new three board

1 Introduction

In the information times, the capital market system shows an essential character in endorsing the coordinated expansion of capital markets at different levels and so long as investment and financing amenities for SMEs [1, 2]. The New Third Board has become the sanctuaries marketplace with the leading number of service companies in the domain. The rapid expansion of the market has also brought many problems. Financial expenses of listed companies increased by 40.5% year-on-year. The growth rate of medium-sized enterprises and minor enterprises reached 65.24% and 42.07%, and the share of financial expenditure in the income share increased by more than 0.2%, which shows that the current indirect financing costs of SMEs can be lower [3].

In recent years, increasing attention has been paid to the financing issues of SMEs, mainly focusing on life cycle, financing efficiency, credit guarantees, financing methods, and cost benefits and so on. Through an evaluation of the collected works, it is found that there are relatively limited readings on the financing efficiency of listed companies, and less research on the impelling aspects of planned companies' financing efficiency

J. MacIntyre et al. (Eds.): SPIoT 2020, AISC 1283, pp. 251–257, 2021.
https://doi.org/10.1007/978-3-030-62746-1_37

[4]. Since the implementation of the new three-tiered layered system, although many scholars have been studying the effectiveness of the layered system during the period, the conclusions of their research are also limited due to the limitation of the availability of annual statement data [5].

This article mainly analyzes the financing efficiency, debt service ability, operating ability and growth ability of listed companies from 2012 to 2018, and analyzes the financing efficiency of companies. This article takes the financing efficiency as the main research line, collects relevant financial data indicators of 50 SMEs listed, and uses SPSS to analyze the factors that affect the financing efficacy of SMEs in the first year after listing. Find out the results of the company's profitability, growth ability and capital structure affecting the financing efficiency of the company, and make recommendations on the results.

2   Theoretical Analysis and Research Hypothesis

2.1   Theoretical Analysis

In the information age, there has been a substantial increase in the level of financing of companies on the New Third Board. But there are still factors affecting and restricting financing:

(1) Enterprise size: The large-scale high-quality asset companies have more financing channels than small enterprises, easier to gain the trust of institutional investors and lower financing costs [6, 7].

(2) Financing costs: It refers to the capital expenditure incurred to pay the owner in order to obtain the right to use the funds to achieve the purpose of completing the transfer of funds, and these paid fees are the costs incurred by the corporate financing [8].

(3) Financing risk: Compared with large enterprises with open information and low cost of access to resources, due to their small size, insufficient human resources for high-tech R&D, imperfect internal management work, small and medium-sized enterprises are more prone to credit crises [9, 10].

(4) Capital Structure: When a company increases its debt financing ratio, it will reduce its financing cost and increase its financial risk. Therefore, an unreasonable capital structure will affect the operating costs of the enterprise, and then affect the internal profitability and play a certain inhibitory role in financing efficiency [11].

(5) Fund utilization: Increasing the utilization of funds can indirectly improve financing efficiency. [9] SMEs can broaden financing channels, strengthen capital management, comprehensively manage corporate fixed assets, and reduce idleness of corporate funds and equipment.

(6) Enterprise profitability: There are many indicators to measure profitability. The stronger the profitability of an enterprise, the more helpful it is to the capital turnover of the enterprise, which will reduce the credit risk of the enterprise and increase its financing income [10, 12].

2.2 Research Hypothesis

This paper believes that profitability mainly takes a huge influence on the financing efficacy of enumerated companies. There are two main indicators to show the profitability of an enterprise: the return on overall assets and the return on net assets. This paper selects the ROE as a measure.

This paper believes that the company's growth ability, operational capacity, capital structure, solvency and profitability may have a certain control on the financing effectiveness. It indicates the progression rate of net assets to reflect the development capability of enterprises. The advanced the net asset development rate, the better the business situation, the stronger market competitiveness and the stronger development. On the contrary, it shows that the business is not in good condition and growth is not good. This paper also chooses equity multiplier to represent capital structure. The higher the equity multiplier, the minor the percentage of capital put in by the vendor of the company's assets and upper the balance level of the company. In terms of solvency, this paper uses liquidity ratio to express the solvency of an enterprise.

2.3 Indicators Selection

This paper uses Guotai'an database to analyze the relevant financial ratios of 50 representative SMEs enumerated on the new third board from 2012 to 2018. Due to the small number of recorded companies in the previous year and the short time for the establishment of the New Third Board, the development is not perfect. Since 2012, the number of itemized companies in the New Third Board has grown hurriedly and the expansion has become more perfect. Therefore, the data analysis is based on 2012.

3 Statistical Description and Correlation Analysis of Samples

3.1 Statistical Description of Sample Variables

From Table 1, we can understand that the ordinary return on net assets representing the financing efficiency of the sample companies is 5.64% and the median value is 10.5%, which is relatively high, indicating that the average financing efficiency of the sample new third board SMEs is better. The transformation among the extreme value and the minutest value is relatively high. However, the standard deviation is 29.88, which is relatively small, indicating that the financing adeptness of different SMEs in the new third board of the sample is relatively small.

As can be perceived from Table 1, the average return on venture capital representing the profitability of the sample company is 26.65%, and the intermediate value is 9.68%. This value is relatively large, indicating that the sample company has higher profitability. The standard deviation is 411.09, which is relatively large, indicating that there is a big difference in the overall profitability of the sample new three-board listed companies.

Table 1. Statistical description of financing efficiency indicators for SMEs listed on the New Third Board

	Roe	Return on invested capital	Net assets growth rate	Business cycle (days/times)	Equity Multiplier	Current ratio
Mean	5.64	26.65	26.57	334.57	1.98	21.24
Median	10.50	9.68	15.55	266.00	1.72	2.07
Maximum	50.22	7786.21	722.77	3516.00	19.88	5561.2
Minimum	−226.18	−288.68	−233.22	0.00	1.02	0.16
Standard deviation	29.88	411.09	66.99	358.75	1.32	299.29
Skewness	−4.52	19.22	3.79	4.52	8.77	199.25
Kurtosis	31.02	336.84	40.02	33.28	109.78	343.26

3.2 Relevance Analysis of Sample Variables

Because there may be collinearity among variables, and multivariate linear regression take up that there is no correlation amongst the elements, under such detection, the test results may not be reliable, and the regression model may lack stability. Therefore, it is necessary to make a correlation analysis of the explanatory variables, as displayed in the resulting Table 2:

Table 2. Relevance analysis of explanatory variables

	X1	X2	X3	X4	X5
X1	1.000	0.018	−0.063	0.056	−0.007
X2	0.018	1.000	−0.136	−0.003	−0.036
X3	−0.063	−0.136	1.000	0.130	−0.063
X4	0.056	−0.003	0.130	1.000	−0.040
X5	−0.007	−0.036	−0.063	−0.040	1.000

In the table: X1 represents the return on investment capital; X2 represents the growth rate of net assets; X3 represents the business cycle; X4 represents the equity multiplier; X5 represents the liquidity ratio.

Through the correlation analysis among the explanatory variables, it can be understood that the selection of descriptive variables is appropriate, and there is basically no multiple collinearity among the explanatory variables.

3.3 Regression Analysis of Financing Efficiency of SMEs in the New Third Board

It is necessary to establish the following models for its analysis:

$$y = b0 + b1x1 + b2x2 + b3x3 + b4x4 + b5x5 + \xi \tag{1}$$

y represents the profit on remaining assets, as a dependent variable, indicating the financing efficiency of the enterprise; b0 represents a constant term; bn represents a regression coefficient of each explanatory variable; ξ is a unsystematic error tenure. Combined with the overhead information, linear regression analysis is carried out by using SPSS software, and the following information is obtained. As shown in the table below (Table 3):

Table 3. Estimation of Multivariate Linear Regression Parameters Coefficients[a]

Model	Unstandardized coefficients		Standardized coefficients			Collinearity statistics	
	B	Std. Error	Beta	t	Sig.	Tolerance	VIF
(Constant)	1.514	1.311		1.181	.241		
Return on invested capital	1.276	.032	.957	45.952	.000	.715	1.456
Net asset growth rate	.024	.008	.058	2.743	.008	.779	1.315
Business cycle	−.002	.002	−.023	−.867	.387	.870	1.154
Equity Multiplier	−2.211	.482	−.091	−4.568	.000	.882	1.141
Current ratio	−.001	.046	−.006	−.302	.774	.889	1.125

a. Dependent Variable: Return on net assets

The VIF values of the selected five constants are less than 10, which indicates that there is no multiple collinearity between them. Therefore, the regression equation can be obtained as follows (Table 4):

$$y = 1.276x1 + 0.024x2 − 2.211x4 + 1.514 \tag{2}$$

This table is the result of multiple linear regression analysis of ROE. According to this table, the goodness of fit test can be performed. Since this is a multiple linear

Table 4. R2 test: model summary[b]

Model	R	R Square	Adjusted R Square	Std. Error of the Estimate	Durbin-Waston
1	0.956[a]	0.881	0.886	7.5641821	2.082

a. Predictors: (Constant), Return on invested capital, current ratio, net asset growth rate, equity multiplier, business cycle (days/times)
b. Dependent Variable: Roe

regression, it is necessary to observe the adjusted decision coefficient of 0.886, and the goodness of fit is high.

4 Conclusion

Listing on the new third board can be an operative approach to develop the financing efficiency of SMEs. It can relief SMEs elucidate the inefficiency of financing problems, endorse the vigorous improvement of SMEs, and utilize the innovation vitality it deserves to indorse sustainable economic growth. However, the capital structure has an undesirable sway on the financing efficacy, while the operational capacity and solvency have little impact on the financing capacity. It can be comprehended that the financing efficiency of SMEs in the new third board market is not high, and the composition of the capital structure, solvency and operational capacity need to be improved. In order to expand the adeptness of financing, enterprises also need to improve their own capabilities in footings of development and cost-effectiveness.

Acknowledgements. This work was supported by the Hubei Innovation Project (NO. T201940).

References

1. Dan, L., Asa, F.-B., Dan, P.: Current and future industry 4.0 capabilities for information and knowledge sharing: case of two swedish SMEs. Int. J. Adv. Manufact. Technol. **105**(9), 3951–3963 (2019)
2. Shankar, K., Elhoseny, M.: Trust based cluster head election of secure message transmission in MANET using multi secure protocol with TDES. J. Univ. Comput. Sci. **25**(10), 1221–1239 (2019)
3. Vidgenab, R., Mortensonc, M., Powellb, P.: Invited viewpoint: how well does the information systems discipline fare in the financial times' top 50 journal list? J. Strateg. Inf. Syst. **28**(4), 101–105 (2019)
4. Kim, H., Yasuda, Y.: Accounting information quality and guaranteed loans: evidence from Japanese SMEs. Small Bus. Econ. **53**(4), 1033–1050 (2018)
5. Xing, R., Zhang, J.: Problems and countermeasures of the application of enterprise management accounting informatization. Agric. Sci. Technol. **18**(8), 1555–1558 (2017)
6. Chit, M.M.: Financial information credibility, legal environment, and SMEs' access to finance. Int. J. Econ. Bus. **26**(3), 329–354 (2019)
7. Matulová, M., Fitzová, H.: Transformation of urban public transport financing and its effect on operators' efficiency: evidence from the Czech Republic. CEJOR **26**(4), 967–983 (2018)
8. Poshan, Y., Zuozhang, C., Jin, S.: Innovative financing: an empirical study on public–private partnership securitisation in China. Australian Econ. Papers **57**(3), 394–425 (2018)
9. Mahfuzur, R., Aslam, M., Binti, I.I., Ruhana, I.C.: Factors affecting the financing cost of microfinance institutions: panel evidence. Enterp. Dev. Microfinance **29**(2), 103–117 (2018)
10. Metawa, N., Elhoseny, M., Kabir Hassan, M., Hassanien, A.E.: Loan portfolio optimization using genetic algorithm: a case of credit constraints. In: Proceedings of 12th International Computer Engineering Conference, ICENCO 2016: Boundless Smart Societies, vol. 7856446, pp. 59–64 (2016)

11. Xun, L., Xiaoliang, Y., Simon, G.: A quantitative study of financing efficiency of low-carbon companies: a three-stage data envelopment analysis. Bus. Strategy Environ. **28**(5), 858–871 (2019)
12. Ke, X., Chengxuan, G., Wei, X.: Research on financing ecology and financing efficiency of strategic emerging industries in China. J. Bus. Econ. Manag. **20**(2), 311–329 (2019)

Application of Virtual Instrument Technology in Electronic Course Teaching

Hui Cheng[✉]

Dalian Ocean University, Dalian, Liaoning, China
15541185675@163.com

Abstract. With the continuous progress of modern information technology and continuous reform of teaching methods, practice has become an important part of students' in-depth understanding of theoretical knowledge. Experience has become an indispensable part of teaching and the most effective way to improve students' creativity and practical ability. In order to improve experimental conditions and reform experimental teaching methods, this article is studying the application of virtual instrument technology in electronic course teaching. Increase students' interest in experiments and improve teaching quality.

Keywords: Virtual instrument · Electronic course · Application

1 Introduction

In recent years, Due to the rapid development of virtual instrument technology and space transmission technology, it is possible to establish an experimental education system through virtual equipment. The virtual instrument teaching system will become an important aspect of student experience and distance education [1]. In addition, the construction of education and scientific research networks, libraries and other departments in Chinese universities and colleges has been basically completed, and the establishment of network labs in universities and universities has become the next major focus of school informatics [2]. With the rapid development of virtual hardware and network technology, a network laboratory can be established through network technology and virtual equipment.

2 Introduction to Virtual Instrument Technology

Virtual instruments are new tools that complement data collection, control, analysis, processing, and display of test results, and recognize that "software and hardware" can assign the functions of the tool to maintain, expand, and update the tool. The virtual instrument "panel" is displayed on the computer screen. The tool is run by selecting different buttons with the mouse and combining keyboard input.

J. MacIntyre et al. (Eds.): SPIoT 2020, AISC 1283, pp. 258–263, 2021.
https://doi.org/10.1007/978-3-030-62746-1_38

3   Composition and Classification of Virtual Instruments

Virtual instruments are generally composed of instruments and solutions, and adopt a unit structure. Figure 1 is a diagram of the overall structure of a virtual instrument. The main function of a virtual instrument is to acquire signals measured in the real world, while the function of the software is to control and integrate data acquisition, analysis, processing, display and other functions. They are in the command environment for instrument operation and operation.

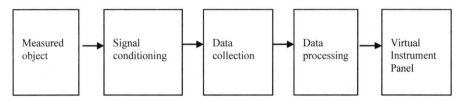

Fig. 1. Block diagram of the overall structure of the virtual instrument

There are many classification methods of virtual instruments, which can be divided into application fields and measurement functions, but the most commonly used are different interface interfaces of virtual instruments. The tool is divided into virtual parallel interfaces, virtual interface tools, virtual VXI tools, virtual PXI tools, Grid tools and the latest IEEE 1394 virtual instrument.

DAQ virtual instruments are widely used in general test and industrial process control systems, and from the previous standard 16 bb to PCI 32 bb cards, thus providing better data acquisition and control capabilities for the design of different test tools. Of course, the virtual device needs to open the host structure connection, which is difficult to use and it is easy to input interference on the computer. Therefore, many external virtual instruments based on the standard configuration interface of public computers will become the direction of development. The external communication plan avoids noise inside the computer, which is especially suitable for low-signal applications, thus providing more space, better isolation and more suitable communications for device design [3, 4].

RS232 / RS422 interface is widely used in various foreign business control tools, supports long-term transmission, has strong interference ability, but the transmission rate is low.

The parallel interface is also a traditional high-speed interface. Generally, printers have a parallel interface installed. Virtual instruments such as digital vibration and logic analyzers have appeared on the market. Of course, the global user chain transport aircraft and high-speed connection IEEEI394 will be more promising in the future.

It becomes a computer's USB and supports hot connect function. The IEEEI394 bus has become common in some advanced desktop and laptop computers. The biggest advantage of USB and IEEEI394 carriers is the high data transfer rate [5]. At present, the virtual instrument based on IEEEI394 has reached a transmission speed of 100 megabytes per second, which fully meets the requirements of high-performance dynamic testing [6].

GPIB, VXT, and PX (bus) are computer interface buses designed for software-controlled tools. Gebb's tools include independent hardware operating interfaces, which can be independent of the computer, or can be connected to the computer through a standard gibb control cable. And PXI does not include a separate tool operation interface, which must rely on the virtual operation interface provided by the tool operation program.

4 Performance Characteristics of Virtual Instruments

Virtual instrument is a new tool developed by computer technology. In the virtual instrument, the computer is at the basic position, and the computer software and test system are more tightly integrated, which has caused a significant change in the tool's structural concept and design perspective. In terms of configuration and functions, the virtual instrument is a novel low-cost tool. It uses existing computers to match the corresponding equipment and special programs to form a sophisticated low-cost tool, which includes the core functions of common documents and common documents [7]. With special features. In terms of use, virtual instruments use a powerful graphical development environment to create a simple, flexible, and fast light board, which effectively improves the efficiency of the tool. The characteristics of virtual instruments can be summarized as follows:

(1) Enrich and improve the functions of traditional instruments. A virtual central tool for analyzing signals, displays, storage, printing and other computer management, making full use of powerful data processing and transmission capabilities, making the system more flexible and simple.

(2) Emphasize the new concept of "software tools". Some traditional tools are replaced by a program in a virtual instrument. Since many vertical analog devices may drift over time and require regular calibration and use of standard transmitters, the accuracy and speed of measurement and the frequency of the instrument are greatly improved.

(3) The instrument is defined by the user. By providing users with source code that can be used for their own tools, virtual instruments can easily modify the functions of paintings and tools, communication design, timing of tool operation functions, connection of terminals, networks, and other applications, and provide users with the full ability And imagination.

(4) Open industry standards. The instruments and software of virtual instruments set open industry standards, increase the possibility of resource reuse, promote employment opportunities, achieve management standardization, and reduce maintenance and development costs. [8]

(5) It is easy to build a complex and economical test system [9]. Virtual instruments can not only be used independently as test tools, but also use software engineering-based virtual software tools to replace traditional hardware engineering tools, which can greatly save the cost of purchasing and maintaining the tool.

In short, the appearance of a virtual instrument breaks the position determined by traditional tool makers and cannot be changed by the user. It uses a lot of software and hardware resources to break the boundaries of traditional data processing, expression, sending, display and storage tools, and has high performance. Compared with traditional tools, virtual instruments have the following characteristics:

(1) This is a function rather than a physical tool. Virtual instruments use existing software and hardware resources, using software technology as a basic content, understand instrument and measurement and control equipment communication through dashboards and dashboards, integrate signal acquisition, analysis, and processing functions into a computer Become a processing center for data signal acquisition, control, and analysis functions, and replace traditional software tools. The concept of "software as a tool" is reflected in the development of modern tools;

(2) The virtual instrument is a graphical user interface that reflects the concept of "what you see is what you get". The traditional tool control panel was replaced with a floppy disk, and replaced with corresponding preparation options and output control of different virtual instruments. A graphical user interface (graphical user interface) makes the use of virtual instruments easier and provides instant online assistance. This is an advantage that traditional electronic tools cannot apply.

(3) Quick update and good maintenance performance of virtual instruments. Users can customize its structure and functions. Because its core is in a specific development environment, users can carry out sub-development, modify and add functions of existing virtual instrument software. Compared with the development of electronic tools, the development cycle is greatly shortened and the cost is greatly reduced.

5   The Specific Application of Virtual Instrument Technology in Electronic Course Teaching

5.1   Research and Teaching

In order to train high-quality electronic talents, many colleges and universities have purchased a large number of measurement and analysis tools to better conduct practical research in teaching and teaching. Traditionally, an instrument cannot meet the learning needs of students and is not conducive to promoting scientific and technological innovation. Virtual instruments can solve the problem of incomplete electronic laboratory equipment and many people participating in the same equipment; it can save equipment maintenance costs to a large extent, so that schools can also transfer more energy to electronic research and electronic education.

5.2   Electronic Curriculum System Aspect

Virtual instruments play an important role in reconstructing the structure of electronic expertise and improving the electronic curriculum system. [10] The emergence and development of virtual instruments has had a significant impact on the teaching of electronic courses in schools. Traditional electronic courses have not paid enough attention to the use and operation of measurement tools, so traditional e-learning has ruled out the use of measurement and analysis tools reasonably. With the continuous development of simulation and computer technology, virtual instruments have gradually replaced traditional tools and become the main tools for detection and measurement. In order to meet market demand, schools must add relevant knowledge and skills from virtual e-learning tools, which will greatly improve the electronic curriculum system and enrich the structure of electronic professional knowledge [11].

5.3   Teaching Methodologies

First, teachers can use virtual instruments to guide students to work in virtual laboratories. This kind of virtualization and simulation can effectively solve the problems of insufficient equipment, equipment damage and limited operating time. [12] The relationship between teachers and students can also be improved to the greatest extent. Teacher training and student practice can be combined to improve teaching efficiency and learning quality.

Second, the advent of virtual instruments requires teachers to adopt typical teaching methods. Although virtual instruments have great advantages, they can shorten the time and time of practice and effectively improve the quality of teaching, but it also means that teachers' teaching methods must be changed. Teachers need to review student learning projects, implement operational principles and methods through stereotypes, enable students to learn some electronic knowledge and skills, and obtain all professional electronic knowledge and skills through measurement.

Third, the use of virtual instrument technology, simulation technology and Internet technology for distance learning activities. On the one hand, virtual instruments and simulation programs have the ability to interact between humans and computers. [13] We can share and share data and build a virtual laboratory based on computer technology to reveal the electronic laws behind electronic phenomena. On the other hand, Internet technology can collect different interface data of virtual instruments for real-time data, waveform and image analysis results, which also provides the basis for remote electronic courses. For example, we can use circuits to design spices, systems, and other simulation programs, and use them in most circuit and system simulation experiments, using ASP, connectors, and other tools to monitor relevant data in real time.

5.4   Teaching Evaluation

The introduction of a virtual instrument greatly changed the preparation of teaching evaluation. Because traditional electronic measurement tools have limited functions and cannot reasonably analyze and process different data, teachers can only evaluate a small number of projects or a small number of students studying simultaneously in the teaching process. [14] Virtual instruments use a variety of technologies. The functionality of a virtual instrument is equivalent to that of many traditional tools. Virtual instruments have powerful storage, analysis, and processing capabilities. Under this influence, the development of teacher teaching evaluation shows a trend of diversification, that is, teachers can evaluate students' ability to use virtual instruments at the same time, and they can combine actual teaching, tool operation and teaching evaluation so that before teaching Play a role in monitoring and diagnosing all education issues during, after and after.

Although virtual instruments have many advantages, it is unrealistic to eliminate completely traditional experimental teaching aids. When designing electronic courses, we must learn from each other's strengths, make up for each other's weaknesses, combine traditional teaching methods with advanced teaching methods, and jointly encourage teaching diversity and maximize teaching results. Facts have proved that the use of virtual teaching tools can improve the level of educational technology, reform teaching conditions, strengthen the impact of classrooms, and improve innovation capacity.

Acknowledgements. Fund Project: Project supported by the "Thirteenth Five-Year Plan" of Educational Science in Liaoning Province: Research on Practical Skills Cultivation and Innovation of Applied Undergraduate Automation Specialty (Subject No. JG18EB033)

References

1. Thakur, Sriti., Singh, A.K., Ghrera, S.P., Elhoseny, M.: Multi-layer security of medical data through watermarking and chaotic encryption for tele-health applications. Multimed. Tools Appl. **78**, 3457–3470 (2019)
2. Zhang, B., Chen, Z., Sun, W., et al.: Application research of virtual instrument in experimental teaching. J. Shandong Electric Power Univ. **13**(5), 87–89 (2017)
3. Wu, X., Xu, W.: Status and prospects of application research of domestic virtual instrument technology. Modern Sci. Instrum. **8**(4), 112–116 (2017)
4. Abdel-Basset, M., Mohamed, R., Elhoseny, M., Chang, V.: Evaluation framework for smart disaster response systems in uncertainty environment. Mech. Syst. Signal Process. **145** (2020)
5. Elhoseny, M., Shankar, K.: Reliable data transmission model for mobile adhoc network using signcryption technique. IEEE Trans. Reliab., June 2019, in Press
6. Zhu, M., Zhang, J., Pan, K., et al.: Virtual instrument technology and its teaching application. Technol. Products **4**, 96–98 (2016)
7. Jiang, H.: Application of virtual instrument technology in college teaching. China Modern Educ. Equipment **15**, 53–54 (2018)
8. Zheng, J., Gong, S.: Reform of circuit principle course. J. Electr. Electron. Educ. **29**(3) (2017)
9. Elhoseny, M., Shankar, K., Uthayakumar, J.: Intelligent diagnostic prediction and classification system for chronic kidney disease. Sci. Rep. **9**(1) (2019)
10. Yu, X., Lu ,W., Wang, S.: Selection and Innovation of Teaching Contents of Professional Basic Courses - Case Study of Circuit Principles Course of Tsinghua University. J. Electr. Electron. Educ. **28**(3) (2016)
11. Li, P., et al.: Integration and optimization of "Principle of Circuits" and "Signals and Systems" course. J. Electr. Electron. Educ. **25**(5) (2015)
12. Yang, L.: Introduction to Virtual Instrument Technology. Electronic Industry Press, Beijing (2015)
13. Liang, G., Dong, H., Zong, W.: Reform and practice of comprehensive teaching of circuit theory courses. China Electric Power Educ. **1**, 89–91 (2017)
14. Ping, Q., Yue, S.: Discussion on "basic analysis of circuits" teaching based on course connections. J. Electr. Electron. Educ. **29**(6) (2017)

A Solution for Internet of Things Based on Blockchain and Edge Computer

Yan Zhou[✉]

College of Artificial Intelligence, Nanchang Institute of Science and Technology Nanchang,
Nanchang 330108, China
770196211@qq.com

Abstract. The resources of Internet of Things devices is limited and some external resources are needed, while the centralized cloud solution can provide sufficient computing storage resources for the Internet of Things. But some constrained factors such as network bandwidth, large amounts of data cannot be timely processed by the cloud. The cloud is based on static password authentication and plaintext storage mechanism, though there are some security problems. The rise of blockchain technology provide a solution method for the Internet of Things security. In addition, the edge computer can shorten the distance of data processing and strengthen the real-time performance. Therefore, this paper provides a solution method based on the blockchain and edge computer for the internet of Things security.

Keywords: Blockchain · Interment of Things · Edge computer · Security

1 Introduction

With the rapid development of sensor, wireless communication, data analysis and processing, the internet of Things devices will become miniaturized and low-cost. It brights more possibilities to solve some problems with internet of Things in family, education and transportation, etc. Because the data of the initial internet of Things devices was very little, it can complete data processing tasks by using the limited resources [1]. However, when the internet of Things technologies have been applied in home appliance, traffic control, there are a large number of data. In 2010, Zhao believed the cloud computer provided a new chance to solve the data processing of internet of Things [2]. In 2014, Biswas thought that the cloud computer could provide powerful resources, but there are some problems, such as heterogeneous connection, dynamic management, etc. [3]. In 2015, Farris used the edge computer and cloud computer to slove a problem of heterogeneous connection [4]. In 2016, Shi thought that the cloud solution could not meet to real-time requirement of transportation processing, but the edge computer could solve these problems [5].

Since the internet of Things is connected to the internet, there are more and more security threats. Sharmeen pointed to the hack act of malicious attackers, who had directly

© The Editor(s) (if applicable) and The Author(s), under exclusive license
to Springer Nature Switzerland AG 2021
J. MacIntyre et al. (Eds.): SPIoT 2020, AISC 1283, pp. 264–270, 2021.
https://doi.org/10.1007/978-3-030-62746-1_39

accessed the internal data of the internet of Things by bypassing the authorization [6]. The solution of cloud-based centralization was hard to guarantee security of widely dispersed internet of Things equirments. The multi-node distributed structure of internet of Things requirements was similar characteristics to the blackchain. Wu proposed a new security methods of double factors based on blockchain technology, so as to guarantee data security of smart devices. Christidis pointed that the blockchain technology could provide the security of an untrusted environment for internet of Things.

This paper put forward a strict access control mechanism to regulate the behavior of internet of Things device and user on the external network, so as to ensure the security of internet of Things devices.

2   The Related Technology

2.1   Blockchain

In 2008, satoshi nakamoto published a paper on cryptograpy forum, which fully elaborated the principle of bitcoin and blockchain technology. The blockchain is compose of block and chain. In blockchain, the data units of the stored permanently are block, the connected timestamps in chronological order are the chains. Therefore, some people believe that the blockchain is distributed ledger, which is recorded in chronological order, and maintained by nodes in the internet through consensus mechanism. The structure of a block is divided into block header and block body. The block header contains the characteristic information of the block itself, such as header hash, parent hash, timestamp, block transaction number and the block size, etc. The header hash is the most critical price of information, and has the same hash value as the next connected block. The blockchain structure is shown in Fig. 1.

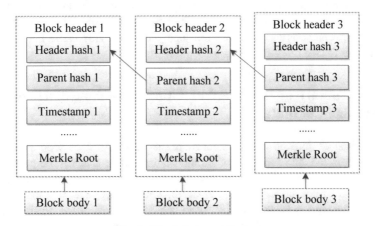

Fig. 1. Blockchain architecture

The block body contains all verified and tradeable record information during block creation. The block examples we will use are shown in Table 1. As long as the newly

Table 1. Comparision in the properties

No	Operation/Event	Merkle root	State
1	Storage	1	Deny
2	Access	2	Allow
3	Query	3	Allow

generated block is generated and added to the end of the block chain, the data of the block cannot be deleted, which will ensure immutability of the data.

According to the range distribution of blockchain nodes, blockchain can be divided into public chain, private chain and alliance chain. The public chain is completely open to the public, so anyone can directly access the blockchain without authorization. The private chain is established by an organization and are accessible only to certain authorized users. The alliance chain is mixture of public and private chains, which are accessible only to alliance users. Considering the problems of public chain in reliability and privacy security, it is difficult to deploy massively the alliance chain, so we chose smaller private blockchain to record information. In addition, Because the private chain is deployed the local plotform, the initial blockchain can be considered safe and reliable.

2.2 Edge Computer

The edge computer is the processing of data at the edge of the network. For example, gateway of smart home can be an edge. For edge computer, some data processing will be completed in home gateway. Of course, if the amount of data is too large, the edge nodes cannot implement data processing timely or completely. The edge nodes must rely on the cloud to complete processing task, the edge nodes will take care of the simple preprocessing. Our solution thinks that the deployed edge node can meet the processing and computer for the internet of Things equirements.

3 The Structure of Interment of Things Based on Blockchain and Edge Computer

The traditional centralized management model is very fragile, as soon as the central node is broken, the entire internet of Things equipment service will be paralyzed. So we must build a system of decentralized regulation. The framework consists of four layers: the perception layer, the edge layer, the data storage layer and the application layger. The perception layer, edge layer and data storage layer together form intranet. The intranet is mainly responsible for collecting, storing, analyzing and processing data. The application layer form the extranet, which is mainly responsible for providing the services for the intranet data.

3.1 Perception Layer

The perception layer includes all kinds of sensor nodes. We can get some information of temperature, humidity, air pressure, light and pressure by these sensors. Because these sensors have limited resources, they can only perform simple data-processing tasks. However, they can transfer complex computational tasks to the edge devices and ask the edge devices for completing processing. In addition, they also receive processing outcomes and complete various responses. Agents are device with strong communication and computing capabilities, whose task is to help nodes with weak communication capabilities or low communication requirement to communicate with devices of the edge and upper layer. But the sensor nodes with strong communication capability do not need agents, they directly communicate with the edge device layer.

3.2 Marginal Layer

The marginal layer is make up of the smart contracts, which mainly runs on edge nodes. The smart contracts is a prewritten and deployed electronic contract, which contains two functional modules. One functional module is responsible for management internet of Things device and the point system of user trustworthiness. Other functional modules is responsible for analyzing the behavior of internet of Things device and giving rules for dealing with them.

3.3 Data Storage Layer

The smart contract will be automatically executed when there are data store operation. First, the smart contract encrypts the data by using an elliptic curve cryptography algorithm. The encryption algorithm and corresponding decrypted private key are automatically selected by smart contract, which does not be leak to other devices or users. Second, the smart contract record the storage event of execution to the block of the blockchain. Finally, the various data of the internet of Things devices are processed by smart contracts and stored on the storage hardware of the edge device. In addition, some the activity events will also be recorded the block.

3.4 Application Layer

The application layer is a platform, which can provide various services for users. In addition, the application layer is also interface to access the data of internet of Things. When the user send a service request to the platform, and the platform will send authenticated user request to the smart contract of the edge layer. After the smart contracts receives the user's request, it first queries the user's credibility score, then decides whether to provide the service after judging the rationality of the request, and feedback it to the application layer. The application layer verifies not only the user's identity, but also the user's request of rationality. The verification process can effectively reduce unreasonable or even malicious behavior, reduce the amount of information processed by smart contract, and avoid wasting system resources.

4   Data Processing and User Access

The internet of Things is facing intrusion, DoS and other attacks. For internet of Things security, we design a security protection mechanism at the edge layer and application layer, where are vulnerable to security threats.

4.1   Data Processing

In ordered to prevent internet of Things devices from stealing and abusing the resource of edge nodes, we design the defense mechanism.

In ordered to improve response efficiency, the blockchain simple the process of IoT devices. The validation information is embedded in the data by requesting the IoT devices, which will be sent to the edge layer. Then the smart contracts checks the validation information and views the binding trust score. After the verification is passed, the smart contracts will allocate resources and process the received data in accordance with the credibility. After the processing is completed, some processing strategies will be triggered by the management rules of the smart contracts.

By simplifying the authentication process, response time can be reduced, but the security will be reduced. Therefore, we can adopt a mechanism for dynamically updating device validation information, which dynamically update some verification information with a certain period. The mechanism can avoid some security attacks.

4.2   User Access

Take the example of querying data, the user firstly requests the application layer to inquire about the data service. Then, the application layer asks the edge layer for the trust score of the identity binding based on the identity of the requesting user. The edge layer will feeds the queried confidence back to the application layer. After the application layer determines the user's credibility, it will further interact with the user to verify the user's legitimacy. In addition, after the verification is passed, the platform of the application layer will make a request to the edge layer to query the data. The decrypted data is transferred to the application platform after some processing by the edge layer. The application layer sends the encrypted data to the user after encrypting data with the user's public key. Finally, the users will obtain the data of the query by using the private key. The processing of user access is shown in Fig. 2.

4.3   Production of New Block

Whether it's internet of Things device or user access, the every data operation will produce a record block. And the record block will be added to the block body. The processing is shown in Fig. 3. When the block record reacher the limit, the smart contract will help the datastore layer to generate a new block, which will be added to the data storage layer. Then, the new block will continue to record the access events.

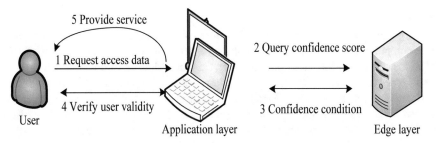

Fig. 2. User access flowchart

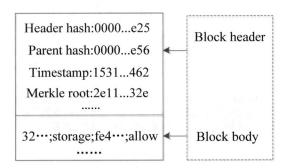

Fig. 3. Block record graph

5 Evolution

This solution is different from the cloud-based internet of Things solution. Some performance estimates is shown in the Table 2. This solution is superior to the cloud-based internet of Things solution in terms of access control and real-time performance, which is realized by block chain and edge computing technology.

Table 2. Comparision in the properties

Property	Solution base on cloud	The solution
Access control model	Static password	Dynamic password
Storage model	Clipetext	Cryptograph
Security mechanism	Static password	Dynamic password, POW
Resource number	More	Medium
Real-time	Ordinary	Strong
Scale	large	Medium

Summary

In this paper, we propose a solution of internet of Things based on the block chain and edge computer. Based on the analysis of the existing internet of Things technology and solutions, we found the shortcoming of the existing solutions and similarities between the internet of Things and structure of bitcoin. In addition, according to the security mode, we summarize the control strategy of strictly controlling the upper layer to access the lower layer and forbidding tampering with lower layer data. Therefore, we use the blockchain technology to record the operation, user and device behavior in the system and control the behavior of users and internet of Things device with the equipment, so as to ensure system security. In addition, the edge computing technology provides a reliable platform for us to process internet of Things data and deploy blockchain. Because the security mechanism of the blockchain WOP takes up a large amount of computering resources, it will cause resource shortage in the short term. However, this work is not the focus of this study, so the solution hasn't further studied it, and only increased the interval of POW.

Acknowledgement. This research was financially supported by Key scientific research projects of Jiangxi Provincial Department of Education (Grant NO. GJJ GJJ191100) and "13th five-year" plan of education science of Jiangxi province (Grant NO. 18YB295).

References

1. Chaudhary, M.H., Scheers, B.: Software-defined wireless communications and positioning device for IoT development. In: International Conference on Military Communications and Information Systems. IEEE (2016). https://doi.org/10.1109/ICMCIS.2016.7496555
2. Zhao, F.: Sensors meet the cloud: planetary-scale distributed sensing and decision making. In: 9th IEEE International Conference on Cognitive Informatics. IEEE (2010). https://doi.org/10.1109/COGINF.2010.5599715
3. Biswas, A.R., Giaffreda, R.: IoT and cloud convergence: opportunities and challenges. In: IEEE World Forum on Internet of Things(WF-IoT). IEEE (2014). https://doi.org/10.1109/wf-iot.2014.6803194
4. Farris, I., Militano, L., Nitti, M., et al.: Federated edge-assisted mobile clouds for service provisioning in heterogeneous IoT environments. In: IEEE 2nd World Forum on Internet of Things (WF-IoT) (2015). https://doi.org/10.1109/WFIoT.2015.7389120
5. Shi, W.S., Cao, J., Zhang, Q., et al.: Edge computing: vision and challenges. IEEE Internet Things J. **3**(5), 637–646 (2016)
6. Sharmeen, S., Huda, S., Abawajy, J.H., et al.: Malware threats and detection for industrial mobile-IoT networks. IEEE Access **6**, 15941–15957 (2018)

Discovery and Advice of Free Charging of Electronic Devices

Yuhang Du[✉] and Chen Wang

Northeast Yucai School, Hunnan District, Shenyang 110000, Liaoning Province, China
83157958@qq.com

Abstract. There are many free charging posts in public places, such as sockets and charging stations for electric vehicles. While users enjoy the convenience of charging anytime and anywhere, it also has certain impacts on public places. Our goal is to create a model to study the impact of these free plug-in charging posts on public places and to quantify these impacts. We will then discuss the extent of these costs and how they will be paid. Finally, we refine the models in different locations and propose two measures to reduce cost.

Keywords: Energy · Cost · Charging

1 Introduction

In recent years, as the living standard of people elevated, the need for electricity has also increased. In our houses, we purchase charging devices, and pay the electricity company for the fee. Many public places offer free charging today. In this article, we will discuss as the time passed by, the free charging devices will have what kind of influences on the public places and whether the users need to pay for it.

2 General Assumption

- We use cell phone to represent the small items that need to charge in public places, and electric vehicles to represent the large items.
- We consider the increase of population, the increase of the number of cell phones, and the increase of the number of electric vehicles as discontinuous process.
- We don't consider the loss cost of the electric vehicle charging stations.
- We assume that the variables are independent of each other and the charging duration is normally distributed.

3 Model Design

3.1 Changes in Energy Consumption

3.1.1 Hypothesis of This Question

To simplify modeling process, we only consider two types of charging devices when establishing the models, cell phone of the small devices and electric vehicle of the large devices.

J. MacIntyre et al. (Eds.): SPIoT 2020, AISC 1283, pp. 271–276, 2021.
https://doi.org/10.1007/978-3-030-62746-1_40

3.1.2   Establishment and Solution of Regression Model

To evaluate the change of the consumption of energy, our calculation consists of two parts, the charging of small devices and that of the large ones.

3.1.3   Small Item

We can calculate the changes in energy consumption caused by cell phones charging, as shown in Table 1.

Table 1. Small item energy consumption change table

Year	Energy consumption (kwh)	Rate of change
2010	1,168,378,998	/
2011	1,170,790,778	0.00206421
2012	1,179,364,587	0.007323092
2013	1,184,695,450	0.004520115
2014	1,188,977,280	0.003614288
2015	1,196,682,010	0.006480132
2016	1,203,992,089	0.006108623
2017	1,203,352,656	−0.000531094
2018	1,209,102,814	0.004778448
*2019	1,215,008,865	0.004884656
*2020	1,220,217,688	0.004287066

3.1.4   Large Item

We can work out the annual changes in electric energy consumption caused by the charging of electric vehicles, as shown in Table 2.

3.1.5   Total Energy Consumption

We draw the energy trends as Fig. 1 shows. The red line is the energy consumption by the small items, the green line is the energy consumption by the large items, and the blue line is the total energy consumption.

3.2   The Cost Model

3.2.1   Hypothesis of This Question

In the cost consideration, we do not consider the additional cost caused by unexpected factors [1–3]. We assume that the depreciation rate and failure rate of all devices are basically the same, and the purchase cost of all devices is calculated according to the items with the highest sales volume on Amazon.

Table 2. Large item energy consumption change table

Year	Energy consumption (kwh)	Rate of change
2010	6,965,166	/
2011	39,635,112	4.69047619
2012	137,737,082	2.475127848
2013	315,920,037	1.293645485
2014	534,788,407	0.692796734
2015	744,609,431	0.392344003
2016	1,038,731,079	0.395001237
2017	1,406,923,030	0.354463209
2018	2,072,680,515	0.473201072
*2019	2,236,833,635	0.079198467
*2020	2,728,316,398	0.219722538

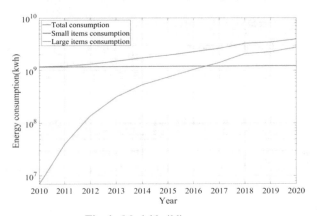

Fig. 1. Model building process

3.2.2 Cost Structure

Continuing with the first question, we divide discussion into the cost of charging small items and the cost of charging large items.

3.2.3 Cost Model Establishment and Solution

Small Item

We calculate the cost and the changes resulting from the charging of small items in public places annually, as shown in Fig. 2(a) (b). We can find that most of the charging costs brought by small items are because of the electricity fee. The purchase of hardware such as sockets and hardware loss only accounts for a small part of the cost, and the cost of purchasing is decreasing year by year [4–6].

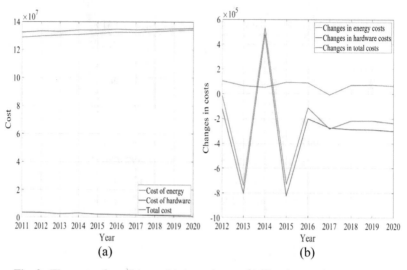

Fig. 2. The costs of small items (a) Annual costs (b) The changes in annual costs

Large Item

We calculate the annual cost of charging large items in public places and the change in cost as shown in Fig. 3(a) (b). It can be found that most of the charging costs brought by large items also come from the electricity fee. The purchase and installation of charging stations only account for a small part of the cost, and the purchase cost is basically stable.

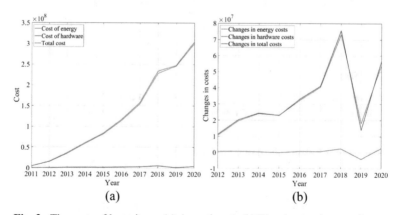

Fig. 3. The costs of large items (a) Annual costs (b) The changes in annual costs

3.2.4   Total Cost (the Users and Governments Burden Costs Are not Calculated)

We can draw a figure as shown in Fig. 4(a), and the changing of the cost as Fig. 4(b) shows, and find that in the overall analysis, whether it is small items or large items charging in public places, the cost of purchasing hardware and maintaining hardware accounts for a very small proportion of the total cost, and the main cost comes from the electricity fee.

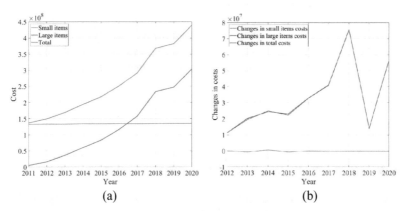

Fig. 4. Total cost (The users and governments burden costs are not calculated) and changes

3.3   Cost Reduction Initiatives

3.3.1   Problem Analysis

In this question, we need to explore the measures to reduce the cost. We can reduce the cost from two aspects: the electricity cost on the one hand, and the hardware cost on the other. We will analyze and adjust our cost model separately from these two aspects.

3.3.2   Reduce Electricity Cost Measures

We can develop a price strategy that is positively correlated with the trend of the load. That is, we can adopt a higher fee when the charging load is high and a lower fee when the charging load is low. In this way, the resources of charging facilities can be better utilized, and can also affect the charging time of users. In addition, more income can be obtained during peak periods, thus, reducing the cost. This strategy of using different methods to collect fee in different time periods can reduce the cost of electricity in public places.

3.3.3   Reduce Hardware Cost Measures

We design a charging device structure that mimic a beehive. We transform the original 1 to 1 charging device into 1 to 6 charging devices. In this case, we only give an assumption, without considering whether it is possible to achieve 1 to 6 charge. If we can achieve this structure, we can reduce the purchase amount of charging devices to 1/6 of before, and also reduce the purchase cost of hardware to 1/6 of before, and keep the electricity charge efficiency unchanged.

4 Free Charging: Discovery and Advice

Based on the data collection and mathematical model establishment of free charging in public places, we get to know that charging for free in public, which we have already used to, is bringing so much cost for public places. We also call on everyone to save as much electricity as possible. After all, coal-fired power accounts for 63.46 percent of the electricity generated in the US, and the energy used to generate electricity is not renewable.

References

1. Mauri, G., Valsecchi, A.: THE impact on the MV distribution grids of the Milan metropolitan area. In: IEEE Energy Engineering Conference, 27–29 March 2012, Shanghai, China, p. 5 (2012)
2. Bayram, I.S, Michailidis, G., Devetsikiotis, M.: Electric power resource provisioning for large scale public EV charging facilities. In: 2013 IEEE International Conference on Smart Grid Communications (SmartGridComm). IEEE (2013)
3. Bendiabdellah, Z., Senouci, S.M., Feham, M.: A hybrid algorithm for planning public charging stations. In: Global Information Infrastructure & Networking Symposium. IEEE (2014)
4. Nijhuis, P.J.: Urban fast charging stations: a design of efficient public charging infrastructure for large numbers of electric vehicles in cities (2015)
5. Gharbaoui, M., Bruno, R., Martini, B., et al.: Assessing the effect of introducing adaptive charging stations in public EV charging infrastructures (2015)
6. Xiong, H., Xiang, T., Zhu, Y., et al.: Electric vehicle public charging stations location optimal planning. Dianli Xitong Zidonghua/Autom. Elect. Power Syst. **36**(23), 65–70 (2012)

Design and Implementation of Tourism Information Management System Based on .NET

Yue Meng, Jing Pu, and Wenkuan Chen[✉]

Sichuan Agricultural University, Chengdu, China
wenkuan_chen@163.com

Abstract. This article relates to the operation mode of the tourism industry to complete the development mode to carry out the design operation. The intention is to use the synergy between many software technologies and corporate market departments to assist the tourism industry in the implementation of many related departments and other affairs to achieve better work efficiency and facilitate high-level implementation decisions deal with. The main contents are: the previous investigation and analysis of the completed tourism development, clarifying its management needs, process information, and model types and making clear analysis and processing in accordance with the actual needs; collecting and mastering software engineering related knowledge. Then, the obtained requirement analysis is actually consistent with the system development, and the performance and functional standards to be performed by the target system are obtained. And use ASP. NET technology to get the architecture and functional composition of the system to complete the process; explore the database design and implementation process and other information to clarify the details of the database design that the system needs to implement. With a targeted explanation of the implementation process and implementation effects of each key module, the overall system implementation effect is analyzed and determined based on the obtained system test situation, and it is determined that the related work of system testing is well implemented necessity.

Keywords: .NET · Tourism information management · Design and implementation · Information flow

1 Introduction

With the continuous progress of material development in recent years, people's spiritual pursuits have also increased, and the corresponding tourism industry has developed rapidly. Leaving the endless hustle and bustle of the city and city, no longer consider the various pressures brought about by work and life, go to a place with beautiful scenery and relax myself [1]. Especially the combination of network technology strength and computer technology strength in recent years has made tourism-related letter and systems optimized and progressed, and tourism-related business implementation has become

J. MacIntyre et al. (Eds.): SPIoT 2020, AISC 1283, pp. 277–284, 2021.
https://doi.org/10.1007/978-3-030-62746-1_41

more convenient and efficient. If the industry is to seek long-term growth and progress, it must be able to meet the needs of users to arrange travel arrangements with more convenient travel paths, better service results, and better experience, and where the users need to perform their work. Seeking a better new model of tourism development.

At present, there is not much investment in tourism [2]. Everyone has the ability to arrange the required travel. Although it is not a necessary part of life, it is indeed a new way for people to seek spiritual satisfaction. In the statistical analysis provided by the country, it is clear that the growth of the existing tourism industry is the key point of the domestic economic growth rate, and the subsequent growth rate will continue to be in the range of 10% [3]. However, the development effect of the existing tourism industry is more different than that of developed countries. Model management and the realization of data sharing cannot meet the needs of the current economic development. The development of intensive and precise tourism is a trend in the future. A comprehensive understanding of the characteristics of the development of tourism can help people better realize the management of tourism development. First, the tourism industry has been able to complete transactions beyond the form of physical exchange, and the two parties to the transaction have realized the theme of interconnecting their information. Therefore, to ensure more effective progress in the tourism industry, it must be clear that the information part is organized; The tourism industry's own development has a strong correlation effect, which involves not only simple food and lodging, but also transportation arrangements [4]. Only when these related departments provide more efficient information support, then the overall tourism development matters will be even more important.

So many tourism development characteristics also make it extremely important to perform efficient information construction on it. The key to completing information construction is to use relatively basic information acquisition methods, and use the acquisition of these information to ensure that more information is effectively applied and processed, so as to achieve the growth of the human spiritual level and promote the development of, and this is also the necessary path for China to follow up with the international community.

2   System Requirements Analysis

Due to the impact of the growing information technology in the tourism industry, people engaged in the industry have begun to focus on the development of information technology in tourism. To this end, various information application systems have emerged. This article briefly describes the current status and future development trends in the field of tourism confidence management from the following three aspects:

2.1   The Promotion of Tourism Products Based on the Internet

Since the development of the tourism industry does not actually require the government to intervene in the spontaneous consumption activities of the people, the intensity of its publicity will be an effective counterfeiting method for businesses engaged in the tourism industry [5]. Due to the advent of computer networks, more businesses are turning their propaganda to the online approach under new media, which is less investment-friendly and widely spread. Most merchants are committed to enriching online resources, so a large number of websites carrying characteristic tourism and related information of their respective merchants have temporarily filled the entire network. Even many travel agencies and tourism agencies in China have begun to focus on online tourism [6]. Resource information. Since such websites can freely support people who will use computer operations to preview and legally use the tourism resource information placed on their web pages, this has also disguised the development of the tourism industry and strengthened the team of "donkey friends".

2.2 Provide Links to Online Booking Services

Relevant businesses engaged in the tourism industry have established their own websites in order to promote their own tourism characteristics and attract people's attention. Through this website, people can not only refer to the tourist information they want to know, but also can perform price consultation, accommodation booking, and experience virtual scenery. In this way, the relationship between the two parties of consumption is brought closer, while the cost of the intermediate intermediary link is reduced, which provides more convenience for tourists. The application that supports tourism information management can reduce the workload for workers in the industry while improving work efficiency, so that it can provide passengers with more efficient, fast, and high-quality service items.

2.3 Commercialization of Tourism Consumption

Due to the development of network technology, the tourism industry has begun to apply it to a number of service projects. In order to meet the needs of people to buy travel products without leaving home, merchants have begun to open online payment models to support people's travel consumption. At the same time, a variety of product payment methods such as credit card channels, online banking channels, and member channels were opened [7]. Unlike other industries, the tourism industry uses this module to make it more convenient for people to operate and use. It only requires the use of a computer to make payment according to the needs to achieve a one-stop service for tourism consumption.

3 System Design

3.1 Overall System Design

As shown in Fig. 1 below: As an open and complex application for the management of travel letters, in order to achieve the management of various subsystems similar to attraction information, hotel information, and shopping information, as well as information interaction with various other systems such as transportation environment to ensure the normal operation of the system, an information flow is established between the subsystems to support service and feedback operations [8].

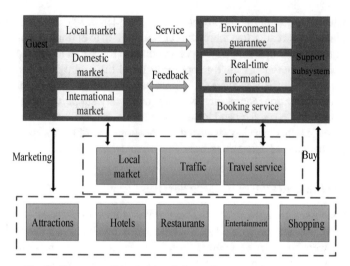

Fig. 1. Structure of the tourism information management system

3.2 Database Design of the System

During the implementation of the required functions of the entire information system, a database needs to provide effective support for it. Therefore, in the process of implementing the database construction, it is necessary to give full attention to the security and reliable effects of its information preservation and prevention and control processing. The specific principles to be followed during construction are: First, operations such as data management and processing are easy to complete. Considering that the original purpose of the database application layer is to better optimize the redundancy of data information content and the efficiency of operation execution under the condition of consistent functions, the implementation of development should be strictly attached to the convenience of the function control processing; second, during the implementation of the program, it is necessary to minimize the occupation of various resources reserved in the system and provide strong guarantees to increase operational efficiency [9]. Third, the log content and the specific data capture database are used to complete independent storage control, which is convenient for accidental intrusion. Just-assisted logs help achieve database recovery.

Here will be combined with the system analysis content given earlier to complete the database analysis and design work. Considering the quality of database construction will affect the overall operating efficiency and implementation effect of the system, so we should focus on it here. In terms of structural design, the design of the conceptual part and the physical part is completed with the help of relatively accurate E-R diagrams [10]. The former focuses on the analysis of the relationship between the entities in the database; the latter focuses on the effective design of the forms involved in the database. In the completion of the conceptual model analysis, the E-R design needs to be done, and the supplementary control of the related content. During the second period, the information content model construction process will be completed with reference to the needs of the user. Use the name and number of the route as the primary key to build an ER map of the tourist route with five attributes including type, price, and time. The details are as in Fig. 2.

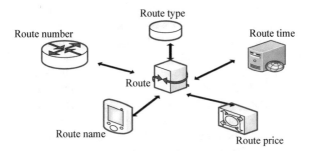

Fig. 2. ER diagram of tourist routes

3.3 Function Modules of the System

The implementation of the six functional modules related to the management information program in the tourism industry is explained in detail here, mainly because there are too many related functional modules in this aspect and it is not easy to explain them one by one [11]. Relying on the information management functions formed by the six aspects of hotel resource information, message information, travel landscape information, travel resource information, and travel route alternatives and travel-related news information.

(1) Implementation of hotel module
      The purpose of building a management module for travel hotels is to provide travelers with convenient and fast real-time hotel information for their choice. Through this module, hotel information containing their respective names, star ratings, prices, and their scenic spots can be distinguished and released according to star ratings and scenic spots. It can also support users to perform multiple operations, such as browsing and booking, for the hotels they need through the Internet. It can also support the staff responsible for managing travel information after interacting with related hotels to delete, modify, and add multiple operations to manipulate travelers' online booking orders and related information. In this module, the system is mainly implemented by adding, modifying, deleting, and querying the four sub-function modules for the hotel information in the scenic spots [12].
(2) Implementation of the message module
      The system also builds a function similar to most program software that guestbooks can cover, which mainly supports users to carry out feedback operations. It is mainly composed of two parts for the administrator to check and respond to the user's message and to support the user to leave a message and display it.
(3) Implementation of tourist landscape map module
      The module is constructed by listing exquisite scenic pictures of various tourist attractions in the form of thumbnails for users to view. The pictures arranged in the

module page are sorted according to their upload time. Users can choose different according to their own preferences. Click on the small landscape image on the search page for the name of the tourist image of the scenic spot. Aiming at the four manipulation forms of adding, modifying, deleting, and searching for tourist attractions and scenery maps, the author's construction of this function is also divided into corresponding four parts for specific implementation. In the construction of this unit, the author pays more attention to the functional design of supporting the response to users' search and browsing of picture resources and the realization of the management function of travel pictures. Among them, the former is mainly responsible for finding out all the relevant pictures stored in the database according to the prompts of the user's specific keywords, and performing thumbnail display. In this way, users can view the pictures they are interested in. Users can view more information and details through the single-frame diagram information; the latter mainly supports the execution of operations such as publishing, modifying, and deleting travel pictures.

(4) Implementation of tourist attractions module

In order to make it easier for users to obtain the details of the scenic spots involved in each scenic spot, the system also builds a number of functional modules that support the scenic spots as a unit to browse and manage scenic spots. The main parts are: 1. Management function modules for adding, deleting, and modifying operations for scenic spots and scenic spots; 2. Supporting the scenic spots module browsing and display functions in two forms: scenic spots list and scenic spot details.

Through the construction of the browse display function, you can search the database for the relevant scenic spot information corresponding to the keywords pointed by the user, and provide the form of display links to help users view the details of the scenic spots they need [10]. The support for the implementation of its management module is implemented for the related operations of saving database data of scenic spots in the system, and mainly implements operations such as adding, deleting, and modifying scenic spots and scenic spots. The part of the source code that can support the execution of related functions is not described in detail.

(5) Implementation of tourism news module

This unit module constructs an implementation platform that supports the release of tourism-related information and is responsible for operation and maintenance tasks by the administrator. The types of news published on this site are mainly based on service guides and travel information. In order to facilitate the management and management of the administrator, the author has built three sub-function modules based on news browsing operations, publishing operations, and management operations. In this way, it is convenient for the operator to add, delete, modify and manage the module and the news released.

According to the structure chart of the travel news system constructed above, we have designed its functions from the aspects of arranging news, browsing news, and managing news: The use mechanism is: the administrator will open the latest travel letter through the relevant control, and the system database It can display and match the corresponding simple sentences or key prompts on the homepage of the news browsing page and supports detailed information reading links [12]. Users can view news related information here according to their interests and needs.

Its management function is mainly to facilitate the system administrator to delete, modify and update the travel news information. The following figure is its operation interface diagram.

(6) Implementation of tourist traffic line module

The system also builds a platform that can support the release of travel routes including accommodation standard information, en route attractions information, and various related information such as travel arrangements [9]. Through the platform's real-time updated travel schedule details, passengers can choose according to their needs. The best travel route can also make it easier for managers to formulate the best travel plan that is more convenient for tourists, based on the differences in time and scenic areas and the adaptation to the crowd. The system also provides an online booking link. With this link, users can realize the operation of booking the travel routes they need without leaving the home. The background manager of the system can not only perform various operations such as adding, deleting, updating, and modifying the line arrangements in the platform, but also control the passengers' line booking orders.

This unit supports multiple operations such as classification, modification, addition, deletion, and display of reservation management for the tourist route: in the display section, it can be divided into two types of display: journey type arrangement and content details; in the management section, it can support background managers various operations such as the addition, deletion, modification and release of each route, as well as the classification of tourism types. According to the structure diagram of the part constructed above, it can be seen that its function realization module mainly includes two parts for browsing and management of the route. The unit function support code is implemented after the construction of its database is completed. The following figure is the page view of its control interface.

4   Conclusion

Based on the B/C model, a management process that can support the tourism industry to perform management of its related information is implemented in the article. In order to show the real-time and accuracy of tourism information, the author has also integrated various new technological concepts to It is simple, practical and easy to operate. The system implemented in the end is not only used to help the tourism management organization to carry out related management operations, but also benefits the tourism provider and people who travel. To a large extent, it has promoted the rapid development of tourism. After focusing on the analysis of the current information management system in the tourism industry, find out its shortcomings. After combining the future development direction of the industry, it summarizes the aspects that need to be improved. In order to better provide people with convenient electronic and intelligent business-oriented tourism information management systems for people's travel, it will be the direction that people who work in this field will work hard. After conducting a full range of exchanges and consultations on the current information management system of the tourism industry and the application of the tourism system, including provider managers and travelers, a system logic function that includes all the performance to be implemented by the system is feasible.

Acknowledgments. Study on the livelihood vulnerability of farmers and its countermeasures in the giant panda national park community, project number: GJGY2019-YB006.

References

1. Schröter, B., Hauck, J., Hackenberg, I., Matzdorf, B.: Bringing transparency into the process: social network analysis as a tool to support the participatory design and implementation process of payments for ecosystem services. Ecosyst. Serv. **3**(4), 206–217 (2018)
2. Schröter, B., Matzdorf, B., Hackenberg, I., Hauck, J.: More than just linking the nodes: civil society actors as intermediaries in the design and implementation of payments for ecosystem services–the case of a blue carbon project in Costa Rica. Local Environ. **23**(6), 635–651 (2018)
3. Hassannia, R., Vatankhah Barenji, A., Li, Z., Alipour, H.: Web-based recommendation system for smart tourism: multiagent technology. Sustainability **11**(2), 323–324 (2019)
4. Min, W., Ku, J.: Tourism information system based on sharing economy using an integrated information communication technology platform. Int. J. u- and e-Serv. Sci. Technol. **9**(5), 279–290 (2016)
5. Xu, B., Xu, L., Cai, H., Jiang, L., Luo, Y., Gu, Y.: The design of an m-Health monitoring system based on a cloud computing platform. Enterp. Inf. Syst. **11**(1), 17–36 (2017)
6. Scheepens, A.E., Vogtländer, J.G., Brezet, J.C.: Two life cycle assessment (LCA) based methods to analyse and design complex (regional) circular economy systems. Case: making water tourism more sustainable. J. Cleaner Prod. **11**(4), 257–268 (2016)
7. Lee, E.K., Shi, W., Gadh, R., Kim, W.: Design and implementation of a microgrid energy management system. Sustainability **8**(11), 1143–1144 (2016)
8. Nesshöver, C., Assmuth, T., Irvine, K.N., Rusch, G.M., Waylen, K.A., Delbaere, B., Krauze, K.: The science, policy and practice of nature-based solutions: An interdisciplinary perspective. Sci. Total Environ. **5**(9), 1215–1227 (2017)
9. Gil, L., Ruiz, P., Escrivá, L., Font, G., Manyes, L.: A decade of food safety management system based on ISO 22000: a GLOBAL overview. Revista de Toxicología **34**(2), 84–93 (2017)
10. Zhang, Z., Xu, G., Zhang, P., Wang, Y.: Personalized recommendation system based on WSN. Int. J. Online Biomed. Eng. **12**(10), 91–96 (2016)
11. Psaltopoulos, D., Wade, A.J., Skuras, D., Kernan, M., Tyllianakis, E., Erlandsson, M.: False positive and false negative errors in the design and implementation of agri-environmental policies: a case study on water quality and agricultural nutrients. Sci. Total Environ. **57**(5), 1087–1099 (2017)
12. Hawkins, S.J., Allcock, A.L., Bates, A.E., Firth, L.B., Smith, I.P., Swearer, S.E., Todd, P.A.: Design options, implementation issues and evaluating success of ecologically engineered shorelines. Oceanogr. Mar. Biol. Annu. Rev. **5**(7), 169–228 (2019)

A Computer Model for Decision of Equipment Maintenance Spare Parts Reserve

Yunxing Wang[✉], Weishan Zhou, Jingjing Zhao, and Mengwei Wang

Qing Zhou High-Tech Institute, Qing Zhou, Shan Dong, China
rzwyx@126.com

Abstract. It is considered that reserving a reasonable quantity of spare parts is an effective measure to increase equipment availability and reduce life cycle cost. Keeping the demand and configuration of equipment maintenance spare parts balanced is an important part to determine a reasonable amount of spare parts reserve. According to reliability theory and maintenance decision theory, under the assumed condition that the lifetime distribution and maintenance time distribution of weapon equipment are subject to exponential distribution, this paper concludes the decision models of two kinds of equipment maintenance spare parts, respectively repairable parts and irreparable parts.

Keywords: Equipment maintenance · Spare parts · Decision · Calculation model

1 Introduction

In general, various devices are composed of numerous parts (or components), and some parts (or components) of the device may break down while using, thus affecting the performance of equipment. Therefore, it is necessary to reserve a reasonable quantity of spare parts during the service period, ready to replace the failed parts, which is an effective measure to improve equipment availability and reduce life cycle cost. With the increasing complexity of equipment, the variety and quantity of spare parts become more and more, and the proportion of the support cost of spare parts in the support cost of equipment use is also increasing. Under the limited funding condition, the optimal configuration of spare parts variety is essential, to ensure that the spare parts required in the use and maintenance of equipment can be timely and adequately supplied [1–3].

To determine the number of spare parts, the calculation model of spare parts must be established first. In the 1970s, the U.S. military issued an order that it was necessary to use mathematical models to determine the number of spare parts. This shows that the mathematical model plays an important role in determining the quantity of spare parts. However, the calculation of mathematical models should be well-reasoned. US military materials also acknowledged: "Calculating the demand for spare parts involves risks and uncertainties. Although mathematical methods are used to predict the demand, it is still common for some spare parts to be stored too many or too few at the moment." The

J. MacIntyre et al. (Eds.): SPIoT 2020, AISC 1283, pp. 285–292, 2021.
https://doi.org/10.1007/978-3-030-62746-1_42

existing research results and spare parts management experience show that the selection of mathematical models to calculate spare parts should follow the following principles [4].

1) Theoretically reasonable, universal and representative;
2) Less computing requirements, and easy operation;
3) The calculation results are similar to that of other models.

2 Analysis on the Decisive Calculation Model of Maintenance Spare Parts Reserve

The reserve quantity of maintenance spare parts is related to the reliability and maintainability of the equipment. According to the general use experience of the equipment, and also for convenience, it is assumed that the life distribution and maintenance time distribution of the equipment are subject to exponential distribution.

In addition, under the condition of modern technology, the structure of the equipment becomes more and more complicated, and the types of the components constituting the equipment also become more and more various. However, from the aspect of maintenance results, it can be attributed to irreparable parts and repairable parts. Irreparable parts mainly refer to consumable spare parts and partially damaged spare parts, and repairable parts are replaceable parts that provide repair turnover. This paper mainly studies the decision analysis of repairable and irreparable maintenance spare parts [5–7].

2.1 Decision Analysis of Irreparable Spare Parts Reserve

When the irreparable parts in the equipment system break down, replacing the parts is a quick and easy method that is often used. If the equipment and spare parts together are regarded as an equipment system, and the time of replacing the failed parts and the failure rate of the spare parts in the reserve period are negligible, the equipment is equivalent to a cold reserve system. If a certain irreparable part of the equipment fails, then the equipment breaks down and the spare part needs to be replaced immediately. There are n spare parts, and the equipment breaks down after n spare parts all have failed.

Suppose the life of $n + 1$ vulnerable parts are random variables $T_0, T_1, T_2, ..., T_n$. T_0 is the component used in the equipment. The total life of the equipment is $T = T_0 + T_1 + \cdots + T_n$. Its task reliability is:

$$R_n(t) = P(T_0 + T_1 + \cdots + T_n > t) \tag{1}$$

Suppose that the random variables are mutually independent, so that the density function $f_n(t)$ of the life distribution of the system is the convolution integral of the density function $f_i(t)$ of the life distribution of $n + 1$ components, namely:

$$f_n(t) = f_0(t) * f_1(t) * \cdots * f_n(t) \tag{2}$$

2.1.1 The Working Life of a Spare Part

In the decision-making of maintenance spare parts reserve, the life distribution of components is exponential distribution. Suppose the failure rate is λ, and it can be obtained by the above Formula (2) that when there is a spare part, the life distribution density function of the equipment is:

$$f_n(t) = \int_0^t \lambda_0 e^{-\lambda_0(t-x)} \lambda_1 e^{-\lambda_1 x} dx = \lambda_0 \lambda_1 \frac{e^{-\lambda_0 t} - e^{-\lambda_1 t}}{\lambda_1 - \lambda_0} \tag{3}$$

Due to the reliability $R_s(t)$ and average life expectancy $\overline{T_s}$ of the equipment are respectively:

$$R_s(t) = \int_t^{+\infty} \lambda_0 \lambda_1 \frac{e^{-\lambda_0 x} - e^{-\lambda_1 x}}{\lambda_1 - \lambda_0} dx$$

$$= \frac{\lambda_1}{\lambda_1 - \lambda_0} e^{-\lambda_0 t} + \frac{\lambda_0}{\lambda_1 - \lambda_0} e^{-\lambda_1 t} \tag{4}$$

2.1.2 The Working Life of n Spare Parts

When the life distribution of n spare parts is the exponential distribution with failure rate λ, the reliability of the system proved by mathematical induction is:

$$R_s(t) = [1 + \lambda t + \frac{(\lambda t)^2}{2!} + \cdots + \frac{(\lambda t)^n}{n!}] e^{-\lambda t}$$

$$= \sum_{k=0}^{n} \frac{(\lambda t)^k}{k!} e^{-\lambda t} \tag{5}$$

The average life expectancy of the equipment is:

$$\overline{T_s} = \frac{1 + n}{\lambda} \tag{6}$$

2.1.3 The Working Life when the Equipment Needs L Components Working at the Same Time

Since the reliability of each component is $e^{-\lambda t}$, the system only works when L components work at the same time, and then the system reliability of such L components is $e^{-L\lambda t}$. If any one of the components in the equipment fails, the part needs to be replaced by spare part immediately, so the reliability of the system remains unchanged until L spare parts are used up [8, 9].

Together with the spare parts, the reliability and average life expectancy of the equipment are:

$$R_s(t) = \sum_{k=0}^{n} \frac{(L\lambda t)^k}{k!} e^{-L\lambda t} \tag{7}$$

$$\overline{T}_s = \frac{1+n}{L\lambda} \tag{8}$$

The two methods of Formula (8) and (9) are mainly used in determining the quantity of irreparable spare parts reserve. If the reliability of the equipment is used as the target (the reliability target value is usually determined by the user and the contractor together), to determine the spare parts reserve, n can be calculated inversely from Formula (8) after the reliability is given; if the utility time of the equipment is given to obtain the minimum reserve of spare parts, Formula (9) can be used for calculation.

2.2 Reserve of Repairable Spare Parts

There would be a situation of the equipment in the work process: there is a fragile part, and another n repairable spare parts for backup. If the failure rate of spare parts at work is λ, the parts are immediately repaired after the replacement, suppose the fault spare parts can only be repaired one by one, the repair rate is μ, and the replacement time is negligible, the failure rate of spare parts during the reserve period is 0, and $K = \mu/\lambda$ is recorded as repair rate.

2.2.1 The Average Time $\overline{T}$ of Spare Parts Lack for the First Time

For the above system with n spare parts, there are $n+1$ states, represented by $S_0, S_1, S_2, ..., S_n$. S_0 represents that the working components are normal, n spare parts are prepared aside for cold storage; S_1 represents that one spare part is in repair, and $n-1$ spare parts are prepared aside; S_2 represents that one spare part is in repair, one spare part is waiting for repair and $n-2$ spare parts are prepared aside ...; S_n represents that only the working components are normal, one spare part is in repair, and the rest spare parts are all waiting for repair. The state transition is shown in Fig. 1.

Fig. 1. Diagram of the system state transition of repairable spare parts

The equipment in S_i state may transfer to S_{i+1} state due to the fault, or S_{i-1} state due to the repair of spare parts, and the state transition probabilities are respectively λ and μ. The average of the system's staying time in S_i state is recorded as T_i, and then the average value $\overline{T}$ of the time before the occurrence of spare parts lack of the system is:

$$\overline{T} = \sum_{i=0}^{n} T_i \tag{9}$$

Obviously, the average of transition times for the equipment transferring from S_i state to S_{i+1} state should be λT_i; and the average of the transition times to S_{i-1} state should be μT_i. One-time transition from S_n state to the lack of spare parts results in the event of spare parts lack,

Therefore,

$$T_n = \frac{1}{\lambda} \tag{10}$$

Although the repair rate of spare parts is often greater than the failure rate ($\mu > \lambda$), that is, the average time between failures is often greater than the average repair time, this is only an average of random events and the event that new fault occurs when the spare parts have not been repaired is not excluded, so that the spare parts would be lacking. Describe this situation by state transition, that is, the number of the transitions from S_i state to S_{i+1} state is one time more than the number of the transitions from S_{i+1} state to S_i state, namely:

$$\lambda T_i - \mu T_{i+1} = 1 \tag{11}$$

$$T_i = \frac{1}{\lambda} + \frac{\mu}{\lambda} T_{i+1} = T_n + K T_{i+1} \tag{12}$$

In the formula, K is the repair rate.

Thus, T_n can be calculated by Formula (11), then T_i ($i = 0, 1, 2, \cdots, n$) are calculated by Formula (13), and $\overline{T}$ is finally calculated through Formula (11). When n is a lot, this calculation is complex. Through the derivation of the above formulas, the average value $\overline{T}$ of the time when the equipment lacks spare parts for the first time is obtained. When $K \neq 1$,

$$\overline{T} = \sum_{i=0}^{n} T_i = \sum_{i=0}^{n} T_i \left(\frac{K^{i+1}-1}{K-1} \right) T_n$$

$$= \frac{T_n}{K-1} \left(\sum_{i=0}^{n+1} K^i - n - 2 \right) = \frac{T_n}{K-1} \left[\frac{K^{n+2}-1}{K-1} - (n+2) \right] \tag{13}$$

When $K = 1$, Formula (13) can be written as: $T_i = (n - i + 1)T_n$
Thus,

$$\overline{T} = \sum_{i=0}^{n} T_i = [1 + 2 + 3 + \cdots + (n+1)]T_n$$

$$= \frac{1}{2}(n+1)(n+2)T_n \tag{14}$$

2.3 Determining the Minimum Spare Parts Reserve According to the Time of the Spare Parts Lack for the First Time

According to Formula (14), n can be calculated inversely with the known $\overline{T}$. However, the time of spare parts lack for the first time in the formula is an expectation value, and the reliability of equipment working to this time is only 36.7%. Therefore, the quantity of spare parts determined by Formula (14) is the minimum spare parts reserve, that is,

290 Y. Wang et al.

the least spare parts reserve for the equipment working for a period of time without lacking spare parts. To facilitate the calculation, the solution can conducted based on four situations.

When $K = 1$, based on Formula (15), for each additional spare part, the average value of the first spare part lack would be increased by $(n + 2)T_n$. Calculate n from known $\overline{T}$, and get a quadratic equation from Formula (15): $n^2 + 3n - 2\lambda\overline{T} + 2 = 0$,

Its positive root is:

$$n_1 = \frac{-3 + \sqrt{9 + 8(\lambda T - 1)}}{2} \tag{15}$$

Use Formula (16) to find a positive root n_1 of n, which is the minimum reserve of spare parts.

When $K \approx 1$, solve the first approximate solution n_1 of n from above formula, and then substitute this value n_1 into Formula (14) to get the second approximation n_2 of n, which is the minimum reserve quantity of spare parts when $K \approx 1$.

$$\overline{T} = \frac{T_n}{K - 1}\left[\frac{K^{n_2+2} - 1}{K - 1} - (n_1 + 2)\right] \tag{16}$$

$$\lambda\overline{T}(K - 1)^2 = K^{n_2-2} - (n_1 + 2)(K - 1) - 1 \tag{17}$$

$$n_2 = \frac{\lg[\lambda\overline{T}(K - 1)^2 + (n_1 + 2)(K - 1) - 1]}{\lg K} - 2 \tag{18}$$

The minimum reserve of spare parts n_1 when $K = 1$ is taken as the first approximation solution, and is substituted into Formula (19) to obtain the second approximation n_2 of n. If the difference calculated between n_2 and n_1 is small (less than 1), the larger one of the two is the required spare parts reserve n; if the difference calculated between n_2 and n_1 is very large, n_2 can be used to replace n_1 in Formula (19), to further get the third approximation solution n_3, until the difference between n_i and n_{i+1} is less than 1, the larger one is taken as the required n.

(3) When $K > 1$, if $K > 1.5$, n is larger, then $K^{n+2} >> 1$, $(n + 2)$ in Formula (11) can be ignored, and $\overline{T}$ approximates to:

$$\overline{T} = \frac{T_n}{(K - 1)^2}(K^{n+2} - 1), \quad \lambda\overline{T}(K - 1)^2 + 1 = K^{n+2} \tag{19}$$

$$n = \frac{\lg[\lambda\overline{T}(K - 1)^2 + 1]}{\lg K} - 2 \tag{20}$$

After using Formula (21) to get the first approximation n_1 of spare parts quantity n, similar to the calculation method when $K \approx 1$, Formula (19) is used to get the approximate solution n_2, n_3, …, until the difference between the adjacent approximate solutions is less than 1, and the greater one is taken as the required spare parts quantity n.

If $1 < K < 1.5$, and $K^{n+2} >> 1$ fails to work, Formula (14) is difficult to be simplified. In this case, two kinds of limit cases are considered first, and Formula (16) and Formula (21) are respectively applied to get the spare parts reserve quantity n_1 and n_2 when $K = 1$, $K > 1.5$ and $K^{n+2} >> 1$, thus through transforming Formula (14), get:

$$f(n) = \frac{T_n}{K-1}\left[\frac{K^{n+2}-1}{K-1} - (n+2)\right] - \overline{T} \qquad (21)$$

Obviously, $f(n_1) > 0$ and $f(n_2) < 0$, so that Formula (22) and dichotomy can be applied to take n_1 and n_2 as the initial points, to conduct stepwise approximation on the minimum reserve of spare parts, and get division points n_3, n_4, ..., until the difference between adjacent points is less than 1, and the one meeting $f(n) > 0$ is the required spare parts quantity n.

(4) When $K < 1$, if K is smaller and n is larger, $K << 1$. Formula (14) can be approximated as:

$$\overline{T} = \frac{T_n}{1-K}(n+2+\frac{1}{1-K}) \qquad (22)$$

That is, for each additional spare part, the time of the first spare part lack will increase by a small value $T_n/(1-K)$. Therefore, we should focus on improving the repair rate of spare parts at the moment. From (22), get:

$$n = [\lambda\overline{T}(1-K) + \frac{1}{1-K}] - 2 \qquad (23)$$

When $K < 1$, find the first approximate solution n_1 of spare parts from Formula (24), the subsequent steps are the same as before, substitute n_1 into Formula (19), successively obtain approximation solutions n_2, n_3, ..., until the difference between the adjacent approximate solutions is less than 1, and the larger one is the required spare parts quantity n.

If $K^{n+2} << 1$ fails to work, Formula (14) is difficult to be simplified. In this case, two kinds of limit cases are considered first as (6), Formula (16) and Formula (24) are respectively applied to get n_1 and n_2 when $K = 1$, $K < 1$ and $K^{n+2} << 1$, Formula (22) and dichotomy are applied to take n_1 and n_2 as the initial points, to conduct stepwise approximation on the minimum reserve of spare parts, and get division points n_3, n_4, ..., until the difference between adjacent points is less than 1, and the one meeting $f(n) > 0$ is the required spare parts quantity n.

3   Conclusion

Maintenance spare parts decision-making is a complex multi-objective decision-making, needing comprehensive balancing and analysis. This paper applies use reliability and working hours as two qualifications, to meet the decision-making requirements of weapon equipment maintenance spare parts. At the same time, the possibility of funding limits and spare parts quantity should also be considered based on the price of spare parts. In general, for equipment spare parts with low requirement for task reliability, it is better to minimize maintenance costs, such as minimizing the quantity of spare parts required, reducing the reliability requirement of the system, appropriately increasing the procurement cycle, speeding up the turnaround and utilization of spare parts and reducing the cost of spare parts backlog. For the equipment requiring higher task reliability, it is necessary to take the reliability of the system completing the task into full consideration.

References

1. Wang, Z.: Rationality of equipment maintenance mode and spare parts management reserve. China Plant Eng. **9**, 7–8 (2001)
2. Yu, J., et al.: Study on the calculation method of missile equipment spare parts quantity. Tactical Missile Technol. **3**, 9–12 (2006)
3. Wang, Q., et al.: Determination of the maintenance spare parts reserve of mechanical equipment. Machinery **4**, 54–56 (2003)
4. Liang, H., et al.: Ordnance maintenance equipment security management. China Science Press, Beijing (1998)
5. Zhao, X., et al.: Determination of the spare parts reserve of electronic equipment of lifetime obeying exponential distribution. J. Naval Aeronaut. Astronaut. Univ. **7**, 426–428 (2006)
6. Li, Y., et al.: Analysis and calculation method of missile weapon equipment spare parts demand. Ordnance Ind. Autom. **8**, 22, 34 (2007)
7. Wu, J., et al.: Study on the determination method of missile support equipment demand. J. Naval Aeronaut. Astronaut. Univ. **6**, 653–656 (2004)
8. Xu, X.: Equipment Maintenance Management. National Defense Industry Press, Beijing (1994)
9. Li, J., Ding, H.: Analysis on the calculation model of spare parts demand. Electron. Prod. Reliab. Environ. Test **6**, 11–14 (2000)

Risk Level Determination of Science and Technology Service Supply Chain with PA-BP Integrated Model

JinHua Sun[✉], Linlin Xu, and Xiang Guo

Chongqing University of Technology, 69 Hongguang Avenue, Chongqing, China
Sjh1009@163.com

Abstract. It is advantageous to predict the risk for scientific and technological services enterprises by making the science and technology service supply chain risk early warning, and help them to take effective actions to reduce or avoid risk timely to improve the decision level. On the basis of analyzing the connotation of science and technology service supply chain risk, this paper finds out the main risk factors influencing the science and technology service supply chain operation, and establishes the model on the science and technology service supply chain risk early warning by making the effective integration of BP neural networks and principal component analyzing method. Finally, the results show that it has high applicability and reliability of the risk early warning model of science and technology service supply chain with BP neural network, which can realize the accurate early warning, detection and analysis for the risk of scientific and technical service supply chain.

Keywords: Scientific · Technological chain · Risk pre-warning model · BP neural network

1 Introduction

Science and technology service industry has become an important driving force of the national economy. How to accurately identify and determine technology service supply chain risk, it's key to the development of the science and technology service industry. Finally, it can improve the decision-making level and ability of participants to cope with risks. Facilitating the corresponding risk prevention measures and reducing the incidence of risk. Thus promoting the promotion of scientific and technological innovation in our country.

2 Literature

E yizhou et al. (2016) constructed the framework of Internet financial risk identification system. Liu haohua (2016) made an in-depth analysis on the supply risk of power lithium battery. Zhuguang et al. (2017) identified the security risks faced by the five stages of

© The Editor(s) (if applicable) and The Author(s), under exclusive license
to Springer Nature Switzerland AG 2021
J. MacIntyre et al. (Eds.): SPIoT 2020, AISC 1283, pp. 293–300, 2021.
https://doi.org/10.1007/978-3-030-62746-1_43

big data flow. Known et al. (2017) comprehensively used functional weight method and difference weight method to identify risks of academic information system. Wang jianhua et al. (2017) used bayesian network analysis method to conduct an empirical analysis on the source of safety risks. Wang H H et al. (2017) analyzed and identified the supply chain risks based on the SCOR model. Zou Xiao-hua et al. (2016) realized the risk grade assessment of the project. Rodriguez A et al. (2016) evaluated and ranked the risk level of information technology development projects.

3   Risk Sources in Science and Technology Service Supply Chain

The risk of science and technology service supply chain refers to the internal and external uncertainties or events. The specific manifestations are as follows [9].

(1)   Risks of technology suppliers. The risks are mainly manifested in four aspects: the risk of science and technology project establishment, the risk of loss of key resources, the risk of investment in science and technology elements, and the risk of imperfect supporting facilities.
(2)   Risk of science and technology service integrato. the instability risk of alliance relationship usually refers to the overall interests are damaged [10].
(3)   Risks of technology users. One is the transformation of scientific and technological achievements and applications of risk, The second is technology outsourcing risk [6–8].

4   Establishment of Risk Level Determination Model

4.1   Design of Risk Level Determination Model for Science and Technology Service Supply Chain

Establish the range of risk factors. This paper constructed the index system as shown in Table 1 below. the range of this kind of index is set as [1–5], and the range of this index is set as [1, 100%] due to the different units and values of the quantitative index.

4.2   Model Selection

With the basic principle of BP neural network, this paper adopts the fuzzy integral principle. Using factor analysis method to extract principal factors. Calculating the score of each common factor and its weight, the comprehensive score of each common factor is obtained after the sum.

Summary analysis, comprehensive training factor score (as shown in Table 2), in order to determine the technology service supply chain risk of grade I, II, III, IV, its represent a high risk, risk, less risk, no risk, among them, the risk is extremely high, less than 0.01 said [0.01, 0.04] said the risk is bigger, [0.04, 0.09] said less risk, minimal risk is greater than 0.09. When risks occur, the indicators set by these four risk levels in the entire judgment model are: high risk (red), high risk (orange), low risk (blue) and low risk (green) (as shown in Table 3 below).

Table 1. Risk level evaluation index system for science and technology service supply chain

Level 1 indicators	Level 2 indicators	Level 3 indicators	Characterization of index
Supply chain risk source Y	Technology service provider risk	Technology project risk X1	Non-project expenditure rate A1 New product development profit growth rate A2 The proportion of non-project participants is A3
		Key resource loss risk X2	The missing degree of invention patent A4 R&D insufficient rate of internal expenditure A5
		Investment risk X3	Proportion of non-R&D personnel A6 Science and technology input growth rate A7
		Poor infrastructure risk X4	Growth rate of expenditure on R&D assets A8 Equipment running turbulence A9
	Technology services integrator risk	Risk of key object loss X5	Employee turnover level A10 High-tech equipment incompleteness A11
		incomplete organization and coordination risk X6	Conflicting A12 Goal inconsistency A13
		Risk of alliance instability X7	Degree of divergence of interests A14 Degree of mutual suspicion A15
		Financing risk X8	Internal expenditure growth rate A16 Technology spending growth rate A17 Main business profit margin growth rate A18
		Patent protection risk X9	Patent applications are inefficient A19 Non-related rate of technological progress A20 Low availability of technology A21
		Node enterprise culture conflict risk X10	Cultural differences A22 Strength of cultural potential A23
		Low quality risk X11	Proportion of staff degree below with master A24 Enterprise exclusion A25

(continued)

Table 1. (*continued*)

Level 1 indicators	Level 2 indicators	Level 3 indicators	Characterization of index
		Risk of information asymmetry X12	Degree of information closure A26 Organizational complexity A27
		Risk of improper supplier selection X13	The proportion of firm lacking R&D activities A28 Technical contract retention ratio A29
		Technology policy barriers risk X14	Degree of policy misinterpretation A30 Policy gap rate A31 Policy change level A32
		Information network vulnerability risk X15	Information technology contribution growth rate A33 Internet penetration rate A34 Supply chain network complexity A35
	Technology user risk	Transformation and application of scientific and technological achievements risk X16	Product batch elimination rate A36 Technical market turnover rate A37 Technology incubator growth rate A38
		Technology outsourcing risk X17	Technical consulting contract elimination rate A39 Technical service contract elimination rate A40

Table 2. Comprehensive score table of training factors

	2000	2001	2002	2003	2004	2005	2006	2007
1	0.381	0.365	0.318	0.27	0.13	0.114	0.051	0.034
2	−0.103	−0.096	−0.076	−0.154	−0.112	−0.102	−0.081	−0.101
3	−0.908	−0.859	−0.724	−0.218	0.243	0.276	0.477	0.619
A	−0.003	0.001	0.007	0.036	0.06	0.061	0.071	0.079
	2008	2009	2010	2011	2012	2013	2014	2015
1	−0.025	−0.019	−0.159	−0.164	−0.054	−0.076	−0.181	−0.234
2	−0.068	−0.034	0.016	0.066	0.234	0.37	0.467	0.528
3	0.754	0.602	1.018	0.855	−0.262	−0.695	−0.617	−0.63
A	0.085	0.076	0.095	0.085	0.019	−0.01	−0.012	−0.016

Table 3. Risk degree table

Level	The degree of risk	Comprehensive range	Warning output	The warning signs
I.	A high risk	$F \in (-\infty, -0.01)$	[1000]	Red
II.	Risk is bigger	$F \in (-0.01, 0.04)$	[0100]	Orange
III.	Less risky	$F \in (0.04, 0.09)$	[0010]	Blue
IV.	Less risky	$F \in (0.09, +\infty)$	[0001]	Green

Through the analysis of the test results, it is found that when the number of hidden layer nodes in the risk level determination model of s&t service supply chain is 11, the actual output of the model has the smallest gap with the expected output, and the accuracy is higher [6–8]. Based on this, this paper established a $12 \times 11 \times 4$ risk level determination model for science and technology service supply chain.

5   Training and Testing

5.1   Model Training

In this paper, Specific parameters of the network model are set as follows, Training function is gradient descent method. Transfer function is log function. Maximum training steps e is 50. Learning rate η is 0.1. Expected error ε Assignment in 0.001. Show time intervals is 1. Other parameters are set to default values.

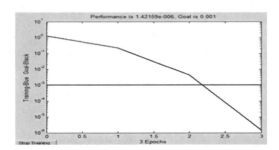

Fig. 1. Variation curve of network training error

As can be seen from Fig. 1, when the risk level determination model of s&t service supply chain is trained to step 3, the performance of the entire risk level determination network reaches the best state. The Fig. 2 and Fig. 3 shows that the technology service supply chain risk level for determining network training error between actual output values and the desired output is small. Specific results (as shown in Table 4), and each phase of the actual output and desired output value changes in smaller change in gap, the fitting degree is higher, consistent characteristics.

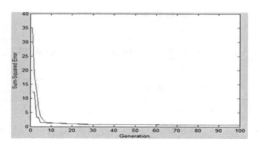

Fig. 2. Error sum of squares curve of network training

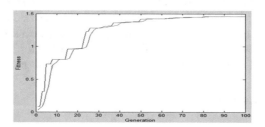

Fig. 3. Fitting curve between expected output and actual output of network training

Table 4. Output results of network training

Year	Expected output value	Actual output value	Risk level	Warning signal
2000	[0010]	0.9983	*III*	Green
2001	[1000]	1.0000	*I*	Green
2002	[1000]	0.0796	*I*	Blue
2003	[0100]	0.5896	*II*	Green
2004	[0100]	0.0339	*II*	Orange
2005	[0010]	0.0687	*III*	Blue
2006	[0010]	0.0067	*III*	Red
2007	[0001]	−0.9464	*IV*	Red
2008	[0010]	−1.000	*III*	Red
2009	[0001]	0.0261	*IV*	Orange

5.2 Model Test

In this paper, the sample data of the last 6 phases of risk level determination index of sci-tech service supply chain were used as test samples to verify the model and output the results of network test (as shown in Table 5 below).

Table 5. Output results of network inspection

Year	Expected value	Actual value	Risk level	Warning signal
2010	[0001]	1.0000	*IV*	Green
2011	[0001]	0.0356	*IV*	Blue
2012	[0001]	−0.0156	*IV*	Blue
2013	[0001]	−0.0004	*IV*	Red
2014	[1000]	−1.0000	*II*	Red
2015	[1000]	−0.0293	*II*	Orange

With the results of Tables 4 and 5, it can be seen that in the selected 16-year sample data, of which 29.4% are at risk, 17.6% of the samples are at higher risk, and 23.5% of the samples are less risky and minimally risky.

6 Conclusion

The level of risk decision model of high precision, good performance, can realize the dynamic tracking of science and technology service of supply chain risk and comparative analysis, not only provide for determining risk level for the whole of science and technology service industry, but also can help decision-makers in a timely manner, and take effective measures to prevent and improve the related enterprise ability to cope with risks.

Acknowledgments. This work is supported by the Humanity and Social Science Youth Foundation of Ministry of Education of China (Grant No. 19YJCZH054), the Science and Technology Research program of Chongqing Municipal Education Commission (Grant No. KJQN201801145) and the Chongqing Social Science Planning Project (Grant No. 2015YBSH051).

References

1. Yi-zhou, E., Yu-yang, Q.I.A.O., Zi-chao, L.I.U.: A study on the evolution and risk identification of internet financial models. East Chin. Econ. Manage. **30**(3), 91–96 (2016)
2. Liu, H.-H.: Study on supply risk identification and assessment of the critical power batteries in NEVs. Sci. Technol. Manage. Res. **36**(24), 189–195 (2016)
3. Guang, Z.H.U., Min-ning, F.E.N.G., Shuo, L.I.U.: Research on security risk identification and countermeasures of big data flow-based on the perspective of information life cycle. Res. Libr. Sci. **9**, 84–90 (2017)
4. Zhou, Z., HU, C.-P.: Research on the risk identification of academic information system based on the comprehensive weighting method. Inf. Sci. **8**, 159–163 (2017)
5. Wang, Z.-H., Liu, Z., ZHU, D.: Safety risk identification and prevention path of the production link in pig supply chain. Chin. Popul. Res. Environ. **27**(12), 174–182 (2017)
6. Wang, H.H., Huang, X., Chen, R.: Research on risk recognition of supply chain for large aircraft collaborative development based on SCOR model. J. Jinling Inst. Technol. (2015)

7. Zou, X.-H., Tian, L.-X.: Comprehensive evaluation of CCS project based on improved integrated cloud model. Stat. Decis. Making **3**, 182–185 (2016)
8. Rodríguez, A., Ortega, F., Concepción, R.: A method for the evaluation of risk in IT projects. Expert Syst. Appl. **45**, 273–285 (2016)
9. Wu, Y., Li, L., Xu, R., et al.: Risk assessment in straw-based power generation public-private partnership projects in China: a fuzzy synthetic evaluation analysis. J. Cleaner Prod. **161**, 977–990 (2017)
10. Jian, L.-L.: Research on risk assessment of logistics strategic alliance based on BP neural network. Sci. Technol. Manage. Res. **24**, 165–167 (2011)

Evaluation of the New Engineering Course System for Cyberspace Security

Xinxin Lin[1], Wei Wang[2(✉)], and Kainan Wang[1]

[1] Cyberspace Security, Changchun University, Changchun 130022, Jilin, China
[2] School of Computer Science and Technology, Changchun University, Changchun 130022, Jilin, China
20017008@qq.com

Abstract. Combined with the construction of the new cyberspace security engineering, in order to solve the problem of the blindness of the curriculum setting in the current curriculum system construction of new cyberspace security engineering, lack of quantitative analysis of course output evaluation and low utilization of traditional industry-academia collaboration resources, based on linear model calculation method in machine learning and use the matplotlib plot in Python, propose an evaluation model that can quantitatively analyze the rationality of curriculum setting and curriculum output, use this method to evaluate the opinions of different numbers of students, companies and experts in the course of "Internet Culture Security", the experimental results show that the algorithm accuracy and reference value of SCP-based curriculum evaluation model are high, the specificity, was eliminated as far as possible to ensure the accuracy rate, which reached more than 93.5%. The evaluation model based on the SCP course setting algorithm and the course output evaluation model can solve the problems of blindness in the course setting of the new engineering course of cyberspace security, effectively reduce the errors in human experience and improve the quality of cyberspace security new engineering construction.

Keywords: Cyberspace security · Evaluation of curriculum system · Evaluation of course output

1 Introduction

With the rapid development of information technology, the content of the Internet-made cyberspace is expanding, according to the new ideas of the construction of "new engineering subjects" pointed out by the Ministry of Education in 2017, combined with the country's new talent needs for cyberspace security, puts forward a new engineering direction of integrating cyberspace technology security and cyberspace culture security, pay attention to the cultivation of comprehensive skills of talents, realize the new professional construction mode of mutual construction of science and engineering and co-construction of arts and sciences, and solve the absolute fault of cyberspace technology security and cyberspace culture security. Many scholars at home and abroad

J. MacIntyre et al. (Eds.): SPIoT 2020, AISC 1283, pp. 301–308, 2021.
https://doi.org/10.1007/978-3-030-62746-1_44

have carried out research and achieved corresponding results for the construction of the curriculum system of emerging disciplines. The all methods have studied the construction of the new engineering course system for cyberspace security from different directions, but both lack a quantitative standard, in order to better realize the integration of technological security and cultural security in cyberspace, the construction of new engineering disciplines, this paper through studying the calculation methods of linear models in machine learning, proposed a model of curriculum system evaluation based on SCP and a model of curriculum output evaluation based on OBE, the combination of the two makes up for the blindness and lack of quantitative indicators of traditional curriculum system construction, teaching students according to their aptitude, building a more reasonable curriculum system.

2   Background and Development of Cyberspace Security

The term cyberspace first appeared in the novels of American science fiction writer William Gibson, Originally known as the "Cyberspace" [1]. US Presidential Decree No. 54 in 2008 officially defines it as cyberspace. With the rapid development of the Internet, the Internet provides new ways for people to communicate, the main body of the network performs various operations through various carriers carrying the Internet to complete the exchange of information resources, cyberspace is a space composed of information constructed by humans using computers and communication equipment. This includes all information transmitted, processed, and stored on the Internet, the Internet of Things, and the network [2]. Academician Fang Binxing defined cyberspace as "an artificial space built on ICT infrastructure to support people's various ICT-related activities in that space." [3] This definition does not explicitly point out the cultural background of information communication in cyberspace, as well as ideological changes and cultural content changes that occur during communication, can increase the content of ideological awareness and ideology. The cyberspace system can be composed of five basic elements: subject, carrier, information resource, operation [3] and culture, take culture as the background and integrate into each basic element.

At present, cyberspace has gradually developed into the fifth largest strategic space, and cyberspace security issues have become more prominent. In 2003, the US government issued the "National Strategy for the Protection of Cyberspace Security" to explain the concept of cyberspace security to the world [4]. Cyberspace security was defined in the "Enhanced Critical Infrastructure Cyberspace Security Framework" issued by the National Institute of Standards and Technology in 2014, that is, "the process of preventing, detecting and responding to attacks and protecting information [5]. Academician Fang Binxing raised the security problems in the application of cyber hardware security to electronic hardware equipment, information systems, and operating data systems at the World Internet Conference 2015, corresponding to equipment, system, data, and application [6]. At the same time, in the face of the current cyber environment, cyberspace security content should pay attention to ideological security in a digital society and cultural content security caused by information collisions in cyberspace.

3   Problems in Traditional Engineering Construction

At present, the traditional engineering construction of colleges and universities has achieved remarkable performance in historical development [7]. However, facing the current social development demand for multi-integrated knowledge system and compound talents, there are some problems in traditional engineering construction:

The traditional teaching model is centered on teachers, students are in a passive position, and teaching activities are mainly based on teachers, forming the weird phenomenon of "teachers in charge and students filling pits", students lack the ability to think independently [8]; The teaching method is mainly based on theory teaching, with few practice links and lack of real cases, deviating from actual combat needs; In the traditional engineering construction, there is a lack of reference standards for courses of various disciplines, teachers and students have an assessment mechanism. However, the course itself lacks evaluation standards, and the course system is not flexible enough; The structure of the teaching team is unreasonable, the knowledge structure between teachers is not complementary, and the knowledge system is single; The traditional engineering construction focuses on technical research and ignores the construction of students' humanistic knowledge system; The evaluation method of the course output is too single, and the reference value of the ability test is not high.

This article proposes a SCP-based curriculum system construction reference model and a OBE-based many-to-many curriculum output evaluation model for the lack of standards for curriculum evaluation, details as follows.

4   Curriculum and Capability Evaluation of the New Engineering Course of Cyberspace Security

4.1   SCP-Based Curriculum System Construction Reference Model

At present, the cyberspace security courses of ordinary universities are mainly based on traditional cybersecurity theories and technologies, lack of popularization of cultural content security in cyberspace security [9]. Cyberspace security itself needs to be supported by knowledge of various disciplines, to build the cultural characteristics of cyberspace, in addition to the traditional cyber security theories and technologies such as attack and defense penetration and information security, the courses involved also need to learn the current popular big data security, Internet of Things security, public opinion analysis, data analysis, and culture, sociology, Communication and other related knowledge, cultivate students' awareness of network culture security, improve cyber cultural literacy, clarify cyber space security related laws and regulations, and gain an in-depth understanding of the development status, trends and research frontiers of network technology, network culture, network society, and network communication [10].

In the construction process of the cyberspace security new engineering course system, multiple factors should be considered during the construction process, including student opinions, social enterprise opinions, and expert opinions. For each course, whether it needs to be established, SCP-based (Student, Company, Professor) can be constructed three Reference model for the evaluation of the course system of the flag value.

In this model, the course evaluation is divided into three parts: unsatisfactory, basic satisfaction and very satisfaction. Second, set the standard mark values of course opening 1–5, which respectively indicate that the course must be cancelled, recommended to cancel the course, recommended to be set as an elective course, recommended to be set as a required course, and required courses. Count the number of students, companies, and experts participating in the evaluation of the course, mark the sum as n, and then count the number of different participants who are not satisfied, basically satisfied, and very satisfied with the course, labeled a, b, c, Refer to Table 1 to record the flag values of different types of participation.

All kinds of marker values calculated by the above method are calculated with reference to Table 2 to get the final marker value marked as p. Refer to Table 3 for the value of p to get the evaluation standard for this course. The reference model's course evaluation cycle is 1 year, and the course evaluation value is recorded. When it is recommended to cancel for 3 consecutive years, its flag value is set to 1.

The reference model provides a numerical standard for new courses and problems in traditional courses, which is helpful for the evaluation of new courses and the update of the curriculum system. According to the reference model, write a program with students as participants to evaluate the opinions of "Internet Culture Security" course for different numbers of students, companies and experts, the results are shown in Fig. 1.

4.2 OBE-Based Many-to-Many Curriculum Output Evaluation Model

The training of talents is the result of curriculum output. The benefit of curriculum output needs to be measured by a quantitative standard. This paper proposes a reference model for evaluating the effectiveness of curriculum output.

The purpose of a course is to develop one or more abilities of students, at the same time, the cultivation of a student's ability is the result of the role of multiple courses. Quantitatively assign 100% of the student's ability to cultivate one or more abilities in the course of the course, record the weighted sum of the assessment scores of all courses supporting a student's ability as SA, the average number of comprehensive SAs for all students in this ability is recorded as MA, record the student's achievement in a single ability as DA, to prevent students from scoring too much in different percentage courses, when SA/MA $>=$ 1.2, DA is marked as high, when SA/MA $>=$ 0.8 and SA/MA $<$ 1.2, DA is marked as qualified, and when SA/MA $<$ 0.8, DA is marked as unqualified.

Add the sum of students' abilities as SB, the average of the average of all abilities of all students is recorded as MB, record the student's total ability achievement as DB, in order to prevent students from becoming too diverse in their abilities, DB is marked high when SB/MB $>=$ 1.2, DB is qualified when SB/MB $>=$ 0.8 and SB/MB $<$ 1.2, and DB is unqualified when SB/MB $<$ 0.8. The output evaluation period of the course is one year, adjust the curriculum system and training plan in time for students' ability changes every year, make the new engineering course system more reasonable.

The specific cases of applying this model are shown in Table 4. The lowercase letters represent the assessment results of courses offered, and different numbers represent the benefits generated by the courses corresponding to different training abilities.

In this case, SA/MA was used as the calculation method, and the redundancy of 0.2 was used as the adjustment, as shown in Fig. 2.

Table 1. Flag value calculation reference table

Type rate	a/n == 1	0.8 <= a/n < 1	0.6 <= a/n < 0.8		0.4 <= a/n < 0.6		a/n < 0.4 and c/n < 0.6		0.6 <= c/n < 0.8		c/n >= 0.8	c/n == 1
–	–	–	c/a <= 0.2	c/a > 0.2	c/a >= 0.7	c/a < 0.7	c/(b+c) >=0.7	c/(b+c) < 0.7	a/c <= 0.2	a/c > 0.2	–	–
Student	1	2	2	3	4	3	4	3	4	3	4	5
Company	1	2	2	3	4	3	4	3	4	3	4	5
Professor	1	2	2	3	4	3	4	3	4	3	4	5

Table 2. Final flag value reference table

Final flag value (p)	$0.1q + 0.5w + 0.4e$

Table 3. Mark reference table

p	Course description
1	Must be cancelled
(1–2]	Recommended to cancel the course
(2–3]	Recommended to be set as an elective course
p	Course description
(3–4]	Recommended to be set as a required course
(4–5]	Required courses

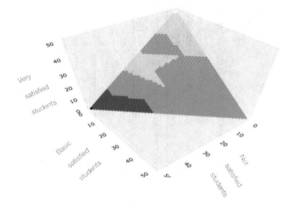

Fig. 1. Student flag value change chart

Table 4. Case reference table

	1	2	3	4	5
a	20%	0	0	30%	50%
b	0	20%	30%	10%	40%
c	0	40%	0	0	60%
d	30%	0	30%	40%	0
Sum	$0.2a + 0.3d$	$0.2b + 0.4c$	$0.3b + 0.3d$	$0.3a + 0.1b + 0.4d$	$0.5a + 0.4b + 0.6c$

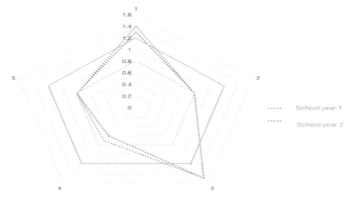

Fig. 2. Student ability index chart

5 Conclusion

The SCP-based curriculum system construction evaluation model combines the opinions of students, companies and experts to give quantitative standards using the linear model calculation method in machine learning, it helps to solve the blindness of the curriculum system construction in the construction of new engineering disciplines and also provides a reference for the adjustment of the curriculum system. The obe-based curriculum output measurement model quantifies curriculum ability development in student tests, convenient to update the training plan in time, effectively solve the problem that the curriculum does not meet the needs of students' ability development. Of these two calculation methods, since it is first proposed, only a simple linear calculation method of weight ratios is used to judge by threshold comparison. The accuracy needs to be improved.

Acknowledgments. This research was supported by teaching Reform of Higher Education in Jilin Province (SJXGK17-04), higher Education Research Project of Jilin Province Higher Education Society (JGJX2015C65) and Project of Vocational Education in Jilin Province (2019ZCY375).

References

1. Huaqin, L.: Cyber space anthropology: Interaction between cyberspace and anthropology. J. Guangxi Univ. Nationalities (Philos. Soc. Sci. Ed.) **02**, 64–68 (2004). (in Chinese)
2. Zheng, G., Yamin, W.: Exploration of practical teaching of cyberspace security personnel training. Comput. Educ. **02**, 113–117 (2016). (in Chinese)
3. Binxing, F.: Defining cyberspace security. J. Netw. Inf. Secur. **4**(01), 1–5 (2018). (in Chinese)
4. Xu, J.: Retrospection and reflection on the construction process of cyberspace security in China. Cyberspace Secur. **7**(08), 4–6 + 23 (2016). (in Chinese)
5. Junzhou, L., Ming, Y., Ling Zhen, W., Xiaodan, W.G.: Cyberspace security system and key technologies. Sci. China Inf. Sci. **46**(08), 939–968 (2016). (in Chinese)
6. Tang Jundong, Yin Hongju, Wang Shuang. An Analysis of Network Culture Security in Colleges and Universities [J]. New West (Last. Theory), 2011 (04): 162 + 207. (in Chinese)

7. Yongjuan, W., Qingjun, Y.: The status quo, characteristics and enlightenment of the training of security personnel in the cyberspace of the world. Wangxin Mil. Civilian Integr. **10**, 51–55 (2019). (in Chinese)
8. Ye, Q.: Research on Self-management of College Students Based on the Cultivation of Innovative Talents. Nanjing University of Aeronautics and Astronautics (2009). (in Chinese)
9. Yan, Z., Guangling, S.: Research on the wisdom classroom teaching mode under the background of new engineering disciplines. Softw. Guide **18**(3), 206–209 (2019). (in Chinese)
10. Huijie, S., Minghua, W.: Research on computer applied talents training. Comput. Educ. **13**, 28–31 (2011). (in Chinese)

Exploration on Teaching Reform of Architectural Design Courses from the Perspective of MOOCs

Qian Yu[✉]

School of Architecture and Engineering, Qing Dao BinHai University, Qingdao 266555, China
448022988@qq.com

Abstract. Scientific and technological innovation has set off a "quality revolution" in China's higher education and has continued to develop in depth. MOOC is a curriculum model based on Internet teaching. It has the characteristics of large-scale, openness, and no time and space limitation. It is very feasible and necessary to introduce the MOOC perspective into the teaching reform of architectural design courses, and it can improve the teaching quality of architectural design courses as a whole, stimulate students' learning interest, and better carry out the architectural reform of teaching reform. The article mainly studies the teaching reform of architectural design courses based on MOOCs, with a view to providing useful reference for teaching reform.

Keywords: MOOC · Teaching reform · Architectural design courses

1 Introduction

Times are developing, technology is advancing, and educational technology changes have had a positive impact on classroom quality revolutions and subtly changed. Mu classes and flipped classrooms have become new terms in the era of educational development. It played a vital role in eliminating the "water course" and the "golden course". MOOC is a teaching method based on the Internet platform using modern educational technology to carry out teaching. Compared to the traditional teaching mode, MOOC has unparalleled advantages. Today, the rapid development of artificial intelligence can better meet the needs of architecture. The teaching needs of design courses and society's demand for talents in architectural design.

2 Conceptual Interpretation and Characteristics Analysis of Moocs

2.1 Concept of MOOCs

In short, MOOC is a course model. It is a large-scale open online course. MOOC does not limit the number of participants and is open to all online users. The "M" in MOOC

J. MacIntyre et al. (Eds.): SPIoT 2020, AISC 1283, pp. 309–314, 2021.
https://doi.org/10.1007/978-3-030-62746-1_45

stands for Massive (large scale). Compared with the traditional curriculum model, there are only tens or hundreds of students [1]. Large-scale students cannot be divided by specific numbers. The two letters "o" stand for open, as long as you are interested in what you want to learn, you can learn through Internet access. The registration and learning process for learners is completely free of charge. After completing the course study, learners can obtain the certificate of competence organized by the MOOC platform by completing assignments and online evaluation. The third letter "O" stands for Oline (online), which completes learning on the network, breaking the limits of time and space. The fourth letter "C" means Course is the course. Course is the core of MOOC. Course indicated by Course is not only the courseware materials shared through the Internet, a single course design, but also the online and offline process from course design to the end of the course.

2.2 Characteristics of MOOCs

Compared to traditional online courses, MOOC not only provides learning video course-ware resources, text materials, and online interactive services, but also provides an inter-active community where students can discuss MOOC learning content with each other. This resource course based on large-scale student learning, mutual communication and participation, and access to Internet-based learning resources has attracted the attention of many students, allowing more people to devote themselves to MOOC learning [2]. The teaching advantages of Mu classes are mainly reflected in the following aspects: First, large-scale. The large-scale Mu classes are mainly reflected in four aspects: student groups, teaching platforms, teacher teams, and course size. For the student group, the number of educated groups in MOOC is huge. Not only are many free learners studying through MOOC, but also a large number of universities participate in MOOC teaching. Abundant online courses are also a large-scale manifestation of MOOC. According to incomplete statistics, the current MOOC courses can provide science, law, and com-puting Machine science and many other fields. Second, openness. The openness of the MOOC teaching mode is reflected in the openness of the educated group.

This idea is truly incorporated into teaching. The openness of MOOC also repre-sents high-quality teaching content. Teachers upload the recorded teaching videos to the MOOC platform. Students can independently motivate teachers to improve the quality of teaching videos by visiting MOOC's website for learning. Third, there is no time and space limitation. No time and space limitation is one of the important characteristics of the MOOC curriculum, which means that learners can break through the constraints of time and space to learn according to their preferences and interests, and obtain learning feedback in a timely manner.

3   Teaching Characteristics of Architectural Design Courses

3.1   Nature of Architectural Design Courses

Architectural design course is one of the most crucial courses in the study of archi-tecture. The key is not only reflected in the importance of the course, but also in the expansion space of the architectural design course [3]. Architectural design courses cover a wide range of topics, including building history, building theory, building components, building mechanics, and building energy efficiency. Therefore, to learn a good architectural design course is not only to design good. The architecture of the show should integrate other subjects of the architecture specialty. Architectural design course contains strict logical thinking and accurate drawing process. It is a kind of engineering design. One of the main tasks of the architectural design course is to cultivate the professional quality and rigorous quality of students in the process of engineering architectural design. Architectural design courses are the basic theory and technical training courses of architectural design courses, which mainly include introduction rendering, architectural modeling foundation, and architectural painting performance techniques. Architectural design courses require students to fully grasp the drawing methods of architectural drawings, students to master the traditional rendering methods, etc., to be familiar with basic architectural design modeling principles, and to pay attention to the three-dimensional and spatial composition of buildings. Master the basic knowledge of building and basic design theory [4].

3.2 Feasibility Exploration of Introducing MOOC into the Teaching Reform of Architectural Design Courses

It is very feasible and necessary to introduce MOOC into the teaching reform of architectural design courses. It is mainly manifested in the following points: First, the teaching of architectural design courses through MOOCs is viewed through videos. Computers and smart phones are now widely used on campus, and almost every student has their own computer and smart phones. Computers and smart phones have become terminal devices for students to study MOOCs. Students can use computers and computers to study architectural design courses anytime, anywhere. Secondly, MOOC is teaching by making videos. Teachers can record videos in a simple way. Teachers can record different teaching content according to teaching tasks and teaching goals. Teachers can use the Internet to upload recorded MOOC teaching content to MOOC teaching platform at any time.

4 Teaching Reform and Practice of Architectural Design Courses for MOOCs

4.1 Theoretical Teaching Reform of Architectural Design Courses Based on MOOCs

The teaching reform of architectural design courses based on MOOCs is not carried out completely according to MOOC teaching styles. Instead, it combines the nature and teaching characteristics of architectural design curriculums to give full play to the advantages of MOOC teaching platform and micro-curricular teaching. Design courses incorporate MOOC instruction. Through the MOOC teaching link, focus on a certain teaching focus of the architectural design course, pointing in detail to specific problems, and exploring at various levels in order to solve problems encountered in classroom teaching reform. Architectural design courses based on MOOCs can not only play the leading role of teachers, but also further propose the subjective initiative of students in the learning process. The teaching reform practice of architectural design courses based on MOOCs can be carried out from the following aspects: First, under the overall grasp of the teaching goals and teaching content, the teacher analyzes the overall teaching goal framework and details the overall teaching goals It is divided into different small teaching goals, and the teaching content of architectural design courses is subdivided into learning tasks that are easier to master, and the overall learning goals are achieved by guiding students to continuously complete the tasks. Secondly, the design of MOOC teaching tasks should pay attention to teaching priorities and difficulties. In the design process of MOOC teaching tasks, it is necessary to consider in detail the difficulty of MOOC teaching tasks, the large number of teachings, and the connection between back and forth, and try to control the duration of MOOC teaching videos to about 10–15 min. Third, in the process of carrying out MOOC teaching, it is necessary to fully analyze the learning situation of students, Taking into account the student's learning age, cognitive characteristics, learning needs, learning status, etc. On this basis, the multimedia form is combined with the MOOC teaching unit to carry out teaching by combining online and offline teaching methods. Fifth, the development of MOOC teaching should take into account the ratio of Internet learning to offline class hours, the platform for students' independent learning and collaborative learning, and the combination with other courses.

4.2 Practical Teaching Reform of Architectural Design Courses Based on MOOCs

Architectural design courses are highly practical courses. Students should master the practical ability to analyze and solve design problems through the study of architectural practice courses. Therefore, the teaching of architectural design courses needs to strengthen practical teaching. Only in this way can we cultivate the talents of architectural design applied to meet the needs of society. Architectural design course teaching needs to pay more attention to the integrity of the teaching process, focus on pre-design investigation and training, and cultivate students' creative thinking. Based on the architectural design practice teaching under the MOOCs, teachers can teach by recording videos of key software operations in architectural design practice teaching. In the teaching of architectural design courses, the use of architectural software is very critical and will directly affect the student's architectural design process. The teaching of architectural design practice courses needs to be carried out in the computer room, and it is difficult for students to independently study after class. The teaching of architectural design practice courses based on the MOOC can be taught by recording small videos using design software. Students can independently study the architectural design software in the dormitory after school and in the computer study room. In addition, the practical teaching of architectural design courses based on MOOC can also be realized through the assignment of architectural design courses. Teachers can arrange assignments of architectural design courses to be done by combining individual and group work. Architectural design practice is a process that requires groups to complete cooperation together. In the teaching of architectural design courses, teachers should guide students to give play to their unique personalities and abilities, while forming a good style of group collaboration.

After arranging the assignment of architectural design assignments, the teacher will give detailed explanations, and may ask the students to collect relevant data by themselves, and do the preliminary investigation of the design. Specific architectural design under the guidance of MOOC videos. In group cooperation, the design emphasis of each student is different. Some students collect relevant knowledge about the principles of architectural design and design specifications, and some students collect related design works through the Internet. Students make resource contributions in the case of group cooperation, and finally upload the design results through the MOOC platform, allowing teachers to comment on the design results. In addition, teachers can also display the architectural design results of each group on the MOOC platform, so that each group can learn from each other and communicate with each other, mobilizing the enthusiasm of the students in architectural design.

4.3 Teaching Achievement Reform of Architectural Design Courses Based on MOOCs

From the traditional instilled teaching method of simple blackboard and chalk teaching to a short and intensive MOOC teaching method that can break through the time and space constraints, students' interest in architectural design courses has also been greatly stimulated. Students can study at any time and any place through the MOOC learning platform. Teachers are not only the imparters of knowledge, but have become the leaders of architectural design courses. They have always centered on the learning needs of students, referring to teaching tasks and teaching goals, and making MOOC instructional videos for students to learn independently after class. MOOC has brought a new look to the teaching reform of architectural design courses, expanded the teaching content of architectural design courses, extended the time and space of architectural design courses, and overturned the traditional teaching mode for schools and classroom definition. The effects of introducing MOOCs in the teaching of architectural design courses are summarized as follows: First, the learning efficiency of students has been significantly improved. Students can take up MOOC food after class to learn without taking up class time. Second, it helps to increase students' motivation to learn. Students can independently complete learning tasks, independently discover problems, think about problems, and solve problems by watching videos after class. In the subsequent learning process, I bring my own problems in the MOOC learning process to study, making the architectural design courses more purposeful. Thirdly, watching MOOC videos is very convenient, which is very helpful for students to review after class. When students have forgotten some knowledge points in the architectural design course, they can use computers or mobilephones to watch at any time, without the need for teachers to demonstrate on the spot again, which significantly improves the practical teaching effect of architectural design courses [5].

5   Research Summary

Generally speaking, for the teaching of architectural design courses, MOOCs is a new teaching method that can solve the current teaching problems of architectural design courses. Through MOOC teaching, the learning efficiency of learners can be significantly improved, so that students are more interested in the study of architectural design courses. They can take the initiative to carry out learning through MOOC teaching videos after class to improve the learning effect. It is believed that MOOC will become one of the important teaching methods in the course of teaching development of architectural design courses, which will help the architectural design courses and the teaching of architecture majors [6].

References

1. Hehai, L., Shuyu, Z., Lilan, Z.: Discussion of essence, connotation and value of MOOCs. Mod. Educ. Technol. **12**(24), 5–11 (2014)
2. How Jeremy Knox: MOOC revolution progressed: three changes in MOOC theme. Chin. Distance Educ. **2**, 53–62 (2018)
3. Guijun, H.: Exploration on the integration of MOOC into architectural courses. Chin. Distance Educ. **3**(36), 68–70 (2018)
4. Yang, J., Wang, Y.: An study of the type of interorganizational defamation risk based on grounded theory——an analysis from the Chinese scenario. Int. J. Front. Sociol. **1**(1), 75–88 (2019)
5. Hualei, J.: Research on the efficiency measurement of Chinese cultural manufacturing industry based on DEA method. Int. J. Front. Sociol. **1**(1), 32–42 (2019)
6. Li, J., Zhao, Y., Huang, L., Yang, Q.: Construction and exploration of innovation and entrepreneurship platform for urban underground space engineering students based on DCLOUD technology. Int. J. Front. Eng. Technol. **1**(1), 48–55 (2019)

Large-Scale Farmland Management on Improving Production Efficiency Under the Background of Smart Agriculture-Based on Panel Quantile Regression

Bangzheng Wu[✉]

School of Economics, Shanghai University, Shanghai, China
wubangzheng1995@163.com

Abstract. With the rise of big data and intelligent manufacturing in recent years, agricultural production in the traditional way of operation has been linked with wisdom gradually. Especially after the rise of the rural revitalization strategy, the key to achieving rural revitalization lies in solving the problems of agriculture, rural areas and farmers. It is inseparable from the intelligent management model and advanced technology to improve the efficiency of agricultural production. It has changed the agricultural production by establishing an intelligent management platform, which can accelerate land transfer and form an intensive agricultural land management model. This paper uses Chinese national panel data from 2002 to 2017 combining quantile regression to examine whether farmland scale can improve production efficiency in the background of smart agriculture. Finally, we will propose relevant policy recommendations based on empirical test results.

Keywords: Smart agriculture · Large-Scale farmland management · Production efficiency · Panel quantile regression

1 Introduction

Traditional agricultural production is based on input of labor and land, which is inefficient and difficult to adapt to increasingly complex changes in the natural environment and social development [1]. The advanced stage of agricultural production is smart agriculture. Smart agriculture is an important part of the smart economy and a concrete manifestation of its form in agriculture. It is conducive to eradicating poverty, achieving latecomer advantages [2], and catching up strategies in developing countries. The application of new information technology achievements in smart agriculture as the use of internet technology. Smart agriculture enables the efficient circulation of rural land management rights. It also can form a large-scale business model and improve production efficiency [3, 4].

J. MacIntyre et al. (Eds.): SPIoT 2020, AISC 1283, pp. 315–320, 2021.
https://doi.org/10.1007/978-3-030-62746-1_46

2 Data

In order to explore the relationship between the scale of farmland management and agricultural production efficiency in the background of smart agriculture. This paper establishes the following econometric models:

$$Production_{it} = \beta_0 + \beta_1 Scale_{it} + \theta X_{it} + \mu_{it} \tag{1}$$

The specific variable setting and calculation methods are shown in Table 1:

Table 1. Variable setting and calculation method

	Variable	Calculation method
Dependent independent	Production	LN (Primary industry per capita output value)
	Scale	LN (Sown area of per capita agricultural products)
	Labor	LN (Employment in the primary industry)
	Eco	LN (Per capita income of rural residents)
	Fertilize	LN (Fertilize usage)
	Power	LN (Total mechanical power)
	Film	LN (Film usage)
	Irr	LN (Effective irrigation area/Total sown area)
	MCI	LN (Total sown area of crops/cultivated area)
	Disaster	Crop affected area/Total sown area

3 Model Construction and Empirical Result

Quantile regression is an extension of the least squares method based on the classical conditional mean model [5, 6]. For a continuous random variable y, if the probability of $y \leq Q(\tau)$ is τ, then we set the τ quantile value of y as $Q(\tau)$, or the τth quantile of y. Let the distribution function of the random variable Y be $F(y) = P(Y \leq y))$, then the τth quantile of Y is $Q(\tau) = \inf\{y : F(y) \geq \tau\}(0 < \tau < 1)$.

For general models:

$$y_i = x_i'\beta_\tau + \alpha_{\tau i} + u_{\tau i} \ (i = 1, 2 \ldots \ldots) \tag{2}$$

Among (2), y_i is the independent variable, x_i is the row vector of k * 1, β_τ is the row vector of k*1, it represents the regression coefficient of each independent variable corresponding to the τth quantile of the explained variable, and $\alpha_{\tau i}$ is the intercept term, $u_{\tau i}$ is a random error term. Now suppose that the conditional quantile of Y is represented by a matrix of k independent variables.

$$Q(\tau|x_i, \beta(\tau)) = x_i'\beta(\tau) \tag{3}$$

In the formula (3), $x_i = (x_{1i}, x_{2i}, \ldots\ldots, x_{ki})'$ is an independent variable vector, and $\beta(\tau) = (\beta_1, \beta_2, \ldots\ldots, \beta_k)'$ is a coefficient vector in the τ quantile.

$$Q(\tau) = argmin_\varepsilon \left\{ \sum_{i, y_i \geq \varepsilon} \tau|y_i - \varepsilon| + \sum_{i, y_i \geq \varepsilon} (1 - \tau)|y_i - \varepsilon| \right\}$$

$$= argmin_\varepsilon \left\{ \sum_i \rho_\tau (y_i - \varepsilon) \right\} \tag{4}$$

When τ fluctuates at (0, 1), solving the following minimization problem can get different parameter estimates for quantile regression:

$$\beta_N(\tau) = \arg min_{\beta(\tau)} \left(\sum_{i=1}^N \rho_\tau \left(\rho_\tau - x_i'\beta(\tau) \right) \right) \tag{5}$$

For the panel data model:

$$y_{it}' = x_{it}'\beta_i + \alpha_i + u_{it} \ (i = 1, 2, \ldots, N; \ t = i = 1, 2, \ldots, T) \tag{6}$$

i represents different sample individuals, t represents different sample observation points, u represents a random error vector, β_i represents a coefficient vector of independent variables, and α_i represents an unobservable random effect vector of different samples.

Panel data quantile regression parameter estimation general linear conditional quantile equation:

$$Q_{y_{it}} \left(\tau_j | x_{it}, \alpha_i \right) = x_{it}'\beta \left(\tau_j \right) + \alpha_i \tag{7}$$

When τ fluctuates at (0, 1), solving the weighted absolute residual minimization problem can obtain the parameter estimates of the quantile regression at different quantile points. Among (7), the parameter β can be solved by the following formula:

$$\hat{\beta} = argmin_{\alpha, \beta} \sum_{j=1}^J \sum_{t=1}^{TT} \sum_{i=1}^N \rho_{\tau_j} \left(y_{it} - x_{it}'\beta \left(\tau_j \right) - \alpha_i \right) \tag{8}$$

After the preliminary quantitative test of the model, this paper uses quantile regression to estimate the impact of the concentration of farmland management scale on different agricultural labor production efficiency, and uses a fixed effect model as compared. The empirical results are shown in Table 2.

Different quantiles in the model represent agricultural production efficiency at different levels of development. The larger the quantile, the higher the level of agricultural production efficiency it represents. The distribution of the per capita output value of the primary industry from small to large is regarded as the development of agriculture. From the low level to the high level, we can observe the dynamic changes of the corresponding influencing factors as the agricultural production efficiency changes from low to high.

According to the quantile regression results in Table 2, the elasticity coefficient of the scale of land management at the other quantile points are all positive except for the 10% quantile point. They are -1.1626, 0.0226, 0.1328, 0.0453, 0.0414. The elasticity coefficient of the scale of land management has gone from negative to positive, gradually increased, and slowly decreased after reaching the maximum at the 50% quantile. On the whole, with the gradual improvement of agricultural production efficiency, the scale of

Table 2. Quantile regression results of various factors on agricultural production

Variable	q(0.1)	q(0.25)	q(0.5)	q(0.75)	q(0.9)	Fixed effects
Scale	−0.1626***	0.0226***	0.1328***	0.0453**	0.0414***	0.3738***
	(−9.02)	(3.17)	(24.34)	(2.55)	(4.10)	(3.95)
labor	−0.6435***	−0.4504***	−0.5148***	−0.5556***	−0.6283***	−0.6909***
	(−31.62)	(−32.96)	(−34.52)	(−36.67)	(−26.47)	(−6.26)
Eco	0.2186***	0.4645***	0.4395***	0.4393***	0.3732***	0.2611***
	(5.13)	(29.02)	(68.06)	(55.39)	(23.61)	(11.20)
Fertilize	0.3484***	0.4243***	0.4624***	0.3652***	0.4186***	0.2670***
	(11.46)	(75.24)	(47.46)	(23.06)	(19.29)	(4.05)
Power	0.0277	−0.1293***	−0.0413***	−0.0715***	−0.0723***	0.1678***
	(0.84)	(−10.72)	(−6.92)	(−23.31)	(−8.10)	(4.62)
Film	0.1514***	0.0988***	0.0040	0.1502***	0.1737***	0.0899***
	(23.57)	(26.91)	(1.39)	(87.46)	(13.83)	(4.29)
Irr	0.0877*	0.2700***	0.2801***	0.1105***	0.0595***	−0.1194
	(1.90)	(26.93)	(16.23)	(25.01)	(3.65)	(−1.13)
Mci	0.5253***	−0.0081	−0.0998***	−0.0396***	−0.073	0.0610
	(4.91)	(−0.61)	(−3.69)	(−4.31)	(−1.15)	(0.92)
Disaster	−0.0186***	−0.0110*	−0.0193***	0.0022*	−0.0725***	−0.1180***
	(−3.47)	(−2.49)	(−3.03)	(0.57)	(−3.66)	(−2.31)

Note: 1) *, **, *** represent the significance level of 10%, 5%, 1%, respectively;

rural land management will have a positive impact on agricultural production efficiency. This shows that with the improvement of the level of agricultural development, smart agriculture has been continuously deepened and applied. The effect of the scale of land production on the scale of production has gradually emerged. The higher the scale, the smarter the agriculture it is, the more conducive to mechanized investment and professional management, and it is more conducive to reducing production costs per unit area and expanding output. Then it will improve agricultural production efficiency at last.

About the results of other variables, the elastic coefficient of rural labor input is always significantly negative at all quantiles, indicating that the increase in labor input will have a negative impact on agricultural output value. Investment should focus on improving the quality of agricultural employees themselves, and improve the comprehensive conditions of agricultural production through multiple channels; The elasticity coefficient of the rural economic development level is positive at all quantiles. The improvement of farmers' living standards has promoted agricultural production and has obvious effects. At the 50% quantile in Table 2, the elasticity coefficient of the rural economic development level is 0.4395, which means that at this level of agricultural production efficiency, for every 10% increase in the level of rural economic development, the level of agricultural production will increase by 4.4%. The level of rural economic development is an important indicator of the degree of regional economic prosperity and infrastructure investment. This shows that the improvement of farmers' income level and the improvement of infrastructure will significantly promote agricultural productivity; Capital and technology inputs represented by fertilizers, mechanical power, and films, they are all 5% significant positive in the fixed effect model, and the elasticity coefficients are divided into 0.2670, 0.1678, and 0.0899, which can all play a positive role in agricultural production. It indicates that the improvement of agricultural production efficiency at this stage will increasingly depend on smart technology and capital on the structure of agricultural production factors. The recultivation index is not significant under some quantile results, and it is also not significant in the fixed effect model. It may be due to the increase in the cost of agricultural production and the attraction of urban non-agricultural employment to migrant workers, which has reduced the annual crops. Finally, the disaster rate is significantly negative in both the quantile model and the fixed effect model, that means the loss of natural disasters in an area will always have an adverse effect on agricultural production efficiency, so we must pay attention to it in the agricultural production process.

4 Conclusions and Recommendations

Based on 2002–2017 Chinese national panel data, in the background of smart agricultural production combining with panel quantile regression, the following main conclusions are obtained:

According to different stages of agricultural production development, the scale of agricultural land plays different roles. When the level of productivity is low, decentralized land management can promote agricultural production more effectively. At this time, attention should be paid to the economic development of various regions and infrastructure, construction and upgrading of agricultural business model. In the process of gradual improvement of productivity, land circulation should be promoted in an orderly and reasonable manner, realizing the scale of land management, increasing the production level. When productivity reaches a certain high level, the government should pay more attention to the comprehensive promotion of smart agricultural production factors.

In the background of smart agriculture, in order to promote the scale, modernization and industrialization of agriculture, this article proposes from the following aspects:

Firstly, the government must bring agricultural science and technology and rural finance into a vital role in agricultural production and also raise the level of smart agriculture. Increase technical investment in agricultural production, and continuously increase the mechanization rate of agricultural production on the basis of large-scale agricultural land; Develop and promote production models such as professional households and family farms according to local conditions. Reduce labor input in agricultural production through changes in agricultural production methods. Finally, we can improve agricultural production technology efficiency.

Secondly, the government should establish and improve land management rights and agricultural product trading markets, and give full play to the regulating role of market mechanisms. Establish and improve the market system including land, labor, agricultural production factors and agricultural products. It must ensure the implementation of the reform of the rural land system, and accelerate the formation of a large-scale management model of rural land by exerting the regulating role of market mechanisms. It must optimize the allocation of production resources, and realize the agricultural product market operating, which guide the rational distribution of agricultural production.

References

1. Qingen, G., Xi, Z., Mingwang, C., Qinghua, S.: Land misallocation and aggregate labor productivity. Econ. Res. J. **52**(05), 117–130 (2017). (in Chinese)
2. Aryal, J.P., Rahut, D.B., Maharjan, S., et al.: Factors affecting the adoption of multiple climate-smart agricultural practices in the Indo-Gangetic Plains of India. Nat. Resour. Forum **42**(3), 141–158 (2018)
3. Yahata, S., Onishi, T., Yamaguchi, K., et al.: A hybrid machine learning approach to automatic plant phenotyping for smart agriculture. In: 2017 International Joint Conference on Neural Networks (IJCNN). IEEE (2017)
4. Arslan, A., Mccarthy, N., Lipper, L., et al.: Climate smart agriculture? Assessing the adaptation implications in Zambia. J. Agric. Econ. **66**(3), 753–780 (2015)
5. Koenker, R., Bassett, G.: Regression quantiles. Econometrica **46**(1), 33–50 (1978)
6. Hallock, K.F., Koenker, R.W.: Quantile regression. J. Econ. Perspect. **15**(4), 143–156 (2001)
7. Lopez-Ridaura, S., Frelat, R., Wijk, M.T.V., et al.: Climate smart agriculture, farm household typologies and food security: an ex-ante assessment from Eastern India. Agric. Syst. **159**, 57–68 (2018)

Impact of Alliance Fitness on Smart Phone Brand Alliance Based on Intelligent Application

Fang Wu[✉], Ying Yang, Qi Duan, and Heliang Song

China National Institute of Standardization, Haidian, Beijing 100193, China
wufang@cnis.ac.cn

Abstract. It has been widely proved that alliance fitness is the key determinant of brand alliance effect, however, little is known about how fitness affects the evaluation of co-branding. Based on Keller's (1993) classic brand association model, this paper identifies the meaning of alliance fitness and puts forward a two-dimensional structure for fitness, then test the effectiveness of the structure and its impact on co-branding evaluation in experimental research. Results indicate that alliance fitness is consist of two dimensions, namely product feature complementary and brand image congruence, both dimension have positive influence on the evaluation of co-branding, while the way and results of the impact differs between them.

Keywords: Alliance fitness · Brand alliance · Product feature complementary · Brand image consistency

1 Introduction

Brand alliance has been widely concerned as an effective way to enhance brand equity, and many brands such as Intel, have achieved success through this strategy. Famous marketing scholar Kevin Keller once pointed out that the most important factor for a successful brand alliance is the logical fit between partner brands. If the two brands do not seem to fit each other, it may even cause adverse queries [1, 2]. A large number of empirical studies show that brand alliance effect is based on consumers' positive evaluation of the alliance brand, and the alliance fitness plays the most important role in the alliance brand evaluation [3]. On the basis previous researches on alliance fitness, this paper analyzes dimensions and connotation of alliance fitness based on Keller's brand association model, and explores the influence of each dimension of alliance fitness on brand alliance evaluation through experimental design [4–6]. The research results of this paper will not only deepen the understanding of brand alliance theory, but also provide new insight for enterprises to select appropriate alliance partners, form effective brand alliance and maximize the effect of brand alliance.

J. MacIntyre et al. (Eds.): SPIoT 2020, AISC 1283, pp. 321–326, 2021.
https://doi.org/10.1007/978-3-030-62746-1_47

2   Theory and Hypothesis

2.1   Dimensions and Connotation of Brand Alliance Fitness

Early research on brand alliance fitness was mainly focusing on product level. Scholars believe that if two products can bring more practical utility to consumers, the alliance is effective. With the deepening of the research, scholars found that the fitness of partnership at brand level also determines alliance fitness. Simon and Ruth (1998) measured alliance fitness from both product and brand perspective. However they measured brand fitness only in general way without any further exploration. Thus this paper will analyze the dimensions and connotation of brand alliance fitness based on Keller's brand association model [7–10].

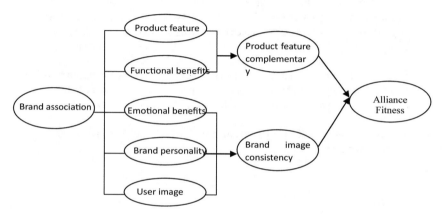

Fig. 1. Brand alliance fitness theoretical model

Brand association is defined as features or benefits which could differentiate one brand from others competitors. Keller (2003) put forward the classic brand association model consist of product attributes, brand benefit (functional and emotional), image, and brand personality. Previous scholars believed that alliance fitness is mainly based on specific or abstract brand associations. Based on specific brand association, when the partner brand complements each other in product attribute or functional benefits, the alliance is considered as a match. Therefore, the first dimension of alliance fitness is defined as product functional complementary. Based on abstract brand associations, it mainly shows whether the partner brands are fitted in emotional benefits, brand personality and user image. According to the theory of cognitive congruence, consumers tend to strive for perceived congruence or harmony when making evaluations. Therefore, we defined the second dimension of alliance fitness as brand image consistency. The theoretical model of brand alliance fitness is shown in Fig. 1.

2.2   The Impact of Alliance Fitness on Brand Alliance Evaluation

Brand alliance effect depends on consumers' evaluation, and alliance fitness is the key factor to determine consumers' evaluation. Early studies mainly focused on the effect of alliance fitness from the product aspect, and believed that product attribute complementary was a necessary condition for a successful brand alliance. Thus, we propose the following hypothesis:

H1: The product function complementary dimension of alliance fitness has a significant positive impact on brand alliance. The more complement of the product function, the higher the consumers' evaluation of brand alliance will be.

Previous studies have not fully discussed the impact of brand image consistency on brand alliance. Previous studies on brand extension also found that the extension is more likely to succeed when the extended brand and the parent brand are consistent in brand image. Therefore we propose that:

H2: The brand image consistency dimension of alliance fitness has a significant positive impact on brand alliance evaluation. The higher the brand image consistency is, the higher the consumers' evaluation of alliance brand will be.

Different dimensions of alliance fitness may have different effects on brand alliance evaluation. Park, Milberg and Lawson (1996) compared the influence of product feature similarity and brand concept consistency on consumer evaluation of brand extension, and the results showed that no matter how similar the product feature is, the extension with high brand concept consistency will get more positive evaluation. Therefore we propose the following assumptions:

H3: Different dimensions of alliance fitness have different impacts on joint brand evaluation. Compared with product function complementary, brand image consistency has a greater impact on brand alliance evaluation.

3   Results

3.1   Experiment Design

The purpose of this study is to reveal the influence of the two dimensions of alliance fitness on brand alliance evaluation, so 2 (product function complementary: high/low) × 2 (brand image consistency: high/low) grouping experiment design is adopted in this paper. A fixed brand is selected as the core brand, and different partner brands are selected to manipulate the two dimensions of alliance fitness.

We choose Chinese smart phone brand MI as the core brand. We chose the lens with the highest complementarities ($M = 5.80$) and the beverage with the lowest complementarities ($M = 1.96, = 3.84, t = 15.82$) as the category of the partner brand. After prior test, in lens products we choose Leica and Zeiss as the partner brands with high and low image consistency. In beverage products, Coca-Cola and Huiyuan are selected as the partner brands with high and low image consistency. (see Table 1).

This study adopts three indicators to measure brand alliance evaluation, which are perceived quality, attitude preference and purchase intention reflecting three aspects of attitude, which are cognition, emotion and behavior.

Table 1. Experiment groups

		Product functional complementary	
		High	Low
Brand image consistency	high	A: MI + Leica	C: MI + Coca-Cola
	low	B: MI + Zeiss	D: MI + Huiyuan

3.2 Hypothesis Testing

ANOVA was used to test the influence of alliance fitness on brand alliance evaluation. Evaluation results of brand alliance under different fit conditions are listed in Table 2.

Table 2. Brand alliance evaluation under different alliance fitness conditions

The evaluation index	Group A	Group B	Group C	Group D	Product functional complementary		Brand image consistency	
	The mean (sd)	The mean (sd)	The mean (sd)	The mean (sd)	F value	Sig	F value	Sig
Perceived quality	5.27 (0.74)	4.78 (0.54)	4.14 (0.67)	3.93 (0.62)	11.98	0.00	1.44	0.11
Attitude preference	4.88 (0.95)	4.53 (0.94)	4.21 (0.97)	3.82 (1.05)	3.60	0.00	2.89	0.00
Purchase intention	5.01 (0.86)	4.45 (1.10)	4.19 (1.02)	4.00 (1.24)	3.14	0.03	1.89	0.02

The results showed that product function complementary had significant main effects on perceived quality ($F = 11.98$, $P < 0.01$), attitude preference ($F = 3.60$, $P < 0.01$) and purchase intention ($F = 3.14$, $P < 0.05$) of brand alliance. H1 has been supported.

Variance analysis results showed that the subjects' attitude preference ($F = 2.89$, $P < 0.01$) and purchase intention ($F = 1.89$, $P < 0.05$) to brand alliance have significant differences under different brand image consistency. That means with the same product function complementary, the improvement of brand image consistency will makes the attitude and purchase intention of brand alliance improved. However, brand image consistency has no significant influence on the perceived quality of brand alliance ($F = 1.44$, $P > 0.10$). H2 is partially verified.

We constructed the SEM to test the difference between two alliance fitness dimensions:

$$QA = \gamma_{11} \cdot PF + \gamma_{12} \cdot BF + \varepsilon 1$$
$$AT = \beta 21 QA + \gamma_{21} \cdot PF + \gamma_{22} \cdot BF + \varepsilon 2$$
$$PI = \beta 31 QA + \beta 32 AT + \gamma_{31} \cdot PF + \gamma_{32} \cdot BF + \varepsilon 3$$

(QA = Perceived quality; AT = attitude preference; PI = Purchase intention; PF = product function complementary; BF = brand image consistency)

The results show that the fitting indexes of the model reach the ideal level and the overall fitting is good (see Fig. 2 for the results). Based on the path load in Fig. 2, we calculated the total effect, direct effect and indirect effect of each dimension of alliance fitness on the three aspects of brand alliance evaluation.

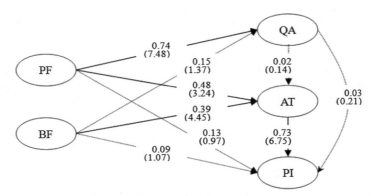

Fig. 2. Fitting results of structural equation model M1

Results show that, in the perceived quality dimension of brand alliance, the total effect of product function complementary is significantly greater than image consistency (PFte = 0.74, t = 7.48). BFte = 0.15, t = 1.37). In the dimension of attitude preference and purchase intention, the total effect of product function complementary is also relatively higher (attitude preference: PFte = 0.49, t = 6.00; BFte = 0.37, t = 4.89; Purchase intention: PFte = 0.49, t = 6.00; BFte = 0.37, t = 4.89), which is contrary to our hypothesis of H3 and the conclusion of Simonin and Ruth (1998). We believe that this may be caused by difference of stimuli in this study and previous studies. Siomin and Ruth (1998) used automobile brands as theirs stimuli. Besides its functional feature as a traffic tool, automobiles have become a symbol of social status. Therefore, brand image may have a greater impact on evaluation in their study.

4   Conclusions

4.1   Alliance Fitness Has Two Dimensions

This paper analyzed alliance fitness from the brand association perspective, verified its composition and dimensions through CFA. Results confirmed that the alliance fitness has two dimensions namely product function complementary and brand image consistency. Product function complementary is based on product attributes, to bring the complementary function to consumer; while brand image consistency means user image and brand personality of partner brands is congruency to consumers. That means when companies are choosing their alliance partner, Partners fit with their own brand in both product and brand perspective will bring more positive effects.

4.2 Two Dimensions of Alliance Fitness Have Different Impacts on Brand Alliance

Two dimensions of alliance fitness have positive impact on brand alliance evaluation, but the impact results, paths and degrees are different. Product function complementary has a significant positive effect on the perceived quality, attitude preference and purchase intention of brand alliance, while the effect of brand image consistency is mainly reflected in the dimensions of attitude preference and purchase intention. The effect of product function complementary on brand alliance evaluation is greater than that of brand image consistency.

Acknowledgement. This work is supported by the National Social Science Fund Major Project (Research on Quality Governance System and Policy to Promote High Quality Development), No.18ZDA079. And the Dean fund project of China National Institute of Standardization under grant No. 552019Y-6660.

References

1. Samuelsen, B.M., Olsen, L.E., Keller, K.L.: The multiple roles of fit between brand alliance partners in alliance attitude formation. Mark. Lett. **26**(4), 619–629 (2014)
2. Keller, K.L.: Strategic Brand Management, 2nd edn., pp. 101–103. Prentice Hall, Upper Saddle River (2003)
3. Veloutsou, Cleopatra: Brand evaluation, satisfaction and trust as predictors of brand loyalty: the mediator-moderator effect of brand relationships. J. Consum. Mark. **32**(6), 405–421 (2015)
4. Lee, J.K., Lee, B.K., Lee, W.N.: The effects of country-of-origin fit on cross-border brand alliances. Asia Pac. J. Mark. Logist. **30**(5), 1259–1276 (2018)
5. Lee, H.N., Lee, A.S., Liang, Y.W.: An empirical analysis of brand as symbol, perceived transaction value, perceived acquisition value and customer loyalty using structural equation modeling. Sustainability **11**(7), 1108–1112 (2019)
6. Retamosa, Marta., Millán, Ángel, Moital, Miguel: Does the type of degree predict different levels of satisfaction and loyalty? A brand equity perspective. Corp. Reput. Rev. **07**, 9–15 (2019)
7. Schiffman, L.G., Kanuk, L.L.: Consumer Behavior, 7th edn., pp. 342–345. Prentice Hall, Upper Saddler River (2000)
8. Washburn, J.H., Till, B.D., Priluck, R.: Brand alliance and customer-based brand-equity effects. Psychol. Mark. **21**(7), 487–508 (2016)
9. Bluemelhuber, C., Carter, L.L., Lambe, C.J.: Extending the view of brand alliance effects. Int. Mark. Rev. **04**, 78–89 (2013)
10. Maiksteniene, K.: Modeling brand alliance effects in professional services. Waset.Org 31–38 (2013)

Effective Control Method of the Whole Process Project Cost Based on BIM

Linlin Yu[(✉)]

Hubei Business College, Wuhan, China
15369229@qq.com

Abstract. Engineering industry is one of the important industries in the process of economic development. In the aspect of project investment, how to achieve the expected effect with the minimum investment under the premise of ensuring the project quality has become the topic of common concern of all stakeholders. In view of the difficulty of the traditional project cost control which is not refined and time-consuming, this paper proposes an effective control method of the whole process project cost based on BIM. BIM technology is used to control the cost of the whole process of the project, so as to realize effective management and avoid out of control investment. Through the actual case, the results show that the method presented in this paper has obvious advantages in time consumption and estimation accuracy, and the satisfaction of management personnel, construction personnel and supervision personnel has reached more than 85%.

Keywords: Whole process engineering management · Cost control · BIM (Building information Modeling) · Effect evaluation

1 Introduction

The cost of construction industry has always been a topic of concern for the contractors, investors and other relevant personnel [1]. For example, the real estate market sales did not show a strong trend, the house did not sell out, the builders have to find a way to control the construction cost of the project. The process from planning to construction to final completion of the building is very long. Once the project cost is not well controlled, it is likely that the follow-up work cannot be carried out [2]. Only by actively strengthening the project cost control of the whole project can we be sure to adapt to the development of market economy, so as to enhance our competitive advantage in the construction field and achieve long-term and healthy development [3]. Therefore, in the process of the project, under the premise of ensuring the quality of the project, the whole process of effective control of the project cost can avoid the bad phenomenon of out of control investment to the maximum extent.

At present, the investment utilization rate of engineering construction is not very good, and the input and output are not positively related, that is to say, the input is very high but the output is poor, and the production efficiency is getting lower and lower

J. MacIntyre et al. (Eds.): SPIoT 2020, AISC 1283, pp. 327–332, 2021.
https://doi.org/10.1007/978-3-030-62746-1_48

with the development of time [4]. Engineering cost control involves many fields, such as housing, highway, subway and so on, which are studied by experts and scholars. In [5], considering the maintenance cost of the overhaul project, the author compares and selects the project design scheme and strictly screens it in the design stage to reduce the pressure of capital control. In [6], in the aspect of urban tunnel construction, the author takes a project of Lanzhou Metro as an example for the cost control of shield machine construction. According to the work process, the author regularly carries out comparative analysis with the target to reasonably control the cost. In [7], aiming at the cost control of EPC project, the author strengthens the cost management from the aspects of design, implementation, procurement and completion to realize the effective control of the whole process project cost. Although the above research has been effective in engineering cost control, there is still room for further optimization. At present, there are some shortcomings in the cost control methods, such as insufficient refinement, low level of project cost informatization, and attaching importance to the results while ignoring the process [8].

In view of the difficulties existing in the project cost, referring to the foreign construction information research, the author intends to introduce BIM (building information modeling) technology into the effective control of the whole process project cost. Based on BIM Technology, this paper analyzes the whole process of engineering cost, in order to provide reference for the long-term healthy development of engineering cost management industry.

2 Relevant Concepts of BIM and Project Cost

2.1 BIM Related Concepts

In the 1970s, a concept of building information model was proposed, namely building information modeling, referred to as BIM [9]. The emergence of BIM Technology has brought good news to the field of engineering construction. The core of BIM Technology is information, which well parameterizes and models the design, construction, management and other relevant information of the project construction process, including all the information and processes in the whole project life cycle [10]. For example, the construction progress, construction process management, related maintenance process, as well as all features, single cost and geometric information of the building, the spatial topological relations of all information are centralized and stored in the database by using the method of 3D Boolean operation, and finally form a digital model. Because of the accuracy of calculation, the comprehensiveness of information storage and the convenience of use, it greatly improves the efficiency and accuracy of project cost management. It has the characteristics of parameterization, visualization, simulation, coordination and exportability, saves a lot of redundant costs, and ensures the realization of project management investment objectives. It has been widely used in the construction field in China and even in the world recognition, and more and more widely used.

2.2 Relevant Concepts of Project Cost Management

The project cost management, in fact, is to ensure the final economic benefits of the project, take the ultimate goal as the principle, use scientific technology, and conduct the behavior of the whole project cost from multiple angles consistent with the policies and laws. There are two kinds of project cost management. The first is construction project investment cost management. The second is project price management. In the first kind of project cost management, it means to measure the project cost in advance or monitor the price change in order to achieve the expected investment goal under the existing project design scheme. As for the second kind of project cost management, from a small perspective, it can be the cost control, bidding and other activities when the production enterprise realizes the management goal again. From a large perspective, it can be the government's price control through economic, legal, policy and other ways to regulate market behavior. Our country is short of natural resources. If we want to maintain a relatively stable development speed, we must invest more funds. However, the amount of capital is too large to complete the fund-raising, so how to achieve the expected goal or even higher than the expected goal with less investment has become a hot topic in the industry.

3   BIM Based Whole Process Cost Control Process Method

(1) Cost control in decision-making stage

The first scientific decision is very important for a project to be completed. In the decision-making stage, it mainly compares various schemes and carries out feasibility analysis, so as to maximize the income result with the minimum investment cost. In the aspect of cost control in this stage, multiple alternative investment schemes are selected, BIM technology is used to compare the cost, and the optimal benefit scheme is finally selected. In this way, the estimation deviation of investment can be minimized.

The BIM model engineering body is built by Revit, the geological model is constructed by Inforsworks and GIS software, and the investment estimation, schedule and other indicators of the project are integrated with the financial software. If there is any modification, the parameter modification operation can be made quickly to guarantee the decision-making.

(2) Cost control in design stage

The goal of cost management in the design stage of a project is to make a construction budget with an error of less than 12%. On the premise of ensuring the project quality, the construction of the project shall be completed with the minimum time consumption and energy consumption. The design unit can use BIM technology to carry out design analysis on energy conservation, sunshine, lighting, ventilation, sound insulation and noise of the project entity, and timely change the information model, analyze the impact on the project cost, and realize the comparison, selection and continuous optimization adjustment of the scheme within the control range.

(3) Cost control in bidding stage

In this stage, it is mainly to complete the control of reasonable bidding methods, and select more capable contractors within a reasonable bidding price. In terms of cost management, it is necessary to follow the principle that, in order to facilitate investment control, the bid price of the project shall not be higher than the bidding control price, but the price shall not be too low, because this may lead to Jerry building and material cutting. In the aspect of cost control, BIM model is used to obtain the information of the whole project, avoid omission, and obtain the price close to the market price from the software, so as to control the bidding price reasonably.

(4) Cost control in construction stage

The cost control here is mainly reflected in comparing the planned cost with the actual cost, making changes, claims, project payment and other links. Depending on BIM design model, the quantity and cost of any part of the project can be obtained. In addition, through the application of BIM related software, it can not only achieve on-time settlement, but also accurately purchase and pick up materials according to the project schedule, add time schedule and cost information to the BIM model, and automatically extract resource use plans such as capital plan, material purchase plan and labor plan in each stage of the project.

(5) Cost control at completion stage

In the completion stage, the main concern here is the possible problems of settlement. For example, the engineering quantity is falsely reported, the charging standard is maliciously increased, and the cost is reported in disorder. By using BIM Technology, all the physical information, geometric information and other attributes of engineering entity can be loaded into BIM model, such as material, factory information, engineering quantity, price and construction progress of each component. When settling accounts, both sides can settle accounts on the same model, effectively avoiding differences, and making the work in the completion stage fast.

4   Effect Analysis of Project Cost Control Based on BIM

In order to verify the effectiveness of the whole process project cost control method based on BIM proposed in this paper, the method proposed in this paper is applied to the actual project. At the same time, the traditional cost control method is used for cost management, which is compared with the method proposed in this paper to evaluate the effectiveness of the method proposed in this paper. In the management decision-making stage, the traditional method and the method proposed in this paper are used to control the cost and make the decision-making scheme. The time spent and the estimated error of the two methods in the resulting scheme are shown in Fig. 1.

From Fig. 1, it can be seen that the BIM based whole process engineering cost control method proposed in this paper has obvious advantages in both time consumed and cost estimation error. In the same project, the traditional method needs 30 days to make decision-making plan, and its estimation error is 31.23%; the cost control method proposed in this paper gives the cost plan in 7 days, and its estimation error is only 10.51%. In contrast, the method proposed in this paper not only takes a short time, but also is more accurate. It can also update the scheme quickly according to the actual changes, and the effect advantage is very obvious.

The progress of the project is interfered by people. Once there is a problem in the project, people of all kinds of work are required to communicate and coordinate with each other to ensure the orderly progress of the project, including management personnel,

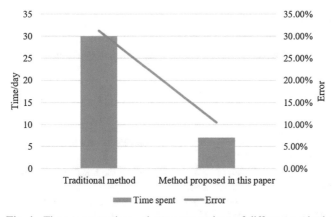

Fig. 1. Time consumption and error comparison of different methods

construction personnel and supervision personnel. In this paper, people from these three types of work are interviewed to evaluate their satisfaction with the BIM based whole process engineering cost control method. The satisfaction is shown in Fig. 2.

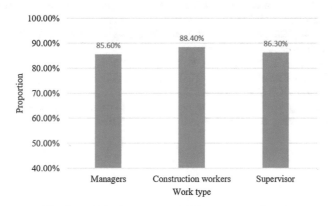

Fig. 2. Satisfaction degree of different types of work

Through the investigation of management, construction and supervision personnel, among the interviewees, the number of people satisfied with the BIM based whole process cost control method proposed in this paper is more than 85%, indicating that BIM based project cost control is very suitable for the effect of relevant personnel in promoting the project.

5 Conclusion

BIM technology can help improve the accuracy of project calculation, arrange resources more reasonably, control design changes and realize data accumulation and sharing. Of course, in practical application, there are also problems such as inconsistent standards and high technical difficulty, which need to be solved by relevant personnel. People's cognition of BIM technology needs to be improved. The potential of BIM Technology is unlimited. In the future, BIM technology will surely receive more attention and recognition.

References

1. Jieying, L.: Key points and control of construction project cost budget control. Build. Mater. Decoration **8**, 140–141 (2019)
2. Jie, Q., Hongcheng, L., Haixia, J.: How to strengthen the construction cost management of salt chemical enterprises. Salt Sci. Chem. Ind. **46**(8), 49–51 (2017)
3. Yang, C.: The importance of cost control in the whole process of construction project management. Green Build. Mater. **137**(07), 214 + 216 (2018)
4. Botong, S., Jie, Z., Yong, W., Yaning, L.: Analysis of the relevance and ripple effect of Beijing real estate industry based on input-output model. Eng. Econ. **26**(9), 44–48 (2016)
5. Yonghong, Y., Wanzhong, Y., Xuancang, W.: Research on cost control of highway overhaul project. Highw. Eng. **42**(2), 76–80 (2017)
6. Pan, Y.: Cost control of shield tunnel construction in urban tunnel construction. Rail. Eng. Technol. Econ. **33**, 56–68 (2018)
7. Dexin, Z.: Analysis of engineering cost control points of EPC general contracting project. Min. Eng. **14**(3), 1–3 (2016)
8. Wei, W., Lejia, L.: Application of BIM technology in engineering cost management. Residential Real Estate **10X**, 104 (2018)
9. Lau, S.E.N., Zakaria, R., Aminudin, E., Saar, C.C., Wahid, C.M.F.H.C.: A review of application building information modeling (BIM) during pre-construction stage: retrospective and future directions. IOP Conf. **143**(1), 012050 (2018)
10. Deguang, Z.: Application of BIM technology in the construction process of prefabricated building engineering. Residential Real Estate **9**, 229 (2019)

Research and Design of FPGA Embedded System

Dandan Jiang[✉]

Zhengzhou University, Zhengzhou, Henan 450000, China
1079605094@qq.com

Abstract. Facing the huge domestic demand, the current relatively backward status of my country's automatic quantitative measurement control instrument technology urgently needs to be changed. How to improve the measurement accuracy and measurement speed at the same time has always been one of the problems to be solved by factories, enterprises and research fields. Under this background, this paper designs a dual processor dynamic measurement system based on FPGA. This system has dual processors to share different control links and work on the same chip. Compared with the currently widely used single processor structure products, it has a higher processing capacity while significantly reducing costs. This paper optimizes the design of the FIR digital filtering algorithm based on FPGA, and uses a combination of software and hardware to carry out system integrity design, design process analysis, application module design of each internal system, and optimized programming of each functional module such as FIR digital filtering. Software design, configure the system NiosII soft core, construct a design scheme based on FPGA and digital filter system, and implement the design scheme with FPGA. Through careful research on the compound sampling filter algorithm and the drop compensation method, this article builds an experimental platform for the system, selects Cyclonell EP2C8 as the main chip, and two NiosIII soft cores as processors, realizing high-precision real-time measurement of powder-like materials, Achieved good results, and laid the foundation for further research in the future. The experimental results show that the 9-bit embedded multiplier used in this paper occupies 4 of the total 36; the global clock resource occupies 6 and the utilization rate is 75%. The above results fully show that the system makes good use of FPGA EP2C8 resources.

Keywords: Dual processor · Embedded system · Feedback control · Dynamic measurement

1 Introduction

With the rapid development of electronic technology, control technology and computer technology, CPLD, FPGA, SOPC, embedded systems and other new technologies continue to develop, based on FPGA and NiosII embedded programmable system-on-chip (SOPC) with its flexible design, hardware and software Programmable, tailorable and

J. MacIntyre et al. (Eds.): SPIoT 2020, AISC 1283, pp. 333–339, 2021.
https://doi.org/10.1007/978-3-030-62746-1_49

other features and advantages, at the same time, it has outstanding characteristics such as small size, strong performance, low power consumption and high reliability [1, 2]. The system can customize the CPU and its peripheral interfaces according to the needs of the software, especially the reconfigurable feature of Nios II brings greater flexibility and adaptability to the design of embedded systems. It proposes a design for embedded systems. The new construction method has changed the traditional system design mode of fixed CPU chips [3, 4].

As China's entry into the WTO accelerates and the investment and trade environment continues to improve, an inevitable trend is that internationally renowned heavy packaging companies will further increase investment and productivity in China, which will inevitably bring certain As a result, it is imperative to comprehensively upgrade the technical content of China's weighing packaging manufacturing industry [5, 6]. In view of the current situation in China and considering that the current export targets are mainly Southeast Asian, Middle Eastern and African markets, in the future, China's quantitative packaging machinery manufacturing industry should focus on the development of fast, low-cost packaging equipment [7, 8]. Make the equipment develop in the direction of smaller, more flexible, multi-purpose and high efficiency [9, 10].

In this paper, the two methods widely used in metrology system, such as composite sampling filter algorithm and drop compensation method, are studied in depth, and a new feedback control method is designed to make the system have online real-time control capability and improve the system's metrology accuracy. Mainly research the design method of digital filters based on FPGA, mainly design and analyze from the aspects of function analysis, structural design, algorithm establishment and optimization, program design, simulation, etc., and finally pass the hardware experimental test. The feasibility and effectiveness of the FPGA-based SOPC system platform design are verified. This paper builds an experimental platform for system verification, and uses a composite sampling filter algorithm, a drop value compensation method, and a feedback control method to control the feeding system. The comparison of different results shows that the feedback control method can work continuously. Under the conclusion of maintaining a good measurement effect.

2 Design of FPGA Embedded System

2.1 Memory Selection

This design uses SDRAM and FLASH to form the embedded hardware storage part. NiosII embedded soft core is built in FPGA, SDRAM is used as data memory, and FLASH is used as program memory. Below I will make a detailed analysis of the above parts.

(1) SDRAM

The SDRAM chip used in this design is MT48LC4M32B2F5-7. The chip is packaged in FBGA, has 90 pins, the clock rate is up to 143 MHz, and the capacity is 128 Mbits. In this system, it acts as a data storage, the actual operating rate is 125 MHz.

(2) FLASH

The FLASH chip used in this design has EPCS4. EPCS4 is used to store the FPGA hardware logic code, which is compiled and generated by QuartusI1; it is used to store the C language program, which is compiled and generated by NiosIIIDE, and is used as the program memory of the embedded system. When the circuit board is powered on or reset, the FPGA downloads the hardware code from EPCS4, constructs the hardware system (including the NiosIICPU inside the FPGA), and then NiosIICPU downloads the solidified C language code into SDRAM to fetch instructions for the CPU to execute.

To realize the embedded hardware system in SOPCBuilder, CPU, memory and peripheral interface are the most basic three elements of the embedded hardware system. Because FPGA itself does not contain CPU and peripheral interface, in order to build the CPU core and peripheral interface needed, Altera's development software SOPCBuilderi is used for design.

2.2 FIR Filter Algorithm Establishment and Optimization

The finite impulse response filter is the most basic unit in the digital signal processing system. It can have strict linear phase frequency characteristics at any amplitude and frequency characteristics, and its unit impulse response is limited, and there is no input to output. Feedback is a stable system.

The expression of the relationship between the output sequence y(n) of the N-order FIR filter and the input time series x(n) is:

$$y(n) = \sum_{i=0}^{n=1} h(i)x(n-i) \tag{1}$$

That is, the output sequence y(n) is the convolution of the unit pulse corresponding h(n) and the input sequence x(n), and the structural relationship diagram can be obtained from the convolution relationship, which is called a direct structure, and a total of N multiplications are required in this structure. Device. Each sample y(n) needs to perform n times of multiplication and $n - 1$ times of addition operation to achieve the sum of multiplication and accumulation.

For linear phase FIR digital filters, the unit sampling response is antisymmetric or symmetric, that is, $h(n) = \pm h(N-1-n)$, then the use of symmetry can simplify the network structure, if h(n) Is even symmetry and N is even.

$$y(n) = \sum_{i=0}^{n=1} h(i)x(n-i) = \sum_{i=0}^{\frac{N}{2}-1} h(i)[x(n-i) + (n-N+m)] \tag{2}$$

The FIR digital filter is a sectioned delay line, which weights and accumulates the output of each section, and finally obtains the output of the filter.

2.3 Dynamic Measurement Signal Processing

The digital filtering method is mainly a digital filter developed on the basis of an analog filter. It is a linear discrete system with frequency selectivity, which can realize hardware functions through software. It is characterized by great flexibility, high precision and good stability. Estimation method is a basic method of dynamic weighing. One of its theoretical foundations is probability statistical analysis. Strictly speaking, it is also a filtering method, but the estimation process is performed in the time domain. The comparison rule is to compare the known mass with the mass to be measured to eliminate the effect of motion acceleration.

3   The Experimental Research of FPGA Embedded System

3.1   Digital Filter Function

Digital filtering is widely used in various fields such as speech processing, noise elimi-nation, harmonic reduction, and pattern recognition. It is a basic signal processing unit. In the process of data communication, temperature and voltage drift, noise interference and other problems can be eliminated. The main function of digital filtering is to filter out some useless frequency components from the signal. After the signal is digitally filtered, it actually multiplies the signal spectrum and the frequency response of the filter to analyze from the time domain. It is the convolution sum of the impulse response of the input signal and the filter.

3.2   Implementation of Mutually Exclusive Hardware Access

This system uses hardware five exclusive modules to achieve access to shared RAM, instead of software to improve access speed through hardware access. This mutually exclusive module can also be regarded as a shared resource, which has the function of "test and set". First, the NiosII processor is tested to see whether the mutually exclusive module is feasible. If it is feasible, it is obtained in a certain operation. When the NiosII processor ends the use of shared resources related to the mutex module, the mutex module is released. At the same time, another processor can obtain the mutually exclusive module and use the shared resources. Without this mutual exclusion module, two separate test and setup instructions are needed to implement the above functions, and a "deadlock" situation may occur.

3.3   Data Acquisition System

The data acquisition system of this system is composed of load cell, AD7730 analog-to-digital converter and NiosIICPU1. Among them, the load cell uses a resistance strain pressure sensor, using elastic sensitive elements and strain gauges to convert the weight of the object into a corresponding voltage signal. This voltage signal is filtered and amplified by the signal conditioning circuit, and the AD7730 is the core. The data acquisition circuit converts the analog signal into a digital signal that can be processed to obtain the required quality value.

4   Experimental Analysis of FPGA Embedded System

4.1   Analysis of System Resource Usage

This article uses the external expansion configuration chip EPCS4 to save the corresponding configuration information. Through the way of programming, the.pof file is programmed into the EP2C8 external expansion configuration chip EPCS4 through a special tool. EP2C8. After power on, it reads the configuration information from the configuration chip. In the actual use process, if the application is not very large, it can be directly written into this configuration chip. Table 1 shows the system's use of various resources on Cyclonell EP2C8Q208C8.

Table 1. System resource usage

Total LABs	346/524 (67%)
I/O pins	112/154 (77%)
Clock pins	2/8(25%)
M4Ks	17/34 (64%)
Total memory bits	76325/165826 (46%)
Total RAM block bits	10557/165842 (64%)

From the resource usage table, it can be concluded that the system's LUT analysis for different input numbers shows that 4233 logic function logic units are 2233, 3-input logic function logic units are 1421, and less than 2-input logic function logic units are 532. The number of occupied registers is 2915, accounting for 34% of the total number of registers 8646, of which 69 I/O registers account for 18% of the total number of I/O registers. 344 logic array blocks are used, accounting for 67% of the total; 106 user I/O ports occupy 77% of the total resources; 9-bit embedded multipliers occupy 4 of 36; global clock resources occupy 6 The utilization rate is 75%. The above results fully show that the system makes good use of FPGA EP2C8 resources. Under the above resource occupancy and constraints, the system clock frequency can be up to 65.21 MHz. The above indicators show that there is still much room for improvement in system efficiency, which provides conditions for further research and improvement.

4.2   Sampling and Filtering Experiment Analysis

Under laboratory conditions, 20 consecutive experiments were carried out on this system, using composite sampling filtering method, drop compensation method and feedback control method to operate. Due to space limitations, only the measurement results obtained in the first 5 times are given here, as shown in Fig. 1. In the experiment, the average metering time of this system is 5.9 s per bag, which is 32% shorter than that of the current domestic metering system, which is 8.7 s per bag. In this experiment, flour was used as the measurement object. In the initial stage of the experiment, the calibration

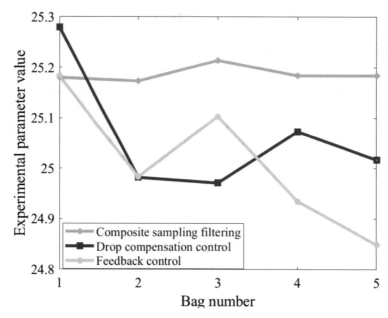

Fig. 1. Comparison of measurement data by different methods

system W was 25 kg, and the initial values of e_1 and e_2 were calibrated to 0.1 and 0, respectively.

As shown in Fig. 1, it can be concluded that the error rate is 0.57% when the composite sampling filtering method is used, the error rate is 0.19% using the drop compensation method, and the error rate is reduced to 0.08% using the feedback control method. It can be seen that the feedback control method maintains a good control effect under continuous operation. It realizes the online adjustment of the switching parameters and has obvious advantages in the measurement process. However, for the feedback control, the measurement deviation value at the first operation of the system needs to be used, which results in the improvement of the measurement accuracy at the first operation of the system, which needs to be further studied and improved in the subsequent work.

5   Conclusions

This paper completed the system design and development work in the laboratory environment, and verified the system functions. From the final experimental results, it can be seen that good results have been achieved. Although some good design ideas were proposed, a series of problems were still encountered in the actual design process. In the process of using the dual-core structure, it is also found that the EP2C8 development board has only 36 M4K RAM blocks and a maximum of 18K of on-chip memory, but the two processors will occupy a considerable part of it. Performance is affected. Since the feedback control needs to use the measurement deviation value at the first operation of the system, the measurement accuracy at the first operation of the system is not significantly improved. Due to limited time and capacity, this system was only tested under laboratory conditions. It did not take into account the influence of interference factors such as vibration of the metering sensor, vibration of the material impact force and uncertain air flying materials on the system. These are the shortcomings of the current system, and need to continue to study and improve in the future study and work.

References

1. Gopi, J., Manjula, J.: Efficient NOC architecture for FPGA based soft core embedded system. Int. J. Appl. Eng. Res. **10**(7), 17029–17036 (2015)
2. Rajasekaran, C., Jeyabharath, R., Veena, P.: FPGA SoC based multichannel data acquisition system with network control module. Circuits Syst. **08**(2), 53–75 (2017)
3. Wu, Y., Kaempf, J.H., Scartezzini, J.L.: Design and validation of a compact embedded photometric device for real-time daylighting computing in office buildings. Building Environ. **148**, 309–322 (2019)
4. Li, A., Kumar, A., Ha, Y., et al.: Correlation ratio based volume image registration on GPUs. Microprocess. Microsyst. **39**(8), 998–1011 (2015)
5. Cuppini, M., Mucci, C., Franchi, S.E.: Soft-core embedded-FPGA based on multistage switching networks: a quantitative analysis. IEEE Trans. Very Large Scale Integr. Syst. **23**(12), 3043–3052 (2015)
6. Perera, D.G., Li, K.F.: A design methodology for mobile and embedded applications on FPGA-based dynamic reconfigurable hardware. Int. J. Embedded Syst. 11(5), 661 (2019)
7. Mcintosh-Smith, S., Price, J., Sessions, R.B., et al.: High performance in silico virtual drug screening on many-core processors. Int. J. High Perform. Comput. Appl. **29**(2), 119–134 (2015)
8. Bangqiang, L., Zhoumo, Z., Jing, D.: A new method of fault tolerance based on partial reconfiguration. Int. J. Inf. Commun. Technol. **9**(2), 232 (2016)
9. Arena, L., Piergentili, F., Santoni, F.: Design, manufacturing, and ground testing of a control-moment gyro for agile microsatellites. J. Aerospace Eng. **30**(5), 04017039 (2017)
10. Liang, W., Wu, K., Xie, Y., et al.: TDCM: an IP watermarking algorithm based on two dimensional chaotic mapping. Comput. Sci. Inf. Syst. **12**(2), 823–841 (2015)

Multi-source Heterogeneous and XBOOST Vehicle Sales Forecasting Model

Fan Zhang[✉], Jing Yang, Yaxin Guo, and Hongjian Gu

China Automotive Research (Tianjin) Automotive Information Consulting Co., Ltd., Tianjin 300000, China
m13502037247_3@163.com

Abstract. With the rapid development of the economy, the automobile industry has developed as the world's no. 1 automobile consumer market and the world's largest consumer potential market. The growth of domestic auto production and sales has an obvious driving effect on auto finance. The penetration rate of auto finance has increased from 13% five years ago to nearly 40% now. The purpose of this paper is to optimize the model algorithm based on multi-source heterogeneous and XBOOST vehicle sales prediction model. In this paper, the application of multi-source heterogeneity and XBOOST algorithm to the auto sales prediction model is discussed by using the sample data of auto sales of A Auto company, and the relevant characteristics of customers are studied to establish the logistic regression model of sales prediction and take it as the standard. Finally, the optimal parameter combination is explored to optimize the model effect. Combined with the evaluation index of machine learning classification model, the performance of each model is compared. The results showed that the difference between forecast and actual sales volume was only 12.1%. This is helpful for the enterprise to provide reference for the forecast of car sales.

Keywords: Multi-source heterogeneity · Vehicle sales · Machine learning · Logistic regression

1 Introduction

With the improvement of the penetration rate of auto finance, the requirements on enterprises' risk control ability are increasingly high, and relevant enterprises also begin to combine scientific and technological means with risk control [1, 2]. Automobile sales forecasting model is a measure for automobile enterprises to prevent risks and forecast market conditions. It can help improve the company's ability to identify customers and reduce the losses caused by auto finance business [3].

Since this year, countries have attached greater importance to new energy vehicles. China's policies to promote the development of new energy vehicles are also constantly introduced. Under the active promotion of relevant national policies, the development of new energy vehicles in China is gradually accelerated. At the beginning of this year, the automobile industry adjustment and revitalization plan principles were adopted, which

© The Editor(s) (if applicable) and The Author(s), under exclusive license to Springer Nature Switzerland AG 2021
J. MacIntyre et al. (Eds.): SPIoT 2020, AISC 1283, pp. 340–347, 2021.
https://doi.org/10.1007/978-3-030-62746-1_50

not only boosted the confidence of the automobile market, but also more reflected the state's support for the development of energy-saving and new energy vehicles. According to the plan, the central government will allocate 10 billion yuan over the next three years to support technological innovation and upgrading of enterprises and the development of new energy vehicles and spare parts. Promote the industrialization of electric vehicles and their key components. The central finance arranges subsidy funds to support the demonstration and promotion of energy-saving and new-energy vehicles in large and medium-sized cities, and the implementation of this policy will certainly have a significant impact on automobile sales [4].

Sun Huifang proposed a decision method based on core and gray level to solve the multi-attribute decision problem of gray multi-source heterogeneous data, and gave the definition of core density and gray level of extended gray number in gray multi-source heterogeneous data sequence. On this basis, the kernel vector and gray vector of the sequence are constructed to whiten the multi-source heterogeneous information, and a grey relational bidirectional projection sorting method is proposed [5]. In the face of massive terminals, the reasonable allocation of limited resources in the system and the fusion of heterogeneous data has become an important research topic of NB-IOT. Liu Yu proposed a nB-iot multi-source heterogeneous data fusion method based on perceptive semantics. Firstly, the advantages and key technologies of NB-iot are introduced. Liu Yu analyzed the centralized mode and distributed mode in NB-iot networks and proposed a semantic perception-based multi-source heterogeneous data fusion to form a unified format [6]. With the help of expert experience information, interval estimation of failure efficiency data under different methods can be obtained. How to make the interval convergent to the new information is an important problem to be solved. Heng Shao USES the concept of generalized standard grey number to describe the multi-source heterogeneous uncertainty failure rate data as a unified framework. Then, the average failure rate is calculated by engineering construction method, and the grey index distribution reliability function is established [7].

This paper makes use of multi-source heterogeneous algorithm and XBOOST algorithm, combined with the sales situation of automobile companies, analyzes related variables, and provides Suggestions for establishing automobile forecast sales model. Use important feature variables to establish corresponding structure algorithm [8, 9]. Then, it introduces the main parameters of the machine learning algorithm, USES the machine learning library tool to explore the optimal parameter combination, and finally, through experiments, provides technical Suggestions for automobile enterprises [10].

2 Proposed Method

2.1 Multi-source Isomerism

In the process of enterprise information management system of construction, due to various business data management system of construction and management of the implementation of the process, data storage and management system and the evolution of technical and other economic and social, and other factors, led to the enterprise in the process of informatization development has accumulated a lot of different business system and data storage and management informatization, including enterprise adopts the distributed data storage and management system is very different, from simple to complex enterprise file management database network management database, Together, they constitute the heterogeneous data system and data source of the entire enterprise.

2.2 Automobile Sales Forecasting Model

Selected history this month sales price index as the dependent variable, explain the variable factors including history of auto sales this month, steel car production, rubber tire car production, monetary supply of consumer goods, network baidu search traffic index, the consumer prices index, etc., assumes that the car sales is not only related to car sales price index for the history of this month, also with the network of steel car production, rubber tire car production, money supply of consumer goods, baidu's search traffic, consumer goods price index many factors, such as k month before you can get under the car sales data of multi-factor nonlinear autoregressive model is defined as:

$$x(t - k) = [x_1(t - k), x_2(t - k), x_3(t - k), x_4(t - k), x_5(t - k)] \tag{1}$$

In the above equation, each factor represents car sales in the first K months, steel production in the first K months, rubber tire production in the first K months, money supply in the first K months, and Internet search index in the first K months. On this basis, assuming that the errors of the base learner are independent of each other, the learning objective of the XBoost algorithm is to find F(t) and minimize the objective function, where $y_i^t = y_i^{(t-1)} + f_t(x_i)$ is the predicted result of the ith sample during the T th model training:

$$L^{(t)} = \sum_{i=1}^{n} l\left(y_i, y^{(t-1)} + f_t(x_i)\right) + \Omega(f_t) \tag{2}$$

3 Experiments

3.1 Experimental Background

In the past two years, the sales growth rate of domestic automobile market has changed from double digit to single digit, which means that blind expansion of production capacity will lead to overcapacity. Due to the capital-intensive, labor-intensive and technology-intensive nature of the automobile industry and its pillar position in the national economy, once there exists the problem of overcapacity in the automobile industry, it will not only cause huge waste of manpower and material resources, but also shake the foundation of the entire national economy. Therefore, the emergence of automobile sales forecasting model is very necessary.

3.2 Experimental Design

The data in this paper are from the sales data and indicators of A automotive enterprise over the years. Taking them as sample values, the importance of each feature in the modeling process to the objective function is plotted through plot_Importance module and Matplotlib module in XBoost. The XBoost algorithm USES the greedy algorithm to segment the training data by recursively selecting the optimal features of the tree structure from the root node. Assuming that IL and IR are the sample sets on the left and right of the segmentation point respectively, I = IL ∪ IR, the information gain of each segmentation scheme is calculated. The segmentation with the maximum information gain is the optimal segmentation of the node. Its model is shown in Eq. (3):

$$y_i = \phi(x_i) = \sum_{k=1}^{k} f_k(x_i) \tag{3}$$

Where $f_k \in F$ and K is the number of CART trees. The XBoost algorithm retains the prediction of the previous T−1 round at each model training, adds a new function F(t) to the model, and then tests the correlation coefficient of the explanatory variables in the model through the model's multicollinearity test. Finally, Eviews6.0 is used to test the correlation of data. In order to avoid heteroscedasticity of the model, logarithmic transformation is performed on the data. Some experimental results are shown in Table 1.

Table 1. Experimental results

	The number of data	MSE	R
Training data	68	4.27629e−17	9.99999e−1
Validation data	14	3.07607e−2	5.10501e−1
The test data	14	6.93728e−2	2.18183e−1

4   Discussion

4.1   Analysis Based on Multi-source Heterogeneous and XBOOST Vehicle Sales Prediction Model

As shown in Fig. 1, MSE is the mean square error between the predicted sales volume and the actual sales volume. The smaller the MSE, the better the prediction effect. When MSE = 0, the predicted sales volume = the actual sales volume. This is because the amount of sample data is small. Only by increasing the sample size can the accuracy of machine learning prediction be improved. Except for the relatively small correlation coefficient between the fuel price index and other variables, the other correlation coefficients are mostly above 0.7, in which the number of passenger vehicles in service is negatively correlated with other factors. For example: the operation of the passenger car and the energy consumption per capita life negative correlation relationship, that is the operation of the passenger car ownership increases, such as the increase in the number of the bus, so more people can take a bus rather than drive, thus reducing the energy consumption, per capita energy consumption is decreased. Therefore, it shows that there is serious multicollinearity among the above explanatory variables.

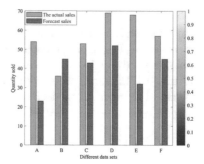

Fig. 1. Comparison of predicted sales volume and actual sales volume in the prediction model

In the figure above, A, B, C, D, E and F respectively refer to the main factors affecting China's car sales: per capita private car ownership, per capita domestic energy consumption, road mileage, operating passenger car ownership, fuel price index and steel production. In the process of building the model, we normalized the data of explanatory variables, so the power index of explanatory variables in the equation cannot be used to judge the degree of influence of a certain factor on automobile sales. The automobile sales forecast model we established contains 6 influencing factors. As long as we know the specific values of these 6 factors and substitute them into the model, we can predict the automobile sales of the corresponding year. This has certain guiding function to the automobile enterprise reasonable formulation production plan. China's automobile sales and fuel price index show a rising trend with the passage of time. Fuel is the driving force of cars. According to statistics, most buyers consider fuel consumption as the primary factor when buying a car less than 10,000 yuan. So a change in the price of fuel inevitably leads to a change in car sales. After Granger causality test, fuel price index is the Granger cause of automobile sales change. In addition, steel production and car sales are increasing over time. After Granger causality test, steel production is the Granger cause of car sales change. After analysis, the above factors have a certain degree of impact on China's car sales. After granger causality test, the above factors are all granger causes of automobile sales, that is, changes in the above factors will cause changes in automobile sales.

As shown in Fig. 2, it is the comparison of grey correlation coefficients of different factors. It can be known that the gray correlation coefficient of steel price is the largest, that is, the driving force of steel price on automobile sales is the largest. The iron and steel industry is the pillar industry of China's national economy and an important symbol of the country's economic level and comprehensive strength. The development of the iron and steel industry directly affects other industries related to it. With the increase of construction and manufacturing demand, China's steel industry has also developed rapidly. Automobile output directly affects automobile sales. For each additional 10,000 automobiles, it is necessary to increase the use of 98,000 tons of steel. In a sense, the steel output restricts automobile production and thus affects automobile sales. Therefore, steel production should be the biggest factor affecting auto sales. In addition, with the rapid development of China's economy, automobile enterprises have also developed. More and more people will buy cars, so that the number of cars will continue to increase, forcing the construction of roads to increase. Especially in some big cities, the urban traffic environment hinders the sales of cars to a great extent. The third driving force is per capita energy consumption, which is an important indicator to measure a country's energy consumption level. Although China is a world energy consumer, its per capita consumption is only 62% of the world level due to its large population. Normal use of cars requires energy consumption, so as per capita energy consumption increases, so does vehicle sales. The fourth driving force is owner-occupied passenger vehicles, which are used as "replacement supplements" for passenger vehicles. An increase in the number of passenger cars will make public transportation more convenient and consumer prices lower, which will reduce consumer demand for passenger cars and affect sales.

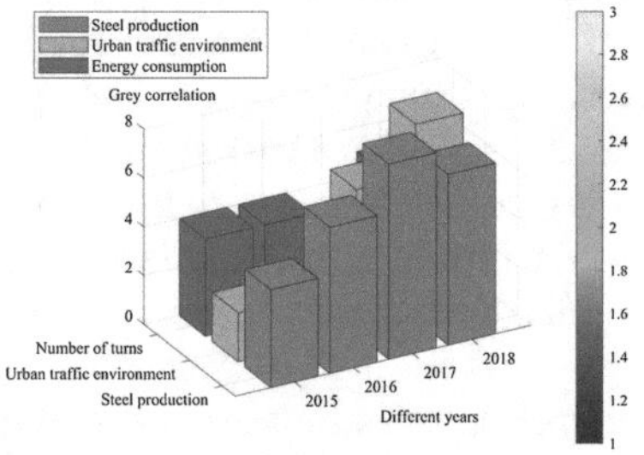

Fig. 2. Comparison of grey correlation coefficients of different factors

Drivers of both the per capita and fuel price indices came in second, suggesting that the impact of these two factors on car sales was less pronounced. With the development of economy, fuel is no longer an important factor hindering automobile consumption. The driving force ranks last is the average price of the automobile market. In fact, it can be observed from the original data that the average price of the automobile market has been on a downward trend for ten years. With the progress of science and technology, the cost of the automobile industry has been reduced, so the price has also been reduced. The price of a car, though a consideration when buying a car, is no longer the dominant factor.

As the sample data size is slightly larger, the calculation speed will be affected by the algorithm. Experiments show that the prediction accuracy of XBoost algorithm reaches the maximum when the number of iterations is 1000, and the modeling process takes about 9 s on average. After the integration of XBoost algorithm, the prediction accuracy reaches the maximum when the number of iterations is 500, and the modeling process takes about 50 s. When the number of decision trees is 500, the prediction accuracy reaches the maximum, and the modeling process takes about 120 s on average. The XBoost algorithm and its integration are significantly faster than other algorithms. To sum up, the user score prediction model based on XBoost algorithm has high prediction accuracy and computing efficiency.

4.2 Recommendations Based on Multi-source Heterogeneous and XBOOST Vehicle Sales Forecasting Model

From the perspective of national policies, the importance and support given by the state to the development of new energy vehicles have a great incentive and guiding role for enterprises. Especially in the initial stage of new energy technology, state support is very important. Many enterprises have regarded new energy vehicle technology as one of the important basis points for future "profit enterprises". The full text of the Auto Industry Adjustment and Revitalization Plan has been released, which puts forward a clear direction for the development of China's automobiles: namely, the production and marketing scale of electric vehicles. We will transform existing production capacity to produce 500,000 new energy vehicles, including pure electric, plug-in hybrid, and ordinary hybrid electric vehicles. Sales of new energy vehicles account for about 5% of total passenger vehicle sales. The main passenger vehicle production enterprises shall have new energy vehicle products that have passed the certification. In order to ensure the ultimate realization of this goal, enterprises need to comb and adjust their new energy vehicle strategies. The industrial process of China's new energy vehicles has just entered the climbing stage, and the Plan undoubtedly depicts a future technology roadmap for China's automobile industry.

For the research of automobile sales forecast, the traditional time series analysis model is mostly established at present. They have two shortcomings: they only use the historical sales data to solve the problem of automobile sales forecasting. In fact, there are many factors affecting automobile sales forecasting, such as raw material factors, consumer factors, network communication factors and macroeconomic factors. It should be assumed that there is a linear relationship between historical sales and sales. In fact, there is a highly non-linear relationship between sales and historical sales and other influencing factors.

The key of integrated learning is how to generate good and different individual learners. The higher the accuracy and diversity of individual learners are, the better the integration effect will be. Compared with LR, NB and other stable classification models, tree model is an unstable classification model which is sensitive to sample disturbance. Decision tree is often used as an individual learner of integrated learning because of its simplicity and intuition and strong interpretability. Compared with ID3 and C4.5, CART USES binary recursive partition to construct a binary tree, and the split feature can be used repeatedly, both for classification and regression. Based on the advantages of integrated learning, a score prediction model was established by using XBoost algorithm based on CART tree as the learner at the bottom of the model. At the same time, in order to increase the diversity of individual learners in integrated learning and improve the generalization ability of the model, data sample perturbations and attribute perturbations are added at the top level with Bagging idea to build multiple good and different XBoost models.

In this paper, the algorithm model is improved on the basis of the traditional algorithm. At the bottom of the model, XBoost algorithm based on CART tree is used to build the scoring prediction model. At the same time, in order to increase the diversity of the integrated study of individual learning, improve the generalization ability of the model, on the top floor by Bagging thought to join the data sample disturbance, disturbance and properties to create a more good and different XBoost model, and USES the ballot for model integration, for a binary classification task, category marker for $\{y_0, y_1\}$, set down the sample x on XBoost model to predict the output is expressed as a two-dimensional vector $\{\phi_n^0(x), \phi_n^1(x)\}$, is on the sample x, by integrated forecast model N XBoost, predicted results can be expressed as the formula:

$$\hat{y}(x) = y \underset{\substack{\arg\max \\ j\in\{0,1\}}}{\overset{N}{\underset{n=1}{\sum}} \phi_n^j(x)} \tag{4}$$

5 Conclusions

This paper makes a preliminary study on the automobile sales forecast of multi-source heterogeneous and XBOOST. This paper makes a brief introduction to some forecasting theories at present, and analyzes the main factors affecting the growth of automobile sales in China. As these factors are constantly changing, automobile sales are also changing accordingly. Then, the correlation between these factors and auto sales is tested, and the existing auto sales forecasting model is optimized.

References

1. Wang Feng, H., Liang, Z.J., et al.: A semantics-based approach to multi-source heterogeneous information fusion in the Internet of Things. Soft. Comput. **21**(8), 2005–2013 (2017)
2. Heng, S., Zhigeng, F., Qin, Z., et al.: Research on an exponential distribution reliability function model based on multi-source heterogeneous data supplements. Grey Syst. Theo. Appl. **7**(3), 329–342 (2017)
3. Gupta, C., Jain, A., Joshi, N.: DE-ForABSA: a novel approach to forecast automobiles sales using aspect based sentiment analysis and differential evolution. Int. J. Inf. Retrieval Res. **9**(1), 33–49 (2019)
4. Junjie, G., Yanan, X., Xiaomin, C., et al.: Chinese automobile sales forecasting using economic indicators and typical domestic brand automobile sales data: a method based on econometric model. Adv. Mech. Eng. **10**(2), 168781401774932 (2018)
5. Sun, H., Dang, Y., Mao, W.: A decision-making method with grey multi-source heterogeneous data and its application in green supplier selection. Int. J. Environ. Res. Public Health **15**(3), 446 (2018)
6. Liu, Yu.: Multi-source heterogeneous data fusion based on perceptual semantics in narrowband Internet of Things. Pers. Ubiquit. Comput. **23**(3–4), 413–420 (2019)
7. Heng, S., Zhigeng, F., Qin, Z., et al.: Research on an exponential distribution reliability function model based on multi-source heterogeneous data supplements. Grey Syst. Theo. Appl. **7**(3), 329–342 (2017)
8. Netisopakul, P., Leenawong, C.: Multiple linear regression using gradient descent: a case study on Thailand car sales. J. Comput. Theoret. Nanosci. **23**(6), 5195–5198 (2017)
9. Yang, L., Qi, Z., Zhiping, F., et al.: Maintenance spare parts demand forecasting for automobile 4S shop considering weather data. IEEE Trans. Fuzzy Syst. **27**(5), 943–955 (2019)
10. Kim, N., Park, Y., Lee, D.: Differences in consumer intention to use on-demand automobile-related services in accordance with the degree of face-to-face interactions. Technol. Forecast. Soc. Chang. **139**, 277–286 (2019)

The Lorentz Component Algorithm for Calculating the Center of the Circular Motion of Charged Particles

Chaoyue Jin[(✉)]

School of Ocean Science and Technology, Panjin City 124000, Liaoning Province, China
jcygold@mail.dlut.edu.cn

Abstract. When the moving point charge enters the magnetic field, it will move in a uniform circular motion without external force. The traditional calculation method can obtain the moving radius and the center position of the circular motion of the moving point charge in the magnetic field according to the charged quantity q of the moving point charge, its own mass m, the moving speed v and the magnetic induction intensity B of the magnetic field. In order to express the accurate position of the center of the circle intuitively and realize the computer programming, a method of using the Lorentz force component formula to orthogonally decompose the speed and calculate separately is proposed here, and the straight line and parameter conditions are determined by the Lorentz formula and the left-hand rule, And finally determine the accurate position of the center of the circle through the intersection of two straight lines.

Keywords: Center of circle solution · Lorentz force component formula · Lorentz force · Electric charge movement

Physics is divided into many different fields. Electromagnetism is an important field of physics. There are applications of electromagnetic theory everywhere in our lives. Electromagnetics mainly involves the two most basic forces: Ampere force and Lorentz force [1]. Among them, the Lorentz force is the force of the magnetic field on the moving charge, which is more abstract [2], and if the moving point charge is only affected by the Lorentz force, the moving point charge will make a uniform circular motion [3]. It is relatively difficult and complicated to use the Lorentz formula to calculate the radius of motion and then use the left-hand rule to determine the position of the center of the circle in the two-dimensional coordinate system. It is relatively difficult and complicated. This also brings a lot of problems to the computer programming software to analyze the movement process of the point charge inconvenient. An algorithm that uses the Lorentz force component formula to determine the coordinates of the center of the circle is proposed below.

1 Algorithm Overview

In 1892, when Dutch physicist Hendrik Antoon Lorentz (1853–1928) established the classical electron theory, as a basic assumption, he gave a formula for the force of a

J. MacIntyre et al. (Eds.): SPIoT 2020, AISC 1283, pp. 348–353, 2021.
https://doi.org/10.1007/978-3-030-62746-1_51

magnetic field on moving charged particles

$$F = q\vec{v} \times \vec{B} \,[4]$$

(1)

Among them, the direction of the vector difference product can be judged by the left-hand rule. Now we expand the componentization algorithm based on this basic formula (essentially the component formula of the particle momentum theorem) [5].

1.1 Parameter Setting

Setting the parameters, assuming that the charge of the moving point is q, its own mass is m, the movement speed is v, and the location is (x, y); the magnetic induction intensity of the magnetic field is B.

1.2 Algorithm Description

First, the orthogonal decomposition along the coordinate axis is obtained v_x and v_y.

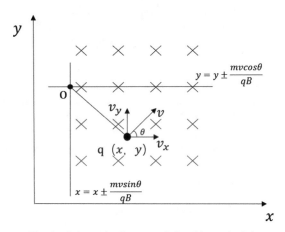

Fig. 1. Schematic diagram of algorithm principle

Among

$$\begin{cases} v_x = v \cos \theta \\ v_y = v \sin \theta \end{cases}$$

(2)

Apply the Lorentz formula and the left-hand rule to the two velocity components, and draw a straight line $y = y \pm \frac{mv \cos \theta}{qB}$. Straight line $(x - \frac{mv \sin \theta}{qB}, y + \frac{mv \cos \theta}{qB})$.

The or-symbol is determined by the direction of point charge movement and the direction of the magnetic field), as shown in Fig. 1 at the intersection of two straight lines $(x \pm \frac{mv \sin \theta}{qB}, y \pm \frac{mv \cos \theta}{qB})$. It is the center of the circle movement made by the moving point charge O.

2   Algorithm Proof

Suppose the velocity of the point charge and the direction of the magnetic field are shown in Fig. 1.

2.1   Two-Line Algorithm

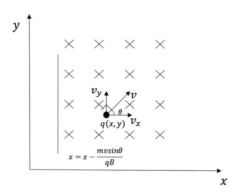

Fig. 2. Schematic diagram of the base line of the circle center

During this movement, the charged particles are, y Sub-velocity in direction v_y Produce edge x Lorentz force of direction f_x in x Sub-velocity in direction v_x Producing edge y Lorentz force of direction f_y [6], Positive cause f_x and f_y The existence of these two centripetal forces can determine that the center of the circle is x and y The corresponding distance between the direction and the charge of the moving point is the source of the two straight lines.

First of all v Obtained after orthogonal decomposition v_x, According to the formula:

$$qBx = m \cdot \Delta v_y \text{ [7]} \tag{3}$$

Available: $R_y = \frac{mv_x}{qB}$ determine with x The distance between the parallel axis and the point charge is $R_y = \frac{mv \cos \theta}{qB}$ Then use the left-hand rule to determine the specific orientation of the line, and finally add the original coordinates of the point charge to get the line shown in Fig. 2. $y = y + \frac{mv \cos \theta}{qB}$ The same goes for v_y Perform the same calculation and the left-hand rule to determine the straight line shown in Fig. 3. $x = x - \frac{mv \sin \theta}{qB}$ The intersection of two straight lines is the center of the circle O coordinate: $(x - \frac{mv \sin \theta}{qB}, y + \frac{mv \cos \theta}{qB})$.

2.2   Verification of the Accuracy of the Obtained Circle Center Position

The trajectory of a charged particle in a magnetic field has three elements, namely the center, the radius, and the central angle [8]. Therefore, the accuracy of the center position must be verified, otherwise the trajectory of the charged particle cannot be

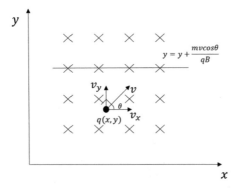

Fig. 3. Schematic diagram of the base line of the circle center

accurately analyzed. According to the characteristics of uniform circular motion and the characteristic that Lorentz force acts as the centripetal force, the conditions that must be met for the position of the center of the circle obtained are:

① The distance between the center of the circle and the point charge is the radius of the particle's circular orbit: $R = \frac{mv}{qB}$ [9].
② The line between the center of the circle and the point charge is perpendicular to the direction of velocity.
③ Satisfy the direction judgment of the left-hand rule.

2.2.1 Radius Verification

According to the distance formula between two points:

$$s = \sqrt{(x_1 - x_2)^2 + (y_1 - y_2)^2} \tag{4}$$

Getting some $x - \frac{mv \sin\theta}{qB}$, $y + \frac{mv \cos\theta}{qB}$ with points (x, y) The distance between is:

$$R = \sqrt{x - \frac{mv \sin\theta}{qB} - x^2 + y + \frac{mv \cos\theta}{qB} - y^2} = \sqrt{\left(\frac{mv \sin\theta}{qB}\right)^2 + \left(\frac{mv \cos\theta}{qB}\right)^2} = \frac{mv}{qB}$$

So it is proved.

2.2.2 Orientation Verification

The geometric relationship is shown in Fig. 4. Needs to be verified that the line between the center of the circle and the point charge is perpendicular to the direction of velocity, that is $\beta = \frac{\pi}{2}$.

According to the geometric relationship:

$$\cos\alpha = \frac{\frac{mv\sin\theta}{qB}}{s} = \frac{\frac{mv\sin\theta}{qB}}{\frac{mv}{qB}} = \sin\theta$$

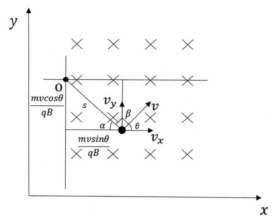

Fig. 4. Schematic diagram of geometric relationship

From this we can see α and θ Mutual redundancy, namely:

$$\alpha + \theta = \frac{\pi}{2} \tag{5}$$

Getting $\beta = \pi - (\alpha + \theta) = \pi - \frac{\pi}{2} = \frac{\pi}{2}$ so it is proved.

2.3 Left-Hand Rule Verification

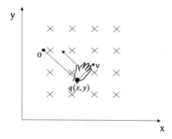

Fig. 5. Schematic diagram of the left-hand rule

It is verified by using the left-hand rule as shown in Fig. 5 that the magnetic field lines penetrate the palm of the hand perpendicularly, the four straight fingers point to the direction of the charge movement, and the thumb points to the center of the circle [10], the result is consistent with the position of the center of the circle obtained by the algorithm.

All three conditions are verified, which proves that the center position calculated by this algorithm has absolute accuracy.

3 Summary

The Lorentz component is used to solve the center of the charged particle in a uniform circular motion in the magnetic field. By orthogonally decomposing the speed of the charged particle in the magnetic field according to the coordinate axis direction, the accurate position of the center is obtained by combining the Lorentz formula and the left-hand rule Coordinates greatly simplify the expression of the center of the circle in the two-dimensional coordinate system, and make it easier to realize the idea of applying computer programming to analyze the point charge movement process in the future.

References

1. Kai, J.: Discuss the ampere force and Lorentz force from the perspective of energy conservation. Sci. Consult. **12**, 45–68 (2019)
2. Su, Y., Cao, Y., Zhang, Y.: Use the Lorentz force demonstrator to explore the movement of charged particles in electric and magnetic fields. Phys. Exp. **9**, 89–96 (2019)
3. Chen, J., Cheng, J.: Explore the method of solving the problem of charged particles moving in a magnetic field. Phys. Bull. **4**, 24–36 (2018)
4. Chen, B., Wang, J., Zhang, R.: How is the Lorentz force formula given? Univ. Phys. **8**, 89–97 (2008)
5. Chen, G.: Using the inference of Lorentz force component to solve physics problems cleverly. Phys. Teach. **6**, 112–125 (2010)
6. Cui, Y., Ma, C.: Using the Lorentz force component to solve the 2018 national volume finale problem. Chin. Phys. (High Sch. Ed.) **12**, 156–167 (2018)
7. Cui, Y., Ma, C.: Using the Lorentz force component to solve the 2018 national volume finale problem. Chin. Phys. (High Sch. Ed.) **12**, 178–183 (2018)
8. Yang, X.: The basic model of charged particles moving in a magnetic field. Phys. Teach. **4**, 64–75(2019)
9. Su, J.: The movement of charged particles. J. Xuzhou Educ. Inst. **3**, 145–158 (2008)
10. Li, C.: Left-hand rule, right-hand rule and right-hand spiral rule. J. Jingmen Vocat. Tech. Coll. **3**, 49–56 (2000)

Controller Design Based on LQR for Rotational Inverted Pendulum

Dong Wang[1]([⊠]), Xinjun Wang[2], and Liang Zhang[1]

[1] Bohai University, Jinzhou 121013, Liaoning, China
wdqn@sina.com
[2] School of Information Science and Engineering, Shandong Normal University,
Jinan 250014, China

Abstract. The rotational inverted pendulum is an unstable system with multiple variables. This study achieve the stability control of rotational inverted pendulum based on the LQR optimal control method. According to Lagrange dynamics equation modeling rotating pendulum. The paper makes a study of the weighted matrix Q and R. Designed controller and simulation according to the mathematical model by the Matlab. Results show the LQR method can achieve the stable control of rotational inverted pendulum.

Keywords: Rotational inverted pendulum · LQR · Optimal control

1 Introduction

Inverted pendulum is a typical nonlinear and unstable system, it is an effective platform for testing control method, which has been considered by many researchers most of which have used linearization theory in their control schemes [2]. Because the inverted pendulum is a typical nonlinear, it is a difficult assignment to control with classical methods. Inverted pendulum has two degrees of freedom and only one control input. At present, most of the research on inverted pendulum is straight line, due to the impact of mechanical structure, influenced the control effect [3, 4]. While the rotary inverted pendulum is not limited by space, which put forward higher requirements on control algorithm. The control method has a wide in some areas. Currently inverted pendulum control methods are mainly PID control, State feedback controller optimal control, Adaptive neural network system, Fuzzy Control and so on [1, 10]. Most of these methods are used to a straight line as the research object, while the rotary inverted pendulum has little research [5].

In this paper, the LQR optimal control method achieves a stable control based on rotational inverted pendulum. In the hardware structure, it reduced the intermediate transmission mechanism, with greater nonlinear, instability and complexity. The state equations of the system shown in (1).

$$\dot{X} = Bu + Ax. \tag{1}$$

J. MacIntyre et al. (Eds.): SPIoT 2020, AISC 1283, pp. 354–359, 2021.
https://doi.org/10.1007/978-3-030-62746-1_52

$$u(t) = -Kx(t) \tag{2}$$

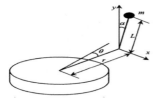

Fig. 1. System model

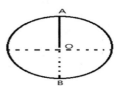

Fig. 2. Analysis of the balance position

Controlled quantity, shown in (2). Through the LQR function of the MATLAB got the best control matrix K, to minimize the performance index function. Compared with other control method, the method has the advantages of a small overshoot and the fast response speed, can realize the stable control of rotary inverted pendulum and finally it can get a good result.

2   Build a Mathematical Model to System

In order to realize the control of the inverted pendulum, building the system model, shown in Fig. 1, m is the mass of the pendulum, L is the pendulum rod to hinge centroid distance, R is the length of the rotary arm, α is the pendulum rod relative to the vertical direction of the zero angular displacement, θ is the angular displacement of rotating arm relative to the horizontal zero.

Based on the Lagrange equation, shown in (3). Where q is the generalized coordinates of the system, H is Lagrangian, V is the potential energy of system, T is the kinetic energy of system. In this system, $i = 1, 2, q = \{\theta, \alpha\}$. Obtained the following equation shown in (4) [4].

$$H(q, \dot{q}) = -V(q, \dot{q}) + T(q, \dot{q}) \tag{3}$$

Simplification and established the nonlinear equation of rotary inverted pendulum, shown in (5).

Where T_{output} is the output torque of the DC servo motor.

$$\begin{cases} \frac{\partial}{\partial t}\left(\frac{\partial H}{\partial \dot{\theta}}\right) - \frac{\partial H}{\partial \theta} = T_{output} - B_{eq}\dot{\theta} \\ \frac{\partial}{\partial t}\left(\frac{\partial H}{\partial \dot{\alpha}}\right) - \frac{\partial H}{\partial \alpha} = 0 \end{cases} \tag{4}$$

$$\begin{cases} (J_1 + mr^2)\ddot{\theta} + mlr[\sin\alpha(\dot{\alpha}) - \cos\alpha(\dot{\alpha})] = T_{output} - B_{eq} \\ \frac{4}{3}mL^2\ddot{\alpha} + mLr\sin\alpha(\dot{\theta}) - mLr\cos\alpha(\ddot{\theta}) - mgL\sin\alpha = 0 \end{cases} \tag{5}$$

Where $T_{output} = \eta_m \eta_g K_t K_g (V_m - K_g K_m \dot\theta)/R_m$, Selecting $\theta, \dot\theta, \alpha, \dot\alpha$ for state variables, considering the inverted pendulum in the initial position, $\alpha, \theta \ll 1\,rad$, $\cos\alpha \approx 1$, $\sin\alpha \approx 0$. Simplification and obtaining the following local linear equation of state.

$$
\begin{bmatrix} \dot\theta \\ \ddot\theta \\ \dot\alpha \\ \ddot\alpha \end{bmatrix} = \begin{bmatrix} 0 & 1 & 0 & 0 \\ 0 & \frac{-4G}{4J_1+mr^2} & \frac{3mgr}{4J_1+mr^2} & 0 \\ 0 & 0 & 0 & 1 \\ 0 & \frac{-3rG}{L(4J_1+mr^2)} & \frac{3g(J_1+mr^2)}{L(4J_1+mr^2)} & 0 \end{bmatrix} \begin{bmatrix} \theta \\ \dot\theta \\ \alpha \\ \dot\alpha \end{bmatrix} + \begin{bmatrix} 0 \\ \frac{4\eta_m\eta_g K_t K_g}{R_m(4J_1+mr^2)} \\ 0 \\ \frac{3r\eta_m\eta_g K_t K_g}{LR_m(4J_1+mr^2)} \end{bmatrix} V_m \tag{6}
$$

Where V_m is the output of the controller, for the armature voltage of DC servo motor. $G = (\eta_m \eta_g K_t K_m K_g^2 + B_{eq} R_m)/R_m$.

3   Integrated Design Based on LQR

The pendulum rod position has two kinds of equilibrium state, including straight up and straight down. The equilibrium analysis is shown in Fig. 2. The vertical downward status is the stable equilibrium point as is shown in B. The vertical upward status is the unstable equilibrium point as is shown in A. In the case of an applied control, just applying slight perturbations, it will make the system deviates from the equilibrium point and diverging shock, and finally returned to the stability of the equilibrium point B. The inverted pendulum control target is to become a stable movement in the unstable equilibrium point.

Quadratic optimal control problem is under the system of linear constraints, selecting the control input to made the quadratic objective function minimum [6]. The purpose is designing linear quadratic regulator for system, referred LQR [9].

For an linear time invariant systems equation

$$
\dot X = BU + AX \tag{7}
$$

$$
Y = DU + CX \tag{8}
$$

This performance index is taken as

$$
J = 0.5 * \int_0^\infty (U^T RU + X^T QX)dt \tag{9}
$$

Where U is r dimensional input vector, X is n dimensional state vector, Y is m dimensional output vector. Q and R is weighting matrix, which is used to balanced between state vector and the input vector. $R > 0, Q \geq 0$.

If the system is out of balance, choosing appropriate controls U^* in order to make the system back to the state of equilibrium and made J to a minimum. U^* is called optimal control.

Based on the optimal control theory, the optimal control is as follows.

$$
U^* = -R^{-1}B^T PX = -KX \tag{10}
$$

Where K is the gain matrix, P is the solution of the equation, then solving the algebraic Riccati equation

$$PA - PBR^{-1}B^T P + A^T P + Q = 0 \qquad (11)$$

$$K = R^{-1}B^T P = [k_1, k_2, k_3, k_4]^T \qquad (12)$$

Based on the modern control theory, optimal control is $U = -Kx$, making the quadratic performance index of system $J = 0.5 * \int_0^\infty (U^T RU + X^T QX)dt$ the minimum. The mechanical parameters of the inverted pendulum device values here shown in Table 1.

Table 1. The system physical parameters table

Symbol	Physical significance	Numerical and unit
L	The pendulum rod to the shaft center distance/m	0.165
M	The swing rod quality/kg	0.105
R	The length of the rotary arm/m	0.175
J_1	Moment of inertia of rotating arm/$kg \cdot m^2$	2.0×10^{-3}
J_2	Moment of inertia between the pendulum and center of mass/$kg \cdot m^2$	9.53×10^{-4}
g	Acceleration of gravity/$m \cdot s^{-1}$	9.8
K_t	Motor torque coefficient/$N \cdot m \cdot A^{-1}$	$7.767 \times 10^{=3}$
Symbol	Physical significance	Numerical and unit
K_m	Reverse potential coefficient/$V \cdot s \cdot rad^{-1}$	$7.767 \times 10^{=3}$
K_g	Transmission gear ratio	5 : 1
R_m	DC motor armature resistance/Ω	2.6
η_m	DC motor efficiency/%	69
η_g	Transmission efficiency/%	90
B_{eq}	Viscous damping coefficient/$N \cdot ms \cdot rad^{-1}$	4.0×10^{-3}

Putting the mechanical parameters shown as Table 1 into the system state equation.

$$\begin{bmatrix} \dot{\theta} \\ \ddot{\theta} \\ \dot{\alpha} \\ \ddot{\alpha} \end{bmatrix} = \begin{bmatrix} 0 & 1 & 0 & 0 \\ 0 & -1.4886 & 48.1672 & 0 \\ 0 & 0 & 0 & 1 \\ 0 & -1.1841 & 82.8603 & 0 \end{bmatrix} \begin{bmatrix} \theta \\ \dot{\theta} \\ \alpha \\ \dot{\alpha} \end{bmatrix} + \begin{bmatrix} 0 \\ 1.5666 \\ 0 \\ 1.2462 \end{bmatrix} V_m \qquad (13)$$

In order to obtain the control matrix K, selecting the appropriate weighting matrix Q and R, the purpose is to balance the sensitivity of the system between output and input.

Choose Q = [2.3 0 0 0; 0 4.5 0 0; 0 0 1.5 0; 0 0 0 0.5]; R = 1. According to LOR optimal regulator theory, calling the statement that the MATLAB provided: [K,S,E] = lqr(A,B,Q,R,N), then obtained control matrix K.K = [−1.5166 −4.3114 175.3191 21.1769]. The eigenvalues of the system is $a_1 = -10.3165$, $a_2 = -8.3517$, $a_3 = -0.6990$, $a_4 = -1.7572$. From the eigenvalues of the system could draw a conclusion that system can achieve the stable at equili- brium point [7].

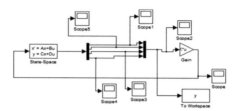

Fig. 3. Simulation model of Simulink system structure diagram

4   System Simulation

First simulation in the Matlab simulink environment, establishing the model shown in Fig. 3. Getting the response curve of the system simulation. Pendulum rod angle and angular velocity curve is shown in Fig. 4. Arm angle and angular velocity curve is shown in Fig. 5.

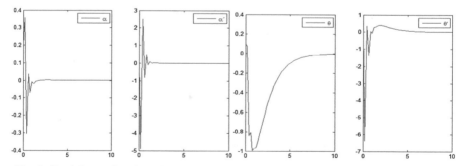

Fig. 4. Pendulum rod angle and angular velocity curve

Fig. 5. Arm angle and angular velocity curve

It can be seen from Fig. 4 that the pendulum angle is not greater than 0.4 *rad*, and tended to 0 in the 5 s, finally stabled at the equilibrium state. It can be seen from Fig. 4 that the maximum angle of rotating arm is no more than 1 *rad*, finally tended to 0. The control system has good control effect and achieves the expected goal.

After many simulation experiments, the inverted pendulum can be stabilized in the equilibrium state, On the one hand, the overshoot of pendulum angle is very small, and it has fast response speed. On the other hand, the inverted pendulum with a little error and could achieve the optimal control.

5 Conclusions

First, modeling for the rotational inverted pendulum by Lagrange equation, achieved the stable control of the inverted pendulum with the LQR method. By the Matlab simulation curve could seen the system with quick respondence, the control effect is satisfactory. The method is simple in principle, could realize the stable control of rotational inverted pendulum. The method is to lay the groundwork for future research.

Acknowledgements. This work was supported by Liaoning Natural Science Foundation under Grant No. 20180550189 and No. 2019-ZD-0491

References

1. Zhao, X., Zhang, L., Shi, P., Karimi, H.R.: Novel stability criteria for T-S fuzzy systems. IEEE Trans. Fuzzy Syst. **21**(6), 1–11 (2013)
2. Bugeja, M.: Non-linear swing-up and stabilizing control of an inverted pendulum system, In: EUROCON Ljubljana, Slovenia (2003)
3. Craig, K., Awtar, S.: Inverted pendulum systems: rotary and arm-driven a mechatronic system design case study. Mechatronics **12**, 357–370 (2001)
4. Shiriaev, A.S., Friesel, A., Perram, J., Pogromsky, A.: On stabilization of rotational modes of an invertedpendulum. In: Proceedings of the 39th IEEE Conference on Decision and Control, Sydney, NSW Australia, vol. 12, no. 5, pp. 5047–5052 (2000)
5. Krishen, J., Becerra, V.M.: Efficient fuzzy control of a rotary inverted pendulum based on LQR mapping. In: IEEE International Symposium on Intelligent Control, Germany, pp. 2701–2706 (2006)
6. Mrad, F., El-Hassan, N., Mahmoud, S.E.H., Alawieh, B., Adlouni, F.: Real-time control of free-standing cart-mounted inverted pendulum using LabVIEW RT. In: Conference Record of the 2000 IEEE Industry Applications Conference, Rome, Italy, vol. 10, no. 2, pp. 1291–1298 (2000)
7. Park, J.I., Lee, S.G.: Synthesis of control inputs for simultaneous control of angle and position of inverted pendulum. In: Proceedings of 1996 4th International Workshop on Advanced Motion Control, Mie, Japan, vol. 3, no. 2, pp. 619–624 (1996)
8. Rotary Inverted Pendulum User Guides & Laboratories. Quanser Inc. (2003)
9. Ang, K.H., Chong, G., Li, Y.: LQR control system analysis, design, and technology. IEEE Trans. Control Syst. Technol. **13**(4), 559–576 (2005)
10. Du, G., Huang, N., Wu, G.: The rotational inverted-pendulum based on DSP controller. In: Proceedings of the 4th World Congress on Intelligent Control and Automation, Shanghai, China, vol. 7, no. 4, pp. 3101–3105 (2002)

Three-Dimensional Scenic Spot Roaming System Based on Virtual Reality Technology

Yujuan Yan[1][✉], Shuo Wang[1], Jialin Li[2], and Suyang Li[1]

[1] Jilin Engineering Normal University, Changchun, Jilin, China
601206679@QQ.com
[2] Changchun Institute of Land Surveying and Mapping, Changchun, Jilin, China

Abstract. With the development of human civilization and the progress of society, some intelligent technologies such as computer software technology and electronic information technology are gradually improved. Data virtual technology has become a new development direction. In this context, virtual reality technology came into being. In this era of rapid economic development, tourism has become a way of life accepted and practiced by people. Therefore, in this information age, many scenic spots began to use the Internet to promote scenic spots, and the research of three-dimensional scenic spot roaming system based on virtual reality technology has started. In this paper, through the method of virtual reality modeling and roaming, combined with the specific situation of the scenic spot, the three-dimensional roaming model of the scenic spot is established, and the real-time three-dimensional roaming of multiple scenes and the humanized design of the system are realized. In the previous research, we found that virtual and augmented reality technology can directly use computer technology to generate a more realistic three-dimensional roaming environment of human-computer simulation, which makes users completely immersed in the virtual environment. The direct interaction between users and virtual environment has been realized through the virtual human-computer interaction interface. Through data analysis, it can be found that the overall success rate of transaction completion and successful transaction reaches 100%.

Keywords: Computer technology · Virtual reality · Three-dimensional scenic spots · Odyssey system

1 Introduction

Virtual reality [1, 2] system is a computer system based on virtual reality technology, which generates virtual three-dimensional scene through computer software and its external equipment. Through the interaction and feedback of visual, auditory, tactile and other senses, everyone can feel the feeling of being in the scene. At present, virtual augmented reality technology has been widely used in military, entertainment, architecture, geology, medicine, geographic information system and other fields and industries with its unique technical advantages. At present, virtual augmented reality technology

J. MacIntyre et al. (Eds.): SPIoT 2020, AISC 1283, pp. 360–366, 2021.
https://doi.org/10.1007/978-3-030-62746-1_53

is widely used in modern computer graphics, multimedia electronic information technology, artificial intelligence [3–5] technology and other research and application fields, further promoting the development of computer vision simulation technology. How to reproduce the real world information more intuitively is the purpose of virtual reality technology and visual simulation technology.

Since the 21st century, with the continuous improvement of modern computer and software technology [6], E-commerce Internet technology, virtualization and reality Internet technology, database management technology and network software and hardware automation technology, etc., it has laid a solid technical foundation for the security and sustainable development of China's digital tourism. Digital tourism and digital earth technology are reflected in the development of Digital Tourism in China. Digital tourism technology is based on three-dimensional simulation technology and network technology, which makes the display of scenic spots or landscapes more vivid, vivid and intuitive. It also has search function, and can make full use of video, panoramic pictures, virtual reality technology and other means to provide tourists with comprehensive tourism information services, such as food, accommodation, transportation, tourism, shopping and entertainment, so as to comprehensively improve the service quality and improve the internationalization level and core competitiveness of Tourism.

This paper starts with the theme of the development of the times, because under the background of the information age, any industry must follow the pace of the development of the times and improve its core competitiveness, because in this era of rapid development, a small outdated problem may be eliminated. Based on virtual reality technology, the 3D roaming system of scenic spot is studied. In this paper, through the modeling and roaming method of virtual reality technology, combined with the specific situation of the scenic spot, the three-dimensional model of the scenic spot is established, and the real-time scene roaming system is designed. In this way, users can better interact with the virtual environment.

2   Related Technologies of 3D Scenic Spot Roaming System

Virtual reality technology is an advanced computer user interface. It provides virtual reality users with a way and means to simultaneously help users realize various intuitive and natural computer real-time visual and perceptual interaction, such as watching, listening, touching, etc., so as to maximize convenience and help users improve the speed and convenience of system operation on the computer, Thus, it greatly reduces the psychological pressure and economic burden of the system users, and improves the network security and work efficiency of the whole system. According to the different application objects between system virtual and reality, the functions of virtual reality can be expressed in different forms, such as designing or conceiving a certain concept.

2.1   Image Feature Point Extraction and Matching Technology

Accurate extraction of image feature points is the basis of camera calibration, image matching, image retrieval and 3D reconstruction. Therefore, image feature point extraction technology has been a hot research topic in computer image processing and computer vision.

In order to make the space of feature points have good scale invariance, scale space is introduced. The theory of scale space was first put forward by Witkin in 1983. It is a Gaussian function space which scales the original image with different feature points $G(x, y, \sigma)$ a mathematical analysis method to map the feature points of original image to scale space by convolution operation. Small scale transform theory of feature points can accurately show the overall details and local features of the original image, while large scale transformation theory can well describe the overall overview and characteristics of the original image. Koendetink proved that Gaussian convolution kernel is the only convolution kernel which can realize the only scale transformation of Gaussian scale linear transformation, and Lindeberg further proved that Gaussian kernel is the only Gaussian linear convolution kernel.

The scale space of a two-dimensional gray image is defined by formula (1):

$$L(x, y, \sigma) = G(x, y, \sigma) * I(x, y) \tag{1}$$

$$G(x, y, \sigma) = \frac{1}{2\pi\sigma^2} e^{-(x^2+y^2)/2\sigma^2} \tag{2}$$

$G(x, y, \sigma)$ It's a scale variable Gaussian function, $I(x, y)$ it is the grayscale image of the original image, symbol $*$ represents convolution operation, (x,y) represents the position of pixels in the image, σ is the scale space factor. σ the smaller the value is, the smaller the corresponding scale is, the more image details are retained. Along with σ gradually, the image is smoothed more and more, leaving only the general picture of the image.

2.2 Virtual Reality Roaming Based on Geometric Model

Virtual reality roaming based on geometric modeling is the basic drawing principle of computer graphics. First of all, we need to digitize and abstract simulate the 3D scene, build it in the virtual 3D landscape model structure (usually including building site, natural landscape effect, etc.), and set the 3D model of material information. Then, according to the requirements of on-site display, in order to achieve a more realistic sensory effect, texture mapping technology usually needs to improve the appearance of the system model. Finally, by setting the control parameter set required for rendering the scene, the observer's position information is obtained through the interactive controller, and the lighting parameters of the system position are rendered to the output device in real time to complete the walkthrough of the whole scene.

3   Experimental Background and Design

3.1   Experimental Background

Accurate extraction of image feature points is the basis of camera calibration, image matching, image retrieval and 3D reconstruction. Therefore, image feature point extraction technology has been a hot research topic in computer image processing and computer vision. Image feature points refer to the points with obvious features in the image. It is usually different from neighborhood points in such attributes as brightness, color, curvature or texture. For example: corner point, line intersection point, discontinuity point, maximum curvature point on contour line, etc. Feature point extraction technology is to detect and describe the points different from neighborhood points in image. The selected feature points should be obvious, easy to extract and have enough distribution in the image. In order to uniquely identify each feature point and the subsequent feature point matching module, a small neighborhood of the feature point is usually selected as the center point, and the description vector of the feature point is generated according to a certain measurement method.

3.2 Experimental Design

In order to find a general feature point extraction algorithm, this experiment will compare SIFT algorithm and surf algorithm from three aspects: the number of feature points extracted by the algorithm, the time required by the algorithm, and the number of correct matching point pairs. The comparison of sift and surf detection feature points is shown in Table 1. The number of feature points extracted from each graph, the time spent and the number of correct matching point pairs are recorded in Table 1. In order to make the data comparable, the same feature point matching algorithm is used to match the feature points extracted by SIFT and surf. From the data in Table 1, compared with surf algorithm, SIFT algorithm can extract more feature points and correctly match point pairs, but it takes a long time. Compared with SIFT algorithm, surf runs faster than SIFT algorithm. In the case of better image quality, the number of effective feature points detected can meet the requirements of estimating model parameters in the process of panorama generation. Therefore, surf can be used as a feature point detection method.

Table 1. Comparison of sift and surf performance

		SIFT			SURF		
		Number of characteristic points (piece)	Time required (seconds)	Match point pairs correctly (yes)	Number of characteristic points (piece)	Time required (seconds)	Match point pairs correctly (yes)
Group 1	(a)	599	1.654	119	264	0.121	65
	(b)	487	0.965		295	0.234	
Group 2	(a)	712	1.414	59	164	0.414	36
	(b)	469	0.885		135	0.144	
Group 3	(a)	255	0.654	15	91	0.103	3
	(b)	295	0.986		116	0.156	

4 Discussion

4.1 System Testing

This paper briefly introduces and demonstrates the function test of some modules of the system. The purpose of testing is to find out the potential errors and defects of the system, and quickly correct the errors and defects, so as to improve the life cycle and quality of the system, and avoid the potential defects caused by various risks. In this test, the actual business scenario is simulated by continuous pressurization. The purpose of this report is to describe the maximum number of users that the server can support in this test, and to show the response time and server resource usage in this case. This test is based on a high concurrency connection, and the number of virtual users is 6000.

Use the pressure mode to set the execution frequency of all users for 3 h. The system is equipped with optical fiber network and better server hardware equipment. After testing, the roaming system can support more than 6000 concurrent connections, and the e-commerce network can support more than 6000 concurrent connections. The overall success rate of completed transactions and successful transactions obtained from system logging is 100%. When the roaming system is concurrent with 6000 users, the average reading time of a single scene is about 8 s; when the roaming system is concurrent with 4000 users, the average reading time of a single scene is about 6 s; when the roaming system is concurrent with 2000 users, the average reading time of a single scene is about 2 s. The specific test situation is shown in Fig. 1.

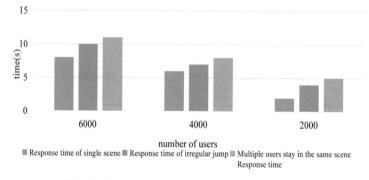

Fig. 1. Concurrent response time of roaming system

In the 3D panoramic roaming system, each scenic spot has its own commentary dubbing. Under the effect of 3D display of real scene, audio interpretation brings users a new display mode, which makes users feel immersive. Visitors can freely shuttle around any scenic spot by operating the mouse to realize real-time scene, move the real scene to users, and fully display the environmental conditions of the scenic spot. Three dimensional panoramic technologies based on multimedia image, audio and programming language can easily show the scenic spot in front of tourists, which is the most powerful weapon for tourism promotion and promotion. After 3 h of stress testing, the CPU utilization is in the ideal value, and the available memory is within the normal threshold. The details are shown in Fig. 2.

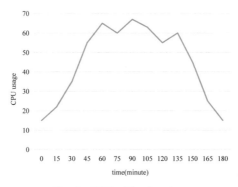

Fig. 2. CPU utilization chart

As shown in Fig. 2, in the first hour, the CPU utilization gradually increases. When it reaches the first peak of 65, the CPU utilization begins to decline. In the second hour, the CPU utilization is basically around 65, and the CPU fluctuates around the first peak 65. In the third hour, the CPU utilization begins to decline. The CPU occupancy peak test shows that the system can basically meet the needs of a large number of users online at the same time, and the system has good stability.

4.2   Development Characteristics of Virtual Reality Technology

Virtual reality technology is a kind of three-dimensional virtual world and real environment established by computer and information technology. In the environment of computer virtual reality, the user's computer can enter the virtual environment through the manual operation and participation of the user, and get the feeling of being in the scene both visually and audibly. The information technology of virtual computer virtual reality and information technology provide our computer and modern industrial users with an interactive three-dimensional virtual world and a real virtual living environment visually. They can interact with any object in the virtual environment world independently, so that users can produce a variety of comprehensive feelings and experiences including vision, hearing and touch. Generally speaking, virtual world and reality technology should have good immersion, autonomy and good interaction in their use. The combination of the two makes users produce a concept similar to reality, that is, the three main characteristics of virtual reality: interactivity, immersion and imagination.

The development of virtual reality technology is very fast, and its application fields and prospects are also very broad. From the current development of information technology, the network has entered people's life, work and learning. At the same time, due to the continuous progress of modern computer image compression and video display processing technology and the continuous improvement of network corresponding software and hardware and equipment, it is not difficult to judge and predict the future development trend of virtual reality network technology. This development trend can roughly include two main development directions: business exhibition, education and training, simulation games and so on. Because the hardware requirement of desktop virtual reality is relatively low, it can be realized by a single PC. At the same time, due to the rapid development of Internet, desktop virtual reality will be more closely combined with it, because in many high-tech fields, such as aerospace, military training, simulation training, due to various special requirements, only simulation experiments can be carried out. Simple desktop virtual reality can no longer meet the requirements, and can only build high-performance immersive virtual reality. At present, many high-tech virtual reality simulation systems have been established at home and abroad.

5   Conclusions

This paper is based on the virtual reality technology of three-dimensional scenic roaming system research, with the development of the times, computer technology and information technology, virtual technology is gradually applied in people's daily life, so, in order to improve their core competitiveness, follow the development of the times, ready to combine virtual reality technology with the three-dimensional model of scenic spots To make it more convenient for users to visit scenic facilities and buildings. In this paper, through the method of virtual reality modeling and roaming, combined with the specific situation of the scenic spot, the three-dimensional model of the scenic spot is established, and the real-time scene roaming system is designed. In this way, users can better interact with the virtual environment.

Acknowledgements. This work was supported by Jilin Engineering Normal University—technologies of Three-Dimensional Scenic Spot Roaming System Based on Virtual Reality Technology.

And this work was supported by the college students innovations special project funded by Jilin Engineering Normal University.

References

1. Bastug, E., Bennis, M., Medard, M., et al.: Toward interconnected virtual reality: opportunities, challenges, and enablers. IEEE Commun. Mag. **55**(6), 110–117 (2017)
2. Nieder, G.L., Scott, J.N., Anderson, M.D.: Using quicktime virtual reality objects in compute-assisted instruction of gross anatomy: Yorick—the VR Skull. Clin. Anat. **13**(4), 287–293 (2015)
3. Jeavons, A.: What is artificial intelligence? Res. World **2017**(65), 75 (2017)
4. Raza, M.Q., Khosravi, A.: A review on artificial intelligence based load demand forecasting techniques for smart grid and buildings. Renew. Sustain. Energy Rev. **50**, 1352–1372 (2015)
5. Bundy, A.: Preparing for the future of artificial intelligence. AI Soc. **32**(2), 285–287 (2017)
6. David, B.: Computer technology and probable job destructions in Japan: an evaluation. J. Japanese Int. Econ. **43**, 77–87 (2017)

Research and Development of Simulation System for Industrial Robot Stamping Automation Production Line Based on Virtual Reality

Songqing Liu$^{(\boxtimes)}$ and Ying Chen

College of Mechanical Engineering, Jilin Engineering Normal University, Changchun 130052, Jilin, China
463499342@qq.com

Abstract. With the increasing demand for automation of production lines, robots are increasingly used in industrial production, making robot production line simulation technology a hot research topic. The virtual test run system of the robot is based on the layout plan of the robot production line to carry out three-dimensional visual simulation of the production line, to verify the rationality of the layout plan, and to avoid unnecessary losses caused by blind production. However, the current research on simulation software is lacking, so the research of robot virtual trial operation system has important significance for domestic industrial development. The purpose of this paper is to use the virtual prototype to solve the problems in the design stage before manufacturing the physical prototype of the stamped product, and contribute to the research of the robotic stamping automation production line. This paper is mainly based on the advantages of high flexibility of robot stamping production line, low mold requirements, simple programming of new workpieces, virtual prototyping technology, computer science and other disciplines, physical simulation, full simulation, and finally the same physical prototype test production method.

Keywords: Stamping production line · Virtual prototype technology · ADAMS · Simulation research

1 Introduction

Since the press needs to wait for the robot to complete the corresponding command, resulting in low production efficiency, the clutches and brakes of the press are frequently turned on and off, reducing the service life. Therefore, it is crucial to optimize the coordination mode of the press and the robot movement to improve the tempo and smoothness of the production line.

In recent years, research on stamping production lines has attracted widespread attention. In 2016, QiuXuesong et al. [1] used a multi-software co-simulation strategy of interface technology to establish a mathematical model of the press line in order to study the large-scale robotic stamping production line. In 2016, Wang Yannian et al. [2]

J. MacIntyre et al. (Eds.): SPIoT 2020, AISC 1283, pp. 367–373, 2021.
https://doi.org/10.1007/978-3-030-62746-1_54

proposed a design scheme of the three-dimensional feeding robot control system based on the joint control of PLC and motion controller in order to improve the automation degree of the press production line. In 2017, Fang Shihui et al. [3] realized the full automatic production of water heater liner molding in order to improve the traditional production mode of water heater liner, through robot substitution and logic timing control. The system has perfect production monitoring, data statistics function and safety. The control mechanism shows that the system has high production efficiency and good product quality. In 2018, Han Xiaoyu et al. [4] used the bending manipulator of the fixture to design the punching manipulator, and used the bending manipulator as an example to analyze the kinematics.

In recent years, the development of virtual prototyping technology has gradually matured, and many researchers have studied it. In 2017, Zhao Ruiwen et al. [5] used the ADAMS software to establish a virtual prototype model of the welding robot in order to study the time-varying force of the components in the actual work of the welding robot. In 2017, Zhang Wenhua et al. [6] used the three-dimensional modeling software UG to establish a wheeled tractor solid model in order to verify the roll stability of the virtual prototype technology, and judge the maximum rollover stability angle of the tractor by analyzing the tire force curve. In 2017, in order to solve the friction and wear problem of the motion gap, Zhang Ruiqiu et al. [7] used the 3D modeling software and the virtual simulation software to establish the plane mechanism model with the motion pair clearance, and carried out the dynamics simulation. In 2018, Ma Jianguo et al. [8] used the hybrid vibration isolation platform for theoretical calculation and modal analysis to obtain the characteristics and design parameters of the active-passive hybrid vibration isolation system. The variable-step method was used to control the active-passive hybrid vibration isolation system. The secondary channel identification is carried out.

In this paper, by proposing a new method of robot and skin machine motion [9, 10], the evaluation index of the program has also been improved. There are also specific principles for the determination of the interference threshold. At the same time, combined with the modal analysis results and the actual motion of the press, three matching schemes are obtained. The new method of robot and press motion coordination has certain practicability for lifting rhythm and provides reference for engineering application [11, 12].

2 Methods

2.1 Define Sports Synergy Evaluation Indicators

Good press and robot motion coordination methods must properly improve production efficiency on the basis of meeting the requirements of motion stability.

(1) Slowest equipment utilization

Equipment utilization is a problem that companies care about and is directly related to productivity. Equipment utilization refers to the ratio of actual equipment operating time to total production time. If the equipment utilization rate is too large, it may cause the production line to break. If the equipment utilization rate is too small, it will cause the production line to accumulate. It is necessary to properly improve the equipment utilization rate based on the coordination of the press and the robot movement to accelerate the production cycle. Defining the slowest device utilization is as follows,

$$E = \frac{T_u}{T_b} \times 100\% \tag{1.1}$$

$$T_b = T_q + T_x \tag{1.2}$$

In the formula, T_u is the time when the press is operated above the interference critical line; T_b is the time required for the crankshaft to run one week; T_q is the feeding time of the loading robot; T_x is the unloading time of the unloading robot.

On the basis of other conditions, the larger the value of E, the better the utilization of the device is proved, and on the contrary, the worse.

(2) Interference incidence

In the press line, the occurrence of dryness can lead to major safety accidents. Checking for interference is a prerequisite for all subsequent work. Interference on the press line mainly occurs between the press and the robot, and the loading and unloading robot. The interference between the press and the robot can be eliminated by the interference height of the press. The ADAMS software dry check function can check for interference between adjacent parts. However, the parameter setting is optimal during the simulation process. In reality, due to the existence of factors such as the production environment, an object that does not interfere during simulation may cause interference during actual motion. The expression of the incidence of interference is:

$$\varepsilon = \frac{R}{H_{min}} \times 100\% \tag{1.3}$$

Where H_{min} is the minimum distance between the end of the robot and the height of the upper slider between the press and the height of the upper slider; R is the stroke of the slider of the press; the larger ε, the safer.

2.2 Virtual Prototype Simulation of Stamping Automated Production Line

The simplified robot is mainly composed of a base, a shoulder, a waist, and a wrist. Convert the virtual loaded press and robot CAD model into a neutral file and import it into ADAMS to add constraints and drivers. The robot adds a rotating pair and a drive between each link, and a total of six rotating pairs and six drives form a six-degree-of-freedom robot. The first three degrees of freedom determine the picking position, and the last three degrees of freedom do the picking attitude.

(1) Geometric drawing method

Because of the six-degree-of-freedom robot, the first three degrees of freedom determine the position of the grabbing material, and the last three degrees of freedom determine the attitude of the robot, so the workspace is determined by the first three degrees of freedom. The base of the robot is 120 mm high, the waist is 230 mm, the shoulder is 136.8 mm, and the elbow is 377.7 mm. Range of robot waist angle θ_1 [−240°, +240°], range of shoulder pitch angle θ_2 [−120°, +120°], range of elbow pitch angle θ_3 [0°, + 161°], robot Workspace boundary curve determination method:

① Set p point as the reference point as shown. The set of all reachable spaces at point p is the workspace of the robot.
② The range of the waist rotation angle θ_1 [−240°, +240°], the base radius is 131 mm, the robot is horizontally oriented to the left position, the clockwise rotation is 240° to the left limit, and the counterclockwise rotation is 240° to the right limit., you can determine the top view of the p point action range.
③ The range of shoulder pitch angle θ_2 [−120°, +120°]. Straighten the shield and elbows, with the vertical up to zero position, the J_3 joints do not move, the J_2 joints turn 120° clockwise to reach the J_2 right limit, and turn from the zero position to the left to 120° to J_2 left limit.
④ The range of the elbow pitch angle θ_3 [0°, +161°]. Taking the left limit as the starting position, the J_2 joints do not move, and the J_2 joints rotate 161° counterclockwise to reach the right extreme position of the J_3 joints, at which time the elbows are contracted in the shoulders.
⑤ Starting from the final position of ④ steps, the J_3 joints do not move, and the J_2 joints turn 240° clockwise. At this time, the J_2 joints reach the left limit position, and the elbows are still in the contracted state.
⑥ Starting from the final position of steps, the J_2 joints do not move, and the J_3 joints turn 161° clockwise. At this time, the J_2 joints reach the left limit position, and the elbows are fully deployed.
⑦ After the above steps, the pattern swept by p point is the boundary curve of the robot workspace.
(2) Numerical analysis

The numerical analysis method for solving the robot workspace is the most commonly used Monte Carlo analysis. The idea of Monte Carlo analysis is to first establish a probability model, the number of parameters is equal to the number of joint variables, then randomly extract the joint variables of the robot, and substitute the values into the position equation of the reference point of the robot end, thus obtaining An approximate solution to the robot workspace. The principle expression is:

$$W(p) = \{w(q) : q \in Q\} \subset R^3 (q_{i,\min} \leq q_i \leq q_{i,\max}) \qquad (1.5)$$

Where W(p) is the robot workspace; p is the reference point; q is the joint variable; w(q) is the joint variable function; Q is the joint space; R^3 is the three-dimensional space.

3   Results and Discussions?

3.1   Analysis of Results of Exercise Coordination Program

According to the set evaluation criteria, different schemes for the two feasible interference thresholds were evaluated. The evaluation results are shown in Table 1. As can be seen from Table 1, the best solution is option 2.

Table 1. Scheme evaluation index results

Evaluation index	Option I	Option II	Option III
Slowest device utilization E	62.5%	62.5%	62.5%
Interference rate ε	23.6%	41.6%	23.6%
Equipment overlap utilization k	40%	20%	40%

The main solution steps of the Monte Carlo analysis robot workspace combined with the software are shown in Fig. 1.

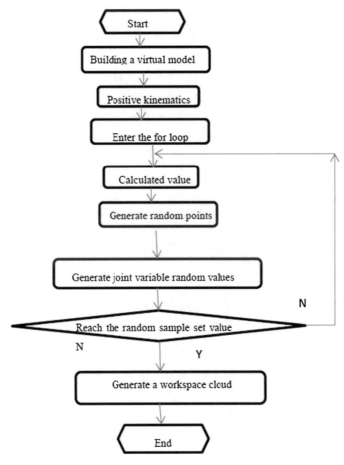

Fig. 1. Robot workspace solution steps

3.2   Robot Trajectory Planning Method

(1) Linear interpolation method

The plane space has two points P_1 and P_2, the coordinates of the starting point P_1 are $[x_1, y_1, z_1]$, the coordinates of the ending point P_2 are $[x_2, y_2, z_2]$, the speed of the linear motion v, the number of interpolations N, and the task of the trajectory is to require the end of the robot to follow the point P_1 at a constant speed. Move straight to point P_2.

Obtained by the linear displacement motion formula;

$$\begin{cases} \Delta x = (x_2 - x_1)/(N+1) \\ \Delta y = (y_2 - y_1)/(N+1) \\ \Delta z = (z_2 - z_1)/(N+1) \end{cases} \tag{1.6}$$

It can be seen from Eq. (1.6) that for any point P_i $(1 \leq i \leq N)$ on the line, the coordinate of P_i is $[x_i, y_i, z_i]$, then

$$\begin{cases} x_i = x_1 + \Delta x \cdot (i-1) \\ y_i = y_1 + \Delta y \cdot (i-1) \\ z_i = z_1 + \Delta z \cdot (i-1) \end{cases} \tag{1.7}$$

(2) Planar circular interpolation

In the plane coordinate system, there is a clockwise arc with a coordinate of $0_0[x_0, y_0, z_0]$, the starting point of the arc is $A[x_1, y_1, z_1]$, the end point is $B[x_{N+1}, y_{N+1}, z_{N+1}]$, the radius is R, the arc center angle is θ, and the starting angle is α (referring to the arc starting point A and The angle between the line connecting the center of the circle and the Y axis), the number of interpolations is N, $\Delta\theta = \theta/N + 1$).

According to the arc calculation formula,

$$\begin{cases} x_a = x_0 + R \cdot \sin\alpha \\ y_a = y_0 + R \cdot \cos\alpha \\ z_a = z_0 \end{cases} \tag{1.8}$$

Then any point on the arc $P_i[x_i, y_i, z_i]$ $(1 \leq i \leq N)$,

$$\begin{cases} x_i = x_0 + R \cdot \sin[\alpha + (i-1) \cdot \Delta\theta] \\ y_i = y_0 + R \cdot \cos[\alpha + (i-1) \cdot \Delta\theta] \\ z_i = z_0 \end{cases} \tag{1.9}$$

4   Conclusion

As the most powerful multi-body dynamics simulation software, ADAMS software is very suitable for studying the kinematics analysis of robots. By studying the motion characteristics of the robot from different directions through virtual prototyping technology, it can provide reference data for the final trajectory optimization.

Establish a complete virtual prototype simulation system for the complete stamping production line for T. Therefore, it is necessary to complete the simulation work of the robot workspace first, and provide the basis for determining the layout of the press and the robot. Then, the virtual prototype model of the stamping automated production line was completed in the ADAMS software, and the dynamic interference check of the virtual prototype model was completed. Finally, the virtual simulation work of the dual material detection system, the robot workspace control system and the grab weight control system on the stamping production line is completed by sensors and measuring tools.

References

1. Qiu, X., Xiao, C., Tan, H., Hou, Y., Zhou, Y.: Multi-software co-simulation of large robot stamping production line. China Mech. Eng. **27**(6), 772–777 (2016)
2. Wang, Y., Ding, H.: Design of automatic feeding manipulator control system for multi-station stamping production line. Forging Technol. **41**(6), 55–60 (2016)
3. Fang, S., Zhao, J., Li, F., Zhang, Z.:. Development of control system for drawing stamping automatic production line. Manuf. Automa. **39**(6), 7–11 (2017)
4. Han, X., Ge, Z., Shen, J.: Design and kinematics analysis of reconfigurable stamping manipulators. Mach. Tool Hydraulics **46**(21), 91–93 (2018)
5. Zhao, R., Tong, Y., Tan, Q., Wu, S., Li, D.: Simulation of welding robot based on virtual prototyping technology. Mech. Des. Manuf. Eng. **46**(3), 36–40 (2017)
6. Zhang, W., Zhang, W., Lu, Z., Wang, B., Bang, T.: Study on roll stability of wheeled tractor based on virtual prototype technology. Mech. Strength **01**, 141–145 (2017)
7. Zhang, R., Zhang, X., Sun, W.: Analysis of the motion accuracy of planar mechanism based on virtual prototyping technology. J. Graph. **38**(2), 278–282 (2017)
8. Ma, J., Shuai, C., Li, Y.: Virtual prototyping technology for active and passive hybrid vibration isolation. J. Naval Univ. Eng. **30**(05), 56–60 (2018)
9. Qian, Z.H., Feng, X., Li, Y., Tang, K.: Virtual reality model of the three-dimensional anatomy of the cavernous sinus based on a cadaveric image and dissection. J. Craniofac. Surg. **29**(1), 1 (2018)
10. Chen, M., Saad, W., Yin, C.: Virtual reality over wireless networks: quality-of-service model and learning-based resource management. IEEE Trans. Commun. **PP**(99), 1 (2017)
11. Abdallah, K.A.A., Namgung, I.: Virtual reality development and simulation of bmi nozzle inspection system for use during regular refueling outage of apr1400 family of reactors. Ann. Nucl. Energy **116**, 235–256 (2018)
12. Overbeck, R.S., Erickson, D., Evangelakos, D., Pharr, M., Debevec, P.: A system for acquiring, processing, and rendering panoramic light field stills for virtual reality. ACM Trans. Graph. **37**(6), 1–15 (2018)

Architectural Roaming Animation
Based on VR Technology

Heng Yang[✉]

Wuxi Institute of Arts & Technology, Yixing, Jiangsu, China
2858093479@qq.com

Abstract. Architectural design, as a kind of art, itself has a strong innovation, and when it found computer animation science and technology to express more ideal and unique artistic effects, architectural performance technology has made rapid progress in the development. As an important means of architectural expression, the virtual technology of architectural animation provides a new digital mode for the promotion of architectural image. Architectural animation virtual 3D film with its unique form of expression to establish a highly technical, humanized virtual space, has become a popular new technology field in the 21st century. The purpose of this paper is to study the application of architectural roaming animation based on VR technology. Based on the background and significance of the application of virtual reality technology in the construction industry, this paper briefly describes the definition of virtual reality, and then introduces two technologies of virtual reality collision detection and automatic pathfinding. Finally, Unity3D and C++ are used to design and implement the building roaming system. Through testing the system's automatic pathfinding performance under different terminals, the pathfinding time on the computer end is the shortest, less than 100 ms.

Keywords: VR technology · Architectural animation · Collision detection algorithm · Automatic pathfinding

1 Introduction

Virtual tour of architecture is a technology with very good practical prospect [1]. At present, virtual tour technology of architecture has been applied in architectural design, public building projects, urban planning and design schemes, virtual landscape [2, 3]. From the perspective of technology, the biggest difficulty in virtual tour of architecture lies in the realistic sense of architectural scene rendering and the efficiency of scene real-time rendering [4]. In order to achieve better user immersion, the scenery presented in the architectural scene needs to be more real, so the model in the architectural scene needs to be constructed very carefully, but it takes a lot of time to build the model [5]. For such a fine and complex model, when it is transplanted to the roaming system, it is often difficult to achieve the effect of real-time roaming due to the restriction of machine hardware performance. Therefore, on the premise of satisfying the sense of reality and

J. MacIntyre et al. (Eds.): SPIoT 2020, AISC 1283, pp. 374–380, 2021.
https://doi.org/10.1007/978-3-030-62746-1_55

immersion of architectural scenes, how to reduce the complexity of scene models, reduce the number of polygons to be processed by the graphics system and realize real-time interaction has become a major topic in the research of computer graphics [6, 7].

Architectural animation is the best way to show the dynamic "visualization" of architecture [8]. In recent years, the development of computer graphics has led to the formation and development of three-dimensional representation technology of architecture, which is a technology that uses computer graphics and image processing technology to transform design data into images displayed on the screen [9]. Dynamic visualization of architectural effects has gathered the strengths of various industries in the field of design, providing designers, customers and the public with ornamental and interactive approaches to understand projects and products in advance [10]. Architectural animation virtual technology endowing people with a simulation, three-dimensional graphics world, using previously unimaginable visual means to obtain information to give play to their creative thinking, saving resources of architectural expression.

Based on the background and significance of the application of virtual reality technology in the construction industry, this paper briefly describes the definition of virtual reality, and then introduces two technologies of virtual reality collision detection and automatic pathfinding. Finally, Unity3D and C++ are used to design and implement the building roaming system. Through testing the system's automatic pathfinding performance under different terminals, the pathfinding time on the computer end is the shortest, less than 100 ms.

2   Architectural Tour Method

2.1   Virtual Reality

Virtual reality VR for short, is a technology to adopt advanced technology of the computer technology as the core to form a kind of very vivid sensory experience a virtual environment, the experience to through the necessary input and output devices and the objects in the virtual environment interaction, mutual influence, and thus gain immersive experience and feelings.

2.2   Collision Detection Technology

The Oriented Bounding Box (OBB) is an directional rectangular OBB with three edges located at arbitrary directions, and its direction depends on the model structure it surrounds. OBB is commonly expressed as the center point, the direction of the three edges, and the length of half of the three edges. The construction of OBB needs to utilize the triangle information of the model grid. If the three vertices of the ith triangle are (p^i, q^i, r^i), the mean μ and covariance matrix C of all triangle vertices are calculated as follows:

$$\mu = \frac{1}{3n} \sum_{i=0}^{n} (p^i, q^i, r^i) \tag{1}$$

$$C_{jk} = \frac{1}{3n} \sum_{i=0}^{n} (\bar{p}_j^i \bar{p}_k^i + \bar{q}_j^i \bar{q}_k^i + \bar{r}_j^i \bar{r}_k^i) \ \ 1 \le j, k \le 3 \tag{2}$$

Where, n is the number of triangles, and $\bar{p}^i = p^i - \mu, \bar{q}^i = q^i - \mu, \bar{r}^i = r^i - \mu$ represents each element in the third-order covariance matrix. C is a symmetric matrix whose three eigenvectors are orthogonal to each other. The three eigenvectors of C were unitized as the three coordinate axes of OBB, and the three side lengths of OBB were calculated according to the maximum interval of the projection of the model on the three coordinate axes.

2.3 Automatic Navigation Pathfinding

Automatic navigation is an important function in the building roaming system involved in this paper. By selecting the target area, the system will display the moving route in the navigation map, so as to facilitate the user's roaming in the building roaming system. Navigation is realized by combining with the preset script of Unity3D. The system firstly determines the target location according to user input, uses the pathfinding component in Unity3D to find the optimal route, and finally prompts the user to move through the friendly display interface. The following three conditions are required for the system to realize the automatic pathfinding function: The target location, which can be found in the virtual scene through the user's input; For scenes that meet the requirements, the scenes in the system need to be set, which scenes can be set on foot and represented by subdividing the grid to meet the input requirements of pathfinding algorithm; Initial location. After the user sends a pathfinder request, the system will automatically find the pathfinder location information.

3 Experimental Design of Roaming System

3.1 Overall Design of Roaming System

The function of the roaming system is to construct an indoor virtual scene with a sense of reality. The observer can change the viewpoint with mouse and keyboard to complete the roaming of the virtual scene. The construction technology of system key Technology Application framework The construction technology of realistic indoor virtual scene real-time interactive tour technology collision detection technology. In general, there are two major modules: the building module of virtual scene and the real-time roaming module.

3.2 Experimental Environment

The hardware and software development environment of the building roaming system designed in this paper are shown in Table 1.

Table 1. Development environment

	Name	Configuration
Software development environment	CPU	Pentiem(R)
	RAM	16G
	Graphics card	NVIDIA GeForce4
	Hard disk	250G
Hardware development environment	Development platform	Windows 10
	Development language	C++
	Modeling software	Unity3D

4 Discussion of Experimental Results of Roaming System

4.1 Analysis and Discussion of Experimental Results

In order to verify the cross-platform, efficient rendering and other characteristics of the system, this section will test the relevant performance of the system.

(1) Render frame rate test

In this paper, a smartphone using Android operating system, a smartphone using IOS operating system and a laptop with Windows7 system are selected to access the system. The test results of average rendering frame rate (unit FPS) of each scene when running the building roaming system with three terminals are shown in Table 2 and Fig. 1:

Table 2. Test results of average rendering frame rate of each scene in the building roaming system

Terminal	Single roaming	Many people roaming	Show scenes	Video playback
Android	50	35	60	24
IOS	50	48	60	24
Computer	60	60	60	60

It can be seen from Table 2 and Fig. 1 that the operating efficiency of the system is low in the process of multi-person roaming and video playing. System due to the need in the process of video playback at the same time rendering 3 d scene and will occupy more resources, at this point does not move as a background, the three-dimensional scene, and a broadcast frame rate of video resources greater than 24 FPS is very smooth, so a run on the mobile side frame rate control at 24 FPS can ensure smooth operation of the system at the same time, effectively reduce the system resource. In the process of multi-person roaming, the system needs to render more 3D content in real time, and the rendering frame rate will be affected by the hardware performance of the system.

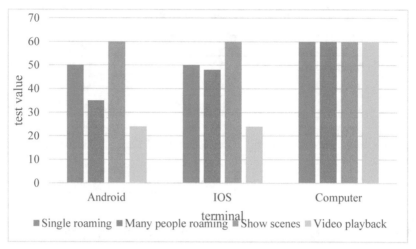

Fig. 1. Test results of average rendering frame rate of each scene in the building roaming system

(2) Pathfinding time test results

In addition, the pathfinding function of the system was tested in this paper. The pathfinding time (in milliseconds) at different distances using three different terminals is shown in Fig. 2.

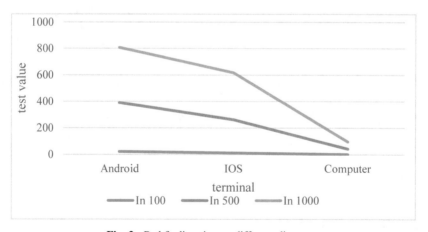

Fig. 2. Pathfinding time at different distances

It can be seen from Fig. 2 that system pathfinding time is closely related to hardware performance and increases with the increase of pathfinding distance. The pathfinding time on the mobile side, though much longer than on the PC side, is still within a relatively short time range. Moreover, the map on the mobile end uses less grid display, and the number of grids occupied by the mobile end is less than that on the PC end in the virtual scene, so the difference in efficiency is not obvious.

4.2 Suggestions on Rational Use of Virtual Reality Technology in Architectural Animation

(1) Reasonable expression of intention

First of all, virtual reality technology, as a new technology, is for the production of services, reasonable use of technology is conducive to express the intention of the designer quickly and accurately. In the process of architectural animation production, it is necessary to avoid the phenomenon of focusing on technology while ignoring animation art. Secondly, the use of virtual reality technology to produce architectural animation requires both "reality" in life and artistic processing. When making urban planning animation and real estate promotion animation, not only should the scene in the planning be completely reproduced, but also the animation should be processed with the lens language, color language and sound to make the audience understand relevant information in a relaxed and joyful audiovisual environment.

(2) Determine the planning scheme

Any architectural animation virtual film works should have a distinctive theme design, so as to attract people, so before the production of the idea of creation is very important, otherwise only rely on excellent 3D rendering effect is far from enough. Conceptual design is primarily determine the key works to express architectural aesthetics performance, for each animation works we will be organized by the project manager, group discussion developed animation production team performance plan, by professional planners to design thinking animation Angle this idea, expressed in the form of written language, will be in the form of a shooting rough sketches design simple first represented in a single frame painting. If you're just making some simple animations, you should just have a rough outline of the script in your head. If you're going to do some big ones, you should do it. A large building project needs multi-field cooperation, so we will draw more detailed sketches for communication.

5 Conclusion

In general, the virtual reality technology is a can bring many convenience to mankind and the benefits of high and new technology, combination of architectural animation performance, makes the building designer to update perspective on all aspects of design, deliberate plan form, materials and space feeling at the same time, attaches great importance to the emerging architectural animation technology performance in aided architectural design aspects of the application and development, and has made great development, in recent years, our country also in vigorously introduce and the development of this technology. Construction of our country animation industry is still in the fledgling experiment exploration stage, the high-tech efficiency of emerging technologies at the same time also have considerable technical level and challenging, still there are many unresolved technical problems and theoretical problems, so we should constantly rich professional knowledge, have the courage to practice, more need our this generation of young people continuously, passion and energy into our personality, talent and enthusiasm, to accomplish very personalized color architectural animation works, bring people a new way of creation and new experience, for our country's construction industry and make great contributions to art.

Acknowledgement. (1) 2018 Jiangsu Province Elite research and training of teachers' professional leaders in Colleges (Visiting scholar of Tongji University)

Project Number: 2018GRFX063.

(2) This paper is 2018 Wuxi Institute of Arts & Technology reform in education projects, (Practice and Research on digital innovation project of ancient buildings around Taihu Lake in 3D animation teaching).

Project Number: 19kt108.

(3) Wuxi modern apprentice and Wuxi Vocational Education excellence course (Three dimensional scene production》

Project Number: 19kc202.

References

1. Lau, K.W., Kan, C.W., Lee, P.Y.: Doing textiles experiments in game-based virtual reality. Int. J. Inf. Learn. Technol. **34**(3), 242–258 (2017)
2. Jinsoo, A., Young, K., Ronny, K.: Virtual reality-wireless local area network: wireless connection-oriented virtual reality architecture for next-generation virtual reality devices. Appl. ences **8**(1), 43 (2018)
3. Mendez, R., Flores, J., Castello, E., et al.: New distributed virtual TV set architecture for a synergistic operation of sensors and improved interaction between real and virtual worlds. Multimed. Tools Appl. **77**(15), 18999–19025 (2018)
4. Elbamby, M.S., Perfecto, C., Bennis, M., et al.: Towards low-latency and ultra-reliable virtual reality. IEEE Network **32**(2), 78–84 (2018)
5. Ji, X., Fang, X., Shim, S.H.: Design and development of a maintenance and virtual training system for ancient Chinese architecture. Multimed. Tools Appl. **77**(22), 29367–29382 (2018)
6. Dash, A., Lahiri, U.: Design of virtual reality-enabled surface electromyogram-triggered grip exercise platform. IEEE Trans. Neural Syst. Rehabil. Eng. **28**(2), 444–452 (2020)
7. Satja, S., Matija, R., Joseph, C., et al.: Collision detection for underwater ROV manipulator systems. Sensors **18**(4), 1117 (2018)
8. Jung, B.J., Moon, H.: Development of collision detection method using estimation of cartesian space acceleration disturbance. J. Korea Robot. Soc. **12**(3), 258–262 (2017)
9. Tang, Y., Hou, J., Wu, T., et al.: Hybrid collision detection algorithm based on particle conversion and bounding box. Harbin Gongcheng Daxue Xuebao/J. Harbin Eng. Univ. **39**(10), 1695–1701 (2018)
10. Gao, L.: Application of collision detection algorithm in 2D animation design. Revista de la Facultad de Ingenieria **32**(14), 467–472 (2017)

The Measurement Method of Housing Affordability

Fang Yang[✉] and Lingni Wan

Department of Business Administration, Wuhan Business University, Wuhan, Hubei, China
1357585034@qq.com

Abstract. According to the data of China's economic blue book in 2010, 85% of the families have difficulties in housing payment. Therefore, the key problem in identifying housing poverty is to determine the purchasing power of commercial housing and the measurement method. This paper introduces several commonly used methods of housing affordability measurement, and analyzes the application scope, advantages and disadvantages of various methods.

Keywords: Housing affordability · Measurement method · Family income

1 Introduction

Generally, there are two ways to determine the purchasing power of commercial housing. One is to measure the proportion of housing consumption in household income, and stipulate a certain proportion as the evaluation standard. If the ratio is lower than this ratio, it will be difficult to pay. Scholars advocate the use of such methods, such as: housing expenditure income ratio, house price income ratio, rent income ratio and so on. The second is the income balance evaluation method to measure the balance between the residual income after deducting the basic living expenses and the cost of housing consumption expenditure. The third method is to identify the income level of the public rental housing applicants by taking the household income measurement unit lower than the division basis multiple or a certain percentage set by the region itself. So, which method is the most effective way to calculate the family income limit is the focus of the public rental housing affordability screening index [1–4].

1.1 House Price Income Ratio (PIR)

House price income ratio refers to the ratio of the total price of a house to the annual income of urban residents. The housing price to income ratio directly reflects the residents' purchasing power to the existing housing. The average housing price income ratio of a country is usually calculated by the ratio of the average price of a house to the average annual household income:

$$PIR = \frac{Ph}{t} = \frac{s \times n \times p}{n \times a} \tag{1}$$

J. MacIntyre et al. (Eds.): SPIoT 2020, AISC 1283, pp. 381–387, 2021.
https://doi.org/10.1007/978-3-030-62746-1_56

In the formula, PIR is the ratio of house price to income, PIR has two variables: the total price of each house Ph and the annual total income of each family, I, s is the per capita housing area, n is the average population of each household, p is the average selling price of houses per unit area, and a is the annual per capita income. The larger the ratio of house price to income is, the weaker the housing payment ability is, and the more difficult it is to buy a house. On the contrary, the opposite is true.

Although PIR is a universal index to evaluate housing affordability in the world, there is no reasonable and strict regulation on its index range. According to the experience of the world bank and developed countries, it is considered that the reasonable ratio of the total house price to the annual income of the residents should be maintained in the range of 3–6 times. If the ratio exceeds 6 times, it is considered that the house price is too high and the residents have difficulties in housing payment [5]. According to the data released by Shanghai E-House Real Estate Research Institute, the price to income ratio of 35 large and medium-sized cities in China in 2015 shows that:

From 1998 to 2015, the trend of China's housing price income ratio was from 1998 to 2003, and the housing price income ratio basically maintained at the range of 6.6–6.9. From 2004 to 2010, the housing price income ratio rose to 9.2, and then fell back to 8.7 in 2015. From the perspective of trend, before 2004, the housing price income ratio basically maintained a stable state, which can reasonably reflect the national housing price income ratio. The report holds that the upper limit of the reasonable range of China's house price income ratio should be kept within 7, and the range deviating from the reasonable value is 3%. 2016 ranking of deviation degree of housing price income ratio in 35 large and medium-sized cities.

1.2 Rent to Income Ratio (RIR)

The biggest difference between the rent income ratio and the housing price income ratio is reflected in the form of the existing housing property rights owned by the residents, whether they own or rent. The ratio of rent to income refers to the ratio between the annual average rent paid by the family for renting housing and the annual disposable income of the family. It is the most direct indicator to measure the affordability of families who apply for public rental housing. The formula of rent income ratio can be expressed as follows:

$$RIP = \frac{R}{A} = \frac{12 \times r \times s}{n \times a} \qquad (2)$$

RIR is the ratio of rent to income, R is the average rent of a year, A is the annual disposable income of the family, r is the unit rent per square meter, s is the rental area, n is the number of households, a is the annual per capita disposable income of tenants. The value of RIR ratio reflects the degree of difficulty in rent payment. The smaller the RIR ratio is, the stronger the ability of rent payment is, and the greater the ratio is, the more difficult it is to pay rent [6–9].

1.3 Residual Income Affordability (RIA)

Income balance evaluation method, also known as residual income method, refers to the difference between the residual income after deducting the basic living expenses and the cost of housing consumption expenditure. It takes the premise that the housing consumption does not affect the basic living standard of the family. It can directly reflect the people's affordability of housing consumption expenditure. The income balance evaluation method consists of three key factors: disposable income, basic living expenditure and housing consumption expenditure. Basic living expenditure refers to the most basic consumption expenditure to ensure the family to meet its own survival and healthy life, including the expenditure on food, clothing, transportation and communication, daily necessities and services. Housing consumption expenditure refers to the cost of buying or renting a house. Therefore, the housing consumption expenditure and living consumption expenditure in the residual income method are generally used to meet the basic housing consumption expenditure and basic living expenditure value.

The income balance evaluation method is expressed as:

$$I = C + H \tag{3}$$

$$RIA = I - C \tag{4}$$

It is the annual per capital disposable income of the family, C is the annual basic living consumption expenditure of the family, h is the annual basic housing consumption expenditure of the family, and RIA is the annual residual income. When RIA $\geq$ H or RI-H $\geq$ 0, it means that the residual income of the family is enough to pay for the basic housing consumption of the family, the housing affordability is strong, and there is no housing payment difficulty. When RI $\leq$ H, or RI-H $\leq$ 0, it means that the non housing consumption of the family is too high, the family housing burden is heavy, the surplus income is not enough to pay for the basic housing consumption of the family, the housing demand can't be solved through their own ability, and there is housing payment difficulty [10].

1.4 Housing Affordability Index (HAI)

According to the upper limit requirement of housing consumption proportion (the proportion of housing consumption expenditure to income), the American Association of real estate agents (APA) proposed to use the housing affordability index to evaluate the affordability of households in the housing market with median income level to live in the median house price. It is a very mature foreign real estate market analysis index, which is used to evaluate the ability of typical income families to afford typical housing prices. The HAI index is released once a month to reflect the change of the family's ability to pay for housing in real time. The specific calculation formula is as follows:

$$HAI = \frac{1}{Ha} \times 100\% = \frac{1}{Pm \times 4 \times 12} \times 100\% \tag{5}$$

$$Pm = Ph \times 0.8 \times \frac{R}{12} = \frac{(1 + \frac{R}{12})^{n \times 12}}{(1 + \frac{R}{12})^{n \times 12} - 1} \tag{6}$$

The median level of family's annual income is expressed by I, and the housing price at the median level is expressed by H; Ha is the monthly income qualification line of the family, that is, the housing affordability of the median price level; the principal and interest of the housing mortgage loan repayable by the family every month is expressed by Pm; the annual interest rate level of medium and long-term housing mortgage loan is expressed by R, and the loan term is n years.

The basic idea of HAI index proposed by the American Association of real estate brokers is: it is usually assumed that the household housing loan is 80% of the housing price, and the down payment is 20%. Generally, the monthly mortgage payment is calculated based on the 30-year mortgage term and equal repayment. The upper limit of the monthly mortgage ratio of housing consumption is 25% of the family's monthly income. If it is lower than 25%, it is considered that there is housing payment difficulty hard. The evaluation benchmark of HAI is 100%. If HAI > 100%, it means that the family can afford a higher housing price than the median family; when HAI = 100%, it means that the family with median income can just afford the house with median house price; when HAI < 100%, it means that the family with median income can only afford the housing with lower price. When Chinese scholars use HAI index, the evaluation standard is basically unchanged. However, due to the incomplete housing statistics in China, the average annual household income is usually used when calculating the median household income of purchasing housing and the average annual selling price of commercial housing is used instead of the median price of commercial housing.

2   Comparative Analysis of Various Calculation Methods

When judging whether there is payment difficulty for the people who buy houses, each calculation method has its own advantages and disadvantages and applicable scope (Table 1).

Table 1. Comparison of calculation methods of household income limit

Calculation method	Advantages	Disadvantages	Scope of application
PIR	1. Directly reflect the changes of house prices in the same country and the same region 2. It is not difficult to obtain the data	1. It does not show the actual economic constraints faced by residents in purchasing housing 2. Lack of pertinence 3. The reasonable range of housing price income ratio has certain subjective color	It can also be used for regional comparison and housing consumption analysis

(continued)

Table 1. (*continued*)

Calculation method	Advantages	Disadvantages	Scope of application
RIR	It can directly reflect the ability of family rent payment	1. The scope of application is narrow 2. It is difficult to define the reasonable range of rent income ratio 3. Ignoring the difference of housing affordability caused by different consumption preferences	It can only be used to evaluate the housing affordability of the rental group, but not for the purchasing group
RIA	It can directly reflect the difference of housing consumption expenditure of families with different income levels and the specific housing affordability of a single family	1. The basic cost of living is difficult to define and data acquisition is difficult. 2. It is difficult to unify the measurement caliber of housing consumption expenditure data	It is more accurate to calculate the housing affordability of low-income families. It can be used to measure the purchasing power of commercial housing and the affordability of affordable housing
HAI	The ability to pay for a house can be measured from the perspective of household housing debt (mortgage loan purchase)	1. Lack of evaluation on the ability to pay for rental housing 2. Lack of consideration of individual housing consumption preferences 3. Ignoring the ability to pay down payment 4. The influence of non housing consumption expenditure on family housing affordability is not considered	Research on housing affordability of typical middle-income families with loans

3 Conclusion

To sum up, through the analysis of the advantages and disadvantages of various assessment methods and their respective scope of application, it can be seen that different methods of measuring household income limit have their own concerns. According to the timeliness of housing payment, it can be divided into immediate payment and delayed payment. For example: the house price income ratio tends to measure the real-time payment ability of the family income used for the full purchase. However, the housing affordability index tends to measure the delayed payment ability of the family income used for loan purchase. The residual income evaluation method measures housing affordability from another perspective of non housing consumption. In the actual measurement of housing affordability, based on the difficulty of data acquisition, the dynamic nature of data, the difficulty of understanding the measurement method and the operability in the implementation process, the identification method of family income limit in policy practice is mainly to measure the public rental housing by a certain measurement unit of family income lower than the division basis multiple or a certain percentage set by the region itself Whether there are difficulties in housing payment for the families who apply for housing. The advantage of this method is that the data is easy to obtain and dynamic and the calculation is relatively simple. However, the factors considered in the calculation are too few, which can't take into account the factors such as individual housing consumption preference and housing expenditure cost. It can only roughly estimate the family housing payment ability, and can not accurately identify the groups with difficulties in housing payment.

Therefore, there are some limitations when choosing the method of calculating the family income limit, if we only stand on one angle (i.e. income or expenditure angle, supply or demand angle), or select only one method to evaluate the housing affordability of the applicant family. Therefore, we should make all-round consideration, combine the government supply with the demand of households for public rental housing, and consider the income and expenditure of the families who apply for the public rental housing, and implement step by step to combine several methods to verify each other, so as to improve the accuracy of the calculation of family income limit and make the calculation method of family income limit more reasonable and scientific.

Acknowledgements. This work was supported by Research on the nature and orientation of social insurance agency, a doctoral research fund project of Wuhan Business University (2017KB010).

References

1. Quigley: Why the governments hold play a role in the house transaction. Econometrica (49), 505–513 (1990)
2. Miyamoto, K., Udomsri, R.: Ananalysis system for integrated policy measures regarding land use transportand the environment in a metro policy. Am. Econ. Rev. **22**, 57–65 (1996)
3. Kutty, N.K.: Derminans of Srucural Adeuay of Dwlingn Joma of Housing Research **10**(1), 27–43 (1999)
4. Kutty, N.K.: A new measure of housing affordability: estimates and analytical results. Hous. Policy Debate. **16**(1), 113–142 (2005)

5. Padley, M., Marshall, L.: Defining and measuring housing affordability using the Minimum Income Standard. Hous. Stud. **34**(8), 1307–1329 (2019)
6. Harry, J.F., Boumeester, M.: The affordability of housing in the netherlands: an increasing income gap between renting and owning? Hous. Stud. **25**(6), 799–820 (2010)
7. Tang, C.P.Y.: Measuring the affordability of housing association rents in England: a dual approach. Int. J. Housing Markets and Anal. 5 (2012)
8. Kuang, W., Ding, Y.: Spatial and temporal distribution of housing affordability of Chinese urban residents: an analysis of rent affordability of 35 large and medium-sized cities. Price Theory Practice (10), 16–19 (2018). (in Chinese)
9. Xin, D.: Dynamic trends and structural differences: a comprehensive measure of the affordability of China's housing market. Econ. Manage. **34**(6), 119–127 (2012). (in Chinese)
10. Stone, M.E.: What is housing affordability? The case for the residual income approach. Housing Policy Debate **17**(1), 151–184 (2006)

Smart Home Security Based on the Internet of Things

Kejie Zhao, Jiezhuo Zhong, and Jun Ye[✉]

School of Computer Science and Cyberspace Security, Hainan University, Haikou, China
yejun@hainanu.edu.cn

Abstract. With the rapid development of Internet of Things, Internet of Things technology has been widely applied in many aspects. However, with the continuous popularization of intelligence and digitalization, the security problem of smart home in the Internet of Things environment is increasingly significant, and there are no specific laws, regulations and standards to protect the security of smart home. Therefore, smart home security has become one of the indispensable research fields of the Internet of Things. At first, this paper describes the development stage of smart home in recent years, then according to the literature at home and abroad in recent years, the Internet of things brought by smart home security hidden danger, respectively from two aspects of device security, communication security for a description, and sums up the harm of security vulnerabilities and the lack of research are summarized. Finally, based on the research status of smart home security, the challenges and opportunities of smart home security in the future are proposed, which is of great reference significance to the development of smart home industry in the Internet of Things.

Keywords: Internet of Things · Smart home · Device security · Communication security · Security vulnerabilities

Strategy Analytics estimates [1] that global consumer spending on smart home devices will grow by 40.9% to a record $62 billion by 2021. Canalys recently predicted [2] that the global smart audio devices (TWS) will grow by 29% in 2020 under the influence of COVID-19, with more than 200 million units shipped. Smart home system is not only within the family of a complete set of facilities, products, and it is a central control center, the auxiliary user management of daily life, the user can use different terminal (such as: mobile phone, computer) to meet the need of remote control of the user requirements, if devices are under attack and will bring unimaginable consequences, because attackers can access and control users to connect all smart devices, So the smart home system is facing serious security problems, security vulnerabilities are not only easy to cause privacy leakage of users, but also cause economic losses. In literature [3–5], the vulnerabilities analysis of smart home security is relatively single, the research scope is narrow, the research value is not high, and there are few reviews in the field of smart home security. Therefore, this paper will provide certain reference opinions for scholars studying smart home security.

J. MacIntyre et al. (Eds.): SPIoT 2020, AISC 1283, pp. 388–393, 2021.
https://doi.org/10.1007/978-3-030-62746-1_57

1   History of Smart Home

Smart home has experienced four development stages [6, 7]. The first stage is mainly based on the coaxial line and two core lines for family networking. The smart home can control lights, curtains, a small amount of security, etc., which is just in a cognitive stage for China. The second stage is mainly based on RS-485 line. The technology of smart home has not only been upgraded, but also has various functions, such as visual intercom, security, etc. The domestic smart home market has begun to improve, and dozens of research and development enterprises have been established. In the third stage, centralized family control is realized, and smart home can realize measurement and security, etc. The fourth stage is the present stage. With the continuous progress of smart home technology, based on full IP technology, Zigbee technology, Bluetooth technology, etc., it starts to move towards personalization and customization.

2   Smart Home Security

Device security and communication security is discussed in this paper. Device security refers to the security problems related to intelligent devices. Communication security refers to the security issues related to the connection of intelligent devices to communication networks.

2.1   Device Security

There are various types of smart home devices, and the number of security vulnerabilities of smart devices is also extremely alarming. In 2020, National Internet Emergency Response Center (CNCERT) released a report titled "Summary of China's Internet network security situation in 2019" [8]. The number of security vulnerabilities recorded by the National Information Security Vulnerability Sharing Platform (CNVD) reached a record high, with a year-on-year increase of 14%, totaling 16,193.

At present, many academics have carried out research on the mining and analysis of security vulnerabilities in smart home devices. In 2018, Muench et al. [9] proposed six heuristic methods that could be used to mine memory damage vulnerabilities. The method could detect the application of memory damage in devices through fuzzy testing, and the experiment proved the effectiveness of the method. Cheng et al. [10] proposed a detection method for device firmware vulnerabilities, which solved the problems of data flow analysis and data structure recovery, and the detected vulnerabilities were significantly higher than Angr in terms of time efficiency. Chen et al. [11] pointed out that most of the smart home devices are controlled by APP, and the generated data is tested in the device by fuzziness, which overcomes the problem of firmware acquisition. In 2019, Zheng et al. [12] analyzed the vulnerabilities mining of device firmware from static analysis, fuzzy test and homology analysis, respectively, and emphasized the urgency of vulnerabilities mining of Internet of Things devices. Mariusz et al. [13] published a paper on device Intrusion Detection System (IDS) distributed between users and ISP, which helps detect large scale security vulnerabilities in the firmware of smart home devices. HaddadPajouh et al. [14] summarized that the increasing destructive use of computing resources in different application fields of Internet of Things devices led to a series of vulnerabilities in the Internet of Things devices, proposed a vulnerabilities classification based on three layer architecture, and provided corresponding solution classification. In 2020, Jaehwan et al. [15] designed a method to apply the effectiveness of SIP to remote proof, which is used for the communication between devices on the home network and external devices, which can prevent the communication of attackers and protect the smart home environment. Sarra et al. [16] proposed a novel and enhanced LoRaWAN framework for remote control of the firmware security of smart home devices.

At present, there are many methods to exploit firmware vulnerabilities of devices, such as fuzzy testing and static analysis, but most of the related researchers are limited to the analysis of a single device. For fuzzy testing technology, the current research cannot simulate lightweight firmware, and it is difficult to monitor and implement the firmware operation. For static analysis technology, although the current research can effectively solve the firmware analysis and vulnerabilities analysis, but specific vulnerabilities can not be carried out efficient analysis, the relevant personnel are worth further research.

2.2 Communication Protocol Security

Communication protocol is the user through a variety of wireless communication protocol the process of the transmission of personal information to the Internet of things devices, and intelligent control of smart home system, users can remote control almost all the household products in the home, when away from home can also be household environment for real-time monitoring, data sharing and help users to manage the family environment [17]. The main popular communication protocols include Zigbee, NFC, Wi-Fi, Bluetooth, etc. These access technologies have become the most common connection methods between devices, but security problems are inevitable in configuration and implementation.

The main solution of communication protocol is the security vulnerabilities when smart devices are connected to the network and the vulnerabilities when data is exchanged between applications. In 2018, Lonzetta et al. [18] analyzed that BLE devices could use the mechanism of generating and exchanging long-term keys, but Bluetooth technology had serious security defects and vulnerabilities [19]. Attackers could infer that the keys could be decrypted to obtain the data sent by smart devices, and even carried out MTM attacks in serious cases. Seyed et al. [20] proved that lightweight RFID is not secure, and implemented DOS and desynchronization attacks of the protocol, through which the attacker can get sensitive information of the user. In 2019, Alshahrani et al. [21] design a based on Zigbee technology device to anonymous mutual authentication and key exchange scheme in intelligent home network, through the AES algorithm to encrypt the payload to ensure data security, only with Internet of Things devices symmetric key controller can connect to other devices and encrypt the message, can effectively prevent a device under threat of attack any security vulnerabilities. Aellison et al. [22] proposed a new BLE injection attack technology for bluetooth low-power devices, which could not force key renegotiation, discussed possible countermeasures against this attack, and provided various defense measures for developers who were aware of vulnerabilities. In 2020, Hamidreza et al. [23] to study the IEEE 802.11x vulnerabilities, namely in the insecure channel frame transmission management, puts forward the performance evaluation,

WIDPS used to evaluate and classify system performance under different conditions, a comprehensive model for wi-fi intrusion detection and defense are analyzed, and the content of the research can help the wi-fi network "cancel the authentication/remove link" in the attack. Mahdi et al. [24] first verified that previous NFC protocols all had security holes and were vulnerable to attacks by attackers, pointed out anonymous and secure key agreement protocols for NFC applications, and used the Real-Or-Random model for simulation experiments to prove security and solve the most likely security vulnerabilities threats.

In the future, relevant researchers can start from lightweight communication protocol and study how lightweight communication protocol can be applied to smart home scenarios, so as to promote diversification of smart home scenarios and bring more convenient life to users.

3 Challenges and Opportunities

After in-depth investigation and research on smart home security [25, 26], the challenges of current smart home security in device firmware vulnerabilities and communication protocol vulnerabilities are proposed, as well as the opportunities to be applied to these challenges.

3.1 Device Firmware Vulnerabilities

With more and more varieties of smart home devices, many smart home manufacturers do not guard against security vulnerabilities in the production of devices. Often, smart devices crash due to security problems and are vulnerable to attacks. The main reason is that a large number of memory vulnerabilities and logic vulnerabilities have long been hidden in the firmware. The main challenge is that device firmware vulnerabilities mining is only in a single architecture, rather than a diversified architecture, and the category of vulnerability mining is small, and device firmware is difficult to obtain.

3.2 Communication Protocol Vulnerabilities

The security of the communication protocol is the main link of the security of the entire smart home system. However, once the vulnerability of the communication protocol is attacked by the attacker, the smart home device will be controlled remotely, thus causing the user data privacy leakage and economic losses. The disadvantage of most smart home devices lies in insufficient communication resources, so the main challenge lies in the need for a large number of lightweight communication protocols to be applied to the smart home environment to protect the security of the smart home. This requires more academics to study the content.

4 Conclusion

In this paper, first introduced the history of smart home and its architecture. Secondly, the device security and communication security in smart home environment are systematically studied. Through the current security problems of smart home, the challenges and opportunities of smart home security are pointed out. Finally, the thesis is summarized. Although a lot of research has been carried out on smart home security, many problems have not been completely and effectively solved, so it is urgent to introduce some laws, regulations and standards to protect smart home security. The next step will mainly study the attacks of vulnerabilities in smart home security and how to make corresponding defenses.

Acknowledgments. This work was partially supported by the Science Project of Hainan University (KYQD(ZR)20021), Hainan Provincial Natural Science Foundation of China(618QN218).

References

1. Strategy Analytics: Global smart home spending is expected to grow by 41% in 2021 [EB/OL]. https://baijiahao.baidu.com/s?id=1672251536385677825&wfr=spider&for=pc
2. Canalys: More than 200 million smart audio devices will be shipped globally in 2020 [EB/OL]. http://www.199it.com/archives/1069574.html
3. Batala, J.M., Vasilakos, A., Gajewski, M.: Secure smart homes: opportunities and chalenges. ACM Comput. Surv. **50**(5), 75:1–75:32 (2017)
4. Zhang, Y., Zhou, W., Peng, A.: Survey of Internet of Things security. J. Comput. Res. Dev. **54**(10), 2130–2143 (2017)
5. Li, D, Gurira, K.: Research review of ZigBee-based intelligent home system. Comput. Age (06), 23–25+30 (2019)
6. Wang, J., Li, Y., Jia, Y., Zhou, W., Wang, Y., Wang, H., Zhang, Y.: Overview of intelligent home security. Comput. Res. Dev. **55**(10), 2111–2124 (2018)
7. Wang, Y., Miao, M., Shen, J., Wang, J.: Towards efficient privacy-preserving encrypted image search in cloud computing. Soft. Comput. **23**(6), 2101–2112 (2017)
8. National Internet Emergency Response Center. Summary of China's Internet network security situation in 2019 [EB/OL]. https://www.cert.org.cn/publish/main/upload/File/2019-year.pdf
9. Muench, M., Stijohann, J., Kargl, F., et al.: What you corupt is not what you crash: challenges in fuzzing embedded devices. In: Proceedings of the 25th Network and Distributed System Security Symposium Reston: Internet Society (2018)
10. Kai, C., Li, Q., Wang, L., Chen, Q., Zheng, Y., Sun, L., Liang, Z.: DTaint: detecting the taint-style vulnerability in embedded device firmware. In: IEEE/IFIP International Conference on Dependable Systems and Networks (DSN 2018), pp. 430–441 (2018)
11. Chen, J., Diao, W., Zhao, Q., et al.: IoTFuzzer: discovering memory corruptions in IoT through app-based fuzzing. In: Proceedings of the 25th Network and Distributed System Security Symposium Reston: Internet Society (2018)
12. Zheng, Y., Wen, H., Cheng, K., Song, Z., Zhu, H., Sun, L.: Research review on vulnerability mining technology of Internet of things devices. J. Inf. Secur. **4**(05), 61–75 (2019)
13. Gajewski, M., Batalla, J.M., Mastorakis, G., Mavromoustakis, C.X.: A distributed IDS architecture model for smart home systems. Cluster Comput. **22**(1), 1739–1749 (2017)
14. HaddadPajouh, H., Dehghantanha, A., Parizi, R.M., et al.: A survey on internet of things security: requirements, challenges, and solutions. Internet Things, 100129 (2019)
15. Ahn, J., Lee, I.G., Kim, M.: Design and implementation of hardware-based remote attestation for a secure Internet of Things. Wirel. Pers. Commun., 1–33 (2020). (Prepublish)
16. Naoui, S., Elhdhili, M.E., Saidane, L.A.: Novel enhanced LoRaWAN framework for smart home remote control security. Wirel. Pers. Commun. **110**(4), 2109–21030 (2020)

17. Chen, W.: Research on internet of things technology and its application in smart home. Sci. Technol. Innov. Appl. **19**, 164–165 (2020)
18. Lonzetta, A., Cope, P., Campbell, J., Mohd, B., Hayajneh, T.: Security vulnerabilities in bluetooth technology as used in IOT. J. Sens. Act. Netw. **7**(3), 28 (2018)
19. Celebucki, D., Lin, M.A., Graham, S.: A security evaluation of popular internet of things protocols for manufacturers. 2018 IEEE International Conference on Consumer Electronics (ICCE), pp. 1–6. IEEE, Las Vegas (2018)
20. Aghili, S.F., Ashouri-Talouki, M., Mala, H.: DoS, impersonation and de-synchronization attacks against an ultra-lightweight RFID mutual authentication protocol for IoT. J. Supercomput. **74**(1), 509–525 (2017)
21. Mohammed, A., Issa, T., Isaac, W.: Anonymous mutual IoT interdevice authentication and key agreement scheme based on the ZigBee technique. Internet Things **7**, 100061 (2019)
22. Santos, A.C.T., Soares Filho, J.L., Silva, Á.Í., Nigam, V., Fonseca, I.E.: BLE injection-free attack: a novel attack on bluetooth low energy devices. J. Ambient Intell. Humanized Comput., 1–11 (2019). (Prepublish)
23. Mahini, H., Mousavirad, S.M.: WiFi intrusion detection and prevention systems analyzing: a game theoretical perspective. Int. J. Wirel. Inf. Netw. **27**(1), 77–88 (2020)
24. Ghafoorian, M., Nikooghadam, M.: An anonymous and secure key agreement protocol for NFC applications using pseudonym. Wirel. Netw. **26**, 4269–428 (2020). (Prepublish)
25. Alaa, M., Zaidan, A.A., Zaidan, B.B., Talal, M., Kiah, M.L.M.: A review of smart home applications based on Internet of Things. J. Netw. Comput. Appl. **97**, 48–65 (2017)
26. Almusaylim, Z.A., Zaman, N.: A review on smart home present state and challenges: linked to context-awareness internet of things (IoT). Wirel. Netw. **25**(6), 3193–3204 (2018)

CS-Based Homomorphism Encryption and Trust Scheme for Underwater Acoustic Sensor Networks

Kun Liang, Haijie Huang, Xiangdang Huang$^{(\boxtimes)}$, and Qiuling Yang

School of Computer and Cyberspace Security, Hainan University, Haikou 570228, China
990623@hainanu.edu.cn

Abstract. Underwater acoustic sensor networks (UASNs) have been exploited in many applications. However, due to the complex, unattended and, worse still, hostile deployment environment of the networks, they are vulnerable to many malicious attacks. Presently, researches on how to cope with these threats are extremely restricted due to the limited capability of the sensors. In this paper, we propose a compressive sensing (CS) based homomorphism encryption and trust scheme (CHTS). To identify several malicious attacks such as eavesdropping attacks, compromising attacks, Sybil attacks, wormhole attacks and selective forwarding attacks, etc., we utilize SVM to train trust model and send to each node so that it can determine whether its neighbor nodes are malicious or not. Also, homomorphism encryption is adopted to ensure data confidentiality. Finally, the security analysis shows that the proposed scheme can effectively ensure data confidentiality and identify malicious nodes.

Keywords: CS · Homomorphism encryption · Machine learning · Trust model · UASNs

1 Introduction

With the development of wireless sensor networks (WSNs), underwater acoustic sensor networks (UASNs), as a promising branch of (WSNs), have attracted much attention of researchers on their applications [1], e.g. underwater environment monitoring, off-shore exploration, auxiliary navigation, and tactical surveillance [2]. Due to the characteristics of underwater communication channels, UASNs utilize acoustic waves to send and receive message.

In general, acoustic sensors are randomly deployed in a complex, inaccessible and, even worse, hostile underwater environment, which makes UASNs vulnerable to many malicious attacks. In addition, these battery-powered sensors cannot mostly be recharged, which requires that the sensors have to only possess constrained working ability in order to prolong the lifetime of UASNs as longer as possible. To design a crypto-system suitable for UASNs, the key issue is how to maintain the trade-off among security, performance and cost [3].

J. MacIntyre et al. (Eds.): SPIoT 2020, AISC 1283, pp. 394–399, 2021.
https://doi.org/10.1007/978-3-030-62746-1_58

In this paper, we propose an integrated security model described as CS-based homo-morphism encryption (HE) and trust scheme (CHTS). SVM-based trust model [4] is utilized to detect the compromised nodes according to several trust evidences. HE can ensure data confidentiality. The aforementioned methods can be combined to better cope with many threats existing in UASNs.

2 Related Work

In this section, we summarized some existing homomorphism encryption and trust model works.

The Okamoto-Uchiyama (OU) algorithm [5] is a public-key cryptosystem as secure as factoring and based on the ability of computing discrete logarithms in a particular subgroup. In [6], authors design an Elliptic curve ElGamal (EC-EG) based additive homomorphism algorithm and cipher-texts from different nodes are collected through addition. Its security is based on the hardness of the elliptic curve discrete logarithm problem (ECDLP). This algorithm is to map plaintext m to the EC point mG, and reverse m from mG. However, EC-EG homomorphism encryption is not suit for large-scale sensor networks.

In [7], the individual trust value is evaluated in each CM by monitoring the commu-nication behaviors of neighbors. Moreover, the cluster head collects the original trust values of CMs and sends them to the base station to calculate final trust values. In [8], a cloud-based trust management system (TMC) was presented. The trust evidence includes link trust, data trust, and node trust, which are generated by analyzing the impact of malicious attacks on the network layer by layer respectively. In [4], authors utilized SVM machine learning to train the prediction model.

3 Description of CHTS

In this section, we provide a comprehensive system scheme for coping with the complex deployment environment of UASNs, which includes measurement matrix, homomorphism cryptosystem, trust model.

3.1 Measurement Matrix

Assuming that the sensing signals are sparse or compressible, the compressive sensing can be utilized to largely reduce the measurements to be transmitted to the next hop. According to the CS theory, the measurement matrix is $\emptyset \in R^{M \times N}$, and the measurement result is $y \in R^{M \times 1}$.

$$y = \emptyset x = \emptyset \psi' \theta = A\theta \tag{1}$$

where $x \in R^{N \times 1}$ is a N-dimension signal vector, $A \in R^{M \times N}$ is the sensing matrix or recovery matrix and θ is the transformation of the signal x. By the way, the measurement matrix should satisfy restricted isometry property (RIP).

A special sparse matrix [9], as a kind of sub-gaussian matrix is available as follows:

$$\begin{cases} P(\emptyset_{IJ} = 1) = 1/2s \\ P(\emptyset_{ij} = -1) = 1/2s \\ P(\emptyset_{ij} = 0) = 1 - 1/s \end{cases} \tag{2}$$

Furthermore, when $\emptyset_{ij} = 0$, the item wouldn't be encrypted. Due to only a few elements have non-zero value, the number of elements that requires encryption and transmission reduces greatly.

3.2 Trust Model

In this paper, OU algorithm [5] can ensure the confidentiality of the sample sets. Therefore, this algorithm could encrypt the sample sets to ensure that the base station receives right training samples. In return, the trust models can identify malicious modes and improve the robustness of the network.

It is assumed that each node requires to maintain a historical window and a current window to store interaction behaviors with its neighbor nodes as Fig. 1. In general, trust evidences are calculated according to the performance of malicious attacks. For example, Dos attacks would consume more energy than normal nodes. On the contrary, selective forwarding attacks would consume less energy than normal nodes. The effect of several attacks on the network is summarized as Table 1. Generally, sink nodes have to consume more energy. So, in order to identify these special nodes, we configure a hierarchic marker for each node. It is assumed that the hierarchic marker of the base station, the sink nodes and the ordinary nodes are 0.1, 0.2, 0.3, respectively.

To integrate with homomorphism encryption, this paper utilizes hierarchic marker, communication trust and energy trust for trust model. The followings are generation of trust value and calculation of prediction model.

Generation of Trust Value. Communication trust: The number of unsuccessful and successful communication between neighbor nodes can partly reflect the probability whether each other is a malicious node in a period. Due to the complex and dynamic deployment environment, the stored interaction information is fuzzy and uncertain. So, the theory of subjective logic is adopted in this part. According to [10], a triplet $\{b, d, u\}$ denotes the impact of successful communication, unsuccessful communication and uncertainty, respectively. If s and f represent the number of successful and unsuccessful communication respectively in the historical time window, the communication trust T_c can be calculated as follows:

$$T_c = b + \frac{1}{2}u \tag{3}$$

where $b = \frac{s}{s+f+1}$, $u = \frac{1}{s+f+1}$.

Energy trust: Here, residual energy E_r and energy consumption rate r_e are utilized to calculate energy trust. Therefore, the energy trust of a node can be calculated as follows:

$$T_e = \begin{cases} 0, & \text{if } E_r < \theta \\ 1 - |r_e - r_a|, & \text{else} \end{cases} \tag{4}$$

where θ denotes the energy threshold and r_a denotes the stable energy consumption rate.

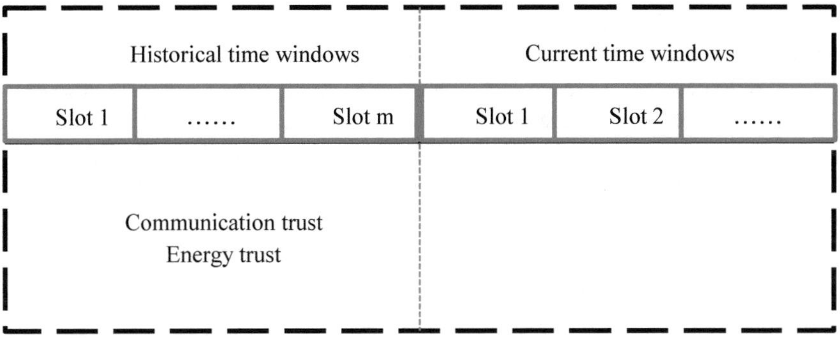

Fig. 1. The sliding time window.

Table 1. The effects of several attacks.

Attack type	The performance of malicious nodes
Dos attack	More energy consumption
Sybil attack	More energy consumption
Wormhole attack	More energy consumption
Fake routing attack	Communication failure and more energy consumption
Selective forwarding	Communication failure and less energy consumption

Prediction Model Calculation. In this scheme, the base station utilizes the k-means algorithm that is relatively simple and quick to classify the initial trust set from all the nodes into two labels like 0 or 1.

According to the labeled data set obtained through the k-means algorithm, the base station utilizes the supervised learning algorithm SVM to classify the labeled data set into several categories, which can achieve relatively good performance in the condition of a sample set. In this algorithm, the means of training the SVM model is similar with the authors in [11]. A Radial Basis Function Kernel is used due to the small number of features of the samples. For improving the accuracy of the SVM model, the regularization technique is adopted. Finally, the base station sends the trained prediction model to all the nodes to identify malicious nodes.

4 Discussion

In this section, the security performance of the scheme is discussed in several attacks. By the way, the security model in this paper is a comprehensive scheme by combining OU with the trust model. So, it can cope with many types of attacks. It is assumed that the base station is secure enough.

4.1 Eavesdropping Attack

In this scheme, measurement results of a sensor node are encrypted using public key technology based OU and aggregated in cipher domain using the property of homomorphism encryption. That is, a node that is responsible for aggregating the data from its child nodes don't have to decrypt the cipher-texts from its child nodes. In addition, the private key is stored in the base station. So, even though an adversary eavesdrops on a transmitted packet, it has no way to decrypt the cipher-text without the private key of the base station. In a word, this scheme can prevent eavesdropping attacks and ensure the end-to-end data confidentiality.

4.2 Selective Forwarding Attack

In this attack [12], malicious nodes behave like legitimate nodes, but selectively drop packets, and refuse to forward the received packets.

This kind of attacks would generally consume less energy than normal nodes and appear the phenomenon of communication failure. So, the base station can cope with this attack by training a prediction model on the trust evidence from all the nodes and send the model to the nodes. The nodes can efficiently identify malicious nodes using the trained prediction model. Better still, the accuracy of the prediction is gradually improved with time due to the advantage of machine learning.

4.3 Dos Attack

Dos attacks can be launched in different ways and at different layers of the protocol stack. They could exist not only in wired networks but also in wireless networks and is hard to defense. The aim of the attacks is to make the resources and services of a network unavailable to the legitimate nodes. Even if utilizing traditional encryption algorithms, UASNs can still be threatened by the Dos attacks.

Fortunately, this kind of attacks generally consume more energy than normal nodes. So, trust model is a feasible solution to Dos attacks. Moreover, our prediction model can adapt to the complex underwater environment of UASNs due to adopting machine learning in this scheme. The prediction accuracy of the model would continuously improve with time.

5 Conclusion

In this paper, we propose a comprehensive security scheme for UASNs for coping with several malicious attacks, which includes a CS-based homomorphism encryption algorithm for guaranteeing data confidentiality and a trust model for coping with malicious nodes. Moreover, a special measurement matrix is selected to largely reduce the energy consumption of encryption. In addition, the SVM and the k-means algorithms are utilized to efficiently train the prediction model, which are suitable for resources-limited, unstable, employment- sparse UASNs. The last section shows that this scheme can effectively prevent several attacks including eavesdropping attack, compromising attack, sybil attack, wormhole attack and selective forwarding attack, etc.

Acknowledgment. This work was supported by the following projects: the National Natural Science Foundation of China (61862020); the key research and development project of Hainan Province (ZDYF2018006); Hainan University-Tianjin University Collaborative Innovation Foundation Project (HDTDU202005).

References

1. Diamant, R., Casari, P., Tomasin, S.: Cooperative authentication in underwater acoustic sensor networks. IEEE Trans. Wirel. Commun. **18**(2), 954–968 (2019)
2. Liu, Z., Gao, H., Wang, W., Chang, S., Chen, J.: Color filtering localization for three-dimensional underwater acoustic sensor networks. Sensors **5**(3), 6009–6032 (2015)
3. Biswas, K., Muthukkumarasamy, V., Singh, K.: An encryption scheme using chaotic map and genetic operations for wireless sensor networks. IEEE Sens. J. **15**(5), 2801–2809 (2015). https://doi.org/10.1109/JSEN.2014.2380816
4. Han, G., He, Y., Jiang, J., Wang, N., Guizani, M., Ansere, J.A.: A synergetic trust model based on SVM in underwater acoustic sensor networks. IEEE Trans. Veh. Technol. **68**(11), 11239–11247 (2019). https://doi.org/10.1109/TVT.2019.2939179
5. Okamoto, T., Uchiyama, S.: A new public-key cryptosystem as secure as factoring. In: International Conference on the Theory and Applications of Cryptographic Techniques, pp. 308–318. Springer, Berlin (1998)
6. Shim, K.A., Park, C.M.: A secure data aggregation scheme based on appropriate cryptographic primitives in heterogeneous wireless sensor networks. IEEE Trans. Parallel Distrib. Syst. **26**(8), 2128–2139 (2015)
7. Shaikh, R.A., Jameel, H., d'Auriol, B.J., Lee, H., Lee, S., Song, Y.-J.: Group-based trust management scheme for clustered wireless sensor networks. IEEE Trans. Parallel Distrib. Syst. **20**(11), 1698–1712 (2008)
8. Jiang, J., Han, G., Shu, L., Chan, S., Wang, K.: A trust model based on cloud theory in underwater acoustic sensor networks. IEEE Trans. Ind. Inf. **13**(1), 342–350 (2017)
9. Zhang, P., Wang, S., Guo, K., Wang, J.: A secure data collection scheme based on compressive sensing in wireless sensor networks. Ad Hoc Netw. **70**, 73–84 (2018). ISSN 1570-8705
10. Jiang, J., Han, G., Wang, F., Shu, L., Guizani, M.: An efficient distributed trust model for wireless sensor networks. IEEE Trans. Parallel Distrib. Syst. **26**(5), 1228–1237 (2015)
11. Jayasinghe, U., Lee, G.M., Shi, Q.: Machine learning based trust computational model for IoT services. IEEE Trans. Sustain. Comput. **4**(1), 39–52 (2019)
12. Yang, G., Dai, L., Wei, Z.: Challenges, threats, security issues and new trends of underwater wireless sensor networks. Sensors **18**, 3907 (2018)

3D Animation Scene Plane Design Based on Virtual Reality Technology

Min Zhang[⊠]

College of Art and Design, Shangqiu Normal University, Shangqiu 476000, China
zm1975.ok@163.com

Abstract. Since it is difficult for the current method to acquire the complete and correct information of the 3D animation scene, and exists the problems of the poor continuity of 3D animation scene and complex computation, a graphic design method of 3D animation scene based on virtual reality technology is proposed. With the method, the grid nodes representing the depth information of animation scene are established in the space coordinates; the camera coordinate system is transformmed into the world coordinate system to get the depth coordinate values of the candidate grid nodes of the animation scene; the preprocessing for the depth image of animation scene is used to smooth the noise to acquire the 3D point cloud of the animation scene image. According to the relationship between the feature points of animation scene image and 3D point cloud, the 3D feature point of animation scene image is obtained, and the ICP method is used to realize the point cloud accurate matching of the animation scene image to establish the 3D animation scene model. The experimental results show that the method can improve the recon struction accuracy of 3D animation scene effectively,and has strong real-time performance and high stability.

Keywords: Virtual reality technology · 3D animation scene · Graphic design · 3D point cloud

1 Introduction

With the development of 3D reconstruction technology, 2D animation scene has been difficult to meet people's visual needs and subjective feelings. At present, most of the 3D animation scene data obtained by the 3D animation scene reconstruction method is part of the information of the reconstructed animation scene surface, and describes the reconstructed animation scene according to the 3D shape and position in the coordinate system with the observer as the center, so it is difficult to present the depth information of the animation scene. "In this case, how to accurately and effectively extract all the information behind the animation scene to complete the 3D animation scene graphic design has become a major problem in the field of computer vision research, which has attracted the high attention of experts and scholars in this field 1459. In reference [1], a 3D animation scene plane design method based on stereo matching is proposed. This method

J. MacIntyre et al. (Eds.): SPIoT 2020, AISC 1283, pp. 400–404, 2021.
https://doi.org/10.1007/978-3-030-62746-1_59

has a good effect on the image reconstruction of the noisy animation scene, but there are some problems such as complicated calculation process and long time consuming. In reference, a graphic design method of 3D animation scene based on corner detection is proposed. The accuracy of animation scene reconstruction using this method is high, but there is a problem of low efficiency of animation scene reconstruction 89 [2].

In view of the above problems, this paper proposes a 3D animation scene plane design method based on virtual reality technology. Experimental results show that the proposed method can effectively improve the reconstruction accuracy of 3D animation scene, and has strong real-time and stability.

2 3D Animation Scene Plane Design Based on Virtual Reality Technology

Suppose (U, V) represents point P, the sub-pixel coordinates in the animation scene image, and (x, y, Z.) represents the three-dimensional coordinates in the world coordinate system of the corresponding points in the animation scene image. Using the sub-pixel coordinates in the animation scene image, the normalized three-dimensional coordinates x in the camera coordinate system under the radial and tangential distortion of the animation scene image can be calculated, and the calculation is carried out by formula (1):

$$\begin{pmatrix} Xd\ (1) \\ Xd\ (2) \\ 1 \end{pmatrix} = \begin{pmatrix} a_x & 0 & u_0 \\ 0 & a_y & v_0 \\ 0 & 0 & 1 \end{pmatrix} \begin{pmatrix} U \\ V \\ 1 \end{pmatrix} \tag{1}$$

In the formula, a represents the scale factor of the x-axis of the animation scene image coordinate system; a represents the scale factor of the Y-axis of the animation scene image coordinate system; (uo, VO) represents the pixel coordinate of the animation scene image center; X (), X. (2) represents the x-direction and Y-direction coordinate points of the animation scene image normalization. Suppose X. Based on the normalized coordinates of the animation scene image, the radial and tangential distortion of the dynamic animation scene image is compensated in the form of multiple iterations. The initial estimate of is expressed by Eq. (2):

$$X_n = \begin{pmatrix} x \\ y \end{pmatrix} = \begin{pmatrix} Xd\ (1) \\ Xd\ (2) \end{pmatrix} \tag{2}$$

Suppose k_{ridial} represents the radial distortion component of the animation scene image, and δ represents the tangential distortion component of the animation scene image, which can be expressed as follows: $k_{ridial} = 1 + k1r2 + k2r4 + k3r6$.

$$\delta = \begin{pmatrix} 2k3xy + k4(r2 + 2x2) \\ K3(r2 + 2x2) + 2k4xy \end{pmatrix} \tag{3}$$

Where R2 = X2 + Y2 represents the relative distortion value of x-direction and Y-direction animation scene image; K., KIR, K, KH represent linear distortion parameters of animation scene image. By solving Eqs. (2) and (4) simultaneously, the image coordinates of dynamic animation scene after domestication can be obtained.

$$X_n = \begin{pmatrix} X \\ Y \end{pmatrix} = \begin{pmatrix} \text{Xd (1)} - \delta\,(1)/k_{\text{ridia}} \\ \text{Xd (2)} - \delta\,(2)/k_{\text{ridial}} \end{pmatrix} \tag{4}$$

In the formula, $\delta\,(1)$ and $\delta\,(2)$ represent the linear distortion value of the tangential distortion of the animation scene image. Combined with the calculation results of X, the radial distortion kridial and the tangential distortion component δ of the animation scene image are iterated for many times, and the converged x is obtained. Because of X. It is the reduced coordinate of the camera coordinate system in the linear model. Suppose (x, y, z) represents the coordinate of the camera coordinate system, then use Eq. (6) to transform the camera coordinate system and the world coordinate system.

$$Z_e \begin{pmatrix} x \\ y \\ 1 \end{pmatrix} = Z_e \begin{pmatrix} \text{Xe}/\text{Ze} \\ \text{Ye}/\text{Ze} \\ 1 \end{pmatrix} = R \begin{pmatrix} \text{Xw} \\ \text{Yw} \\ \text{Zw} \end{pmatrix} \tag{5}$$

Where R and t represent the rotation matrix and translation vector of camera external parameters respectively. By solving Eq. (6), the depth coordinates of candidate mesh nodes in the depth direction are obtained [3].

3 Experimental Results and Analysis

In order to verify the accuracy of the scene, the CCD image sequence with a resolution of 380i/s was used to verify the accuracy of the scene. Through the windows platform, the experimental results of 3D animation scene plane design based on the improved method and binocular vision method are described. The improved method and binocular vision method are respectively used for 3D animation scene plane design experiment, and the point cloud computing time (unit: ms) of 3D animation scene under the two methods is compared, and the comparison results are shown in Fig. 1. Through the analysis of Fig. 1, it can be seen that the computing time of point cloud for 3D animation scene with improved method is shorter than that of binocular vision method, which is mainly because in the process of 3D animation scene plane design with virtual reality technology, the grid nodes representing the depth information of animation scene are established in space first, According to the relationship between the feature points of animation scene image and the 3D point cloud, the 3D feature points of animation scene image are obtained. The ICP method is used to complete the accurate registration of the point cloud of the animation scene image, and the 3D point cloud of the animation scene image is obtained, which makes the calculation time of the point cloud of the 3D animation scene plane design using the improved method shorter.

The improved method and binocular vision method are respectively used in 3D animation scene plane design experiment, and the two different methods are used in 3D animation scene plane design experiment.

The mean square error H (4) of the location distance of the performance index parameters of the surface design is compared, and the formula (12) is used to calculate

$$H(li) = 1/2 + mi(2\| \, li \, \| - mi) \tag{6}$$

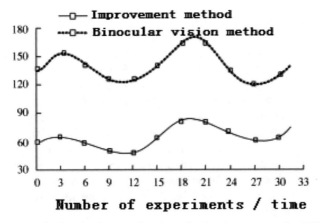

Fig. 1. Comparison of point cloud computing time of animation scene under different methods

Where: l; represents the distance difference between corresponding points in the i-th group of animated scene images after transformation; m represents the preset value. This is mainly because in the process of 3D animation scene plane design using virtual reality technology, the normalized three-dimensional coordinates of camera coordinate system under radial and tangential distortion of animation scene image are firstly calculated, and then compensated in the form of multiple iterations The radial and tangential distortion of the dynamic animation scene image is combined with the bilateral filtering method to filter the animation scene image, and the spatial and gray domain weights of the animation scene image are obtained, which makes the position distance mean square error of the improved 3D animation scene plane design lower.

The improved method and binocular vision method are used to carry out 3D animation scene plane design experiments. The reconstruction accuracy of two different methods for 3D animation scene plane design is compared. The reconstruction accuracy of 3D animation scene plane design under different methods is compared by the number of point clouds. The comparison results are shown in Table 1.

Table 1. Comparison of point cloud numbers under different methods

Number of experiments/time	Improvement method	Binocular vision method
5	56 437	23 615
10	54 672	26124
15	55 437	24381
20	53 416	23746
25	52 741	25736

Through the analysis of Table 1, it can be seen that the number of point clouds in 3D animation scene plane design using the improved method is more than that of binocular vision method, which can effectively improve the reconstruction accuracy of animation scene, mainly because in the process of 3D animation scene plane design using virtual reality technology, Firstly, the normalized three-dimensional coordinates of the camera coordinate system under the radial and tangential distortion of the animation scene image are calculated, and the radial and tangential distortion of the dynamic animation scene image is compensated in the form of multiple iterations to obtain the normalized animation scene image coordinates, On this basis, combined with the point cloud registration method, the minimum rotation matrix and the objective function of the minimum translation vector of the animation scene graph are calculated, which makes the 3D animation scene plane design using the improved method can effectively improve the reconstruction accuracy.

4   Conclusions

In the process of 3D animation scene plane design, it is difficult to obtain the complete and correct depth information of 3D animation scene, which has the problems of complex calculation and poor continuity of 3D animation scene. A 3D animation scene plane design method based on virtual reality technology is proposed. Experimental results show that the proposed method can effectively improve the accuracy of 3D animation scene reconstruction, and has strong real-time performance and reliability.

Acknowledgements. Henan philosophy and social science planning project "research on the integrated creation of film and television animation based on the central plainsculture" (project no.: 2018BYS010).

References

1. Li, X., Yang, A., Qin, B., et al.: 3D reconstruction of monocular vision based on optical flow feedback. Acta optica Sinica **35**(5), 228–236 (2015)
2. Yu, S., Fu, J., Fan, C., et al.: Gear design and precise modeling based on virtual reality. Korean J. Ceram. **37**(2), 200–204 (2016)
3. Ye, F.: Effect optimization of 3D animation scene lighting rendering algorithm. Sci. Technol. Bull. **31**(8), 262–264 (2015)

Design of VR Computer Network and Maintenance Experiment System

Fancheng Fu$^{(\boxtimes)}$, Mei Yang, and Xiaofang Tan

School of Computer Information Engineering, Nanchang Institute of Technology, Nanchang 330044, China
ffc2009@126.com, maynni@163.com, txfcoffee@163.com

Abstract. This paper mainly studies how to apply virtual reality technology in multimedia experiment teaching. On the premise of using many kinds of technology, the virtual teaching platform of multimedia experiment with imagination and interactivity is established by means of constructing virtual space and other methods and means. It is known that: v r technology mainly relies on the hardware and software resources of virtual reality, So as to realize human-computer interaction, and provide users with visual, auditory and tactile personal experience by virtue of virtual space, so that users can interact and perceive the three-dimensional objects in the actual teaching application of the technology, which can enhance students' operation ability, optimize the experimental teaching environment, and improve the teaching management level.

Keywords: Virtual reality technology · Teaching experiment · Multimedia teaching system

1 Introduction

The practical training of networking and maintenance of computer network is a practical technical basic course that information majors must master. Through the practical training course offered by experiments, students can not only deepen their understanding of the course of computer network composition principle, but also master the assembly and maintenance technology of computer network and the use of common tools and software. From the aspect of hardware, we can further cultivate students' practical ability, analyze and solve problems independently. However, in the process of training, the equipment is disassembled many times, which is easy to cause equipment damage. At the same time, due to the limited number of experimental equipment, most students can not carry out effective training in person, which makes it difficult for students to understand the working mechanism of computer network and the principle of networking and maintenance. Especially when learning the principle of computer network, because the students have no professional work experience, they have relatively few opportunities to "see" in school [1–3].

J. MacIntyre et al. (Eds.): SPIoT 2020, AISC 1283, pp. 405–409, 2021.
https://doi.org/10.1007/978-3-030-62746-1_60

Therefore, the problem of students' perceptual cognition should be solved first. In this regard, we created a virtual model library, and used UG and Pro/engineer software to create the virtual elements of some typical instruments in the computer network networking and maintenance training, and used the way of virtual assembly and dynamic demonstration to build the virtual experiment teaching mode of computer network and maintenance testing. Through the motion simulation of virtual components, students can clearly understand the composition and working principle of the computer network hardware components, more easily understand its working principle and maintenance, improve learning interest, and lay the foundation for the following courses [4–7].

2 Function Design of Multimedia Teaching System Simulation Experiment

T Through practical research, the key of simulation design in this paper is to develop two major equipment of teaching system, namely projector and video display platform. In order to provide convenience for users, this paper includes the following modules: resource link, knowledge introduction, system description and the platform of actual communication [8, 9].

First, system description: It is convenient for users to understand the function and operation of the system.

Second, knowledge explanation: This paper briefly introduces the contents of virtual experiment and all media equipments of multimedia teaching system.

Third, virtual experiment: This part is the focus of the design. It mainly includes projector, multimedia system, virtual laboratory and video stage experiment.

Fourth, system help: The operation steps of virtual experiment system are explained reasonably to solve various problems in the experiment.

Fifth, online communication: Create a platform for students to communicate with each other. Students can communicate online with the help of system module content, answer various questions with each other, and improve their knowledge reserve.

3 Overview of Multimedia Teaching System Development Platform

First, hardware platform overview
2 GB system memory;
Intel (R) core (TM) 2 Duo E7200 @ 2.53 GHZ-PC;
Windows XP operation;
BS_Contact_VRML_ 61 browser plug-in;
NVIDIA Geforce 8400se video card.
Second, overview of development tools
Dreamweaver MX, net - website development 3DS MAX8.0;
Vrmipad 2.1, JavaScript, X3d-Edit 3.2-program development;
Cool Edit Pro 2.1 - voice recognition and processing;
Photoshop CS - image recognition processing;
Premiere pro2.0-video identification and processing.

VR algorithm is equivalent to neural network, which is composed of neurons. Neurons are composed of dendrites and axons. Dendrites are used to receive signals from the previous neuron, and then act on the next neuron from the axons. Artificial neural networks are similar. As shown in the figure, X_i represents the input of the previous neuron, W_i represents the weight, and Y represents the output. Output y can be written as:

$$Y = W_1 X_1 + W_2 X_2 + W_3 X_3 + \cdots + W_n X_n \qquad (1)$$

(1) The weight matrix, equivalent to the memory of neural network, adjusts the appropriate weight through continuous training.
(2) Activation function, equivalent to the calculation equation of neural network, commonly used activation functions are:

$$\sigma X = \frac{1}{1 + e^{-x}} \qquad (2)$$

Because it's a continuous function, and it's bounded.

4 Realize the Modeling and Development of Virtual Experiment Environment

The modeling method is analyzed as follows: First, code by hand. According to the actual geographical location of the model in the entity space, the modeling statement of the program language is used to show the location relationship of the model in the virtual environment scene by virtue of the three-dimensional coordinate points. The main characteristics of the virtual scene are simple, short and fast. In VRML, special geometric nodes are set to construct the three-dimensional model, At the same time, it gives the model a certain material and appearance. This paper adopts this method. VRML geometry node refers to shape node, three-dimensional node, two-dimensional space modeling node and other related nodes. In the actual modeling, the above nodes should be reasonably selected to make the object shape clear. Second, modeling software coding, including Maya, 3dmax and other related software. This method can effectively simulate the more complex three-dimensional shape, at the same time, it has the corresponding plug-in support of exporting to the virtual reality software, and can export a number of different formats of three-dimensional geometry

The main models include two types, the irregular model and the rule model. The irregular model is usually constructed with the help of 3DMAX software, and then the post processing is carried out with VRML. The rule model is best handled directly with VRML software, so the file size can be effectively reduced

Generally speaking, the laboratory scene consists of two parts, namely, outdoor doorway, aisle and indoor (see Fig. 1). To build an outdoor scene is usually to provide a virtual reality entrance for the user. When the user approaches, the door of the laboratory will open itself, so that the user can enter into the experimental scene. The construction of this scene is mainly based on language modeling, By using Box. IndexedFaceSet of geometry basic geometry node in VRML software and Transforon space transformation node, the regular shape of laboratory scene, including ceiling and floor, is constructed.

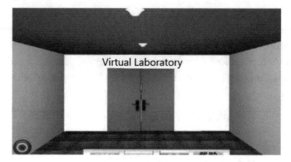

Fig. 1. The scene of laboratory corridor

5 The Ability to Reasonably Design the Interaction of Video Display Platform

5.1 Meet the Interactive Function of Moving Parts of Video Display Platform

The camera lens and arms of the video display platform can move in a specific area. The main activities include: first, fixed rotation axis point; second, rotation angle in a fixed range. The design of the interaction function can be developed by VRML language modeling, or visual design can be developed by 3DMAX software. With VRML language modeling, first of all, it should be clear in the space. The three-dimensional coordinate points of this part include not only the transformation of spatial displacement coordinates, but also the transformation of spatial rotation direction. Therefore, the code of VRML function to realize the interactive ability of video display platform is complex, and at the same time, it is very easy to generate jumping conditions. With 3DMAX technology, it has the ability of animation design, and can set the animation of active points in 3DMAX in advance, Build the movement function at a specific time, and then import it into VRML software to set it up.

5.2 Meet the Interactive Function of Moving Parts of Video Display Platform

Generally speaking, both the base plate and the arms of the video display stand are equipped with light bulbs. If the image projected by the object on the video display stand is relatively dark, the two arm lights can be turned on and the brightness can be improved. The base plate light bulbs of the display stand are usually used in slide projection to improve the clarity of the image. Use the buttons on the control panel to control the light bulbs, When both bulbs are turned on, you can turn on the double arm bulb first, then press it, turn on the floor light when the double arm lamp is turned off, and then press it once, that is, both lights are turned off. In the actual design, you should do a good job in setting the state in advance: the state of the bottom lamp, the state of the arm lamp and the state without lamp. In the actual operation, you can use 3DMAX technology to pre condition the light range, intensity and position of the above states.

In the above three conditions, the color of the base plate lamp and the double arm lamp will also change. The value of emissivity of the self illumination attribute of the object can be controlled by VRML software. In the off state, the value is O, i.e. no light. In the on state, the value is 1. The light is white. After the light is set completely, the JavaScript code of VRML software is used for interactive control. The final operation of the virtual experiment is completed, and its effect is as follows (see Fig. 2, Fig. 3).

Fig. 2. Effect of no light video display table   **Fig. 3.** Video display effect of bottom light on

6   Conclusions

This paper mainly studies how to realize the interactive function of virtual equipment and how to construct virtual scene. With the help of many kinds of technologies, the multimedia virtual experiment teaching with imagination and interaction is established by appropriate means.

Acknowledgements. The Design of Computer Network and Maintenance Experiment System Based on VR Subject No.: GJJ180998.

References

1. Wanning: Design and Development of Experimental Teaching Courseware Based on Virtual Scene. Sichuan Normal University, Chengdu (2007)
2. Wang, Y.: Application of Virtual Experiment in Middle School Physics Teaching. Sichuan Normal University, Chengdu (2008)
3. Zhang, X., Hu, L.: Experimental Course of Modern Educational Technology. Electronic Industry Press, Beijing (2008)
4. Zhang, J., Zhang, J., Zhang, J.: X3D Virtual Reality Design. Electronic Industry Press, Beijing (2007)
5. Tang, Y.: Application of Virtual Reality Technology in Education. Science Press, Beijing (2007)
6. Burdea, C.C., Coiffict, P., Wei, Y., et al.: Virtual Reality Technology. Version 2. Electronic Industry Press, Beijing (2005)
7. Yu, M.: Design and Development of Virtual Multimedia Classroom Based on EON Studio. Shaanxi Normal University, Xi'an (2008)
8. Feng, W.: Design and implementation of medical virtual learning situation. Master's thesis of the Fourth Military Medical University, Xi'an (2008)
9. Hu, F.: Virtual reality simulation of blast furnace distribution process. Master's degree thesis of Wuhan University of Science and Technology, Wuhan (2008)

The Quantum State Engineering by Adiabatic Decoherence-Free Subspace

Yue Gao, Yifan Ping, and Zhongju Liu$^{(\boxtimes)}$

School of Science, Shenyang University of Technology, Shenyang
110870, People's Republic of China
Agaoyue@163.com, 15735151479@163.com, liuzhongju10@163.com

Abstract. Decay effect for open quantum system is investigated in this paper. After comparing the fidelity with Adiabatic evolution of Decoherence-Free Subspace, Dissipated Engineering Subspace and Adiabatic Evolution Subspace. We show that, for open quantum system, Adiabatic evolution of Decoherence-Free Subspace has its advantages for preparation of steady state.

Keywords: Open quantum system · Adiabatic evolution · Decoherence-Free Subspace · Steady state

1 Introduction

The adiabatic theorem is the important principle of quantum theory [1, 2], which reflects the time-dependent Hamiltonian operator changes very slowly [3, 4]. For a long time the adiabatic theorem was almost exclusively concerned with closed systems. In the actual physical environment, the quantum system are inevitably effected by the environment [5]. The evolution equation of the open quantum system [6, 7] is very similar to that for quantum closed systems, except that the evolution equation has non-unitary terms. The dynamic effects described by non-unitary terms are called decoherence [8]. In order to eliminate the effect of decoherence, there is an approach that a subspace immune to decoherence is found in the Hilbert space of the system, which makes the quantum system unitary evolution to ensure the integrity of quantum information [9].

On the other hand, quantum adiabatic theorem has been widely used in physical chemistry, quantum field theory and NP-complete problems [10]. In order to study couplings of qubits to environments into account, there is a method of combining adiabatic dynamics with decoherent-free subspace in some heuristic methods, which is called Adiabatic evolution of Decoherent-Free Subspaces (ADFSs) [8]. S.L. Wu et al. have discussed the adiabatic theorem and shortcuts of time-dependent quantum open systems [11]. In this paper, S.L.Wu et al. based on the definition of dynamically stable decoherent-free subspaces, an adiabatic theorem for open systems in time-dependent decoherent-free subspace (t-DFS) is given, and the adiabatic conditions for quantum systems in time-dependent decoherent-free subspace are proposed [12, 13].

J. MacIntyre et al. (Eds.): SPIoT 2020, AISC 1283, pp. 410–414, 2021.
https://doi.org/10.1007/978-3-030-62746-1_61

In this paper, we will investigate the preparation of steady state due to dissppation in ADFS. Meanwhile, we study Dissipated Engineering Subspace (DES) and Adiabatic Evolution Subspace (AES). We will discuss the advantage for decay effect of ADFS by the fidelity of the system.

2 Model

We consider a two-level system, with and ground state $|0\rangle$ and excited state $|1\rangle$ coupled to a coherent control field $\Omega(t)$ and squeezed vacuum field. The Hamiltonian of the system can be written as

$$H_0 = \Omega(t)|0\rangle\langle 1| + \text{H.c.}, \tag{1}$$

and can be described by the master equation under the Markov approximation

$$\partial_t\rho(t) = -i[H_0(t), \rho(t)] + L_D\rho(t), \tag{2}$$

where the L_D is a dissipator which coupled to the squeezed vacuum. If the vacuum squeezing field is perfect, the LDcan be transformed into the Lindblad form as follows,

$$L_D\rho(t) = \frac{\gamma}{2}\left(2L\rho(t)L^\dagger - \left\{L^\dagger L, \rho(t)\right\}\right), \tag{3}$$

Where γ is the decay of the system and L is the decoherence operator which can be described as

$$L = \cosh(r)\exp\left(-\frac{i\theta}{2}\right)\sigma_- + \sinh(r)\exp\left(\frac{i\theta}{2}\right)\sigma_+ \tag{4}$$

where r and θ is the squeezing strength and squeezing phase respectively, and σ_+ (σ_-) is the raising(lowering) operator. Here, we focus on that the adiabatic theorem of open systems can be expressed as two independent conditions [14, 15]. It is shown that by adjusting the coherent evolution, the quantum state can be retained in t-DFS without any purity loss. Under the theorem for Adiabatic evolution of Decoherence-Free Subspace (ADFS), we can obtain the coherent control field as

$$\Omega(r, \theta) = \frac{i\gamma\exp(-r - i\theta)\sqrt{\sinh(r)\cosh(r)}}{2} \tag{5}$$

where the squeezing parameters can be described as $r = r_0 + \mu t$ and $\theta = \theta_0 + \upsilon t$. The Hamiltonian of the system is

$$H_{\text{eff}}^{\text{ADFS}} = H_0 + \frac{i}{2}\sum_\alpha\left(C_\alpha^* L - C_\alpha L^\dagger\right) \tag{6}$$

where C is eigenvalue of L. In the next section, we will discuss the fidelity for Adiabatic evolution of Decoherence-Free Subspace (ADFS).

For comparision, we consider the system reach steady state by the dissipation of the environment. In the Dissipated Engineering Subspace (DES), the Hamiltonian of the system is time-independent. The Hamiltion can be writen as $H_{\text{eff}}^{\text{DES}} = H_0 + \Gamma$, where Γ is the dissipation of the system. Meanwhile, we study the Adiabatic Evolution Subspace, in which, $L = 0$ and $H_{\text{eff}}^{\text{AES}} = H_0 + \omega_0|\phi_1\rangle\langle\phi_1|$, where $|\phi_1\rangle$ is the eigenstate of L.

3 Numerical Analysis

In this section, we will investigate the fidelity of system by the numerical analysis. As shown in Fig. 1, the dash line represents the Adiabatic evolution of Decoherence-Free Subspace (ADFS). The change rate is chosen to be $\mu = \upsilon = 0.1\,\omega_0$, meanwhile, the final state is prepared on the time $\omega_0 t = \frac{\pi}{2}$. The solid line represents the Dissipated Engineering Subspace (DES). The system reach steady state depending on the dissipation of the environment. The dot line represents the Adiabatic Evolution Subspace (AES). The results illustrates that, the fidelity versus the decay γ reach to 1, with the decay increase in the ADFS and DES. However, it is almost constant in the AES due to no dissipation. Moreover, the system reach to the steady state quickly in ADFS.

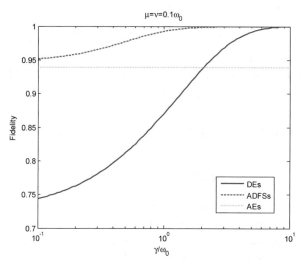

Fig. 1. Fidelity versus γ (in units of ω_0) with $\mu = \upsilon = 0.1\,\gamma$ in the courses of DEs (solid line), ADFSs (dash line) and AEs (dot line).

In Fig. 2, we make the change rate to be $\mu = \upsilon = 0.1\,\gamma$, and the final-sate is prepared on the time $\omega_0 t = \frac{\pi}{2}$. In this part, we investigate the fidelity versus the frequency ω_0 in ADFS (dash line), DES (solid line) and AES (dot line). The results show that, the system is on the steady state from beginning to end in ADFS and DES. However, the system in the AES cannot reach steady state with small frequency ω_0.

In Fig. 3, we make the change rate to be $\omega_0 = \gamma$, and the final-state is prepared on the time $\upsilon t = \frac{\pi}{2}$. In this part, we investigate the fidelity versus change rate μ in ADFS (dash line), DES (solid line) and AES (dot line). We can obtain that, the system is on steady state in all subspace with sm-all change rate μ. The system begin to damp due to the increasing of change rate μ. However, for ADFS, the system remain has a larger fidelity than the other subspace.

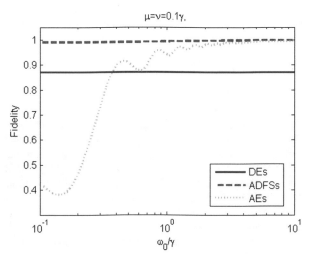

Fig. 2. Fidelity versus ω_0 (in units of γ) with $\mu = \upsilon = 0.1\,\gamma$ in the courses of DEs (solid line), ADFSs (dash line) and AEs (dot line).

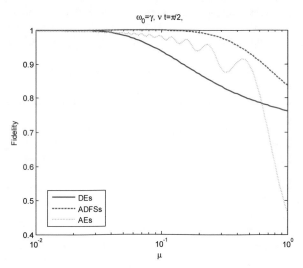

Fig. 3. Fidelity versus μ with $\omega_0 = \gamma$ and $\mu T = \frac{\pi}{2}$ in the courses of DEs (solid line), ADFSs (dash line) and AEs (dot line).

4   Conclusion

In this paper, we investigate preparation of steady state for open quantum system due to the Decay effect. We study the influence on the fidelity in ADFS, DES and AES against decay, frequency and change rate. We can obtain that ADFS has the advantage in preparation.

Acknowledgments. This work is supported by Natural Science Foundation of Liaoning province under Grant No. **2019-ZD-0199**.

References

1. Born, M.V., Fock, Z.: Proof of the Adiabatic theorem. Physics **51**,165 (1928)
2. Kato, T.: On the Adiabatic theorem of quantum mechanics. Phys. Soc. Jpn. **5**, 435 (1950)
3. Gosset, D., Terhal, B.M., Vershynina, A.: Universal Adiabatic quantum computation via the space-time circuit-to-hamiltonian construction. Phys. Rev. Lett. **114**, 140501 (2016)
4. Wild, D.S., Gopalakrishnan, S., Knap, M., Yao, N.Y., Lukin, M.D.: Adiabatic quantum search in open systems. Phys. Rev. Lett. **117**, 150501 (2016)
5. Nielsen, M.A., Chuang, I.L.: Quantum Computation and Quantum Information. Cambridge University Press, Cambridge (2000)
6. Bacon, D., Flammia, S.T.: Adiabatic gate teleportation. Phys. Rev. Lett. **103**(12), 120504 (2009)
7. Cong, S., Yang, F.: Control of quantum states in decoherence-free sub-spaces. J. Phys. A Math. Theor. **46**(7), 075305 (2013)
8. Carollo, A., Palma, G.M., Lozinski, A., Santos, M.F., Vedral, V.: Geometric phase induced by a cyclically evolving squeezed vacuum reservoir. Phys. Rev. Lett. **96**, 150403 (2006)
9. Prado, F.O., Duzzioni, E.I., Moussa, M.H.Y., Almeida, N.G., VillasBoas, C.J.: Nonadiabatic coherent evolution of two-level systems under spontaneous decay. Phys. Rev. Lett. **102**, 073008 (2009)
10. Nenciu, G.: On the adiabatic theorem of quantum mechanics. J. Phys. A Math. Gen. **13**, L15–L18 (1980)
11. Wu, S.L., Huang, X.L., Li, H., Yi, X.X.: Adiabatic evolution of decoherence-free subspaces and its shortcuts. Phys. Rev. A **94**, 042104 (2017)
12. Berry, J.M.V.: Transitionless quantum driving. Phys. A Math. Theor. **42**, 365303 (2009)
13. Luo, D.W., Pyshkin, P.V., Lam, C.H., Yu, T., Lin, H.Q., You, J.Q., Wu, L.A.: Dynamical invariants in a non-Markovian quantum-state-diffusion equation. Phys. Rev. A **92**, 062127 (2015)
14. Song, X.K., Zhang, H., Ai, Q., Qiu, J., Deng, F.G.: Shortcuts to adiabatic holonomic quantum computation in decoherence-free subspace with transitionless quantum driving algorithm. New J. Phys. **18**, 023001 (2016)
15. Beige, A., Braun, D., Tregenna, B., Knight, P.L.: Quantum comput-ing using dissipation to remain in a decoherence-free subspace. Phys. Rev. Lett. **85**, 1762 (2000)

Design and Implementation of Training System for Broadcast Hosts P2P Network Video Live

Yuzao Tan[(✉)]

China Three Gorges University, Yichang 444300, China
tanyuzao@163.com

Abstract. With the continuous popularization and development of network technology and network platform, network live broadcasting has gradually become a common way of entertainment in the society. Network live broadcasting brings development opportunities for broadcasting and hosting professionals, with some negative impact. This paper will briefly analyze the change trend of information dissemination mode in the new media environment, and discuss the influence of live broadcast on the cultivation of broadcasting and hosting professionals and the solutions. Through the analysis and research of this paper, the purpose is to promote the cultivation quality of broadcasting and hosting professionals.

Keywords: Live broadcast on the internet · Major of broadcasting and hosting · Personnel training

1 Introduction

In the era of new media division, with the continuous deepening of media integration, traditional media and emerging media have a good space for development. Every university in China has set up broadcasting and hosting professional courses, but both the teaching quality and teaching mode are deficient. In recent years, the employment effect of graduates majoring in broadcasting and hosting can be found that it can not meet the needs of the industry. The emergence and development of network live broadcast brings innovation opportunities and negative effects to the development of professional talent training of broadcasting and hosting. Therefore, it is of great value and significance to analyze the influence of network live broadcast on the cultivation of broadcasting and hosting professionals and the countermeasures.

2 The Change of Information Transmission Mode Under the Influence of Network Live Broadcast

Change of information production mode. The rapid development of network communication technology provides people with a mobile, convenient and diverse mode of information communication interaction. With the help of Internet technology and network

J. MacIntyre et al. (Eds.): SPIoT 2020, AISC 1283, pp. 415–419, 2021.
https://doi.org/10.1007/978-3-030-62746-1_62

platform, supported by mobile terminal devices and programs, network live broadcasting is a network information interaction mode that people are interested in. With the help of network media platform, the acceptance and dissemination of information will effectively break the time limit and space limit of information interaction. Traditional information production and dissemination are limited by fixed time and platform. The new network live broadcast mode will accelerate the integration of various media and innovate the way of information generation. Therefore, under the social background of the prevalence of live network, the way of information production has changed.

Changes in the main body of information dissemination. With the advent of the mobile Internet era, people's information exchange and interaction appear the trend of mobility, fragmentation and interaction. With the help of the network platform, everyone can become the publisher and the reprinter of information content, and the information dissemination presents a diversified development trend. In traditional information communication, TV, newspaper, radio and magazine are the main body. The main body of information communication is fixed and has certain authority. With the popularization and development of the network platform, the main body of information dissemination has changed, and live network has become a popular way of entertainment and information interaction.

3 The Influence of Network Live Broadcast on the Cultivation of Broadcasting and Hosting Professionals

3.1 Positive Impact

The network live broadcast has a positive impact on the cultivation of broadcasting and hosting professionals. First of all, the popularization and development of network live broadcasting provides the opportunity for students majoring in broadcasting and hosting to exercise their language expression ability and on-the-spot response ability with the help of network platform. For example, the teacher can select a traditional cultural content as the theme of the activity, stipulate that the students will broadcast live at the designated time, and introduce the relevant cultural content to the netizens. Students can make clear their own shortcomings according to the interactive reactions and opinions of the audience in the live broadcast. Teachers can also make more targeted teaching plans. Secondly, in the classroom teaching, teachers can guide students to establish correct professional quality and moral quality by taking some incorrect comments and values of network anchors as examples.

In the live broadcast on the Internet, students are constantly instilled with the sense of competition, sharpened their minds, and stimulated their desire for success and passion for success. In this jungle age full of all kinds of competition, students need to have the quality of self-improvement and self-reliance, and more importantly, have a strong sense of competition. In particular, the current family education, is generally the indulgence of students. In the teaching of higher education, it should not be all the protection of ivory tower. Only with a strong sense of competition and the cultivation of a strong sense of competition, can we face a more brutal employment environment and have more core competitiveness.

Arouse students' interest and cultivate their independent consciousness.

Today's students, in the family, their parents will take care of them wholeheartedly. Many parents would like to help them with all their children's work if they did it all by themselves. This kind of parents' love for their children is a great and selfless love without any superfluous feelings. But in the middle school campus, because of the heavy pressure of entering a higher school. Teachers can't help but join in the parent industry and try their best to clear the obstacles in campus life for students. Students only need to bury their heads in books and complete their studies wholeheartedly. But everything is too much, come to the university campus, many students are at a loss, doing nothing all day: in the dormitory, boys only know to play games every day, girls only know to hold mobile phones every day, holding tablets to catch up with the never-ending Korean dramas. These are the lack of a good mentality. But in the campus life, some students have not even seen the washing powder, do not know how to use, it is inevitable to make people laugh. From small to large, being arranged to study and live in the name of love, we have to say, is to make students become such a culprit. The emergence of network live broadcast plays a good role in linking up. In the live broadcast room, students can't help but contact those unfamiliar things and learn the skills they are not good at. For example, some anchors are cooking and sewing clothes live Naturally, students can learn how to be independent and gradually develop the sense of independence.

3.2 Negative Effects

The popularization and development of network live broadcasting has shaken the position of broadcasting and hosting profession. With the rise of new ways of communication, the traditional way of media based on TV has been seriously impacted. The network live broadcast has attracted a huge attraction, and the development prospect of the broadcasting and hosting profession is not good. At the same time, the network live broadcast will have an impact on the traditional media form, and the values and hosting style of the students majoring in broadcasting and hosting will be affected to some extent.

In the network live broadcast, although has the merit. But there are still many vulgar live broadcasts. Many illegal businesses spread bad values through live webcasts, making huge profits, regardless of the moral bottom line. The network live broadcast is full of many vulgar live broadcasts. Some anchors are doing all kinds of shocking and boundless things, which are just for the audience's cheers and rewards. In this regard, college students in the new era should have a sincere heart, abstain from anxiety, arrogance and impetuosity, and resolutely resist the bad values, rather than being trapped in them.

4 Solutions to the Influence of Network Live Broadcast on the Cultivation of Broadcasting and Hosting Professionals

(1) Cultivate correct values. The emergence and development of network live broadcast has a certain impact on the development of professional talent training, among which the urgent problem is how to guide students to establish correct values. The development of network information technology in China started late, so the supervision and management of network environment is insufficient. In the network

environment, the wrong values are rampant. In order to get more attention, and then get economic benefits, the live broadcast content is relatively vulgar. This kind of blind pursuit of economic interests is easy to affect the establishment of correct values of college students. In the process of training broadcast and host professionals, we should pay attention to the cultivation of students' correct values and professional quality to ensure that broadcast and host professionals establish correct values and professional beliefs.

(2) Clear professional development orientation. In order to solve the problem of the influence of network broadcast on the cultivation of broadcasting and hosting professionals, we should find out the development orientation of broadcasting and hosting professional law and cultivate professional talents with professional competitiveness. Colleges and universities should adhere to the guidance of the employment market, combine the activities of personnel training with the new media industry, and cultivate professional talents with innovation ability and the spirit of the times. For example, in 2016, Nanguang College of China Media University cooperated with pepper platform and announced to be the first live broadcasting college in China. Based on the accurate positioning of talent training objectives, improve the effectiveness and accuracy of broadcasting and hosting major.

(3) Reform teaching methods and improve teaching level. In view of the positive side of the network live broadcast, colleges and universities can take advantage of this opportunity to learn from its beneficial aspects, so as to improve the teaching skills of the broadcasting and hosting major, so as to reform the teaching mode with a new attitude and perspective, and not to lose the trust of their families and countries. In today's new era of socialist development, in today's reform and opening up development, we can first "do not break", but we can "first establish". For example, in the daily teaching of colleges and universities, the education of information "armed" enables teachers to make use of "Internet plus" to achieve "linkage teaching mode". We should change the traditional classroom teaching method which is dominated by teachers. With the form of network live broadcast, teachers and students can communicate with each other. For the broadcasting and hosting major, not only teachers need to give lectures, but also students need to "give lectures", so as to change the past teacher led teaching into teacher assisted teaching, so that students can play their own role in the classroom, better absorb knowledge by virtue of their own "what they see and what they hear", so as to rely on the network directly Broadcast to better improve the teaching level.

(4) Build a good supervision mechanism to protect the growth of students. Compared with foreign education, domestic students generally live in the "greenhouse", from small to large, students are growing up under care. Domestic education pays little attention to social phenomena, not to mention how to prevent these malicious spread. At the same time, with the popularity of the Internet platform, the network is full of more and more vulgar, kitsch, vulgar and other phenomena. Therefore, it is necessary and the general trend to build a good supervision mechanism. School, parents, society and other three parties work together to establish an effective supervision mechanism to protect the growth of students.

5 Conclusions

In the era of self media, network live broadcasting has made great achievements and progress in the cultivation of broadcasting and hosting professionals. For individuals, we must adjust our mentality in time, correct our attitude towards learning with a positive outlook on life and a correct methodology; for society and schools, in addition to the indispensable positive and correct guidance, we should also pay attention to the refinement of teaching content, the diversification of business models and the improvement of teaching models in the future, constantly learn advanced experience at home and abroad, and improve our awareness, so as to better accomplish the great tasks assigned by our country.

Broadcasting and hosting professional talent training activities will face certain development opportunities and difficulties. Teachers and students need to actively adjust their mentality and state, keep up with the development of social media, and accurately locate the talent needs of the new media environment. Through the integration of broadcasting and hosting education resources, we should actively expand the content of the training of broadcasting and hosting professionals, and cultivate broadcasting and hosting professionals in line with the spirit of the development of the times.

References

1. Qu, Y.: The influence of online live broadcast on the cultivation of broadcasting and hosting professionals and countermeasures. Cradle Journalists (10), 90–91 (2018)
2. Liang, Y.: The influence and Enlightenment of online live broadcast on the development of broadcasting and hosting art. News Lovers (12), 50–53 (2017)
3. Ye, C.: The fog of the anchor and the sadness of the host – theoretical association caused by the network anchor in the first year of live broadcast in 2016. South. TV J. (06), 58–63 (2016)
4. Liu, Q.: Innovation of professional personnel training mode broadcast and presided over in the era of mass media. Educ. Teach. Forum (21) (2016)
5. Jia, X., Li, X.: The impact of live network on contemporary college students. Mod. Econ. Inf. (3) (2017)
6. Miao, J.: Research on broadcasting and hosting change in the era of new media. New Media Res. (2) (2017)
7. Li, Y.: On the cultivation of broadcasters and hosts in the new media era. China Media Technol. (11) (2017)

Application Research of Enterprise Production Logistics System Based on Computer Simulation Technology

Zhou Gan Cui[✉]

Department of Business Management, Laiwu Vocational and Technical College, No.1, Shancai Street, Laiwu, Jinan 271100, Shandong, China
zhougancui2001@163.com

Abstract. The popularization and development of computer technology drives the development of this era and society, and has a more and more profound impact on people. With the computer technology coming into the market and life is the intelligent technology and simulation technology, and these advanced technologies are also very effective in the enterprise competition. At present, the competition of enterprises is more and more fierce. If enterprises want to get further development, they must optimize their own production logistics system. Therefore, this paper discusses and analyzes the optimization of the enterprise production logistics system based on the computer simulation technology from many angles and aspects, introduces the specific steps of the computer simulation technology, and compares the enterprise production logistics system under the computer simulation technology with the traditional enterprise production logistics system, and then obtains the computer simulation technology The purpose of application impact is to provide theoretical suggestions for enterprises to carry out system optimization.

Keywords: Computer simulation technology · Enterprise production · Logistics system · Optimi-zation work

1 Introduction

The good development of an enterprise is directly related to the management of the enterprise. At present, many enterprises do not seem to attach importance to the management of the production logistics system. Most enterprises have not been able to make clear the importance of the production logistics. In the current market competition process, if the enterprise wants to get further development, it must optimize its production logistics system, which includes production, sales and transportation [1]. In the process of enterprise management, the management of the logistics system is generally in a weak stage, but after data research, the production logistics system accounts for 30% to 40% of the cost of the enterprise, so the enterprise must optimize the production logistics system at this stage. The optimization of production logistics is not only beneficial to the production logistics system, but also to the overall cost and efficiency of the enterprise.

J. MacIntyre et al. (Eds.): SPIoT 2020, AISC 1283, pp. 420–425, 2021.
https://doi.org/10.1007/978-3-030-62746-1_63

At the present stage, there are obvious deficiencies in the production process of enterprises in China, specifically reflected in the confusion of logistics facilities layout, and the lack of clear system work, which seriously affect the development and production of enterprises. Using computer simulation technology to strengthen the management of enterprise production logistics, to optimize the enterprise production logistics system, and to improve the efficiency and management mode of the enterprise.

2   The Importance of Simulation Technology in the Optimization of Production Logistics System

The application of simulation technology in the current stage has been more mature. In our society, simulation technology has gradually taken on the simulation work of many industries, and has achieved good results. Since the middle of the twentieth century, people have tried to deal with simulation work in computer technology, but they have not made very obvious progress. It was not until the late 1970s that many scientists developed the computing technology that can improve the efficiency of simulation. At present, computer simulation technology has been applied more mature, and can be applied in various industries. Therefore, computer simulation technology has become one of the important means to adapt to the development of the times. It has a very important computer significance for enterprises. If enterprises want to get stronger competitiveness, they need to use computer simulation technology to improve their management level. Only in this way can enterprises better keep up with the pace of the times and improve their competitiveness and efficiency [2].

3   Optimization Analysis of Enterprise Production Logistics System Based on Computer Simulation Technology

3.1   The Concept of Computer Simulation Technology

The computer simulation technology is one of the modern processing technologies based on the computer technology, which is interdisciplinary and multi-faceted. In short, the main operation process of computer simulation technology is to use computer technology for system theory and various data collection and analysis, and then through the establishment of the model to simulate the data, so as to ensure the feasibility and accuracy of the simulation system. In recent years, a large number of computer simulation software have appeared in China's technology market. Arena software is used in the optimization of enterprise production logistics system in this paper. Because Arena software is more commercial, and its function is more powerful than other software, and it has higher evaluation in the industry, with the advantages of animation display. Therefore, Arena software is mainly used in this study to optimize the simulation work in the field of enterprise management and production logistics management (AEMA simulation model is shown as follows) (Fig. 1).

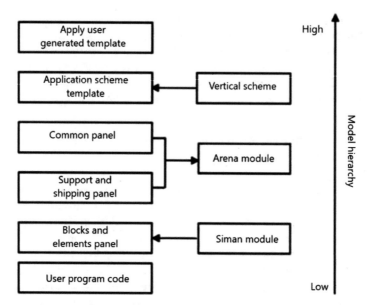

Fig. 1. AEMA simulation model brief view

3.2 The Steps of Computer Simulation Technology in the Optimization of Enterprise Production Logistics System

AEMA software is used for simulation optimization, and finally a systematic simulation model is established. In this paper, nine specific steps in the optimization work are briefly described and explained, as follows:

(1) Problem clarity

In the optimization simulation work, the software first needs to make clear the simulation problem of the enterprise, the main contents of which are the transportation path and the characteristics of the transportation method in the production process of the enterprise, which human resource scheme is better and which scheme can bring the enterprise the saving of funds and the improvement of efficiency [3].

(2) Confirm objectives

After clarifying the problem, confirm the goal of the problem. Before the simulation, first determine the goal of the simulation. The goal of the simulation is the production logistics system of the enterprise. The main goal is to explore the most reasonable and efficient path or logistics mode for the production logistics system of the enterprise.

(3)   Collect data

Simulation technology is mainly to establish simulation model through data, and data is the main basis of simulation process. Before the establishment of the model, the first step is to carry out logarithmic simulation. The main data collected in this simulation are transportation time, transportation distance, transportation funds, etc.

(4)   Model building

Modeling step is one of the most important steps in the whole simulation process. Simulation is to build a model through data, and then get the desired results more clearly and intuitively. In the process of building the model, through the accumulation of data and related knowledge modeling, and then through the comparison of mathematical software and computer technology, the most reasonable simulation model is obtained.

① Define data

After getting the model, in the data definition of the model, the model data is classified, the input data is divided into one category, the output data is divided into one category, and the different functional data in the input data are separately classified.

② Programming

Through these data, the program is written by professional technicians, and then the simulation model is established into a program, so that the paper can be turned into reality.

③ Simulation debugging

Simulation debugging is used to detect the reliability of the final model and data. In the process of simulation debugging, some problems in the simulation model can be found and solved as soon as possible.

④ Simulation run

Simulation operation is the last operation step of simulation. After the most reasonable path calculated by the simulation model, the enterprise inputs the data, obtains the corresponding parameters, and finally defines the most reasonable way and content of the path.

⑤ Statistics of simulation results

After the completion of all the simulation work, the enterprise needs to make statistics of simulation results, that is, summarize and analyze the whole simulation work, and make a report to facilitate the application analysis of the enterprise in the future.

4   Real Time Simulation Algorithm

The real-time simulation is shown in Fig. 2. At this time, the input and output of the computer is a set of data columns with a fixed sampling period T.

Real time simulation integration algorithm

Let the computer solve the following nonlinear equation of state

$$X = F(X, \ U(t)) \tag{1}$$

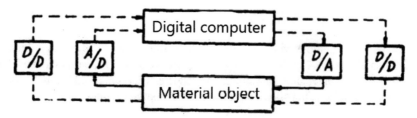

Fig. 2. Real time simulation block diagram

Where x is the state variable and U (t) is the input vector.

The actual input to the computer is the sequence $\{U_n\}$, where $U_n = U\ (_aT)$, n ≈ 0.1,....... According to different integration methods, the computer uses $U_n\ X_n = x\ (_aT)$ or $T = {_aT}$ to calculate $X_{n+1} = X\ [(n + 1)\ T]$ of course, calculate X_n +1. The actual time required must be equal to or less than T s, so that the computer can output $_{Xn+1}$ at $t = (n + 1)$ T. at the same time, it also requires real-time integration algorithm.

5   Research on the Influence of Computer Simulation Technology in the Optimization of Enterprise Production Logistics System

The application of computer simulation technology in the optimization of production logistics system is a very step to adapt to the development of the times and improve their competitiveness. However, in the specific optimization of the enterprise production logistics system, it is found that the computer simulation technology has the following two obvious advantages.

1. Simplified data processing

In the past, enterprises need to apply a large number of human resources in the production and logistics work. In the process of production and transportation routes, there are also a lot of data to measure. Using computer simulation technology and building data model can effectively simplify the data processing process. In addition, the computer simulation technology can also reduce the probability of error, effectively reduce the cost of enterprise management, and effectively improve the competitiveness of enterprises.

2. Simplify the transportation management of enterprise production logistics system

Production logistics transportation management is an important part of enterprise logistics. Traditional enterprise production logistics management does not use intelligent equipment and methods, so it has high requirements for operators. The production logistics system based on computer simulation technology is easy to operate, with high efficiency and low error rate, which can save unnecessary losses and reduce production costs.

6   Conclusions

All in all, at this stage, due to the continuous development and progress of social economy and science and technology, the market competition of enterprises in our country is becoming more and more serious. Therefore, if enterprises want to enhance their competitiveness in the process of competition, and always remain invincible, they need to carry out management rectification and optimization of various systems. Therefore, the reasonable use of computer simulation technology to optimize the production logistics system has become one of the important research issues at this stage. This paper discusses how to optimize the production logistics system of an enterprise by computer simulation technology in many aspects, and briefly explores the important influence of computer simulation technology in optimizing the production logistics system of an enterprise, hoping to play a theoretical role in the development of each enterprise.

Acknowledgements. 2019 teacher research fund project of Laiwu vocational and Technical College (No. 2019jsky13): Research on e-commerce logistics distribution based on Internet of things;

2020 horizontal project of Laiwu vocational and Technical College (No.2020hxky35): logistics and distribution scheme research of small and medium-sized e-commerce enterprises.

References

1. Biaohong, T., Zhiqiang, Z.: Production logistics optimization analysis of small and medium-sized packaging enterprises. Old Area Constr. **12**, 7–9 (2017)
2. Yukun, C., Yunfeng, L., Chen, W.: Production logistics simulation system of equipment manufacturing enterprises. Appl. Comput. Syst. 12–16 (2016)
3. Xingguo, W.: On the improvement strategy of production logistics in manufacturing enterprises. Logistics Eng. Manag. **2**, 15–16 (2016)

Atomic Decay-Based Clustering for Operation Log

Ren Guo[1(✉)], Jue Bo[1], Guoxian Dou[2], Yang Bo[3], and Qiong Wang[3]

[1] State Grid Liaoning Electric Power Co. Ltd., Shenyang 110005, Liaoning, China
guoren_dianli@163.com, bojue_dianli@163.com
[2] Douguoxian Anhui Jiyuan Software Co. Ltd., Hefei 230000, Anhui, China
douguoxian_jiyuan@163.com
[3] State Grid Gansu Electric Power Co. Ltd., Lanzhou 730000, Gansu, China
yangbo_gansu@163.com, wangqiong_gansu@163.com

Abstract. The so-called data clustering refers to dividing the data into some aggregated clusters according to the intrinsic properties of the data. The elements in each aggregated cluster have the same characteristics as much as possible, and the differences between different aggregated clusters are as large as possible. With the advent of the information age, as early as the 1950s, the concept of clustering has been proposed. The purpose of data clustering is not only limited to the integration of data, but also widely used in data classification, feature extraction and other key areas of data mining. However, the research on clustering methods aiming at complex sources, massive items and operation and maintenance systems is relatively few. This paper focuses on the operation logs, and propose an atomic decay-based clustering (ADC) method to handle the operational logs from State Grid IT data center, which can be used to effectively discover the changes of the topology structure of State Grid IT data center. Through the analysis of the operation logs of State Grid IT data center with time series characteristics, the proposed ADC method can effectively outperform the existing approaches on the operation logs.

Keywords: Clustering method · ADC · Operational log

1 Introduction

In the era of big data, with the continuous expansion of IT architecture of State Grid, the number of servers and storage devices is increasing, and the network is becoming more complex, which brings huge challenges to the operation and maintenance work. In order to ensure a good user experience and data timeliness, the operation and maintenance work is very difficult. Its monitoring system needs to collect tens of thousands of data every minute, which is not easy, but it is more difficult to process and analyze the massive data collected. If the data is not processed, it has no significance and value for operation and maintenance. Therefore, in the present and future when the trend of big data centralization is becoming more and more obvious, the analysis and mining technology

J. MacIntyre et al. (Eds.): SPIoT 2020, AISC 1283, pp. 426–438, 2021.
https://doi.org/10.1007/978-3-030-62746-1_64

of big data should be comprehensively used to realize the comprehensive awareness of information communication risk situation, the comprehensive analysis of operation and maintenance data, and the real-time warning of operation and maintenance risk, so as to achieve the active operation and maintenance effect of solving problems before the occurrence of faults, and improve the overall ability of information communication fault monitoring and risk warning of the company.

In particular, the state grid information system is complex, and the operation and maintenance information generated is diverse, involving a wide range of aspects. If the alarm log of the operation and maintenance system is used to analyze the root cause of the fault, it becomes a key problem. Many operation data generated by the operation and maintenance system, such as alarm, fault, log, performance, and other data, contains valuable information for experts to analyze faults and solve problems. So, for this kind of data, how to effectively integrate the classification becomes the key problem of the later data association analysis, and the key of the integrated classification technology is the data clustering technology. At present, the lack of clustering technology for complex source data, such as its infrastructure of State Grid, leads to data redundancy, "data explosion, lack of information", which hinders the stable operation of IT infrastructure of State Grid.

To solve this problem, this paper proposes an atomic decay-based clustering (ADC) method to handle the operational logs from State Grid IT data center, in order to perceive the network topology changes based on expert knowledge. In contrast, the traditional clustering algorithm cannot mine the operation log. Traditionally, the similarity measurement between texts determines the clustering effect dramatically [1], which is improper here as there are too many similar texts in operation logs. The proposed method makes full use of the log timing for operation log analysis, and utilizes the clustering target to justify clustering process in order to provide more reliable data. We have compared and validated our results in experimental section with other existing approaches.

This paper is organized as follows. Section 2 reviews existing studies on clustering. Section 3 presents our atomic decay-based clustering (ADC) method to handle the operational logs from State Grid IT data center. Section 4 discusses the empirical studies on our proposed ADC method. Finally, Sect. 5 concludes our work and further work in the near future.

2   Related Works on Clustering

Lloyd's first proposed K-means clustering algorithm based on the idea of partition in 1957 [13], and Macqueen also studied k-means algorithm in 1967 [12]. Because K-means clustering algorithm usually stops when obtaining a local optimal value, and is only suitable for clustering of numerical data. Particularly, only data sets with convex clustering results can be found. [14] propose k-center algorithms PAM and Clara respectively; Huang [15] also proposed a K-Modes algorithm suitable for classifying attribute data to improve the shortcomings of K-means algorithm; Ester, kriegel, sander and Xu [16] abandon the concept of distance, and propose a new clustering algorithm DBSCAN based on the idea of density; Agrawal, Gehrke, gunopulos and Raghavan [17] proposed a high-dimensional growth subspace clustering algorithm called clique based on Apriori idea for clustering high-dimensional data.

Due to the difference of applications, the traditional clustering algorithms have significant differences on the discovery results of different clusters, and the data cluster of the operation log is often different from other applications [3], which has high requirements for the clustering algorithms.

Taking the famous K-means algorithm as an example, it firstly determines K centroids as the beginning points of clusters. This brings a big issue to the operation log clustering. First of all, it is obviously impossible to define the value of K randomly without any background knowledge. Hence, there are some existing improvements for this point of the K-means algorithm such as the elbow method and the contour coefficient method [4–6].

Besides above-mentioned K-means algorithms, although there some density-based clustering algorithm such as DBSCAN, which can avoid K-value selection during the clustering process, this also brings in two new parameters, Eps and MinPts, which should be determined firstly. To adopt these density-based clustering algorithms, we need to ensure that the entries in the same cluster in the operation log share a high density. But this may be true or not for different logs. The work in [7] uses the additional conditions to accelerate the clustering process, and it can locate the clusters under the combination and/or the separation of clusters and reduce the impact of low-density regions on cluster discovery.

In addition to the above-mentioned parameter predetermination problems, existing clustering algorithms [8] are also difficult to handle the operation log clustering for another reason, i.e., how to propose a similarity measurement to make the distance between clusters is large enough and the one among clusters is small enough. There are some existing works aiming to solve this problem. For example, in [9] a local weighting strategy can adopt a local weight matrix to cluster similarity ones dynamically between clusters. Specifically, [9] introduces the weights of existing algorithms, compares their performance, and combines them together. In contrast, the channel-based clustering algorithm in [10] can evaluate segmentation scores of clusters to clustering within clusters. However, there is still no suitable way to solve the clustering process with temporal conditions.

According to the analysis of the above existing works, based on the properties of State Grid IT operation logs, it is necessary and important to introduce a novel clustering method to process the operation logs. Hence, the clustering in this paper is to discover the dynamic modules in operation logs, and an atomic decay-based clustering (ADC) method is proposed to handle the operational logs from State Grid IT data center. Experimental results show that the proposed ADC approach can handle the operational logs effectively.

3   Operation Log and ADC

3.1   Properties of Operation Log

Operation log is composed of many entries, which have both high dimensions to describe the properties of operation logs and temporal relationship between entries. At the same time, operation log often contains a lot of noisy information, such as frequent services, events, and warnings, which are unnecessary for the clustering in our proposed method.

As for reducing the high dimensions of entries, i.e. the high feature dimensions, we can classify different module behaviors described by different entries of operation logs. The classic dimension reduction methods including Linear Discriminant Analysis (LDA), Principal Component Analysis (PCA), Locally linear embedding (LLE), Laplacian Eigenmaps, and etc. [2]. With the dimension reduction process for the operation logs, the clustering process be optimized, including accelerated clustering process and improved clustering precision.

Except for the above-mentioned characteristics of operation logs, the State Grid IT operation logs have more specific properties. For example, we can find that the operation logs corresponding to different modules have almost indistinguishable features in similarity for traditional similarity measurement. This makes it quite difficult to select an effect similarity measurement method for operation log clustering. As illustrated in Fig. 1, different modules produce different threads, and the last column in Fig. 1 represents the behavior label of the module. As for these operation logs, traditional clustering algorithms are difficult to divide it into different clusters when they are almost the same.

```
[2018-04-1310:24:11,213]  pool-2-thread 21  RefreshingRootWebApplicationContext:start
[2018-04-1310:24:11,213]  pool-2-thread 16  RefreshingRootWebApplicationContext:start
[2018-04-1310:24:11,213]  pool-2-thread 23  RefreshingRootWebApplicationContext:start
[2018-04-1310:24:11,214]  pool-2-thread 13  RefreshingRootWebApplicationContext:start
[2018-04-1310:24:11,214]  pool-2-thread 14  RefreshingRootWebApplicationContext:start
[2018-04-1310:24:11,220]  pool-2-thread 26  RefreshingRootWebApplicationContext:stop
```

Fig. 1. An operation log example of state grid

As for the temporal relationship between entries in operation logs, due to the properties of the thread, we can observe that the modules often start at the same time, which makes the time series analysis method such as Dynamic Time Warping (DTW) not perform well for clustering which leads to a poor similarity accuracy. For example, in Fig. 2, the behaviors generated by the same module is depicted with the same abscissa, and the ordinate represents the corresponding time.

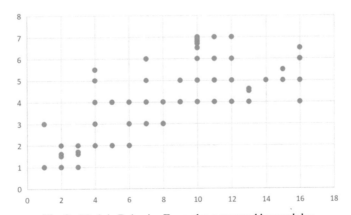

Fig. 2. Module Behavior Examples generated by modules

3.2 Data Sets from State Grid IT Operation Log

The data set adopted in this paper is from the operation logs generated by certain State Grid IT Data Center, which mainly includes a part of the data center module startup and a part of the alarm data generation.

Table 1. operation log example from a State Grid IT Data Center

Source	Type	Thread	Event type	Description
UE-ALARM_CENTER	INFO	N\A	60	Policy changes
UE-ALARM_CENTER	INFO	Pool-2-thread 14	109	Change com.sgcc.uap.kernel.policy
UE-ALARM_CENTER	INFO	N\A	60	Policy changes
UE-ALARM_CENTER	INFO	N\A	60	Policy changes

In this paper, the former one is used for operation log clustering, which has the properties of 6 dimensions, as shown in Table 1. The data set used in this experiment has a total of 3,000 log entries.

In the operation log, the thread part may share some default values. In addition, from these data of operation logs, we can observe that the behaviors generated by the same module should belong to the same thread. Therefore, in this paper, the thread is used as a class label to evaluate the effectiveness of the experimental results.

3.3 Operation Log Preprocessing

In order to reduce the noisy data in operation logs, especially the redundant data, here we introduce a data preprocess. As shown in Fig. 2, the log entities which are hard to be handled by the traditional clustering algorithms are the focus of the ADC method. As for the part of the operation log which is the event description that directly contains the module name, this have the longest texts compared with other parts of the operation log, as shown in Table 2. Here, we directly omit this event description, so as to reduce the noise for the similarity measurement, which makes a little easier for the similarity measurement, so as to improve the overall clustering performance.

Table 2. Event behavior with module name description

Time stamp	Thread	Description
[2018-04-1515:34:12,126]	Pool-2-thread 6	Dynamic-Module:com.sgcc.uap.execption changes to event starting
[2018-04-2115:34:12,126]	Pool-2-thread 14	Dynamic-Module:com.sgcc.uap.policy changes to event starting
[2018-04-2115:34:12,219]	N\A	Dynamic-Module:com.sgcc.uap.execption changes to event started
[2018-04-2115:34:12,239]	N\A	Dynamic-Module:com.sgcc.uap.policy changes to event started

This paper deals operation logs with the dates to separate clustering entities due to the fact that the operation of the data center is counted in days.

3.4 ADC Algorithm

In order to improve the performance of operation log clustering, ADC should make good use of the temporal information of operation logs. More specifically, as the clusters overlap on the scale of time, the ADC algorithm should be able to handle the time information well during the clustering process. In order to satisfy the above requirements, the ADC uses an atomic decay-based clustering process to improve the performance of operation log clustering.

3.4.1 Algorithm Parameters

The parameters of this ADC algorithm are defined as follows:

- Nuclear mass M: is the time span of modules (i.e. clusters).
- Nuclear attraction A: It is proportional to M, and the nuclear has an infinite attraction to the data points with the same time within a certain error.
- Electron e: is a data point which is attracted to the nuclear.
- Repulsive force R: The repulsive force R of e on another data point, which is defined by the module behavior. Technically, this can be evaluated as the cosine similarity between module behaviors.
- Electronic gravity F: The attraction of e to another data point, as defined above.
- Generation time t: is the time when M appears.
- Annihilation time T: is the time when M becomes zero.

3.4.2 Algorithm Framework

Algorithm 1: ADC Algorithm

Input: Operation Log, g is the mass of the nucleus, gravitational coefficient f(g);
Output: Cluster sets
1.Iinitialize the nuclear set K;
2.for each sample in log p do
3. for each k in K do
4.   if (quality M of k!=0) then
5.   Calculate the initial gravity of nucleus Fk=f(g);
6.   If (Fk!=INF) then
7.   For each e surrounds k do
8.   Fk = Fk + F(e,p)-R(e,p);
9.   End for
10.   End if
11.   if (F is max) then
12.   divide peripheral electrons where p is k,
13.   update the quality of k;
14.   End if
15.   End if
16. End for
17.End for
18.End

As for ADC algorithm, firstly, according to the properties of the operation logs, the initialization of the nuclear set should be different from each other. In addition, the mass of each nucleus can be determined according to the temporal properties of the operation logs, and the mass of different clusters can determine the attractiveness of the cluster to the data points. The mass of nucleus k updated in line 2 of Algorithm 1 should be consistent with the initialization method in line 1. More specifically, when electron e is assigned to nucleus k, the mass M of nucleus k should be equal to the result of the nuclear annihilation time T minus the generation time t of electron e.

In line 5 of Algorithm 1 the function $f(g)$ can be defined as linear one or non-linear one, and it can be determined according to the size of the target cluster. Particularly, in this paper the longer the module survives in State Grid IT operation logs, the larger the cluster size. Meanwhile, the proportion of the cluster size to the nuclear mass is 1:1, which makes $f(g)$ always equal to 1. But there are often some special electrons in State Grid IT operation logs which are often combined with the generation of the nucleus, which make $f(g)$ s unable to be fit for such data points very well. Therefore, we define gravitational threshold $MinG$ and time difference between electron e and nucleus t, which are adopted by $f(g)$.

$$f(g) = \begin{cases} INF, & (g \leq MinG or \Delta t \leq m) \\ g, & other \end{cases} \quad (1)$$

Where m can be the defined as a constant, and $m = MinG$ is taken in the experiments due to the fact that the proportion of the mass of the nucleus to the time span is 1:1.

The gravitational function F and the repulsive function R in line 2 can be defined as follows:

$$F(e, p) = \begin{cases} INF, & (1 > similar > (e, p) \geq max\_s) \\ 0. & other \end{cases} \quad (2)$$

Where max_s is the similarity threshold, and when some property similarity between electron e and electron p is larger than the threshold, the gravitation is set as infinite at this time, i.e.

$$F(e, p) = \begin{cases} INF, & (similar(e, p) = 1 \\ 0, & other \end{cases} \quad (3)$$

It should be also noticed that clusters in State Grid IT operation logs usually have the data points with the same time, and these data should not be assigned to one cluster at the same time according to their practical applications. Therefore, we define the repulsion between some electronic components with the same characteristics as infinity.

4 Experiments

4.1 Experimental Settings

This experiment adopts the dataset from the module startup logs of a data center of State Grid from March 22 to May 24, 2018, which are 9106 pieces of operation logs in total. According to the data preprocessing as mentioned above, we have a total of 353 pieces of operation logs, which can be divided into two parts: 181 pieces and 172 pieces.

Each State Grid IT operation log contains 6-dimensional features, and the experiment only uses 3-dimensional features of them, i.e., time, event codes, and event descriptions.

4.1.1 Evaluation Measurements

The experiment takes the thread information of the module in State Grid IT operation logs as the label of the cluster, and takes the event description in the cluster to evaluate the specific function of the module. Hence, the normalized mutual information NMI, Purity and Accuracy are introduced to evaluate the performance of clustering results for State Grid IT operation logs.

- NMI is a popular evaluation standard adopted for the similarity evaluation between clustering results and benchmarks, whose value should stay between 0 and 1. The larger NMI value, the better clustering performance.
- Purity indicates the proportion of the number of correct classified clusters to the total number; the higher Purity value, the better clustering performance.
- Accuracy represents the proportion of data points in the cluster which are assigned with the correct class label to the total number. The higher Accuracy value, the stronger the ability to correctly predict the class label.

4.1.2 Comparison Methods

The comparison algorithm adopted in experiments is the improved SC-kmeans algorithm proposed by Nuwan Ganganath et al. in [11]. In order to be consistent with the ADC algorithm proposed in this paper, the time span can be taken as the prior knowledge of cluster size. Meanwhile, in order to keep the fairness of comparisons in the experiments, the centroid of the algorithm used in the experiment is predetermined.

Mainly, this experimental section consists of 3 parts:

- Performance Comparison between ADC and SC-kmeans;
- Performance Comparison of the impact of tagged data on ADC and SC-kmeans;
- Time performance comparison between ADC and SC-kmeans.

Due to the properties of State Grid IT operation logs in the experimental dataset, the SC-kmeans algorithm as the comparison algorithm has been improved to be fit for the properties of the State Grid IT operation logs, including:

- Data points are only to be selected to the cluster that is most likely to be assigned in the actual applications.
- Following the local weighting strategy proposed in [9], the algorithm weights the similarity measurement locally.

- As the centroid in the dataset is predetermined, the convergence of SC-kmeans in the iterative process on the experimental data set is turning worse, so this experiment only uses the first iteration result as the best result.

Performance Comparison *between ADC and SC-kmeans*

Table 3. Performance comparison with dataset 1

Algorithms	Accuracy	NMI	Purity
ADC	0.7	0.79	0.69
SC-kmeans	0.49	0.77	0.6

Table 4. Performance comparison with dataset 2

	Accuracy	NMI	Purity
ADC	0.82	0.87	0.79
SC-kmeans	0.41	0.73	0.54

From Tables 3 and 4, it is obviously that ADC outperforms SC-kmeans in all three evaluation indicators on the two datasets. Particularly, as for Accuracy, SC-kmeans can cluster some data points together correctly, but with the centroid predetermined, it can assign some points to the wrong centroids. In addition, its Purity is significantly lower than that of the ADC algorithm, which indicates that the purity in the cluster is not good enough, and there are many data points mis-clustered.

4.2 Performance Comparison of the Impact of Tagged Data on ADC and SC-Kmeans

See Figs. 3, 4, 5, 6, 7 and 8.

4.3 Time Performance Comparison Between ADC and SC-Kmeans

Due to the fact that the datasets adopted in experiments are too small, the time performance is mainly discussed theoretically here.

Practically, in the experiments, the operation time of ADC and SC-kmeans algorithm is both about 15 ms, as there is no iterative process in the proposed ADC algorithm in this paper and we only use the first iteration result as the best result in SC-kmeans algorithm.

Theoretically, the overall time complexity of ADC algorithm is $O((n^2 - k^2)/2)$; and the time complexity of the SC-kmeans algorithm is $O(nkt)$ for the first iteration result only.

All in all, we can conclude that the proposed ADC algorithm is better in time efficiency. This is due to the fact that there is no iteration process for ADC algorithm, and the iterations required by the traditional clustering algorithms are much more than that.

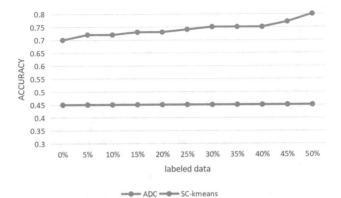

Fig. 3. Comparisons on Accuracy with dataset 1

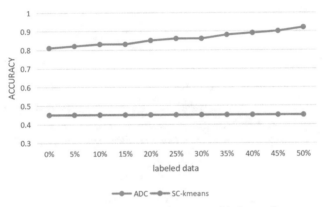

Fig. 4. Comparisons on Accuracy with dataset 2

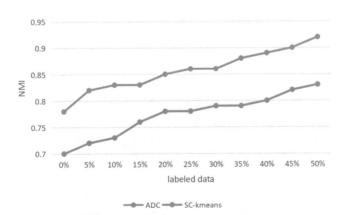

Fig. 5. Comparisons on NMI with dataset 1

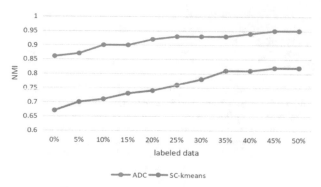

Fig. 6. Comparisons on NMI with dataset 2

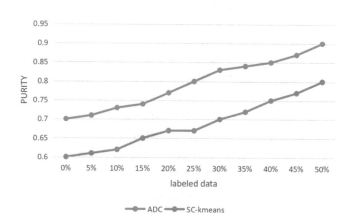

Fig. 7. Comparisons on Purity with dataset 1

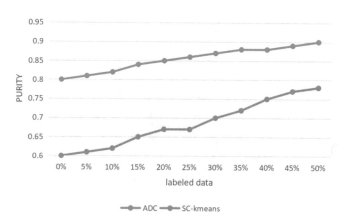

Fig. 8. Comparisons on Purity with dataset 2

5   Conclusions

The proposed atomic decay-based clustering (ADC) method performs well on State Grid IT operation logs. More Specifically, the introduction of atomic decay-based model can classify the data points with high confidence into the centroid, which improves the effectiveness and efficiency of clustering process. According to the properties of State Grid IT operation logs, the iterative process of the traditional clustering algorithm can be transformed into the process of atomic decay, which accelerates the convergence speed of the clustering algorithm. Experimental results have been presented to validate the effectiveness and efficiency of our proposed ADC algorithms. With the clustering results of the ADC algorithm, reliable clustered data can be provided for the subsequent module behavior analysis for State Grid IT operation logs.

Acknowledgements. This research was financially supported by the Science and Technology projects of State Grid Corporation of China (No. 500623723).

References

1. Canlun, Z., Miyamoto, S.: Text clustering using fuzzy neighborhood and evaluation of clusters. In: IEEE International Conference on Granular Computing, pp. 19–24. IEEE Press, Noboribetsu (2014)
2. Huang, X.: Research and development of feature dimensionality reduction. Comput. Sci. **45**(Z6), 16–21 (2018)
3. Steinbach, M., Karypis, G., Kumar, V., et al.: A comparison of document clustering techniques. KDD Workshop Text Min. **400**(1), 525–526 (2000)
4. Liu, G., Tingting, W., Yu, L., Li, Y., Gao, J.: The improved research on k-means clustering algorithm in initial values. In: International Conference on Mechatronic Sciences, pp. 2124–2127. IEEE Press, Shengyang (2013)
5. Na, S., Xumin, L., Yong, G.: Research on k-means clustering algorithm: an improved k-means clustering algorithm. In: International Symposium on Intelligent Information Technology and Security Informatics, pp. 63–67. IEEE Press, Jinggangshan (2010)
6. Wang, H., Shi, Y.: K-means clustering algorithm based on initial center optimization and feature weighted. Comput. Sci. **44**(Z11), 457–459 (2017)
7. Zhao, W.-Z., Ma, H.-F., Li, Z.-Q., Shi, Z.-Z.: Efficiently active learning for semi-supervised document clustering. J. Softw. **23**(6), 1486–1499 (2012)
8. Huang, D., Wang, C., Lai, J.: Locally weighted ensemble clustering]. IEEE Trans. Cybern. **48**(5), 1460–1473 (2018)
9. Nock, R., Nielsen, F.: On weighting clustering. IEEE Trans. Pattern Anal. Mach. Intell. **28**(8), 1223–1235 (2006)
10. Mishra, R.K., Saini, K., Bagri, S.: Text document clustering on the basis of inter passage approach by using K-means. In: International Conference on Computing, Communication & Automation, pp. 110–113. IEEE Press, Noida (2015)
11. Ganganath, N., Cheng, C., Tse, C.K.: Data clustering with cluster size constraints using a modified K-means algorithm. In: International Conference on Cyber-Enabled Distributed Computing and Knowledge Discovery, pp. 158–161. IEEE Press, Shanghai (2014)
12. Mcqueen, J.: Some methods for classification and analysis of multivariate observations. In: Proceedings of the Fifth Berkeley Symposium on Mathematical Statistics and Probability, pp. 281–297 (1967)

13. Lloyd, S.P.: Least square quantization in PCM, Bell Telephone Laboratories Paper, 1957, Published in journal much later: Lloyd, S. P. (1982)
14. Kaufman, L., Rousseeuw, P.: Finding Groups in Data: An Introduction to Cluster. Wiley, New York (1990)
15. Huang, Z.: Extensions to the v-means algorithm for clustering large data sets with categorical values. Data Min. Knowl. Discovery **2**, 283–304 (1998)
16. Ester, M., Kriegel, H.-P., Sander, J., Xu, X.: A density-based algorithm for discovering communities in large spatial databases with noise. In: Proceedings of 2nd International Conference on Knowledge Discovery and Data Mining (KDD-96) (1996)
17. Agrawal, R., Gehrke, J., Gunopulos, D., Raghavan, P.: Automatic subspace clustering of high dimensional data for data mining applications. In: Proceedings of SIGMOD Record ACM Special Interest Group on Management of Data, pp. 94–105 (1998)

Hybrid Mac Protocol Based on Security in Clustering Topology

Wei Dong[1,2], Yanxia Chen[2], Xiangdang Huang[2,3(✉)], and Qiuling Yang[2,3]

[1] Information and Communication Engineering of Hainan University, Haikou 570228, China
[2] Computer Science and Cyberspace Security of Hainan University, Haikou 570228, China
990623@hainanu.edu.cn
[3] State Key Laboratory of Marine Resource Utilization in South China Sea, Haikou 570228, China

Abstract. Underwater wireless sensor networks use acoustic waves as the transmission medium which have shortcomings such as long propagation delay, limited bandwidth and time-varying. Because of the openness of underwater acoustic channels, underwater nodes are also vulnerable to various types of attacks. These characteristics bring great challenges to the design of MAC protocol and the safety of underwater network. This paper presents an idea of a hybrid MAC protocol based on security in a clustered topology. In the proposed idea, the nodes are first clustered according to the existing clustering algorithm for energy saving, and then the trust management mechanism is used in the network to resist internal attacks. Then the security based hybrid MAC protocol is used in intra-cluster communication, the main ideal is use an asymmetric encryption mechanism to encrypt the symmetric keys, the cluster head allocates time slots based on the information of the member nodes in the cluster, and encodes with CDMA to eliminate inter-cluster communication interference. The proposed protocol can resist multiple attacks and achieve secure communication.

Keywords: Underwater wireless sensor network · Hybrid Mac protocol · Cluster · Trust management

1 Introduction

Underwater acoustic communication has the characteristics of long and variable propagation delay, limited bandwidth, multipath fading, etc. [1]. The underwater acoustic channel is an open channel, nodes are vulnerable to various forms of attacks underwater, which can be divided into active attacks and passive attacks. The existing cryptography mechanism can effectively guarantee the confidentiality and integrity of the data and resist passive attacks. At present, the cryptography mainly include two kinds of symmetric encryption mechanisms and asymmetric encryption mechanisms, The characteristics of symmetric encryption algorithm are open algorithm, fast encryption speed and high encryption efficiency. Compared with the symmetric encryption mechanism, the asymmetric encryption mechanism is more secure, but the energy consumption is

J. MacIntyre et al. (Eds.): SPIoT 2020, AISC 1283, pp. 439–444, 2021.
https://doi.org/10.1007/978-3-030-62746-1_65

also higher. Clustering structure is good for balancing the energy consumption of nodes and for network security management. This paper adopts the two-dimensional clustered network topology. Reasonable, efficient and secure clustering algorithms are of great significance to network performance and network security. Most of the existing clustering algorithms are based on reducing the energy consumption of nodes and prolonging the life cycle of the network as a starting point, but rarely consider the security aspects and cannot identify malicious nodes. As shown in Fig. 1, once the malicious node is selected as the cluster head node, the malicious node can steal the data in the network at will, or even destroy the network.

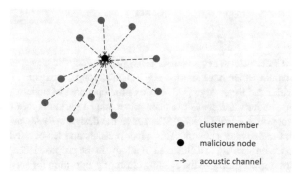

Fig. 1. A malicious node that becomes a cluster head

Trust management mechanism is fundamentally to identify malicious nodes by monitoring the behavior of nodes, and take corresponding security measures for defense. The core issue of trust management is to monitor the behavior of the evaluated person and determine their trust based on it.

The MAC protocol is a channel access control mechanism that allows multiple nodes to communicate through a shared channel. MAC protocol has an important impact on the performance of network throughput, end-to-end delay, fairness and so on. MAC protocols can be roughly divided into non-competition MAC protocols and competition-based MAC protocols [2].

The rest of the paper is organized as follows: Sect. 2 first briefly introduces various attacks faced, then introduces the traditional energy-based clustering algorithm, and finally analyzes the existing research on MAC. Section 3 proposes a secure clustering idea based on energy and trust management. Section 4 introduces a idea of a hybrid MAC protocol based on encryption mechanism in a clustered topology. Section 5 analyzes the proposed content.

2 Related Work

Due to the particularity of the underwater environment, the research of underwater wireless sensor networks is facing more severe challenges [3].

Among all kinds of security attacks, there is also an attack called internal attack [6]. The internal attack is mainly an attack launched by the enemy by capturing nodes or injecting malicious nodes into the network. Compared with other attacks, internal attacks are more difficult to detect and difficult to defend.

LEACH [4] is a distributed clustering algorithm proposed for energy saving purposes. In this algorithm, in order to balance the energy consumed by network nodes, all nodes are periodically used as cluster head nodes. LEACH-C [5] does not allow nodes with energy lower than the average remaining energy of all surviving nodes to participate in the competition of cluster head nodes. These algorithms can extend the life cycle of the network, but they do not involve security issues and cannot identify malicious nodes.

MAC protocols can be roughly divided into non-competition MAC protocols and competition-based MAC protocols. The contention-based MAC protocol can be divided into a random access MAC protocol and a MAC protocol based on a handshake mechanism. ALOHA, slotted ALOHA and MACA are some typical random access protocols [6], but these protocols will have data collisions. In order to avoid data packet conflicts, a MAC protocol based on a handshake mechanism was proposed to avoid data conflicts by using RTS/CTS/DATA/ACK handshake mechanisms before transmitting data. Non-competition MAC protocols mainly include FDMA [7], CDMA, and TDMA [8]. CDMA is code division multiplexing technology. The CDMA system has the advantages of strong anti-multipath interference capability, high frequency band utilization rate, and no inter-modulation interference. TDMA technology is a simple and mature mechanism for channel allocation. It divides the time period into time slots, each node is assigned a time slot for data transmission/reception, the nodes only communicate in their own time slots.

3   Secure Clustering Algorithm Based on Energy and Trust Management

In this paper, the source node extracts the packet DATA by hash algorithm, then we can get $H = \text{Hash}(DATA)$, and then encrypt it with the original packet DATA, we can get $E_S(DATA|Hash(DATA))$, and sends it through the relay node. The gateway node or the destination node decrypt the $E_S(DATA|Hash(DATA))$, and then get the $DATA'$, and then extracts the summary using the same hash algorithm $H' = \text{Hash}(DATA')$. If $H' = H$, it is considered that the relay node j forwards the packet in time, and the packet has not been tampered with. The gateway node or the destination node replies a confirmation message ACK_{DATA}, and the relay node and the source node participating in the routing think that the node forwards the packet after receiving the message, then the trust value of node j is increased. If the packet is not received after a certain period of time, node j is considered to have failed to forward the packet, and its trust value is reduced. In this paper, each node maintains an integer t about the neighbor node locally. When the network is initialized, t = 0. If the relay node and the source node receive ACK_{DATA}, then $t = t + 1$, or t will not change. The relay node or source node i calculates the trust value of node j according to $T_{i,j} = \frac{t}{n}$. According to Formula 1 the cluster head node collects the trust value and takes the average value as the final trust value of node j.

$$T_j = \frac{\sum\limits_{\substack{i=1 \\ i \neq j}}^{n} T_{i,j}}{n-1} \tag{1}$$

3.1 Clustering Algorithm

Initial Phase. After the node is deployed, the gateway node generates a clustered message, encrypts it with the master key S stored in advance, and broadcasts it. After receiving the message, the node decrypts it and starts autonomous clustering. The nodes in the network first use the energy-based LEACH clustering algorithm for clustering. Note that all messages in the clustering process must be encrypted with the master key S.

Cluster Head Security Re-selection Phase. Since the cluster head node needs to perform data aggregation and key management on the member nodes in the cluster, the cluster head node consumes more energy than the member nodes. The nodes within the cluster periodically report their residual energy to the cluster head node.

E_{ri} represents the residual energy of node i, then the average residual energy in the cluster is defined as $E_r = \dfrac{\sum\limits_{i=1}^{n} E_{ri}}{n}$, where n represents the number of member nodes in the cluster and E_{rc} represents the residual energy of the current cluster head. When $E_{rc} < E_r$, the cluster head broadcast message informs the member nodes in the cluster to start the cluster head re-selection. In the cluster head re-selection phase, the current cluster head node as the management center takes the trust value and residual energy of member nodes as the comprehensive evaluation factor, and selects the node with the highest evaluation value as the new cluster head, and the node with the second trust value as the alternative cluster head. The main function of the alternative cluster head is to act as the new cluster head when the new cluster head fails to work normally due to emergency.

Evaluation value of node j: $T_{lj} = \rho_1 T_j + \rho_2 \frac{E_{ri}}{E_r} + P$, among which ρ_1, ρ_2 is constant, which can be adjusted according to the needs of the network. P is the default packet loss rate.

4 Hybrid Mac Protocol Based on Security in Clustering Topology

4.1 Intra-cluster Communication Key

The Intra-cluster key is used for encryption when the members in the cluster transmit data to the cluster head node. The security of the asymmetric encryption mechanism is better than that of the symmetric encryption, but its computational load is large. If the asymmetric encryption is used for communication within the cluster, it will increase a large amount of energy consumption and shorten the life cycle of the network. Although the implementation of symmetric encryption is simple, fast, and consumes less energy, the security of symmetric encryption depends on the security of the symmetric key. Once the symmetric key is obtained by the enemy, the security of the network cannot be guaranteed. Therefore, in the process of key generation and distribution within the cluster, an asymmetric encryption mechanism is used for encryption to ensure the security of the symmetric key.

4.2 Intra-cluster Broadcast Key

After the symmetric communication key distribution between each node in the cluster and the cluster head node is completed, the cluster head node generates an Intra-cluster broadcast key, which is used to encrypt the broadcast message in the cluster, and the broadcast key is encrypted by the cluster head's private key and sent to the member nodes in the cluster.

4.3 Secure Hybrid Mac Protocol

Intra-cluster Secure Hybrid Mac Protocol. After the generation and distribution of the inter-cluster key and the inter-cluster broadcast key, the cluster head node collects the information of all cluster member nodes and allocates a transmission time slot for each member node, and then generates a time division multiple access TDMA message, then use the inter-cluster broadcast key to encrypt. In order to avoid the interference between the nodes in the cluster, the generated cipher text message is sent to the member node after CDMA encoding, and the member node decrypts after receiving the cipher text, and the data is transmitted according to the time slot obtained by the decryption Information, all data must be encrypted with the symmetric encryption key with the cluster head node, and then sent to the cluster head node using CDMA encoding.

Inter-cluster Security Hybrid Mac Protocol. The key generation and distribution methods of the cluster head node and the gateway node are the same as those in the cluster. The distance between the cluster head and the gateway node is one hop or even multiple hops. Therefore, TDMA is not suitable for inter-cluster communication, so some common MAC protocols based on handshake can be used for inter-cluster communication. Of course, all messages must be encrypted.

5 Analysis

This paper presents an idea of a hybrid MAC protocol based on security in a clustered topology. In the proposed idea, under the energy-based clustering strategy, a trust management mechanism is added. The node performs trust evaluation on the behavior of the neighbor node, and sends the trust evaluation to the cluster head, and the cluster head performs corresponding processing. After adding the trust management mechanism, it can effectively resist the attack of the compromised node, combined with the trust management for cluster head re-selection, it can avoid the compromised node becoming a cluster head and threatening the security of the network. Then the article proposes a idea of a secure hybrid MAC protocol in the cluster. The main point is that the cluster head node allocates time slots based on the information of the member nodes in the cluster, and encodes with CDMA to eliminate inter-cluster communication interference. The cluster head node generates and distributes symmetric keys for communication within the cluster, and use an asymmetric encryption mechanism to encrypt this process, which can ensure the security of the symmetric encryption key, and reduce network energy consumption and extend the network life cycle. Because the inter-cluster communication adopts the multi-hop communication mode, the inter-cluster communication can adopt some common competitive MAC protocols based on handshake.

References

1. Liu, L.: Prospects and problems of wireless communication for underwater sensor networks. Wirel. Commun. Mob. Comput. **8**(8), 977–994 (2008)
2. Chen, K.: A survey on MAC protocols for underwater wireless sensor networks. IEEE Commun. Surv. Tutor. **16**(3), 1433–1447 (2014)
3. Lopez, J.: Analysis of security threats, requirements, technologies and standards in wireless sensor networks. In: Foundations of Security Analysis and Design, pp. 289–338 (2009)
4. Heinzelman, W.R.: Energy-efficient communication protocol for wireless sensor networks. In: Proceedings of the 33rd Annual Hawaii International Conference on IEEE, pp. 201–207 (2000)
5. Heinzelman, W.B.: An application specific protocol architecture for wireless microsensor networks. IEEE Trans. Wirel. Commun. **1**(4), 660–670 (2000)
6. Liao, Z.: A handshake based ordered scheduling MAC protocol for underwater acoustic local area networks. Int. J. Distrib. Sens. Netw. (2015)
7. Shao, C.: Performance analysis of connectivity probability and connectivity-aware MAC protocol design for platoon-based VANETs. IEEE Trans. Veh. Technol. **64**(12), 5596–5609 (2016)
8. Rasheed, M.B.: Delay and energy consumption analysis of priority guaranteed MAC protocol for wireless body area networks. Wirel. Netw. **23**(4), 1249–1266 (2017)

Experiments, Test-Beds and Prototyping Systems for IoT Security

The Transformation and Remodeling of Literary Education Concept Under Mobile Internet Era

Xiaowei Dong[✉] and Jie Zhao

Sichuan University Jinjiang College, Meishan, Sichuan, China
yinxing_tx@163.com

Abstract. Literature education occupies an important position in the strategic training of college talents, and the concept of literature education determines the true effect of literature education. Under the guidance of advanced literary education concepts, literary education can not only improve students' literacy in basic aspects such as characters, words, sentences, and texts. Moreover, it enables students to absorb rich spiritual food and cultivate their own sentiments in literary works, and constantly improve their own literary taste and literary literacy, thereby enhancing the core competitiveness of college talents.

Keywords: Mobile internet · Internet age · Literature education · Education concept · Transformation and remodeling

1 Introduction

With the universal application and full coverage of network and multimedia information technology, open courses and smart classrooms have gradually emerged, and a huge transformation have taken place in the field of education and teaching. The mobile, digital, and informatization of teaching has become the university literature teaching in the context of the "Internet +" era. The new theme of model is under development. Education informatization should pay attention to three aspects: education-oriented innovation-oriented, high-quality educational resources and the construction of information-based learning environment, and innovation in teaching models and learning methods as the core [1]. The virtualized and intelligent teaching platform and high-quality teaching resources have built a good learning environment for the mixed teaching mode of university literature. As shown in Fig. 1, the mobile teaching mode is for the majority of literature lovers and learners, especially non-literary majors, who are interested in literature. The missing college students have created a vivid and effective learning model, realized flexible and diverse learning methods, and laid a foundation for completely breaking the limitations of the traditional learning model of university literature and realizing the interaction between teachers and students [2].

Today is the world of mobile Internet. Under the rendering and influence of this large environment, college students carry the mobile Internet with them, which brings great convenience to students to study all kinds of literature. The literary learning method

J. MacIntyre et al. (Eds.): SPIoT 2020, AISC 1283, pp. 447–452, 2021.
https://doi.org/10.1007/978-3-030-62746-1_66

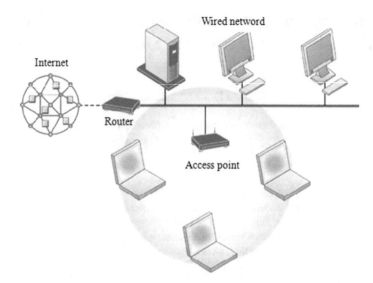

Fig. 1. Virtualized and intelligent teaching platform

of college students is no longer limited to classrooms and libraries, but to learn on mobile clients anytime, anywhere. This has a huge impact on the traditional concept of literary education in colleges and universities, and must be highly valued by schools and teachers [3]. This article starts by analyzing the problems of traditional literary education concepts, then clarifies the advantages of online literature, and then studies how to transform and reshape the literary education concepts of colleges and universities to adapt to the learning of college literature in the era of mobile Internet.

2 The Opportunities Faced by Literary Education in the Era of Mobile Internet

Literature itself is a concept that only covers a wider area. It includes two major pieces of domestic literature and foreign literature, and it is even more comprehensive under the subdivision [4]. Therefore, to do a good job of literature education is a huge project. Contemporary college students living in the era of the mobile Internet have important opportunities for the learning methods of literature.

2.1 The Era of Mobile Internet has Broadened the Horizons of College Students

In the traditional literary education model, teachers have absolute right to speak. In the era of mobile Internet, students can use mobile devices such as mobile phones to study anytime, anywhere [4]. While gaining a lot of knowledge, they can use Baidu mode to answer questions encountered in the learning process. Compared with the traditional one-word learning process, with the emergence of mobile terminals, students tend to learn independently, which broadens their horizons.

2.2   The Mobile Internet has Changed the Literary Reading Mode of College Students

There are three main traditional literary reading modes; one is textbooks; the other is library borrowing; the third is bookstore buying or just reading in the store. What they have in common is that they all rely on paper as a carrier, so their limitations are self-evident [5]. First, the space occupied by books is large. The mobile internet has changed the literary reading mode of college students. As shown in Fig. 2, a mobile phone can be downloaded in the mobile Internet era, and at most one charging treasure is added. Secondly, it occupies a large amount of students' time and wastes a lot of fragmented time. Whether in a library or a bookstore, students need to spend a lot of time reading.

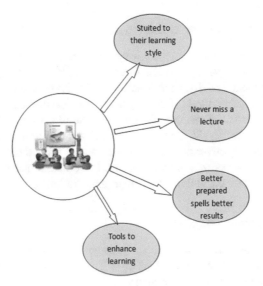

Fig. 2. Virtual reality system architecture

2.3   The Rise of Network Literature

The rise of network literature has an important position in college literature education. The advantage of network literature is mainly reflected in the literature education of colleges and universities. Students can personally try a small knife in literary creation. In addition to reading literature anytime, anywhere, online literature allows students to show their literary talents [5].

3   Transformation and Remodeling of Literary Education Concepts in Universities

The concept of literary education in colleges and universities in the era of mobile Internet needs to be transformed and reshaped. It must be in line with the times and applicable to current students.

3.1 Educational Thinking Should Keep Pace with the Times

Quality education requires schools to cultivate talents with all-round development of virtue, intelligence, physical fitness and labor. Under the employment pressure of college students, some colleges and universities have some phenomena of quick success in talent cultivation. They pursue employment rate one-sidedly, only pay attention to professional knowledge education, and ignore the cultivation of humanities [6]. Therefore, major colleges and universities should pay attention to students. Literary education, affirming the important position of literary education in the cultivation of talents in colleges and universities, while cultivating professional skills of students, increase the intensity of literary education training.

3.2 Improve the Relevant Curriculum System

Under the examination-oriented education system, Chinese literary education is only manifested in language learning, so the literacy of college students is generally low [7]. After entering the university, he will no longer be suppressed by the exam-oriented education. With the requirement of quality education, literary education is becoming more and more important. It can completely break the limitation of literary education, which is Chinese education [8]. For example, colleges and universities can add more literary courses focusing on aesthetics and appreciation during their freshman year, and select some literary works that embody the truth, goodness and beauty to stimulate students' interest in learning and improve their appreciation ability.

3.3 Start with Teacher Teaching

In the era of mobile Internet, the authority of teachers in the process of literary education has been challenged. To reshape the concept of literary education in universities, it is necessary to regain the authority of teachers.

1) Teachers should establish the concept of lifelong learning and constantly enrich themselves and improve themselves [7]. In daily life and work, we constantly update our knowledge reserves, innovate our own teaching methods, and use our hard knowledge to capture the respect and love of students.
2) Teachers should keep pace with the development trend of the times and keep pace with the times. Today, with the development of the mobile internet, teachers should also take the initiative to contact and learn these new high-tech products. This poses a severe test especially for some older teachers. Mobile internet can arouse students' interest in learning. Teachers should take the initiative to conquer major mobile internet platforms and apply them to actual teaching activities [8]. The mobile internet can provide students with a lot of information and many advantages, but it cannot classify and integrate all the students' problems, nor can it intuitively convey knowledge points.

3) Classroom management should be tight and loose. University classrooms are different from other stages of classroom learning. Most students have their own mobile Internet, so classroom management is too tight, which can easily arouse students' resentment [9]. If it is too loose, some students' self-control ability is too poor, easy to indulge in games, etc., so be relaxed.

In the era of mobile Internet, after reshaping the authority of teachers, teachers can play their unique role in literary education [9]. Use your own knowledge and cultural literacy and personal charm and teaching management ability to guide students to read literary works in the literary education class, taste the theme and ideology of literary works, and encourage students to create their own, thus forming a people-oriented interactive learning classroom, as shown in Fig. 3.

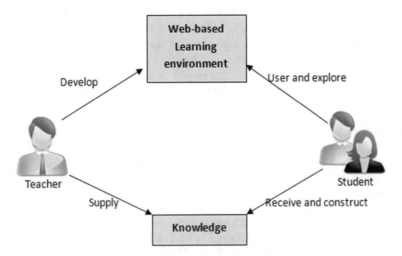

Fig. 3. Interactive learning classroom

3.4 Start from Students

Students have always been the main body in the learning process, which is particularly prominent under the requirements of the new curriculum standards and quality education. In the era of the mobile Internet, the reshaping of college literary education concepts is inseparable from the active play of students [10].

1) Encourage students to use mobile internet to improve their reading of literary works. One of the foundations of literary education is the necessary reading of literary works. Freshmen who have just entered the university have no time and conditions to read a large number of literary works except for some excerpts of literary works in the Chinese language class during the exam-oriented education stage [11]. In the era of mobile Internet, college students basically have a mobile phone, and some classmates even have a tablet computer, which provides the conditions for reading a large number of literary works.

2) Encourage students to actively perform literary creation. Schools can organize related creative competitions, and the subject matter can be poetry, small talk, chicken soup, etc., to stimulate students' creative desire [11]. In this process, it can not only increase the reading volume of students' literary works, but also stimulate students' interest and love for literature, and cultivate students' interpersonal communication skills and improve students' comprehensive quality.

4 Conclusion

With the continuous development of the Internet, it has gradually changed traditional industries and people's living habits. Universities should change their thinking in a timely manner, fully understand the influence of the Internet, change traditional thinking based on the new changes brought by the Internet, and gradually form Internet thinking. According to the development direction of the industry and industry, the literature teaching method should be changed in time, the training program of the literature major should be updated, the Internet thinking should be integrated with the teaching of literature, and the new development vitality of literature should be given, which will provide references for the reform and development of literature teaching.

References

1. Li, Y.: Thinking about the model of college talent training under the background of global automation. J. Zhangjiakou Vocat. Techn. Coll. **1**, 17–19 (2015). (in Chinese)
2. Wang, L.: Reflection on literary transformation and cultural research. J. Jiangxi Electr. Power Vocat. Tech. Coll. **27**(3), 88–91 (2014). (in Chinese)
3. Gan, F., Ma, L.: Reflections on several basic theoretical issues of the transformation of contemporary literature. Contemp. Chin. Lit. Art Res. **5**, 103–107 (2016). (in Chinese)
4. Chen, J., Yan, T., Zhang, Z.: Exploration of new teaching models in the Internet age. J. Yulin Univ. **04**, 108–111 (2019). (in Chinese)
5. Wang, S.: Rain Classroom: Smart teaching tools in the context of mobile Internet and big data. Modern Educ. Technol. **05**, 26–32 (2017). (in Chinese)
6. Xu, J., Fan, M.: Practice and exploration of wisdom teaching in the field of educational informatization 2.0. Jiangsu Educ. Res. (15), 41–43 (2019). (in Chinese)
7. Nan, Zhang: Research on the wisdom classroom teaching mode under the background of "Internet + education". J. Jilin Inst. Technol. **01**, 119–122 (2019). (in Chinese)
8. Wang, G.: A preliminary study on the teaching status and countermeasures of the introduction to Chinese literature in local universities. J. Ankang Univ. **12**, 103–108 (2018). (in Chinese)
9. Liu, H.: Text reading in college English and American literature teaching. J. Chin. Lit. **07**, 121–122 (2009). (in Chinese)
10. Huang, S.: Research on the variation of the "new criticism" reading method in college literary criticism textbooks. Chongqing Normal Univ. **11**, 111–114 (2015). (in Chinese)
11. Chen, J.: Mobile Internet and college literature teaching. J. Guangzhou Radio TV Univ. **03**, 45–48 (2013). (in Chinese)

Teaching Design Based on Network Red Blue Confrontation

Guofang Zhang[✉] and Lili Yan

Hainan College of Software Technology, Qionghai, China
guofang_zhang@haoxueshu.com

Abstract. The "network attack and defense competition for college students" held by the Ministry of education and the office of industry and information technology every year attracts more and more colleges and universities to participate in it. In order to achieve good results in the competition every year, colleges and universities are actively preparing for training. Network technology security application, as a professional technology course of computer network in Colleges and universities, whether from the perspective of competition or from the professional positions students will be engaged in the future. Designing this course is very important for students to participate in competitions and work in professional posts. This paper analyzes the characteristic of this course from the perspective of students future professional work, puts forward the scene teaching concept based on the network red blue confrontation with the aim of improving students' professional skills and stimulating students' interest in learning. Constructs the course module according to students' cognitive law, designs the teaching mode of the network red blue confrontation from the reality, and calculates for colleges and universities, provide reference for the reform of machine course.

Keywords: Red blue confrontation · Teaching model · Security

1 Introduction

The "network attack and defense" competition jointly held by the Ministry of industry and information technology of the people's Republic of China and well-known IT enterprise in Colleges and universities across the country is attracting more and more colleges and universities and various industries to actively participate in it. This kind of competition has aroused the curiosity and curiosity of middle-aged young scholars in Colleges and universities. According to the school running tenet of "training skill application talents", vocational colleges pay more and more attention to this kind of competition. It is the goal of vocational colleges to encourage teachers and students to participate in the competition and win prizes. The competition award is not only the affirmation of student's skills, but also a strong basis for improving the ranking order of colleges and universities in higher vocational colleges. How to organization student's training, improving practical skills and stand out in the competition is a subject that vocational colleges are constantly exploring. How to design classroom teaching mode is the key point [1, 2].

J. MacIntyre et al. (Eds.): SPIoT 2020, AISC 1283, pp. 453–458, 2021.
https://doi.org/10.1007/978-3-030-62746-1_67

2   Problems Faced by Curriculum Teaching

The course of network security technology is different from the traditional culture course, which is characterized by theory + practice. For teachers, after explaining the theory of a technology, they need to practice on the computer and network equipment to prove the correctness of the theory. From the student's point of view, after listening to the teacher's explanation of the principle of a network technology, there is an urgent need to see the teacher's practical operation process and results, which also belongs to the learning process. Then students also need to imitate the teacher's operation on the computer and network equipment for experiments, and verify their understanding and mastery of this lesson based on the experimental results [3]. The simultaneous interpreting of this theory and practice of information technology classroom teaching is facing new problems different from traditional classroom teaching.

2.1   The Cognitive Process of Dichotomy

The classroom learning of theory + experiment practice is a two-part learning process for students, which is the key to design a good class. It should not only consider the integrity of the content and the time limit stipulated in the classroom, but also consider the reality of student's learning and acceptance. The key to solve this problem is to reasonably allocate the theoretical explanation time and experimental operation time [4]. How to grasp this scale and simplify the complex and boring theory? Even sentence should be carefully considered. It is the goal of curriculum design to use words properly and inspire students to achieve artistic conception. Each student has its own characteristics, and there are differences in learning and understanding ability. Classroom teaching is designed according to the cognitive law of students. By simplifying the boring and complex computer network security theory, it can not only improve the classroom teaching effect, but also promote the ability of teachers to control the classroom. It is also an effective way to improve students' quality through classroom teaching. After key to design classroom teaching is the practical operation [5]. In the limited time, on the one hand, students should observer the practical operation process operation of teachers, on the other hand, they should also simulate the practical operation of teachers to conduct experiments and judge the experimental results. The practical operation is a recovery that signals stimulate the brain of students to respond to the brain of students and guide the limbs to operate Miscellaneous process, which needs to be designed according to students' cognitive psychology and cognitive law, is a practice that aims to improve students' professional ability [6].

2.2   Scene Design of Practice Link

According to the job requirement of network security engineer, to be able to deal with unexpected network security incidents is the focus of the job requirements. The forms of network security incidents are various, and the losses caused to users are not the same. Some network intrusions are mainly to steal user information, while others are catastrophic. Most network security intrusions do not have early warning, which are often found after the occurrence of network security incidents. Network security personnel need to spend time to find out the mechanism of security events and the loopholes in the network, and then decide the security measures to be taken. There is a time difference between network intrusion and later remedial measures [7]. It is the goal of the practical teaching design of this course to bring many network security events into the classroom for restoration, design the real network security events scene, and make the students experience the working situation of the professional post in the future in advance.

2.3 Design Network Countermeasure Training Project

How to design the project of red blue network confrontation is the key point of realizing the teaching of red blue network confrontation in the course of network security technology application, which is the also the difficulty of this paper. In many network security events, not all network security events can be designed against each other. It is necessary to select typical network security events for design according to specific standards. According to the node of network communication experience, terminal computer, switch, border gateway, network communication link are selected in order to design teaching module [8]. First, in order to protect the security of node equipment, the encryption of user accounts of each node equipment is designed as a training project. Second, the boundary firewall configuration is set as a special project. Third, the configuration project training of the system firewall of the intranet node host is designed. Fourth, the VPN training project is designed in the gateway. Fifth, the reinforcement training project of the intranet server is designed. Sixth, design the information training project of scanning system. Seventh, design the ARP cache attack training project. Organize students to carry out purposeful training in various projects, stimulate students' desire for knowledge through a certain amount of training, and achieve the teaching and training goal of improving professional skills in training [9].

3 Build a Virtual Confrontation Experiment Platform

No matter in normal teaching or competition training, the primary problem of network security course is the lack of suitable training platform and high simulation training place. The high cost of network security products and the fast upgrading of network security products often make the construction funds, school running costs and the actual situation of learning and training, the existing equipment resources and virtualization technology are used to build a virtualization experiment platform. There are two ways to build a virtualization lab: virtual machine and private cloud platform.

3.1 Using Virtual Machine to Set Up the Experiment Platform

Virtual machine technology is a popular way of network construction at present. The advantage of using virtual machine technology is not only to save the cost of equipment, but also to quickly build a variety of experiment equipment that meeting the needs. What is sacrificed is only some CPU time and memory resources of physical computer, but this is insignificant compared with the benefits brought by virtual software [10]. Now there are two mainstream virtual machine software, VMware workstation and virtual box. VMware workstation is a commercial software, which is not opensource and needs to be paid, but there are many free versions or cracked versions, which can be used to build the required experimental environment. This software has three ways of network connection, which can simulate different scenes of the network and users. Virtual box is an open-source free software, which takes less resources and requires less installation environment. The two virtual software are both graphical interfaces, which are easy to operate. They can manage and transplant virtual machine snapshot, and realize dynamic migration of virtual laboratory. Therefore, in the training scenario of building a virtual laboratory, any software can be selected according to the needs and user's personal preferences to meet the needs. In this paper, VMware workstation is selected to build the virtual countermeasure experiment platform [11].

3.2 Using Virtualization Technology to Build a Private Cloud Platform

The virtualization technology has been quite mature. The full virtualization technology based on KVM has been widely used in various modern network cloud platforms. Some of the more famous cloud technologies use virtualization technology. For example, Alibaba cloud, baidu cloud, Huawei cloud and other large-scale enterprises have built cloud platforms. On the other hand, cloud platform also makes business development very easy. Virtualization technology requires high hardware resource [12]. As a cloud platform, the physical machine memory must be large enough. The CPU supports virtualization technology, but this is negligible compared with the benefits of virtualization. It is a shortcut to build network security training platform with virtualization technology. There are many kinds of optional virtualization software, and Xen and KVM are common virtualization software. Xen software is an early semi open source software, which is fast and has been applied in the initial stage of virtualization. However, Xen software has complex configuration and is semi virtualization software, which is not widely used. Later, KVM virtualization software with superior performance appeared. KVM is an virtualization software based on the Linux kernel, which is a module conforming to the open source software standard. It has been recognized by the Linux open source organization and added to the Linux operating system, which brings great convenient to the use of virtualization. In addition, there are some popular virtualization software, such as open stack, docker, VMware esxi, etc. These virtualization software have their own strengths when implementing virtualization, and users can choose according to their preferences.

4   Network Attack and Deployment of Network Defense

Network attack and network protection are the theme of network red blue confrontation. They are two aspects of one thing, the relationship between spear and shield. How to divide their boundaries and how to set up their teaching methods are the key points of classroom teaching. If we don't draw a clear line between cyber attack and defense, classroom teaching can't grasp the key points and sort out the clue. Classroom teaching design will be more chaotic, teachers and students will be affected in the classroom, and eventually students will lose interest in learning. In order to solve the problem of dividing the boundary between network attack and network defense, the following rules and standards are set up in this paper:

(1) students study in groups with three students in each group, and each group uses the method of draw lots to decide the red blue pair of two confrontation, and each group takes into account both red and blue team identity, and when the group attacks the other part's system target, it is the red team identity.

(2) each group system platform is the same, but each group's systems are independent of each other. The system includes attack platform and system target. The attack platform adopts the most cutting-edges Kali Linux operating system, which is an open-source operating system and integrates the latest and most complete network tools on the internet, which a number of thousands. The system target machine uses the Windows XP, Windows server 2003 and Linux operating system that students are familiar with. On the Windows Server 2003 platform, IIS server and MySQL database DHCP server can be built.

(3) Determine the score of the confrontation game, and set a score for the preset vulnerability on the system target. The size of the score can be set according to the degree of vulnerability being attacked. When a vulnerability is successfully attacked and obtained from the target system, the score set for the corresponding vulnerability will be obtained.

(4) In the first stage, each team builds its own system, including building a system target and a system attack platform; in the second stage, each team strengthens its own system target, and the strengthened rules shall ensure that the system target function of the team can be accessed by the other team normally.

5   Conclusion

The internet, which is made up of various kinds of hardware and software, has such and such loopholes, which creates opportunities for network intruders. In this territory where is no smoke of gunpowder, crime is inevitable. Various cyber security incidents happen from time to time. The forms of cyber attacks are changing with the progress of technology. The quiet internet is always full of swords and swords. The teaching content of network security technology application course also needs to be updated with the form of network security incidents, some outdated network confrontation training projects need to be discarded, and new network security incidents need to be studied and designed into classroom learning training projects, which is the characteristics of this course and the key point to promote the development of network security teaching.

Acknowledgement. This paper is supported by "General research project of education and teaching reform in colleges and universities of Hainan Province in 2020: Hnjg2020-144".

References

1. Red blue confrontation. https://he1m4n6a.github.io
2. Red blue confrontation. https://www.secpulse.com/archives/107194.html
3. Brief introduction of red blue confrontation of security. https://www.cnblogs.com/he1m4n6a/p/10057473.html
4. Practice drill: Green Alliance Technology red blue confrontation service. http://www.ijiandao.com/2b/baijia/184585.html
5. Wang, D.: Teaching reform of computer network security course. Chin. Foreign Entrepreneurs **6**, 35–36 (2019)
6. Zhang, G.: The web penetration based SQL injection. Lecture notes in electrical engineering, vol. 551, December 2019
7. How to improve the efficiency of classroom teaching? Teachers should do "three taboos" and "three demands". https://baijiahao.baidu.com/s?id=1631597347538496274&wfr=spider&for=pc
8. Network security. https://wenku.baidu.com/view/8b0f4da34793daef5ef7ba0d4a7302768e996fa4.html
9. Zhang, G., Li, W.: A security reinforcement scheme for the server. WOP Educ. Soc. Sci. Psychol. **14**, 92–92 (2018)
10. Teaching design of network information security. http://blog.sina.com.cn/s/blog_14063191e0102v5cv.html
11. China cyber security technology competition. https://baike.baidu.com/item/2018%E4%B8%AD%E5%9B%BD%E7%BD%91%E7%BB%9C%E5%AE%89%E5%85%A8%E6%8A%80%E6%9C%AF%E5%AF%B9%E6%8A%97%E8%B5%9B/23380041?fr=aladdin
12. Introduction of various information security and network confrontation competitions at home and abroad. https://blog.csdn.net/weixin_34342992/article/details/92304115

Construction and Implementation of the "Pyramid" Online Learning Platform for Experimental Teaching

Bingjie Zhu, Tongna Shi, Zhenjiang Shi, and Wenhua Wu[✉]

National Demonstration Center for Experimental Materials Science and Engineering Education,
Donghua University, Shanghai, China
clkx2491@163.com

Abstract. This paper introduces a "pyramid" online learning platform for experimental teaching. The platform provides basic courses, interest courses and advanced courses to students with different demands and learning abilities; graduate and undergraduate students are divided into different groups and can receive education in accordance of their aptitude. The platform effectively integrates online and offline learning, and provides a model of carrying out online experimental education for universities and colleges around the country.

Keywords: Pyramid online learning platform · Experimental teaching · Teaching informatization

1 Introduction

On April 11, 2018, the Ministry of Education issued the *2.0 Education Informatization Action Plan*. The education informatization gradually shifted from "point" breakthrough to "surface" development, making the teaching more personalized, the education more balanced, the management more refined and the decision-making more scientific. The new era endows education informatization with a new mission. During the epidemic, the popularization and application of "Internet plus education" is the touchstone of educational informatization achievement. Meanwhile, it will inevitably lead education information technology to develop towards the 2.0 era.

2 Present Situation of Experimental Teaching Informatization

2.1 Students' Autonomous Learning

The real significance of education not only includes the acquisition of knowledge, but also includes the mastery of learning methods; students should learn how to learn, and ultimately achieve the autonomous learning through education. In order to truly cultivate students' autonomous learning ability, we need to constantly explore in the process

J. MacIntyre et al. (Eds.): SPIoT 2020, AISC 1283, pp. 459–467, 2021.
https://doi.org/10.1007/978-3-030-62746-1_68

of education reform. Only by breaking through and subverting traditional education methods, can we cultivate high-quality talents who can adapt to the changing needs of the society [1]. In recent years, universities began to reform the traditional education mode through the new mode of "Internet plus" experimental teaching, and constantly develop innovative teaching methods to achieve the goal of cultivating high-quality talents. In the development of the Internet plus teaching mode, how to provide differentiated and individualized courses is also a hot topic.

2.2 The Importance of Experimental Teaching

Experiment is an important way to cultivate students' operation skills, practical abilities and scientific literacy; it cannot be replaced by other teaching contents or teaching links [2]. Experimental teaching can effectively improve students' understanding of knowledge, enhance their exploration ability, develop their innovative thinking, cultivate their cooperation consciousness, and provide strong support for the overall development of college students [3].

2.3 Problems of Existing Information Platforms for Experimental Teaching

In 2012, the Massive Open On-Line Course (MOOC), a new online education method, gradually changed the world. Under the background of the new era, online open teaching has become an important "field" in the new development and reform of today's education system. However, MOOC is more suitable for the teaching of theoretical and knowledge-based courses; it is not suitable for experimental teaching courses, which needs to combine theory with practice [4].

First, existing MOOC cannot meet the needs of experimental courses. The MOOC online learning platform is built mainly for theoretical courses. Many experimental courses need to be operated by students to ensure the learning effect, which cannot be achieved by online MOOC learning. Second, the existing platform is not personalized enough. In the existing online platform of experimental teaching, basic experimental online courses are set up according to existing courses; it fails to consider the differences of students in personalities and learning abilities. Third, the platform level construction is not clear. In the existing platform, most of the courses are mixed, and there is no clear hierarchy, which is not conducive to students' independent upgrade learning. Students cannot quickly find the advanced learning channel.

Therefore, how to further focus on the "learner's" perspective, find tools to stimulate students' interest in autonomous learning, and set up precise targeted courses, are important problems to be solved in the platform construction.

3 Construction of the Student-Based "Pyramid" Online Learning Platform for Experimental Teaching

The National Demonstration Center for Experimental Materials Science and Engineering Education of Donghua University (hereinafter referred to as "center") is a large open service platform integrating teaching, scientific research and management. Relying on key materials disciplines, the center is jointly constructed by the College of Materials Science and Engineering and the State Key Laboratory of Fiber Materials Modification. Through this platform, the online learning and offline experiments are combined organically. It changes the traditional experimental teaching mode, realizes the cooperation and co-compensation of online teaching platform and real experiments in the laboratory, and establishes the "Internet plus" experimental teaching mode [5].

3.1 Student Oriented and Individualized Learning Environment

The center adheres to the overall development of students, takes the cultivation of students' practical ability and innovation ability as the core, and cultivate high-quality innovative talents of materials sciences and engineering with strong innovation consciousness and practical ability. Based on the needs of users, the center has gone through three stages of "diagnosis and positioning", "strategy making" and "evaluation tracking" to construct the "pyramid" online learning platform for experimental teaching.

3.1.1 Diagnosis and Positioning

Starting from "diagnosis and positioning", we study the needs and interests of different students, so as to accurately locate and explore the direction of constructing the platform and related resources. From the two major directions of improving the professional and technical ability and the autonomous learning ability, undergraduate and graduate students are divided into four categories. The first group includes most basic undergraduate students hope to complete the study, get credits, and meet the requirements of undergraduate graduation. The second group includes undergraduate students who have completed basic learning and hope to acquire more knowledge and have certain autonomous learning ability. The third group includes graduate students who need to learn professional skills and hope that learning experimental technology can help them to publish professional articles and patents. The fourth group includes graduate students who not only know how to do experiments, but also want to acquire in-depth knowledge, so as to develop themselves and make innovation.

3.1.2 Strategy Making

Then through "strategy making", we can analyze the behaviours, abilities and habits of different groups, and set up relevant courses. For example, students in the first group need a "step-by-step" learning process; the learning contents should focus on required courses that can help students to obtain credits. Undergraduate students of the second group needs courses which can help them to increase their professional knowledge and improve their skills; students' initiative should be fully considered in the design of learning process. The third group of students pay more attention to the improvement of professional and technical ability; courses can be chosen by students according to their own needs. Students of the fourth group hope to learn more. This part of students should focus on online learning. After they have mastered online learning, they can also apply for offline experiment separately.

3.1.3   Evaluation Tracking

Finally, through "evaluation tracking", we can evaluate the use condition of the platform and the quality of each courses in the trial stage, in order to gradually optimize and continuously improve this platform. Through the three stages of construction, the "pyramid" experimental teaching online learning platform is constructed. Finally, all courses on the platform are divided into "Basic Courses", "Interest Courses" and "Advanced Courses". From basic learning to advanced learning, a large online learning platform is created. The courses on the platform are useful and can provide some offline support.

The basic course module includes general and universal courses; they are mainly aimed at students of group one. The compulsory experimental course for undergraduates can be found in this module. Under the guidance of teachers, students can learn anytime and anywhere according to the requirements of the course.

Interest courses are professional promotion courses; they are mainly aimed at students in group two and most students in group three. According to the characteristics and needs of these students, the center has set up some independent and selective experimental courses, such as large and precise instrument courses, professional technology experimental courses required in the field of materials, and so on. The canter also holds regular offline training sessions to cooperate with these courses. In addition, there are some experimental courses on foreign language, which can be applied for by all international students in the school. In this stage, the teacher plays the role of promoter. In the course, students can improve their learning according to their own learning progress anytime and anywhere.

Advanced courses are in-depth and extensive courses; they are mainly aimed at students in group four and a small number of students with higher requirements in group three. This part of course has certain difficulties and requirements, which requires learners to have strong autonomous learning ability. There are some virtual courses, science and technology lectures and large-scale competition training. In this module, the teacher only plays the role of companion. More importantly, students can explore, research and innovate on their own. If they encounter problems in the learning process, they can consult the teacher and solve their problems.

3.2   Platform Technology

Based on modern information technology such as big data, multimedia and virtual reality, the "Pyramid" online learning platform for experimental teaching was built, and the Internet plus learning system was established. Combined with the intelligent laboratory management system developed by the center [6], the online learning platform and offline entity platform were integrated based on existing technology of campus network and campus card.

3.2.1 Multimedia Technology

Multimedia videos integrate text, image, sound, animation and other elements, so that students can maintain their interest in learning and watching. Based on the needs of students, the center uses AE, flash, Premiere and other video production software to integrate animation, pictures and sound effects into the courseware to enrich forms of experimental teaching, and increase the perceptual knowledge in teaching, so as to strengthen and expand students' understanding of experimental knowledge.

3.2.2 Virtual Reality Technology

Virtual Reality (VR) technique, Augmented Reality (AR) technique and Mix Reality (MR) technique are used to provide learners with personalized and self exploring learning space. The center, in cooperation with virtual software development companies, makes experiments with professional characteristics into well-made virtual courses, so that students can learn by playing. It does not only increase students' interest in autonomous learning, but also enables students to access more professional knowledge and technology.

3.2.3 Big Data Technology

In the "pyramid" online learning platform for experimental teaching, the center has functions of data collection, data storage, data statistics, data analysis and data feedback. It collects and stores data on students' learning records, learning reports, assessment papers, and so on. Through the big data visualization technology and graphical means, relevant data can be clearly communicated and effectively analyzed. Students' learning situation can be mastered in time, which can be used as an important basis for the teaching reform and promote the continuous improvement of experimental teaching contents and methods.

3.2.4 Internet of Things Technology

Internet of things technology means, based on existing Internet technology, it can achieve the real-time control and management of sensing objects through the RFID technology, the intelligent sensing technology, the satellite positioning technology and other techniques [7]. In the "pyramid" online learning platform, the center adopts the Internet of things technology. Through the laboratory intelligent management system, the offline entity platform is activated. After online learning, students can apply to enter the offline entity platform to carry out the experiment in corresponding laboratory and with corresponding experimental equipment, so as to further improve the management level of intelligent laboratories in the center (Fig. 1).

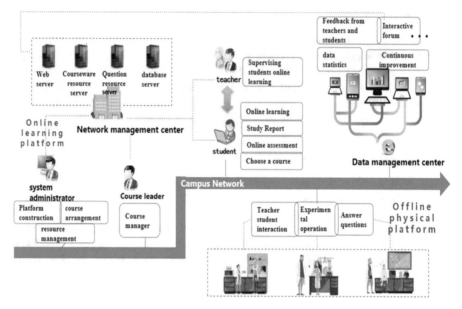

Fig. 1. Architecture of the "pyramid" online learning platform for experimental teaching

3.3 Platform Process

In the process of platform construction, the user needs are the fundamental starting point. We need to design the platform using process from the perspective of different roles. In basic courses, teachers organize students to learn. In interest courses, teachers promote students' autonomous learning, and the learning process is no longer fixed. In advanced courses, teachers accompany students to study independently. Students independently choose the right time to study, and the learning process is completely autonomous.

4 Implementation of the Student-Based "Pyramid" Online Learning Platform

Since 2013, the center began to build an online learning platform. Through several years of application and implementation, it has gradually formed a "pyramid" online learning platform which is composed of "Basic Courses", "Interest Courses" and "Advanced Courses". The situation of implementing the "pyramid" online learning platform in 2019 is shown in Table 1.

Basic course is provided for all undergraduate students in the school. It covers compulsory experiments, and has features of wide coverage, great number of students participate, and high use frequency. Students can get corresponding credit after completed the course. It is located at the "solid foundation" part of the "pyramid" online learning platform. Only with solid foundation, can students complete other courses better.

Table 1. Implementation of the "pyramid" type online learning platform in 2019

Type of courses	Title	Project	Number of students	Person-time of study
Advanced courses	Training for national polymer material experiment and practice competition	1	33 universities	858 person-time
	Virtual course on large analytical instruments	12	17 students	30 person-time
	Virtual course in organic chemistry	6	10 students	10 person-time
Interest courses	Equipment training course - scanning electron microscopy	3	104 students	385 person-time
	Equipment training course – cell culture	1	5 students	53 person-time
	Instrument training course-1	11	23 students	984 person-time
	Instrument training course-2	9	23 students	426 person-time
	Instrument training course-3	3	24 students	273 person-time
	Professional elective course	6	6 students	114 person-time
	Basic elective course	4	44 students	499 person-time
Basic courses	Material science experiment	10	242 students	23043 person-time
	Physical and chemical experiments of composite materials	12	61 students	5642 person-time
	Physical and chemical experiments of polymer materials	12	70 students	6429 person-time

Interest course includes elective experiments of interest for all students in the school. Students choose these courses completely based on their needs and interests. Basic elective courses and professional elective courses are provided for students who have studied the basic courses of material science experiment, polymer physical chemistry experiment and composite material physical chemistry experiment. After completing required experiments, they can selectively study according to their own needs.

Advanced courses are mainly provided for students who have special needs and can have more autonomous learning ability. One school can select the course as one unit, and then provide the course to a lot of students in the school. In addition, there are many virtual experiment courses offered, so that students with higher ability can have more learning opportunities and access to more high-precision instruments and technologies.

As can be seen from the Fig. 2, in terms of the overall number of learners, the number of people in basic courses is far greater than that in interest courses, and the number of people in interesting courses is greater than that in advanced courses. Students of advanced courses account for 7% of basic courses, and students of interest courses account for 17% of basic courses. At the same time, most of the students who have spare time to study interesting courses and advanced courses are students with good academic performances in basic courses; most of them are students with top 50% academic scores.

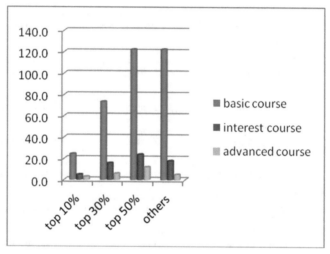

Fig. 2. Students who select basic courses, interest courses and advanced courses and their grades

5   Conclusion

Through dividing students into different groups and providing targeted courses for them, the pyramid online learning platform overcomes problems in existing platforms of experimental education. It effectively integrates online and offline learning, and provides a model of carrying out online experimental education for universities and colleges around the country.

References

1. Li, Y.L.: Research on strategy developing students' autonomous learning ability under the mixed teaching mode. Sci. Educ. Article Collects **493**, 46–47 (2020)
2. Fan, H.X., Chai, C.W., Gu, C., et al.: Learning from the concept of MOOC, accelerating the construction of network platform for university chemistry laboratory. Exp. Technol. Manage. **31**(11), 26–28 (2014)

3. Song, Q.F.: On the importance of experimental teaching in modern education mode. Road Success **595**(3), 12 (2019)
4. Cui, G.X., Xiong, J.P.: Development of MOOE practice teaching platform based on virtual simulation technology. Exp. Techno. Manage. **33**(4), 103–107 (2016)
5. Wu, W.H., Zhu, B.J., Yang, W., et al.: Construction and effect of national experimental teaching demonstration center of materials science and engineering. Res. Explor. Lab. **38**(10), 151–201 (2019)
6. Wu, W.H., Yang, Q., Shen, X.Y., et al.: Laboratory open management under intelligent laboratory management system. Exp. Technol. Manage. **28**(2), 172–197 (2011)
7. Zhao, J.: Application of internet of things technology in smart campus: taking tibet university as an example. Sci. Inf. **6**, 105–108 (2020)

The Research on the Strategy of College English Teaching Mode Based on Mobile Terminal Under the Background of Internet+

Rongjian Li[✉]

Jilin Animation Institute, Changchun, Jilin, China
haoxue163@eiwhy.com

Abstract. With the development and progress of the times, people have new ways to explore knowledge. With the advent of the information age, people can learn the knowledge they want more and more conveniently on the Internet, and through Internet-based big data methods, people can also effectively reduce the burden of learning. Through mobile phones, computers, tablets and other electronic people can learn all kinds of knowledge without leaving home, and get rid of the shackles of traditional preaching learning. At present, the study of college English is also adapting to the pace of social trends. The emergence of mobile terminal platforms has provided students with more possibilities for learning college English, not limited by time and place, and has multiple means and comprehensive customized learning programs. The advantage has been liked by more and more college students. The following is a research and analysis on the mode of college English teaching on the mobile terminal platform.

Keywords: Internet+ · Mobile terminal · College English · English teaching mode

1 Introduction

The development of the information age provides more and more conveniences for people's daily lives. Nowadays, society has stepped into 5G, and the communication technology based on 5G Internet will further promote the development of society. In this context, mobile terminal technology has also been unprecedentedly developed and applied in many fields. In the context of the mobile Internet, there are more and more teaching platforms based on mobile terminals [1]. People can learn on their mobile phones without leaving their homes. Based on Internet big data technology, these platforms focus on multimedia video teaching and cooperate with more vivid videos, sounds, and teaching courseware allow learners to easily learn knowledge. Currently, common mobile terminal teaching platforms in China include Netease Cloud Classroom, Cocoa English, and Tencent Classroom. Increased, more and more people began to study on mobile phones and tablets, making learning a part of life and promoting personal development [1].

© The Editor(s) (if applicable) and The Author(s), under exclusive license
to Springer Nature Switzerland AG 2021
J. MacIntyre et al. (Eds.): SPIoT 2020, AISC 1283, pp. 468–473, 2021.
https://doi.org/10.1007/978-3-030-62746-1_69

2   Overview of "Internet+"

"Internet+" refers to the use of information and communication technologies and Internet platforms to achieve the deep integration of the Internet and traditional industries; through Internet optimization and integration of social resource allocation, Internet innovations are applied to all areas of society, driving the vitality of social and economic entities and promoting the formation of new forms of social and economic development. Affected by the role of "Internet+", the "Internet + education" model has gradually developed. The essence of "Internet + education" is mainly on the basis of education consumption needs, in accordance with the basic attributes of the Internet, in the allocation of production factors, the integration and optimization of the Internet are fully exerted, so that a new educational ecology and form can be created [2]. College English teaching under the "Internet+" thinking mode is the understanding and practice of the concept and mode of "Internet + education" [1].

3   Definition and Characteristics of Mobile Learning

3.1   Definition of Mobile Learning

Mobile learning refers to a new type of learning model in which learners obtain learning resources through Internet technology and mobile terminal devices, so that learners can carry out autonomous learning regardless of time and place. The mobile terminal refers to a mobile computer device, including a smartphone, a tablet computer, a notebook, etc., and also includes an APP in a smartphone frequently used by learners [2], as shown in Fig. 1.

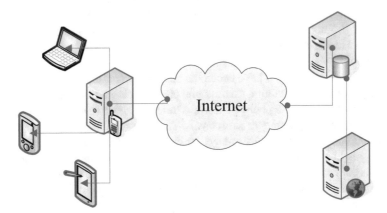

Fig. 1. Internet model of mobile learning

3.2 Features of Mobile Learning

There are many types of mobile terminals and clever design, which not only makes the learner's learning break through the limitation of time and space, but also stimulates the learner's interest in learning. Compared with the traditional teaching mode in which teachers teach mainly, the mobile terminal-based learning greatly improves students' participation in the courses, and teacher-student interaction and student- student interaction are well reflected [3]. As a supplement to the formalization of extracurricular teaching, mobile learning can allow students to freely arrange a reasonable time for learning, so that the teaching and learning time outside the classroom can be continued.

4   Analysis of the Status Quo of Mobile Terminal Education Development

The development of informatization has made technology more and more close to people's lives and integrated into life, providing more and more unexpected conveniences to people's lives. As an important medium for people to communicate, mobile phones are no longer just used to make calls as before. The development of technology has given mobile phones more possibilities [3]. Mobile phones have also become more intelligent, convenient, and functional. Mobile phones have become people. An indispensable part of daily life. Modern mobile phones are different from previous mobile phones. It is more like a portable multimedia computer device. It can realize mobile phone Internet access based on wireless network technology. It can also provide people with more diverse photographs, videos, music, etc. Customized functions, so the development of mobile phones is also towards more integrated development [4].

According to the latest 2018 "Mobile Internet Industry Data Research Report", the scale of China Mobile Internet users has risen to 1.132 billion in the past year, which is an extremely large number, and the total daily use time of China Mobile terminals reached 3.172 billion Hours, and compared with 17 years, it has increased by 17% points. With the passage of time, these data are still slowly rising, and it is not difficult to see from these data that mobile phones have become a very important tool in people's lives [5]. Due to the accelerated development of mobile phones, a large number of learning apps have emerged on the mobile phone platform. These apps involve different departments, different types, and different ages at all levels. They are suitable for students and office workers, and it also has many Advantages, such as portability, ready-to-learn, and low prices are sought after by more and more people. A lot of high-quality English learning software has appeared on the English mobile terminal platform, such as our common Youdao English, Cocoa English, Palm English, voa slow English, BBc online, scallop words, hundred words, Edo English, etc. English learning software. Mobile phones have gradually replaced computers and become important assistants for people to work and study [5].

5 Advantageous Application of Mobile Terminal Platform in College English

Mobile phone English teaching has changed the traditional teaching concept. After breaking through the teaching restrictions, it has been sought after by more and more people. Here is a relevant analysis of its advantages:

5.1 Not Limited by Time and Region

Mobile phone English learning is based on Internet big data technology, providing students with a portable learning platform. Through mobile phone software, students can learn online, and the content is rich and diverse. It can be watched online or downloaded offline to provide students with learning [6].

5.2 Personalized Customized Learning and Review Plan

English is a subject that requires long-term review. The mobile software platform will customize a learning plan that suits the students' learning according to the habits of many students, such as vocabulary memory. The mobile phone can help students remember in the form of animations, pictures and sounds. The method of rote memorization in the past; in addition, the platform provides more customized review programs, such as a certain grammar learning, such grammar-related exercises will appear to help students understand [7].

5.3 Efficient Multi-dimensional Evaluation Tool

Intelligent today, English learning is no longer constrained by traditional teaching. Students can take English tests through the mobile terminal platform test, and the platform also presets related levels. Only students who pass the level can get the qualification for the next level [7]. Through this interesting learning mode, students can improve their English learning ability in multiple dimensions.

6 The Use of College English Teaching Based on Mobile Terminals

6.1 Integrate Traditional Teaching Methods and Network Teaching Methods

Make full use of the advantages of the mobile terminal teaching platform, combined with traditional teaching methods, give full play to the advantages of different teaching methods, learn from each other's strengths and weaknesses, so as to achieve English teaching. In traditional teaching classrooms, teachers are often the main supplement to the students. Teachers teach knowledge to students, and they get little or no feedback from students on the knowledge they teach, which leads to students' incomplete understanding of classroom knowledge [8]. The college English teaching model can improve this phenomenon with the supplement of the network environment. Before the class, students can fully prepare for the preparation work through the mobile terminal, and they can ask questions in the classroom if they don't understand. In this way, the full play of the student's leading role can be brought into full play, and the teacher can also understand the students' preparation through the questions [9].

6.2 Break Through the Limitations of Time and Space

Traditional teaching is limited by time and space, which hinders students from acquiring more knowledge after class. However, college English teaching based on mobile terminals has benefited from the network environment, which has broken through the limitations of time and space. In this environment, teachers can optimize the teaching mode, such as publishing the relevant content of the text on the App in advance, so that students can learn the content of the next class through preview; teachers can also release micro-classes and admiration on the teaching platform in advance teaching resources such as lessons are released in advance to students, and the students' completion can be directly checked in the classroom [10]. This can not only consolidate students' knowledge at any time before, during and after class, but also improve classroom efficiency. In addition, because of the diversity of network resources, it can also increase students' interest in learning.

6.3 Reasonable Use of Mobile Terminal Teaching Software

In addition to tools such as QQ and WeChat, which are often used for communication, various teaching software also appears with the popularity of mobile devices. Major institutions have also developed various teaching platforms to facilitate students to learn English better and more conveniently. These softwares and platforms are highly interactive, which is convenient for teachers to check, review, and supervise student learning on computer or mobile terminals, as shown in Fig. 2. For example, teachers can publish composition or translation exercises on the platform, and the system will intelligently evaluate the scores and give suggestions for modification [11].

Fig. 2. Teaching software and platform for mobile terminals

7 Conclusion

With the continuous improvement of network technology, traditional classroom teaching can no longer meet the needs of teachers, especially students. Compared with traditional teaching, the teaching of mobile terminals based on Internet + background is more innovative and time-sensitive. The combination of mobile learning and college English teaching is a product of the development of the times, which is conducive to promoting the reform of college English, stimulating students' interest in learning, and achieving efficient college English classroom teaching.

Acknowledgments. This paper is the outcome of the study, Construction and Practice of College English Interactive Learning Mode Based on Mobile Terminals, which is supported by the Foundation for Higher Education Reform Projects of Jilin Province; the project number is SY1904.

References

1. Jin, Y.: A new model of college English teaching under the background of "Internet + Education". Yangtze River Series. Theoretical Research **8**, 109–110 (2017). (in Chinese)
2. Wei, Z.: Research on college english teaching under the "Internet +" mode of thinking. Anhui Literature **2**, 130–132 (2016). (in Chinese)
3. Xie, C.: Innovative research on college English teaching model from the perspective of "Internet + education". Anhui Lit. Monthly **8**, 121–122 (2017). (in Chinese)
4. Zheng, W.: Research on the interactive learning path in and out of college English classes based on mobile terminals. J. Jiamusi Vocat. Coll. **02**, 228–231 (2019). (in Chinese)
5. Jiang, W.: A comparative experiment on the learning effect of micro mobile terminals and traditional methods. Overseas English **24**, 147–148 (2018). (in Chinese)
6. Meng, S.: Research on mixed college English teaching mode based on mobile network terminal platform. J. Chengdu Polytech. Univ. **04**, 64–66 (2018). (in Chinese)
7. Zhang, C.: A review and prospect of research on mobile English-based college English mobile learning. J. Jiangxi Vocat. Tech. Coll. Electricity **11**, 45–47 (2018). (in Chinese)
8. Zhen, W.: Design and empirical study of English mixed learning based on mobile terminals. Educ. Informatization China (10), 10–14 (2018). (in Chinese)
9. Li, S.: Research on the promotion mechanism of mobile APP in oral English teaching in higher vocational education. Contemp. Educ. Practice Teach. Res. **6**, 212–214 (2016). (in Chinese)
10. Wang, Y.: Innovative research on hybrid English teaching model based on mobile APP. Overseas English **6**, 65–68 (2017). (in Chinese)
11. Bao, Q.: Application of mobile Internet environment in college English mixed teaching. International PR (7), 58–62 (2019). (in Chinese)

The Research on College Physical Education Innovation Based on Internet and New Media

Bin Dai[⊠]

Chongqing College of Architecture and Technology, Chongqing, China
debbracn@eiwhy.com

Abstract. In the education system of universities, physical education is one of its important contents. The purpose of physical education is to improve the physical fitness of students and promote the comprehensive development of students. In the context of the rapid development of China's social economy, the new Internet media is gradually applied in various disciplines, which can guide students to actively participate in physical education and ensure the effective physical education development. With the promotion and application of new media, teaching can be carried out with the help of visual video and micro video, providing rich and diverse teaching content and cultivating students' lifelong sports awareness. In this regard, from the aspects of optimizing teaching concepts, implementing micro-teaching and promoting sports culture, this article analyzes the strategies of innovative college sports teaching models based on the Internet and new media.

Keywords: Internet · New media · College physical education · Physical education · Teaching innovation

1 Introduction

In Internet era, sports, as a discipline to improve the physical fitness of students, face the new media environment, and need to use information technology scientifically and rationally, take advantage of information technology, enrich classroom teaching content, effectively use physical education resources, and make students proactive Participate in sports learning and feel the fun and charm of sports. Under the new media environment, college physical education should break the previous single teaching model and create a good physical education atmosphere to ensure the smooth development of physical education [1].

2 The Internet Impact on College Physical Education

2.1 Internet + Overview

The so-called Internet+ is actually a brand-new format for the Internet development. In short, it is "Internet+ various traditional industries". It penetrates and integrates the

J. MacIntyre et al. (Eds.): SPIoT 2020, AISC 1283, pp. 474–479, 2021.
https://doi.org/10.1007/978-3-030-62746-1_70

Internet into various industries and fields, and uses Internet concepts, technologies, and equipment to build new developments [2]. The Internet+ era has six characteristics, namely cross-border integration, innovation-driven, restructuring, respect for human nature, open ecology everything and connecting everything. At present, China has just entered the Internet+ era, these six characteristics have not been fully reflected, and continuous innovation and development of information technology are still needed as an important thrust. It is foreseeable that with the gradual the Internet+ era development, the entire human society will undergo earth-shaking changes, and various fields will inevitably present a completely different trend from the current, forming a new normal [3].

2.2 Optimizing Teaching Ideas and Improving College Physical Education Methods

Optimizing the physical education teaching concept teachers are an effective way to improve the physical education quality of classrooms in colleges and universities. As a school, we should strengthen the propaganda work, promote the new media advantages in teaching, and use the special activities to promote the characteristics of new media in classroom activities, so that physical education teachers recognize the value of new media in teaching and fully understand their teaching reform [4]. At the same time, the new media application can stimulate students' enthusiasm for sports participation and ensure the effective physical education development.

2.3 Use Micro-classes to Innovate Classroom Teaching Content

With the development of the new media environment, micro-classes have been gradually proposed. As a product of the Internet era, it has the characteristics of short, sophisticated, and interesting. It can fully arouse the students enthusiasm and deepen the learning and mastering of sports knowledge [5]. The usage of micro-classes can make the shortcomings and deficiencies up for the traditional teaching model, ensure the effectiveness of students' classroom learning, and improve the quality of teacher teaching. Before the beginning of classroom teaching, teachers can copy the created micro-classes on the computer and show them to students with the help of projection mode, so that students can clearly understand the key points and difficulties.

3 Advantages of New Internet Media Applied in College Physical Education

3.1 Enriching the College Physical Education Environment

The physical teaching environment in universities generally includes track and field, professional training rooms, necessary teaching equipment, outdoor venues, etc. Physical education courses in universities are compulsory public subjects. On the one hand, physical education courses are designed to improve the physical fitness of students, and on the another hand, it is easy to students. The traditional college physical education environment includes all indoor and outdoor teaching venues and teaching resources [6]. Under the influence of the new Internet media, college physical education is no longer restricted to traditional field space teaching, and instead it will be a competition on the new Internet media on-site and teaching videos are moved into college physical education classrooms. Pictures, videos, real-time games and sports star classrooms are used to create a vivid and three-dimensional teaching atmosphere and improve college physical education environments [5], as shown in Fig. 1.

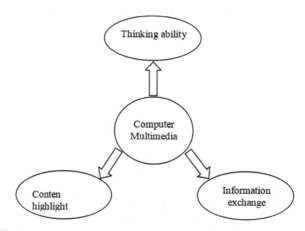

Fig. 1. Advantages of computer multimedia application in teaching

3.2   Strengthen the Exchange of Students and Teachers in College Physical Education

The traditional one-to-many teaching format has severely lacked teacher-student communication. With the development of new Internet media, teacher-student communication is no longer to limit the classroom, but appears in the teaching process and at any time after the teaching, giving students more. The combination of new Internet media and college physical education resources will not only improve the teaching quality, but also ensure the healthy development of college students' physical and mental health by giving sufficient free communication space for teachers and students [7].

3.3   Help Students Improve Their Innovation Ability

Physical education in universities, while enhancing the physical fitness of students, cultivates students' comprehensive strength. The addition of new media on the Internet has brought new teaching methods and technologies to physical education in universities, and the realization of college students by combining the latest teaching resources of new media on the internet improvement of innovation ability. The teaching materials of physical education teachers can be more abundant and diverse. Only on the basis of the richness and interest in physical education can be promoted students' learning interest, and students should be more actively involved in physical education and thinking innovation [8]. The new media on the Internet allows college physical education to keep up with the trend of the times.

4 Innovative Forms of College Physical Education from the New Internet Media Perspective

4.1 Internet New Media Sports Knowledge Enriches Teaching Resources

The theoretical knowledge and practical knowledge of physical education in universities are combined with each other. When teaching theoretical knowledge in class, you can get a lot of online resources through the Internet to gather sports-related social events and social sports news [9]. Injecting fresh blood into sports theory knowledge can not only improve the interest of classroom teaching and student participation, but also enable students to have a more comprehensive and deeper sports understanding, as shown in Fig. 2.

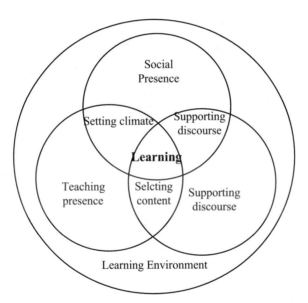

Fig. 2. Internet sports online education

4.2 Internet New Media + Sports Practice Three-Dimensional Teaching Form

The main purpose of physical education in universities is to promote the comprehensive development of physical fitness of college students. In the teaching process, there will be relatively complex activities and sports skills. In addition to regular physical training. Physical education teachers will have certain difficulties in teaching. In the learning process, students are also prone to problems such as non-standard movements, excessively fast demonstration movements, and excessive difficulty [10].

478   B. Dai

4.3 Internet New Media + Intelligent Teaching and Student Exchanges in Physical Education

There is a general lack of comprehensive communication between teachers and students in colleges and in universities, especially in physical education. The interaction is the key to affecting the realization of physical education skills [11]. The new Internet media can improve the communication between students and teachers in physical education. First, the information storage function of the new Internet media in physical education, and store the key points of the classroom to store information; Second, using the interactive platform of new media on the Internet, you can use intelligent teacher-student interactions such as barrage, WeChat and QQ in the visual teaching stage to connect the intelligent social platform to physical education, as shown in Fig. 3; Third, the massive sports information and teaching resources on the new Internet media can provide richer information for teachers and students' communication [12]. The development of new media on the Internet can strengthen the interaction in physical education.

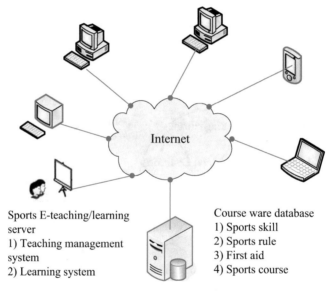

Fig. 3. Three-dimensional sports multimedia teaching platform

5 Conclusion

In short, In college physical education, with the going of new Internet media technology, the classroom teaching concept and classroom teaching model are innovated, so that students can actively participate in physical learning. At the meantime, the transformation of the role of teachers and students, highlight the main position of students in the classroom, fully implement the physical education in universities, ensure the quality of physical education, promote the physical education development in Chinese colleges and universities, and achieve the goal of physical education.

References

1. Chen, T., Zhang, H., Hu, X.: Internet+ times college physical education laboratory construction and teaching reform. Exp. Technol. Manage. **33**(06), 233–235 (2016). (in Chinese)
2. Xiao, D.: Exploring the mixed learning model of college physical education under the background of "Internet+". China Audio-Visual Educ. **11**(10), 123–125 (2019). (in Chinese)
3. Chen, Q.: Research on the creation of a new physical education model in the Internet+ era. J. Huainan Vocat. Tech. Coll. **5**, 95–96 (2018). (in Chinese)
4. Zhang, L.: Research on the innovation of physical education models in local colleges under the "Internet+" environment. Mod. Vocat. Educ. **1**, 78–80 (2018). (in Chinese)
5. Ji, N.: A preliminary study on the innovative thinking of college physical education under the background of "Internet+". Cult. Sports Supplies Technol. **19**, 124–126 (2018). (in Chinese)
6. Wang, Z., Zhang, Y., Wang, Y.: The road to the reform of university public physical education under the new media environment. J. Hengshui Univ. **19**(04), 32–35 (2017). (in Chinese)
7. Kuang, X., Feng, G.: Using new media to improve the effectiveness of physical education in colleges and universities. Inf. Recording Mater. **19**(07), 163–165 (2018). (in Chinese)
8. Duan, Y.: Research on the teaching reform of public physical education in colleges and universities under the background of new media. Caizhi **8**(27), 110–112 (2017). (in Chinese)
9. Liu, C., Li, Y.: The difficulties and countermeasures in the construction of physical education teaching resources in colleges and universities under the background of "Internet+". Teach. Educ. People (High. Educ. Forum) **12**, 15–16 (2017). (in Chinese)
10. Jiang, Y., Wang, D.: Research on the influence of mobile internet on college physical education. J. Nanjing Inst. Phys. Educ. Nat. Sci. Ed. **5**, 99–101 (2018). (in Chinese)
11. Li, X., Yu, H.: Discuss the research on the construction of networked physical education in colleges and universities based on the "Internet+" background. Youth Sports. **5**, 74–75 (2017). (in Chinese)
12. Zou, W., Ma, L.: Influential factors and countermeasures of students' physical health in Jiangxi colleges and universities under the background of "Internet+" era. Tomorrow Fashion **4**, 193–195 (2017). (in Chinese)

The Development Strategy of College English Education Under the Language Education Planning in the Age of Artificial Intelligence

Xiangying Cao[✉]

Department of Foreign Languages, Nanchang Institute of Technology, Nanchang, Jiangxi, China
104379295@qq.com

Abstract. At home and abroad, how to apply artificial intelligence technology to education in the context of the development of artificial intelligence has been given great attention. In 2019, the Ministry of Education put forward requirements on the integration, application, and promotion of artificial intelligence technology and the education industry. Based on the introduction of related concepts, this article describes the main problems of college English teaching in the context of language education planning based on the background of artificial intelligence. Then it analyzes the development strategies of college English education under language education planning in order to improve college English in the new era.

Keywords: Artificial intelligence · Language education planning · College english education · Development strategies

1 Introduction

With the introduction and launch of the "Education Informationization 2.0 Action Plan", the era of artificial intelligence in education has officially begun, and college English teaching has ushered in a major change. In recent years, major breakthroughs have been made in artificial intelligence technology. Big data, cloud computing, speech recognition, and deep learning have become new hot spots. Driven by related information and communication technologies, artificial intelligence appears in all aspects of people's lives with new algorithms, new technologies, new experiences and new applications. Among them, image recognition, speech recognition, language translation, language testing and other fields that are closely related to language services have developed rapidly. In popular education stage, how to use advanced artificial intelligence technology to achieve personalized and precise education is the main problem. New applications of teaching and learning intelligent identification technology, learner learning analysis technology, and artificial intelligence technology based on personalized learning are driving changes in teaching and learning. Cloud computing and big data have enabled artificial intelligence, interactive lessons such as micro-lessons, flipped classrooms, and personalized teaching to achieve "teaching" and "learning" in real-time, personalized feedback, and

J. MacIntyre et al. (Eds.): SPIoT 2020, AISC 1283, pp. 480–486, 2021.
https://doi.org/10.1007/978-3-030-62746-1_71

effective communication. In February 2017, the Ministry of Education issued the Guide to College English Teaching, which pointed out the direction for the construction and reform of college English courses. In terms of teaching methods and means, the Guide states: "College English teaching should follow the rules of foreign language learning, according to the characteristics of teaching content, fully consider individual differences and learning styles of students, use appropriate and effective teaching methods, and reflect teacher-led the teaching concept of taking students as the main body enables the transformation of teaching activities from "teaching" to "learning," and forms a teaching normal to feature teachers' guidance and inspiration and students' active participation.

2 Language Education Planning Concept

Language planning can be divided into ontology planning, status planning, reputation planning, and acquisition planning. Language education planning is a language planning acquisition plan, which refers to "the act of obtaining the specific language ability of an educated person through educational means, which is the concrete realization and implementation of the general goal of the national language plan in the field of education." [1].

3 The Main Problems of College English Education Under the Language Education Planning in the Era of Artificial Intelligence

In the past 40 years, public foreign language teaching in Chinese universities has been fully restored and developed. College foreign language teaching is accompanied by national economic construction, scientific and technological progress, social change, and the development of higher education. It has pioneered and experienced the path from reform of curriculum system to teaching model, integration of curriculum to technology integration, and development from scale to content enhancement. Professor Cai Jigang (2017) pointed out that the failure of our foreign language education policy is the failure of the orientation of college English teaching [2]. Shen Qi (2018) also believes that the fundamental disadvantage of the current reform of college English teaching lies in "unclear positioning and value orientation of college foreign language teaching" [3]. Therefore, to measure the gains and losses of college English teaching reform, it is necessary to re-examine the value proposition of college foreign language teaching planning from the perspective of language planning and language policy.

4 Analysis of the Contribution of Artificial Intelligence to English Education

4.1 Artificial Intelligence Helps to Implement in-Depth Education

Teaching students according to their aptitude has always been an important goal of high-quality and top-quality development of vocational education teaching and it is also an ideal state of vocational education teaching process. Due to the limitations of various subjective and objective factors in traditional vocational education, it is difficult for teachers to truly implement teaching according to their aptitude. However, with the help of information technology and artificial intelligence technology such as big data and cloud computing, teachers can collect student's personality information, quality information and learning information more systematically, efficiently and conveniently, and develop personalized teaching based on intelligent learning terminals [4].

4.2 Artificial Intelligence Improve the Scientific Nature of College English Education Management

Regardless of whether it is regional group education management or individual education management, scientific decision-making needs to rely on comprehensive and accurate information. Only with sufficient information can it be possible to achieve "all starting from reality" [5]. By establishing regional and campus big data centers, creating an artificial intelligence management service platform, and using technical methods such as big data analysis, business modeling, and data visualization, education managers can be provided with far-reaching information support, thereby significantly improving higher education management.

4.3 Artificial Intelligence Drives Innovation in Teaching Models

Artificial intelligence has changed the mode that teachers are the main knowledge imparters and information carriers in the traditional college English teaching system. Under the background of artificial intelligence, students' learning objects and interactive objects can be transferred from real people to intelligent machines. Various artificial intelligence-based teaching modes can be effectively used in college English teaching classrooms to increase students' language [6]. Therefore, the emergence of artificial intelligence has contributed to the reform and innovation of college English teaching models.

5   Analysis of College English Education Development Strategy under the Language Education Planning in the Age of Artificial Intelligence

5.1 College English Education Deeply Integrates Artificial Intelligence

Under the impact of high-tech development such as artificial intelligence, the reform of foreign language teaching in universities must also strengthen the integration and innovation with technology, and explore a teaching mode that deeply integrates learning science. Although the current artificial intelligence technology is still in a weak artificial intelligence stage, it can only provide a supplementary means for foreign language teaching, and it cannot completely replace the traditional foreign language classroom teaching model [7]. However, from the perspective of the development of language technology, the science of learning that has a disruptive effect on traditional foreign language teaching is unstoppable, and the learning science system that deeply integrates people and machines is the general trend. University foreign language teaching must respond to and carry out related research in advance. Foreign language education and research in the era of artificial intelligence will face a profound change and innovation, which means that foreign language teachers will be freed from the traditional language knowledge and simple listening and speaking teaching, but also the "gold content" of university foreign language teaching. Traditional foreign language classrooms may face a technological revolution. In future foreign language classrooms, foreign language teachers will actively guide students to engage in autonomous learning and exploration, allow classroom interaction and dialogue, and integrate teaching guidance and research and analysis into a single teaching resource space. This will bring huge challenges to the existing teaching model, curriculum system [8].

5.2 Establish a One-Stop System for English Teaching, and Plan Overall English Teaching in Primary and Secondary Schools and Colleges

The "one-stop" plan for English teaching in universities, primary and middle schools is a new orientation for the development of China's education in the new century. It is scientific and forward-looking. It refers to starting from a macro perspective and standing at the level of national English teaching to plan English teaching in all stages of our country, that is, to carry out unified design, planning and operation of English teaching in schools across the country as a whole. "One-stop" system of teaching advocates to fully reflect the teaching characteristics of "students as the main body", teachers, students, teaching materials are the three elements of classroom teaching. Obviously, the element of students is even more critical. Therefore, in the future, one-stop system design of English teaching in colleges, middle schools, and primary schools must be carefully observed and analyzed to understand their actual needs and foreign language skills. Based on this, we set appropriate textbooks and teaching methods. The foreign language quizzes and examinations of universities and middle schools should be different, but they must be well connected, otherwise it is difficult for college students to get out of the shadow of the exam. Only when these aspects are effectively connected, can we build a practical and adaptable one-stop system, and avoid unpracticed action to a certain extent [9].

In order to better develop in the future, education departments and schools at all levels must truly recognize the importance of good English teaching in middle and primary schools. In addition, college and middle school English teachers need to further strengthen scientific research and earnest exploration in teaching practice, strengthen exchanges, in order to understand the current situation of each other's education in a timely manner, find out the problems existing in each stage, and work out effective connection methods.

5.3 English for Special Purposes as the Teaching Direction, and Train Professional Composite Talents with Strong English Proficiency

English for Special Purposes (ESP for short) is the development direction of public English teaching reform in Chinese universities. This teaching mode of setting college

English courses according to specific majors changes a single language skills training, avoids the waste of teaching resources, promotes the cultivation of compound talents, and has great significance for the current reform of public English teaching. We should pay attention to needs research, clarify the curriculum teaching goals and teaching outlines, formulate a scientific and comprehensive curriculum system and novel teaching methods, and promote language extension to various professions. ESP teaching in colleges and universities is an important part of language teaching in terms of its fundamental properties. English courses, which are guided by specific goals, should strictly observe the actual language requirements and instrumental guidelines. According to the analysis of students' internal learning motivations, learning goals, and future professional language requirements, ESP teaching can implement targeted language training, establish a comprehensive classroom with knowledge, vividness, and applicability, and promote the cultivation of talented language. The formulation of the teaching goals and teaching plans of the ESP curriculum must be based on the actual needs of the students, and a comprehensive analysis must be made. Not only should we focus on the professionalism and standardization of English knowledge, but also strengthen the practicality and pertinence of language skills. In the curriculum arrangement, we must strengthen the comprehensive use of language as the core of teaching, attach importance to the organic combination of English and other majors, and also include training in cross-cultural communication to ensure the professionalism of the English curriculum system for special purposes, diversity and instrumentality. In addition, the course design must implement the stable start of the combination of language and major in accordance with the students' real language ability and the actual difficulty of the major, and carry out the teaching in each link step by step, and carefully observe and counsel the dynamic process of student learning to ensure students can gradually migrate from the consolidation of professional knowledge to interdisciplinary, interdisciplinary learning and knowledge integration in their course learning. Colleges and universities must, in accordance with their own school-running philosophy, based on needs analysis, proceed from the actual situation of students and the development of disciplines, and reform and develop aspects such as training goals, teaching theory, curriculum settings, teaching methods, textbook compilation, teacher construction, and evaluation methods to improve ESP teaching, and gradually form a systematic, scientific and developmental teaching system to cultivate practical senior winter talents [10].

5.4 Optimize the Evaluation System, Focus on Formative Evaluation, and Establish a Comprehensive Evaluation System Based on Artificial Intelligence

Evaluation is part of the teaching process and runs through it. According to evaluation methods, teaching evaluation can be generally divided into two types: "summative evaluation" and "formative evaluation". Most teachers mainly use final evaluation in their daily college English teaching, and occasionally use formative evaluation. A few teachers rarely use formative evaluation. With the continuous development of artificial intelligence technology, most of the final evaluation models in the form of examinations or assignments can be replaced by artificial intelligence evaluation. The assessment of objective questions in the college English test can be completely reviewed by artificial intelligence, and it also shows the superiority that manual evaluation such as efficient, accurate and objective cannot reach. Some artificial intelligence software has also implemented the review of subjective questions such as translation and composition. However, there are still some shortcomings in the subject analysis, ideological cognition, and context understanding. However, the auxiliary evaluation of artificial intelligence can still judge many objective language problems [11].

Intelligent detection provides students with a convenient and efficient platform and means for independent English assessment. Students can test their English learning anytime, anywhere through their smart phones, computers or other digital devices with smart detection software installed. This not only saves teachers' working time, but also provides timely feedback for students' learning. Teachers can use the intelligent platform to post assignments and ask students to complete them within a specified time. At the same time, they can also complete assessments automatically through smart devices.

Although artificial intelligence evaluation has many advantages, it still cannot completely replace teacher evaluation. The evaluation of college English teaching aided by artificial intelligence must involve teacher evaluation in the review of subjective questions; otherwise the evaluation results may not be fair and comprehensive [12].

6 Conclusion

In the current period of important opportunities for the strategic development of the country and universities, university foreign language teaching reform and planning cannot be simply positioned as a basic public course, but its value should be repositioned. Through the top-level design of university language education planning, actively innovate the curriculum system and follow policies. From the system innovation, language and culture, science and technology and discourse planning, etc., the state carry out university foreign language teaching planning, to create a distinctive feature in the new era, to connect and serve the national strategy and university development overall university foreign language teaching system to create favorable conditions. College English teaching under the background of artificial intelligence + big data will produce great changes. The application of these two technologies in English teaching can restore students dominant position in the classroom, stimulate students' interest in learning, and help students find ways to adapt to themselves. At the same time, through the use of network platforms, software, etc., it can break the constraints of time and space, enable students to learn knowledge in more scenarios, and improve the teaching quality of English courses. Facing such changes, college English teachers must also strive to improve their overall quality and keep up with the development of education.

Acknowledgments. This study was supported by the "Thirteenth Five-Year Plan" 2020 Project of Jiangxi Provincial Education Science (University Series): Research on the Countermeasures for the Development of College English Education from the Perspective of Language Education Planning (20YB221).

References

1. Qi, S.: University foreign language teaching reform from the perspective of language planning. Foreign Lang. Teach. **2018**(06), 49–53 (2018)
2. Cai, J.: Reflections on 40 years of English education in Chinese universities: failures and lessons. J. Northeast Normal Univ. (Philos. Soc. Sci. Ed.) **000**(005), 1–7(2017)
3. Qi, S.: Chinese discourse planning: a new task for language planning in the construction of a community of human destiny. Appl. Lang. Writ. **112**(04), 35–43(2019)
4. Wang, S.: Change concepts and deepen reforms to promote the new development of foreign language teaching in universities. China Univ. Teach. **000**(002), 59–64(2017)
5. Shen, Q., Wei, H.: China's foreign language strategic planning from the perspective of building a community with a shared future for mankind. Foreign Lang. Circles **000**(005), 11–18 (2018)
6. Shu, D.: Thoughts on the planning and layout of foreign language education in my country. Foreign Lang. Teach. Res. **45**(003), 426–435(2013)
7. Zhao, S., Zhang, D.: The internationalization trend of language planning: a new field of language communication and competition. Foreign Lang.: J. Shanghai Int. Stud. Univ. **2012**(04), 2–11 (2012)
8. Lulu, H., Lin, C., Mengmeng, S.: Research on artificial intelligence to promote English learning reform. Modern Distance Educ. **2019**(6), 29–33 (2017)
9. Bin, C., Zhongshan, T.: Analysis of the main problems existing in the teaching of english for specific purposes in universities and the countermeasures. J. Inner Mongolia Normal Univ. (Educ. Sci. Ed.) **230**(03), 104–107 (2018)
10. Luo, A., Cao, Y.: The construction of "one-stop" system for english teaching in primary and middle schools. J. East China Univ. Sci. Technol. (Soc. Sci. Ed.) **2008**(01), 77–80 (2008)
11. Wei, F., Ma, Q.: Foreign language education from the perspective of language education planning. Nankai Linguist. J. **2010**(1), 151–159 (2010)
12. Jiang, X.: Research on College English Education Development Strategy from the Perspective of Language Education Planning. Jilin University (2016)

Construction of Smart City Needed for a Better Life Under the Background of the Times—A Perspective of Holistic Governance Theory Based on Governance and the Structure of Local Political Institutions

Dan Li[(⊠)] and Kexin Yang

Faculty of Management, Wuhan Donghu University, Wuhan, China
`50444936@qq.com`, `245077962@qq.com`

Abstract. What we face now is the contradiction between unbalanced and inadequate development and the people's ever-growing needs for a better life. Technology changes life. Smart cities have a important theoretical significance and practical value in meeting and satisfying people's needs for a better life, improving the sense of gain, happiness and security. In view of this, the article harnesses the methods of literature, logic analysis, and field-participatory observation to follow the main line of "what, why, and what", and uses Perry Hicks's theory of overall governance as a theoretical tool. It is proposed that smart cities are oriented to meet people's differentiated needs, with responsibility, information and technology as the core elements, and integration and coordination as the operating path as the background, and proposed that the SMART city construction should be in the analysis of the city PEST and SWOT, following the concept of SMART law. they followed the concept of the SMART rule in order to provide reproducible and generalizable theoretical experience and operation mode for the construction of smart cities.

Keywords: Smart city · Integrity · Governance

1 Introduction

The theme of the 2010 Shanghai World Expo is "Better City, Better Life". It is conceivable that why cities make life better is technology changes life. So what are smart cities, what are their characteristics, and how to build them [1]?

Perry Six's holistic governance theory points out that we should take meeting the public's needs as the guide, optimize the information system, improve the responsibility system, enhance the trust relationship, integrate the resources and institutions of all departments, and coordinate the relations of all aspects to achieve integrated and holistic governance.

The article harnesses the methods of literature, logic analysis and field-participatory observation to follow the main line of "what, why, and what",and explores the theory of holistic governance in the construction of smart cities.

J. MacIntyre et al. (Eds.): SPIoT 2020, AISC 1283, pp. 487–493, 2021.
https://doi.org/10.1007/978-3-030-62746-1_72

2 Password Mining: The Meaning of the Smart City

For the definition of a smart city, scholars at home and abroad have defined it from different angles [2]. The article refines its public elements and tentatively proposes that a smart city is the use of big data information, with science and technology as the core carrier, in order to meet the differentiated needs of people [3]. Big data information makes people's lives simple and convenient through integration and coordination.

3 The Status Quo Sketch: Problems Existing in the Current Smart City Construction

The first person to eat crabs is a hero. However, after the smart city was proposed, there are several problems in its current construction.

3.1 Construction of Artificial Fengshui Mechanically Copied Suit

There is no one-size-fits-all approach, which means that the construction of it requires attention to its ecology. However, in the construction of smart cities, some cities imitate or directly copy other urban construction paradigms, without considering whether the city has the prerequisites for building a smart city [4]. For the sake of political achievements, the project was built blindly, and the project was launched blindly, and even three shots of smart city construction appeared. Head shot—brain storming decision-making to build a smart city, shot chest—blindly confident the construction of a smart city, buttocks—Irresponsible, make the construction of smart cities end and shirk responsibility.

3.2 Fragmentation Lack of Integrated Construction

When the government departments make decisions, there are often many political outlets among the relevant departments. The construction of a smart city needs the unified coordination, overall layout and integration of multiple institutions and departments [5]. However, it is difficult to form an integrated force because there are many political departments, information is not shared between them, and they fight alone. There is no unified paradigm, which leads to the diversification of supply mode, supply subject and supply channel when a smart city provides people with relevant information or services.

3.3 Game Imbalance Between Public Interest and Self-interest

However, according to Buchanan's theory of government failure, the government will rationally pursue the maximization of its own institutions or personal interests based on the assumption of economic humanity, and the phenomenon of bloated institutions and even rent-seeking will occur. Some projects are not conducive to the construction of smart cities. However, the pursuit of the public interest of most people has been neglected due to the increase in political performance or the number of posts in the department, the increase in ranks, and the increase in funding [6]. Failure to fulfill the original intention and responsibility of making people's lives better, abiding by the spirit of the contract, seeking happiness for the people, and seeking development for the national rejuvenation.

3.4 Deregulation of Processes Leads to Bad Ends

The construction of smart city must follow the reasonable process. For example, some projects need to carry out corresponding feasibility analysis, estimates, estimates and budgets, but they are not launched before they are launched. Some projects require land acquisition and construction in accordance with the "first survey, Post-design, and re-construction" while applying for construction-related procedures. For the construction of smart cities, some need to invest a large amount of money in the construction of hardware and software. However, project financiers and investors may not be satisfied with the investment ecology in the construction, which leads to the withdrawal of investment. For all sorts of reasons, there was a break at the end-the tail was bad.

4 Optimization Mechanism: The Construction of Smart Cities Requires External PEST and Internal SWOT Ecological Analysis

The ancients said: orange planted in the north is sweet while it is bitter if planted in the south. This shows the importance of the ecological environment. To build a smart city, we should take into account the above situation of Rote Copying, fragmentation, ignoring public interests and procedures, etc. [7]. It should be optimized from the ecological point of view, aiming at the cities that need to build smart cities to carry out the internal and external ecological interpretation.

4.1 External Ecology—Exploration of the Proposed Smart City PEST

PEST theory is an abbreviation of the four English words of politics, economics, society, and technology. It means that we should pay attention to the analysis of political system, economic atmosphere, social culture and technical support when we analyze the external environment of things.

1) *The Political system excavation of the proposed smart city*

Whether a city's policy is permitted and whether there is a corresponding policy inclination for the construction of a smart city. It must be carried out to the extent permitted by the rule of law. Because politics is the division of interests, and the construction of a smart city will have different impacts on the relevant interests, it is necessary to confirm whether the proposed city has the system of smart city construction.

2) *Analysis of the economic atmosphere of the proposed smart city*

First of all, the construction of smart city needs to invest a lot of money to build hardware and software. Therefore, whether the city to be built as a smart city has a certain economic strength is the so-called "strength without advancement". And after making corresponding investment estimates, design estimates, construction budgets and investment payback periods, investment returns and other trade-offs, make correspond-ing decisions. Secondly, analyze the existing economic development status, industry characteristics, industrial economic conditions, and economic prospects of the city, and whether it is necessary to make a smart city construction in this city, and whether the smart city construction will give the city a better positive direction, externalities, etc. Finally, the question of supply and demand needs to be considered. If the smart city is built and put into use, the services provided by this smart city need to take into consideration the income and consumption levels of qualified cities, otherwise there will be no price, no service, or Equipment and facilities become furnishings [8].

3) *Socio-cultural research of the proposed smart city*

Generally speaking, cities have their own cultural heritage, which is a historical and cultural heritage. The proposed smart city needs to include the city's customs, consumption habits, etiquette and local customs, total population and distribution to the feasibility analysis elements. For example, this city has a better traditional cultural background and is worthy of preservation and protection. It should not cost much time to the break ground and start building a smart city, which has affected the original traditional cultural heritage.

4) *Analysis of technical support for the proposed smart city*

It is true that science and technology are the primary productive forces, which change the means of production [9]. The construction of a smart city needs certain technical support and guarantee. It is just the so-called "no fine steel drill, no porcelain work". For a proposed smart city, it is not to say that high-rise buildings are built on the ground, but to comprehensively study various factors, especially in terms of technical support, whether it is only necessary to upgrade and optimize the existing technical level of the city [10–12]. If it costs a lot of money to introduce and build a new City, it must take into account the input-output ratio. And, after the completion, whether the general public accepts the smart city, this acceptance refers to the ability to operate with this technology, such as Alipay interface operations, and related services of smart cities are embedded in Alipay related operations. But while some older people hold smartphones, some are still too weak to operate Alipay. Therefore, the construction of smart cities also need to take into account China's aging population operating technology.

4.2 Internal Ecology—Consideration of SWOT for Smart City

Does the city have the advantages to build a smart city, what are its advantages, disadvantages, opportunities and threats?

1) *Explore the advantages of the proposed smart city*

What is the core competitiveness of this proposed smart city, and whether it has the icing on the cake, taking it to the next level for smart city construction and promoting its more competitiveness. This requires a complete understanding of the city's signature and core strengths.

2) *Disadvantages of the proposed smart city*

It is the so-called confidant knowing one after another. Where is the shortcoming of the proposed smart city, and whether the construction of the smart city will bring optimization and improvement to the shortcoming, or will it directly step in the pit or adopt an avoidance mechanism? This makes the construction of smart cities more feasible.

3) *Opportunity analysis of the proposed smart city*

In view of the smart city project in this city, we need to consider the smart city construction in this city, what development opportunities will be obtained, what is the probability of success of the smart city construction, and what is the cost of the opportunity to spend a lot of money to build the smart city. We have to consider these opportunities. According to the theory of

$$M = E * V \tag{1}$$

"(1)" is clear that there are large opportunities or more opportunities, high expectations, and optimistic benefits, so the motivation for smart city construction is stronger.

4) *The threat of the proposed smart city*

According to the jungle law, the city will build a smart city based on the survival of the fittest. From its own analysis, what threats will the city face, the competition of the next city, what changes will be made to the industry of the city itself, whether there will be technical unemployment, whether there will be resistance or opposition from people, etc.

5 Operation Strategy: SMART Concept Should Be Followed in the Construction of a Smart City

The principle of SMART means that we should pay attention to the specific of performance index, measurable, attainable, relevant and time-bound of performance appraisal. I think that smart city construction may take smart rules into consideration.

5.1 The Construction of a Smart City Needs to Integrate Resources and Be Clearly Positioned

We don't pursue one city, one product, but we must integrate the city's own resource advantages to build a smart city for this city, and its positioning needs to be clear. For example, the core resource of Xianning City in Hubei Province is the hot spring, so the construction of the Xianning smart city needs to consider and locate the optimization and integration of hot spring related resources, "put intelligent wings on the hot spring", so that tourists can better enjoy the hot spring service.

5.2   The Construction of Smart Cities Must Be Quantifiable

For the construction of a smart city, then, how much capital needs to be invested, how long the construction period is, how much risk probability is faced, how much the investment payback period and investment return rate are, how much the popularity and reputation of the city will be improved, how much the investment environment will be improved, etc. It must be supported by data, and can't be carried out blindly, purposelessly and in a hot way. It can be concluded that, the construction of it needs to quantify human, material, financial and investment income.

5.3   Implementation of the Construction of a Smart City

Adhere to the principle of hop-by-hop, which means that some goals are within their ability, and a little bit more breakthrough can be achieved. Similar to a peach tree on the side of the road, without a front condition that is not suitable for outside tools, a person jumps hard and can pick a peach that is taller than his own height. By analogy, the construction of it should pay attention to the feasibility of the operation in the city and the construction capacity of the city.

5.4   The Construction of Smart Cities Must Coordinate the Relationship Between All Parties

The construction of a smart city involves coordination and cooperation in politics, economy, culture, technology and other aspects. Many government departments, such as land and resources administration, development and Reform Commission, transportation, education, finance, public security, Taxation Bureau and Water Affairs Bureau, can be said to be a big action to launch the whole body. All departments should coordinate with each other, take unified actions and pay attention to their own Role, to prevent dislocation, offside, absence of the phenomenon. In this interconnected and interacting related parties, we need to control the overall governance, communicate information smoothly, and integrate the communication platform with science and technology.

5.5   The Construction of a Smart City Should not Neglect the Guarantee of Time

The last rule of cleverness is to have a sense of time. First of all, the development of science and technology is changing with each passing day. Then, the sooner the construction of a smart city is completed, the sooner people can enjoy the fruits of development, the advancement and convenience of science and technology. From this perspective, when the construction of a smart city needs to be completed, this must be clarified. Second, for project financing parties, efficiency is life and time is money. Investors also consider the time value of funds. Finally, for the governors, there is also a time limit for completing the corresponding performance projects during the term of office. Therefore, when it is completed is a ballast stone for the construction of smart cities.

6   Reflective Discussion

The construction of a smart city is a new thing. In the process of urbanization and the process of Chinese people's pursuit of a better life, theoretical exploration of it has certain value. The article incorporates Perry Six's overall governance theory into research on the construction of smart cities, and tries to propose that smart cities are oriented to meet people's differentiated needs, supported by science and technology, integrate big data, and pursue a responsibility system. The enhancement of trust relationships, the optimization of information systems, the integration of resources from all parties and the coordination of various relationships make it easier and simpler for people's lives. In response to the emergence of rigidity, fragmentation, neglect of public interests, and irregular processes, we constructed an internal and external PEST and SWOT ecological analysis of the proposed smart city, and further promoted the construction of smart cities under the SMART concept of smart rules. Provide certain theoretical support and practical direction for the construction of smart cities.

References

1. Saracevic, M., Adamović, S., Macek, N., Elhoseny, M., Sarhan, S.: Cryptographic Keys Exchange Model for Smart City Applications. IET Intelligent Transport Systems (in Press, 2020)
2. Kumar, H., Singh, M.K., Gupta, M.P., Madaan, J.: Moving towards smart cities: Solutions that lead to the Smart City Transformation Framework. Technological Forecasting & Social Change (2018)
3. Yan, T., Lu, Z., Jing, L.: A comparative study on participatory microregeneration of old communities: from the perspective of social capital. China City Plann. Rev. **28**(04), 50–58 (2019)
4. Liu, J., Chaowen, C.A.O.: The application of old city planning based on the concept of smart cities: a case study of Wanli District, Nanchang. J. Landscape Res. **10**(06), 17–20 (2018)
5. Ok, J.A., Yo, S.: Directions and improvements of the future smart city development: a case of Gyenggi province. Spatial Inf. Res. **25**(2), 281–292 (2017)
6. Breeden II., J.: How cities are getting smarter. Nextgov.com (2019)
7. Dong, J.: Operation mode of smart cities: a comparative study. Ecol. Econ. **14**(03), 190–199 (2018)
8. Hurrah, N.N., Parah, S.A., Loan, N.A., Sheikh, J.A., Elhoseny, M., Muhammad, K.: Dual watermarking framework for privacy protection and content authentication of multimedia. Future Gener. Comput. Syst. **94**, 654–667 (2019)
9. Axelrod, J.: Roadmap to a smart city master plan. The American City & County (2019)
10. Shah, J., Kothari, J., Doshi, N.: A survey of smart city infrastructure via case study on New York. Procedia Comput. Sci. **160** , 702–705 (2019)
11. Alaa, A., Vangalur, A., Zaki, C., Nematollaah, S.: Characterization and efficient management of big data in IoT-driven smart city development. Sensors (Basel, Switzerland) **19**(11) (2019)
12. Anonymous. Stantec report hails smart cities for many benefits. Daily Commercial News **92**(193) (2019)

Investigation on the Growth Mechanism of Small and Medium-Sized Agricultural Leading Enterprises from the Perspective of Intelligent Agricultural Niche Theory

Haohao Li[✉] and Nan Zhou

JiangXi Institute of Economic Administrators, Nanchang, China
37246509@qq.com, 185698223@qq.com

Abstract. Small and medium-sized agricultural (SMSA) leading enterprises in the economic development and agricultural development more and more prominent position. Especially in recent years, with the increasing policy support, SMSA leading enterprises are playing an extremely important role in promoting agricultural development and protecting agricultural ecology. The purpose of this paper is to find out the problems existing in the development of SMSA leading enterprises from the perspective of intelligent agricultural niche theory, and summarize the scientific and efficient growth mechanism of agricultural leading enterprises. In this paper, the related theoretical concepts of smart agriculture and niche theory are elaborated, and then the definition, function and ecosystem of agricultural leading enterprises are briefly explained. Then, by means of experimental investigation, this paper analyzes the growth and development status of China's SMSA leading enterprises from the perspective of intelligent agricultural niche theory, and finds out the problems existing in the development of China's SMSA leading enterprises. Finally, combining with the ecological niche theory of intelligent agriculture, a scientific growth mechanism is proposed to promote the development of SMSA leading enterprises in China. The experiments of this paper show that there are many problems in the development of SMSA leading enterprises in China, which are contrary to the ecological niche theory of smart agriculture.

Keywords: Smart agriculture · Niche theory · Agricultural leading enterprises · Growth mechanism

1 Introduction

SMSA leading enterprises occupy an important position in China's. According to the data, during the 12th five-year plan period, the number of households of SMSA leading enterprises increased by about 8%, and the economic contribution rate increased by about 9.27%. However, due to the economic reform in China and the limited development capacity of enterprises, most of the SMSA leading enterprises in China have gone through the phenomenon of ecological niche decline, which hinders the sustainable and healthy

J. MacIntyre et al. (Eds.): SPIoT 2020, AISC 1283, pp. 494–499, 2021.
https://doi.org/10.1007/978-3-030-62746-1_73

development of agricultural leading enterprises. Therefore, it is of great significance to study the growth mechanism of SMSA leading enterprises from the perspective of intelligent agricultural niche theory.

The deepening of ecological concept, the theory of intelligent agriculture and ecological niche has been the focus of domestic and foreign scholars in recent years. In [1], the author firstly introduces cloud computing into agricultural analysis based on Internet of things and Internet technology, and establishes a cloud platform of intelligent agricultural Internet of things, which promotes the realization of digitalization of agricultural information, production automation and intelligent management. In [2], taking the acceptability and enforceability of enterprises as the starting point, combined with the UTAUT model, the author deeply analyzed the main factors influencing the development of smart agriculture, and based on this, built a smart agriculture transformation path model suitable for SMSA enterprises. In [3], the author first sorts out the development history of niche theory, puts forward the enterprise ecology theory.

Because SMSA leading enterprises occupy an important position in economic development, Chinese scholars have also conducted a series of studies on the development of SMSA leading enterprises. In [4], the author puts forward that the sustainable development of SMSA leading enterprises should start from the organization, management and operation, so as to continuously promote the improvement of enterprise scale and market share, and enhance the acquisition capacity of enterprise resources. In [5], the author firstly analyzes the main factors influencing the development of SMSA leading enterprises by means of factor analysis, and then puts forward the development strategy of constructing the growth of agricultural enterprise(AE) clusters based on the current situation and development background of China's agricultural development.

In order to summarize the scientific and efficient growth mechanism of agricultural leading enterprises and realize the sustainable development of AEs in China, this paper firstly elaborated the relevant theoretical concepts of intelligent agriculture and niche theory, and then briefly explained the definition, function and ecosystem of agricultural leading enterprises. Then, by means of experimental investigation, this paper analyzes the growth and development status of China's SMSA leading enterprises from the perspective of intelligent agricultural niche theory, and finds out the problems existing in the development of China's SMSA leading enterprises. Finally, combining with the theory of intelligent agricultural niche, a scientific growth mechanism suitable for the development of leading enterprises in SMSA in China is proposed [6, 7]. The research in this paper not only promotes the continuous improvement of the growth mechanism of the leading enterprises of SMSA in China, but also lays a theoretical foundation for the related research in the future [8].

2   Method

2.1   Perspective of Intelligent Agriculture and Ecological Niche Theory

With the continuous improvement of Internet technology, traditional agricultural production has fallen behind. The combination of agricultural production and information technology has given birth to smart agriculture. Smart agriculture is the inevitable result of agricultural production and scientific and technological progress, and belongs to the agricultural production in an advanced stage [9]. Intelligent agriculture information such as the Internet, the Internet of things, cloud computing technology as support, placed at the scene of the agricultural production sensor nodes, and with the help of wireless communication technology, agricultural production condition of intelligent monitoring, early warning and decision-making and analysis, experts according to the monitoring results of the analysis provide online guidance of agricultural production, and finally realize the precise planting of agricultural production, greatly improve the efficiency of agricultural production [10, 11]. In essence, it is a biological concept, which is generally used to discuss various relations between organisms and the environment, and is the embodiment of the competition and cooperation phenomena existing in organisms and their growth space. Niche theory mainly refers to the interaction between the spatio-temporal position occupied by a fixed population in an ecosystem and other related populations. Niche theory covers a comprehensive range of contents, including niche theory overlap and separation, construction, schematic measurement and situation theory. There is a close relationship between smart agriculture and niche theory. On the one hand, niche theory is the theoretical basis for the development of smart agriculture; On the other hand, the development of smart agriculture promotes the continuous improvement of niche theory [12, 13].

2.2 Leading AEs and Their Ecosystem

According to the relevant documents of our country AEs, agricultural leading enterprises mainly refers to processing and circulation of agricultural products, with the help of interest to establish the contact between the farmers, realize the integration of development of agricultural production, processing, marketing, and management scale and the indexes reached the standard of stipulating in the whole AEs and relevant government departments. Agricultural leading enterprises have played an important role in the overall social and economic development, mainly in the following aspects: Second, the promotion of agricultural science and technology level, so that the practical application of agricultural technology is expanding; Thirdly, the sustainable development of rural economy can be realized by promoting the continuous improvement of farmers' income level. This paper attempts to introduce the niche theory from the natural ecosystem to the AE. From the perspective of niche theory, research on agricultural leading enterprises is conducive to in-depth analysis of the relationship between agricultural leading enterprises and the production environment, in-depth grasp of the development law and trend of agricultural leading enterprises, and provide positive and beneficial guidance for the exploration of future growth mechanism of SMSA leading enterprises.

3   Investigation and Experiment on the Development of SMSA Leading Enterprises

In order to study the growth mechanism of SMSA leading enterprises, it is necessary to grasp the current development status of SMSA leading enterprises. First, the collection of raw data on AE development. This paper visited the website of China's agricultural economy, the database of Chinese enterprises and other websites, and also consulted relevant research data in the databases of cnki and wanfang. Secondly, the question-naire survey about the development of SMSA leading enterprises. In order to grasp the development status and growth mechanism of SMSA leading enterprises, this paper collected data by means of questionnaire survey. The respondents included personnel from relevant departments of the state, department leaders of SMSA leading enterprises, and experts and scholars in this field. The contents of the survey involved the enterprise development mode, development effect, fit with the ecological niche theory of smart agriculture and other aspects. A total of 1000 questionnaires were issued and 974 valid questionnaires were collected. After the questionnaire survey, SPSS software was used to conduct statistical analysis on the questionnaire survey data. Finally, the comparative analysis of the data, the conclusion of the development of SMSA leading enterprises.

4 Discuss

4.1 Experimental Results and Analysis

The following experimental data can be obtained according to the above investigation experiment on the development of SMSA leading enterprises. The specific experimental data are shown in Fig. 1 and Table 1 below. The data in the chart are the results of the author's investigation and arrangement.

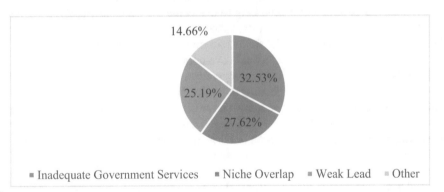

Fig. 1. Problems existing in the growth mechanism of SMSA leading enterprises from the perspective of niche theory

As can be seen from the data in Table 1, the development level of leading enterprises in SMSA in China is relatively low at present. The development efficiency and speed are all around 50%, which is only half of the desired effect. Moreover, the development speed remains around 15%, which belongs to low-speed development. Most importantly, the fit between the development of leading enterprises in SMSA and niche theory is low at the present stage, which is only 45.12%. As can be seen from Fig. 1, there are many problems in the growth mechanism of SMSA leading enterprises from the perspective of niche theory, mainly focusing on three aspects: insufficient government services, ecological niche overlap and weak leading role of leading enterprises, all of which involve the

Table 1. Development data table of China's leading SMSA enterprises at present stage

Project	Data	Development trend	Niche theoretical fit
Development speed	54.17%	16.3%	45.12%
Development model	49.62%	12.4%	
Development effect	53.17%	10.9%	

*Data were collected from questionnaires and documents

enterprise ecological environment system theory in niche theory. The emergence of these problems hinders the sustainable development of SMSA leading enterprises in China.

4.2 Improvement Strategies for the Growth Mechanism of Small and Medium-Sized Leading AEs

First, we should promote the transformation of government functions and strengthen services for leading AEs. To this end must do the following points: first, to promptly implement the relevant AE development policy, and pay attention to the leading enterprises. On the one hand, AEs should be provided with a good farm environment. On the other hand, the government should help to overcome the unstable factors in the environment. Second, the government established a system to ensure the development of leading AEs. For example, special funds for industrial operation and risk funds should be established according to local actual development conditions, and loans should be provided for leading AEs. Third, build cooperation platforms. For example, establishing agricultural cooperatives, strengthening the connection between farmers and enterprises and forming a community of development interests can better promote the development of enterprises. Secondly, avoid the excessive overlap of ecological niche and promote the continuous optimization of agricultural industry pattern. The excessive overlap of enterprise niches is to increase the competition of enterprises and cause the disorder of agro-enterprise ecosystem. In order to reduce the occurrence of this phenomenon, on the one hand, it is necessary to conduct dislocation management of enterprise ecological niche. Enterprises should make full research before agricultural product production to avoid the overlap between the products produced and the existing products in the market. On the other hand, enterprises should highlight the focus of production and form an industrial pattern. Finally, in order to give full play to the leading role of leading AEs, enterprises must strive to improve the level of science and technology, efforts to expand the scale of production, promote the continued development of export-oriented economy.

5   Conclusion

The growth and development of SMSA leading enterprises are related to the development of the overall social economy. Based on the research on the growth mechanism of SMSA leading enterprises from the perspective of smart agricultural niche theory, the following conclusions are drawn:

(1) The development of leading AEs plays an important role in the improvement of the distribution of agricultural industry, the improvement of agricultural science and technology and the promotion of farmers' income.

(2) From the perspective of smart agricultural niche theory, the problems in the growth mechanism of SMSA leading enterprises mainly focus on three aspects: insufficient government services, ecological niche overlap and weak leading role of leading enterprises.

Acknowledgement. This work was supported by Soft Science Research Base Project of County-level Economic Competitiveness in Jiangxi Province: A Study on the Growth Mechanism of Leading Small and Medium-sized Enterprises from the Perspective of Niche—Based on the Research of Agricultural Development in Jiangxi Province (NO. 2019RKXYB001)

References

1. Mao, Y., Wang, K., Tang, C.: Application practice analysis of Internet of things technology in modern agriculture at home and abroad. Jiangsu Agric. Sci. **11**(4), 412–414 (2018)
2. Gao, L., Gao, L.: The practice and enlightenment of developing "smart agriculture" in the United States to promote the transformation of agricultural industry chain. Econ. Vertical Horizontal **14**(12), 120–124 (2017)
3. Yang, D.: Development strategy of China's smart agriculture industry. Jiangsu Agric. Sci. **13**(4), 1–2 (2017)
4. Zhang, J.: Development path and guarantee of smart agriculture in China. Reform Strategy **15**(6), 104–107 (2017)
5. Wang, X., Deng, C.: Development strategy and path of smart agriculture in China based on "Internet+". Jiangsu Agric. Sci. **24**(16), 312–315 (2017)
6. He, H., Zhang, Y.: Evaluation model of smes' transformation and upgrading capability based on niche theory. Enterp. Econ. **11**(5), 117 (2017)
7. Yanan. Research on enterprise niche and its dynamic selection. J. Southeast Univ. **14**(7), 62–64 (2017)
8. Zhu, J.: Research progress of niche theory and its measurement. J. Beijing Forest. Univ. **14**(1), 116–119 (2017)
9. Xu, F., Li, J.: Research on enterprise niche principle and model. China Soft Sci. **9**(5), 130–139 (2017)
10. Qian, Y., Ren, H.: Research on the competitive relationship of enterprises based on ecological niche. Financ. Trade Rese. **13**(2), 123–127 (2017)
11. Liu, Z.: Theoretical basis and path selection of enterprise niche optimization. Enterp. Econ. **15**(3), 117–119 (2017)
12. Cao, Y., Liu, Z.: Research on the basic connotation, prominent problems and countermeasures of enterprise niche optimization. Future Dev. **3**(6), 142 (2017)
13. Liu, K., Liu, Y.: Analysis of "leading enterprise + peasant household" model in agricultural industrialization. South. Agric. Ind. **31**(12), 148–150 (2018)

The Optimization and Construction of College English Teaching by Information Technology

Jun Chen[✉]

Fuzhou Melbourne Polytechnic, Fuzhou, China
chenjun85info@163.com

Abstract. With the popularization and application of modern technology, teachers from colleges and universities integrate various information technology into the English teaching in college. The teaching methods which are based on the modern technology have the advantages of rich teaching resources, strong interaction, convenient and rapid data processing, etc. By the use of all sorts of technology, like the mobile terminals and various online teaching software and platforms, the whole teaching process can truly be optimized. Through the optimization and construction of teaching goals, tasks, contents, forms, methods and ways of evaluation, the classroom teaching effect is improved. It also improves the comprehensive application ability of students in English and promotes the reform of English teaching in college.

Keywords: Information technology · Optimization strategy · Teaching process · College english

1 Introduction

In China, many people learn English from primary school, but there are still problems in the communication and use of English in real life after years of learning. College English teaching reform has been actively carried out by the universities in China. In recent years, teaching modes such as hierarchical teaching, flipped classrooms, Moocs, micro-classes, etc. have attracted wide publicity from college English teachers. Teachers have actively worked on the effective ways to improve the effect and efficiency of College English teaching.

With the in-depth study of teaching modes such as flipped classrooms, Moocs, micro-classes, teachers have found that these innovative teaching modes are inseparable from the help of modern technology and the comprehensive and creative use of online and offline materials. Therefore, how to build an information technology platforms, how to tap the strengths of the technology in teaching, and set up the most reasonable and optimal teaching process is the key to promote the effects of English teaching.

J. MacIntyre et al. (Eds.): SPIoT 2020, AISC 1283, pp. 500–506, 2021.
https://doi.org/10.1007/978-3-030-62746-1_74

2 Problems Existing in the Teaching Based on Information Technology

The modern technology has been widely applied to the teaching in universities. But there are some issues that cannot be ignored, such as emphasizing technology over practical application, ignoring teaching content itself, etc. [1] Teachers are busy using various and mixed information technology and often overlook the teaching effects. However, using information technology itself is not equivalent to better teaching effects. The following will analyze the existing problems from four aspects.

2.1 The Software and Hardware Facilities Applied in Information Technology-Based Teaching Can not Keep up with Each Other

Many colleges and universities' information technology-based teaching started late, failing to provide teaching and learning with the optimal environment required for the blended teaching. Information technology-based teaching needs to combine information technology, Internet, multimedia, etc. with practical English course teaching [2]. If the school cannot create the optimal conditions for information technology-based teaching and learning, cannot build the hardware environment such as networks, servers, and mobile terminals, and cannot provide teachers and students with information technology-based resources such as online courses, as well as the online teaching software and platforms for auxiliary teaching and automatic feedback of various data, then the information technology-based teaching mode will be difficult to continue.

2.2 More Attention of Teaching Is Paid for Using the Information Technology Than the Teaching Content Itself

During the teaching, teachers attach more importance to the use of various teaching software and platforms. However, due to the simple mixture and blindly piling up of information technology, many problems have been caused, such as the emphasis on technology over application, the emphasis on form over content, and the emphasis on skills over application abilities, etc. Teachers often ignore the actual teaching effects, the delivery of optimal teaching content, and the comprehensive development of students [3].

2.3 Lack Capacity for Teaching Resource Production and Integration

The teaching by using modern technology has greatly tested teachers' capability to reconstruct course materials and content [4]. Many teachers lack the awareness or experience of using modern technology while teaching, lack the ability to utilize modern technology to produce, integrate or process teaching materials, and fail to tap the strengths of existing information technology platforms. For example, traditional paper-based textbooks cannot be produced and integrated into flexible and easy-to-use e-books for students.

2.4   Lack Capacity to Collect, Analyze and Process Data

Most of the college teachers still fail to use online teaching software or other information technology to collect, process, and analyze the feedback. Nowadays, most of the data collected by teachers are the students' phased or final test results, lacking evaluation data on students' learning attitudes and learning processes. The acquisition of these data requires the help of various information technology software. Without the data collected, teachers can neither assess the learning progress of students objectively and comprehensively, nor can they further adopt practical, effective and optimized teaching methods to improve students' academic performance.

3   The Optimization of Teaching Process by Using Information Technology

The teaching process includes teaching conditions, teaching goals and tasks, teaching contents, forms, and methods, as well as teaching assessment, etc. [5]. With the aim of achieving better effects of teaching, teachers should tap the strengths of modern technology, optimize and construct the teaching process of College English courses.

3.1   Actively Create Internal and External Conditions Required for Information Technology-Based Teaching

The information technology-based teaching should rely on mobile terminals and various teaching software and platforms. It requires the cooperation from schools, teachers and students to construct and improve together. 1. For colleges, they should actively build the environment, resources and platforms required for information technology-based teaching for teachers and students, including hardware environments such as the Internet, servers, mobile terminals, smart classrooms, as well as software environments such as basic software, application software, and online teaching resources. Meanwhile, colleges need to update and improve the existing equipment, and create a high-quality external environment for the teaching activities. 2. For teachers, they need to break the traditional way of thinking, continuously strengthen their capacity to apply information technology in teaching, and bring the existing software and platforms into full play, such as Mosoteach, Superstar Learning Apps, So Jump Apps, quizlet Apps, and Pigai Apps [6]. Teachers should also enhance their practical capabilities, such as resource production and integration capabilities, data collection and processing capabilities, etc. 3. For students, they need guidance from teachers to help them learn to use the software to broaden the width and breadth of learning, and enhance the ability of self-learning, research and exploration. During the teaching, teachers should refrain from overusing various information technology, which will not achieve the optimal effect of teaching, but will greatly increase the burden on teachers and students.

3.2   Use Data Feedback from Teaching and Learning Software to Plan Teaching Goals and Tasks Rationally

Scientific and rational planning of teaching goals and tasks is the prerequisite and basis for achieving optimal teaching. By using modern technology, teachers can analyze the teaching goals and students' knowledge reserves more objectively, and then establish the teaching goals and tasks more accurately. Teachers can set questions and issue questionnaires to students through the online teaching and learning software. After students receive the questionnaires, they complete and submit them. The data will be automatically generated and analyzed by the software. Teachers can also design questions similar to brainstorming through the software and encourage students to actively participate in answering questions. Through the software platform, teachers can quickly view the answers of different students. Through the feedback of backstage data, and according to the views of students, teachers can optimize the overall planning of teaching goals and tasks.

3.3 Optimize and Refine Teaching Contents by Using Software and Internet Platforms

Use information technology to select and optimize teaching contents, achieve better learning effects, and reduce the burden on students. 1. Apply information technology to reconstruct and integrate teaching contents. Teachers can make the best of the online teaching and learning software to make e-books, which carry out in-depth excavation and processing of the contents and knowledge points of the original paper-based textbooks. The e-books can visually present the textbook content from multiple angles and dimensions, which is easy to understand for students [7]. Teachers can also integrate and intercept important contents in e-books according to actual teaching needs, and send them to students for learning through the teaching and learning software. The functions of bookmarks, notes and labels in e-books facilitate students' learning. 2. Utilize the online resources, optimize and innovate the teaching content of English courses by tapping the strengths of rich teaching resources, wide knowledge, huge amount of information and convenient search. College teachers can make the best of high-quality online resources, such as Chinese University MOOC, College Online Education Alliance, etc., to select high-quality learning resources for students to help students understand and overcome the difficulties in learning [8].

3.4 Set up Information Technology-Based Teaching Forms in a Reasonable Way

In view of the fact that English courses pay more attention to cultivating students' practical communication and application ability, a combination of small class teaching, group cooperation, and independent inquiry is more likely to achieve better teaching results by using modern technology [9]. 1. Small class teaching is more conducive to carry out the in-class activities. 2. On the basis of small class teaching, teachers make the best of the grouping function on the online teaching and learning software to divide the whole class into several groups for practice or competition by random. Then, the students complete the tasks assigned and upload them to the software groups by groups. After teachers and students check groups' tasks completed online, mutual evaluation between groups is carried out, or teachers give the incentive scores to groups according to their performance, with the aim to inspire students' interest and motivation in learning. 3. The online teaching and learning software can also help students carry out independent inquiry learning. Teachers arrange learning tasks through the software.

Students receive, complete and submit tasks assigned by teachers, and then carry out self-evaluation and mutual evaluation on the software. Based on the outcomes submitted by the students, the software gives overall evaluation, modification suggestions, specific scores and class rankings [10]. Students can learn about the existing problems at the first time, modify and improve in time, or refer to the outcomes of other students to learn from each other. Combined with the overall data feedback from the software system, teachers analyze and comment. In the age of modern technology, the reasonable use of information technology-based teaching aids to conduct the teaching will mobilize and maintain students' learning interest and motivation more effectively.

3.5 Effectively Integrate Various Information Technology Means to Optimize the Teaching Methods Applied in College English Courses

On the basis of observing universal teaching rules and principles, taking the entire teaching process into consideration, teachers should choose the best and most effective methods to meet the needs of teaching goals, teaching tasks, and teaching contents [3]. Relying on the information technology, the optimization and upgrading of teaching methods will further promote the achievement of teaching goals and achieve the maximum teaching effectiveness. In the College English course teaching, teachers can release the learning tasks before, during and after class through the online teaching and learning software, and organize students to carry out learning activities step by step. 1. Before class, teachers make or select micro-class videos, integrate and edit core knowledge points and upload to the software for students to watch and learn, open discussion forums to discuss, exchange and solve doubts during the learning process. Teachers can also set up brainstorming, quizzes and other activities to check students' understanding. With the help of the software system, teachers can check the student's learning progress, learning duration and the difficulties emerged during the learning process in real time, which help determine the language points of the lesson, making the teaching more targeted. 2. During the class, teachers use information technology to set up different situations for teaching, release tasks through the interactive area of the online software, and guide students to complete the tasks to learn knowledge points and overcome key and difficult points. Teachers can also release mini quizzes through the software to check students' understanding during the class. 3. After class, students can draw mind maps to record learning gains and learning experience and upload them to the software for teachers and students to review and comment. They can also set up mini quizzes themselves to challenge other students.

3.6 Utilize the Functions of Data Collection and Analysis in the Information Technology Software to Comprehensively Assess Teaching Effects

Use the testing and questionnaire functions in information technology software to comprehensively evaluate and collect teaching results, including changes in student performance during different phases, students' feedback on teachers and teaching effects, students' self-evaluation of learning effects, and students' time and energy consumption, teachers' feedback on the information technology-based teaching model, etc. Teachers can obtain the feedback from all aspects through the software at the first time, and then analyze and gain lessons. Teachers should focus on analyzing whether the teaching effects have achieved the teaching goals, the reasons why the teaching effect may deviate from the optimal standard, and the reasons for the difference between the good learners and poor learners [3]. On the basis of data analysis, teachers should also actively reflect on how to optimize and upgrade the teaching process. All the feedback from the software system should be served as a reference for the next round of teaching.

4   Conclusion

Under the background of economic globalization, China's education is moving towards the direction of internationalized development, pushing the realization of education modernization. The information technology-based teaching model promotes the reform of College English courses teaching and achieves better results in teaching. In China, more and more colleges and universities have introduced information technology-based teaching to English classroom teaching. In essence, information technology is a teaching tool. College English teachers should tap the strengths of the technology, organize and guide students to learn English with its help. Information technology-based teaching optimizes the entire teaching process, highlights the student's subjective status, increases their learning motivation, and improves the effectiveness of classroom teaching. Furthermore, it enables teachers and students to harvest the best results with the least time and energy, and effectively improves students' comprehensive English application ability, providing experience for College English teaching reform in China.

Acknowledgment. This study was funded by the Education and Scientific Research Project for Young and Middle-aged Teachers of Fujian Province, China (Project No.: JAT191664).

References

1. Zeng, Z.: Teaching promotion strategy of information technology course in secondary vocational school based on information literacy training. Educ. Forum **02**, 340–341 (2020)
2. Yang, M.: Study on the construction of evaluation index system for information teaching ability of primary and secondary. Southwest University, pp. 17–18 (2016). (in Chinese)
3. Babanski, Optimization of Teaching Process: General Teaching Theory. People's Education Press, Beijing (2007). (in Chinese)
4. Lu, C.: The research and practice of flipped classroom teaching in Journalism under the background of media integration. West China Broadcast. TV **24**, 84–85 (2019)
5. Li, F., Jin, J.: Teaching of engineering thermodynamics based on MOOC platform. Technol. Wind **06**, 81–82 (2020)
6. Lai, W.: Research on Teachers' development based on the concept of flipped classroom. Theory Practice Innov. Entrepreneurship **17**, 58–59 (2019)
7. Xu, Y., Wang, D.: Curriculum construction based on Internet plus. Educ. Modernization **30**, 96–99 (2018)
8. Peng, X., Xu, Q.: Investigating learners' behaviors and discourse content in MOOC course reviews. Comput. Educ. **143**, 101–103 (2020)
9. Zheng, F., Chen, S.: The application of MOOC + SPOC mode in the teaching of domestic universities. J. Liuzhou Vocat. Tech. Coll. **20**(01), 57–61 (2020)

10. Shao, J.: Research on the design of college oral English course assisted by network resources and mobile technology. Nanjing University of Posts and Telecommunications, pp. 46–50 (2019)

The Application of Computer Technology in Chinese Teaching for Foreign Students from the Perspective of Internet

Bingjie Han[✉]

Shandong Vocational College of Light Industry, Zibo, Shandong, China
Bingjie_Han81@haoxueshu.com

Abstract. As we all know, language must be taught and learned in a certain environment. In recent years, international exchanges have been deepened year by year, and the popularity of teaching Chinese as a foreign language has become higher and higher internationally. There has been an upsurge in learning Chinese throughout the world. The Chinese as a Foreign Language major is a hot specialty in China's universities. With the rise of international Chinese teaching, the teaching model of Chinese as a foreign language is also constantly reforming. With the rapid development of science and technology represented by computer technology, it has become possible to construct a virtual language environment based on virtual reality technology, which has changed the difficult situation of traditional Chinese teaching as a foreign language. This article first explains the concept of virtual reality technology, and then analyzes the theoretical basis and important value of virtual reality technology in teaching Chinese as a foreign language, and then puts forward the current idea of using virtual reality technology to reform the teaching mode of Chinese as a foreign language in order to promote foreign students. Improve the quality of Chinese teaching.

Keywords: Virtual reality technology · Chinese as a foreign language · Teaching reform · Theoretical basis

1 Introduction

Chinese as a Foreign Language is an important language course offered by colleges and universities in China. Its main teaching target is people from countries or ethnic groups whose mother tongue is other languages. It is more difficult for them to teach Chinese because they have a poor foundation in Chinese, or even zero, and Chinese itself is a vast and profound language, so many students say that learning Chinese is difficult. In the context of quality education, more and more attention is being paid to the cultivation of students' comprehensive practical ability [1]. Many universities in China continue to practice the concept of quality education, carry out teaching reforms in teaching concepts, teaching models, teaching management, and so on. The comprehensive training of students and the effective use of diversified teaching material resources will

J. MacIntyre et al. (Eds.): SPIoT 2020, AISC 1283, pp. 507–513, 2021.
https://doi.org/10.1007/978-3-030-62746-1_75

help to improve the level of teaching Chinese as a foreign language. In the information technology era, the application of various computer technologies and communication technologies has made the teaching of Chinese as a foreign language more and more convenient, and has also expanded the learning channels for students. Distance education technology is a brand-new teaching mode that can greatly improve students learning level. Virtual Reality (VR) technology comes along with the continuous development of computer technology. It combines system simulation, computer graphics, digital image processing, sensing and measurement and other disciplines into one. People have created a perceptible, highly simulated virtual reality environment. At present, many universities at home and abroad have applied virtual reality technology to classroom teaching [2]. In September 2015, Google announced that it would cooperate with California's top public schools to promote virtual reality classroom systems for free. However, these application cases are more focused on professional courses in science and engineering such as architecture, physics, medicine, and biology, and the implementation methods are mostly limited to relatively simple applications such as experimental teaching. New Oriental CEO Yu Minhong said that the panoramic teaching mode of virtual reality technology can enable students to achieve "immersive learning" and improve the efficiency of foreign language classroom learning [3]. As a new technology, virtual reality technology will create a new foreign language learning model, build a real language environment, help students experience traditional foreign language conventions and rules of use in virtual foreign language communication scenarios, and then master foreign language knowledge in scene-based experiences. This article will mainly analyze the concept and characteristics, theoretical basis of virtual reality technology, and its role and application in Chinese teaching.

2 Concepts and Characteristics of Internet-Based Virtual Reality Technology

2.1 The Concept of Virtual Reality Technology

The narrow sense of virtual reality technology generally refers to the use of computers and sensor gloves, 3D controllers, stereo glasses, sensors and other supporting equipment for data exchange, so that people can get sensory information such as hearing, force, vision, and even touch to build a The perceivable, highly simulated virtual reality environment produces a 3D realistic feeling relatively close to reality [4]. With the continuous development of technology, the meaning of the broad sense of virtual reality technology includes not only the narrow content, but also refers to all kinds of software and hardware, use technology, and implementation methods related to virtual reality technology that can implement virtual simulation. It should be said that virtual reality technology is the cross-fusion of different disciplines such as system simulation, visualization technology, computer graphics, digital image processing, software engineering, sensing and measurement, and artificial intelligence [5]. Therefore, virtual reality technology is a more effective advanced human-computer interaction technology that simulates human behaviors such as sight, movement, smell, force, hearing, and touching in the natural environment. It uses computer technology to create a perceivable, highly simulated virtual reality environment, so that operators in this virtual environment can directly operate this virtual environment through various sensor interaction devices, and get real-time display and exchange feedback. When the external world and virtual environment form a closed loop of feedback through sensory interaction devices, under the control of the operator, the interaction between the operator and the virtual environment will have a corresponding reaction on the external world (see Fig. 1) [6].

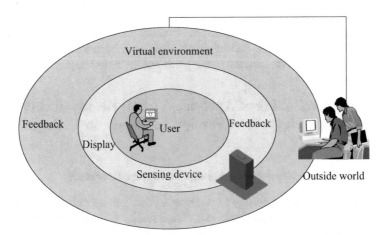

Fig. 1. Virtual reality system

2.2   Features of Virtual Reality Technology

French scholar Grigore C. Burdea and American scholar Philippe Coiffet published in 1993 the book "Virtual Reality Technology". The famous "triangular theory" proposed that virtual reality technology has the following three typical characteristics: immersion, interactivity and imagination.

(1)   Immersive

Immersion, also called telepresence, is a core feature of virtual reality technology [7]. The immersiveness of virtual reality technology means that the operator can be immersed in the virtual environment generated by the computer system, and then can perceive the virtual environment through more special equipment, thereby generating an immersive feeling and "immersing himself" In a virtual environment [8].

(2)   Interactivity

Interaction means that participants are no longer just passively participating in the virtual environment established by virtual reality technology, but can actively change or choose the content they feel by operating some special equipment themselves, that is, participants are no longer limited For the traditional keyboard and mouse for information processing, you can use data gloves, haptic feedback systems, sensors and other new devices and virtual objects to interact and operate [6].

(3) Imagination

Imagination emphasizes that virtual reality technology has a very wide imagination space, and thus continuously expands the human cognitive space. Virtual reality technology can not only reproduce the objectively existing real environment, but also create an unlimited imagination space for the participants, thereby returning the initiative to the participants, and then making the participants boldly try to imagine and create [7].

3 The Important Value of Applying Virtual Reality Technology to Teaching Chinese as a Foreign Language in Colleges and Universities

3.1 Help Students Better Acquire Foreign Language Knowledge

Applying virtual reality technology to teaching Chinese as a foreign language in colleges and universities can put students in a virtual reality system, thereby breaking the limits of traditional foreign language classroom teaching, enabling them to learn, use, and feel languages in virtual classrooms, and then integrate them into Into their own language learning ability and language application ability [9]. With the support of virtual reality technology, a virtual cognitive environment similar to the real environment is constructed, so that scenes that are difficult for students to observe in reality create corresponding virtual objects in the virtual environment, making foreign language learners more intuitive In the cognitive environment, it is better to recognize concepts and construct knowledge [10]. At the same time, virtual reality technology can transform traditional foreign language learning methods. Learners can choose corresponding foreign language learning resources from the master according to their own learning needs, and carry out exploratory learning according to their own learning method and speed, so that each foreign language learning Anyone can turn passive into active, which is conducive to enhancing students' creativity and initiative.

3.2 Strengthening Students' Foreign Language Communication Skills Training

Applying virtual reality technology to the teaching of Chinese as a foreign language in colleges and universities can create a rich and colorful foreign language communication environment, allowing students to "play" roles in the virtual environment, allowing them to effectively communicate with their favorite characters in an immersive environment. Help foreign language learners to learn the daily idioms and spoken expressions of foreign languages in a real and rich language situation. Through imitation and reuse in the communication process, master the fixed collocation and flexible application of the language to fully master the foreign language communication ability of college students [11]. At the same time, the application of virtual reality technology can also enable foreign language learners to practice a variety of language listening, speaking, reading and writing at any time and place, cultivate a good sense of autonomous learning, greatly enhance students' language application ability, and improve foreign language learning Comprehensive language skills. At present, multiple foreign language virtual worlds have begun to run with the support of relevant countries and governments, such as the "Second Life" launched and operated by Linden Laboratories in the United States in 2003, and 3DMAX-based software development in Europe "Virtual Reality Language Learning Network" (VIRLAN), a "three-dimensional multi-user Russian virtual world" designed and developed by the Russian Ministry of Education and State Petersburg University [12].

3.3   Effectively Motivate Students' Verbal Communication Motivation

It should be said that "mechanical drills" and "meaningful drills" are not enough for foreign language learners to acquire language and must carry out real verbal communication activities. Therefore, in order to internalize the foreign language listening and speaking of foreign language learners, they must also only through a lot of long-term verbal communication practice can it be realized smoothly. As the intrinsic motivation for foreign language learners to develop speech communication, speech communication motivation requires not only explicit external incentives, but also specific internal needs. In traditional Chinese language teaching classrooms, when carrying out speech communication training with specific functions, students and teachers often need to imagine through their own episodic memory to virtualize the communicative subject, environment, and events, and the foreign language virtual world constructed based on virtual reality technology can immerse in the language environment of virtual daily life, creating the effect that the foreign language learner exists in the native language environment, thereby stimulating his strong desire for exploration and the specificity of verbal communication driven by his curiosity [10]. Internal requirements, and the virtual environment itself has many colorful "external incentives." Therefore, in the virtual speech communication environment, the "external inducement" and "internal demand" of speech communication motivation can be organically combined, so as to better stimulate the language communication motivation of foreign language learners.

3.4   Reduce Students' Foreign Language Anxiety

Anxiety, as a psychological phenomenon, plays a certain beneficial role in people's survival and development, but if it evolves into a strong emotional state, it will become a psychological obstacle that affects individual actions [11]. Situational language anxiety is a kind of negative psychological emotion most easily generated by foreign language learners. It refers to the negative emotions such as tension, doubt, fear, anxiety, and fear that foreign language learners have during the process of learning and applying the target language. Excessive anxiety will inevitably have a corresponding inhibitory effect on foreign language learning. On the one hand, it is easy for foreign language learners to be forced to abandon speech communication motivation due to lack of confidence during language training. Language information is difficult for foreign language learners to understand and absorb smoothly. Students' foreign language anxiety mainly depends on three factors: speech scene, communication level, and time limit. In the traditional foreign language classroom teaching, the above three factors may cause students to lose their language expression opportunities in foreign language situational anxiety. The foreign language virtual world based on virtual reality technology can reduce the negative impact of speech context because it removes the interference of many miscellaneous elements. Moreover, foreign language learners can be more casual and free of time when making speech expressions in the virtual world of foreign languages. When language communication obstacles occur, they can pause to make students feel more secure without generating excessive foreign language anxiety.

3.5 Breaking New Ground in Extracurricular Language Acquisition

The famous American language educator Stephen Krashen (SD) proposed unconscious foreign language learning teaching concepts, and called this foreign language learning process the "language acquisition process", emphasizing that foreign language learning must be the same as children learning their mother tongue. A relatively natural way is to use language unconsciously to carry out social communication, so that foreign language learners can naturally acquire the language [8]. However, in reality for foreign language beginners, natural language acquisition is relatively difficult, because beginners' foreign language level is not high, they have not mastered the basic skills of direct language communication in foreign languages, and they have acquired the target language. The internal and external environment of children's acquisition of mother tongue is completely different, and it is difficult to obtain rich and effective speech input from natural speech materials. To some extent, virtual reality technology has created a verbal communication environment for foreign language learners, such as life situations such as travel, shopping, learning, entertainment, etc. that may be encountered in the target language country, so as to exercise and improve foreign language learning.

4 Conclusion

Using virtual reality technology and 3D streaming technology to provide real-time teaching scenes of outstanding foreign language teachers to users of distance learning in real time, not only can achieve online learning and synchronization of on-site teaching, but also can realize one-to-many real-time teaching. At the same time, you can also use virtual reality technology to make excellent teaching videos for teaching Chinese as a foreign language. For example, you can virtualize the resources of excellent courses at the municipal, provincial, and national levels, which can not only help outstanding teachers share their teaching experience, but also use them as teaching resources for permanent preservation. As an emerging technology, virtual reality technology has not yet been widely used in teaching Chinese as a foreign language in colleges and universities. However, as the technology related to virtual reality technology matures, colleges and universities teaching Chinese as a foreign language based on virtual reality technology will inevitably flourish. It will inevitably inject fresh blood into traditional foreign language teaching, which will bring a benefit to everyone in Chinese as a foreign language.

References

1. Teo, T., Sang, G., Mei, B., Hoi, C.K.W.: Investigating pre-service teachers' acceptance of Web 2.0 technologies in their future teaching: a Chinese perspective. Interact. Learn. Environ. **27**(4), 530–546 (2019)
2. Chen, M., Zhou, C., Meng, C., Wu, D.: How to promote Chinese primary and secondary school teachers to use ICT to develop high-quality teaching activities. Educ. Tech. Res. Dev. **67**(6), 1593–1611 (2019)
3. Xu, S., Yang, H.H., MacLeod, J., Zhu, S.: Interpersonal communication competence and digital citizenship among pre-service teachers in China's teacher preparation programs. J. Moral Educ. **48**(2), 179–198 (2019)
4. Gao, S.: Integrating multimedia technology into teaching chinese as a Foreign language: a field study on perspectives of teachers in Northern California. J. Lang. Teach. Res. **10**(6), 1181–1196 (2019)
5. Hong, J.C., Hwang, M.Y., Tai, K.H., Lin, P.H.: Improving cognitive certitude with calibration mediated by cognitive anxiety, online learning self-efficacy and interest in learning Chinese pronunciation. Educ. Tech. Res. Dev. **67**(3), 597–615 (2019)
6. Oakley, G., Pegrum, M., Xiong, X.B., Lim, C.P., Yan, H.: An online Chinese-Australian language and cultural exchange through digital storytelling. Lang. Cult. Curriculum **31**(2), 128–149 (2018)
7. Zhou, M.: Chinese university students' acceptance of MOOCs: a self-determination perspective. Comput. Educ. **9**(2), 194–203 (2016)
8. Teo, T., Zhou, M., Fan, A.C.W., Huang, F.: Factors that influence university students' intention to use Moodle: a study in Macau. Educ. Tech. Res. Dev. **67**(3), 749–766 (2019)
9. Tseng, J.J., Lien, Y.J., Chen, H.J.: Using a teacher support group to develop teacher knowledge of Mandarin teaching via web conferencing technology. Comput. Assist. Lang. Learn. **29**(1), 127–147 (2016)
10. Lai, C., Li, X., Wang, Q.: Students' perceptions of teacher impact on their self-directed language learning with technology beyond the classroom: cases of Hong Kong and US. Educ. Tech. Res. Dev. **65**(4), 1105–1133 (2017)
11. Sang, G., Liang, J.C., Chai, C.S., Dong, Y., Tsai, C.C.: Teachers' actual and preferred perceptions of twenty-first century learning competencies: a Chinese perspective. Asia Pacific Educ. Rev. **19**(3), 307–317 (2018)
12. Ho, W.Y.J.: Mobility and language learning: a case study on the use of an online platform to learn Chinese as a foreign language. Lond. Rev. Educ. **16**(2), 239–249 (2018)

The Application of Computer "Virtual Simulation" Experimental Teaching in Basic Football Tactics

Xiaowei Di[(⊠)] and Huhu Lian

Sichuan Agricultural University, Yaan, Sichuan, China
lubianinfo@foxmail.com

Abstract. Football tactics are complex and play an important role in football match. Based on analyzing the characteristics and advantages of the virtual simulation teaching platform, this paper uses virtual simulation technology to help students improve their abilities of making tactical decisions in the football game. In this experiment, real competition scenes are reduced to virtual competition scenes, which enables students to judge and make decisions from the first perspective. The method ensures the reality in teaching and effectively improves the efficiency of teaching and learning. It also adapts to the new trend of teaching reform.

Keywords: Virtual simulation · Football tactics · Experimental teaching

1 Introduction

In 2004, China introduced the virtual simulation experiment teaching. Now the project has been supported by the state and the Ministry of Education; relevant policies on the construction of virtual simulation experiment projects have been introduced successively. Colleges and universities are promoted to actively explore a new experimental teaching mode which is personalized, intelligent, universal, and combines online and offline education, so as to form a new information-based experimental teaching system with reasonable professional layout, excellent teaching effects as well as open and shared information, and support the overall improvement of the higher education quality [1]. At the same time, on the basis of the traditional teaching mode, there is an urgent need for new and auxiliary teaching methods to help football education goes out of the difficulties of single teaching method, few teaching hours and insufficient actual combat experiences. As a new teaching method, virtual simulation experiment teaching is supported by the national policy and can solve the problem of insufficient actual experiences. Therefore, it is urgent to combine the basic football tactics teaching with the virtual simulation technology [2].

J. MacIntyre et al. (Eds.): SPIoT 2020, AISC 1283, pp. 514–522, 2021.
https://doi.org/10.1007/978-3-030-62746-1_76

2 The Necessity of Applying Virtual Simulation Technology in the Teaching of Basic Football Tactics

2.1 Virtual Simulation Technology Is the New Development Trend in the Teaching of Basic Football Tactics

Football is the first sport in the world. The development of football tactics is changing with each passing day. In colleges, it is difficult to achieve certain effects only by using traditional teaching and training methods [3]. The teaching of basic football tactics in virtual simulation experiments has become a new trend. In the virtual simulation experimental teaching, scenes of real competitions are simplified into virtual tactical scenes. In this way, the virtual environment can be connected with the objective reality; students can learn and use basic tactics of football attack and defense through the computer system.

2.2 The Reality and Effectiveness of Virtual Simulation Technology in the Teaching of Basic Football Tactics

In the virtual simulation experiment, students first identify the scene of the game, including the time of the game, the location of the field, team members and so on. After identification, students quickly enter into the process of judging basic attack and defense tactics, including the types and behaviors of basic tactics [2]. After that, students need to make quick and effective decisions and choose reasonable tactics. The system will display the tactical actions selected by students in the competition scene, and give objective evaluation according to the situation of scene recognition, decision-making time and action rationality. Through this system, students can learn and apply basic tactics of football attack and defense [4]. When they acquire knowledge, they can not only deepen their understanding, but also master the knowledge application. At the same time, the efficiency of teaching and learning is also improved.

2.3 The Extension and Expansion of Traditional Teaching

The teaching of football basic attack and defense tactics in virtual simulation not only breaks the limit of learning space, but also realizes the free learning time and content, which greatly improves the efficiency of teaching and learning. Students can learn in the context of virtual scenes, and enjoy the experience of substitution [4]. At the same time of acquiring knowledge, it further increases students' understanding on the application of knowledge, and at the same time of theoretical learning, it further exercises the personal practical ability. At the same time, the traditional teaching method is one-dimensional explanation of tactics through the board or videos; students do not have deep understanding of the scene. The virtual simulation experiment enables students to learn in the three-dimensional environment and lays the foundation for the consolidation of theoretical knowledge and the transition in the real scene.

3   Advantages of Virtual Simulation Experiments in the Teaching of Basic Football Tactics

3.1   Overcome Shortcomings in the One-Dimensional Teaching of Basic Attack and Defense Tactics

The traditional method of learning basic attack and defense tactics is realized through one-dimensional explanation of tactics on board or in videos. Students do not have deep understanding on the scene. When students make decisions, they need to consider the positions, distances and angles of teammates and defenders. The one-dimensional explanation can't achieve these effects; students can't recognize the scene and make correct decision effectively in the real competition. The virtual simulation system helps students to observe the positions, distances and angles of teammates and defenders through the first perspective of the ball holder [5]. They can have more sense of substitution and achieve better learning effects. This way can encourage students to carry out independent inquiry learning, and produce positive effects.

3.2   Help to Improve Students' Learning Enthusiasm

Virtual simulation technology has the function of improving students' learning ability. It enables students to gain more profound cognition on image and intuition, so as to deepen and consolidate knowledge and cognition. Students can learn online according to unified requirements of the teacher, and arrange the learning time, place, teaching contents, as well as preview before class and review after class according to the schedule [6]. The combination of online and offline experimental teaching methods greatly stimulates students' learning motivation, breaks the barriers of time and space in learning, improves the efficiency of teaching and learning, and plays an important role in developing students' learning potentials and cultivating their ability of combining theory with practice.

3.3   Observation from Multiple Perspectives Can Provide Students Various Choices

In football games, players often make decisions based on their own observation from the first perspective. But they may ignore a lot of important information due to the pressure of the game or the angle of observation. Virtual simulation system provides students with a variety of angles of view in decision-making, which is more convenient for students to correct their own choices. Using this system not only enriches the way of experiencing football tactics, but also improves the learning efficiency, and lays a foundation for the understanding and application of football games [7].

3.4   Help to Improve Students' Ability of Quickly Identifying Competition Scenes

The virtual simulation system can accurately restore the practice scene. The scene of football match changes rapidly, and students usually make decisions in an instant. Through the means of virtual simulation, we can present the instantaneous competition scene to students more clearly and accurately, which is more conducive to cultivate students' ability of identifying competition scenes.

3.5 Help to Improve the Efficiency of Football Teaching

Due to the large number of students in traditional teaching, teachers can not guide every student. Through the auxiliary practice of virtual simulation, every student can have the opportunity to obtain accurate teaching guidance, which greatly improves the teaching efficiency.

4 The Application of Virtual Simulation Experiments in Football Tactics Teaching

4.1 Setting Virtual Learning Scenes According to Real Competitions

The virtual simulation experiment uses the virtual simulation technology to simplify the real game scene in football matches through three-dimensional animation. Through a variety of perspectives, it can help students identify the scene. After that, the system will return to the perspective of the player holding the ball, and then the ball holder will make tactical decisions. In the "basic attack and defense module", students should have basic football skills, including passing and catching the ball, handling and controlling the ball, as well as shooting and other skills [8]. At the same time, they should learn the application principle and key points of this technology. Through the fixed module of virtual simulation exercise, students will master the ability to identify the scenes of "one person attack and defense", "two persons attack and defense", as well as "three persons attack and defense", and understand the key points of decision-making in such scenes, so as to make the best decision and take action when encountering similar scenes in the real competition. In the "complex situation decision-making module", students should have professional tactical knowledge, including shooting, passing, dribbling as well as other theoretical points and skills. In addition, they should have good theoretical basis for keeping the three lines in attacking and defending and maintaining the formation of the team.

4.2 Building an Effective Teaching Mode

The virtual simulation experiment in the teaching of basic football attack and defense tactics has created a new teaching mode combining online and offline teaching. It combines distance teaching based on network, guided teaching based on classroom theory and practical teaching based on field training (as shown in Fig. 1). The use of the imagery training teaching method and the simulation training teaching method greatly improves the teaching efficiency and promotes students to learn professional skills. Through the experiment, they can recognize the match scene quickly and accurately and make reasonable decisions. The virtual simulation experiment teaching enriches the traditional mode of teaching football techniques and tactics, in which students acquire knowledge in the theory class but lack the intuitive feeling of the match. In addition, in the real football match, the scene is fleeting; it is difficult to achieve 100% reproduction. Through the virtual simulation experiment, students can consolidate the knowledge they have learned in theory class, and lay the foundation for identifying different scenes in real matches. After returning to the experiment after comparing with the teaching match, they can improve the scene recognition ability and make reasonable decisions [9].

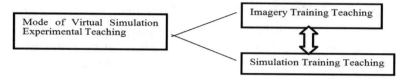

Fig. 1. Virtual simulation experimental teaching mode.

4.2.1 Recovery of Real Game Scenes

The teacher uses the virtual simulation system and asks students to watch video clips of high-level football matches. After that, the system will simplify the complex game scene into the theme scene, and impart theoretical knowledge corresponding to the scene.

4.2.2 Immersive Scene Recognition

After entering the virtual link, students identify the scene through the first perspective, make decisions and take actions. At the same time, they can switch to the God perspective, which provides more intuitive, three-dimensional and vivid perception of the game scene to assist their decision-making. Through the accurate representation of the game scene, it solves problems caused by one-dimensional scene explanation through videos or the tactical board [10].

4.2.3 Interactive Exercises

After entering the practice phase, students make basic decisions according to the progress of the practice, and adjust decisions according to the feedback. The system realizes the trinity of practice, learning and adjustment; mistakes and deficiencies can be corrected and supplemented in time.

4.2.4 Feedback Evaluation

In the process of the experiment, the system will make timely feedback and evaluation according to students' decisions. According to the feedback, teachers know whether students have mastered the offensive and defensive tactics.

4.3 The "Trinity" Evaluation Mode

The evaluation system of virtual simulation experiments of basic football attack and defense tactics realizes the new evaluation mode which combines system evaluation, theory evaluation and practice evaluation.

4.3.1 System Evaluation

After studying in the system, the system will objectively evaluate the rationality of the tactical action according to the selection of the student in each scene. At the same time, the learning process of the student will be recorded and sent to the administrator (the teacher). Some teachers will evaluate the learning process according to the performance of students.

4.3.2 Theory Course Evaluation

The evaluation of the theory course mainly focuses on students' mastery of basic theoretical knowledge, such as the tactical knowledge and the strategy of basic attack and defense tactics. The students are assessed by the way of answering questions; the teacher scores according to students' answers.

4.3.3 Practice Course Evaluation

There are two parts in the assessment of practice course. The first part is to evaluate students' mastery of basic attack and defense tactics according to their techniques. The other part mainly tests the organization and guidance ability of students. The role of students is changed to "coaches" or "teachers". The assessment process is that, the student is required to design a training lesson according to the assessment theme, and implement the designed training course in the venue. The assessment criteria mainly include appearance, practice organization, guidance position, language expression, key points in guidance, method selection and other aspects; the teacher scores according to the comprehensive performance. The teacher will systematically and comprehensively evaluate students according to their comprehensive performances in the system, in the theory class and in the practice course

5 Analysis on the Effects of Virtual Simulation Experiment in Football Tactics Teaching

5.1 Effects of Learning Basic Football Tactics

In order to clarify the effects of virtual simulation technology in the teaching of basic football tactics, this study selected two elective courses of different majors in Dujiangyan campus of Sichuan Agricultural University as the experimental group and the control group. After 64 h of football teaching and training, students of the experimental class and the control class were investigated by questionnaire to analyze the learning effect. The results of show that, students in the experimental group have better understanding on the teaching content and practice more often on basic football tactics after class than the control group, which fully shows that the virtual simulation experimental teaching can help the students achieve good learning effects (as shown in Tables 1, 2 and 3).

Table 1. Students' understanding and reflection on football tactics after class. Unit: %

Group	Frequent	Normal	Occasionally	No
Experimental group	9.72	63.40	15.89	10.99
Control group	5.85	55.38	23.57	23.57

Table 2. The consolidation of teaching contents by students after class. Unit: %

Group	Frequent	Normal	Occasionally	No
Experimental group	15.92	68.82	12.76	2.50
Control group	10.65	50.99	29.41	8.95

Table 3. Students' practice of tactical teaching contents after class. Unit: %

Group	Frequent	Normal	Occasionally	No
Experimental group	11.42	50.48	27.81	10.29
Control group	6.47	37.93	29.50	26.10

5.2   Effects of Teaching Basic Football Tactics

Students in the experimental group are tested from three aspects: operation in the system, theoretical knowledge and practical skills. Students in the control group only need to test their practical skills. The practical skills assessments of the two classes are carried out on the football field respectively. The assessment contents include, basic attack and defense (30% for one person attack and defense, 30% for two person attack and defense, and 40% for three person attack and defense), complex situation decision-making (30% for scenario one, 30% for scenario two and 40% for scenario three, 100 points in total). The test results are compared by the independent sample t-test. The situation (football training) is assessed through organizing and guiding theme training courses, and scored by teachers (with the full score of 100 points).

The data shows that there are significant differences ($t > 1$, $P < 0.05$) between the experimental group and the control group in the evaluation results of teaching basic attack and defense skills and making decisions in the complex situation. It is proved that after the traditional football training course, the virtual simulation technology is more effective in teaching. (Results are shown in Tables 4, 5 and 6).

Table 4. Independent sample T test of football skills and tactics teaching effect.

Item	Group	N	Average value	Standard deviation	t	P
Theory	Experimental group	37	80.36	8.456	3.233	0.017
	Control group	38	77.45	7.567		

Table 5. Independent sample T test of football skills and tactics teaching effect.

Item	Group	N	Average value	Standard deviation	t	P
Skills	Experimental group	38	86.57	7.352	3.326	0.026
	Control group	39	82.46	7.237		

Table 6. Independent sample T test of football skills and tactics teaching effect.

Item	Group	N	Average value	Standard deviation	t	P
Tactics	Experimental group	6	88	11	3.357	0.012
	Control group	6	79	6		

6   Conclusion

On the virtual simulation platform, students can choose corresponding scene and carry out online learning according to theoretical knowledge they have learned in the classroom and the practical operation in the field. It has realized the "four in one" learning and practical teaching mode which takes students as the center, and includes the links of preview before class, learning in the class, review after class and the assessment. Through the combination of online and offline learning, students can organically combine the theoretical learning, the practical operation and the feedback improvement. Through the imagery training teaching method and the simulation training teaching method, students can understand basic attack and defense tactics quickly, make correct tactical decisions in the game more effectively, and learn in the real scene. The method can effectively improve the effect of learning special skills.

This project has been applied to the football elective course of sophomores and got good feedback. It has greatly aroused students' interest in learning and greatly improved the efficiency of teaching and learning. To popularize this project can greatly improve the comprehensive ability of students to apply knowledge, and greatly promotes the development of teaching contents, teaching methods and teaching evaluation ways in football courses of colleges and universities.

References

1. http://www.moe.gov.cn/srcsite/a08/s7945/s7946/201707/t20170721_html
2. Guo, Y.X.: Application of Virtual Simulation Experiment in Tennis Referee Teaching. Wuhan Sports University
3. Xing, Z.Y.: Research on the application of virtual simulation technology in the classroom teaching of sports majors. J. Xuchang Univ. **27**(8), 84–88 (2018)
4. Li, C.M, Ren, D.M.: Virtual Simulation Experiment of Basic Attack and Defense Tactics in Football Match. National Projects of Virtual Simulation Experiment Teaching Platform (2019)

5. Zheng, Y., Wu, Y.: Design and management of virtual simulation experiment for TD-LTE base station installation. Exp. Technol. Manage. **35**(10), 125–128 (2018)
6. Guo, T., Yang, S.G., Jiang, Y.S., et al.: Research on construction and application of virtual simulation experiment teaching projects. Exp. Technol. Manage. **36**(10), 215–217 (2019)
7. Yi, Y.C., Li, K.F., Zhan, C.J.: Application of virtual reality technology in the teaching of sports art courses: a case study of multigen creator software application. J. Shandong Agric. Eng. Coll. **34**(6), 79–81 (2017)
8. Chen, Y.P., Li, Z.H., Xu, Y.M.: The application of virtual simulation technology in physical ability evaluation experiment teaching. Sports Res. Educ. **32**(01), 68–71 (2017)
9. Xu, J.P.: Research on college physical training based on virtual technology. J. Xi'an Univ. Arts Sci. Natural Sci. Edition **19**(5), 93–96 (2016)
10. Xu, L.J., Zhang, A.J., Wu, T.: Virtual reality technology and its application in volleyball teaching. J. Tonghua Normal Univ. **29**(4), 95–97 (2008)

Application of Campus Happy Running APP in the Reform of University Physical Education Integration

Yuanhai Liu[✉] and Yong Liu

School of Physical Education, Hubei University of Science and Technology, Xianning, Hubei, China
67929662@qq.com

Abstract. Objective: In view of the current decline in college students' physical fitness and the Ministry of Education's basic requirement of "effectively ensuring that students exercise for one hour a day", the reform of universities is carried out by using the campus trail Lepao APP software to promote the integration of physical education in and out of universities. Methods: Literature review, questionnaire survey and logical analysis were used. Results: The introduction and APPlication of footpath running app incorporated extra-curricular running exercise into the evaluation system of physical education curriculum, realized the diversified evaluation of college physical education, and promoted the integration of inside and outside physical education classes. Conclusion: The campus trail running APP effectively supervises college students' after-school exercise, realizes the integration of college physical education inside and outside the class and the diversification of performance evaluation, which is of great significance to mobilize the enthusiasm of students' after-school exercise, cultivate lifelong physical education awareness and habits, and perfect the school physical education work. At the same time, it has important reference value for the construction of the integrated evaluation mode of physical education inside and outside the class.

Keywords: College sports · Integration inside and outside the class · Trail fun run APP

1 Introduction

Under the trend of the reform of integration inside and outside physical education classes, how to effectively guide and supervise students' extracurricular exercises and how to incorporate extracurricular physical exercises into the evaluation of physical education courses are the problems that need to be solved urgently. The campus trail Lepao APP software developed by Lepao Sports Internet (Wuhan) Co., Ltd. can learn about students' extracurricular running from the campus intelligence platform starting from the campus sports management plan. With the help of this software, students' after-class directional running can be effectively supervised, which provides ideas and reference for the integrated evaluation inside and outside the class.

J. MacIntyre et al. (Eds.): SPIoT 2020, AISC 1283, pp. 523–528, 2021.
https://doi.org/10.1007/978-3-030-62746-1_77

1.1   Research Background

According to a survey released by the Scientific Papers Report of the 9th National University Games, the physical quality of contemporary college students has a continuous downward trend [1]. Speed quality and strength quality in physical fitness quality have declined for 20 consecutive years, while endurance quality has declined for 30 consecutive years. Poor life style and lack of physical exercise are the main reasons that affect the decline of physical fitness quality of college students [2]. It is imperative to strengthen physical exercise of college students.

1.2   Research Status

On June 11, 2014, the Ministry of Education issued the "Basic Standards for Physical Education in Colleges and Universities" with the number 4 of "Teaching Sports Arts [2014]" stating that college physical education must ensure a certain exercise intensity, and the exercise content to improve students' cardiopulmonary function must not be less than 30%. And will reflect the students' cardiopulmonary function quality exercise items as the examination content [3], its weight shall not be less than 30%; Extra-curricular sports activities will be incorporated into the school's teaching plan, and students will be organized to take part in at least three extra-curricular sports exercises per week to ensure one hour of sports activities per day [4]. It can be seen that it is an urgent task and requirement to promote the integration of physical education in and out of college and to strengthen the exercise of cardiopulmonary function. In foreign countries, college physical education reform is mainly embodied in three aspects: first, taking physical exercise, strengthening physical fitness and improving health as the ultimate goal of college physical education [5]. For example, the goal of physical education teaching in American colleges and universities is to improve the physical health level of participants, develop their sports ability, enable them to master certain physical sports skills [6], promote them to form a healthy lifestyle, and cultivate the habit and ability to participate in physical exercise for life. The second is to emphasize the respect and cultivation of students' personality [7], so that students can happily participate in the sports learning process. For example, the leading direction of physical education in Japanese colleges and universities is happy physical education, which emphasizes that physical education in schools should meet students' physical needs, improve students' enthusiasm to participate in physical exercises, and respect and develop students' personalities. Third, lifelong physical education has become the leading goal of physical education teaching in colleges and universities in various countries [8]. Whether it is Japan's physical education, the best physical fitness curriculum in the United States, or Britain's college physical education, it has made clear the idea that physical education lays the foundation for lifelong physical education.

1.3   Deficiencies

The campus trail running APP effectively supervises college students' after-class exercise, realizes the integration of college sports inside and outside the class and the diversification of performance evaluation, which is of great significance to mobilize the enthusiasm of students' after-class exercise, cultivate lifelong sports awareness and habits, and improve the school sports work. It promotes the integration of college sports inside and outside the class, but there are still some deficiencies. For example, the campus trail running APP is a software for effectively monitoring the campus running. Intelligent settings can be made for route setting, step number control and running speed, but no effective judgment can be made for other extracurricular exercises (not running according to the established route), and no reasonable judgment can be made for exercise quality and attitude (cheating behavior).

2 APPlication of App Software for Campus Trail Lepao

The campus trail running APP is a professional online running platform, which can record running speed, distance, number of steps and evaluate energy consumption through mobile phones. Also can carry on the theory study and the exchange, shares the splendid instantaneous, participates in the interesting interaction, but also can exchange the score, stimulates participates in the running enthusiasm with the exquisite gift; what is more important is that the third party (managers and teachers) can know the running situation through the day after tomorrow. The software was introduced to our school for trial use in the fall semester of 2017, requiring freshmen and sophomores to take part in not less than 50 rounds of the campus every semester, with boys not less than 2 km and girls not less than 1.6 km each time, with a pace of 3–9 min. They must go through random clock-in points, effectively record the number of times once a day, and include 20% of the results of the round-robin in the physical education curriculum assessment. Extra-curricular campus running will be implemented to urge students to take extra-curricular exercises. Boys run for more than 100 km and girls run for more than 80 km each semester to develop cardiopulmonary function and endurance quality. The campus trail run brings extra-curricular exercise into the examination of physical education results, which increases the examination of conscious exercise attitude, promotes the integration of physical education courses in and out of class, and is conducive to the improvement of students' physical education work. The main significance lies in: first, to strengthen the supervision of students' after-school physical exercises so that the Ministry of Education's requirement of "one hour of exercise per day" can be implemented; The second is to use the footpath music running APP software to push college students to the playground and promote the formation of lifelong exercise habits, which is beneficial to students' physical and mental health. The third is to integrate students' after-class exercise into the evaluation of physical education performance, accelerate the integration process of physical education curriculum inside and outside class [9], and promote the reform and development of college physical education teaching. Fourth, it enriches college students' campus life, is conducive to creating a new situation in school physical education, and promotes the great development of school physical education.

3 Construction and Promotion of Integrated Evaluation Model in and Out of Class

3.1 Construction of Integrated Evaluation Model Inside and Outside Class

From the above analysis, it can be seen that the APPlication of campus trail running app software, which includes college students' extracurricular running into the examination

system of physical education curriculum, has promoted the integration of college physical education both inside and outside the class. However, the software itself has great
disadvantages, mainly reflected in the inability to effectively supervise all extracurricular physical exercises. Therefore, its evaluation is imperfect and needs to be improved
and improved. In line with the principle of strengthening students' consciousness and
habits, improving physical quality and developing students' personality, the integrated
evaluation of inside and outside the class is fully realized. After consulting the management department, the evaluation model of university physical education in-class and
out-of-class integration has been reconstructed. summative evaluations such as the "three
basics" test and physical health standard test results, as well as process evaluations such
as classroom performance, extracurricular physical exercises (including extracurricular free exercises, sports club activities and extracurricular training and competitions)
have been incorporated into the evaluation of university physical education curriculum,
and the integration of in-class and out-of-class integration has been comprehensively
promoted (Fig. 1).

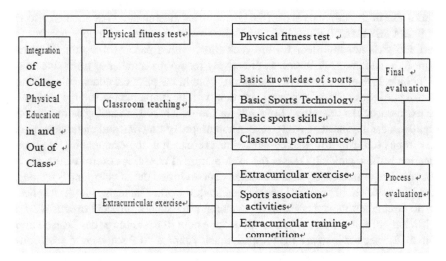

Fig. 1. Integrated evaluation of college physical education in and out of class

3.2 Promotion of Integrated Evaluation Model in and Out of Class

The APPlication of app software for campus trail running integrates after-school exercise into classroom evaluation, improves the single classroom evaluation mode of university sports, and promotes the diversified evaluation mode of university sports scores.
Although there are many deficiencies, its idea of effective supervision of students' after-
school running through campus sports management as the breakthrough point with the
help of campus smart software is worth learning and promoting. The implementation of
the integrated evaluation mode inside and outside the class is both a supervision and a

promotion to the students' after-class exercise. It "drives" the students out of the dormitory from their after-school time, pushes them to the campus, walks towards the nature, enjoys the happiness and happiness brought by sports, mobilizes the students' consciousness and enthusiasm, and realizes the Ministry of Education's decision to include "extracurricular sports activities into the school teaching plan". Its reform method is worth popularizing.

4 Evaluation Effect Analysis After Introducing Trail Race

The introduction and implementation of APP software for campus trail running has strengthened the supervision of students' running after class, conscientiously implemented the national policy on college sports, and established the guiding ideology of "health first". Actively deepen the reform, improve the single classroom evaluation mode of college physical education, include students' exercise after class into the evaluation of physical education results, and speed up the integration process of physical education curriculum inside and outside class; Highlight the student's dominant position, students can flexibly use extra-curricular time and promote a lifestyle of sports. The effect of the campus trail running has been good for one year, mainly in two aspects:

First, the students have conscientiously implemented extracurricular running exercises, and their endurance quality has been significantly improved. From the smart campus platform, it can be seen that the qualified rate of students in the spring semester of 2018 is 97.3%, the average time for boys is 4' 09 "for 1000 meters, the average time for girls is 4' 16" for 800 meters, the qualified rate of students in the autumn semester of 2018 is 98.3%, the average time for boys is 4' 06 "for 1000 meters, and the average time for girls is 4' 07" for 800 meters. The endurance quality of students is obviously improved.

The second is that the students highly approve of the trail race, and the number of people taking part in the race has increased significantly. In addition to completing their tasks, students have been trained physically and mentally through two semesters of hiking [10].. The students' approval rate for trail running is as high as 94.6%. After class, running exercises become the norm of their life. The atmosphere for campus running activities is strong. The habit of consciously running exercises after class is being developed.

5 Conclusion

The Campus Trail Lepao APP fully integrates extra-curricular exercise into the examination of physical performance, transforms compulsory physical education into a conscious behavior, sets up an activated lifestyle for physical education students, forms a habit of lifelong exercise, drives college students out of dormitories from their extra-curricular time, pushes them to the campus and moves towards nature, and realizes the Ministry of Education's decision to incorporate "extra-curricular physical activities into the school teaching plan". Its reform method is worth promoting.

Acknowledgments. This research was supported by Teaching Research Project of Hubei Institute of Science and Technology (2018-XB-016).

References

1. Hu, G., Zhang, L., Wang, H., Xing, Z.: National fitness" and "national health" under the background of the new era of public sports teaching reform in colleges and universities. Chin. School Sports (High. Educ.) **5**(07), 25–30 + 35 (2018)
2. Xiang, L.: Reality and outlet of school sports in China from the perspective of rule of law. Shandong Sports Sci. Technol. **41**(02), 59–63 (2019)
3. Xia, Q.: Research on the Current Situation and Development Countermeasures of Extracurricular Sports Activities and Sports Competitions in Universities in Northern Anhui. Huaibei Normal University (2016)
4. Weihua, Hu: Research on the development trend of physical education teaching objectives in colleges and universities at home and abroad. Educ. Occup. **02**, 190–191 (2010)
5. Li, Y.: Research on the Current Situation and Influencing Factors of Sunshine Sports in Some Universities in Heilongjiang Province. Beijing Sports University (2012)
6. Ren, L.: Research on the Construction of Integrated Teaching Mode of Volleyball in and out of Class in Changchun Universities. Changchun Normal University (2018)
7. Dai, X., Xie, D., Qin, C., Wang, D., Xie, H.: Optimization of the curriculum environment of university physical education "in-class and out-of-class integration"-taking Shenzhen University as an example. J. Phys. Educ. **22**(06), 75–79 (2015)
8. Zhang, X.: On the coercion and freedom in school physical education-thoughts triggered by "compulsory physical education" in schools. J. Beijing Sports Univ. **40**(12), 78–83 (2017)
9. Liu, Y., Zuo, T., Zhu, X., Tian, C.: Research on the current situation and countermeasures of the second class of physical education in colleges and universities from the perspective of health promotion-taking a university as an example. J. Univ. **39**(02), 85–88 (2019)
10. Cao, X., Cao, W.: Research on exercise motivation of app in college students' campus trail-taking Shaanxi Normal University as an example. Sports World (Academic Edition) 2019(06), 65–66 + 47 (2019)

Data Mining of Educational Service Quality of Professional Degree Masters Based on Improved Expectation Confirmation Model

Yu Xiang[1,2] and Jianmin Liu[1(✉)]

[1] Guangxi University of Finance and Economics, Nanning 530007, China
jianminliu2007@163.com
[2] National Institute of Development Administration, Bangkok 10240, Thailand

Abstract. Based on the theory of Expectation Confirmation, this article discusses the formation mechanism of graduate education satisfaction from the perspective of graduate students' expectation and perception quality, constructs a research model of graduate education satisfaction. We conduct data mining for graduate students of Guangxi University of Finance and economics, analyzes the influencing factors of graduate education satisfaction, and confirms the expectation, perception performance and expectation through the theory of Expectation Confirmation. The satisfaction is verified to provide theoretical basis and decision-making reference for improving the quality of graduate education in research universities, strengthening the cultivation of innovative talents and promoting the development of graduate education in China. The research in this article shows that the establishment and promotion of the school image through various methods has a significant positive impact on improving graduate satisfaction.

Keywords: Expectation confirmation theory · Graduate education · Service quality

1 Introduction

Postgraduate education is an important part of higher education in China. Since the postgraduate system was restored in 1979, after four decades of development, a relatively complete system has been formed. Therefore, under the social background of comprehensively improving the quality of teaching in China's higher education and building a harmonious campus, it is of great significance to carry out research on the service quality of the professional degree and master's degree education process. How to improve the quality has also become the focus of attention of all sectors of society.

Graduate satisfaction, as one of the important indicators to measure the quality of graduate education, has also become the focus of research in graduate education. Based on the theme of "Graduate Education + Satisfaction", a search was performed in the China CNKI, and it showed a total of 205 articles from 2000 to 2019.

J. MacIntyre et al. (Eds.): SPIoT 2020, AISC 1283, pp. 529–535, 2021.
https://doi.org/10.1007/978-3-030-62746-1_78

By retrieving previous literature, we can see that research on the quality of graduate students mainly focuses on the curriculum setting, teaching mode, and learning quality of professional graduate students. Few researchers have studied the graduate students' expectations, perceived quality, and perceived performance. Analysis with the degree of expectation confirmation degree, and the research results based on expectation confirmation theory are even rarer.

Therefore, this article is based on the theory of expectation confirmation, which specifically studies the influencing factors of graduates' satisfaction with schools, makes graduates the subject of higher education service quality evaluation, and explains the determinants of graduates' satisfaction.

2 Design of Quality Model of Professional Degree Master's Education Service

2.1 Theoretical Model

Expectation confirmation theory is mainly based on the research of consumer satisfaction. This theory was proposed by American marketing scientist Richard L. Oliver in 1980. The theory believes that whether a person is willing to buy a certain product or service repeatedly is mainly composed of expectations, Factors such as perceived performance, degree of expectation confirmation, and satisfaction are jointly determined [1].

2.2 Research Hypothesis

Sun Youran, Yang Miao, and Jiang Ge (2016) believe that the image of universities has a significant direct positive effect on student expectations [2]. Li Rui, Ni Chuanbin, Xiao Wei, Su Qiujun (2016) believes that the platform's interactive functions should be used reasonably to improve learners' listening and speaking skills [3].

Therefore, from previous research conclusions, we can see that postgraduates' factors such as school image, student expectations and quality perception can all have a significant positive impact on the degree of expectations confirmation. Based on the above literature and inferences, the following hypotheses are therefore proposed: Hypothesis 1: University image has a significant direct positive effect on student expectations. Hypothesis 2: Student expectations have a positive effect on the degree of expectation confirmation. Hypothesis 3: Students expect positive effects on quality perception.

Hui Liu and Zhengnan Lu (2012) believe that on the basis of customer satisfaction and other related theories, a higher education student satisfaction model is constructed. Studies have shown that among the factors that affect student satisfaction, the most direct influence is quality perception [4]. Su Shengqiang (2012) believes that perceived quality has a significant impact on student satisfaction. If TVU wants to improve student satisfaction, it must be the top priority of all work [5]. Zhou Xiaogang, Chen Xiao, Liu Yuemei, Fan Tao (2017) believe that the service quality of Didi Chuxing includes five dimensions: security, tangibility, reliability, responsiveness and empathy; service quality and conversion costs are positive This affects passenger satisfaction and loyalty, and the effect of service quality is most obvious [6]. Wang Qin and Luo Jianchao (2018) believe that rural Internet finance farmers' loyalty is not high, and their satisfaction with financial services and financial products are the most basic factors affecting farmers' loyalty [7]. Li Wu (2017) believes that user satisfaction with e-book reading clients significantly affects their loyalty [8]. Yang Tao (2016) believes that satisfaction and continued use intention may be transformed into usage habits [9]. Gong Qifeng (2011) believes that the quality of teaching services is undoubtedly the most important [10]. Many studies have shown that the more positive a student's perception of the quality of a school is, the stronger his satisfaction with the school is.

Based on the above literature and inferences, the following hypotheses are therefore proposed: Hypothesis 4: Quality perception has a positive effect on the degree of expected confirmation. Hypothesis 5: Quality perception has a positive impact on satisfaction. Hypothesis 6: The degree of expectation confirmation has a positive and significant effect on satisfaction.

Zhang Wenqin, Shi Jintao, Liu Yun (2010) believe that the learning goal orientation of team members has a significant positive impact on their innovation behavior [11] . Therefore, the more positive and positive a student is towards his or her learning goals, the stronger the degree of confirmation of his expectations. On the other hand, the less ambiguous his attitude towards learning, the lower his degree of confirmation of learning expectations.

Hypothesis 7: Goal setting can have a significant positive impact on the degree of expectation confirmation.

Figure 1 shows Improved Expectation Confirmation model described above.

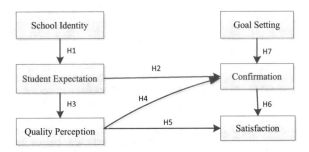

Fig. 1. Improved expectation confirmation model

3 Measurement and Evaluation

3.1 Questionnaire Design

This research focused on the 2017 and 2018 MPAcc graduate students of Guangxi University of Finance and Economics. The data comes from field surveys of graduate students of Guangxi University of Finance and Economics. The survey was conducted in the form of questionnaires. The survey period was from November 1, 2019 to 2019. December 1 for one month. The surveyed area is mainly for all MPAcc graduate students of Guangxi University of Finance and Economics.

The questionnaire contains two parts: The first part surveys the background of students, mainly the basic information of graduate students of Guangxi University of Finance and Economics, including gender, age, grade, and undergraduate graduation school; The second part mainly investigates the relevant dimensions of the education satisfaction of MPAcc graduate students in Guangxi University of Finance and Economics. Related entries and sources are shown in Table 1.

Table 1. Dimension, Code and Entry

Dimension	Code	Entry
Quality Perceived	QP	1. I am very satisfied with the quality of graduate teaching in the school 2. I am very satisfied with the quality of the classroom teaching of graduate training 3. I am very satisfied with the school's practical teaching base 4. I am very satisfied with the setting of the school thesis
Student Expectation	SE	1. I have slightly higher expectations for school 2. I think the current status of the school is in line with my expectations 3. I think that through postgraduate studies, 4. I bring great utility
Goal Setting	GS	1. I set learning standards for my learning tasks. 2. I set short-term (daily or weekly) and long-term (monthly or semester) learning goals. 3. I maintain a high standard of study quality. 4. I set goals to help me manage my study time.
School Identity	SI	1. I think the school is a "high-level university with distinctive characteristics" 2. I think that the school's emphasis on teaching quality can guarantee the gold content of graduate education 3. Do you think schools can provide high-quality teaching services
Expectation Confirmation	EC	1. I think the experience I got during my postgraduate study was better than expected 2. I think the benefits of postgraduate studies are greater than I expected 3. Overall, my expectations for the graduate level are met during the learning process
Satisfaction	Sa	1. I would like to recommend a school to more students 2. I think this school is my best choice 3. I would like to publicize the school to friends and family 4. I am willing to publicize the school more and better 5. If possible, I would like to support the construction of my alma mater

3.2 Research Method

Based on the expectation confirmation model, this study draws on the seven dimensions proposed by Liu Hui and Lu Zhengnan (2012) in the study of Chinese higher education student satisfaction evaluation as the measurement dimensions of student satisfaction, combined with our university's professional graduate students The current status quo is a subjective measurement of the quality perception, student expectations, school image, value perception, degree of expectation confirmation, and satisfaction of higher education students. The Likert seven-point method is adopted to measure (the higher the score, the higher the degree of identification of the person being tested on the dimension. The students are required to choose the option that they think is the most suitable and complete the questionnaire independently.

3.3 Statistical Analysis

This study uses SPSS statistical software to analyze the data. The analysis methods mainly include analysis of variance, regression analysis and "expectation confirmation" analysis. The analysis of variance is mainly used to explore the individual differences in education satisfaction of professional degree masters. The regression analysis is designed to examine the factors that affect the education satisfaction of professional degree masters. The "Expectation Confirmation" analysis explores the students' degree of professional degree masters. The degree of expectation confirmation and satisfaction evaluation of various service indicators of education.

4 Measurement Results and Analysis

A total of 200 questionnaires were sent out in the graduate students of MPACC of Guangxi University of Finance and Economics. Incomplete questionnaires and over-consistent forms were excluded. Finally, 195 valid questionnaires were recovered. The effective questionnaire recovery rate was 95.5%. The model of influencing factors obtained through comprehensive analysis of data is shown in Fig. 2.

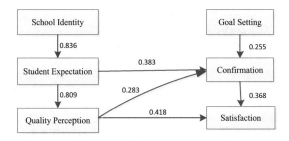

Fig. 2. Influencing factor model

The path map and path coefficients after analysis are shown in the above figure. All path effects p-values are less than 0.01, reaching a significant level. The correlation coefficient between the two exogenous variables of school image and goal setting is 0.503, and the p value is 0.000, which is less than 0.05, reaching a significant level, indicating that there is a positive relationship between the two. The indirect effect of school image on satisfaction through the intermediate variables of student expectations and quality perception is equal to 0.283, and the indirect effect of student expectations through the intermediate variables on the degree of satisfaction is 0.118, and the indirect effect of student expectations, quality perception level, and expectation confirmation level on satisfaction through intermediate variables is 0.07, so the total impact of school image variables on satisfaction variables The effect is equal to 0.471, which shows that the establishment and promotion of the school image through various methods has a significant positive impact on improving graduate satisfaction. The indirect effect that goal setting affects satisfaction through the degree of expectation confirmation of intermediate variables is equal to 0.094, so the total effect of goal setting variables on satisfaction variables is equal to 0.094, which indicates that graduate students have conducted various methods. Self-targeting has a positive impact on improving graduate satisfaction.

5   Conclusion

This article proposed an Improved Expectation Confirmation model. The research in this article shows that the establishment and promotion of the school image through various methods has a significant positive impact on improving graduate satisfaction. Moreover, Self-targeting has a positive impact on improving graduate satisfaction.

Acknowledgments. This research was supported by Innovation Project of Guangxi Graduate Education, Innovation Project of Guangxi University of Finance and Economics Graduate Education (Grant No. XWJG201903) and Phd Research Foundation of Guangxi University of Finance and Economics.

References

1. Oliver, R.L.: Satisfaction: A Behavioral Perspective on the Consumer: A Behavioral Perspective on the Consumer. Routledge, New York (2014)
2. Sun, Y.R., Yang, M., Jiang, G.: Research on the construction of a college equation practical teaching satisfaction model. High. Educ. Explor. **1**, 74–81 (2016)
3. Li, R., et al.: Research on the influencing factors of the continuous use of interactive english platforms under the ubiquitous learning concept. Distance Educ. China **10**, 72–78 (2016)
4. Liu, H., Lu, Z.N.: Research on Chinese higher education student satisfaction evaluation based on PLS path modeling technology. High. Educ. Explor. **2**, 30–36 (2012)
5. Su, S.Q.: Empirical research on distance learner satisfaction from the perspective of structural equation model. Distance Educ. China **5**, 49–55 (2012)
6. Zhou, X.G., et al.: An empirical study of the impact of 'didi travel' service quality on customer satisfaction and loyalty. Stat. Inf. Forum (32), 117–122 (2017)
7. Wang, Q., Luo, J.C.: Research on the influencing factors of rural internet finance user loyalty. J. Northwest A&F Univ. (Soc. Sci. Edition) (6), 16 (2018)

8. Li, W.: Research on the impact of perceived value on user satisfaction and loyalty of E-Reading clients. J. Libr. Sci. China **43**, 35–49 (2017)
9. Yan, T.: Research on eBook users' continuous use behavior: expansion of expectation confirmation model. Res. Libr. Sci. **22**, 76–83 (2016)
10. Gong, Q.F.: Educational service quality, student satisfaction and loyalty: SERVQUAL or SERVPERF? China Soft Sci. **2**, 21–26 (2011)
11. Zhang, W.Q., Shi, J.T., Liu, Y.: Two layers of influencing factors on team members' innovation behavior: personal goal orientation and team innovation atmosphere. Nankai Bus. Rev. **5**, 22–30 (2010)

Intelligent Development of Urban Housing

Tingyong Wang$^{(\boxtimes)}$, Yuweng Zhu, Fengjuan Zhang, and Wei Zhao

Strategic Research Center for Rural Revitalization in Ethnic Areas Qiannan,
Normal University for Nationalities, Duyun, China
wwll202014@qq.com

Abstract. The urban housing intelligence relying too heavily on the Internet with low security, and a disorderly market seeing false propaganda of some products with low compatibility, insufficient practicality and comprehensiveness is found based on the analysis and prospect of the definition, composition, current status and problems of urban housing intelligence. Given that, these situations can be addressed by strengthening algorithms and firewalls, offering in-person experience and careful thinking by consumers, establishing a unified technical standard with focus on scientific development. Finally, it is concluded that housing intelligence is the inexorable trend of social development, which will gradually integrate into and change people's lifestyle.

Keywords: Urban housing · Intelligent residential area · Community intelligence · Smart home · Development

The Internet came into being with the success of sending two letters "LO" from University of California, Los Angeles (UCLA) to Stanford Research Institute, 500 km away on October 29, 1969. With the increasing number of hosts access to ARPA network and NSF network, originally serving for scientific research, the Internet has also been commercialized and eventually becoming globalized, evolving into the Internet today, deriving big data and cloud network, as well as intelligent industries. Taking smart watch and artificial intelligence, etc. as an example, such intelligent products are gradually influencing our life and lifestyles, building a safe, convenient, comfortable, energy-efficient living environment for people [1–4].

1 Concept of Urban Housing Intelligence

The current study on urban housing intelligence in China has made progress to some extent, but lacking a complete consensus on the specific concept of urban housing intelligence. It was not until 2001 that the national relevant authorities formally proposed the basic concept of what is an intelligent residential area. The so-called intelligent residential area is a kind of modern intelligent residential area integrating property, fire safety, service and management through modern advanced 4C network transmission and control technology, thus achieving a more comfortable and high-quality family living environment [5].

J. MacIntyre et al. (Eds.): SPIoT 2020, AISC 1283, pp. 536–543, 2021.
https://doi.org/10.1007/978-3-030-62746-1_79

2   Structural Composition of Urban Housing Intelligence

(1) Control center system

The control center plays a role in managing and controlling the internal and external inter-action of the entire intelligent residential area, which thereby can also be called a bridge for effective communication between the residential area and the surrounding. Mean-while, in respect of structural composition, this system consists of five parts, namely information management module, parking lot management module, fire safety man-agement module, smart home management module and property management module [6].

(2) Network cabling system

An efficient and convenient cabling system plays an irreplaceable role in improving the intelligence level of the community. First of all, the cabling system, as a neces-sary hardware supporting platform for all information transmission, offers a high-speed information transmission channel, guaranteeing the interconnection between intelligent modules of the entire residential area. Secondly, people's living standards has increas-ingly improved with the sustained economic and social development, therefore a more and more efficient and complete information transmission and service quality is required.

(3) Smart home system

It can be said that the smart home system is a multi-field technology integration system. It includes not only modern computer technology, communication technology, and elec-trical circuit design technology, but also the basic treatment technology of the medical industry. Meanwhile, the smart home system also satisfies the user-defined settings of people's living habits, such as the control of lighting, fire safety, health and epidemic prevention, and the control of kitchen natural gas. It can effectively integrate multiple different subsystems, so as to greatly improve quality and level of people's lives [7].

Residence is the carrier of home, and home is the composition part of residence. People spend at least half of their life at home. Therefore, people have higher require-ments for the family environment. The home system can optimize the comfort of the family environment.

3   Problems in Housing Intelligence

The housing intelligence in China at present is a market startup period, making the investment amount and technology of the industry steadily rising with the "Internet plus". The continuous development of social economy, the achieved overall well-off level, the increased middle class group, and the user's particularly strong demand make the market pay more attention to the pursuit of quality of life. There are also intelligent products with different prices in the market for people to select. However, it is far from enough compared with the popularity of smart phones. In addition to few attentions paid to electronic products by the older generation, the poor existing technology and practical experience of products are also a major reason. In practical use, there are several problems as follows:

(1) High dependence on the Internet with low security
Internet plays an important role in information transformation. It is an important method to control smart home, and is the basis of smart home, which highlights the importance of the Internet, and also reflects the dependence severity of smart home on the Internet. It means that once leaving the Internet, smart home will fall into a state of paralysis, it cannot be used even if there is electricity. Secondly, the use logs and monitoring records of smart home contain a large number of user privacy. The ordinary users can do nothing about the stealing of hacker, even don't know the stealing occurs, leading personal privacy disclosure.

(2) Mixed market with intelligence concept confused by businesses
Quite a quantity of products is with the label of intelligence. In essence, they only add logic system or technology on the basis of traditional home appliances. However, this is not the real smart home. For example, the common smart phones, which is the traditional phones with Android or IOS system added in, and becomes to our smart phones. However, we can't feel smart in practical use.

(3) "Intelligence" stunt is more significant than reality, and even becomes a scam
In general, most people considered that intelligence is an attribute that makes all kinds of things take the initiative to meet people's daily needs by the application of modern network technology. In fact, a lot of intelligence is just to add a touch screen to a stuff, or to set some time points, or even under the cloak of intelligence, with essential of ordinary home appliances to deceive consumers.

One of the most typical cases is the intelligent fraud of Juicero juicer. Juicero is a scientific research company established in San Francisco in 2013, with the intelligent juicer as one of its main products. When it was first put into the market, its selling price was raised to US$ 699 all the way. The juicer boasts a series of good qualities-it does not need fresh fruits and vegetables, but only needs to buy matching material bags, from which the freshest fruit and vegetable juice in the world are available, and it is free of washing for life-which is exactly this product advertised. However, when the Company's founder sold the product, he put forward a mandatory condition: to buy the material package, consumers must first buy a juicer. Finally, the myth about the material bag and juicer becomes completely clear. The so-called material bag turned out to be filled with suppressed fruit and vegetable juice. It is such a juicer with extremely low technology content that it has completed four rounds of risk financing totaling US$ 120 million before.

(4) Low compatibility
The poor compatibility of the equipment reflects the non-connection in the industry. In the actual application of technology, each brand manufacturer has different research and development capabilities and technical levels, and different emphases, which leads to the production equipment not necessarily compatible. Even in the same brand, there are incompatible cases. For example, Huawei Honor A1 Watch and other watches in Huawei cannot be recorded and bound via the same APP.

(5) Lack of practicality and comprehensiveness

Taking the intelligent sweeping robot as an example, the invention solves the tedious daily cleaning, but its shortcomings are also obvious: a. It will sweep pet excrement all over the floor; b. If the gate is not marked with a virtual wall, it will run out of the house; c. the route planning algorithms of some brands are not scientific enough; d. Battery capacity; e. Wet mop needs to dry and the rag need to wash manually, which are more troublesome than using mops. f. Collision with furniture; g. Noise, etc. Each product is bound to have its own advantages and disadvantages. Only through practice can we continuously make improvements. Each generation of products needs to be gradually improved. However, for families with pets, owing the sweeping robot is not a worry-free experience. Wet drag mode is actually practical, but failed experiences will reduce consumers' trust in such products.

4   Solution to the Problem of Housing Intelligence

(1) Adding Bluetooth, AI and data algorithms to strengthen firewalls

Firstly, in the absence of the Internet, wireless technologies such as Bluetooth and NFC can be used in short distance, while algorithms and artificial intelligence in long distance to calculate, estimate the control time and method, and implement them. Secondly, when designing intelligent products, network security protection issues should be taken into account. For example, the firewall system should be written into the mainboard module or the network protection should be modularized, which can be sold as a separate product, with some of the obtained funds used to maintain and update the protection system to form a virtuous circle.

(2) Self experience and evaluation observation of consumers

What consumers really need is the humanized experience, just like the rapidness, high efficiency, safety and comfort required by the definition of housing intelligence. The actual experience of the consumers should be the standard to judge whether a product is intelligent and those products that haven't been used can't arouse the natural trust of consumers. Different measures to local conditions are of great importance. The condition and building design of each family are different and the installation design shall meet the personalized requirements, which avoids the waste of money and space and guarantees the high quality life experience.

(3) Careful thinking and bold demonstration

With Juicero juicer as an example, it has very low intelligence content, but the accessories used are quite expensive. The panel is very hard and thick and proved that the juicer will not fall accidentally and the brushed metal patterns are decorated on the inner wall. However, the juicing space is very narrow, so it is needed to buy the special material bag produced by the manufacturer. The circuit board provides WIFI internet function, and has camera and scanner to scan the barcode on the material bag. The core is an ARM processor with USB interface for the convenience of upgrading of juicer system. The transformer is provided by the company specialized in producing electric drills. The juicing capacity is the same with the promotion. The technology firm in Silicon Valley might have been cheated by the liner with almost perfect integration of the hardware in engineering science. Therefore, think twice is necessary faced with this kind of product. For example, whether it is necessary to apply high technology in the juicer, how to apply the technology, how to operate it and how to realize the capacity as it has promoted, etc. The conclusions after thinking can't equal the truth, but can be demonstrated one by one, to obtain the truth and conclusion finally.

(4) Constructing unified technical standards and improving the compatibility and universality of products

The final target of production is to sell the products to consumers. The low compatibility of products means to force the consumers to buy products of the same brand. However, the research, development and promotion of a brand take a long period and require a large quantity of resource and money consumption, which will force companies to select one direction for special research and development. When the consumers need the functions and products developed by another company and fail to use them due to the incompatibility, they will be dissatisfied with the original brand, but the cost to change it is high and not cost-efficient. So, it seems to have guaranteed the consumers' loyalty to the brand and user viscosity in a short term, but is not beneficial to the long-term development of the brand. What's more, consumption plays the guiding role for the adjustment and upgrading of production. Who has mastered the demand and pain point of consumers will be the final winner.

As the intelligent market grows bigger and affects people's life style more deeply, a set of standardized technical requirements and rules and regulations are inevitable and in need. The compatibility is the necessary trend after the technical requirements are unified.

(5) Science as the primary productive force

Practicability and comprehensiveness are mostly restricted by the technological development level, e.g. the intelligent floor cleaning robot spreads the waste of pets on the floor because of its working mode and principle, the route planning is not scientific due to its algorithmic fault and the same reason applies to its hitting the furniture. All the similar circumstances happen because technology has restricted the imagination and development power of human beings. Only with the technological development, can these problems be solved. Furthermore, also due to the technological development, can the output cost of advanced technologies decrease and only with the reduced cost, can the corresponding technological means be commercialized and put into use.

5 Development Trend in Prospect

(1) The popularity of smart home

Seen from the Market Scale Change Chart of China's Smart Home Industry from 2012 to 2018 released by Chinabgao, the smart home market has steadily increased with fantastic room for improvement. It is estimated that the total size of China's smart home market is expected to reach RMB 225 trillion in 2018, and the overall growth rate of China's smart home market is expected to be about 13% in the next 3–5 years (Fig. 1).

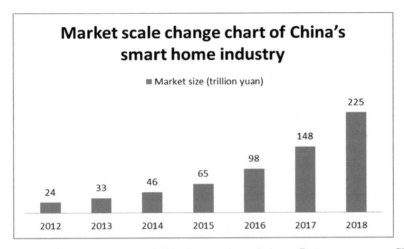

Fig. 1. Market scale change chart of China's smart home industry **Data source:** www.Chinab gao.com

It can be foreseen that with the development of the times and society, the continuous improvement of technology and the gradual maturity of intelligence, the intelligent electronic equipment market is poised for take-off, and a storm of electronic equipment reform is preparing. The smart home market is promising and is expected to become the next "winner" (Fig. 2).

category	Inventory of traditional products (100 million yuan)	Assuming a reasonable price for intelligent products (yuan)	Scale of potential intelligent products (100 million yuan)	Market growth	Industry concentration	Market share
Intelligent lighting	2300	Dozens to dozens	6742	25 % – 30 %	Lower	12 %
Intelligent air conditioning	11025	More than 5000	16538	slower	high	28 %
Intelligent refrigerator	9788	More than 4000	14682	slower	high	25 %
Intelligent washing machine	7623	More than 3000	11434	slower	high	20 %
Intelligent door lock	700	Hundreds to thousands	3300	20 %	Lower	6 %
Sun Shade	1997	250	4992	slower	Lower	9 %
Sports and health monitoring		Hundreds to thousands	89	50 % – 100 %	Lower	0 . 2 %
Home camera	?	250	538	Lower	Lower	0 . 9 %
Total			58315			

Fig. 2. Statistical table of classified data of China's smart home industry **Data source:** www.Chinabgao.com

In view of the table, smart air conditioners, refrigerators, and washing machines rank among the top three, with a total market share of 73% and a slower market growth. Intelligent lighting, door locks, and sunshading are all in the midstream in terms of traditional product inventory, market growth, and market size. In comparison, domestic cameras and sports and health testing have the fastest market growth, while the market size is only 1.1%.

From the perspective of industry concentration, smart air conditioners, refrigerators, and washing machines rely on the advantages of companies and brands to occupy the market, but due to their high prices, big body sizes, and low rates of replacement, the market growth is slow. Intelligent lighting, intelligent door locks, and intelligent sunshading are also low-consumables with slow rates of replacement, but because of their moderate price, families with the ability to purchase or those who like supporting smart homes have a high purchase rate. However, owing to their low prices and small body sizes, compared with other types of products, domestic cameras and sports and health testing can better research and develop aiming at the pain points of consumers, so the market growth is relatively faster.

(2) Service intelligent robot

Robots are divided into industrial robots and service robots. As for service intelligent robots, they are composed of three major parts: chips, OS, and AI. At present, most intelligent robots adopt common, with Android operating system is as the mainstream. Artificial intelligence technology still needs to be explored, and some sectors are still immature. What's more, after the robot has its own "thought", it will act autonomously. For example, the Russian robot named Promobot IR77 tried to escape from the laboratory twice in a week, although both ended in failure. An Austrian cleaning robot Irobot Roomba turned on the power by itself and climbed onto an electric baking tray to commit suicide. In addition, Facebook robots create their own language to chat. It makes sense for us to believe that these robots with independent thought, are the same as their creators longing for freedom – you can create me, but you cannot restrict me.

(3) Artificial intelligence

The assertion that "machines can act intelligently (they act as if they are intelligent)" is called weak artificial intelligence hypothesis by philosophers, while the assertion that "machines capable of doing so are indeed thinking (not just simulate thinking)" is called strong artificial intelligence. With regard to these two kinds of artificial intelligence, various technology companies are still exploring and trying to put them into commercial use, most of which are weak artificial intelligence, and the commercialization of strong artificial intelligence is limited by productivity, technology, research and development, and consumer purchasing power. As technology and engineering gets mature, consumer habits develop, and the penetration into traditional industries deepens, strong artificial intelligence has the tendency to become a pervasive item like mobile phones.

(4) People's emotional sustenance

The pace of life in cities is getting faster and faster, and the development of electronic products such as mobile phones is getting more and more rapid, which has made communication between people less and less, and social circles narrower and narrower. However, communication on the Internet cannot fully meet the human's social nature, and smart devices are increasingly regarded as the nourishment for the mind and objects for emotional sustenance. For example, people like to personify the floor mopping robot, and even considers it as a pet. With the development of technology, artificial intelligence will become more and more mature, and the days when the robot can truly become people's emotional sustenance and companionship is approaching.

(5) Energy conservation and emission reduction

While providing convenient and efficient services for people, intelligent housing reduces the waste of resources. Those is difficult for the human brain to calculate and for human power to reach will become simple and easy through data calculation and intelligent distribution. Under the premise of resource scarcity, smart products must make full use of resources, reduce waste of resources, and even achieve fewer resources required by themselves. Energy conservation and emission reduction is a topic that the whole world needs to strive to realize, and it is also the trend of history.

6   Conclusion

There are still many technical problems in intelligent housing, but its development and expansion is an inevitable trend in society. In the process of its development, we shall adhere to the principles of energy conservation, environmental protection and sustainable development, to energetically develop the intelligent industries, with the goal of benefiting mankind, and the task of promoting the progress of the times. We shall apply intelligence to real life, making it close to life and integrated into life, thus changing people's lifestyle like electronic payment. At the same time, we shall also pay attention to its humanized design, for the real intelligence is thinking based on the people and serving for the people.

References

1. Gu, C.: Practice of Intelligent Property Management in Urban Residential Area. Southwest University, Chongqing (2014)
2. Jia, X., Han, W.: The current situation and development trend of intelligent urban residential buildings. Heilongjiang Sci. Technol. Inf. (2), 206–206 (2014)
3. Li, C.: Research on planning and design of intelligent system in urban residential area and its implementation. Intell. City (3), 10–11 (2016)
4. Wang, Z.: intelligent system design in urban residential area. Dig. World (9), 76–77 (2016)
5. Intelligent Equipment. Losberger's sharing of intelligent warehousing solutions in the "Internet Plus" era. Eng. Mach. (9), 150–150 (2017)
6. Cui, M.: Design of electrical and intelligent control system for residential area buildings. Build. Mater. Decor. **26**(3) (2016)
7. Lv, J.: Application of intelligent design in residential buildings. Urban Archit. (9), 28–28 (2016)

Effective Approach to Cultivating Physical Education Teaching Skills of Preschool Majors in the Context of Big Data

Baolong Wang[✉]

Qing Dao BinHai University, Qingdao 266555, China
yiyiyayawbl@163.com

Abstract. Under the guidance of the major national education strategy of building healthy China and realizing the modernization of education, strengthening P.E. teaching in school has become the main theme of current school physical education. However, children's physical education is still a weak link in school physical education. With the incomplete awareness of the importance of children's physical education and the shortages of professional teachers, the development of children's physical education lacks sufficient scientific guidelines which deviating from the requirements of the Guidelines for Kindergarten Education. By literature review and qualitative research, this paper takes the curriculum system of preschool education as the research object to find an effective approach to cultivating physical education teaching skills of preschool majors. Moreover, through standardized physical education skills training and with the help of the Big Data, the study noted that the curriculum training system and the skills of students of this major can be enhanced from 5 dimensions: teaching design, sports technique, teaching practice, teaching evaluation and research.

Keywords: Big data · Early childhood sports · Physical education teaching skills · Pre-school education · Curriculum

1 Introduction

Regular involvement in physical activity brings a variety of psychological, social and physical benefits for children . The time on children's daily physical activities has been clearly stipulated in the "Health" section of the Guidelines for the Learning and Development for Children Aged 3–6 issued by the Ministry of Education, that is "The time on Children's daily outdoor activities should generally be no less than 2 h, in which the physical activity time should be more than 1 h despite of the seasons change" [1, 2]. In 2016, to strengthen the physical and mental fitness for the all-round development of students in schools, the General Office of the State Council stated that "kindergartens' education should take into consideration of the age characteristics and the physical and mental development patterns of children, and accordingly carries out various types of sports activities." How to scientifically guide preschool children to carry out activities

J. MacIntyre et al. (Eds.): SPIoT 2020, AISC 1283, pp. 544–550, 2021.
https://doi.org/10.1007/978-3-030-62746-1_80

effectively and safely is the ultimate duty of the PE teachers. It is an indisputable fact that the obesity rate and myopia rate of preschool children have been rising year by year [3, 4]. There is neither "magic device nor magic doctor "to prevent obesity, which calls for the combination of sports and outdoor sports. The development of PE teaching skills of preschool teachers should optimize the P.E. curriculum of preschool education major based on school PE curriculum. Physical teaching skills refer to a series of behaviors of PE teachers and methods they use to accomplish certain teaching tasks in the course of classroom teaching, which are the basic skills of PE teachers and the essential qualities for teachers as well as other professional skills [5–8]. Data shows that the professional theories and artistic qualities are the principle directions of the pre-primary education program in colleges and universities, however, few of them have been taken P.E. teaching abilities of students seriously. As early as 1976, the Madrid International Federation of PE attached the importance to the PEDAGOGY of PE teachers, which makes the argument of cultivating P.E. teaching skills available. Guided by the orientation of occupation, we can enrich the teaching contents of PE in preschool education and cultivate the PE teaching skills of preschool education students with the help of big data technology. Therefore, the health development of children can be empowered by optimizing the preschool physical education faculties [9, 10].

2 Training and Improving the Teaching Design

2.1 Constructing the Online Classroom and Implementing Theoretical Framework

Teachers are expected to master all fields of knowledge in PE and establish a reasonable knowledge structure. They can guide students to learn the basic knowledge, the principles and rules, the main contents and methods, values and significance of various sports activities of pre-school children in kindergartens. Different from students majoring in PE, students of pre-school major have less PE class, so their theoretical study should be concise and practical. But this problem can be solved via online PE courses offered on the network platform with stressing rapid and effective teaching methods. Teachers should also be capable of distinguishing all kinds of PE teaching phenomena, grasping the correct basic laws of PE teaching and guide PE teaching practice. The improvement of the teaching theories based on the enhancement of the teaching skills of PE.

2.2 Reversed Teaching Design

The teaching design aims at formulating the teaching plans and objectives of each class in accordance with the requirements of the Syllabus. It is the principle for teachers to follow and determines the quality of the class. Students of this major should be trained to learn the result-oriented curriculum design and master the reversed design of teaching contents and teaching methods. They should learn to set scientific, specific, measurable, realistic and time-limited "SMART" teaching objectives, and design sports activities based on the development patterns of preschool children and their own characteristics scientifically and safely, meanwhile strictly modulate the amount of exercise, see Table 1.

Table 1. Design of teaching process of Preschool Physical Education

Teaching process	Teaching content/Selection principle	Time allocation percentage
At the beginning	Small and medium amount of exercise is the key. Such as: formation, warm-up gymnastics; dance; Small amount of running, jumping, climbing; Fun games, etc.	10%–20%
Main Body	High volume of activity. The content may be the games, the exploration, the rules, the synthesis types,etc. The small-size class teaching mainly adopts the imitative practice and follows the teacher's activities. Medium -size class teaching begin to emphasize the combination of plot and cognition of the action, and gradually increased the regularities The large- size class apply fewer teaching contents,in which creative sports are encouraged, sports rules are standardized, and low-intensity competition activities are tried to cultivate the spirit of unity and cooperation for preschool children	70%–80%
The end	Small amount of games, massage, dance, collective equipment, breathing and emotional adjustment	10%

2.3 Research on Teaching Materials

Teaching material is an important reference for teachers' teaching and research. Unfortunately, previous literature measurement analysis has consistently demonstrated that in nearly past 10 years, from 2010 to 2019, the number of PE books for children in China is less than that of other subjects, and the content of those teaching materials is mainly imitating methods and experience abroad and focus on games. They are not consistent with characteristics of children in China at all. Based on the qualitative research, We may conclude that features of planning teaching materials for preschool education mainly focus on two aspects: First, a small number of textbooks on designing and guidance of children's physical activities; Second, an over-emphasis on the cultivation of literary and artistic skills in planning teaching materials which does harm to the all -round development of preschool children's personalities. To counter this disturbing trend, it has been assumed that formal teachers should be able to study and compose school-based teaching materials, and instruct students to use those textbooks critically so as to enrich the preschool teaching materials of PE.

3   Standardize Technical Movements and Improve Sports Skills

Physical exercises are the basic means of sport activities, and for children PE, imitation is their main approach to learning. The standard of the teacher's own sports skills will directly affect the preschool children's learning. Meanwhile, skills qualities are also the basic skills of PE teachers. The teachers need master relative skills, techniques and knowledge and output in class, not only demonstrate but also explain. As to the characteristics of preschool children's physical activities, it is also necessary to train the students of this major to create gymnastics. The combination of improving and training sports skills can appropriately elevate the assessment criteria of physical education courses for students of this major and enhance their sports skills.

4   Construct Comprehensive Physical Education Curriculum

4.1   Provide Demonstrations

In order to make the students of preschool education to clearly apply the PE teaching skills to the classroom, it is necessary to provide the students with typical and exemplary demonstration of teaching skills and specify them in details. By observing the application of organizational teaching skills in the classroom, students can acquire perceptual knowledge, establish distinct classroom concepts, take advantages and deepen understanding and produce exemplary effects. In this process, the big data technology plays an important role which helps to assign online teaching video tasks and answer questions and solve puzzles offline to improve learning efficiency. In choosing demonstration subjects, positive and negative cases are both required to be taken into consideration so as to guide students to find problems, propose solutions, and master the correct norms of teaching.

4.2   Innovate PE Teaching Methods for Children

Physical education teaching method refers to the ways or means to accomplish the teaching tasks in the teaching process. As far as its principle is concerned, the P.E teaching method of preschool children is the same as that of school-age children, but with its own characteristics. It is impossible to apply the PE teaching method in primary and secondary schools to preschool education, especially the method of sports training. In the process of children's physical education, the teaching principle of inspiration should be carried out from beginning to end, to stimulate children's interest in learning and participating in physical activities, and to cultivate their ability of active exercise and creative development. The children's physical education teaching methods should be applied in professional physical education curriculum of the preschool education major, to help students to learn the appropriate methods of children's physical education teaching, such as to use explanation and demonstration, combination the language tips and specific help and so on.

4.3 Development of Curriculum Resources

School programs reach the vast majority of children, and can affect physical activity both directly and indirectly. In order to cope with the current situation that students of this major have few PE curriculums, we adopt online micro-class to teach simple theory courses in classroom teaching organization and teaching practice to promote learning efficiency. It is an effective supplement to preschool students' learning of physical education theory with guide students to learn to make use of network resources. In view of the shortage of sports mooc resources, we should integrate the resources and fusion development to build online sports courses for preschool education.

4.4 Teaching Practice

Teaching practice is the independent starting period of students' career as teachers. In this stage, physical education teachers are required to participate in guiding interns to carry out physical education practice with children. In this stage, students change their role from learning-oriented to teacher-oriented, apply the theoretical knowledge to the practice and also have emergent problems.

Although gap exists between the original abilities and requirements of teaching, they can still manage to get familiar with basic teaching process under the guidance of professional teachers, complete the formulation of teaching plan independently, write teaching design and teaching plan and organize teaching.

In this stage, professional teachers should pay special attention to the performance of preschool education students. Although a preliminary teaching ability has formed in the teaching practice, their teaching skills have not been fully trained. In the process of teaching, interns tend to imitate their teachers in the classroom and pursue an active classroom atmosphere. However, Children in the early childhood stage are very active which makes such teaching skills in vain. It is necessary to have interns to master the psychological characteristics of children at this stage and improve their abilities to observe the children's emotions, response to and manage classes.

5 Learn to Evaluate

5.1 Evaluation of Learning Habits

The evaluation of preschool children must put non-intellectual factors in the first place, mainly through the development of their emotions, wills, attitudes and interests, Observe attention (concentration, indifference), interest (decline, inactivity, inactivity, hyperactivity), motivation, discipline, and the quality of the completed action to judge the child's behavior. Big data technology allows teachers to master the learning behavior habits of different students, so as to carry out diversified learning evaluation, and it is easier for them to pay attention to the differences of students and teach students according to their aptitudes.

5.2 Sports Load Evaluation

There are two methods commonly used to assess children's exercise intensity: observational and physiological measurement (pulse or heart rhythm measurement). Observation method refers to that teachers need observe the changes of children in physical activities at any time, such as the facial and skin color (red, purple, white),sweating, breathing conditions (with the nose, mouth, asthma) and know the changes of heartbeat to judge their mental state and amount of exercise. Physiological measurement is more objective and accurate than observational method which includes the measurement of pulse, blood pressure, respiratory rate, lung capacity, temperature changes in urine protein and so on. Pulse measurement is easier to operate in the kindergartens, and the rationality of sports load arrangement can be analyzed through multiple measurements of the whole course of physical education. Teachers should carefully study children's health status and check children's health cards and materials, and timely master children's health status through sports data accumulated in physical education. The basic purpose of sports load evaluation for preschool children is to monitor the physiological load of students, reasonably adjust the reasonable process of physical education, and finally achieve the teaching goal. Because of the differences load indicators of various sports, it is necessary for teachers to master and evaluate the exercise load indicators of different items, monitor the physiological load of preschool children during the course, at the end of the course and at the recovery stage after the class, conduct data analysis after the class, and improve the teaching system.

6 Developing Teaching and Research Abilities

From a development perspective, the preschool physical education curriculum is still in its initial stage, which needs all the preschool physical education teachers to keep enthusiastic and consistent in developing and improving teaching theories and practices. In this Big Data era, the ever-changing educational theories, teaching contents and educational objects require lifelong learning and research. Only by conducting discipline research can the PE curriculum of children become more effective in guiding students with the most cutting-edge knowledge and network technology and cultivate the habit of lifelong PE. The method of physical exercise will be more scientific and perfect, if PE research integrated with the physical education curriculum, so that the students of preschool education can master the initial methods and means of sports research, and cultivate the consciousness and abilities of doing research."

7 Conclusion

In order to cultivate the teaching skills of PE for pre-school students, it is necessary to establish a wholesome education and training system to improve the students' knowledge, renew the concepts of physical education, improve the health quality, and make them actively involved in sports. In this way, students can gradually learn to practice and teach. We should equip the students of this major with the above mentioned abilities, brim kindergarten with vigor of sports, consolidate the foundation of physical development, improve the quality of children's physical education, and last but not least balance "status" of art education and physical education in the training system of preschool education students, so that all the courses children will develop healthily and happily.

Acknowledgments. This work was financially supported by the social science project of Qingdao in 2019: A study on the construction of Preschool Physical Education from the perspective of dynamic quotient – a case study of kindergarten in Qingdao (QDSKL190123) fund

References

1. Guidelines for Learning and Development of Children Aged 3–6. Ministry of Education, Beijing, PRC (2012)
2. Mao, Z.: Teaching Theory of Physical Education, p. 106De. Higher Education Press, Beijing (2005)
3. Partment of Basic Education, Ministry of Education. Interpretation of Guidelines for Kindergarten Education. Jiangsu Education Press, Nanjing (2018)
4. Xie,W., Xi, M.: Research on the construction of vocational practical physical education curriculum model in higher vocational colleges. Vocat. Educ. Forum (32), 60–64 (2016)
5. Ding, G.: The impact of industrial transformation and upgrading on Shanghai Higher vocational colleges and the countermeasures of the reform. Vocat. Educ. Forum (12), 72–74 (2015)
6. Cui, G.: Research on Preschool Teachers' Physical Literacy, p. 9. Nanjing normal university, Nanjing (2015)
7. Wang, C.: Design and Guide of Children's Sports Activities, pp. 15–31. Fudan University Press, Shanghai (2018)
8. Pang, J., Liu, Q.: Health Education for Preschool Children, p. 27. Hua Dong Normal University Press, Shanghai (2008)
9. Zhang, Y.: Research on the cultivation of physical education abilities of preschool education majors. Educ. Teach. Forum (2012)
10. Liu, X.: Construction of Preschool Physical Education Curriculum for Preschool Education Majors in Higher Vocational Colleges in Sichuan Province, p. 35. Chengdu Institute of Physical Education, Chengdu (2016)

Coordination Game in Online Ride-Hailing Market of Smart City

Qi Duan[✉]

China National Institute of Standardization, Beijing 100191, China
duanqi@cnis.ac.cn

Abstract. Objective: analyze the modes of interest coordination and development of online ride-hailing market of smart city. Method: systematic and comprehensive analysis was made in this paper about the coordination game between smart city's government, enterprise and car owner in the online ride-hailing market to establish a system evolutionary game model, say the government - enterprise evolutionary game model and enterprise - car owner evolutionary game model. Results: the use of this model enabled the study of dynamic government- enterprise game evolution process under four situations and dynamic enterprise-car owner evolution process under six situations, sequentially to obtain a condition model, where the online ride-hailing market system converged to a non-supervision/non-monitor/coordination mode. Conclusion: the government must make proper regulation in the evolutionary process to develop appropriate regulations and policies, and guide the market participants to adopt healthy and sustainable strategies.

Keywords: Online Ride-Hailing car · Evolutionary game · Dynamic evolution process

1 Introduction

In 2018, with the murdering of two online ride-hailing passengers successively, safety problem on online ride-hailing operation aroused wide public concern and discussion in China. Actually ever since the operation of online ride-hailing service, supervision problem has been rather serious due to random parking, inconsistent car information, poor service quality and attitude when online ride-hailing service has become one of the important expressions of sharing economy. Smart cities should be linked with security, order and harmony.

Now some scholars have used classical game theories to analyze supervision problem in the field of sharing economy. For example, benefit relationship between enterprise, government and citizen [4, 5] was analyzed from the perspective of economy participated government evolution behaviors [1], the management mode based on the platform reward and punishment mechanism [2], the trip mode evolution model [3] and reverse logistics etc. Regarding the supervision of online ride-hailing service, some scholars have collected the various policies made by the local Chinese governments on supervising online ride-hailing cars and based on this, strict indexes for vehicle supervision

J. MacIntyre et al. (Eds.): SPIoT 2020, AISC 1283, pp. 551–557, 2021.
https://doi.org/10.1007/978-3-030-62746-1_81

have been built and regional difference [6] in it has been discussed. Also "regulatory capture", simpler administration and the relationship between the central and local governments have been studied [7] in supervising online ride-hailing cars. The summary of the various problems about it included the offside, vacancy and dislocation of roles [8] for the reasons including government [9], the twisted market relationship [10], the shortsightedness of the government sectors [11] and the lack of laws and regulations [12] etc.

It's a new research area regarding the system evolutionary game in supervising online ride-hailing market. In this paper, comprehensive and systematic study was made on the supervision over Chinese online ride-hailing market from government, enterprise and car owner. By analyzing the game relationship between government, enterprise and car owner in online ride-hailing market, a system evolutionary game model had been established. The analysis on the dynamic system evolution process under different situations reveals that the system evolution finally converges to a non-supervision/non-monitor/coordination mode and so suggestions on supervising online ride-hailing market can be raised from the perspective of government.

2   Analysis on the Benefit Game in Online Ride-Hailing Market

The benefits involved in the online ride-hailing market mainly include government, enterprise and car owner. During the supervision over the online ride-hailing market, the government generally guides, supports or gives aid to the online ride-hailing market via laws, regulations or investments. Targeting the maximum social and economic benefits for optimal strategies, the enterprise has dual properties. On one hand, as an independent entity, the enterprise tends to maximize its self-interest. But on the other hand, it must take social responsibility such as tax revenue, employment and safety when it's pursuing economic benefits. Therefore in the face of government supervise, some enterprises might fail to handle properly the relationship between economic interest and social interest without a positive attitude and measures taken to perform laws and regulations. But at the same time, from the perspective of car owner, the supervision over its service provided by the enterprise will reduce the number of online hailing cars, which in turn will deteriorate the problem of no car available in rush hours, resulting in less revenue made by the car owner in the short run, but do protect the rights and interests of the consumers regarding safety and service quality in the long term. Such a coordination game has played a significant role in online ride-hailing market.

Based on the above analysis, a coordination game model for the relationship between enterprise and the other parties has been established, as indicated in Fig. 1:

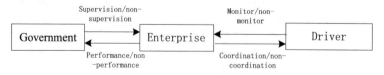

Fig. 1. The game model between enterprise and the others in online ride-hailing market

3 Coordination Game Model in Online Ride-Hailing Market

3.1 Research Hypothesis

Hypothesis 1: The government is under the pressure from economy and social concern, while the enterprise is facing compliance supervision from the government and public opinion supervision.

Hypothesis 2: Enterprises in online ride-hailing market have bounded rationality.

Hypothesis 3: In the government-enterprise evolutionary game, the strategy set includes {supervision, non-monitor} and {performance, non-performance}. But in the enterprise-user evolutionary game, the strategy set is {monitor, non-monitor} and {coordination, no-coordination}.

3.2 Analysis on the Government (Government) -Enterprise (Enterprise) Evolution Model

In the government-enterprise evolutionary game, assume that G_E is the gain of the enterprise when he doesn't perform laws and regulations and $G_E > 0$. C_E is the input cost arising from his performance of laws and regulations, and $C_E > 0$. P is the punishment imposed by the supervision department to the enterprise for his non-performance, and $P > 0$. G_G is the gain of the government after the performance of the enterprise, and $G_G > 0$. C_G is the input cost of the government for the supervision work and $C_G > 0$. R is the rewards given by the government to the enterprise for his performance. L_G is the opportunity loss for the non-supervision of the government when the enterprise fails to perform laws and regulations, and $L_G > 0$. x is the probability for the enterprise to perform policies and regulations on the market, and $0 < x < 1$. y is the probability for the government to implement the supervision on the market, and $0 < y < 1$. The payoff matrix is shown in Table 1.

Table 1. The pay-off matrix between government and enterprise

Enterprise / Government	performance	Non-performance
Supervision	$G_G - C_G, -C_E - G_E + R$	$P - C_G, -P$
Non-supervision	$G_G, -C_E - G_E$	$-L_G, 0$

According to the payoff matrix as above, then: The average government revenue $u_G = x[(G_G - C_G)y + (P - C_G)(1 - y)] + (1 - x)[G_G y - L_G(1 - y)]$. The average enterprise revenue $u_E = y(Rx - C_E - G_E) + (1 - y)(-Px)$. The theory of replicator dynamics applied in the evolutionary game reveals that the participants in online ride-hailing market will adjust their strategies in real time according to the gains and then turn to copy the high yield strategy. Therefore the probability proportion of adopting a certain strategy will be changed in real time, and the proportion growth rate of a certain strategy is actually the difference between the yield brought by this strategy and the average total revenue. For this reason, the proportion growth rate for the government to adopt supervision strategy is: $\frac{dx}{dt}\big/x = (1-x)[(P+L_G)(1-y) - C_G]$, Then the replicator dynamics equation for government is $\frac{dx}{dt} = x(1-x)[(P+L_G)(1-y) - C_G]$. Similarly, the replicator dynamics equation for enterprise is $\frac{dy}{dt} = y(1-y)[(P+R)x - G_E - C_E]$.

Assume that $\frac{dx}{dt} = 0$ and $\frac{dy}{dt} = 0$, then five equilibrium points for the dynamic government- enterprise game evolution system can be obtained, which are separately $(0,0), (0,1), (1,0), (1,1), (\frac{G_E+C_E}{P+R}, 1-\frac{C_G}{P+L_G})$. The analysis on the evolutionary stability of the government - enterprise game strategy reveals that the evolution system can be grouped into the following four situations:

1) Infinite loop: When $0 < \frac{G_E+C_E}{P+R} < 1$ and $0 < \frac{G_E+C_E}{P+R} < 1$, there are totally five equilibrium points in the dynamic government - enterprise game system, where the dynamic evolution trend at this moment is indicated by Fig. 2. Due to the divergence of the system, all of these equilibrium points are not ESSs. That's to say, this dynamic game system won't converge to any point and will enter into an infinite loop.

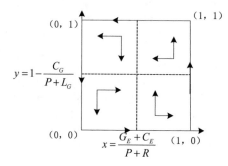

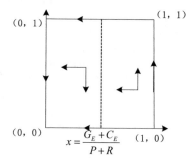

Fig. 2. Infinite loop   **Fig. 3.** (0,0) as an ESS

2) (0,0) as an ESS: When $0 < \frac{G_E+C_E}{P+R} < 1$ and $\frac{C_G}{P+L_G} \geq 1$, there are totally four equilibrium points, say $(0,0), (0,1), (1,0)$ $(1,1)$ in the dynamic government - enterprise game system, where the dynamic evolution trend for the change of government and enterprise strategies is shown in Fig. 3. Among all of these four points, only (0,0) is an ESS. Therefore the system will finally converge to a non-supervision/non-performance mode.

3) (1,0) as an ESS: When $\frac{G_E+C_E}{P+R} \geq 1$ and $0 < \frac{C_G}{P+L_G} < 1$, there are also four equilibrium points, say $(0,0), (0,1), (1,0)$ $(1,1)$ in the dynamic government- enterprise game system, where the dynamic trend for the change of government and enterprise strategies is shown in Fig. 4. Among all of these points, only $(1,0)$ is an ESS. Hence the system will gradually converge to a supervision/non-performance mode finally.

4) (0,0) as an ESS: When $\frac{G_E+C_E}{P+R} \geq 1$ and $\frac{C_G}{P+L_G} \geq 1$, the dynamic evolution trend for the change of government and enterprise strategies is indicated in Fig. 5, where only (0,0) is stable as an ESS. It means that the system will finally converge to a non-supervision/non-performance mode in any case.

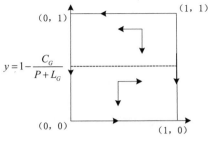

$$y = 1 - \frac{C_G}{P + L_G}$$

Fig. 4. (1,0) as an ESS

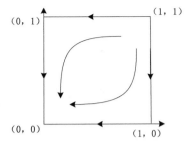

Fig. 5. (0,0) as an ESS

3.3 Analysis on the Enterprise (Enterprise) - Car Owner (User) Evolution Model

In the enterprise - car owner evolution game, the enterprise-car owner strategy set includes {supervision, non-supervision} and {coordination, non-coordination}. Assume that G_U is the extra income earned by the car owner when he doesn't coordinate than that when he does, and $G_U > 0$. C'_E is the cost of an enterprise for his supervision over a car owner. Take note that C'_E is different from C_E, which is the input cost for the supervision over the whole market and $C'_E > 0$. L_{UA} is the loss of car owner caused by any accident when he doesn't coordinate and $L_{UA} > 0$. P_U is the punishment imposed by the enterprise to the car owner for the non-coordination and $P_U > 0$. P_{EA} is the punishment made by the government to the enterprise for any accident caused by the non-coordination of the car owner, and $P_{EA} > 0$. p is the probability of an accident caused by the violation of laws and regulations by the car owner. The payoff matrix is showed in Table 2.

Table 2. The payoff matrix between enterprise and car owner

Enterprise / Enterprise	Coordination	Non-coordination
Supervision	$-C'_E, 0$	$P_U - C'_E, G_U - P_U - p(L_{UA} + P_{EA})$
Non-supervision	$0, 0$	$-pP_E, G_U - p(L_{UA} + P_{EA})$

The dynamic enterprise-car owner game evolution process can be concluded into six circumstances, where the first four situations are quite similar to those in the government-enterprise evolutionary game, including infinite loop, gradually converging to a non-supervision/non-coordination mode ((0,0) as an ESS), converging to a supervision/non-coordination mode ((1,0) as an ESS). The other two circumstances are described as below:

1) (0,1) as an ESS: When $P_U + pP_E > C'_E$ and $P_U + pP_E > C'_E$, the dynamic evolution trend for the change of enterprise and car owner strategies is shown in Fig. 6, where the enterprises tend to choose the non-supervision strategy, as the car owners are inclined to coordinate in all circumstances. In this case, the system will finally converge to (0,1), which also means a non-supervision/coordination mode.

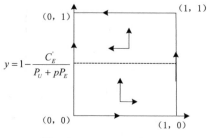

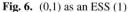

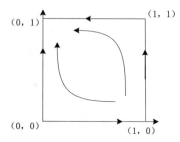

Fig. 6. (0,1) as an ESS (1)   **Fig. 7.** (0,1) as an ESS (2)

2) (0,1) as an ESS: When $P_U + pP_E \leq C'_E$ and $p(L_{UA} + P_{EA}) \leq G_U$, the dynamic evolution trend for the change of enterprise and car owner strategies is shown in Fig. 7, where the enterprises tend to choose non-supervision strategies in all circumstances, as the car owners are prone to the coordination strategy. In this case, the system will converge to (0,1), which also means a non-supervision/coordination mode.

4   Conclusion

The above analysis reveals that when $-P \geq L_G - C_G$ and $p(L_{UA} + P_{EA}) \geq G_U$, the whole system will converge to a healthy, sustainable and stable state, which is also a non-supervision/non-monitor/coordination mode expected by all of the participants in the online ride-hailing market. Therefore in order to make the system converge to this expected mode, the government must make proper regulation in the evolutionary process to develop appropriate regulations and policies, and guide the market participants to adopt healthy and sustainable strategies. The measures that can be taken by the government include:

1) Punishment of the enterprises for their non-performance but not higher than the input cost arising from the supervision works. In other words, the punishment must be appropriate and not too harsh.
2) Increase of the punishment on car owners for any accident caused by their failing to coordinate during the supervision.
3) Sufficient protection on the legitimate income of car owner, who is also a legal holder, with an attempt to reduce the income gap between the car owner's selection of non-coordination strategy and coordination strategy.

Acknowledgements. This research was financially supported by China Central Public-interest Scientific Institution Basal Research Fund (Grant No.: 552018Y-5930 & 552019Y-6660).

References

1. Wang, H.: Behavioral analysis of shared economic participants based on evolutionary game. Econ. Manag. **32**(2), 75–80 (2018). (in Chinese)

2. Zhang, Y., Zhang, J.: Evolutionary game between government supervision and shared bicycle platforms. Stat. Decis. **23**, 64–66 (2017). (in Chinese)
3. Chen, X., Zhou, J., Zhu, Z.: Evolutionary game analysis of choice of urban transportation travel modes. J. Manag. Eng. China **2**, 140–143 (2009). (in Chinese)
4. Du, M., Tao, B., Zhu, Y.: Research on incentive mechanism of reverse logistics of waste appliance recycling based on tripartite game. Soft Sci. **12** (2014). (in Chinese)
5. Wen, L., Chen, G., Wang, X.: The problem of reverse logistics in the development of e-commerce: game analysis of different interest relations of e-commerce enterprises, government and consumers. Ind. Econ. Rev. **1** (2017). (in Chinese)
6. Ma, L., Li, Y.: How the government supervises the sharing economy: an empirical study of china's urban internet car booking policy. Electron. Gov. Aff. **4**, 17–28 (2018). (in Chinese)
7. Wang, X.: The reform of the online car invitation in struggle. People's Forum **17**, 63–65 (2016). (in Chinese)
8. Xu, M., Liu, H.: Reform and improvement of local taxi market legal supervision under the background of online car appointment. Guangdong Soc. Sci. **5**, 249–256 (2016). (in Chinese)
9. Cai, C.: The rise of the sharing economy and the innovation of government supervision. South. Econ. **3**, 99–105 (2017). (in Chinese)
10. Zhang, X.: Government regulation innovation of "internet-based car-riding" from the perspective of experimental regulation. E-Gov. Aff. **4**, 32–41 (2018). (in Chinese)
11. Zhang, H., Li, M.: The standardized development path of the sharing economy in the new era: a new perspective of innovation governance. J. Hehai Univ. (Phil. Soc. Sci.) **5**, 41–49 (2018). (in Chinese)
12. Shi, X., Chen, D., Li, Y.: P2P network lending platform guarantee issues and legal regulations. J. Harbin Commercial Univ, **6**, 118–130 (2018). (in Chinese)

Construction of English Learning Community in the Smart City Environment

Liying Gao[✉]

Shandong Women's University, Jinan 250001, Shandong Province, China
397877@qq.com

Abstract. Smart city is an emerging concept mentioned in urban construction in recent years. It creates efficient, fast, flexible and clear intelligent responses to the needs of different aspects and different levels, creating a person and society, people and people, people and things A harmonious environment. Adhering to the people-oriented concept, the English learning community places English learners at the center of English learning, and relies on the platform of lifelong education and lifelong English learning to improve the lifelong English learning ability of community members, thereby driving the continuous development of the entire community. Therefore, it is becoming more and more important to build English learning communities in smart cities. This article elaborates the basic strategies of constructing English learning communities in smart cities in China. The guiding ideology and overall goals and basic principles of the construction of smart city English learning communities are clarified, and the four aspects of constructing a modern education system, social service system, individual culture system, and characteristic resource system are used to build a smart city English learning community in China. The scientific design of the construction provides a feasible path for the research of this article from theory to practice. Finally, in the form of questionnaire surveys and interviews, the status quo of the construction of English-speaking communities in smart cities in China was investigated. Based on the information and data of the survey items, the problems in the construction of English-speaking communities in smart cities were raised. It also analyzes and summarizes key issues such as basic cognition, system architecture, and resource guarantee, providing a realistic basis for the further development of this thesis. The survey results show that about 19% of the community residents think that the construction of a learning community has little to do with themselves, and they should actively mobilize the enthusiasm and enthusiasm of community residents to participate in community activities.

Keywords: Smart cities · English-learning communities · Building problems · Building strategies

1 Introduction

In recent years, in the context of advocating lifelong learning and creating a learning society, the issue of the construction of English learning communities has received

J. MacIntyre et al. (Eds.): SPIoT 2020, AISC 1283, pp. 558–565, 2021.
https://doi.org/10.1007/978-3-030-62746-1_82

increasing attention from domestic academic circles. As a basic unit of China's current advocacy of a learning society, learning communities have recently It has become a hot issue in the social and educational fields. Therefore, how to study the development status and development direction of the learning community in China from the dynamic and complex interdisciplinary development situation, and to build a Chinese-oriented community in the new century based on the actual situation in China to build a learning community, has become a professional scholar and Important issues of common concern to community managers and workers.

The construction of a learning community can not only start from the micro level, its learning scope is no longer confined to traditional schools, but it is extended to the entire society and its various cells. Learning is obtained both in connotation and extension. With a huge leap, it has also been extended in depth and breadth, and the learning process has become a unity of individual lifelong learning and social lifelong education. This inevitably requires continuous reform and innovation of the city's grassroots management system and community operation mechanism, and the coordination and integration of the interests of all parties. The construction of a learning community enhances the common participation of all members of the community, promotes mutual exchange and communication, unites the sense of community, strengthens the sense of community identity and belonging, builds a harmonious community, moves the community toward a healthy track, and ultimately builds a harmonious society Lay a solid foundation.

This article focuses on exploring the construction of a new community development situation that is in line with the three concepts of "lifelong education, lifelong learning, and a learning society" and the development requirements and construction laws of a harmonious society and a well-off society. The smart city English learning community seeks a construction path that has both commonality and personality, and is both scientific and effective. It provides theoretical support for fundamentally solving the problems that arise in the implementation of smart city learning English communities. And practical guidance.

2 Proposed Method

2.1 Construction of Smart City Learning Community

The construction of a smart city learning community, as a social practice closely linked to politics, economy, culture, and education, undoubtedly requires the infiltration and support of a modern theoretical system [1]. The theoretical foundations supporting the development of elderly education in smart city communities in China mainly include four aspects: the theory of a harmonious society, the theory of lifelong learning, the theory of self-organization, and the theory of learning organizations.

(1) Theory of a harmonious society

The theory of a harmonious society elaborates the social value of the construction of a smart city learning community in China. The essence of a harmonious society lies in the word "harmony", that is, to realize the harmonious development of man and nature, man and society, and human self. This emphasizes the dominant position of man in the construction of a socialist harmonious society and also shows the comprehensiveness of man. The importance of development. This fully shows that improving the overall quality of people and promoting "all-round development of people" with "all-roundly developed people" will become the task of building a harmonious socialist society. Deeper and deeper, the all-round development of people is actually the process of individuals continuously acquiring new knowledge and new skills in learning, and their potentials are fully explored and applied. Therefore, this is also the essential characteristic of the construction of a smart city learning community in China. And historical mission. In the process of constructing a learning community, it is necessary to take the realization of "all-round human development" as the starting point and end point. It attaches importance not only to the development of community members 'intellectual skills, but also to the development of community members' personality, personality, cultural connotation and life value. Enhancement, by advancing the three-phase harmonious learning concept of "human and nature", "human and society", and "human and self", so that community members can improve their quality and provide a solid foundation for the construction and exploration of a harmonious society Protection [2].

(2)   Lifelong learning theory

The theory of lifelong learning provides a direct basis for the construction of the learning community of smart cities in China. Lifelong learning is a lifelong education. It is a learning society that puts the lifelong and comprehensive nature of learning and the autonomy of learners in a prominent position. It believes that learning is a way of survival. "Without lifelong learning, there is no human being. The existence of a life-long society does not matter the quality of life of a person without life-long learning. "It is advocated that learning activities should run through each person's life. Learning activities should be" individuals choose the means and methods suitable for them according to their needs. The main task is to stimulate community members' subjective consciousness, development consciousness, self-worth realization consciousness, and learning motivation. It is not only beneficial for each community member to change existing learning concepts, and to make autonomous learning concepts rooted in the minds of each member. In the study, learning can be regarded as "a self-conscious responsibility for the realization of self-worth and sustainable development of the community", and all resources in the community (public, private, formal, and informal) (Informal, informal) are effectively integrated to build multiple sequences, multiple levels, multiple types, and diversity The "accessible learning platform" provides theoretical guidance, thus providing strong theoretical support for the realization of community-wide participation in learning services and the formation of a community lifelong learning culture within the entire community [3].

(3)   Self-organization theory

Self-organization theory provides theoretical reference for the construction of a smart city learning community in China. The theory of self-organization points out that the stronger the self-organization function, the stronger its ability to maintain and generate new functions. In the current era of social transformation and economic transition, in the process of constructing a learning community, how to give full play to the autonomy of the community is given priority. This autonomy requires not only that community members can have The willingness to participate in community learning activities is more important how the community members acquire the endogenous development ability of knowledge through learning, and in the process, through continuous integration of the internal resources of the community, gradually improve the self-education, self-regulation, and self-management of community members Level of self-restraint and self-renewal in order to achieve community autonomy in the true sense. Based on such development requirements, self-organization theory can help community members find the resources they need within the community, develop their spirit of cooperation, and put this spirit into practice, in order to stimulate community members' self-reliance to solve problems. In order to effectively solve the internal complex problems encountered in the process of constructing a learning community, it also provides a theoretical basis and internal mechanism for the effective organization of community members with a common willingness to learn and common interests [4].

(4) The theory of learning organization

The theory of learning organization has the most fundamental supporting significance for the construction of English learning communities in smart cities in China. A basic sign of the formation of a learning community is that "community learning organizations have been fully developed. It is not difficult to see that the construction of learning communities must be attributed to the question of "how to cultivate various organizations in the community into learning organizations". Come up. According to the theory of learning organization, a scientific learning organization "is able to stimulate new, forward-looking, systematic thinking and promote organization through a common, conscious, and interactive learning for a common ideal goal. The collective and system of continuous innovation and continuous development of the whole and each member". Therefore, how to support the various types of office organizations (such as schools, institutions, enterprises, etc.) and non-governmental organizations in the community based on the theory of learning organization. (All kinds of associations, associations, etc.), and other community autonomous organizations, establish "common vision", form common values, promote "interactive" learning, create "team spirit", achieve "intelligent sharing", and achieve "common transcendence" It has also become an important link in the construction process of smart city learning communities [5, 6].

2.2 Strategies for Constructing English Learning Communities in Smart Cities

(1) Building a learning community with a common vision

The first step in the construction of an English learning community is not the construction of many learning facilities. The most important thing is to create a lifelong learning atmosphere in the community so that all community members have the endogenous motivation to build a common learning home and reach agreement. The goal of building a learning community has the cohesion and appeal that make the entire community flourish [7, 8].

(2) Horizontal dimension of English learning community construction in smart cities

Achieve community and school resource sharing. Numerous colleges and universities provide unique educational resources for the construction of a learning community. The flow and sharing of educational resources can fully realize the value of resources and achieve a win-win situation between the community and the school. Judging from the role of universities in educating communities, it can be divided into hardware resources and software resources [9]. Colleges and universities can provide activities and facilities for community education, such as libraries, museums, sports fields, laboratories, etc. Various types of community activities organized by colleges can also attract the participation of community members, infecting everyone with learning with the educational and cultural atmosphere of colleges and universities. Willing person [10].

3 Research Method

(1) Literature method

On the basis of summarizing, collating, analyzing, and identifying related books and discussions on books and materials, this paper will summarize and summarize the research status of smart city English learning communities in terms of connotation, signs, operating mechanisms, and evaluation index systems at home and abroad. Have some knowledge, find out the lack of research and imperfections, and initially grasp the new problems and new contradictions in the process of the construction of English learning communities in smart cities at this stage, in order to study the construction of English learning communities in smart cities with regional characteristics The necessary literature was prepared.

(2) Questionnaire survey method and interview method

In-depth investigation of the actual development status of English-speaking communities in smart cities, a comprehensive understanding of the basic situation of the construction of English-speaking communities in smart cities through questionnaires and conversations, mastering first-hand information, focusing on the analysis of English-speaking communities in smart cities in the province The experience and problems in the construction process lay the foundation for targeted solutions.

(3) Statistical analysis method

Comprehensively analyze and demonstrate the collected data and various types of information materials. On the basis of understanding the current status of the operation of smart city English learning communities, analyze the internal factors affecting the construction of smart city English learning communities, and build a smart city for our country. Innovations in English-speaking communities provide factual evidence.

4   Discussion

4.1   Lack of Awareness of Community Individual Participation

The direct target of the English learning community should be all members of the community. Does it "meet the lifelong, full-length English learning needs of all community members, guarantee their basic learning rights, comprehensively improve the quality and quality of life of all community members, and realize human "All-round development" is an important criterion for measuring a learning community. It can be seen that the participation of community members is the final point for the construction of a learning community. Therefore, the level of community members' awareness of participation in the construction of an English learning community is an important prerequisite to ensure the construction and development of urban learning communities. However, it can be seen from the questionnaire survey of community members, as shown in Table 1 below.

Table 1. Survey of community members

	Close relationship	Little relationship	It doesn't matter	Unclear
Male resident	15%	40%	20%	25%
Female resident	16%	42%	18%	24%

As shown in Table 1 above, about 19% of residents think that the construction of learning communities has little to do with themselves, 42% of residents think that the construction of learning communities has little to do with themselves, and even 24% of residents think that learning The construction of a type community has nothing to do with itself.

4.2   Unity of Community Activity Forms

Most of the activities held by English-speaking communities are mainly entertainment and other recreational activities. However, some learning activities related to learning communities such as reading activities and special lectures are held infrequently, and even some English-speaking communities are almost No events of any kind have been held. As shown in Fig. 1 below.

As shown in Fig. 1 below, several learning activities that community residents are most willing to participate in are related vocational skills training, family education for infants and young children, law education, elderly education, urban adaptation of migrant workers, and social and cultural life education. It can be seen that the singularity of the existing forms of English community activities cannot meet the learning needs of the diversity of community residents. In addition, it is affected by external factors such as work, study, and family. As a result, "old residents" have appeared in community activities in most communities. It is a regular customer, young and middle-aged residents are visitors, school students are tourists, and preschool children are accompanying guests. The enthusiasm and enthusiasm of community residents in participating in community

564   L. Gao

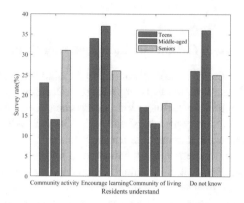

Fig. 1. Community residents' perception of the connotation of learning community

activities has not been fully mobilized, let alone being part of the construction of a learning community. The subject power is up.

5 Conclusions

Subsequent research should aim at the main characteristics of the construction of smart city English communities in China, and strive to bring the theoretical and practical elements obtained through previous research into the practice of building smart city English learning communities within a reasonable range Provide timely feedback on practical results, continuously deepen the practical and theoretical significance of conclusions, and provide data with high reliability and validity for the construction and development of smart city English learning communities. At the same time, various communities in the city can also find new problems in the construction of learning communities through horizontal comparisons, and analyze and compare the improvement direction, so as to promote in the process of continuous comparison and continuous improvement Smart city learning English communities are moving in depth.

References

1. Anderson, T.: Thinking collaboratively - learning in a community of inquiry. Int. Rev. Educ. **62**(1), 1–3 (2016)
2. Wu, W.C.V., Hsieh, J.S.C., Yang, J.C.: Creating an online learning community in a flipped classroom to enhance EFL learners' oral proficiency. J. Educ. Technol. Soc. **20**(2), 142–157 (2017)
3. Rapchak, M., Cipr, A.: Standing alone no more: linking research to a writing course in a learning community. Portal Libr. Acad. **15**(4), 661–675 (2015)
4. Deogade, S.C., Naitam, D.: Reflective learning in community-based dental education. Educ. Health **29**(2), 119 (2016)
5. Hall, B., O'Neal, T.: The residential learning community as a platform for high-impact educational practices aimed at at-risk student success. J. Sch. Teach. Learn. **16**(6), 42 (2016)
6. Delmas, P.M.: Using VoiceThread to create community in online learning. Techtrends **61**(3), 1–8 (2017)

7. Preece, J.: Negotiating service learning through community engagement: adaptive leadership, knowledge, dialogue and power. Educ. Change **20**(1), 104–125 (2016)
8. Fahara, M.F., Bulnes, M.G.R., Quintanilla, M.G.: Building a professional learning community: a way of teacher participation in Mexican public elementary schools. Int. J. Educ. Leadersh. Manag. **3**(2), 113 (2015)
9. Miles, J.M., Larson, K.L., Swanson, M.: Team-based learning in a community health nursing course: improving academic outcomes. J. Nurs. Educ. **56**(7), 425–429 (2017)
10. Friedrichsen, P.J., Barnett, E.: Negotiating the meaning of next generation science standards in a secondary biology teacher professional learning community. J. Res. Sci. Teach. **55**(7), 999–1025 (2018)
11. Moore, J.A.: Exploring Five online collaboration tools to facilitate a professional learning community. Techtrends **62**(4), 1–6 (2018)
12. Palatta, A.M., Kassebaum, D.K., Gadbury-Amyot, C.C.: Change is here: ADEA CCI 2.0—a learning community for the advancement of dental education. J. Dental Educ. **81**(6), 640–648 (2017)
13. Ludy, M.J., Morgan, A.L.: Using a learning community approach to improve health and wellness among university faculty and staff: 1564 board #217 June 2, 8: 00 AM - 9: 30 AM. Med. Sci. Sports Exerc. **48**(5S Suppl 1), 428–429 (2016)
14. Cox, M.D.: Four positions of leadership in planning, implementing, and sustaining faculty learning community programs: four positions of leadership in planning. New Direct. Teach. Learn. **2016**(148), 85–96 (2016)

Automatic Dispensing Technology of Traditional Chinese Medicine Formula Granules in Pharmacy Under the Background of Artificial Intelligence

Wenjing Wang[✉]

Shaanxi University of Chinese Medicine Xi'an, Shaanxi, China
SUCM200101@126.com

Abstract. At present, the research and application of artificial intelligence technology in the field of traditional Chinese medicine has become one of the important methods of modern inheritance strategy of traditional Chinese medicine. In this paper, under the background of artificial intelligence, the application of automatic dispensing technology of traditional Chinese medicine formula granules in pharmacy is studied. In this paper, two hospitals in a city were compared. Hospital a was the control group, and hospital B was the experimental group. Through experimental observation and data collection, it was found that the data of hospital B pharmacy using artificial intelligence automatic dispensing technology was better than that of hospital A in all aspects during the experiment. During the experiment, hospital A pharmacy processed 1056 prescriptions in total, which took 2110 min, and the experimental evaluation was 80 points, while hospital B pharmacy processed 1056 prescriptions During the experiment, 2317 prescriptions were processed in our hospital, which took 1853 min, and the experimental evaluation was 98 points. Moreover, the data of pharmacy of hospital B in dispensing accuracy, storage environment, drug control and other aspects were better than that of hospital A.

Keywords: Artificial intelligence · Dispensing granules of traditional chinese medicine · Automatic dispensing technology · Pharmacy taking medicine

1 Introduction

With the rapid development of science and technology, artificial intelligence technology is also developing rapidly, many industries are involved in the application of artificial intelligence, pharmaceutical industry is no exception [1, 2]. As the cultural treasure of the Chinese nation, traditional Chinese medicine has gone through thousands of years of development, and it should be protected, promoted and inherited no matter from the perspective of cultural heritage or medical resources [3, 4]. In recent years, the traditional Chinese medicine formula granule is a new dosage form with rapid development. It has the advantages of rapid action, small size, convenient use, convenient transportation and

J. MacIntyre et al. (Eds.): SPIoT 2020, AISC 1283, pp. 566–572, 2021.
https://doi.org/10.1007/978-3-030-62746-1_83

so on, so it is welcomed by the majority of patients [5, 6]. However, due to the influence of traditional Chinese medicine formula granule seed, curative effect and traditional medication habits, it brings certain difficulties to the extensive application of traditional Chinese medicine formula granules [7, 8].

Through the application of modern mechatronics and information technology, the automatic dispensing work of traditional Chinese medicine formula particles is carried out. The amount of drug required is automatically grasped as much as it is needed. A complete set of intelligent solutions is provided for the dispensing process of Chinese pharmacy, so as to improve its operation efficiency, modernization and informatization level, It is helpful to solve the problems of high error rate, bad environment, low automation and low efficiency in traditional Chinese pharmacy, reduce manual operation, ensure the quality and safety of patients' medication to a certain extent, and improve the standardization, standardization and modernization level of the whole Chinese medicine industry [9, 10].

This study first introduces the application of artificial intelligence in the dispensing of traditional Chinese medicine, and then expounds the automatic dispensing technology of the dispensing of traditional Chinese medicine. Then, under the background of artificial intelligence, this paper applies the automatic dispensing technology of the dispensing of traditional Chinese medicine to the pharmacy. In order to study the advantages of the automatic dispensing technology of the dispensing of traditional Chinese medicine, two hospitals in a city are selected as the research objects The object of this study is to compare and analyze the experiment data, and analyze the advantages of the automatic dispensing technology of traditional Chinese medicine formula granules in the pharmacy.

2   Proposed Method

2.1   Application of Artificial Intelligence to Traditional Chinese Medicine Formula Granules

With the rapid development of science and technology, the application of artificial intelligence has penetrated into all aspects of our life, including traditional Chinese medicine. The application of artificial intelligence in traditional Chinese medicine granules involves servo control and servo control, which usually refers to closed-loop control, that is, to measure the change of controlled object through feedback regulation, so as to modify the control technology of motor output. In this process, PID algorithm is often used. The expression of PID control algorithm to realize integral separation by computer is as follows:

$$U(i) = K_p \left\{ e(i) + \frac{T}{T_i} \beta \sum_{j=0}^{i} e(j) + \frac{T_d}{T} [e(i) - e(i-1)] \right\} \tag{1}$$

$$\beta = \begin{cases} 1 & |e(i)| \le \Delta X_{ma} \\ 0 & |e(i)| > \Delta X_{ma} \end{cases} \tag{2}$$

Where, i is the sampling sequence number, $e(i)$ is the position deviation input by the error calculator at the time of subsampling, $U(i)$ is the control value output by the computer at the i-th sampling time, T_i is the integral coefficient, T_d is the differential coefficient, K_p is the proportional coefficient, T is the sampling period, and β is the common switching coefficient of the integral term.

2.2 Automatic Dispensing Technology of Dispensing Granules of Traditional Chinese Medicine

Traditional Chinese medicine dispensing refers to the operation technology of accurately preparing traditional Chinese medicine for patients according to the traditional Chinese medicine drugs listed in the Clinical Prescriptions of traditional Chinese medicine doctors. Generally, there are six links such as drug inspection, pricing, dispensing, review, packaging and distribution. The pharmacy personnel take the medicine according to the doctor's prescription. The method is to take the ingredients in the order of prescription medicine, weigh them, and then mix them. Most prescriptions have at least a dozen kinds of drug materials, and the dosage requirements of each drug material are very accurate. Therefore, pharmacy personnel need to focus on and operate carefully to avoid taking the wrong or misweighing the dosage, so the traditional way of drug delivery has great disadvantages.

Automatic dispensing technology of traditional Chinese medicine formula granules is used for dispensing of traditional Chinese medicine formula granules. It can complete the actions of drug identification, weighing measurement, compatibility taboo reminder, etc. according to the dosage, taste, dosage and other formula parameters of doctor's prescription. The equipment has high precision and accuracy, fully reflecting the theoretical essence of "syndrome differentiation and treatment, plus and minus with syndrome, flexible medication" of traditional Chinese medicine, It has realized the real automatic regulation of traditional Chinese medicine formula granules.

3 Experiments

(1) Research object

In this paper, two hospitals in a city were tested and compared. Hospital a was the control group, the hospital pharmacy management was mainly based on the traditional way of taking medicine, hospital B was the experimental group, the hospital pharmacy used the artificial intelligence automatic dispensing technology, the research scope was the traditional Chinese medicine formula granule prescription.

(2) Evaluation index

In this paper, through a week-long experimental study, data collection is carried out through video monitoring, personnel observation and questionnaire visit to hospital patients. The scope of data collection mainly includes: the number of prescription of traditional Chinese medicine granules, the time consumed in the prescription of traditional Chinese medicine granules, the number of lost traditional Chinese medicine granules drug bags, the staffing of hospital pharmacy. After the end of the experiment, the following aspects are discussed Evaluation: dispensing accuracy, storage environment, dispensing convenience and drug control.

4   Discussion

4.1   Key Technology Analysis of the Application of Automatic Dispensing Technology of Traditional Chinese Medicine Formula Granules in Pharmacy Under the Background of Artificial Intelligence

Under the background of artificial intelligence, the realization of automatic dispensing technology of traditional Chinese medicine granules depends on the following key technologies:

(1)   High precision and high efficiency particle blending technology

The dispensing accuracy of traditional Chinese medicine formula granules is measured by weight. Because the dispensing task of a prescription is carried out continuously and completed in a very short time and in a very small range, there is little difference between boxes (bags). Therefore, the weight check data of the whole prescription can be used to monitor, modify and adjust the dispensing process and results of a single box (bag), while improving the dispensing accuracy, However, it will affect the adjustment efficiency.

(2)   High reliability drug identification technology

The traditional Chinese medicine dispensing system usually uses barcode and RFID (radio frequency identification) technology for drug identification. In particular, RFID technology can identify high-speed moving objects and multiple tags at the same time, which is fast and convenient to operate. Under the condition of being covered, RFID can penetrate non-metallic or non transparent materials such as paper, wood and plastic, and carry out penetrating communication, with strong anti-interference, so it is very suitable for particle identification on dispensing equipment. RFID tags are water-proof, oil-proof, dust-proof and pollution-proof, supported by highly reliable information encryption and verification algorithm, which can solve the problem of wrong drug delivery.

(3)   Closed loop control precision transmission technology

The result of the dispensing is to send the drug particles to the corresponding containers. The container is positioned by its position. If the container is transported and positioned incorrectly, the medicine will be loaded in the wrong place, causing the wrong medicine and leakage. In order to prevent the accidental errors of software and hardware in the control system and improve the anti-interference ability of the system, the equipment can adopt the precise transmission of closed-loop control, and add multiple redundant detection points to ensure the accurate and reliable movement. And according to the relevant management regulations, in the key links of man-machine dialogue, multiple reviews, to ensure that the dispensing is accurate, not wrong medicine.

(4) The equipment adopts closed-loop control precise transmission technology for positioning, box by box (bag) automatic drug distribution and automatic packaging, and man-machine dialogue for multiple reviews, so as to provide complete drug type correctness inspection and high reliability drug distribution and packaging function, and avoid drug dispensing errors in the process of manual dispensing.

(5) Rapid and automatic dispensing of various particles

The intelligent dispensing system of traditional Chinese medicine dispensing granules is controlled by the computer in the whole process, which can be linked with his system of the hospital to realize intelligent dispensing in the whole process, automatic downloading of prescriptions, automatic measurement, automatic sub packaging, automatic sealing and information recording in the whole process. The whole process voice and information prompt, the equipment can also be installed with an automatic detection module, which can automatically detect the status of all modules, motors, sensors, etc. of the whole machine, and automatically alarm when a fault is found, and recover the coordination work after the fault is removed.

4.2 Analysis of Application Results of Automatic Dispensing Technology of Traditional Chinese Medicine Formula Granules in Pharmacy Under the Background of Artificial Intelligence

In this paper, two hospitals in a city, a hospital and B hospital, experimental data collection is shown in Table 1.

Table 1. Application of automatic dispensing technology of traditional Chinese medicine formula granules in pharmacy under the background of traditional methods and artificial intelligence

Research object	Number of prescriptions	Dispensing time (min)	Medicine loss (bag)	Staffing (PCs.)
Hospital A	1056	2110	51	2
Hospital B	2317	1853	4	1

It can be seen from Table 1 that hospital a took 1056 drugs in the pharmacy during the experimental study period, which took 2110 min in total. The average time for each drug was about 2 min, and the total loss was 51 bags. The proportion of pharmacy staff was 2 people, while hospital B took 2317 drugs in the pharmacy during the experiment period, which took 1853 min in total, and the average time for each drug was about 0.8 min, the total loss is 4 bags, and the proportion of pharmacy staff is 1 person.

After the experimental observation, the application of the automatic dispensing technology of traditional Chinese medicine formula granules in the pharmacy was evaluated by comparing the traditional method with the artificial intelligence background. The experimental evaluation results are shown in Fig. 1.

After the end of the experiment, by comparing the four aspects of dispensing accuracy, storage environment, dispensing convenience and drug control of the two hospitals,

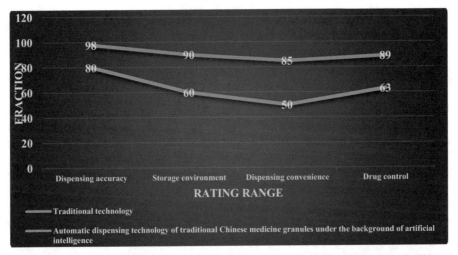

Fig. 1. Application evaluation of automatic dispensing technology of traditional Chinese medicine formula granules in pharmacy under the background of traditional methods and artificial intelligence

it can be seen from Fig. 1 that the score of a hospital pharmacy in dispensing accuracy is 80, while the score of B hospital in dispensing accuracy is as high as 98, in the pharmaceutical storage environment, the score of a hospital pharmacy is 60, while the score of hospital pharmacy is 90 Compared with the convenience of dispensing, hospital a pharmacy scored 50 points, hospital B pharmacy 85 points, hospital a pharmacy 63 points and hospital B pharmacy 89 points in terms of drug control. From the above scores, hospital B scored higher than hospital a in all aspects, which shows that the automatic dispensing technology of artificial intelligence Chinese medicine formula particles used in hospital B pharmacy is more scientific than the traditional pharmacy management method Effective.

From the above experimental data, it can be seen that the automatic dispensing technology using artificial intelligence Chinese medicine formula particles has more advantages in pharmacy management:

(1) The dispensing is accurate. In the dispensing process, the barcode scanning method is used to recheck the drugs to ensure the accurate distribution.
(2) The storage environment of drugs is good, the drugs are stored in a sealed environment, with waterproof and moisture-proof measures to ensure the quality of drugs.
(3) With the function of modern network management, the electronic formula can be generated on different computers in the hospital, and transmitted to the central database, then automatically classified and finally dispensed through the control platform of the dispensing machine.
(4) The operation is simple and reliable. The matching automatic dispensing system of traditional Chinese medicine has a good man-machine interface, which can operate

traditional Chinese medicine, prescription medicine and dispensing machine. The work intensity is greatly reduced. Only one person can complete all the procedures.

5 Conclusions

The development of artificial intelligence has been more than ten years, and the technology and application are gradually becoming mature. But at present, the hospital in our country still takes medicine by artificial. This paper studies the application of automatic dispensing technology of traditional Chinese medicine formula granules in pharmacy under the background of artificial intelligence, and with the development of artificial intelligence, In the future, the automatic dispensing technology of traditional Chinese medicine granules will further develop to the direction of higher automation, faster speed, networking, more intelligent and easy to use, and the pharmacy will gradually develop to the direction of intelligent and unmanned.

References

1. Chen, J., Sun, S., Zhou, Q.: Direct and model-free detection of carbohydrate excipients in traditional Chinese medicine formula granules by ATR-FTIR microspectroscopic imaging. Anal. Bioanal. Chem. **409**(11), 2893–2904 (2017)
2. Xiaodong, L., Peng, W., Zhe, L., et al.: Evaluation of a granulated formula for the nerve root type and vertebral artery type of cervical spondylosis: a multicenter, single-blind, randomized, controlled, phase III clinical trial. J. Trad. Chin. Med. **37**(2), 193–200 (2017)
3. Cao, X.J., Huang, X.C., Wang, X.: Effectiveness of Chinese herbal medicine granules and traditional Chinese medicine–based psychotherapy for perimenopausal depression in Chinese women: a randomized controlled trial. Menopause **26**(10), 1 (2019)
4. Yang, H., Liu, J.X., Shang, H.X., Lin, S., Zhao, J.Y., Lin, J.: Qingjie Fuzheng granules inhibit colorectal cancer cell growth by the PI3K/AKT and ERK pathways. World J. Gastroint. Oncol. **11**(05), 33–48 (2019)
5. Peng, W., Zhe, L., De, L., Feng, H., Jinwen, L., Xinyu, C., Huajian, Z.: Evaluation of a granulated formula for the nerve root type and vertebral artery type of cervical spondylosis:a multicenter, single-blind, randomized, controlled, phase III clinical trial. J. Trad. Chin. Med. **37**(02), 51–58 (2017)
6. Zhang, H., Chen, Y., Wang, J.N., et al.: Application of fingerprint technology in quality evaluation and process control of traditional Chinese medicine formula granules. China J. Chin. Materia Medica **43**(19), 3822–3827 (2018)
7. Kohl, S.: European directorate for the quality of medicines: automatic drugs dispensing report. Eur. J. Hosp. Pharm. **25**(3), 169–172 (2018)
8. Hamon, M., Capelle, F., Passemard, R., et al.: Assessment of an online training tool for the automated unit-dose dispensing system (ADS) process. Pharm. Technol. Hosp. Pharm. **4**(1), 41–46 (2019)
9. Deng, Z., Zhang, J., He, T.: Automatic combination technology of fuzzy CPN for OWL-S web services in supercomputing cloud platform. Int. J. Pattern Recogn. Artif. Intell. **31**(7), 1759010.1–1759010.27 (2017)
10. Ge, Y., Ploetner, M., Berndt, A., et al.: All-printed capacitors with continuous solution dispensing technology. Semicond. Sci. Technol. **32**(9), 095012.1–095012.6 (2017)

Exploration and Practice of Coaching Technology in College Career Guidance Under the Internet + Era

Zenglu Meng[✉]

Suzhou Institute of Technology, Jiangsu University of Science
and Technology, Suzhou, Jiangsu, China
Mengzenglu2020@163.com

Abstract. With the innovation and upgrading of Internet technology, the concept of Internet + has been widely used in all aspects of society. Today, with the rapid development of society, employment guidance is the primary task facing colleges and universities, and it is also the difficulty in the work of colleges and universities. The employment guidance of colleges and universities has also made new breakthroughs due to the rise of the Internet +. In the context of Internet +, the application of coaching technology in college career planning can help improve students' self-understanding, tap their potential, continuously enhance their ability to adapt to society, and promote the practical significance of higher quality employment. This paper takes the current situation of college students' career planning as the research background, and further explains the feasibility of coaching technology in college career planning education under the background of Internet + through the investigation of the current situation. The research results show that, compared with the traditional employment guidance courses, the Internet + supported coaching technology has the advantages of accurately positioning itself, developing career planning ideas, and improving the ability of target decomposition and execution. Therefore, as an emerging educational concept and technology, coaching technology in college employment guidance in the Internet + era enriches and develops the theories of employment planning education, which may completely change the existing employment guidance education mode and open up a new education mode.

Keywords: Employment guidance · Coaching skills · Career planning · Quality employment

1 Introduction

In the era of Internet +, college students' career guidance courses can be supported by a variety of assistive technologies [1]. Coaching technology is a popular new management tool in the information age, which is of great research value among relevant workers in psychology and education [2]. This technology originated from the business management industry and has been applied in the education industry, which is undoubtedly

J. MacIntyre et al. (Eds.): SPIoT 2020, AISC 1283, pp. 573–580, 2021.
https://doi.org/10.1007/978-3-030-62746-1_84

an innovation. It can be said that coaching technology is the most revolutionary and effective management concept in the 20th century [3].

The rise of coaching technology is relatively short, and its application history in all walks of life is relatively short [4]. In European and American countries, coaching technology is regarded as the most powerful weapon to improve the productivity of enterprises, and is greatly favored by all enterprises [5]. Although developed in the field of business management, coaching technology has changed into a new profession and is one of the fastest growth points of service industry in many western developed countries [6]. Western scholars tend to explain the development direction of coaching technology from various perspectives. In literature [7], the author explains the development of coaching technology from the perspective of psychology. In literature [8], the author analyzes the introduction of coaching technology for daily teaching from the perspective of pedagogy, and believes that the application of coaching technology in the field of education is still a concept of description, lacking of relevant case studies. In literature [9], the author studies the application of coaching technology from the perspective of enterprise management, and also conducts research and analysis combining with specific cases.

Although the existing research on the promotion and application of coaching technology in the field of real economy has made a lot of elaboration, the application results are fruitful, but the research on the application of coaching technology in college employment guidance is still very lack. On the official WEBSITE of CNKI, with "coaching technology + College Guidance for Employment" as the key word, only 124 relevant documents can be found, and even with "coaching technology + education" as the theme key word, only 582 relevant documents can be found [10]. Based on the current research status, the bold attempt of this research will coach technology combined with colleges and universities career guidance, the coach technology as the center of the guiding ideology of university employment guidance course, through effective incentive method to improve the ability of being educators take the initiative to create, to excavate the potential of their maximum, with the best state to create results.

2 Related Concept Description and Research Methods

2.1 Coaching Skills

"Coaching" was introduced to the public as a management technique. The international coach federation coach technology is defined as: "the coach to reach an agreement and the two sides agreed to by the coach approved targets, coach through guided by coach on the analysis of the surrounding environment and to build, as well as help by coach to sort out their existing knowledge experience, can help them develop a good ability to adapt to social and then to be flexible in order to achieve vision technology and coach method".

Coaching technology is the science of coaching. It is the product of different disciplines or behaviors. It is both an effective skill and a practical tool. The acquisition of coaching technology is not only the surface of teaching, but more of the understanding of the spirit of the instructor, so as to better explore and improve their own purposes. After years of development, coaching technology has become one of the most effective means to stimulate students' potential and improve training efficiency. Today's coaching technology has become a new management technology, and has achieved good results in the relevant fields such as employee training. It has developed into a systematic and scientific theoretical system by virtue of its systematic and scientific theory, which enables more trainees to realize the methods and approaches to accomplish tasks.

2.2 Career Planning for College Students

The purpose of employment guidance training for college students is to assist college students to make a career plan in line with their own characteristics. In the opinion of the researchers in this paper, career planning is that an individual's career planning can be called as any activity, arrangement or expectation related to his career, which includes his design of career planning and his understanding of ideal life, value orientation and other aspects.

College students are different from other groups in society. College students are a special group who have received higher quality education, but have not fully entered the society and know the social situation. On the one hand, college students are in the stage of semi-socialization and have strong plasticity. On the other hand, there is great uncertainty. In the process of social development, college students bear great responsibility for promoting the development of the country and society. Therefore, the career planning of college students has its own particularity, it is not easy to do well the career planning of college students, it is possible to make a forward-looking career plan only by mastering the methods and approaches to realize value.

2.3 Research Ideas and Methods

Under the background of Internet + era, it is necessary to summarize and organize the advanced and beneficial practical cases at home and abroad for the employment guidance of college students, and explore the educational influence of college students' career planning caused by coaching technology through the cases. This study designed the form of investigation and interview to analyze the development status of career planning for college students in a well-known university in China, analyzed the feasibility of the application of coaching technology in college employment guidance education, discussed the development status of coaching technology in college students' employment guidance, and put forward corresponding countermeasures and Suggestions.

The research in this paper combines the methods of literature research and investigation. In the link of literature research, the author summarizes and sorts out the domestic and foreign coaching technology, the status of college students' career planning and the practical achievements of coaching technology. In the application of the survey method, the research adopts the forms of questionnaire survey and semi-structured interview to compare and analyze the effects of coaching technology before and after the intervention by understanding the current situation, and then analyze the practical application effects of coaching technology in college students' employment guidance.

3 Applied Research Cases

3.1 Respondents

Experiment research survey link (entrust the third party) is random sampling method to extract the five from the domestic well-known colleges and universities a total of 250 students to participate in the survey. The specific information about the subject samples is shown in Table 1. 250 subjects were equally divided into experimental group and control group, with 125 subjects in each group. There were 65 boys and 60 girls in the experimental group. A total of 68 boys and 57 girls were in the control group. In the sample, in terms of grade composition of students, the number of senior students is relatively small, which is caused by the fact that some senior students are doing internships abroad and the overall number of students on campus is relatively small. The overall composition of the professional distribution of the control group and the experimental group is relatively symmetrical, and the overall has a reasonable comparison of sampling.

Table 1. The overall situation of the survey samples

Catalogs		Experimental group	Control group
Gender	Male	65	68
	Female	60	57
Grade	Freshman year	37	35
	Sophomore year	31	37
	Junior year	36	34
	Senior year	21	19
Professional	Literature and History	58	62
	Science and engineering	67	63

3.2 Experimental Design

The experiment adopted the intervention of career guidance with coaching technology classroom in the experimental group. The intervention took the form of novel Internet teaching mode. The intervention course lasted for eight weeks. In the control group, only pre-test and post-test questionnaires were used, without other forms of guidance intervention. The dependent variables of the experiment were the change of self-efficacy and the dynamic change of career decision-making difficulty of college students before and after the intervention. The collected data were analyzed and collated, and SPSS 23.0 statistical software was used for statistical analysis.

4 Experimental Results and Discussion

4.1 Pretest Results and Analysis

Taking groups as independent variables and the difference in pre-test self-efficacy and career decision-making difficulty as dependent variables, a statistical independent sample T-test was conducted to explore whether there were differences in self-efficacy (M) and career decision-making difficulty (SD) between the two groups before intervention. The results are shown in Table 2 and Fig. 1.

Table 2. Differences in self-efficacy and career decision-making difficulties between the experimental group and the control group before intervention

Variable	Experimental group		Control group		t	p
	M	SD	M	SD		
Self-efficacy (A)	2.242	0.566	2.864	0.534	−0.873	0.374
Lack of preparation (B)	2.251	0.483	3.239	0.486	−0.738	0.351
Information search difficulty (C)	2.237	0.657	3.171	0.514	2.263	0.055
Plan conflict (D)	2.268	0.454	3.59	0.501	2.242	0.163
Career decision score (E)	2.256	0.678	3.573	0.696	0.371	0.773

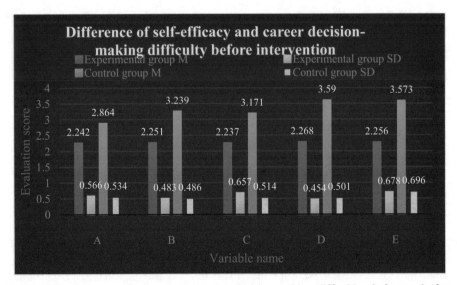

Fig. 1. Differences in self-efficacy and career decision-making difficulties before and after intervention

As can be seen from the experimental results in the figure, the level of self-efficacy of the experimental group and control group before intervention was slightly higher than that of the experimental group, but there was no significant difference between the experimental group and the control group (T = 0.873, P = 0.374). In other words, statistically speaking, the difference of self-efficacy between the two groups was not significant. Subjects of the two groups felt generally in control of their own actions and generally in confidence in dealing with things. In addition, both the experimental group and the control group scored slightly higher in career decision-making difficulties, which indicated that college students in general showed certain characteristics of career planning difficulties. In addition, before the intervention, there was no significant difference between the coaching group and the control group in terms of self-efficacy and career decision-making difficulties. College students had a general sense of self-efficacy, which caused certain career decision-making difficulties. To a certain extent, the differences in the post-test were completely caused by the intervention of coach-style career planning, not including the differences between the two groups of subjects themselves.

4.2 Post-test Results and Analysis

With the group as the independent variable and the difference in post-test self-efficacy and career decision-making difficulty as the dependent variables, a statistical independent sample T-test was conducted to explore whether there were any differences in self-efficacy (M) and career decision-making difficulty (SD) between the two groups after coaching intervention. The results are shown in Table 3 and Fig. 2.

Table 3. Differences in self-efficacy and career decision-making difficulties between the experimental group and the control group after intervention

Variable	Experimental group		Control group		t	p
	M	SD	M	SD		
Self-efficacy (A)	2.938	0.508	2.122	0.562	1.273	0.042
Lack of preparation (B)	2.322	0.815	3.639	0.84	−0.358	0.044
Information search difficulty (C)	2.856	0.423	3.171	0.604	−2.252	0.058
Plan conflict (D)	2.697	1.188	3.294	0.457	−1.452	0.023
Career decision score (E)	2.551	1.269	3.273	0.995	−2.315	0.05

Feedback the information from the chart, through online coaching intervention guidance, the experimental group of self-efficacy of the students, a certain increase, have more confidence to deal with all sorts of things, in life than to participate in the career planning before the class more active, more active, presented the bigger difference compared with control subjects (t = 1.273, p = 0.042). Through online coach technical guidance, the experimental group members also improved awareness of career planning, started to carry out career planning related work, and through the use of coaching and members

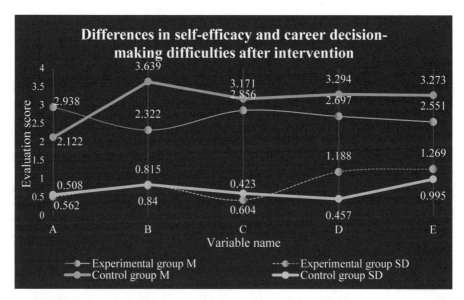

Fig. 2. Differences in self-efficacy and career decision-making difficulties after intervention

of the group sharing and discussion, some of the original irrational beliefs and ideas to some extent, in the face of career planning yes hesitation will be reduced. Therefore, after the online coaching technique course intervention, the experimental group members obtained a certain degree of improvement in the sense of self-efficacy, while their scores in the difficulty of career decision-making were reduced. The coaching intervention played a very good role.

5 Conclusion

In the era of Internet +, the combination of coaching technology and employment guidance for college students is an innovative move. It implements the coach by mastering the practical problems in college students' employment, provide targeted guidance advice, to establish trust relationship with students, help graduating students search for the inner real demand, not only enhance the value of the students' ability of self affirmation, also inspired the students' individual potential, promotes the school career guidance work efficient and orderly.

References

1. Beasley, H.L., Ghousseini, H.N., Wiegmann, D.A., et al.: Strategies for building peer surgical coaching relationships. Jama Surg. **152**(4), e165540 (2017)
2. Jones, H.P., Mcgee, R., Weber-Main, A.M., et al.: Enhancing research careers: an example of a US national diversity-focused, grant-writing training and coaching experiment. BMC Proc. **11**(S12), 16 (2017)

3. Deiorio, N.M., Carney, P.A., Kahl, L.E., et al.: Coaching: a new model for academic and career achievement. Med. Educ. Online **21**(1), 33480 (2016)

4. Brooks, B.A., Skiem, P.T.: career coaching: preparing for what's next. Nurse Leader **16**(3), 190–192 (2018)

5. Hur, Y., Cho, A.R., Kwon, M.: Development of a systematic career coaching program for medical students. Korean J. Med. Educ. **30**(1), 41–50 (2018)

6. Brooks, B.A., Skiem, P.T.: Career coaching 101. Imprint **64**(1), 26–29 (2017)

7. Dawson, A., Dioth, T., Gastin, P.B.: Career facilitators and obstacles of Australian football development coaches. Int. J. Sports Sci. Coaching **11**(2), 255–269 (2016)

8. Dahlstedt, M., Vesterberg, V.: Portrait of authority: a critical interrogation of the ideology of job and career coaching. Pedagogy Cult. Soc. **27**(2), 199–213 (2019)

9. Stachiu, M., Tagliamento, G.: Career coaching and community social psychology: analysis of an intervention. Temas Em Psicologia **24**(3), 791–804 (2016)

10. Yoon, M., El-Haddad, C., Durning, S., et al.: Coaching early-career educators in the health professions. Clin. Teach. **13**(4), 251–256 (2016)

Research on Enterprise Data Governance Based on Knowledge Map

Xiaoying Qi[✉]

School of Management, Hebei University, Baoding 071002, China
2814970041@qq.com

Abstract. Data governance is a system integrating data management, improvement of data quality and data application and the whole process is a closed-loop procedure that is mutually associated, mutually promoted and improved, which serves as a solution to problems ignorance of data, uncontrollable of data and invalid data. With the development and application of big data, data governance plays an important role in improving utilization of enterprise data, enhancing enterprise performance and competitiveness. Based on the literature on corporate data governance, this paper analyzes the status quo, hot topics and research frontiers of corporate data governance at home and abroad based on the knowledge map. It is found that research objects of corporate data governance mainly focus on multinational enterprises, family enterprises, high-tech enterprises and smart cities. The research is conducted with focuses on data warehouse, data governance framework, methods, models and risk management. The future research of enterprise data governance is mainly on determining the scope of master data objects and metadata management.

Keywords: Data governance · Enterprise data governance · Knowledge map

1 Introduction

Data science has promoted a new paradigm of scientific research, which has led to significant changes in the perspectives, processes and methods of data processing and management. With the growth of the organizational business, a large number of high-value multi-structure data sets are produced, which pose new challenges to the traditional data management tools and methods, and increase the importance and urgency of the organizational data governance. Data governance is a set of management behaviors that involve the use of data in an organization. Related research institutions issued all kinds of definitions about data governance, proceeding from different perspectives, some foreign scholars believe that data governance is the definition of series of policies and rules based on the bills [1], and some scholars emphasize that data governance is decision-making and division responsibility of the relevant organizations data assets [2–4]. There are also many scholars holding that data governance is data management process and and methods integrating people, process and information technology by taking the process,

© The Editor(s) (if applicable) and The Author(s), under exclusive license
to Springer Nature Switzerland AG 2021
J. MacIntyre et al. (Eds.): SPIoT 2020, AISC 1283, pp. 581–586, 2021.
https://doi.org/10.1007/978-3-030-62746-1_85

technology and responsibility in the process of data management and control [5, 6], which can ensure the rational use of organizational data assets. The logic of enterprise data governance takes governance structure as the starting point with interaction based on good governance mechanism so as to pursue the ultimate goal of good governance performance [7].

2 Data Governance

Data governance was initially used in the field of enterprise management, and then it is introduced and studied in the fields of medical treatment and government. In enterprises, data governance is used to solve the problem of enterprise management and control, which is the key to improving the level of data asset management and application. In the telecommunications field, data governance system is positioned as a crucial step in the information architecture integrating data, technology, organization and application. Informatica defined data governance as "Data governance is used for asset management to coordinate and define the procedures, policies, standards, technology and personnel within the scope of the organization, so as to realize th available management and controllable growth for accurate, consistent, safe and timely data. As a result, improved decisions on businesses can be made to reduce risks and improve business processes". It can not only promote the requirements for enterprises to connect and share data resources within the organization, but also drive the enterprises to make scientific decisions and enhance their competitiveness by integrating data resources. In 1993, IBM began its exploration of data governance. In 2004, H. Watson started its trial on "data governance" in enterprise management by exploring the best practices in Blue Cross and Blue Shield of North Carolina. In 2005, J. Ghriffin L. Chheong, D. Ower and others discussed the model, framework, factors and mechanism of enterprises in the data governance environment [8–12].

3 Visualized Analysis

3.1 The Status Quo of Research on Data Governance in China

With the theme of "Enterprise Data Governance", 262 Chinese literature were retrieved on China National Knowledge Infrastructure (CNKI), and 234 valid foreign literature were obtained after filtering and deleting the literature and removing the articles such as soliciting contributions and annual catalog. In terms of the number of published articles, the publication volume on data governance in China showed a fluctuated trend from 2006 to 2013, and has increased rapidly since 2014. Therefore, it can be predicted that the number of published articled on data governance in the next few years will maintain a steady growth or in downward trend. The key words with quotation frequency for more than three times in the study of Chinese data governance are respectively data management, big data, data quality, data standard, enterprise data, metadata, data assets, data architecture, data warehousing, enterprise management, big data technology, open data, data sharing, data standardization, government management, data integration, data value. Through cluster analysis, it is found that the hot topics of research on data governance in China mainly include enterprise data, data standardization, big data governance, data architecture and government data governance.

3.2   The Status Quo of Research on Foreign Data Governance

With the theme of "Enterprise Data Governance", data governance related literature were searched in databases and websites such as Wos, Google Scholar, The Data Governance Institute and DGI, etc., and the literature were screened and deleted, and 904 valid literature in foreign languages were obtained in addition to articles such as journal subscription and annual catalog.

In terms of countries to be studied, The main countries to studied on data governance abroad mainly include China, the United States, the United Kingdom, Australia, France, Germany, Canada, Spain, India and Taiwan, etc., as shown in Fig. 1. The research on data governance conducted by foreign researchers are relatively diversified with keywords mainly including data governance, collaborative governance, enterprise performance, corporate management, enterprise innovation, governance strategy, governance model, master data, etc., as shown in Fig. 2 and Table 1. According to cluster analysis (Fig. 3), hot topics of research on data governance abroad mainly include ownership structure, public election, service reuse, collaborative governance, state power, enterprise risk management, smart city and institutional diversification. According to the analysis of time line (Fig. 4), the theme evolution of ownership structure includes enterprise performance, emerging enterprises, etc. The theme evolution of public election includes information cost, capital structure, government-dominated enterprises, etc. The theme evolution of service reuse includes enterprise ability, enterprise social responsibility, etc. The theme evolution of collaborative governance includes enterprise innovation, system development and utilization, etc. The theme evolution of state power includes political connection and management development. The theme evolution of enterprise risk management includes absorbing ability, enterprise knowledge, globalization, family business, etc. The theme evolution of smart city includes collaborative governance, diversified enterprises and international development. The evolution of institutional diversification includes economic development, enterprise data system, internationalization, big data, cities, etc.

Fig. 1. Analysis on countries with research on data governance

Fig. 2. Analysis on keywords cited by foreign countries in research of data governance

Table 1. Statistics of using frequency of foreign researchers in studying data governance

Keyword	Freq	Centrality
governance	175	0.12
corporate governance	122	0.08
performance	94	0.09
china	68	0.11
firm	64	0.07
ownership	62	0.04
firm performance	52	0.15
management	51	0.10
determinant	40	0.07
enterprise	39	0.11
impact	38	0.02
perspective	37	0.07
innovation	36	0.03
model	33	0.12
market	30	0.03
institution	26	0.05
organization	25	0.05
privatization	25	0.06
business	24	0.05
strategy	24	0.01
corporate social respo…	19	0.05
growth	19	0.06
corruption	18	0.03
social enterprise	18	0.01
financial performance	18	0.05
challenge	16	0.03
panel data	16	0.00
state-owned enterprise	15	0.03
entrepreneurship	15	0.02

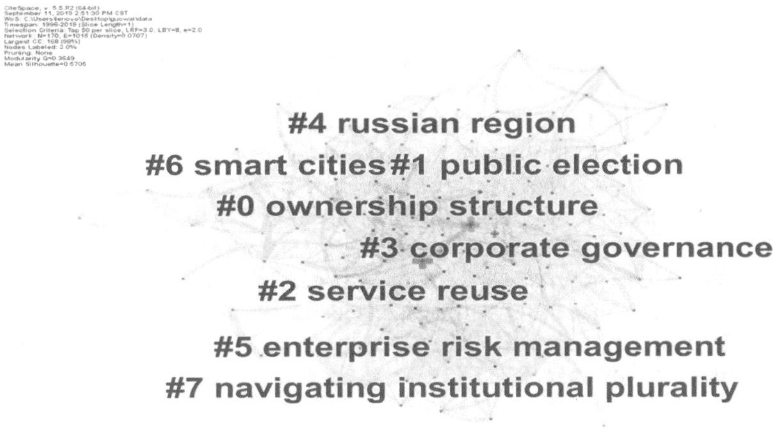

#4 russian region
#6 smart cities #1 public election
#0 ownership structure
#3 corporate governance
#2 service reuse

#5 enterprise risk management
#7 navigating institutional plurality

Fig. 3. Cluster analysis on data governance conducted by foreign researchers

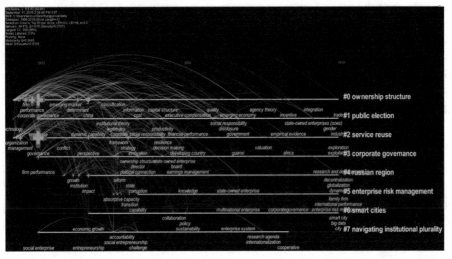

Fig. 4. Visualized and time line based analysis on data governance conducted by foreign researchers

4 Conclusion

Research objects in foreign countries are made with focuses mainly on transnational enterprises, family enterprises, high-tech enterprises, smart cities, etc. The research is designed with focuses on data warehouse, data governance framework, methods, models, enterprise governance, and enterprise risk management. The research objects of data governance in China mainly focus on information technology enterprises, power enterprises, banking and finance enterprises and social media enterprises. Research contents are designed with focus on data quality, metadata, data standards, data security, master data and data life cycle management, business process integration and data architecture management, data warehouse, data assets, big data processing platform, etc. For example, in 2012, China Everbright Bank made many explorations in data governance, such as the development of enterprise-level basic data platform. In 2015, the state issued the white Paper on Data Governance, and the enterprises in banking, insurance, power, telecommunications and other industries have carried out the work in data governance with research objects mainly focusing on information technology enterprises, power enterprises, banking and finance, and social media enterprises. The research focuses on the description of data governance, including data quality, metadata, data standards, data security, master data, data life cycle management and data application. China's Huawei Company expressed its vision on data governance at the 2016 Analysts Conference, and proposed that data governance was conducive for enterprises in reshaping the value of data.

Informatica believes that the key to the success of data governance lies in metadata management, that is, to endow a reference framework that gives context and meaning to data. Therefore, the breakthrough of future research is to determine the scope of master data objects and metadata management based on visualized analysis. In general, there is still more space to be explored in research of data governance, but it is still difficulty in measurement in enterprise data processing network, metadata management, master data management, etc. This study is conducted with the aim to promote the research and practice of enterprise data governance.

References

1. Donaldson, A., Walker, P.: Information governance—a view from the NHS. Int. J. Med. Inf. **73**(3), 281–284 (2004)
2. Khatri, V., Brown, C.V.: Designing data governance. Commun. ACM **53**(1), 148–152 (2010)
3. Thomas, G.: Alpha Males and Data Disasters: The Case for Data Governance. Brass Cannon Press, York County (2006)
4. Putro, B.L., Herbert, S.K.: Leadership and culture of data governance for the achievement of higher education goals (Case study: Indonesia University of Education). In: International Seminar on Mathematics, Science, and Computer Science Education (MSCEIS), vol. 1708, pp. 1–12 (2015)
5. Fernandes, L., O'Connor, M.: Data governance and data stewardship. Critical issues in the move toward EHRs and HIE. J. AHIMA Am. Health Inf. Manag. Assoc. **80**, 30–39 (2009)
6. Zhang, J.: Research on the Relationship between network organizational governance structure and governance performance. Shanxi University of Finance and Economics (2015)
7. Yao, X., Tian, Z.: Strong relationship and weak relationship: a study on social relationship dependence of enterprise growth. J. Manag. Sci. China **11**(1) (2008)
8. Zhang, Y., Gang, T.: Design of agent behavior rules for multi-agent simulation model. Soft Sci. **22**(3) (2008)
9. Discussion on enterprise data governance and its unified process. China Manag. Inf. **19**(16), 57 (2016)
10. Huang, H.: Huawei released the grand vision of data governance. Commun. World **15**, 52 (2016)
11. Ning, Z., Yuan, Q.: Research review on data governance. J. Inf. **36**(05), 129–134+163 (2017)
12. Sun, Z.: Data quality management under enterprise-level data governance framework. Electron. Finan. 06, 57–60+6 (2011)

Website Design and Implementation of Xunyang Lion's Head Orange

YongHui Ma[✉]

College of Engineering and Technology, Xi'an Fanyi University, Xi'an 710105, China
23272435@qq.com

Abstract. With the rapid development of Internet technology, the development and maturity of e-commerce platforms has been promoted, and online shopping has become a trend. Its fast, convenient, and simple operation has an important impact on daily life and makes people's choices on product consumption. It is also becoming more personalized and diverse. The development of e-commerce has provided a broad market for the marketing and promotion of agricultural products, provided convenient sales channels, reduced transaction links and losses, improved product circulation, and met user shopping needs. The market awareness of Xunyang Shitou is not high. I hope that through the form of a website, I will use the advantages of Internet promotion to break through the geographical restrictions, improve the popularity, and efficiently and quickly send it to each consumer. Demand has developed rapidly, driving local economic development.

Keywords: Xunyanglion head orange · Web design · Database design

1 Introduction

1.1 Research Background and Significance

1.1.1 Research Background

With the development of network technology and social economy, the improvement of people's living standard has promoted the rapid development of online shopping, the network platform has changed the traditional consumption way and the consumption habit. With the maturity of logistics technology, breaking the limitation of time and region, e-commerce platform has already entered a brand-new mode to promote the sale of network goods. With the increasing demand of regional fruits, Shitou Mandarin Orange is one of the characteristic agricultural products in Xunyang County. Due to the lack of attention to the training of e-commerce talents, the lack of network marketing talents has affected the sales and market expansion of its products. Xunyang Shitou Mandarin Orange website through the integration of resources, with the help of Internet marketing means to break the sales constraints of Shitou Mandarin Orange, expand the marketing channels of Shitou Mandarin Orange, solve the problem of sales difficulties, promote the income growth of Shitou Mandarin Orange and better promote the advantages of

J. MacIntyre et al. (Eds.): SPIoT 2020, AISC 1283, pp. 587–592, 2021.
https://doi.org/10.1007/978-3-030-62746-1_86

Shitou Mandarin orange. Through the large-scale cultivation of the industrial base for consumers to provide convenient, fast, safe and effective product services, cultivate loyal users, enhance market competitiveness, good word-of-mouth products, and promote the prosperity of the Xunyang County economy.

1.1.2   Research Significance

With the rapid development of the network economy, people's living standards have gradually improved, more and more agricultural products have been moved onto the network platform to promote the rapid growth of the local economy, and the Shitou orange industry is the key supporting industry in Xunyang County, and an important part of Xunyang County's economic development. In order to promote the coordinated development of Xunyang County's economy, Shitou Mandarin Orange (citrus sinensis osbeck) is featured. In the context of the Internet, the market competitiveness of Shitou Mandarin Orange (citrus sinensis osbeck) is enhanced by integrating the industrial base of large-scale cultivation in Xunyang County, through the promotion of the business model to create the Shitou citrus electric brand, to adapt to the different consumer pursuit.

1.2   Research Main Content

The rise of online shopping has changed people's consumption concept, do not leave home to understand the required product information details, through product specifications comparison to choose to buy, convenient and efficient completion of the shopping process. Online online and offline network sales model to expand the Shitou Mandarin orange sales channels, through a full range of product display, to meet the needs of users about Publicis Tangerine. This paper describes the functional requirements and feasibility analysis of the Shitou Orange Web site in Xunyang County, and combines the existing technical support and the further development of the web site framework to come up with a conceptual design of the web site to gradually improve the functions of the web site, xunyang Shitou Orange website through product display and promotion, to provide users with the best quality service attitude and product services.

1.3   Research Methods

(1) Research and analysis method through market analysis on the product influence of Xunyang County Shitou Mandarin Orange, the product classification and price fixing can meet the needs of different users as far as possible, by combining the products with the local orchard, [9] users can experience the growth situation of Xunyang County Satsuma in different time periods, book in advance and enjoy different discounts. (2) the case analysis method initially forms the function module of the Xunyang Shitou Orange Net by referring to the demand characteristics of the agricultural products and fruit wholesale website, reasonably guides the user in the content and the function classification, catches the user's eyeball through the high-quality product, stimulate the idea of purchasing and promote the development of Shitou orange.

2   System Analysis

2.1   Functional Requirements Analysis

In order to enable users to intuitively understand the Xunyang County Shitou Orange Network related product information, provide products while giving users more choice of comparative space, according to the different needs of users to adjust the function to facilitate the fastest way to complete the user's shopping experience [3]. Design the site with the following functional requirements.

2.1.1   The Function Structure

To understand the origin of the citrus, the growth cycle of citrus, users find products, compare hot products, special products, interested in the purchase of products. Part of the functional design is as follows: (1) product information display: according to the user's customary product specifications set product classification, so that users choose product quantity, optimize product arrangement while launching star products, special offers, combination of products. (2) member Login and registration: users can browse the products, select the products, store the shopping cart temporarily, check the visit foot-print, check the perfect personal information and follow-up feedback of order details. (3) shopping cart: add products to the cart to compare product quality, price differences, and maintain the status of the product. Remove the comparison product, empty the cart, or make a purchase. (4) order operation: After the shopping cart product information is checked, the order is paid. The order matches the delivery address according to the registration information. The user can also cancel the order and modify the order information, and check the delivery time through the order inquiry.

2.1.2   The Background Function Structure

Which is convenient for the administrator to modify and adjust the content of the website, to inquire and feed back the user information, and to evade and eliminate the system faults and problems that occur in the system effectively, routine maintenance reduces the potential risk to the system. Edit new product information, publish news and marketing activities. Part of the functional requirements are as follows: (1) administrator LOGIN: In order to improve the user's shopping experience, set up a number of administrators for background data management and Monitoring, to ensure the normal operation of the site, timely resolution of system problems and feedback. (2) product management: In order to strengthen the management of products, different specifications of products are set up, product information is released according to classification, the prices of products are revised, and the characteristics and selling points of products are supplemented. Timely adjust the status of products, do a good job inventory tips.

2.2 Analysis of Feasibility

2.2.1 The Overall System of the Website

Making great efforts in website construction technology, combining relevant technology and professional knowledge, the overall system of the website is planned and designed, in the function and database on the detailed design [8].

2.2.2 Economic Feasibility

With the rise and popularity of online shopping, more and more e-commerce platform, through the combination of online and offline sales mode, reduce the overall operating costs, reasonable human and material expenditure control human costs. At the same time, with the help of the network influence and the word-of-mouth of the users, many wholesalers came to buy in advance to solve the problem of overstock.

2.2.3 Operational Feasibility

Provide users with a wealth of products to choose from, and effective search is also a measure of user experience. The management of User's personal information and order form is the core of website design. To prevent user's information from being stolen, we need to provide technical support, ensure the website's security, reliability, and secure transaction payment [5].

2.3 Database Analytics

To be more user friendly by collecting user information, and to use the collected data for analytics, adjust product recommendation and set a reasonable price.

3 System Design

3.1 System Project Planning

Through functional requirements analysis, through the front to achieve product display and origin of cultural output, close interaction with users, users better choices, complete the shopping experience. In the development of the website, we will enrich our products according to the needs of users, manage the relevant data of the website effectively with the help of the database, and adjust the website function module and website style design with the increase of the number of users, optimize the shopping process of the website, upgrade the service of the system, simplify the process in operation and payment to improve the user experience and shopping satisfaction.

3.2 Functional Module Design

The first step should be to improve functional module independence and to test whether some functional modules should be extracted or merged, strive to reduce coupling and improve cohesion. For example, a subfunction common to multiple functional modules can be invoked by a single functional module, sometimes by decomposing or merging functional modules to reduce the transmission of control information and references to global data, and reduce the complexity of the interface. An application module is the sum of the modules that provide services to the entire user, including user login, online evaluation, information browsing (including evaluation news, evaluation results, system help, evaluation indicators), user message, Change Password, information query (including user information and evaluation records). Xunyang Shitou Orange Network has carried on the analysis to the different demand function, after the system front stage demonstration and the backstage management reasonable plan design, conceives how to design realizes the user's user's shopping flow, the user completes the transaction through the product choice fast payment. Through the display of product information and background data management settings, modification, constantly improve the site content to maintain the normal operation of the site, accumulation of users, increase product sales.

3.3 Database Design

3.3.1 Database Conceptual

Through the constraints of the database data processing, to clarify the entity and the relationship between the main body and the relationship [2], in data processing, there will be separate entities and entities, entities corresponding to the set of tables. From the actual requirements, grasp the design core of the database, separate the corresponding entity table, combined with the table structure analysis description.

3.3.2 Datasheet Design

The induction, the restriction related data, corresponds to each entity, the datasheet is the database important constituent. Now for the associated data to make a brief data analysis, the data table description [4]. (1) user registry, in order to achieve the effective management of user personal information and order details, set different fields to constrain user data, in order to call data information to feed back to users [10, 11], and deal with user queries and problems in a timely manner.

The product category information table sets up the different information parameters of the product, and the classification of the product makes it easy for users to quickly understand the product and compare the price difference between the product and the product.

This paper summarizes the use of PHP [6] and MySQL [1] database website building technology, from the site function analysis, feasibility analysis of the site has a full understanding of the functional design. In the system design and the realization always take the user demand as the core, in the test process simulates the user to eliminate the potential risk, causes the website the overall style and the content layout to adapt with the user, give users a comfortable, convenient and quick shopping experience. In practice, we constantly combine what we have learned with our proficiency in developing technical software. During this time, we encountered many thorny problems. We analyzed and solved the problems through trial and error, also for the future management and maintenance of the website has accumulated some experience. The completion of the website design so that I gained a lot of setbacks in learning to analyze the problem, deal with the problem, attention to detail. At present, the website is still to be adjusted and there are still some deficiencies [7]. In the future, the system functions and technologies will be upgraded to meet the users' browsing, query, product order service, and enhance the users' experience, attract a large number of users to participate in marketing activities.

References

1. Tong, X.T., Rui, Q., Feng, G.: Development and implementation of Nongjiale extension network system based on PHP + MySQL. Comput. Prod. Circ. **05**, 52 (2020). (in Chinese)
2. Chinese Journal of gerontology has been included in several international databases and retrieval journals. Chin. J. Gerontol. Chin. **4008**, 1735 (2020). (in Chinese)
3. Li, Y.: Design and implementation of Traction Power Supply Management Information System based on Big Data Platform. East China Jiaotong University (2020). (in Chinese)
4. Lee. D.Y.: The implementation of online voting system based on complex rules. Comput. Program. Skills Maintenance **20**, 41–43 + 58 (2004). (in Chinese)
5. Tian, W., Deng, M., Wang, X.M.: Research and implementation of key technologies of h 5 and P P 5 in patient management system for catheterization. Comput. Program. Skills Maintenance **04**, 86–90 (2020). (in Chinese)
6. Moutaouakkil, A., Mbarki, S.: Generating a PHP metamodel using Xtext framework. Procedia Comput. Sci. **170**, 838–844 (2020)
7. Hongbo, W.: Video-on-demand website design based on PHP Technology. J. Integr. Circ. Appl. **3704**, 68–69 (2020). (in Chinese)
8. Xiaona, Q.: The application research of PHP technology in dynamic web page form control extraction. J. Comput. Literacy Technol. **1606**, 217–218 (2020). (in Chinese)
9. Guo, C.: Implementation of MySQL database operation based on MySQLi in PHP. J. Hunan Vocat. Tech. College Posts Telecommun. **18**(04), 28–30 + 34 (2019). (in Chinese)
10. Ip, P.L.: Teaching reform and practice of student-centered PHP technology course. J. Comput. Literacy Technol. **1528**, 188–190 (2019). (in Chinese)
11. Wang, F.: Mysql Database Application from Beginner to Proficient. China Railway Press, Beijing (2014). (in Chinese)

Vocational Students' Academic Self-efficacy Improvement Based on Generative PAD Teaching Mode

Yaya Yuan[✉]

Zhejiang Business Technology Institute, Ningbo, Zhejiang, China
23704645@qq.com

Abstract. Generative PAD teaching mode breaks through the restriction of traditional "presupposition-execution" teaching mode, emphasizes the generative feature of curriculum standard, teaching resources and teaching process on the basis of limited presupposition, and effectively changes the traditional passive teaching. It is widely applied in higher vocational colleges. In view of the common phenomenon of learning burnout and learning procrastination among the vocational college students, this research, based on the academic self-efficacy theory, constructs a model of generative PAD teaching to promote students' academic self-efficacy, by influencing three factors, like students' successful direct experience, high level of learning motivation and positive verbal evaluation. Through collecting the data of normal and experimental groups' post-test of questionnaire, by analyzing the two independent T test sample with SPSS software, the results showed that students' learning interest, learning confidence and learning attitude were significantly improved after the experimental teaching. The generative PAD teaching mode provides valuable references for the vocational teachers to practice teaching reform, by analyzing learners' needs, setting teaching goals, building teaching resources, creating learning environment and improving teaching evaluation, which is helpful to improve classroom teaching quality in higher vocational colleges.

Keywords: Higher vocational education · Generative PAD teaching mode · Academic Self-efficacy

1 Introduction

Higher vocational education provides a great number of applied talents and plays an important role in the country's economic construction and development. With the industrial upgrading and the substantial adjustment of economic structure, higher vocational education is becoming more and more urgent to cultivate highly skilled talents for all walks of life. However, the current teaching quality of vocational colleges is worrying. It finds that traditional teaching mode can no longer effectively stimulate students' learning drive. The phenomenon of learning burnout and learning procrastination has become

J. MacIntyre et al. (Eds.): SPIoT 2020, AISC 1283, pp. 593–600, 2021.
https://doi.org/10.1007/978-3-030-62746-1_87

quite common, which leads to students' low academic self-efficacy. Therefore, it is of great significance to carry out generative PAD teaching mode reform in higher vocational colleges for improving students' academic self-efficacy, overcoming students' learning burnout and learning procrastination, and thus improving the quality of talent cultivation.

2 Relevant Research on the Academic Self-efficacy

2.1 Concept of Academic Self-efficacy

The concept of academic self-efficacy originated from Social Foundations of Thought and Action, which was written by American psychologist A. Bandura. According to Bandura, self-efficacy is "a fairly specific expectation about people's belief in their ability to perform a particular behavior or perform the behavior required to produce a certain outcome" [1]. Self-efficacy is different from the skills or abilities that an individual actually possesses. It emphasizes the confidence that people show in subjectively judging that they can accomplish a particular task. Self-efficacy is specific and is associated with behavioral goals, such as physical self-efficacy, interpersonal self-efficacy, professional self-efficacy, etc. Similarly, academic self-efficacy is "the judgment of an individual's ability to organize and implement an action process to achieve a desired educational achievement" [1]. Pintrich and DeGroot divided academic self-efficacy into "learning ability self-efficacy and learning behavior self-efficacy" [2]. Strong academic self-efficacy directly affects students' judgment of their learning ability and behavior, and even their confidence in successfully completing learning tasks and achieving learning goals.

2.2 Relationship Between Academic Self-efficacy and Academic Performance

Jinks and Morgan's research shows that academic self-efficacy is significantly positively correlated with academic performance, that is, students with high academic self-efficacy have better academic performance, while students with low academic self-efficacy have poorer academic performance [3]. Other studies have found that students with low academic self-efficacy show higher learning burnout, and even have doubts and distrust towards their own learning ability and learning methods, which eventually led to poor academic performance. On the contrary, students with high self-efficacy tend to be willing to meet new situations and challenges [4–6].

3 The Basic Structure of the Generative PAD Teaching Mode

The traditional teaching mode ignores the students' cognitive level and individual learning ability. The difficulty to achieve learning goals will lead to students' low level of academic self-efficacy, thus further lead to their learning burnout and learning procrastination. The author believes that generative PAD teaching mode should be adopted to reform the traditional teaching mode.

3.1 The Concept of PAD Teaching Mode

In order to create good interaction between teachers and students, the PAD teaching mode converts the traditional lecture-based learning to one with both teachers' lecture and students' discussion, and it should be executed strictly in accordance with the order of presentation (P), assimilation (A) and discussion (D) [7]. As for the time allocation, the teachers give a lecture in the second half of class, and then after several days' assimilation, the students make discussion in the first half of the next class. In the P stage, the teacher gives a framed lecture and assigns different-level tasks for students' choice. In the A stage, students carry out independent study and complete the corresponding task, absorb the internalized knowledge and make full preparation for the group discussion. In the D stage, students themselves lead group discussions and carry out intra-group, inter-group and whole-class communication.

3.2 The Generative Features of PAD Teaching Mode

Generative teaching mode is different from traditional "presupposition-execution" teaching mode. Ye Lan puts forward that the core of teaching process reform is to change "presupposition and implementation" into "presupposition and generation" [8]. Chen Ruifeng believes that the PAD teaching mode is a generative one [9]. In the P stage, the teacher adjusts the curriculum standard, only makes framed presentation, leaving space for the generation. In the A and D stages, students become independent explorers and experiencers, making it possible to create generative resources.

4 Three Factors to Effect Academic Self-efficacy

Previous studies have shown that student's successful direct experience, high-level learning motivation and positive verbal evaluation will highly improve their academic self-efficacy. Therefore, we will discuss the effect of generative PAD teaching mode from those three factors.

4.1 Successful Direct Experience

Successful experience helps students enhance self-efficacy. In the generative PAD teaching mode, the teacher is no longer the role of infusing all knowledge to students while ignoring their learning level, but the role of helping students obtain the basic knowledge framework and setting up tasks of different levels. In the P stage, it is easy for the students to understand the content. In the A stage, the students have plenty of time to absorb the knowledge and choose the corresponding task, fully adaptive to the students' cognitive load. It is easier for students to obtain direct successful experience.

4.2 High-Level Learning Motivation

High-level learning motivation can constantly strengthen students' academic self-efficacy. As for the generative PAD teaching mode, in the A stage, students conduct independent learning and form their own opinions. In the D stage, students will share their learning outcomes with others. They will not only "highlight" their own opinions, "test" their doubts, but also "help" their peers' difficulties. Their diverse learning experiences and perceptions can be used as generative teaching resources in the discussion. When students' learning results are displayed or even discussed as in-class teaching content, it will greatly stimulate students' learning interest.

4.3 Positive Verbal Evaluation

Generative PAD teaching mode emphasizes the process evaluation and diversified evaluation. It includes teacher-student evaluation and student-student evaluation. The forms of evaluation include learning achievement sharing, group cooperation and participation in discussion, etc. Positive evaluation can improve students' academic self-efficacy by affecting their learning attitudes. The encouragement and support from teachers and peers can effectively affect students' academic self-efficacy.

5 Model Construction of Generative PAD Teaching

Academic self-efficacy is correlated with students' academic performance. The author believes that generative PAD teaching mode can effectively improve students' academic self-efficacy. By combining the information-based teaching platform and means, a model of academic self-efficacy improvement is built, as shown in Fig. 1:

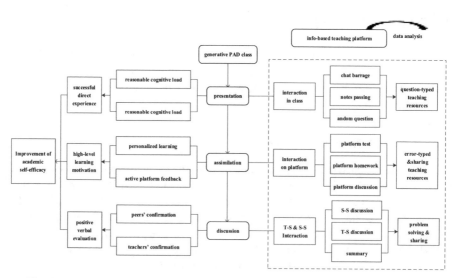

Fig. 1. Model of generative PAD teaching to improve students' academic self-efficacy

This model will take good effect on the students' academic self-efficacy, which provides a new way for teaching reform.

5.1 Analyze Learners' Needs

The cognitive load equal to students' level can help students get successful direct experience. Therefore, teacher needs to analyze learners' learning needs and sets personalized learning tasks. Generative teaching design not only focuses on the generation of knowledge, but also on the generation of students' emotion and ability [10]. For example, the questionnaire survey and individual interview help understand students' learning interest, confidence and attitudes, which helps make the design of PAD teaching mode more reasonable.

5.2 Define Teaching Objectives

The teaching objectives are classified into presupposed teaching objectives and generative teaching objectives. The design of the three stages of PAD can better achieve the presupposed objectives like improving students' abilities of data search, information screening, independent learning and teamwork. The valuable generative teaching objectives are also set up during the PAD teaching process.

5.3 Set up Teaching Resources

The teaching resources include presupposed teaching resources and generative teaching resources. Presupposed teaching resources include PPT courseware, video, audio, graphic webpage and others for students to carry out independent learning. Generative teaching resources is a kind of valuable learning resources, including question-typed resources generated from the classroom interaction in the A stage, the error-typed and sharing resources generated from platform learning in the D stage.

5.4 Create a Learning Environment

The info-based teaching platform and means help create an open and interactive learning environment, which is the premise of generative PAD teaching mode. In the P stage, for example, through the barrage function of teaching app, students can express viewpoints at any time in the classroom, through notes-passing function of teaching app, students can convey to the teachers' own doubts. In the A stage, tests are carried out on the questionnaire webpage and tasks are assigned through the teaching platform. Error-typed learning resources are generated on the basis of data statistics. Sharing learning resources are generated through discussion on the teaching platform.

5.5 Perfect Teaching Evaluation

Generative PAD teaching mode can meet the needs of students' independent and personalized learning. More emphasis should be placed on the evaluation of students' independent learning effect, especially the evaluation of effect of generative resources learning. In the process of evaluation, positive verbal evaluation among peers and teachers is encouraged, which can improve students' academic self-efficacy.

6   The Evaluation of Effect on Generative PAD Teaching Practice

As for the model of the generative PAD teaching mode, the author carries out the generative PAD teaching practice in business English, and then adopts questionnaire survey and comparative research methods to explore the teaching practice effect. The teaching experiment subjects are freshmen from classes 1922, 1923, 1924 and 1925, majoring in international trade. The average academic levels in four classes are almost the same. The four classes are all taught by the author, in which classes 1922 and 1923 are normal group using the teacher-centered lecturing mode while classes 1924 and 1925 are experimental group using generative PAD teaching mode.

6.1   Reliability Analysis of Questionnaire

In order to test the effectiveness of this teaching experiment, the questionnaire survey was conducted on the two groups at the end of the experiment. The validity of the questionnaire was verified to meet the experimental requirements. This survey collected a total of 98 valid questionnaires, covering three dimensions of students' learning interest, learning confidence and learning attitude. SPSS software was used to analyze the reliability of the questionnaires. By SPSS software analysis, the Cronbach α of three dimensions is greater than 0.8, so it can be said that the reliability of the post-test questionnaire is very good.

6.2   The Post-test Analysis of Normal Class and Experimental Class

SPSS analysis was used to conduct T test of two independent samples on the questionnaire results, and a comparative analysis was conducted on the post-test situation of two groups from the three dimensions.

6.2.1   Learning Interest Dimension

As can be seen from Table 1, the mean value of post-test learning interest in the normal group is 3.8368, which is smaller than that of the experimental group, which is 4.6895. The P value obtained by T test of SPSS is 0, less than 0.05. It shows that, after the experiment, there are significant differences between the two groups in learning interest. Therefore, combined with the analysis of the mean value and Sig. (bilateral), it can be concluded that after the experiment, although the interest of students in the normal group in learning business English has been improved somewhat, students in the experimental group have improved to a greater extent.

6.2.2   Learning Confidence Dimension

As can be seen from Table 2, the mean value of post-test learning confidence in the common group is 3.5660, which is smaller than that of the experimental group, which is 4.3526. The P value obtained by T test of SPSS is 0, less than 0.05. It shows that after the experiment, there are significant differences between the two groups in learning confidence. By combining the mean value with the analysis of Sig. (bilateral), it can be

Table 1. The Post-test analysis of the learning interest dimension between two groups

	Group	N	Mean value	Standard deviation	Standard error of mean value	Sig.
Interest dimension	Normal	96	3.8368	.83455	.08518	0
	Experimental	95	4.6895	.37078	.03804	

concluded that after the experiment, although the general group has increased in business English learning confidence, the students in the experimental group have increased to a greater extent.

Table 2. The post-test analysis of the learning confidence dimension between two groups

	Group	N	Mean value	Standard deviation	Standard error of mean value	Sig.
Confidence dimension	Normal	96	3.5660	.92195	.09410	0
	Experimental	95	4.3526	.50813	.05213	

6.2.3 Learning Attitude Dimension

As can be seen from Table 3, the mean value of post-test learning attitudes in the common group is 3.7951, which is smaller than that of the experimental group, which is 4.6632. The P value obtained by T test of SPSS is 0, less than 0.05. After the experiment, there are significant differences between the two groups in learning attitude. By combining the mean value with the analysis of Sig. (bilateral), it can be concluded that after the experiment, although students in the general group have improved their business English learning attitude, students in the experimental group have improved to a greater extent.

Table 3. The post-test analysis of the learning attitude dimension between two groups

	Group	N	Mean value	Standard deviation	Standard error of mean value	Sig.
Attitude dimension	Normal	96	3.7951	.82840	.08455	0
	Experimental	95	4.6632	.37895	.03888	

Acknowledgements. This work was supported by The First Batch of Teaching Reform Projects in the 13th Five-Year Plan for Higher Education in Zhejiang Province. Practice of PAD Teaching based on Motivation and Cognitive Theory in Higher Vocational Colleges – A Case Study of Business English Teaching for International Trade Major (Project No. jg20180631)

References

1. Bandura, A.: Self-efficacy: toward a unifying theory of behavioral change. Pshchol. Rev. **84**(2), 191–215 (1977)
2. Pintrich, P.R., DeGroot, E.V.: Motivational and self- regulated learning components of classroom academic performance. J. Educ. Psychol. **82**(1), 33–40 (1990)
3. Morgan, V.L., Jinks, J.L.: Self- efficacy and achievement: a comparison of children' beliefs from urban, suburban and rural schools. In: Divine, J.H., Tompkins, R.S. (eds.) Interdisciplinary Studies. Proceedings of the 17th National Conference of the Society of Educators and Scholars, pp. 216–224. The University of Southern Indiana Press, Evasville (1994)
4. Ma, Y.: Study on the relationship between college students' learning burnout and learning self-efficacy. Mod. Educ. Sci. **01**, 84–86 (2010). (in Chinese)
5. Shan, P.: Study on the relationship between learning burnout and academic self-efficacy of higher vocational college students. Commun. Vocat. Educ. **17**, 24–28 (2017). (in Chinese)
6. Zhang, W., Zhao, J.: Relationship between burnout and academic self-efficacy in college students. Psychol. Res. **5**(02), 72–76 (2012). (in Chinese)
7. X, X.: PAD classroom: a new exploration of classroom teaching reform in universities. Fudan Educ. Forum **12**(05), 5–10 (2014). (in Chinese)
8. Ye, L.: Presupposition and generation in classroom teaching reform. Fund. Educ. Forum **10**, 6–7 (2012). (in Chinese)
9. Chen, R.: PAD classroom: exploration of generative classroom teaching mode. Shanghai Res. Educ. **03**, 71–74 (2016). (in Chinese)
10. Zhou, H.: Research on generative teaching design in information environment. China Educ. Inf. **08**, 20–23 (2011). (in Chinese)

An Improved K-Means Clustering Algorithm Based on Density Selection

Wenhao Xie[1,2(✉)], Xiaoyan Wang[1], and Bowen Xu[1]

[1] School of Science, Xi'an Shiyou University, Shaanxi, China
xwhaoxwhao@163.com
[2] School of Management, Northwestern Polytechnical University, Shaanxi, China

Abstract. *K-means* clustering algorithm is an unsupervised learning method with simple principles, easy implementation, and strong adaptability. Aiming at the disadvantages of this algorithm that the clusters' number is difficult to determine, sensitive to the initial cluster center, and the clustering result is easily impacted by the outliers, this paper proposes an improved clustering algorithm based on density selection, which compares the neighborhood density of each sample and the average density of all the samples, treats the samples with lower density as the outliers or isolated points, and then deletes them. After data pre-processing, the cluster validity index is modified to obtain the optimal clusters' number by minimizing the cluster validity index, and then optimizes the initial cluster center by density selection strategy. Finally, it is verified by the experiment that the improved algorithm has better accuracy than the traditional *K-means* algorithm, and it can converge to the global minimum of SSE faster.

Keywords: Clustering analysis · *K-Means* algorithm · Cluster validity index

1 Introduction

In data mining, the clustering algorithms belong to the unsupervised learning methods. In the absence of classification labels, clustering is such a process that a set of samples are divided into different categories by different similarity. The clustering process divides the data objects in the sample space into different clusters, so that the similarity between data in the same cluster is high, and the similarity between data in different clusters is low, which is the essence of clustering. As one of the commonly used methods in traditional machine learning algorithms, clustering analysis is widely favored due to its practical, simple and efficient characteristics, and it has been successfully applied in many fields.

For the *K-means* algorithm which is a classical clustering method, that users assign randomly the number of the clusters *K* and the initial cluster center. First, the samples are divided according to the randomly allocated centers, and then the cluster center is determined again by the average value of the data in the same cluster, and the next iteration starts until all the cluster centers don't change, and then the iterations end. The

J. MacIntyre et al. (Eds.): SPIoT 2020, AISC 1283, pp. 601–607, 2021.
https://doi.org/10.1007/978-3-030-62746-1_88

Sum of Squared Error (SSE) usually is taken as the cost function, and the minimum SSE is taken as the training objective to obtain the final clustering result. The smaller the SSE, the smaller the sum of the distances between all the samples from their own cluster center, and the better the clustering effect. To minimize the SSE, it is necessary to repeatedly adjust the centers and repeatedly change the samples in each cluster.

The *K-means* algorithm has the advantages such as simple principle, fast clustering speed, and better clustering effect, but it also has some defects [1–3]. For example, the algorithm is sensitive to the number of clusters and the initial cluster center, the clustering effect depends on the setting of the initial values [4, 5]. In addition, the algorithm is liable to fall into the locally optimal solution and the inability to find non-convex clusters.

2 Improved Clustering Algorithm Based on Density Selection

As mentioned above, the number of clusters, the confirmation of the initial center, and the detection of the outliers can affect the effect of the clustering. In particular, if the outliers are selected as the initial cluster centers, the algorithm will be falling into the local optimal solution, and the clustering effect is affected. In order to solving the above problems, an improved algorithm based on density selection is proposed.

2.1 Cluster Validity Index

In order to improve the sensitivity of *K-means* algorithm to the number and center of clusters, some scholars put forward the cluster validity index [6–8], which can provide reasonable K values under certain conditions. A reasonable cluster partition needs not only to measure the compactness within the cluster, but also needs to consider the degree of separation between different clusters. The literature [8] defines the compactness index and the inter-cluster separation index, but its inter-cluster separation index only considers the distance between the different clusters' centers, and does not consider the separation degree of each sample in one cluster relative to other clusters. So this paper improves the inter-cluster separation index proposed by the literature [8].

Definition 1(Cluster compactness index): In the formula (1), n represents the number of all the samples; x_j represents the data object and μ_i is the center of the cluster C_i; *compactness* reflects the average of the sum of the squared distance between all samples of C_i and its center μ_i.

$$compactness = \frac{1}{n} \sum_{i=1}^{k} \sum_{j=1}^{n} \|x_j - \mu_i\|^2, \ (x_j \in C_i) \tag{1}$$

Definition 2 (Inter-cluster separation index) [7]: The separation degree between the sample x_i of class C_k and the class C_m is:

$$\sum_{x_i \in C_k} sep(x_i, C_m) \bigg/ (n_k \cdot q) \tag{2}$$

$$Sep(x, C) = \sum_{j=1}^{q} \alpha(x, x_j) \cdot \|x - x_j\|^2, \alpha(x, x_j) = \begin{cases} 1, x_j \in C, x_j \in N_q(x) \\ \\ 0, x_j \notin C, x_j \in N_q(x) \end{cases} . N_q(x) \text{ rep-}$$

resents q neighbors of object x. The smaller the value of formula (2), the greater the degree of separation between object x_i and class C_m.

Definition 3: The separation degree between class C_k and class C_m is:

$$separation\,(C_k, C_m) = \max\{ \sum_{x_i \in C_m} sep(x_i, C_k) \Big/ (n_m \cdot q), \sum_{x_i \in C_k} sep(x_i, C_m) \Big/ (n_k \cdot q)\}$$

(3)

n_k, n_m respectively represent the number of samples of class C_k and C_m.

Definition 4: The inter-cluster separation index of the whole is:

$$separation = \max_{k \neq m}\{separation(C_k, C_m)\} \qquad (4)$$

Definition 5 (Cluster validity index): In summary, the smaller the cluster compactness index, at the same time, the smaller the inter-cluster separation index, the better the clustering effect. Therefore, the improved cluster validity index V is defined as follows:

$$V = compactness \cdot separation \qquad (5)$$

The value of K that makes V has the minimum value is the optimal solution [8].

2.2 Selection of the Cluster Centers

In the *K-means* algorithm, compared with randomly selected K cluster centers, if it considers using the maximum distance method to confirm the initial cluster center, it can maximize the separation between clusters. However, because of the inevitable existence of the isolated points or the outliers in the data set, the method of determining the cluster centers only by the farthest distance is unreasonable for the data set with many outliers, and these outliers are easy to be selected as the cluster centers, so that a good clustering effect cannot be obtained.

2.3 The Improved *K-Means* Algorithm Based on Density Selection

In order to overcome the influence of the outliers on clustering effect, this paper proposes an improved algorithm to select the optimal number of clusters and the initial cluster center. First, the average density method is used to delete the outliers to ensure that the centers of the clusters don't include the outliers. The average density algorithm is as follows [9]:

(1) Calculation of the average density of the whole samples and the neighborhood density of a single sample x_i.

Suppose n is the number of the samples; $\bar{d}$ indicates the average Euclidean distance of the all samples; $d_{\max}$ is the maximum in all distances; the dimension of the samples is m.

$$\bar{\rho} = n \left/ t(\frac{\sqrt{3}}{2}d_{\max})^m, \; \rho(x_i, r_i) = n_i \right/ (\frac{\sqrt{3}}{2}a\bar{d})^m \tag{6}$$

Where, $\bar{\rho}$, $\rho(x_i, r_i)$ respectively represent the average density and the neighborhood density. t is the constant coefficient; n_i refers to the number of samples contained in a circular area with x_i as the center and a specified positive number r_i as the radius; $a \in (0, 1]$ is the adjustable coefficient.

(2) Deletion of the outliers and the isolated points

For a sample whose neighborhood density is much smaller than the average density, that is, $\rho(x_i, r_i)$ of a sample is much smaller than $\bar{\rho}$, is regarded as the outlier, and it should be deleted before using the algorithm training sample.

(3) Determination of the initial cluster center based on density selection

According to the relationship between the neighborhood density and the average density, the samples whose neighborhood density is much smaller than the average density are regarded as the outliers, and deleted [10]. For the selection of cluster centers, the following strategy is adopted: the sample with the largest neighborhood density in the remaining samples is used as the first center μ_1, and then the sample with the largest distance from μ_1 is chosen as the next center μ_2, $\cdots\cdots$ and the mth center μ_m is the data object x_i that satisfies the following formula (7) [10]:

$$\max(\min(d(x_i, \mu_1), \, d(x_i, \mu_2), \cdots\cdots d(x_i, \mu_{m-1}))), i = 1, 2, \cdots n; m = 1, 2 \cdots K \tag{7}$$

(4) The improved K-means clustering algorithm based on density selection

The improved algorithm which is called *Den-K-means* algorithm, and its process is:

Step 1. Preprocessing of data: delete the samples whose neighborhood density is much smaller than the average density;

Step 2. Confirm the value of K^* by minimizing the improved cluster validity index, and K^* is the optimal number of clusters;

Step 3. Choose the cluster centers u_i based on the density selection method;

Step 4. Perform *K-means* clustering algorithm on the basis of optimization the value of K^* and $u_i (i = 1, 2 \cdots K^*)$.

3   The Experimental Results

In order to verify the clustering effect of the *Den- K-means* algorithm, we selected three data sets in the UCI database: Iris, Wine and Diabetes, and the clustering is performed using the traditional *K-means* algorithm and the *Den-K-means* algorithm. In the *K-means* algorithm, for data set Iris and Wine, they are shown their average accuracy of 10 tests, and for data set Diabetes, the 10 initial centers, the accuracy of 10 tests, and the average

Table 1. The clustering results of dataset Iris and Wine

Algorithm	Data set	Iris	Wine
K – means	Average accuracy rate	86.15%	74.67%
Den – K – means	Initial Center	(80,110,73)	(70,140,8)
	Accuracy rate	88.65%	78.38%

Table 2. The clustering results of dataset Diabetes

Data set Algorithm		Diabetes	
		Initial Center	Accuracy rate
K – means	1	(81,347)	76.01%
	2	(84,365)	71.21%
	3	(360,220)	65.78%
	4	(21,749)	74.56%
	5	(65,785)	88.28%
	6	(178,255)	80.88%
	7	(78,640)	88.98%
	8	(189,430)	65.24%
	9	(102,222)	71.87%
	10	(22,436)	73.69%
	Average Accuracy rate	75.65%	
Den – K – means	Initial Center	(72,812)	
	Accuracy rate	76.93%	

accuracy of all the tests are demonstrated. The clustering results are shown in Table 1 and Table 2:

Table 1 and Table 2 show that the *Den- K-means* algorithm has a relatively high clustering accuracy. Taking a two-dimensional data set as an example, when the number of iterations is 5, it can be intuitively seen that the improved algorithm has a better clustering effect than the traditional algorithm (See Fig. 1).

SSE is used to evaluate the stability of the algorithms. Under the same number of iterations, the smaller the SSE, the better the algorithm and the smaller the loss. Taking the data set Iris as an example, Table 3 shows that the SSE value for the traditional algorithm at the 10th iteration is 268.92, which converges to the global minimum, while the SSE value for the improved algorithm can reach to 268.92 at the 5th iteration. So

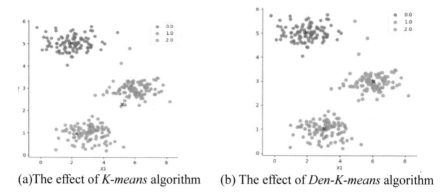

(a)The effect of *K-means* algorithm   (b) The effect of *Den-K-means* algorithm

Fig. 1. The clustering effect of the two algorithms on Iris

the improved algorithm not only obtains better clustering accuracy, but also has higher efficiency that can make the overhead smaller.

Table 3. SSE value of the two algorithms for dataset Iris

Number of iterations	SSE ($K - means\,algorithm$)	SSE ($Den - K - means\,algorithm$)
2	903.25	276.49
4	879.23	270.17
5	801.66	268.92
8	490.13	268.92
10	268.92	268.92
11	268.92	268.92

4   Conclusion

The traditional *K-means* clustering algorithm is sensitive to the number of clusters and the initial center setting. At the same time, the clustering effect will also be affected by the outliers and the isolated points. This paper proposes an improved algorithm based on density selection, improves the cluster validity index, and optimizes the number of clusters and the initial cluster center on the premise of getting rid of the influence of the outliers to the greatest extent. The experiments have proved that the improved clustering algorithm proposed can obtain a relatively stable initial cluster center, has a better clustering accuracy, and can converge to the global minimum of SSE faster.

Acknowledgements. This work was supported by Shaanxi Higher Education Teaching Reform Research Project (19BZ030) and the Project of Education Department of Shaanxi Province (JK190646).

References

1. Yang, S., Li, Y., Hu, X.: Optimization study on k value of k-means algorithm. Syst. Eng.-Theory Pract. **26**(2), 97–101 (2006). (in Chinese)
2. Yang, J., Zhao, C.: Survey on K-means clustering algorithm. Comput. Eng. Appl. **55**(23), 7–13 (2019). (in Chinese)
3. Hung, C.H., Chiou, H.M., Yang, W.N.: Candidate groups search for K-harmonic means data clustering. Appl. Math. Model. **37**(24), 10123–10128 (2013)
4. Wang, J., Ma, X., Duan, G.: Improved K-means clustering k-value selection algorithm. Comput. Eng. Appl. **55**(8), 27–33 (2019). (in Chinese)
5. Gan, G.J., Ng, M.K.P.: K-means clustering with outlier removal. Pattern Recogn. Lett. **90**, 8–14 (2017)
6. Zhu, W., Wu, N., Hu, X.: Improved cluster validity index for fuzzy clustering. Comput. Eng. Appl. **47**(5), 206–209 (2011). (in Chinese)
7. Mao, C.: Automatic three-way decision clustering approach based on K-means. Chongqing Univ. Posts Telecommun., 1–51 (2016). (in Chinese)
8. Xie, J.,Zhang, Y.,Jiang, W.: A k-means clustering algorithm with meliorated initial centers and its application to partition of diet structures. IN: International Symposium on Intelligent Information Technology Application Workshops, pp. 98–102. IEEE (2008)
9. Shi, H., Zhou, S., et al.: Average density-based outliers detection. J. Univ. Electron. Sci. Technol. China **6**(36), 1287–1288 (2007). (in Chinese)
10. Yuan, F., Zhou, Z., X, X.: K-means clustering algorithm with meliorated initial center. Comput. Eng. **33**(3), 65–67 (2017). (in Chinese)

Calculation and Analysis of Family Care Burden Coefficient

Lingni Wan[1,2]([✉])

[1] Department of Public Administration, Zhongnan University of Economics and Law, Wuhan, Hubei, China
317135275@qq.com
[2] Wuhan Business University, Hubei, China

Abstract. The family care burden can be identified by measuring the family support burden coefficient, the support burden coefficient and the family special disability, the patient population care coefficient three quantifiable numerical indicators to identify the most urgent housing needs.

Keywords: Family care burden · Support burden coefficient · Support burden coefficient · Family special disability and patient population care coefficient

1 Introduction

In the mainstream studies of western sociology, family burden focuses on the measurement of micro level, which refers to the care, economic and mental difficulties, problems or adverse effects brought to the family due to long-term illness of family members, including mental illness. There are some special surveys and index systems about family burden in foreign countries [1–3].

1.1 Research Trends at Home and Abroad

These surveys are also designed and studied according to the negative impact of diseases and other problems on families. The main research scales include:

Caregiver strain index (CSI; Robinson 1983), Cost of care index (CCI; kosberg & cairl 1986; Kosberg, Cairl & Keller 1990); Caregiver burden measures(Siegel, Raveis, Houts & Mor 1991); The burden interview (BI; Zarit, reever & Bach Peterson, 1980), The burden scale (Schott Baer 1993); The family impact survey (covinskytal 1994).

Domestic related research is also seen in the field of medical research, such as the family burden of patients with mental illness, chronic disease or cancer (Wang Lie et al. 2006; Su Chunyan et al. 2008; Xiong Zhenzhen Yuan Liji 2008; Li Hong et al. 2009).

These studies have analyzed in depth the various burdens brought about by the illness of family members, but their limitations mainly lie in the fact that the research and design are based on the special situation that some members of the family suffer from illness, which is difficult to be used in the study of bearing the burden in normal family life [4–6].

J. MacIntyre et al. (Eds.): SPIoT 2020, AISC 1283, pp. 608–613, 2021.
https://doi.org/10.1007/978-3-030-62746-1_89

1.2 Identification of Family Care Burden

Family burden research usually divides family burden into three different categories: economic burden, labor (care) burden and spiritual burden. This paper mainly studies the burden of family care. The burden of family care mainly refers to the burden that other family members need to take care of as a result of illness, loss of physical function of family members, or loss of working ability under age or old age [7].

From the perspective of the feasibility of family burden measurement and the universality of general families, this paper focuses on the direct measurement of the burden of family support and family support, through the support coefficient, dependency coefficient and family special disability, patient population care coefficient three quantifiable numerical indicators to identify the most in need of public rental housing products. Therefore, this paper does not consider the family economic burden and family spiritual burden [8].

2 Measurement of Family Care Burden Coefficient

The measurement of family care burden is based on four principles: effectiveness, practicability, convenience and reliability. First of all, the effectiveness is reflected in the determination of the family burden, care burden should fully consider the number of family members and the basis of composition, design the simplest family care burden ratio. Secondly, the practicability can reflect the difference of family care burden of different family structure types and the change of family burden in different family life cycle stages. Thirdly, it must be convenient to use, and can directly use the existing census data or social survey data to calculate. Finally, reliability must reflect the real burden of family care [9].

Based on the above factors, there are three main indicators of family care burden measurement, namely: family support burden, family support burden and family special disability, patient population care, respectively. The larger the value of this index, the more urgent the demand for housing is, the more necessary it is for housing [10].

2.1 Burden Coefficient of Family Support and Care

The burden coefficient of family support and care, also known as the support coefficient of the elderly population or the dependency ratio of the elderly population, refers to the ratio of the number of middle-aged and elderly population in a certain family to the total number of family members. It is used to show how much burden each family has to bear to support and care for the elderly. The calculation formula is as follows:

$$B_o = \frac{P_{65+}}{P_1} \tag{1}$$

Among them: Bo is the dependency ratio of the elderly population; p65+ is the number of elderly population aged 65 and above; P_1 is the total family population.

For example, if a couple of working age couples have two elderly people over 65 years old and the couple has no children, the supporting burden coefficient of this type of family should be 0.5:

$$B_o = \frac{P_{65+}}{P_1} = \frac{2}{4} \tag{2}$$

2.2 Burden Coefficient of Family Support and Care

The coefficient of family support and care burden is that of minors. It refers to the ratio of the number of minors in a family to the total number of family members. This indicator is used to reflect how much responsibility the family applying for public rental housing should bear for the maintenance and care of minors. The calculation formula is as follows:

$$B_m = \frac{P_{0-17}}{P_1} \tag{3}$$

Among them, BM is the dependency ratio of minors, p_{0-17} is the number of minors aged 0–17;

For example, for a second-generation family with a working age husband and wife with one minor child, the non labor age population in the support and care burden coefficient is one child, i.e., $P_{0-17} = 1$, while the total family population should be both husband and wife plus one minor child, i.e., $P_1 = 3$, then the support and care burden coefficient of this type of application family should be 0.33:

$$B_m = \frac{P_{0-17}}{P_1} = \frac{1}{3} \tag{4}$$

If there are two minor children in the family, then by analogy, the burden coefficient of maintenance and care of the family is 0.5.

2.3 Family Special Disability, Patient Population Care Coefficient

The family special disability and patient population care coefficient refers to the ratio of the number of family members who are suffering from serious diseases and disabilities and are unable to take care of themselves due to serious diseases and disabilities, and the total number of family members. This index is used to reflect the number of families who apply for public rental housing to bear the care responsibilities of the disabled and patients. The calculation formula is as follows:

$$B_d = \frac{P_i}{P_1} \tag{5}$$

Among them, B_d refers to the dependency ratio of the family with special disabilities and patients; P_i refers to the number of people whose families are suffering from serious diseases and disabilities and can't take care of themselves and completely lose the ability to work.

2.4   Total Burden Coefficient of Family Care

The total burden coefficient of family care refers to the ratio between the number of non labor force population and the total number of family members, which can be expressed by formula:

$$B_1 = \frac{P_2}{P_1} \tag{6}$$

Among them: B_1 is the total burden coefficient of family care; P_2 is the number of non labor force population in family members; P_1 is the total number of family population. It can also be expressed by the sum of specific sub index coefficients, and the formula is as follows:

$$B_1 = \frac{P_{0-17} + P_{65+} + P_i}{P_1} \tag{7}$$

Among them: P_{0-17} is the number of minors aged 0–17; $P_{65}+$ is the number of elderly people aged 65 and above; P_i is the number of people whose families are suffering from serious diseases and disabilities and can't take care of themselves and completely lose the ability to work. For example, If a couple of working age have two minor children, and there are two elderly people over 65 years old in the family, and one of them is paralyzed and incapacitated in bed, the total family population P_i is 6, If the family has minor children ($p_{0-17} = 2$), the number of elderly people aged 65 years and above who need to support $P_{65}+$ is 1, and the family suffering from serious diseases and disabilities can't take care of themselves, $p_i = 1$, then the total burden coefficient of family care is 0.83,

$$B_1 = \frac{P_{0-17} + P_{65+} + P_i}{P_1} = \frac{2 + 1 + 1}{6} = \frac{5}{6} \tag{8}$$

It needs to be explained here: if the family is suffering from serious diseases or disabilities, and the population who has completely lost the ability to work is the elderly or minors over 65 years old. When calculating the family special disability and patient population care coefficient, it can't be calculated repeatedly with the family support and care burden coefficient or family support and care burden coefficient.

If there is only a couple of working age couples, there are no minor children in the family, there are no elderly people over 65 years old who need to be supported and disabled members who need to be cared for, then the total coefficient of family care burden is 0,

$$B_1 = \frac{P_{0-17} + P_{65+} + P_i}{P_1} = \frac{0}{2} \tag{9}$$

If a 65 year old couple living alone has no children, and the total family population should be two husband and wife, that is 2, and the elderly who need family support $P_{65}+$ is 2, then the family burden coefficient of this type of family is 1:

$$B_1 = \frac{P_{0-17} + P_{65+} + P_i}{P_1} = \frac{2}{2} \tag{10}$$

By analogy, as long as the age structure of the population in a specific family can be divided into different family types, and the total number of family members in the corresponding family type and the number of people who have lost the ability to work can be calculated.

3 Conclusion

The family burden care coefficient based on the age structure of family members and the social responsibility shared by families has four advantages: first, the data are easy to obtain. As long as the content related to family is investigated, the age of family members and family members is usually considered as an important variable and survey content. Second, the indicators are simple. Only two indicators are considered in the design of family care burden coefficient, one is the number of population and the other is the age structure of population. Only these two indicators are needed to calculate the family burden coefficient. Third, it can reflect the real situation of family care burden of public rental housing application.

The main limitations are: the difference of family life is not considered enough. In fact, each family member of the same age group is regarded as an undifferentiated statistical unit based on the size of population. However, in real life, different families may face different family burdens. In fact, it ignores the difference of different types of family burden in individual life cycle and family life cycle. For example, at the beginning of a family, if a couple has not given birth to children, they will not have the pressure to raise children or educate children. However, the housing pressure they are facing may be higher than that of families that have already given birth to children and need to be educated. Similarly, the medical burden increases with the age of family members and the increase of disease risk. Therefore, the relationship between family care burden coefficient and family life needs to be further improved.

Acknowledgements. This work was supported by Research on the nature and orientation of social insurance agency, a doctoral research fund project of Wuhan Business University (2017KB010).

References

1. Schott-Baer, D.: Dependent care, caregiver burden and self-care agency of spouse caregivers. Cancer Nurs. **16**(3), 230–236 (1993)
2. Zarit, S.H., Reever, K.E., Bach-Peterson, J.: Relatives of the impaired elderly: correlates of feelings of burden. Gerontol. **20**, 649–655 (1980)
3. Robinson, B.: Validation of a caregiver strain index. J. Gerontol. **38**, 344–348 (1983)
4. Sikora, S.A.: The University of Arizona college of medicine optimal aging program: in the shadows of successful aging. Gerontol. Geriatr. Educ. **27**, 59–68 (2006)
5. Marchegiani, F., Marra, M., Spazzafumo, L., et al.: Paraoxonase activity and genotype predispose to successful aging. J. Gerontol. Ser. A Biol. Sci. Med. Sci. **61**, 541–546 (2006)
6. Bartres-Faz, D., Junque, C., Clementel, C., et al.: Angiotensin i converting enzyme polymorphism in humans with age-associated memory impairment: relationship cognitive perfomance. Neurosci. Lett. **290**, 177–180 (2000)

7. Hu, P., Bretsky, P., Crimmins, E.M., et al.: Association between serumbeta - carotene levels and decline of cognitive function in high-functioning older persons with or without apolipoprotein E4alleles: MacArthur studies of successful aging. J. Gerontol. Ser. A Biol. Sci. Med. Sci. **61**, 606–620 (2006)
8. Newson, R.S., Kemps, E.B.: Cardiorespiratory fitness as a predictor of successful cognitive ageing. J. Clin. Exp. Neuropsychol. **28**, 949–967 (2006)
9. Chunyan, X.: A study on the burden reduction mechanism of the support pressure of the only child family under the individual income tax reform system. Econ. Res. Guide **19**, 49–50 (2019)
10. Taona, L., Zhang, S.: Research on family support ability based on Grey GM (1,1) - Markov forecasting model. Stat. Decis. Mak. **20**, 97–100 (2012). (in Chinese)

A Study on the Impact of Formal Social Support on the Mental Health of the Elderly in Urban and Rural China: Based on the 2015 CHARLS Data

Dongfang Li[✉]

School of Public Administration, Zhongnan University of Economics and Law, Wuhan 430073, Hubei, China
lidongfang@zuel.edu.cn

Abstract. In the context of implementing the "Healthy China" strategy, the health problems of the elderly in urban and rural China cannot be ignored.

Compared with other age groups, the elderly have a series of obvious changes in mental, physical and social functions, and they need care and help from family, society and government in all aspects. This article used the 2015 CHARLS micro data, selected the elderly over the age of 60 group as the research object, using Stata statistical software for data statistics and analysis, the Probit regression model and OLS regression model analysis of social support on urban and rural elderly health effects. It is found that formal social support also has a significant impact on the degree of depression and cognition in the elderly, indicating that formal social support also has a significant impact on the mental health of the elderly.

Keywords: Formal social support · The elderly · Mental health

1 Introduction

In the context of China's aging population, whether the health status of the elderly can be improved or promoted directly determines whether China can successfully cope with the challenges brought by the aging population. According to the definition of health as defined by the World Health Organization (1947), the health of the elderly is not only physical health but also including important aspects such as mental health [1].

Social support refers to a certain social network by means of certain material and spiritual help free for the social vulnerable groups, using social support intervention in the social network resources are insufficient or lack of the ability of using the social network of the elderly, to help them expand social network resources, improve its ability to use social networks. Social support can be divided into formal social support and informal social support. The difference between the two is that the resource providers of formal social support network include government agencies, regional organizations and professional organizations, etc. Formal social support has strong characteristics

J. MacIntyre et al. (Eds.): SPIoT 2020, AISC 1283, pp. 614–620, 2021.
https://doi.org/10.1007/978-3-030-62746-1_90

of institutionalization, contract and stability. Informal support, on the other hand, is a social support network based on geography, blood relationship and kinship. Its resource providers mainly include family members, relatives and friends, and neighbors, etc., with strong "emotional" color and "differential pattern" characteristics. The formal social support mentioned in this research mainly includes social security, social assistance, etc., which is mainly measured by indicators such as medical insurance, endowment insurance and other social subsidies [2–4].

Social support is particularly important for the elderly, can help the elderly to prevent the negative effect on the mental health physical disability (Wan et al. 2013). People with high levels of social support enjoy higher levels of health and well-being (Cohen & Wills 1985; Pierce et al. 1996; Sarason et al. 1990). Social support can improve physical health and reduce the occurrence of depression (Cohen & Wills 1985; Cutrona & Russell 1987; Roberts & Gotlib 1997; Lynch et al. 1999). Social support can reduce the loneliness of the elderly (Jones & Moore 1987; Russell 1996). In terms of the impact of formal social support on the mental health of the elderly, Tao Yuchun and Shen Yu (2017) believe that formal social support, such as medical insurance and endowment insurance, has a "buffer model" effect on the physical and mental health of the elderly in rural areas. Based on this, we propose the research hypothesis that formal social support promotes the improvement of mental health status in the elderly [5–8].

2 Data, Models and Analysis

The data used in this paper are microscopic data from The 2015 CHARLS of Peking University. The elderly population over 60 years old is selected as the research object, and the mental health status of the elderly is measured by the degree of mental depression and cognitive status [9, 10].

In this paper, the degree of depression and cognition were used to measure the mental health of the elderly. When the influence of social support on the degree of depression was analyzed, the binary probit model was used to estimate the degree of depression since the degree of depression was 0–1 dummy variable (1 means no depression, 0 means depression). When analyzing the impact of social support on cognitive level, this paper adopts the least square method (OLS) to estimate the cognitive level as it is a continuous variable. The specific method is as follows:

2.1 Binary Probit Model

Set $x^T = (x_1, \cdots, x_h)$ as a set of independent variables and $x^T = (x_1, \cdots, x_h)$ as dependent variables. When not depressed, write as *depression* $= 1$; When it is depression, write it as *depression* $= 0$. The probit regression model is established as follows:

$$Probit(depression = 1) = \alpha_j + \beta_1 x_1 + \beta_2 x_2 + \cdots + \beta_h x_h$$

2.2 Classical Linear Regression Model

$$cognition = \beta_{61} isocial\ support + \beta_{62}X + \varepsilon$$

Where, cognition means the cognitive level, is the continuous value. social support means the formal social support. X is the control variable, including gender, age, marital status, whether to live in the town. β_{61} is the coefficient of social support. β_{62} is the coefficient of the control variable. ε is the residual term.

The endowment insurance, medical insurance and other social subsidies all play a role in the degree of depression of the elderly at the significant level of $P < 0.1$, among which the endowment insurance and medical insurance have a positive influence, while other social subsidies have a negative influence. Endowment insurance and medical insurance enable the elderly to enjoy life security, health security and a sense of peace of mind. Other social subsidies are ACTS of free assistance, which bring certain influence and psychological burden to the dignity of the elderly.

The control variables including age, town and income all had a significant effect on the degree of depression of the elderly at the significant level of $P < 0.01$, among which age and income had a negative effect, while town had a positive effect. Older people have a higher tendency of depression, mainly because with the increase of age, the elderly face greater psychological pressure of death, and thus prone to depression. The degree of depression among the elderly in urban areas is relatively low, mainly because the causes of depression in rural areas mainly come from physical health, while the elderly in urban areas are not only affected by physical health factors, but also affected by social and economic pressure. Older people with higher incomes have higher levels of depression, possibly because they have higher levels of work stress and intensity than people with lower incomes. Other control variables including married, separated, divorced, widowed and cohabiting all had positive effects on the degree of depression in the elderly, but not significant in the total sample regression (Table 1).

Endowment insurance, medical insurance and other social subsidies all play a role in the cognitive status of the elderly at the significant level of $P < 0.1$. Among them, endowment insurance and medical insurance have a positive influence, while other social subsidies have a negative influence. The reasons are largely the same as above.

Among the control variables, gender, age, town and income all had a significant effect on the cognition degree of the elderly at the significant level of $P < 0.01$, among which gender and town had a positive effect, while age and income had a negative effect. The cognitive level of elderly men is better than that of elderly women, mainly due to the influence of traditional ideas. The education level of the previous generation of men is higher than that of women, which has an impact on their cognitive level. The cognitive level of the elderly in urban areas is better than that in rural areas, because the level of economic development in urban areas is better, their education level and medical level are higher than that in rural areas, and they are prone to understand and recognize the mental health of the elderly in urban areas. The cognitive status of older people on mental illness is relatively poor, mainly because older people have less knowledge of mental illness, which is not conducive to their cognition. The higher the income level is, the lower the cognitive status of mental illness of the elderly, the more they pay attention to physical health and tend to neglect mental health. Marriage, separation,

Table 1. Influence of formal social support on the degree of depression in the elderly

	(1)	(2)	(3)
	Depression	Depression	Depression
Pension	0.134***	0.129***	0.101***
	(0.0188)	(0.0190)	(0.0192)
Medins		0.0628*	0.0719**
		(0.0326)	(0.0329)
Other subsidy			−0.625***
			(0.0253)
Gender	0.361***	0.360***	0.382***
	(0.0183)	(0.0183)	(0.0185)
Age	−0.0107***	−0.0106***	−0.00626***
	(0.000910)	(0.000910)	(0.000937)
Urban	0.159***	0.160***	0.184***
	(0.0198)	(0.0198)	(0.0200)
Married	5.020	5.029	4.191
	(187.5)	(187.5)	(86.28)
Separated	4.517	4.529	3.660
	(187.5)	(187.5)	(86.28)
Divorced	4.689	4.702	3.878
	(187.5)	(187.5)	(86.28)
Widowed	4.795	4.806	3.999
	(187.5)	(187.5)	(86.28)
Cohabitated	4.387	4.394	3.632
	(187.5)	(187.5)	(86.28)
Single income	−0.0943***	−0.0942***	−0.0942***
	(0.0154)	(0.0153)	(0.0154)
_cons	−4.431	−4.498	−3.826
	(187.5)	(187.5)	(86.28)
N	9785	9785	9785

*means $p < 0.1$,**means $p < 0.05$,***means $p < 0.01$.

divorce, widowhood and cohabitation all had positive effects on the cognition degree of the elderly, but not significant in this regression (Table 2).

The endowment insurance, medical insurance and other social subsidies all play a role in the degree of depression of the rural and urban elderly at the significant level of $P < 0.1$, among which the endowment insurance and medical insurance have a positive

Table 2. Influence of formal social support on the degree of cognition in the elderly

	(4)	(5)	(6)
	Cognition	Cognition	Cognition
Pension	0.132***	0.119***	0.0785***
	(0.0161)	(0.0162)	(0.0164)
Medins		0.166***	0.188***
		(0.0277)	(0.0278)
Other subsidy			−0.890***
			(0.0218)
Gender	0.314***	0.312***	0.345***
	(0.0157)	(0.0157)	(0.0158)
Age	−0.0234***	−0.0233***	−0.0177***
	(0.000785)	(0.000785)	(0.000802)
Urban	0.474***	0.477***	0.525***
	(0.0174)	(0.0174)	(0.0176)
Married	6.221	6.246	5.475
	(126.8)	(126.8)	(98.45)
Separated	5.898	5.928	5.114
	(126.8)	(126.8)	(98.45)
Divorced	6.072	6.110	5.366
	(126.8)	(126.8)	(98.45)
Widowed	6.058	6.088	5.363
	(126.8)	(126.8)	(98.45)
Cohabitated	5.904	5.924	5.270
	(126.8)	(126.8)	(98.45)
	−0.1212***	−0.1215***	−0.1218***
	(0.0132)	(0.0132)	(0.0132)
	(126.8)	(126.8)	(98.45)
N	9785	9785	9785

effect, while other social subsidies have a negative effect. Endowment insurance and medical insurance enable the elderly to enjoy life security, health security and a sense of peace of mind. Other social subsidies are ACTS of free assistance, which bring certain influence and psychological burden to the dignity of the elderly.

Endowment insurance, medical insurance and other social subsidies all have an effect on the cognitive status of the rural and urban elderly at the significant level of $P < 0.1$.

Among them, endowment insurance and medical insurance have a positive effect, while other social subsidies have a negative effect.

From the analysis of the degree of depression in the elderly, as far as the rural elderly are concerned, the controlling variables of age, separation, widovil and cohabitation have a negative effect on the degree of depression in the elderly. Marriage and divorce had a positive effect on the degree of depression in the elderly. Towns and unmarried women are deleted because of collinearity. For the urban elderly, the control variables including age, marriage, separation, divorce, widowhood, cohabitation and unmarried all had a positive influence on the degree of depression of the elderly. Towns were deleted for collinearity. None of the control variables were significant in the rural sample regression or the urban sample regression.

The endowment insurance enables the elderly in urban areas to obtain health security and a sense of peace of mind, and promotes positive cognition. Other social subsidies worsen the cognitive status of the elderly in urban and rural areas, and the elderly receiving social subsidies are generally in poor physical condition. The older the rural elderly, the poorer their cognitive status, which is associated with physical deterioration. The reverse is true for the elderly in urban areas, but not significantly.

For the rural elderly, the control variables of married, separated, divorced and widowed all had a positive effect on the cognitive level of the elderly. For the urban elderly, marriage, divorce, widowhood, cohabitation and unmarried also had a positive impact on the cognitive level of the elderly. The above control variables are not significant in the two sample regressions.

3 Conclusions

3.1 Formal Social Support Had a Significant Effect on the Degree of Depression in the Elderly

The endowment insurance, medical insurance and other social subsidies all play a role in the degree of depression of the elderly at the significant level of $P < 0.1$, among which the endowment insurance and medical insurance have a positive influence, while other social subsidies have a negative influence. Endowment insurance and medical insurance enable the elderly to enjoy life security, health security and a sense of peace of mind. Other social subsidies are ACTS of free assistance, which bring certain influence and psychological burden to the dignity of the elderly. Sample regression shows that the above conclusion is still valid.

3.2 Formal Social Support also Had a Significant Effect on Cognition in the Elderly

Endowment insurance, medical insurance and other social subsidies all play a role in the cognitive status of the elderly at the significant level of $P < 0.1$. Among them, endowment insurance and medical insurance have a positive influence, while other social subsidies have a negative influence. The reasons are largely the same as above. Sample regression shows that the above conclusion is still valid.

References

1. Hwang, K., Hammer, J., Cragun, R.: Extending religion-health research to secular minorities: issues and concerns. J. Relig. Health **50**, 608–622 (2011)
2. Benjamins, M.: Religion and functional health among the elderly: is there a relationship and is it constant? J. Aging Health **16**(3), 235–274 (2004)
3. Green, M., Elliott, M.: Religion, health and psychological well-being. J. Relig. Health **49**, 149–163 (2010)
4. Ware Jr., J.E.: Standards for validating health measures: definition and content. J. Chronic Dis. **40**(6), 473–480 (1987)
5. Chen, F., Short, S.E.: Household context and subjective well-being among the oldest old in China. J. Fam. Issues **29**(10), 1379–1403 (2008)
6. Cobb, S.: Social support as a moderator of life stress. Psychosom. Med. **38**(5), 300–314 (1976)
7. Chen, X., Silverstein, M.: Intergenerational social support and the psychological well-being of older parents in China. Res. Aging **22**(1), 43–65 (2000)
8. Ren, G., Wang, F., Luo, Y.: The impact of income and individual income deprivation on the health of urban and rural residents: an empirical study based on cgss2010 data analysis. Nankai Econ. Res. **6**, 3–18 (2016). (in Chinese)
9. Deneve, K.M., Cooper, H.: The happy personality: a meta-analysis of 137 personality traits and subjective well-being. Psychol. Bull. **2**, 90–197 (1998)
10. Jinling. Influencing factors and comparative scores of life satisfaction of the elderly analysis. Pop. Econ. (2), 85–91 (2011). (in Chinese)

Bi-Apriori-Based Association Discovery via Alarm Logs

Feng Yao[1(✉)], Ang Li[2], and Qiong Wang[1]

[1] State Grid Gansu Electric Power Co. Ltd., Lanzhou 730000, Gansu, China
yaofeng_gansu@163.com, wangqiong_gansu@163.com
[2] Anhui Jiyuan Software Co. Ltd., Hefei 230000, Anhui, China
liang_jiyuan@163.com

Abstract. The state grid information system is complex, the operation and maintenance information are diverse, involving a wide range of aspects. It becomes a key problem that how to use the alarm log of the operation and maintenance system to analyze the root cause of the fault. At present, there is a lack of alarm correlation technology for operation and maintenance system, and the traditional method lacks mature application systems for new scenarios. The reliability and efficiency of operation and maintenance system not only depends on the accurate collection of equipment fault information, but also depends on how to analyze massive fault information effectively and quickly, so as to grasp the key, handle the key, solve the key, three "keys". In this context, a Bi-Apriori algorithm is proposed, to solve the problem of frequently scanning the dataset and generating a large number of candidate set. It adopts the vertical data structure to mine frequent items and then as for alarm logs the proposed Bi-Apriori can be used to mining frequent items from alarm logs. Experimental results show that the proposed Bi-Apriori algorithm is superior to the existing association rule discovery algorithms on several datasets in real applications, especially for the dataset of alarm logs.

Keywords: Association discovery · Bi-Apriori · Alarm logs

1 Introduction

At present, the operation and maintenance of the system in State Grid is mainly carried out by the system operators, but in the case of problems in the information system, it can only be handled by the operation and maintenance personnel manually, which is time-consuming, laborious and inefficient, and poses a greater threat to the safe and stable operation of the information system. These work that takes up considerable workload of operation and maintenance management needs to realize the automation of information system operation and maintenance. How to liberate the operation and maintenance personnel from the mechanical and boring work that can be completed automatically, effectively improve the operation and maintenance efficiency of information system, reduce the operation and maintenance cost, improve the quality of operation and maintenance services, and reduce the risk of information system operation? This is an urgent

J. MacIntyre et al. (Eds.): SPIoT 2020, AISC 1283, pp. 621–632, 2021.
https://doi.org/10.1007/978-3-030-62746-1_91

problem we need to solve, so as to solve the problem of low level of resource alloca-tion information perception and maintenance automation, incomplete management and unable to match the operation and maintenance quality requirements.

Traditional association analysis is to find interesting association and correlation between item sets from a large number of data. A typical example of correlation analysis is shopping basket analysis. In this process, we can find the relationship between different products that customers put into their shopping basket, and analyze customers' shop-ping habits. By understanding which products are frequently purchased by customers at the same time, this kind of association discovery can help retailers to develop market-ing strategies. Other related applications include price list design, product promotion, product emission and customer segmentation based on purchase mode.

With the advent of the information age, there are more and more scenarios of associ-ation analysis application. Its prominent feature is no longer for small-scale data, but for the knowledge discovery of massive data, to find the association, correlation, or causal structure. This feature of association analysis is consistent with the real demand from alarm logs to solve the "Big Bang" problem. Since the 1990 s, there have been many technical schemes domestic and abroad to conduct research on association analysis, but there are few researches aiming at complex sources, massive items and operation and maintenance system. This paper is just from the perspective of association analysis to study and establish data association points, and analyze the technical model of knowl-edge base. Specifically, the comprehensive application of big data association analysis technology can realize the comprehensive awareness of information communication risk situation, comprehensive analysis of operation and maintenance data, and real-time early warning of operation and maintenance risk, achieve the active operation and main-tenance effect of solving problems before the fault occurs, and improve the State Grid company's overall information communication fault monitoring and risk early warning ability.

In this study, based on the knowledge base of association rules, through the research of related technologies of association rules, on this basis, according to the characteristics of alarm logs from the IT infrastructure of State Grid, this paper proposes a Bi-Apriori algorithm. It takes advantage of vertical data structures to compute the support degree quickly. Hence, it is unnecessary to scan the dataset totally to have the frequent item set. This leads to a significant improvement on the time performance, which has been validated with a series of experiments on some datasets from real applications, especially on the alarm logs from State Grid IT data center.

2 Related Works

2.1 Association Discovery

When we talk about the association discovery, we have to mention Apriori [1] algorithm firstly, which is the most popular and widely adopted algorithm for association discovery. Simply speaking, it aims is to obtain the frequent item sets through scanning the dataset with certain support degree. However, its disadvantages are also obvious, i.e., candidate sets are generated by self-joining, which leads to many unnecessary items. In addition, it requires scanning of the dataset back and forth to calculate the support degree, resulting in significant input and output overhead.

Recently, in order to solve the above-mentioned problems in Apriori algorithm, there are many researches aiming to solve them.

As for reducing the times of scans, [2] proposes an algorithm with storing transaction collections with matrices, and it significantly reduces the times of scans with datasets. On the top of this algorithm, [3] proposes a compression strategy of the matrix to further improve its performance. In addition, [4] introduce orthogonal lists to accelerate the computational time of the Apriori algorithm.

As for the pruning step, [5] introduces a hash technology to compute the support degree, and this significantly improve the computational time in the pruning step. [6] introduces the support matrix with frequent 2-item sets to achieve the fast pruning;

As for the support degree calculation, [7, 8] partite the transaction set into several smaller datasets, and then the frequent item set of the dataset can be gotten, which combines the frequent item sets of smaller datasets. This can improve the time efficiency, but not too much, due to the fact that it only reduces the single-step computational memory with the transaction set partition. In contrast, [9, 10] reform the data structure into the vertical data, and this can help to have the support degree directly from the length of the transaction item set Tidset, and the frequent items by the intersection of the Tidset sets. It significantly improves the computational efficiency with the support degree.

Although the above algorithms can discover associations to a certain extent by adopting different data storage structures, special pruning methods or special support degree calculation methods, but they do not make full use of the prior knowledge well. It is necessary and important to introduce a novel approach to improve the Apriori algorithm.

2.2 Process of Apriori Algorithm

As for the association discovery, the Apriori algorithm adopts a layer-by-layer iterative process to get frequent k-item sets, which is the most idea in the algorithm. More specifically, the process of Apriori algorithm can be illustrated as following steps in detail:

Step 1: Process the dataset, in order to get the support number of each item in the dataset, and obtain the candidate 1-item set C_1;

Step 2: With the support degree of each item in C_1, we need delete the items which is less than the support degree threshold *Min_Sup* * *Data.size*, and obtain the frequent 1-item set F_1;

Step 3: With frequent $(k-1)$-item sets F_{k-1}, we can have kth candidate set by $F_{k-1} \times F_1$ or $F_{k-1} \times F_{k-1}$; Process the dataset, in order to count the support degree of each item, and delete the set of items which is less than the support degree threshold;

Step 4: Repeat Step 3 until a new frequent item sets cannot be obtained.

The prior knowledge which has been adopted in the above algorithm is as follows:

Property 1: If an item set is frequent, all its subsets should be frequent.
Property 2: If an item set is not frequent, all its super sets should not be frequent.

2.3 Disadvantages of Apriori Algorithm

Although Apriori algorithm can solve a lot of problems in applications, there are still some disadvantages which makes it hard to be applied in some cases:

Case 1: Processing the dataset too frequently, resulting in too much input and output overhead.

Case 2: When the frequent item sets are to many, the self-connection will lead to a large set of candidate items. The pruning steps can help to alleviate the situation, but it cannot solve this problem.

Case 3: When the connection mode is $F_{k-1} \times F_1$ or $F_{k-1} \times F_{k-1}$ this can lead to a redundant set of candidates.

Case 4: The computational time to obtain the support degree is usually too large, as it has to process the whole dataset.

With the listed cases about disadvantages, the Apriori algorithm can be improved with the following directions:

Case 1: We need propose a data structure, which can not only store transaction sets, but also reduce the time for processing the dataset.

Case 2: We have to present novel strategies for the connection method, in order to avoiding the generation of redundant candidate sets.

Case 3: If it is possible, the size of the dataset should be reduced in order to improve the time efficiency.

Therefore, with all the above-mentioned directions about improvements, we can propose our Bi-Apriori algorithm.

3   BI-APRIORI Algorithm

3.1 Preliminary Concepts

With the data sets consisting of n items and m transactions, for any item $I_j \in \{I_1, I_2, \ldots, I_n\}$, its vector representation is:

$$I_j = \{T_{j1}, T_{j2}, \ldots T_{jk}\}, \ 0 \le k \le m,$$

where T_{jk} is a transaction including item I_j.

The support degree of item I_j is the size of the transaction set including it:

$$\sup(I_j) = k \tag{1}$$

With this property, there is no need for the traditional methods to process the whole dataset to compute the support degree of item I_j, and only to process the transaction set including item I_j is enough.

The intersection $\{I_j, I_i\}$ of item I_j and item I_i is the intersection of the transaction set including I_j and the transaction set including I_i:

$$\{I_j, I_i\} = \{T_{j1}, T_{j2}, \ldots T_{jk}\} \cap \{T_{i1}, T_{i2}, \ldots T_{ik}\}$$

With this property, there is no need to process the whole dataset to obtain the multiple transaction sets. In addition, with the above-mentioned property for support degree computation, these two properties can significantly accelerate the associate discovery process.

3.2 Equalization in Association Discovery

Traditionally, the association discovery methods usually only introduce the convenient insertion into memory [7] to improve the time efficiency. [8] has tried to divide the dataset to the pruning, however it only determines whether the current k-item set is redundant and should be deleted.

In contrast, here in this paper, we introduce the equalization during association discovery. More specifically, the data set is partitioned into two 2 parts D_1 and D_2. items J of D_1 and D_2 are named as I_j^1 and I_j^2. For each item $\{, I_2, \ldots, I_k\}$ of a frequent k-item set, the Cache is added with the following value:

$$Cache(I_1, I_2, \ldots, I_k) = \begin{cases} 0 & \sup(I_{1,2,\ldots K}^1) \geq Min\_\sup * D.size/2 \,\&\&\, \sup(I_{1,2,\ldots K}^2) \geq Min\_\sup * D.size/2 \\ 1 & \sup(I_{1,2,\ldots K}^1) \geq Min\_\sup * D.size/2 \,\&\&\, \sup(I_{1,2,\ldots K}^2) \geq Min\_\sup * D.size/2 \\ 2 & \sup(I_{1,2,\ldots K}^1) \geq Min\_\sup * D.size/2 \,\&\&\, \sup(I_{1,2,\ldots K}^2) \geq Min\_\sup * D.size/2 \end{cases}$$
(2)

With the above-mentioned Properties 1 and 2, we can have the conclusion as follow:

Conclusion 1: As for any set in the frequent (k−1)-item set F_{k-1}, during the $F_{k-1} \times F_{k-1}$ connection process, if

$$Cache(I_1, I_2, \ldots, I_k)! = Cache(I_1, I_2, \ldots, I_q)! = 0,$$

then item set $\{I_1, I_2, \ldots, I_k, I_q\}$ cannot be a frequent item.

Proof: We know that the support degree of item $\{I_1, I_2, \ldots, I_k\}$ in data set D_1 is less than $Min\_Sup * Data.size/2$.

From

$$Cache(I_1, I_2, \ldots, I_k)! = Cache(I_1, I_2, \ldots, I_q)! = 0$$

We can obtain that the support degree of item $\{I_1, I_2, \ldots, I_q\}$ in data set D_2 is also less than $Min\_Sup * Data.size/2$. Hence, we have

$$\sup(I_1, I_2, \ldots, I_k, I_q) < Min\_\sup * Data.size.$$

That is what we need prove here.

Conclusion 2: If $Cache(I_1, I_2, \ldots, I_k)$ in dataset D, for data set $\bar{D}$, if

$$\sup(I_1, I_2, \ldots, I_q) < Min\_\sup * Data.size - \sup(I_1, I_2, \ldots I_k)$$

then item $\{I_1, I_2, \ldots, I_k, I_q\}$ is not a frequent item.

It is obviously that we can use the similar process as above to prove this conclusion.

Algorithm 1: Bi-Apriori Algorithm

Input: Transaction dataset D, Support threshold min_support（%）；
Output: All frequent item sets satisfying support threshold；

---

1. Equivalent transaction dataset D into D_1 and D_2 ;

2. For each T in D_1 :

3.   For each I_i in T:

4.   If(I_i *not in* c_1):

5. $c_1^1 = c_1^1$ .add(I_i);

6.   *Else:*

7. $I_i = I_i$ .add(T);

8.   End for;
9. End for;

10. D_2 do the same(line 2-6), get c_1^2 ;

11.   For each I_i in c_1^1 & c_1^2 :

12.   If(sup(I_i^1)+sup(I_i^2)<D.size*min_support):

13.   Delete(I_i) in both D_1 & D_2 ;

14.   Else:

15.   Assign a value to $Cache(I_i)$;

16.   End for;

17.while(F_k != $\varnothing$) do

18. For each $\{I_1, I_2, ..., I_{k-1}, I_q\}$ in F_{k-1} :

19.   For each $\{I_1, I_2, ..., I_q\}$ in F_{k-1} :

20.   If(satisfy self-connection conditions):

21.   If ($Cache(I_1, I_2, ..., I_{k-1})$! = 0) :

22.   If($Cache(I_1, I_2, ..., I_q) = 3 - Cache(I_1, I_2, ..., I_{k-1})$):

23.   Continue;
              // Conclusion 1, the best case
24.   Else:

25.   New_sup = $min\_sup* D.size - sup(I_1, I_2, ..., I_{k-1})$;

26.   P = $Cache(I_1, I_2, ..., I_{k-1})$;

27.   If(sup($I_1^p, I_2^p, ..., I_{k-1}^p, I_q^p$) < New_sup):

28.   Continue;
                // Conclusion 2, the better case
29.   Else:

30.   If(sup($I_1, I_2, ..., I_{k-1}, I_q$) $\geq$ min_ sup* $D.size$)

31. F_k .add($I_1, I_2, ..., I_{k-1}, I_q$);

32.   Assign a value to $\{I_1, I_2, ..., I_{k-1}, I_q\}$ by property 2;

33.   Assign a value to $Cache(I_1, I_2, ..., I_{k-1}, I_q)$;

34.   Else:
35.   do step 30-33；
36.   Else:
37   break;
38. End for;
39.End for;

40.return F_k ;

41.End;

With the above conclusions, we can observe that during the process of $F_{k-1} \times F_{k-1}$ connection, we only have the three following cases:

The best case: Any two items in D_1 and D_2 to be connected have the support degree less than $Min\_Sup * Data.size/2$. This leads to these items to be deleted without considering the intersection of the two items.

The better case: There are two items $\{I_1, I_2, \ldots, I_k\}$ in D_1 and $\{I_1, I_2, \ldots, I_q\}$ in D_2 to be connected with the support degree less than $Min\_Sup * Data.size/2$. And these two items should be deleted directly. During this process, only one of the two datasets has been taken into account, and the computation has been shortened dramatically. This is similar to increasing the support degree threshold, as a higher support degree threshold usually leads to fewer frequent item sets and faster process.

The worst case: Like the above case, with the increased support degree threshold, item set $\{I_1, I_2, \ldots, , I_k, I_q\}$ may still need intersection process. Meanwhile, if there is no $Cache! = 0$, it requires intersection process as well.

3.3   Algorithm Framework

The process of our proposed Bi-Apriori algorithm is as follows in detail, which is also shown in Algorithm 1:

Step 1: We divide the dataset into two part D_1 and D_2 equally, and then these two datasets are processed to form the Item-Tid set with the vertical data structure as C_1^1 and C_1^2.

Step 2: The support degree of these two candidate sets can be computed. Then we can have the Cache table, delete the items with less support degree, and obtain frequent 1-item sets F_1.

Step 3: As for F_{K-1}, we can use self-join method $F_{K-1} \times F_{K-1}$ to get candidate items according to the above mentioned method. Meanwhile, F_k can be obtained by tailored as mentioned above.

Step 4: Repeat Step 3 until there is no more frequent item sets obtained.

3.4   Examples of Bi-Apriori Algorithm

To illustrate our proposed Bi-Apriori algorithm, we take the transaction table listed in Table 1 as an example, where the support degree threshold is 40%, and the process is based on frequent 3-item set F_3:

The transaction set is first divided into 2 parts and converted into vertical data structures D_1 and D_2 in the form of Item-Tid, as shown in Table 2:

After the transformation into a vertical data structure, we can observe that the support of the item becomes the size of the Tid set, i.e. $\sup(I_1^1) = 4$, $\sup(I_1^2) = 1$. As for the total transaction sets, $\sup(I_1) = 4 + 1 = 5$, in addition,

$$\sup(I_1^2) < 4/2 \text{ and } \sup(I_1^1) > 4/2,$$

so $Cache(I_1) = 1$. Scanning the item sets, as there is no item with support degree less than 4, there are only frequent 1-item, and have the cache table of F_1, as shown in Table 2:

As for $F_1 \times F_1$ connection step, for I_1, $Cache(I_1)= 1$, so I_1 is less than the local support degree threshold in D_2. Meanwhile, in D_1, for any two containing I_1, the local support degree is 3, so $\{I_1, I_2\} = \{1, 2, 3, 4\} \cap \{1, 3, 5\} = \{3\}$ does not satisfy. With

Table 1. Transaction sets D

Tid	Item
1	I_1, I_2, I_4
2	I_1, I_3, I_5, I_6
3	I_1, I_2, I_5
4	I_1, I_3, I_5
5	I_2, I_3, I_5
6	I_1, I_3, I_4, I_6
7	I_2, I_3
8	I_3, I_4
9	I_2, I_3, I_5, I_6
10	I_2, I_3, I_4, I_6

Table 2. Transaction sets and frequent 2-item sets

Item	Tid (D_1)	Tid (D_2)	Cache
I_1	1, 2, 3, 4	6	1
I_2	1, 3, 5	7,9,10	0
I_3	2, 4, 5	6,7,8,9,10	0
I_4	1	6,8,10	2
I_5	2, 3, 4, 5	9	1
I_6	2	6,9,10	2
I_2,I_3	5	7,9,10	2
I_3,I_5	2,4,5	9	1
I_3,I_6	2	6,9,10	2

Conclusion 2, $\{I_1, I_2\} \not\subset F_2$, $\{I_1, I_3\} \not\subset F_2$. Hence, the 2-item set containing I_1, and we only need compute D_1. Therefore, the computation process is halved, and the pruning is improved due to the larger of local support degree threshold; Particularly, due to Cache(I_1)!= Cache(I_2)!= 0, $\{I_1, I_4\} \not\subset F_2$, it is only based on the Cache table to get results. The obtained frequent 2-item sets are shown in Table 5. In this table, we can observe that $\{I_2, I_3\}$ is not a subset of frequent 3-item. $\{I_3, I_5, I_6\}$ can be deleted directly.

4 Experiments

To validate the effectiveness and efficiency of our proposed Bi-Apriori algorithm, four existing algorithms are compared in this paper, which includes: classic Apriori algorithm, the MC_Apriori algorithm based on matrix in [3], improved Apriori algorithm based on vertical data structure in [10] and the proposed Bi-Apriori algorithm in this paper;

4.1 Retail DataSet

In Retail dataset, we have a total of 88,162 transactions, 16,469 items, and the average transaction length is 10. Comparing the running time of four algorithms under different support degree threshold, the experimental results are shown in Fig. 1:

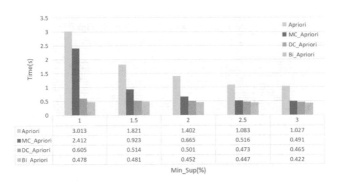

Fig. 1. Performance comparison in Retail Dataset

The Bi-Apriori algorithm outperforms the other three algorithms under different support degree thresholds. Particularly, Apriori performs the worst; the DC_Apriori and Bi-Apriori algorithms with vertical data structures both share better performance time. In contrast, the retail dataset's frequent item sets are less than 200 with 1% support degree threshold, so the difference between DC_Apriori and these algorithms is not significant.

4.2 Mushroom DataSet

As for Mushroom dataset, we have a total of 8,124 transactions, 119 items, and an average transaction length of 23. Comparing the running time of four algorithms under different support degree threshold, the experimental results are shown in Fig. 2:

With a lot of frequent items set, our proposed Bi-Apriori algorithm outperforms the other three algorithms, and with the increase of the support degree threshold, the frequent item sets are becoming smaller, and the difference of these algorithms becomes smaller.

4.3 T10I4D100K DataSet

As for T10I4D100K dataset, we have a total of 100,000 transactions, 1000 items, and an average transaction length of 40. Comparing the running time of four algorithms under different support degree threshold, the experimental results are shown in Fig. 3:

The performance of our Bi-Apriori algorithm is 10–100 times that of the other three algorithms when the number of transactions is large enough and frequent sets are not small enough. When the support degree threshold becomes larger, the performance of several algorithms tends to be the same.

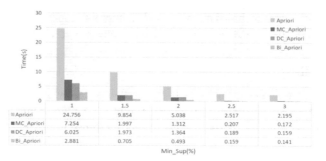

Fig. 2. Performance comparison in Mushroom Dataset

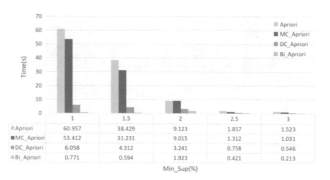

Fig. 3. Performance comparison in T10I4D100K Dataset

4.4 Alarm DataSet

Alarm dataset is collected from a IT data center of the State Grid, 2018/3-6, with a total of 8,574 transactions, 966 items, and the average transaction length of 23. The corresponding experimental results are shown in Fig. 4:

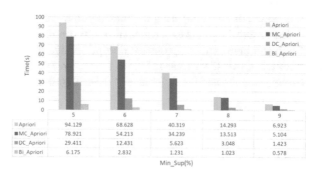

Fig. 4. Performance comparison in Alarm Dataset

In the alarm dataset, with lower support degree threshold, the Bi-Apriori algorithm is significantly better than other algorithms, specifically it is 5 times than that of the

DC_Apriori algorithm. Comparing with other dataset, it seems like our proposed Bi-Apriori algorithm works particularly well for alarm logs, due to several reasons as follows:

1) First of all, the alarm information is produced in temporal, and for some alarms, the former of the dataset appears, and the latter of the number of times appear little or no. For example, I_1 this alarm appears in the former, I_2 alarm out of the latter, then $\{I_1, I_2\}$ cannot be a frequent item set.

2) Even if I_2 as noise which may occasionally appear in the former, and happens to meet the local support degree threshold, but I_1 appears in the latter and satisfies the local support degree is almost impossible, so $\{I_1, I_2\}$ only need to process the former of the dataset, and the local support threshold has also been improved.

5 Conclusions

This paper proposes a Bi-Apriori algorithm for association discovery. Particularly, with the divided data set, we can determine the cache table, accelerate the connection and pruning process, and reduces the intersection dramatically. Experimental results show that the proposed algorithm obviously outperforms existing algorithms.

This method is helpful to construct the knowledge base of alarm association rules of state grid through tools, and improve the operation and maintenance evaluation and repair mechanism, operation and maintenance resource scheduling mechanism and operation and maintenance assistant decision mechanism of state grid information system.

Acknowledgements. This research was financially supported by the Science and Technology projects of State Grid Corporation of China (No. 500623723).

References

1. Borgelt, C., Kruse, R.: Induction of Association Rules, Apriori Implementation. In: Compstat, pp. 395–400. Physica-Verlag HD (2002)
2. Cheng, M., Zhou, X.: Optimization of apriori algorithm based on matrix. Comput. Modernization **12**, 5–7 (2008)
3. Miao, M., Wang, Y.: Research on improvement of apriori algorithm based on matrix compression. Comput. Eng. Appl. **49**(1), 159–162 (2013)
4. Zhao, X., Sun, Z., Yuan, Y., Chen, Y.: An improved apriori algorithm for orthogonal linked table storage. J. Chin. Comput. Syst. **37**(10), 2291–2295 (2016)
5. Yanyan, Yu., Li, S.: Improved aprioritid algorithm for correlation rules based on hash. Comput. Eng. **34**(5), 60–62 (2008)
6. Ji, H.: Apriori, improved algorithm based on frequent 2 item set support matrix. Comput. Eng. **11**, 183–186 (2013)
7. Lin, Z.: A divide-and-conquer apriori algorithm for generating frequent itemsets directly. J. Comput. Appl. Softw. **4**, 297–301 (2014)
8. Zhang, Y., Xiong, Z., Geng, X., Chen, J.: Analysis and improvement of eclat algorithm. Comput. Eng. **36**(23), 28–30 (2010)

9. Huseyinov, I., Aytac, U.C.: Identification of association rules in buying patterns of customers based on modified apriori and eclat algorithms by using R programming language. In: 2017 International Conference on Computer Science and Engineering (UBMK), Antalya, pp. 516–521 (2017)
10. Du, J., Zhang, X., Zhang, H., Chen, L.: Research and improvement of apriori algorithm. In: 2016 Sixth International Conference on Information Science and Technology (ICIST), Dalian, pp. 117–121 (2016)

Research on Perceived Information Model in Product Design Under Cloud Computing

Qiangqiang Fan[1](✉) and SungEun Hyoung[2]

[1] College of Arts, Northeastern University, Shenyang 110819, Liaoning, China
fanqq@mail.neu.edu.com
[2] College of Arts, South China Agricultural University, Guangzhou 510642, Guangdong, China
hsungeun@hanmail.net

Abstract. In product design, it is an important ability for modern design to find the intersection of the image that the designer wants to express and the perceptual meaning of the product audience. There have been a lot of research results about the approach method of perceptual design, but there is no saying that which approach method is the standard. As a new approach method, this study puts forward the structural modeling process of sensibility. By introducing a new concept of sensibility, this approach to perceptual design can improve the creativity of analysts and make more use of it by proposing a rational structure. The modeling process of perceptual structure is based on rational and perceptual evaluation. It is a method to improve the accuracy of model construction and to explain perceptual evaluation objectively.

Keywords: Product design · Perceptual information model · Model establishment

1 Introduction

1.1 Research Background

In the twenty-first century, perceptual design no longer only exists in the field of aesthetics, but also appears in the trend of daily life through scientific analysis and trend orientation. "Sensibility" is widely used in marketing, commodity planning, advertising, design and other fields; Especially when the market is mature and the development of new products becomes difficult, in order to meet the needs of consumers, enterprises have tried to improve the product characteristics from the perceptual level of consumers.

It is necessary to recognize the influence of structural patterns on the variables of sensibility elements and conduct in-depth research on them. However, only based on the understanding of perceptual elements and structure, it can not clearly explain perceptual. If we want to use sensibility to express design in the knowledge we have experienced, we need to change many possibilities into hypotheses, and combine common attributes and relevance.

© The Editor(s) (if applicable) and The Author(s), under exclusive license
to Springer Nature Switzerland AG 2021
J. MacIntyre et al. (Eds.): SPIoT 2020, AISC 1283, pp. 633–639, 2021.
https://doi.org/10.1007/978-3-030-62746-1_92

Therefore, in order to extract the perceptual design, it is particularly important to analyze and understand all the elements, structures, evaluations, etc. that appear, so as to transform the knowledge that is mastered into perceptual information. In addition, attention must be paid to the relevance of the things exhibited by the researcher of the evaluation object. As a process of understanding information, it is necessary to conduct research that can understand "sensibility based on known information or consciousness."

1.2 Research Methods and Objectives

In order to understand how to build perceptual information into perceptual model in perceptual design, this study defines perceptual as knowledge or information that everyone knows. In the research, in order to divide the perceptual technical knowledge into tacit and explicit knowledge to determine the perceptual range, the perceptual extraction method uses dichotomy. In addition, the sample of the actual object is extracted from the information model, and the results of the experimenter are analyzed and discussed by using KJ method, quantitative theory, cluster analysis and so on (Fig. 1).

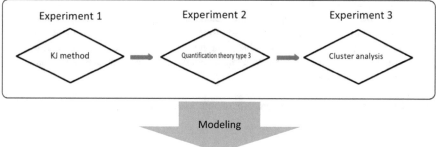

Fig. 1. Experimental method

So far, in the research of perceptual design, many theories and experimental methods have realized the importance of perceptual, but in the practical research, for how perceptual is applicable, the current development and accurate data are not satisfactory. The purpose of this study is to focus on cooking utensils. In product design, through experimental research, the perceptual elements are used to construct the design process with a more objective and convincing perceptual information approach method.

2 Literature Review

In the design process, planning is the most important factor to determine the design success. In terms of design, before product information is sorted out, conceptualization and visualization of product information as design objects are interrelated. On the other hand, in the process of planning to concept, it also uses the methods of design thinking divergence and collective consciousness decision, as well as the methods of thinking map and brainstorming to integrate information and build a framework.

2.1 Perceptual Information of Intuitive Understanding

Sensibility can lead people to be induced by experience. Designers analyze and add factors that can cause this kind of sensibility reaction in product design, so that it can be reborn into a commodity with wider market. Imagine the two knowledge states that affect the perceptual activities of the masses as explicit knowledge and tacit knowledge. Sensibility can be regarded as the characteristics given in the process of conversion between the two.

Here it is, experience is transformed into tacit or normative knowledge rather than consciously used, which is perceived as a state of "known knowledge". In addition, "known knowledge" can be used as tacit or normative knowledge obtained as a result. Therefore, although sufficient conditions are needed, it cannot be said that this expression method will be must succeed.

2.2 The Meaning and Operation Method of Dichotomy

In terms of design, the phased divergence method mainly carries out information collection activities through brainstorming, confirmation lists, etc., but in order to be able to organize and conceptually grasp the information, the combination generation under the "similar" of the KJ method is introduced. The concept of segmentation.

The dichotomy is different from the multi-division method and the parallel method. Its advantage is that the analysis is very clear and it is the simplification of concept classification. According to the analysis of dichotomy, when the perceptual elements of design are extracted, only the concepts with high correlation between the elements are extracted. When the design concepts or conclusions are drawn, the scope and correlation are relatively high. Of course, although Multiple segmentation method and Coordination have the nature of free analysis, they may lead to inconsistent level of segmentation concept, or low correlation between classification elements, which is not conducive to the conceptualization of sensibility.

3 Experimental Methods

3.1 Experiment

In this study, to establish a perceptual model, experiments were conducted on common cooking utensils in daily life. The experimental method is based on the use of cooking utensils, extracting potential languages, using the extracted language as the object, and the image of keyword extraction and usability by the dichotomy method, etc., with the subject as the object experiment.

The constructed perceptual model is to extract keywords by compression and segmentation to select product samples for experiments. In the experimental method, the cluster analysis of the key attributes of the sample was carried out, and the significance was analyzed through the analysis of product classification and quantification theory type 3 to establish the sensitivity model for cooking utensils. Based on the above process, the perceptual approach to perceptual patterns is explained and concluded.

In the parsed categories, the elements with high contribution are represented by an intuitive and easy-to-understand tree diagram. At the same time, the sampling score is analyzed by group analysis as a homogeneous factor to supplement the results, through the process of these perceptual information To build a perceptual design process.

3.2   Construction of Perceptual Information

The perceptual model experiment was carried out using the dichotomy and KJ method mentioned in the literature investigation. In order to establish the experimental object-a sensual model of cooking utensils, we classified the characteristics of gourmet food.

74 items extracted from perceptual vocabulary are classified by similarity. The content of the classification includes related items for ease of use, divided into 7 items: cleanliness, functionality, convenience, stability, new materials, cooking information, and dietary impact. The projects related to the ample external space are divided into 7 projects: aesthetic, comfort, coordination, high-end sense, lighting, uniqueness, and tradition.

The perceptual key words are respectively 'functionality' related to the convenience of operation, 'dietary impact degree' related to the freshness of food, whether the information about cooking is provided in the product function, whether the product has a new form to bring users' curiosity 'uniqueness', the' high-end 'degree of the product itself, the' comfort 'of cooking, and the' aesthetic 'of the product modeling elements. To whether can get along well with the people around as a "harmonious" project was sorted out.

3.3   Analysis of Cooking Utensil Cluster

Based on the extracted perceptual keywords, experiments were carried out on the perceptual information and design methods of perceptual design of actual products. The experiment selected 34 food-related samples.

In order to classify the characteristics of products, the experimental samples applied 8 perceptual keywords to each product. The applicable method is to investigate whether each sample has classified data. For each item, use "Yes" and "No" to classify, convert it into quantitative data in the form of "0/1", and list the responses of the continuous variables of items and categories. Count and make charts.

The analysis of the samples is centered on the classified charts. In order to grasp the homogeneity of the products, cluster analysis is implemented. The results of statistical analysis using spss show that cooking utensils can be divided into 5 groups (Fig. 2).

The characteristics of each classification group are as follows:

Group A is mainly classified as supplies used on the table, Group B is classified as functional supplies used for cooking, Group C is classified as cooking aids, Group D is classified as auxiliary tools, and Group E is classified as cooking Products for heating. In order to grasp the characteristics of keywords, quantification theory type 3 analysis was carried out on the basis of the five characteristics of such classified cooking utensils.

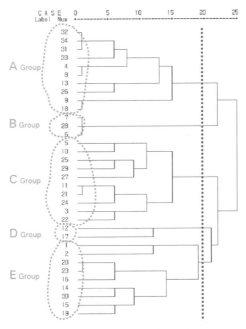

Fig. 2. Cluster analysis results

3.4   Quantitative Analysis of Perceptual Keywords

In order to grasp the characteristics of perceptual keywords, quantification theory type 3 analysis was performed on 8 items, and the category coordinates were classified. The analysis results show that in order to find the classification characteristics and the validity of the data in 8 items, the intrinsic values and the cumulative contribution rate were investigated, and two axes with higher correlation coefficients were selected.

Data contribution rate (10.0 40.07%, 2 axis 30.27%, 3 axis 22.44%), plus the cumulative contribution rate of the third axis up to 92.78%, but if the third axis is also explained, it is difficult to classify and analyze the data. In the analysis of data, the data can be clearly analyzed using the first axis and the second axis. Therefore, the first axis is selected as the X axis, and the second axis is selected as the Y axis. In addition, the correlation between the first axis and the second axis is that the first axis is 0.5711 and the second axis is 0.6567 (Fig. 3).

As the modeling elements of the product, it should contain unique, high-end, rare value, want to own, collect value, etc. As a functional element, it needs to be durable, easy to operate, easy to use, simple in shape, and economical in material. As an information element, must be able to know the freshness of the content, be able to measure the unit's label and measure the temperature, and explain the method of use. As an element of atmosphere, it must look delicious to conform to the trend, and it must conform to each other. There must also be exoticism. Therefore, the perceptual design of cooking utensils needs to meet the four conditions and details of shape, function, information and atmosphere.

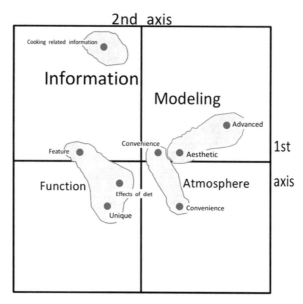

Fig. 3. Category scatter diagram

4   Conclusion

In actual experiments, through the analysis of quantification theory type 3 (Quantification theory type 3) to explain the meaning of attributes, the perceptual model of cooking supplies was tested. Through experimental results, through category analysis items, high-end and aesthetic can be defined as items related to product styling, and comfort and coordination can be defined as "atmosphere during cooking" items. Functionality, influence of diet, and uniqueness are the manifestation of product functions. The information about cooking is defined as cooking information.

Therefore, the way to analyze the new trends in social phenomena and consumer demands and approach sensibility is to prioritize the development of new technologies according to the requirements of the times, introduce designs based on changes in lifestyle, and guide marketing sensibility through continuous advertising and promotion. However, this kind of perceptual field sometimes dominates in a specific field, and sometimes cooperates in mutual fields according to the situation, such as the field of new technology and design, cooperation in the field of design and marketing, etc. It is the result of joint efforts in these various fields. Therefore, enterprise designers must cultivate the design ability to manufacture perceptual products, thereby improving the ability to visualize, conceptualize, and perceptualize, as well as the ability to learn knowledge in various fields, in order to succeed in the market.

References

Hou, G.: The establishment and application of gene intention model of product modeling. Mech. Des. (03) (2014)

Guo, L., Ji, X., Hu, G., Chu, J.: Product modeling design method based on RFID multi-channel deformation. Mech. Sci. Technol. (11) (2013)

Liu, G., Cao, H., Cui, F.: User-oriented product modeling method. J. Henan Univ. Sci. Technol. (Nat. Sci. Ed.) (02) (2012)

Fan, Y.: Research on product image modeling design system based on perceptual engineering and neural network. Lanzhou University of Technology (2011)

Wang. K.C.: A hybrid Kansei engineering design expert system based on grey system theory and support vector regression. Exp. Syst. Appl. (7) (2011)

Yang, C.C.: Constructing a hybrid Kansei engineering system based on multiple affective responses: application to product form design. Comput. Ind. Eng. (4) (2011)

Smith, S., Fu, S.H.: The relationships between automobile head-up display presentation images and drivers' Kansei. Displays (2) (2010)

Karana, E., Hekkert,P., Kandachar, P.: Meanings of materials through sensorial properties and manufacturing processes. Mater. Des. (7) (2008)

Early Warning Management of Enterprise Human Resource Crisis Based on Fuzzy Data Mining Algorithm of Computer

Hua Wu[✉]

Business School, Macau University of Science and Technology, Macau 610041, China
wuhua1218@163.com

Abstract. Since the 21st century, human resources, as an important force to promote the healthy and rapid development of enterprises, has been attached great importance to by all walks of life, which has effectively promoted the rapid development of human resources in China's enterprises in recent decades. However, with the increasingly fierce market competition and the penetration of highly developed information internet, the current enterprise human resource management has gradually produced many crises, which constitutes a serious constraint on the development of enterprises. A data mining method is designed, which can cluster and analyze the existing talents in the organization, so as to discover the types of talents in the organization and judge which type an employee belongs to. How to deal with the human resource crisis effectively and change the crisis into a turning point has become the most urgent problem for many enterprises. Combined with practice, taking effective early warning management is an important way to solve this problem. Based on the author's many years of enterprise human resource management practice, this paper will first summarize the human resource crisis of the enterprise, and then discuss the current difficulties faced by the crisis early warning management, and on this basis explore some corresponding countermeasures.

Keywords: Enterprise human resource crisis · Early warning management · Dilemma · Counter-measures · Data mining

1 Introduction

In the 21st century, human society will be an information age with network technology as the core, which has been recognized by the world. Digitalization, networking and informatization are the characteristics of the 21st century. Modern human resource management requires high efficiency, high processing ability and high analysis ability. The application of computer technology in human resource management will promote the improvement of modern human resource management level and meet the basic needs of modern human resource management. In modern human resource management, by giving full play to the high processing and storage capacity of computer, we can master a

J. MacIntyre et al. (Eds.): SPIoT 2020, AISC 1283, pp. 640–645, 2021.
https://doi.org/10.1007/978-3-030-62746-1_93

lot of information, which is convenient for the meticulous and scientific human resource management. Therefore, at present or in the future, computer technology will be more widely used in human resource management, and it will also improve the position of computer in human resource management [1].

2   An Overview of Human Resource Crisis in Enterprises

1. The meaning of enterprise human resource crisis: according to the research of relevant literature and the practice of the author, the enterprise human resource crisis refers to the events that have a serious impact on its normal operation and production order when the enterprise is suffering from some unexpected or out of control management. These human resource crises generally include natural disasters, wars, major defects in technology and deterioration of employees' mentality, which will bring extreme risks or difficulties to the operation and production of enterprises, so it is very important to adopt crisis early warning management.
2. Early warning management of enterprise human resource crisis: in view of the crisis caused by human resource, many enterprises have adopted early warning management in response, which means that the enterprise forecasts the possible human resource crisis according to the relevant methods, then finds out the source of the crisis through the way of tracing back to the source, and finally formulates the corresponding measures to control and manage it on this basis, so as to protect the enterprise to the greatest extent under the condition of effectively reducing the probability and damage degree of the human resource crisis The industry develops healthily and rapidly.

3   The Predicaments in the Early Warning Management of the Current Human Resource Crisis in Enterprises

According to the author's research on human resource crisis early warning management in enterprises, although many enterprises have implemented some crisis early warning management measures, due to many factors, the current overall human resource crisis early warning management in enterprises still faces the following difficulties:

1. The enterprise human resource crisis early warning system is not perfect: a complete enterprise human resource crisis early warning system is an important support to ensure that the great role of crisis early warning management can play, which mainly consists of four aspects: information collection, processing, decision-making and alarm, and none of them is indispensable. According to the research, when enterprises use the system to carry out human resource crisis early warning management, they should first process and make certain decisions based on the collection of relevant crisis information according to relevant methods, and then the system will send out corresponding alarms in this case. Through the research on the operation of the system, it is found that once there is a problem in any link in the process, it will inevitably lead to the failure of the early warning management of human

resources crisis due to the failure of the alarm In this way, the normal operation and production order of enterprises will be seriously affected by human resource crisis, which shows that the perfect early warning system of human resource crisis of enterprises is of great significance. However, the actual situation is not optimistic. At present, many enterprises' human resource crisis early warning systems are affected by many reasons, resulting in the lack of one of the above four links in the system construction, which leads to the failure of enterprise human resource crisis early warning management.

2. Lack of staff capacity in early warning management of enterprise human resource crisis: in the current early warning management of enterprise human resource crisis, in addition to the imperfect early warning system, the lack of staff capacity is also one of the difficulties. As the executors of crisis early warning management, their ability to a large extent determines the effectiveness of this work and the role of play. However, in practice, many human resource crisis early warning management staff are generally lack of ability, such as lack of awareness and awareness of crisis early warning, lack of professional knowledge and skills, etc., which will inevitably lead to human resource crisis early warning management unable to play a huge role.

4   Fuzzy Mining Algorithm

In the process of data mining, data mining algorithm is the most important. Using the fuzzy mining algorithm and the data extracted from the data warehouse, we can find the existing cadre types in the current organization. In addition, we can judge which type an employee belongs to.

4.1   Set up the Sample Set u to Be Classified on All Data Records of the Data Warehouse

The objects to be classified are called samples. For example, $u_1, u_2,..., u_n = \{u1, u2, ...,un\}$ is the sample set. In order to achieve a reasonable sample classification, its specific attributes should be quantized. The quantized attributes are called sample indicators, with m indicators, which can be described by m-dimension vector, namely [2]:

$$ui = (u_{i1}, u_{i2}, \ldots\ldots, u_{im}) \, (i = 1, \, 2 \ldots, n) \tag{1}$$

Because in the actual data, the collected data is often not the number of [0,1] closed intervals, so these original data should be standardized, and the average value should be calculated first, for example, there are n samples in the sample set, for a certain index k of the sample, you can get n data u'ik, u'2 k,,, u'nk, where u'; it means the data obtained by the I sample for the K index, and their average value is calculated according to the following formula:

$$u'_k = \left(u'_{1k} + u'_{2k} + \Lambda + u'_{nk}\right)/n = \frac{1}{n}\sum_{i=1}^{n} u'_{ik} \, k = 1, 2, \cdots m \tag{2}$$

Then, the standard deviation S_k of these original data can be calculated as follows):

$$S_k = \sqrt{\frac{1}{n} \sum_{i=1}^{n} (u'_{ik} - u'_k)^2} \tag{3}$$

Then calculate the standardized value of each data in one form u''_{ik}:

$$u''_{ik} = \left| \frac{u'_{ik} - u'_k}{S_k} \right| \tag{4}$$

At this time, the obtained standardized data u''_{ik} is not necessarily within the [0,1] closed range, and the following extreme value standardization formula must be used:

$$u_{ik} = \frac{u''_{ik} - u''_{mink}}{u''_{maxk} - u''_{mink}} \tag{5}$$

Here u''_{maxk} and u''_{mink} represent $u''_{1k}, u''_{2k}, \cdots u''_{nk}$, respectively... Maximum and minimum values in ().

4.2 Establish Modulus Similarity R

R can be expressed as a similar matrix in the following general form:

$$R = \begin{pmatrix} r_{1j} & r_{2n} & \Lambda & r_{1n} \\ r_{21} & r_{22} & \Lambda & r_{2n} \\ \Lambda & \Lambda & \Lambda & \Lambda \\ r_{n1} & r_{n2} & \Lambda & r_{nn} \end{pmatrix} \begin{pmatrix} 0 \le r_{ij} \le 1 \\ i = 1, 2, \Lambda, n \\ j = 1, 2, \Lambda, n \end{pmatrix} \tag{6}$$

There are many methods to calculate r_{ij}. Here, the maximum minimum method is used, that is:

$$r_{ij} = \sum_{k=1}^{m} \min(u_{ik}, u_{jk}) \Big/ \sum_{k=1}^{m} \max(u_{ik}, u_{jk}) \ (i, j \le n) \tag{7}$$

4.3 Cluster Analysis

The maximum tree method is used to construct a special graph. When ($r_{ij} \ne 0$), the vertex i and the vertex J can be connected to one edge. The specific method is to draw a certain I in the vertex set first, then connect the edges in the order of r_{ij} from large to small, and require no loop until all the vertices are connected, so as to get a maximum tree. To be precise, it is a "weighted" tree, each edge can be given a certain weight, namely r_{ij}. However, due to different connection methods, the largest tree can not be the only one.

Then the λ cut set is taken for the maximum number, that is, the edges with weights $r_{ij} < \lambda$ are removed, $\lambda \in [0,1]$. In this way, a tree is cut into several disconnected subtrees. Although the largest tree is not unique, after taking the cut set, the resulting subtree is the same. These subtrees are the patterns of inductive discovery in data warehouse (Fig. 1).

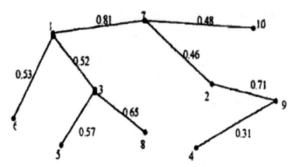

Fig. 1. Cluster analysis: (using the maximum tree method)

5 On Strengthening the Early Warning Management of Human Resource Crisis in Enterprises

1. Establish a perfect enterprise human resource crisis early warning system: in view of the situation that the enterprise production and operation are threatened due to the imperfect human resource crisis early warning system of some enterprises, it is urgent to speed up the establishment of a perfect crisis early warning system. In this regard, the author believes that enterprises can achieve this through the following two ways:

① Establish human resource crisis management department. As mentioned above, information collection, processing and decision-making are the most important basic support in the four links of enterprise human resource crisis early warning system. However, in practice, at present, in most enterprises, the crisis early warning system is generally in the charge of the human resource management department, but often the Department also takes into account other aspects of the human resource work of the enterprise, which is very easy to make it difficult to have sufficient manpower and energy to carry out the work of the crisis early warning system, which will inevitably lead to the failure of early warning management. Therefore, enterprises must actively change this situation and establish a special crisis management department. First of all, enterprises should fully combine their own human resource management staff, the operation requirements of the crisis early warning system and capital, and prepare enough personnel for the crisis management department by means of transfer or recruitment. Secondly, select the personnel with rich human resource management experience and strong sense of responsibility as the management personnel of the Department. Finally, in order to ensure the crisis management department to play a role, enterprises should not only provide financial support, but also formulate the corresponding management, work rewards and punishments and supervision system [3].

② Improve the early warning system. After the enterprise establishes its special crisis management department, the next step is to perfect the human resource crisis early warning system. First of all, according to the four links of the system operation, the corresponding staff and responsible person are assigned to it, so as to lay the foundation for the normal operation of the system. Secondly, the crisis management department formulates specific work flow, formulation and supervision methods according to the different work content of each link. For example, in the information collection work, it is necessary to make relevant collection methods according to the possible sources of human resource crisis.

2. Strengthen the capacity building of human resource crisis early warning management staff: first, in view of the lack of awareness and awareness of crisis early warning of human resource crisis management staff, the core of crisis early warning is to prevent in advance, so enterprises should strengthen the construction of this aspect. For this reason, the human resource crisis management department of the enterprise must collect the relevant knowledge of human resource crisis through the Internet, learning from excellent companies and other ways, and then convene the staff for education. Second, staff professional knowledge and skills training. In the process of training, enterprises should pay attention to the conformity of training methods and contents, and carry out crisis warning training according to the particularity of enterprise development. In addition, in order to effectively ensure the ability of human resource crisis management staff to be effectively improved, the enterprise must also develop a corresponding reward and punishment combined assessment system for the above education or training, so as to ensure their ability to be improved under the premise of fully mobilizing their enthusiasm for education and training.

6 Conclusions

Computer technology is the most popular technology in modern times. Its great advantage is that it can replace the traditional human resources to carry out human resource management. It can also promote the development of human resources in enterprises and provide necessary help for the construction of human resources in enterprises. In the future human resource management, we should introduce more computer technology and innovate its application. Promote the optimization and improvement of human resource management. Data mining can discover the patterns and laws of talents in an organization. In this paper, a data mining method is designed to cluster the existing talents in an organization, so as to discover the types of talents in an organization, which is very useful for decision makers to make talent decisions and talent planning.

References

1. Fan, L., Lian, S.: Early warning management of enterprise human resource crisis. Ind. Eng. Manag. 8(4), 10–14 (2003)
2. Ye, L.: Thoughts on early warning management of human resource crisis in enterprises. Commod. Qual. Consumption Res. 6, 72 (2015)
3. Liu, Y.: Early warning management of enterprise human resource crisis. Orient. Enterpr. Cult. 4, 157 (2014)

New Media Creative Writing and Computer Application Technology

Xiaoci Yang$^{(\boxtimes)}$

Xi'an Fanyi University, Xi'an 710105, China
yangxiaoci1978@163.com

Abstract. Under the current "Internet plus development" mode, creative writing has changed in the context of university training. The advent of the media era has brought opportunities and challenges to creative writing. The development of media and media has changed the background and carrier of creative writing, and has been gradually refined by the classification of people and the way of training is more diversified. Taking Jiangsu Normal University as an example, this paper analyzes the current situation, opportunities and challenges, training needs and other aspects of College Students' creative writing under the background of media convergence, so as to explore the methods of innovative media training for college students' creative writing. With the continuous development of information carriers, the value of new media has been gradually developed; and the new era education focusing on new media is particularly important; especially in the teaching of computer science, with the help of network and digital advantages, we can fully feel the significance of new media technology; in this paper, the application of new media in computer teaching is combined Analysis of teaching methods and put forward relevant views, hoping to promote the improvement of modern education.

Keywords: Creative writing · Media integration · Personality identity · New media · Computer teaching · Application

1 Introduction

The concept of new media convergence was put forward in the United States in the 1980s. Its original definition is to combine different types of media for new development; creative writing also originated in the United States, the original intention is to cultivate writers. However, under the background of "Internet plus" mode in China, media convergence has developed rapidly, and media convergence has gradually moved towards cross industry convergence from simple cross media convergence. Under this media background, the creative industries relying on traditional media for writing have also changed [1]. Demand and development. How to balance the network education of creative writing with traditional classroom education and how to create a new creative writing education environment by using the combination of the two has become a new problem worthy of discussion. Computer science is a compulsory course in Colleges and

J. MacIntyre et al. (Eds.): SPIoT 2020, AISC 1283, pp. 646–651, 2021.
https://doi.org/10.1007/978-3-030-62746-1_94

universities. With the development of the times and the updating of network technology, the traditional teaching mode has a serious contradiction with the modern network teaching theory. The new media teaching is to add new content to the traditional teaching, so that computer teaching can make use of the advantages of the Internet, Constantly expand students' vision, enrich students' imagination; make their students more proficient in the operation and application of computer network technology, so as to vigorously promote the development of new media diversification.

2   Definition of New Media

New media is a direction of future media development, and also the change of traditional media constitution. Traditional media refers to the process of directional information dissemination by means of TV, newspaper and other forms; while new media is based on traditional media, using computer digital technology to process the original information, and finally transmit it to people's advanced equipment in different and most convenient ways; In this way, people with different lifestyles can see the information they want to understand at different times and places. At present, the Internet is the core of the network technology. The popularization and application of this technology has established a spatial communication mode for the masses, sending information to everyone in the form of minimum cost, but it has brought the value of information to the extreme; Network reflects not only the spiritual era, but also the science and technology that many fields of life want to achieve and apply. For example, the teaching of computer science combined with new media technology will make the system design in the computer more personalized and diversified [2].

3   Explore the Significance of the Application of Computer Technology in the New Media Era

With the rapid development of science and technology, the media industry and settlement and technology are also constantly updated and revolutionized. With the rise of new media, the relevance between the two industries is becoming stronger and stronger. With the development of the new media era, China's media industry has entered a new era. The collection and dissemination of information in the media industry, including text, pictures and videos, need the help of computer technology. Compared with the traditional media, these are essentially changes. In the new media era, all media practitioners can not escape the step of changing to new media. The traditional way of information processing is also changing, the application of computer technology instead of traditional media tools has become the core of the application of new media technology. This transformation has a promoting effect on social progress. Nowadays, various computer technologies such as digitization and big data cloud computing are widely used in various industries, including the media industry. The application of computer technology in the media industry can fully reflect the drawbacks and problems existing in traditional media, And help new media through technological innovation to change the problems of traditional media, especially in image and video processing, the progress of computer technology also drives the improvement of this technology.

4   New Media Creative Writing Environment and Current Situation

"Creative writing is a creative activity in the form of writing and the carrier of works. It is the most important and basic work link in the cultural and creative industry chain." The need of media convergence promotes the integration of media. In this trend, the era of mass media is coming. The gap between "people" and "media" has been broken. There is no essential difference between media audience and creator, and the way of information dissemination tends to be diversified. The limitation of the creative subject is gradually disappearing, and the fast-food culture which uses fragmented time and exaggeration content is becoming popular. The advent of the era of mass media has brought new opportunities to the new way of creative writing based on the Internet. For college students, creative writing ability becomes new [3].

5   Abstraction of New Media Model

The functions of these new media models can be abstracted into two levels: control level and data level. As the control level, it is mainly responsible for the dynamic maintenance of the relationship between nodes and other nodes. Nodes exchange control information with other nodes through the control level. The data layer is responsible for receiving and sending data. Nodes exchange media data with other nodes through the data layer. Different new media models are mainly different in the control level. In the tree based model, the main role of the control layer is to find a parent node for the node to provide data, such as the PeerCast model we mentioned earlier. Or look for multiple parent nodes, like the splitstream model mentioned earlier. Another important function is to quickly repair the multicast tree when the parent node leaves. In zig zag model, in order to maintain the balance of multicast tree, the control layer needs to maintain a logical cluster structure management. In the DONet model based on gossip and the hybrid structure new media model proposed in this paper, the main role of the control layer is to find the appropriate data provider for each piece of data required by the node. The main difference between the model based on multicast tree and the model not based on multicast tree is that in the model based on multicast tree, all the data fragments required by nodes are always obtained from one or more specific parent nodes until the parent node changes due to some situation. However, in the model based on non multicast tree, the provider of data fragment always changes dynamically. With the above understanding, at the end of this paper, I propose a layered new media framework (Fig. 1).

The layered new media framework can be divided into three layers: application layer, node abstraction layer and transport layer. The application layer is the playback layer of media data, and the application layer obtains the media data through the node abstraction layer. The node abstraction layer is responsible for maintaining the logical relationship between nodes, that is to realize the function of the control layer in the new media model. The node abstraction layer can abstract different new media models and provide a unified interface for the application layer. Transport layer plays the function of encapsulating data layer, and is also responsible for the transmission of media data and control messages.

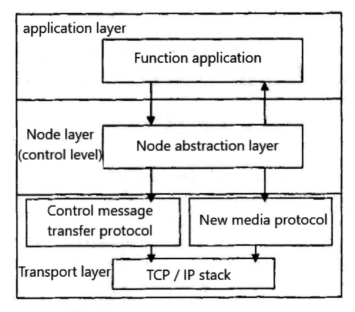

Fig. 1. Layered streaming media framework

6 Cache Replacement Algorithm

The design of cache replacement algorithm generally needs to pay attention to two issues, that is, what content needs to be cached (what to cache); what content needs to be eliminated from the cache (replacement rules). Aiming at the first problem, the general method is to cache every download and play stream media block. The disadvantage is that it is easy to replace the stream media with high popularity and insufficient cache copies by caching the stream media with low popularity and enough copies in the system. To solve the second problem, a general method is to use a simple LRU algorithm. This algorithm eliminates the media blocks with the least number of visits in this node. It only focuses on the current node's statistics of the popularity of streaming media, and does not consider the situation of other nodes in the system. A better replacement algorithm should not only consider the popularity of streaming media, but also pay attention to the caching of the streaming media block in other nodes. We believe that if the number of copies of each streaming media block cached in the whole system is proportional to its popularity, the number of media content that nodes download directly from the server can be reduced as much as possible. Therefore, according to formula (1), the ranking basis for the use value of streaming media block in the replacement algorithm is defined. Where f is the popularity of a streaming media block, CN is the number of copies of the streaming media block that has been cached in the system, and R is the use value of the streaming media block. The essence of replacement algorithm is to replace the one with smaller value and the one with higher value.

   R = popularity of streaming media blocks (F)/number of copies of streaming media blocks cached by the system (CN)

Here is a new replacement algorithm called F/CN algorithm. The algorithm requires the server to count the access frequency of each media block to estimate the popularity of streaming media. The specific algorithm is as follows, the meaning of the symbol in the algorithm is given later.

When the peer requests the streaming media block a and receives a response, the following algorithm is executed:

```
if (Empty_ _Bufer_ Size ≥ sizeof(A)) cache A
else
(1) Download fA and NA from the server;
(2) For all Bi ∈ φ, Download FBi and NB from the server
(3) RBi = FBi/nbi-1 and RA = f/NA-1 were calculated;
(4) (RBi < HA)
(5) if(C > sizeof (A)
{
while (Empty_ Bufer_ Size < sizeof (A))
Delete BJ, where RBj min {RS}; (Bi ∈ φ)
Cache A;
}
else
Do not cache a
```

The meaning of parameters in F/CN algorithm:

A represents the streaming media block to be downloaded

N_A represents the number of online peers with a cache

B_i represents the i-th streaming media block that has been cached by this node

N_{Bi} represents the number of peers that the cache has online

F_{Bi} represents the recent Bi request frequency counted by the server, and approximately represents the popularity of Bi

φ represents the collection of all streaming media blocks cached by this node

Empty_ Buffer_ Size indicates the free cache space size of this node

7 Conclusions

Combined with this article, with the development of society and the progress of science and technology, the application of computer in the new media era promotes the development speed and popularity of new media, and also affects all aspects of people's life, greatly facilitating people's life and improving people's living standards. This paper focuses on the application of computer technology for the new media era of the positive role, in order to explore the application of computer technology in the new media era, the role of computer technology in the development of new media. In recent years, the relevant literature of creative writing has been constantly introduced in China, and the "writer class" of creative writing colleges and universities has gradually formed in Colleges and universities across the country. However, the introduction of creative writing training mode can not meet the needs of creative writing talents in our country. With the rapid development of China's economy and the rapid change of the industry market, the current training mode of creative writing should be combined with China's current situation In recent years, the results of the development mode of "Internet plus" have been revolutionized to adapt to the development of China, while enhancing cultural guidance and forming a "creative writing training mode with Chinese characteristics".

Acknowledgements. Drama adaptation research (XFU18KYTDC01).

References

1. Ge, H.: Subject orientation of creative writing. J. Xiangtan Univ. (Phil. Soc. Sci.) (5) (2011)
2. Hu, Z., Liu, J.: how to amplify the mainstream voice of traditional mainstream media in the era of media convergence. North. Media Res. (1) (2017)
3. Li, D.: Application of computer technology in the new media era. Comput. Technol. Appl. (2018)
4. Liu, Y.: Research on the application of computer technology in the new media era

Short Paper Session

Application of Automatic Obstacle Avoidance Unmanned Vehicle Based on Beidou and Inertial Navigation

Pan Li$^{(\boxtimes)}$, ZhengJiang Pang, LiZong Liu, Wei Meng, and LiPing Zhang

Beijing Zhixin Microelectronics Technology CO., Ltd., Beijing 100085, China
scoourgeall@163.com

Abstract. Autonomous navigation unmanned vehicles based on Beidou+DR. The mature outdoor positioning system is used to control the patrol car movement, and the laser radar obstacle avoidance adaptive adaptive road environment and emergencies. Using RFID technology to add key patrol points for unmanned vehicles, planning the selection of special routes or multiple routes, RFID readers can count materials or key building signs during unmanned vehicle movement, and the maximum measurable distance is 10 m. Using a combination of various technologies to improve the navigation accuracy and practical application value of the unmanned vehicle; the positioning accuracy of the Beidou+DR navigation system can reach ± 10 cm in the outdoor environment, and the positioning data is integrated with the actual vehicle chassis crawler speed through PID. The algorithm compensates for the error of the two tracks so that the unmanned vehicle can move exactly according to the route.

Keywords: Unmanned vehicle · Beidou+DR · Rfid · Laser radar

Preface

At home and abroad, the UAV navigation technology has invested a lot of material and manpower, tracking navigation, slam visual navigation, multi-sensor data fusion navigation, inertial navigation and so on. But no matter which technology has some shortcomings, it needs to be optimized by continuous precipitation iteration. Unmanned vehicles have extremely high application value in the context of the Internet of Things. They can not only replace personnel in patrolling and inventory materials in harsh environments; they can also help factories improve production efficiency and collect data. Unmanned vehicles can monitor power plants and parks in real time by carrying cameras and various sensors. In case of unexpected situations, they can alarm and upload data in time, which can provide more support for the construction of strong power grid.

1 Beidou + Dr

The unmanned vehicle navigation and positioning is composed of a Beidou receiver and an inertial navigation device. The Beidou receiver needs to continuously receive signals

J. MacIntyre et al. (Eds.): SPIoT 2020, AISC 1283, pp. 655–659, 2021.
https://doi.org/10.1007/978-3-030-62746-1_95

from four or more visible Beidou satellites outside the receiver before positioning can be achieved. However, the electromagnetic wave signal of the satellite is easily interrupted or weakened by natural obstacles such as buildings on the ground, and the receiver itself is also interfered and blocked by electromagnetic waves; these factors can easily cause the positioning accuracy to drop or even interrupt the positioning. The inertial navigation system has good concealment and robustness [1–3]. The measured data error and noise are small. The motion acceleration and rotational angular velocity of the carrier are very accurate in a short time, but the error and noise will follow the integral calculation process. The time in the accumulation is getting bigger and bigger. In summary, the method of Beidou+dr can be mutually corrected to reduce the accuracy of navigation data. The system composition is shown in Fig. 1.

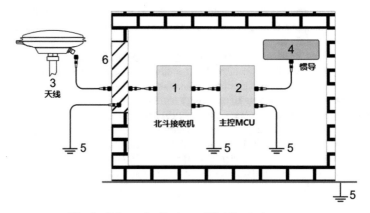

Fig. 1. Schematic diagram of Beidou + dr system

The measurement equation of the Beidou double-sub-sub filter

$$Z_{BI} = \begin{bmatrix} L_I - L_B \\ \delta_I - \delta_B \end{bmatrix} = \begin{bmatrix} \gamma L + V_{LB} \\ \gamma \delta + V_{\delta B} \end{bmatrix} = H_{BI} \dot{X}_{BI} + V_{BI}$$

In the formula: $H_{BI} = \begin{bmatrix} 0_{2\times6} & \text{diag}(1\ 1\ 0) & 0_{2\times11} \end{bmatrix}$, $V_{BI} = \begin{bmatrix} V_{LB} & V_{\delta B} \end{bmatrix}^T$.

Due to the nonlinearity of the terrain, the inertial part leads to the nonlinearity of the measurement equation. The terrain randomization linearization algorithm can obtain the slope of the terrain in real time, and obtain the linearized measurement equation:

$$Z_T = H_T X_T + \gamma$$

among them $H_T = \begin{bmatrix} -h_x & -h_y & 1 & 1 & 0 \end{bmatrix}$, h is the slope of the terrain x, y direction;

$X_T = \begin{bmatrix} \delta_x & \delta_y & \delta_h & \delta_{v_x} & \delta_{v_y} \end{bmatrix}^T_{5\times1}$, δ is the three-dimensional position and the two-position speed;

$\gamma = -\gamma_m - \gamma_r - \gamma_i$, γ is the measurement noise.

Here, the first-order Taylor expansion method is used to simplify the nonlinear measurement equation to a linearization equation. Then two local filters are designed for the two subsystems, Beidou, Inertial Guide (dr), and local estimates of the error state of the integrated navigation system can be obtained by the local filter. Fault detection and isolation methods are used in each local filter to determine if a subsystem has failed. A main filter is designed to combine the local state estimates from the local filters to obtain global estimates of the error state of the combined navigation system, and then use these estimates for feedback correction.

2 RFID Assisted Positioning

RFID tag reading and writing is realized by transmitting and receiving RF signals. The RFID reader can send the relevant information of the read tag to the unmanned vehicle through the serial port. A Received Signal Strength Indication (RSSI) can be used to indicate the strength of the RF signal at a certain location. This article uses RFID tags as beacons. According to the RFID reader, the label can be penetrated, and the label can be placed underground or placed at a specific location such as a building wall. When the reader reads multiple tags, the unmanned vehicle selects the 4 closest distances for positioning based on the distance of each tag learned by RSSI. The RFID device works at 900 MHz, and its transmit power setting can be set to 4 levels of read/write distance. This article sets the maximum readable distance to about 8 meters. It is known that the coordinates of the four labels are $(x1, y1), (x2, y2), (x3, y3), (x4, y4)$, and their distances to the unmanned vehicle R are $d1, d2, d3, d4$, respectively. Assume that the coordinates of the unmanned vehicle R equipped with the reader are (x, y). The coordinates of the mobile robot R can be obtained using the standard minimum mean square error estimation method:

$$\hat{X} = (A^T A)^{-1} A^T b$$

The distance between the reader and the four tags is obtained by simulation. The corresponding deviations of each coordinate point are: 0.1416, 0.0775, 0.0124, 0.0473, 0.9536, 0.0067, 0.136, 0.0619, the unit is meter, and the maximum deviation is less than 15 cm. The minimum deviation is less than 1 cm and the positioning is more accurate. When the unmanned vehicle identifies the building through the map, it can locate its position by RFID, and assist in calibrating Beidou + DR navigation.

The independence of the rfid tag can provide specific motion identities in unmanned vehicle movements, such as in-situ rotation, special location stop taking photos, changing scheduled routes, and the like [4, 5].

3 Obstacle Avoidance, Drive System

Relying on the Beidou+dr+rfid system unmanned vehicle can achieve accurate route planning and navigation, but the actual road conditions are complex; moving pedestrians or fixed obstacles will have a great impact on the unmanned vehicle movement. Increase the terrain of the unmanned vehicle by laser radar to ensure that the unmanned vehicle can break or avoid obstacles when encountering abnormal road conditions [6].

With the 16-line laser radar, the feature points can be extracted to judge the road condition, such as the movement of the pedestrian generated by the pedestrian and the vehicle-driven unmanned vehicle. Figure 2 below is a schematic diagram of the road conditions drawn by the laser radar.

Fig. 2. Road diagram

The movement of the unmanned vehicle relies on the two crawler drives of the motor control chassis, and the motor is controlled by pid as shown (Fig. 3):

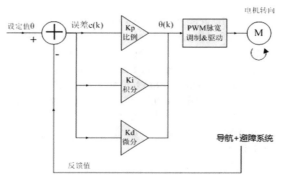

Fig. 3. PID control

The control algorithm uses an incremental pid control algorithm with the formula:

$$\triangle \theta_k = K_p \cdot [e(k) - e(k-1)] + K_i \cdot e(K) + K_d[e(K) - 2e(K-1) + e(K-2)]$$

Here $\triangle \theta_k$ indicates the turning angle increment of the cart, $e(K)$. For system control deviations, K_p. for system control deviations, $K_i = K_p \cdot \frac{T}{T_i}$. for the integral coefficient,

T_i for the integration time, T for the sampling period, $K_d = K_p \cdot \frac{T_d}{T}$ for the integral coefficient, T_i for the integration time, T_d for the differential time, T for the sampling period [7–11].

Because the integral part is mainly to eliminate the static error, the effect of the motor control with large interference is not obvious, but it will reduce the response speed. Therefore, the PD control algorithm with the integral link is removed. Through the test, the lateral deviation of the unmanned vehicle when moving in a straight line is ±30 cm, and the correction of the navigation system can basically ensure the correctness of the predetermined route.

4 Conclusion

With the rapid development of unmanned vehicle technology, navigation with multi-system fusion technology has become mainstream. The Beidou+dr+rfid data fusion proposed in this paper obtains more information including positioning, and has higher accuracy in outdoor navigation and more accurate positioning of unmanned vehicles. However, the prior art still has problems such as the occlusion of the Beidou signal, the cumulative error of the inertial navigation, and the environmental requirements of the RFID tag. In addition to the upgrade of the technology itself, there is still much room for improvement in data fusion and data filtering.

References

1. Tian, J., Chen, G.: Beidou navigation and positioning technology and its application, 40–45 (2017)
2. Ge, K.: Lidar-based mobile robot positioning and navigation technology. Shanghai Jiaotong Univ. Thesis, 48–51 (2006)
3. Xibin, C., Dongfa, Z.: Development and application of inertial navigation technology. Sci. Technol. Guide **26**(9), 21–22 (2011)
4. Kelly, A.: A 3D state space formulation of navigation kalman filter for autonomous vehicles. The Robotics Institute, Carnegie Mellon University, vol. 3 (1994)
5. Abbot, E., Powell, D.: Land-vehicle navigation using GPS. Proc. IEEE Pos. **87**(1), 145–162 (1999)
6. Yan, G.M.: Quaternion method of inertial navigation, 1–10 (2010)
7. Guo, T.Y.: Robotics and its intelligent control, 20–25 (2017)
8. Meng, Q.X., Wang, X.D.: Fundamentals of robotics, 55–65 (2006)
9. Martinez.: ROS robot programming, 70–80 (2012)
10. Xia, L.Y.: Application of Beidou in high precision Positioning, 88–90 (2016)
11. Lv, Z.: Observations and reflections on electric automobile industry developments. Int. J. Front. Eng. Technol. **1**(1), 98–104 (2019)

Application of "Computer Network Teaching Platform + Flipped Teaching Model" in Online Education-Taking "Information Technology Teaching Method" as an Example

Qingchuan Gu[1], Yao Zhang[2(✉)], and Haiyan Yang[2]

[1] College of Physics and Information Engineering, Zhaotong University, Zhaotong, China
[2] Zhaotong University, Zhaotong 657000, China
237763661@qq.com

Abstract. With the development of Internet and its application in education, "Computer Network Teaching Platform + teaching model" is gradually highly praised by colleges and universities. Combining the advantages of advanced teaching mode, this paper creatively proposes "Computer Network Teaching Platform + Flipped Teaching Model". Taking "Information Technology Teaching Method" as an example, this paper expounds the advantages of this mode in teaching by combining with teaching cases, which provides practical ideas for the reform of college curriculum teaching, as well as practical accumulation for promoting the construction of disciplines.

Keywords: Computer network teaching platform · Flipped teaching model · Dingtalk platform

1 Introduction

With the Internet applied to teaching, various Computer Network Teaching Platforms have emerged one after another, which has changed the single teaching model in which teachers and text books are the only teaching materials. In the teaching process, teachers can use the platform to pass on the prepared course content. The Computer Network Teaching Platform is taught to students in the form of micro-classes (recording and broadcasting classes), dynamic videos and other teaching methods. From teacher centered teaching mode to learning centered teaching mode, students form a student-centered teaching mode through various teaching resources provided by teachers in the Computer Network Teaching Platform. Moreover, the platform adopts a teaching method combined with the Flipped Teaching Model, which is conducive to giving full play to the initiative of students. Especially in this year when China is facing the COVID-19 virus epidemic and schools cannot start on time, "Computer Network Teaching Platform + Flipped Teaching Model" can effectively carry out online teaching, and do a good job of non-stop classes. This paper is based on the Information Technology Teaching Method course.

J. MacIntyre et al. (Eds.): SPIoT 2020, AISC 1283, pp. 660–664, 2021.
https://doi.org/10.1007/978-3-030-62746-1_96

2 The Overview of "Computer Network Teaching Platform + Flipped Teaching Model"

Computer network education is a new form of the Internet in the information era. It is the transformation of Internet application in education. It has really formed a change in the education center from "teaching" to "learning". Focusing on "learning" means that teachers use various educational methods and teaching methods to enable students to learn more knowledge, emphasizing that students construct knowledge actively. College stage is one of the key stages of students' learning career, and it is also an important stage to acquire comprehensive knowledge and ability. Computer Network Teaching Platform can provide teachers and students with a lot of teaching resources, which is helpful for teachers' teaching and students' learning.

Taking DingTalk network platform as an example, it is a free communication and collaborative multi-destination platform specially designed for Chinese companies by Alibaba Group. It supports PC version, Web version and mobile version, and can realize the transfer of files between mobile phones and computers. Dingtalk provides "class group" and other scene groups, which are suitable for different industries. This platform can provide high-definition live videos, with functions such as check-in, online classrooms, courseware playing, homework assignment. It can also establish learning circle, share learning experience and exchange learning experience, which is suitable for online teaching. Besides, the platform also has the function of "Statistics", where teachers can check the number of students' visits at any time to supervise them. Flipped Teaching Model first appeared in 2011 when American educator Salman Khan proposed "the Flipped Classroom" [1] 46 in his speech report "Recreate Education with Video". It has been popular in the United States and in recent years, this teaching model has been appreciated by many educators in China. The so-called "flip" or "reverse" refers to the roles of teachers and students, which means that in this teaching model, the roles of teachers and students are actually "reversed". Combining the [2] 128Computer Network Teaching Platform and this new type of flipped classroom teaching mode, students can log in to the teaching platform before class to watch learning videos provided by teachers, other excellent online videos, courseware and so on, and acquire knowledge through educational resources before class. Besides, students can discuss the preview confusion in the class and on the basis of solving the puzzles, in which teachers guide and help students to master the major points, so that students' knowledge can be expanded. After class, the students can consolidate the new knowledge by re-watching the videos. In this way, "Computer Network Teaching Platform + Flipped Classroom Model" can create a highly efficient virtual classroom.

Compared with the traditional teaching model, "Online Education Platform + Flipped Teaching Model" has obvious advantages. The traditional teaching model is teacher centered, in which the teacher is the main role of the whole classroom while the students passively accept knowledge. Also, the teaching resources provided by teachers in the classroom are limited to textbooks or limited reference books. However, "Online Education Platform + Flipped Teaching Model" helps students to complete the preview preparation, complete learning procedures such as teaching key and difficult points, and review to consolidate after class through network platform, which can improve students' ability of independent learning and knowledge exploration.

3   Advantages of Using "Online Education Platform + Flipped Teaching Model" in Information Technology Teaching Method

3.1   Redesigning the Teaching Process Is More Conducive to Students' Knowledge Construction

Nowadays, most colleges and universities emphasize the cultivation of practical personnel, and "Online Education Platform + Flipped Teaching Model" is one of the best teaching methods to cultivate applied talents. Information Technology Teaching Method is a compulsory course for undergraduate computer teachers majors in colleges and universities and it is also a practical course combining theory and practice. The overall learning goal is to learn to be a new information technology teacher under the new curriculum reform. [3] 101 Traditional teaching methods generally have two stages: "transfer information" and "absorb internalization". "Transmitting information" means that the teacher systematically teaches the knowledge points of the course in the classroom, and the students listen to the lesson and accept new knowledge. At this stage,

Teachers are taught in groups, and they cannot teach according to students' aptitude. As a result, students with better reception ability cannot get higher promotion, which inhibits the development of top students. On the contrary, students with poor receptive ability are afraid of difficulties because they can't keep up with the teacher's progress, and even give up listening. The second stage as "absorption and internalization" is the consolidation of knowledge by students through after-school exercises and other forms. At this stage, the traditional teaching model often leads to the students with poor acceptance ability unable to complete their homework and creating a feeling of loss. Due to the lack of solid theoretical knowledge, many students in the practice class do not have the ability to combine teaching theory and practice, and cannot control the teaching of junior middle school information technology courses. However, using "Online Education Platform + Flipped Teaching Model" to a certain extent, considering the cognitive structure and cognitive ability of students, can cultivate students' ability of automatic learning and active participation in teaching, and give full play to their enthusiasm and initiative, so as to better learn and master the course.

Using "Online Education Platform + Flipped Teaching Model" requires reasonable planning and design of teaching process. The "information transmission" stage should be set before the class, so not only can students learn through the vivid interactive short-term videos provided by the teacher, but also can tutors answer questions online through the network platform. The "absorption internalization" stage is set in the class through the interaction between teachers and students, students and students, which reverses the traditional learning process. Through this learning mode, students can better grasp the knowledge of the subject. "Information technology teaching method" also carries out teaching practice courses, which involve students' lectures. Teachers can provide high-quality teaching videos or lecture videos for middle school teachers through network platform, which is of great help for students to have an intuitive understanding of the teaching of middle school information technology courses. In addition, in the teaching practice of students' lectures, teachers can also record video of student teaching practice, and put the video of representative students on the learning network platform and students can use the network platform group summarize and discuss the students' practice videos. And through the discussion and summary, they can find out the advantages and disadvantages of the student in practical teaching, and then help the student to improve the teaching level, which in turn allow the class to achieve the purpose of learning from each other.

3.2   Provide a Variety of Teaching Media, Which Helps to Cultivate Students' Interest in Learning

First of all, teachers should make teaching courseware suitable for the use of network platform. The teaching courseware should take into account the characteristics suitable for students' autonomous learning as far as possible. For example, in the chapter about the teaching method of information technology courses, in addition to presenting the basic concepts and main teaching content of the chapter, the courseware is aimed at important teaching methods, and teachers should also provide detailed teaching cases, so that students can learn the theory of teaching methods on the basis of specific combination to understand and digest knowledge, and to achieve the purpose of flexible use of teaching.

Secondly, teachers should record in advance to help students to solve the teaching difficulties and predict the possible teaching problems before class. The advantage of instructional video is that it can enable students to feedback the learning situation in time. And the teachers can summarize and process before class to help them understand students' learning situation [4] 114. Another example is in the teaching work of teaching information technology courses, which involves the teaching content of "how to teach lessons". Through the network platform, teachers let students watch the videos of high-quality courses of information technology, organizing students to comment and exchange on the network, which can cultivate students' interest in learning and improve their enthusiasm and effectiveness of teaching.

3.3   American Scholars Ference Marton and Roger Saljo Proposed in Essential Differences in Learning

Results and Processes that "Tend to focus on what may be tested is shallow learning, but tend to think in conjunction with your own profession and try to grasp the core idea of the entire article is deep learning." Obviously, deep (level) learning is a more active, exploratory and comprehensible way of learning, which requires learners to pay attention to critical understanding, emphasize information integration, promote knowledge construction, focus on transfer applications and solve problems. [5] 102 College students have relatively strong ability to construct and transfer knowledge and also have strong ability to learn independently, so they are more suitable for "Online Education Platform + Flipped Teaching Model". The role of the teacher has changed from a knowledge lecturer or presenter to a classroom director, a leader and mentor of student learning. The main responsibility of the teacher in the class is to control the class, so in this teaching mode, each student is the subject of knowledge learning and can expand his own knowledge. For example, when learning the chapter on teaching evaluation of information technology courses, as the evaluation of teaching is an important but difficult part, the teacher can play the video of the class's students trying to talk about the middle school information technology course through the network video based on the students' theoretical knowledge, which allows students to combine the knowledge learned in this chapter and evaluate the teaching video of the student from the aspects of the teaching process, learning activities and learning effects, and finally the teacher summarizes. With the help of the network platform and the combination of theory and practice, students can truly master the theoretical knowledge of information technology teaching evaluation and use it flexibly through students' discussion and teachers' summary.

3.4 Break the Space-Time Restrictions of Class and Improve the Interaction of Teaching

Network platform provides diverse teaching media. College students belong to a kind of group with strong autonomous learning ability, so they can strengthen the interaction and communication in learning on the basis of autonomous learning via various teaching media. Whether it is in class or after class, under the guidance of teachers, the communication between students or teachers and students can be carried out on the network platform. Teaching interaction not only helps to improve students' understanding of knowledge and teaching skills, but also helps to train and cultivate students' interpersonal skills and teamwork spirit, and promote students' social development, so as to achieve the cultivation of applied talents in line with social needs.

4 Conclusion

"Online Education Platform + Flipped Teaching Model" is a teaching mode that meets the learning needs of students in this information era, which is an innovation and attempt from the traditional teaching model. Besides, the integration of this model with the "Information Technology Teaching Method" course not only greatly enriches the content of the course learning, but also allows "knowledge transfer" and "ability development" to keep pace with each other. Furthermore, it enables students to fully experience the process of autonomous learning and to build knowledge with the help of teachers. [6] 141 Finally, this teaching mode will play an active role in the teaching reform of this course and other subjects.

References

1. Zhang, J., Wang, Y., et al.: Flip classroom teaching mode. Distance Educ. J. 08 (2012). (in Chinese)
2. Tan, Q.: Research on the application of flipping teaching mode in mandarin teaching. J. Hezhou Univ. 02 (2015). (in Chinese)
3. Yang, Y.: A discussion on the mixed teaching mode of the "information technology teaching method" course. J. Huainan Teach. Coll. 03 (2011). (in Chinese)
4. Shen, S., Liu, Q., Xie, T.: A flipped classroom teaching model based on electronic schoolbags. Chin. Audio-Vis. Educ. 12 (2013). (in Chinese)
5. Qiao, R.: Problems in the application of flip-type teaching in ideological and political courses in colleges and universities. Sci. Teach. J. (Late) 04, 2018. (in Chinese)
6. Zhu, J.: The research and practice of the "dual platform + flipping" teaching model based on MOOC and Pan Ya SPOC - Taking the course CPA financial management as an example. Mod. Mark. 09 (2018). (in Chinese)

Design and Implementation of an Information Service System for Scientific and Technological Achievements Transformation Based on Spring Boot

Jiayin Song[✉]

Science and Technology Industries Division, Northeast Electric Power University,
Jilin City 132012, China
15144206620@139.com

Abstract. Through the mobile cloud service platform of Heyuan University Town, a system is built to improve the level of community informatization of Heyuan University Town, improve everyone's lifestyle, to facilitate the clothing, food, housing, and transportation of community residents, businesses and college students in the university town. The server side of the platform adopts spring boot technology, the mobile side adopts Android+HTML5 technology, and the front and back end data communication is carried out by using JSON key value to transmit value. The platform mainly provides such functional modules as cardio University Town, University mall, University Town BBS, my university, etc., to meet the information needs of different users in the university town community.

Keywords: Spring boot technology · Cloud services · Mobile internet · HTML5 technology

1 Summary

With the rapid development of mobile Internet and the popularization of smartphones, the use of mobile terminal has gradually surpassed the use of PC, and the use of mobile terminal has become an important part of people's life. Heyuan university town community includes the Heyuan campus of Guangdong Normal University, Heyuan vocational and technical college, Heyuan Polytechnic School, Heyuan technician college, Heyuan health school, etc. at the same time, there are many businesses and tourist attractions around the school. Everyone wants to get information about food, accommodation, travel, shopping, and entertainment around the university town, To facilitate everyone's life. To serve the users of Heyuan university town well, the mobile cloud service platform of Heyuan University Town Based on spring boot is designed and developed, through which tourists and users can easily access the relevant information of the university town [1].

J. MacIntyre et al. (Eds.): SPIoT 2020, AISC 1283, pp. 665–669, 2021.
https://doi.org/10.1007/978-3-030-62746-1_97

2 Significance of Platform Research

In 1999, At present, there are many platforms based on mobile Internet, which are generally applied to a certain field, such as e-mail, campus forum, etc. as a service platform, this technology is applied to the surrounding communities of University Town, which is less to meet the needs of different users, and there is no mature application platform.

Therefore, the research significance of this platform is as follows:

(1) Help tourists to understand Heyuan University Town faster and better. For those who want to or have come to Heyuan to play, the platform will provide a brief introduction of the University Town, the latest consultation of the University Town, and the cultural and natural landscape around the university town.

(2) Help the freshmen who are about to enter the university town school know their campus life in advance. For the freshmen of different universities in University Town, after they register and submit the admission notice, the administrator of their school will pass the examination of their application members. Through this platform, the freshmen not only know the living environment around the university town of Heyuan City in advance but also get relevant campus information of the school, such as campus culture, distribution of teaching buildings, distribution of canteens, distribution of dormitories, community organizations, etc., so that they can integrate into campus life faster and better [2].

(3) Help school students enrich their campus life. It provides the school community information, entrepreneurial park information, convenient tools, University City Mall, University City BBS, curriculum management, and other modules for the school students. By using the University City Mobile cloud service platform, you can easily query and arrange your own life and study.

(4) Provide an online shopping mall platform for University City merchants. The platform has a University Town Mall, where businesses can register as merchants to provide convenience for people living in the university town.

(5) It can enhance the project practice experience of information majors in our university. As the project has great commercial promotion value, the early stage of demand analysis has been completed. At present, the professional teachers are leading the students of information major to code. The design and development of the platform can not only improve the team cooperation ability of students but also accumulate practical project experience for students to get employment today.

(6) The research results of this platform are easy to be converted into teaching. The platform is jointly developed by front-line teachers and students, using the enterprise's advanced Java framework, which is convenient for teachers to decompose into modules of appropriate size for teaching. Based on the above significance, it is not only necessary but also very practical to build a mobile cloud service platform of Heyuan University Town Based on spring boot.

3   System Design

3.1   Platform Topology

By studying the research significance and demand analysis of the platform, the overall design topology of the mobile cloud service platform of Heyuan University Town Based on spring boot is obtained, as shown in Fig. 1. As can be seen from the topology map, the servers of the platform are mainly in the cloud, and Android mobile clients communicate with the cloud server through switches. Developers deploy the tested system to the cloud server. Administrators can connect to the cloud server through a PC or mobile terminal to manage and maintain the platform. Tourists or members can access the cloud server through the mobile terminal to log in and use the platform [3–5].

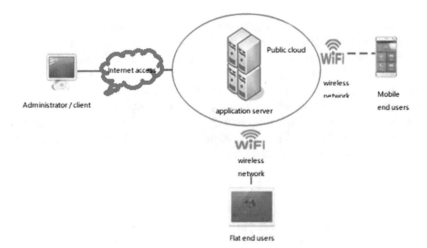

Fig. 1. Mobile cloud service platform topology of University Town

3.2   Platform Function Module Diagram

The mobile cloud service platform of Heyuan university town is mainly used for the residents of Heyuan University Town, including the students, teachers, businesses, and residents in the university town. Therefore, the platform mainly consists of four roles: administrator, tourist, student user, and nonstudent user. Administrators are divided into system administrators, school administrators, merchant administrators, forum administrators, etc., which respectively conduct user registration review and module information maintenance for each management module; tourists mainly browse information consultation, transportation, entertainment, and other information near the university town; nonstudent users can not only have the rights of tourists, but also visit Purchase University mall goods, and can go to the University City BBS to post and reply; student users not only have the rights of nonstudent users but also have their unique functions, such as access to my campus, convenient tools, campus mall, and other modules.

4　Key Technologies of System Implementation

4.1　Spring Boot

Spring framework has the characteristics of aspect-oriented Pro programming (AOP). It is an open-source framework based on the Java development platform. Its core feature is an inversion of control (IOC), which is realized by dependency injection (DI), To realize the management of the object life cycle. Spring boot is developed by the pivotal team. It not only inherits the advantages of the spring framework but also simplifies the process of setting up and developing configuration of the spring framework environment and solves the problems of version conflict and unstable reference. At present, the spring boot framework is the preferred framework for developing the Java platform.

4.2　HTML5

HTML5 is a programming language description method to construct the content of web pages. It is improved based on html4.01. It provides a specification for HTML, the core language of the web, and makes HTML more in line with the requirements of modern network development. HTML5 is not only a simple list of HTML, CSS, and javascript technology, but also has the characteristics of semantics, local storage, device compatibility, connectivity, web multimedia, and other features, so it has its advantages in web network standards, adaptive web design, multi-device cross-platform and so on. The emergence of HTML5 not only reduces the technical threshold of programming developers but also improves the interactivity and usability of the website, thus enhancing the user experience.

4.3　Data Exchange

The mobile terminal and server of Heyuan University City Mobile cloud service platform based on spring boot adopt JSON mode to exchange data. JSON is a text format used to store and represent data. This format has a clear and concise hierarchical structure, which is not only beneficial to developers to write and read, but also to improve the efficiency of machine parsing and generation, to shorten the time in network transmission. In data communication, JSON generally uses {key: value, key: value} Key value pair

It is more readable than the XML document model parsing and traversing node transmission mode.

5　Conclusions

The platform has such functional modules as heartbeat University City, University City Mall, University City BBS, my university, etc., which can meet the needs of different customer groups such as community merchants, students, ordinary residents, etc. By using this platform, tourists visiting Heyuan university town can also learn about shopping, accommodation, transportation, scenic spots, and other information in advance, to arrange their itinerary. The mobile cloud service platform of Heyuan University Town Based on spring boot can effectively serve the community residents of Heyuan University Town, which is relatively high for other communities

Therefore, the platform has a certain promotion value.

References

1. Chang, X., Wang, Y.: Design and implementation of a campus assistant service platform based on Android. Comput. Knowl. Technol. (2018)
2. Roger, et al.: Design and implementation of a campus assistant app based on the android platform. Softw. Guide (2016)
3. Yu, W.: Design and implementation of a mobile cloud storage system based on data security and privacy protection. Univ. Electron. Sci. Technol. (2018)
4. Xiao, D.: Design and implementation of a campus light blog system based on spring boot. Huazhong Univ. Sci. Technol. (2018)
5. Lei, Z., Yue, W.: Research on MVC model based on spring boot microservice architecture. J. Anhui Inst. Electron. Inf. Technol. (2018)

Design of Oral English Learning System Based on Cloud Computing

Tong Shen[✉]

Ji Lin Justice Officer Academy, Jilin 130062, Changchun, China
cindyandsteven@126.com

Abstract. With the maturity and popularization of cloud computing and mobile network technology, mobile learning develops rapidly in many fields, and the demand for mobile learning system is expanding, especially in oral English learning. A cloud computing based oral English learning system is designed. Users can learn oral English knowledge through a mobile terminal, simulate oral English tests and establish a special dictionary. Managers can manage learning resources, user information, and word base through the cloud server. The system can realize mobile learning based on cloud computing, cloud sharing of learning resources and other functions, improve the user's learning experience, and effectively promote the user's aural learning.

Keywords: Cloud computing · Spoken english · Learning resources

1 Introduction

Cloud computing is based on the increase, use, and interaction mode of Internet-related services. It has a powerful computing ability, can handle a large number of data, and more efficiently realize the functions of traditional computers. It also supports users to obtain application services at any location, any time, using different terminals, making it more convenient to use. In recent years, with the rapid development of cloud computing technology, cloud computing technology is widely used in various fields of people's lives, and people increasingly rely on cloud computing technology. At present, the communication between our country and other countries is more and more frequent, so we need high-level English talents urgently. However, English Teaching in schools tends to focus on writing rather than oral English. In the market, there are few learning software for oral English, and there are some problems such as the lack of function of the mobile terminal, resulting in limited computing. This system combines cloud computing technology with spoken English mobile learning, enabling users to learn spoken English anytime, anywhere, making full use of scattered time, experiencing fast and effective learning methods, maximizing learning benefits and realizing real mobile learning [1].

J. MacIntyre et al. (Eds.): SPIoT 2020, AISC 1283, pp. 670–674, 2021.
https://doi.org/10.1007/978-3-030-62746-1_98

2   The Overall Design of the System

The main purpose of an oral English learning system based on cloud computing is to enable users to learn oral English on mobile devices anytime and anywhere. Users are users of the system, and system managers are only responsible for updating and maintaining the data in the system. The user is the main body of the system, which mainly realizes two functions: learning and testing. The local client is used by users, mainly including pronunciation tests, data reading, and vocabulary customization. The cloud server is used by managers, mainly including editing and publishing the latest learning materials, managing word banks, updating standard voice, and other functions.

2.1   System Architecture

Cloud computing is the core of the oral English learning system. The system is based on B/s multi-layer distributed system design, without the need to install the software in the terminal, all operations are completed through a web browser, with convenient, fast, and efficient user experience. The server is based on the cloud server provided by the cloud computing service provider, which is used to deploy the system, application, and data center. The multi-layer distributed architecture is divided into the device layer, data layer, application layer, and presentation layer; the device layer mainly includes PC, mobile terminal, and other local physical equipment; the data layer is the system database, which is responsible for the calculation and storage of data such as users, word library and learning progress; the application layer is the integrated management and Application module of the system platform, including the functions of fetching oral data, updating data and establishing word library; the performance layer is the information presented to users, including testing oral scores and checking learning progress Degree and learning materials display, etc. [2].

3   Function Design of the System

This system refers to the functions of the relevant app on the market, according to its characteristics for reference and innovation, get the following functions. The pronunciation evaluation function is the basic function of the system. It can evaluate the pronunciation of words and provide different pronunciation training. Pronunciation evaluation is divided into two sub-functions: first, repeated, and intensive training for a single word. After the user completes the recording, it is uploaded to the cloud storage. The cloud computing system gives a score for the pronunciation of each phonetic sign of the user, and gives standard pronunciation for the user's reference, and records the wrong English phonetic sign of the user as the basic data for the continuous analysis of the word library module. The second is the pronunciation training of the word bank. Users can choose the word bank provided by the system or the custom word bank for the overall pronunciation training. In this module, the system embeds a single word in the database into a complete sentence and displays it to the user to detect the user's overall understanding of the word. After each training, the system will recommend the training form of the next stage to the user according to the pronunciation of the user and carry out single word training or long sentence training.

The data display module contains original English articles and news from all over the world. Users can choose the English articles they are interested in to play and can control the speed of playing and the display of Chinese. Each article is equipped with a corresponding video, which can be used for users to relax and watch while learning. At the same time, it can strengthen users' understanding of the article and improve users' sense of language. The system also has the function of collection and sharing. Users can collect their favorite articles so that they can watch them next time, share the articles with their friends, and discuss and learn with them [3, 4].

The Thesaurus management module mainly manages the word resources of the platform, including user word usage information, all word databases, etc. The default classification mode of the thesaurus is to classify by grade. The thesaurus will automatically record the words learned by the user according to the user's usage, and give a reminder when the user learns next time. At the same time, the thesaurus automatically enters the words whose pronunciation score is lower than the standard score into the user's thesaurus. The standard score is customized by the user. The user can also manually enter the words into the personal thesaurus according to personal needs.

The user message module is the main communication channel between users and managers. Users who find system bugs or have good suggestions for the system can feedback to the manager at any time. The manager can also reply to the user through the user message module and give a certain reward to the active user.

4   System Learning System Design

An app is developed on the Android platform to realize a spoken English learning system based on speech recognition technology. This system can help users learn and practice spoken English better and more conveniently. The main function modules of the system include user login, voice acquisition and recognition, voice evaluation, voice broadcast, and oral dialogue. Database design is to store the data generated in the system. The original intention of the design is to meet the needs of users, to construct simple and reasonable data tables and database patterns.

4.1   Design of Speech Acquisition and Recognition Module

The special interface responsible for the top-level audio application in Android is audio, and all classes in the audio local framework are part of lib media. So. Audioflinger is responsible for calling the local library libaudioflingerso. In the audio system, only the interaction between the application layer and the underlying hardware is managed. The I/O format is PCM. JNI and Java language provide an interface for the application layer. On the Android platform, the class responsible for voice data collection is Audio Recorder and the voice reading is completed by the audio data reading function read. The class that supports the application framework layer is an audio record.

(1)   Voice collection mainly uses Audio Recorder like parameters, such as sampling rate, audio data format, audio source Mundi record, and storage format. Doacquindsave() indicates the start of voice data collection, and stopacquiandfile() indicates the stop

of voice data collection. Writedatatof file() transforms the original voice signal into a wav uncompressed pure waveform file and is Card Exist() detects whether an SD card is inserted into the Android terminal. The above is based on the methods provided by Audio Utils.

In the voice recognition module of the Android platform, firstly, the method of issdcardexist() in audiofileutils class will be used to detect whether the SD card is inserted into the Android terminal, then the file directory of the SD card will be obtained through getwavfilepath() function, and it will be used as the storage location of the new input original voice data. Final audio. Wav file is used to save the voice collected each time.

The operation steps of the speech recognition module are as follows: first, read the best codebook and template file provided by MATLAB; then, collect the speech data in the speech acquisition module, preprocess the input speech data, extract the feature parameters of MFCC, and then vectorize them; Finally, the parameter probability is calculated and the template file is matched to get the most consistent text information and output. Speech recognition library lib speech.so It is a dynamic link library developed by NDK and C + + , which is called by JNI speech.so The recognize () method of the speech recognition interface in. It is the core library of the speech recognition module on the Android platform. When using, first load the speech recognition library locally, and then call recognize. The specific implementation code is shown in Fig. 1.

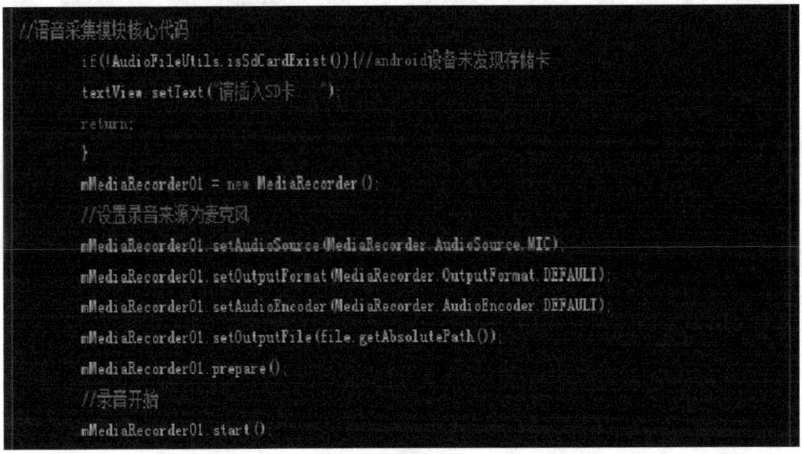

Fig. 1. Voice recognition module program

5 Conclusions

Spoken English learning system is an Android application developed based on speech recognition technology to meet the needs of users to learn and practice spoken English accurately anytime, anywhere. The system selects some functions that are more suitable for the mobile terminal to develop, and provides users with basic learning and practicing modules of spoken English, including voice recognition, voice evaluation, oral broadcast, and oral dialogue. The overall design of the system and detailed description of the functional module structure of the system, and the database applied to the system in the concept and logic aspects of the analysis and design, completed the database table building work; In the detailed design stage, the input and output items of the system are analyzed briefly, the interaction interface is designed, and the business process of each functional module is analyzed and described in detail. Finally, a lot of coding work will be carried out, and the system design will be completed according to the requirements: through the test and analysis work after development, the system can be normally used by users and is stable in function and performance.

This paper designs a cloud computing-based oral English learning system. The main purpose of this system is to realize the two parts of oral English learning and testing. The feature of this system is to use cloud computing technology to increase the efficiency of learning. The pronunciation evaluation module of the system can meet the needs of users for self pronunciation detection, and the word library module can provide word library for users to learn. This system provides a convenient, fast, and efficient way for the majority of users to learn spoken English.

References

1. Jie, Z., Zhitong, H., Xiaolan, W.: The selection principle and research of state number of hidden Markov model in speech recognition. Comput. Eng. Appl. **1** (2000)
2. Zuoying, W., Xiao, X.: HMM, speech recognition model based on segment length distribution. Acta Electron. Sin. **32**(1), 46–49 (2004)
3. Guilin, L.: Research on Robot Embedded Speech Recognition System, vol. 66. China University of Petroleum (East China), Beijing (2009)
4. Haoyi, H.: Computer speech recognition technology and its application. Popular science and technology, June 2005

The Current Situations of Mobile Assisted Language Learning

Guihua Ma[✉]

Xi'an International University, Xi'an 710077, China
605244593@qq.com

Abstract. The pervasive use of mobile devices has profoundly influenced people's life style and leaning style. Currently, it is pretty popular to use smart phones, IPADs, or PCs to have language learning. Mobile assisted language learning (MALL) makes it possible for language learning to occur at any time and place, and it can also compensate the interactive, cooperative and authentic learning environment which is often lack in tradition foreign language learning classroom. The paper mainly discussed the current situation of MALL and its application in the foreign language learning and teaching by summarizing the definition, characteristics, merits and flaws of MALL. It concluded that foreign language teachers should fully consider the functions and advantages of different mobile devices and design appropriate tasks according to learners' cognitive level and foreign language proficiency in order to simulate learners' interest in MALL and also encourage learners to develop their own learning strategies in MALL so as to promote their independent and autonomous learning ability.

Keywords: Mobile assisted language learning · Foreign language learning · Mobile devices · Current situations

1 Introduction

Mobile assisted language learning (MALL) is a comparatively new idea in the field of language learning in 21 Century, which is a means of language learning supported by smart phones, PDAs or tablet PCs. MALL has the characteristics of mobility, efficiency, universality, interaction, sharing and individuality, and it breaks through the conventional network dependence on "line" or "computer" and brings a brand-new learning experience for learners.

MALL has further improved E-learning and allowed learners to learn online almost anywhere if they are connected online. This makes it possible for learners to stay connected to learning during any free time students may have.

MALL devices compose of mini-computers, smart phones and IPADs. MALL focuses on the good use of learners' spare time and free time. Using mobile devices to create learning opportunities is one of the very important parts of mobile learning.

MALL is very fast and convenient because it can be connected to virtual learning space anywhere. Sharing opinions or having discussion are almost simultaneous among

J. MacIntyre et al. (Eds.): SPIoT 2020, AISC 1283, pp. 675–679, 2021.
https://doi.org/10.1007/978-3-030-62746-1_99

those who are using the same learning materials, which can also make timely feedback possible. MALL can also replace some books by making online notes during learning.

2 The Characteristics of Mobile Assisted Language Learning

Because MALL is a new learning mode by the use of mobile devices, so it is different from the traditional ways of language learning. These features mainly include three aspects: mobility of learning environment, variability of learning content and extendibility of learning time [1–3].

Firstly, MALL environment is variable, and is likely to be always changing. It can be realized in a wide variety of places which usually cannot be known in advance, and it can occur on the way from home to school, outdoors or any leisure place, which requires the learning devices with mobility. The flexible mobile learning mode increases more chances for language learners. Real learning contexts can also encourage language learning behaviors by using mobile devices. For instance, if a language learner in a foreign country has a smartphone or IPAD walking in the street and has doubts for some traffic indication terms or store advertisement, so he/she can use the mobile device installed with automatic translation software or image recognition software, and the corresponding language shows quickly. He/she can also mark and store it in the device and prepare himself for future study. As Reinders pointed out that, mobile devices like smart phones are social interaction tools, so they can promote real authentic language communication, and makes it the best means for authentic learning. It is believed that when information and situation are related, and can be immediately put into application by language learners, so real learning is more prone to occur.

Secondly, MALL can achieve close link between information acquisition and physical environment, that is, when a learner lives in a different environment, the learning material is also different. For example, English learners can use mobile devices in a bar learning English names of a variety of beers, or listening to the introductions of different exhibits while visiting in a museum. The learning can be realized via the connectivity of the mobile devices, which not only includes the connection to the Internet, but also the ability to locate the user position via GPS connectivity. When a learner's mobile device has location aware learning software and enters a room, mobile vocabulary learning software will automatically display foreign names of the items and related information of the room, the learner can use the information to learn new vocabulary and knowledge. The learner can also use two-dimensional code recognition learning software through phone camera and the corresponding items of foreign language vocabulary and relevant information will be displayed for the learner to learn. Therefore, different environmental perception technology had profound influences on the effects of foreign language vocabulary learning. If language learning can be related to the actual context, learning effects will be greatly improved, and this is also one of the distinctive characteristics between mobile learning and the traditional learning mode [4–6].

Thirdly, MALL can be well used both in and out of classroom learning. Wong and Looi's study (2010) explained how to use mobile learning to combine formal study in classroom and informal learning out of classroom, and ultimately promote comprehensive learning effect. In their experiment, subjects in the classroom learned some phrases and the usage of prepositions, then the students were given smart phones with network connectivity and imaging features to extend their learning, they used phones to look for inside and outside the campus to learn these new words in context, took photos, and uploaded them to the online learning platform to share and learn these words further. They found out that the mode of combining inside and outside classroom learning enhanced contextual learning and produced creative output. To link classroom learning and extracurricular learning by using mobile devices can further expand the depth and breadth of learning, avoid the separation between learning and application in traditional teaching mode, and is also conducive to the development of learners' ability to learn autonomously, and this is quite crucial for their language learning in the future.

3   The Merits of Mobile Assisted Language Learning

MALL is gradually changing the way of learning and teaching. The wide use of mobile devices makes language learning is no longer limited to desktop computers and school language laboratory, it is gradually becoming "mobile", which is independent of time and space, so learning can occur at any time and place. As long as learners take portable mobile devices, they can have some language learning. Using the mobile device to present language learning material can make learning process more flexible, and it can also provide a two-way wireless communication channels for students and teachers. Mobile assisted language learning greatly expands the continuity and real-time of language learning. The use of mobile devices increases the time and space of learning, and is also conducive to promoting learner-centered and individualized learning process.

In the real foreign language learning environment, there is no target language interaction environment outside the classroom, which has become the most headache for foreign language teachers. With the pervasive use of mobile technology in the foreign language learning, it can solve the problem quite well. Using the connectivity of mobile devices, students and teachers can easily create their own virtual language application environment and increase interaction outside the classroom. Students can also take advantage of the mobile device's network connection function to interact with foreigners to participate in real language interaction anywhere and anytime on line.

In a word, with the development of wireless communication and mobile technologies, like phones, IPDAs, PCs and other mobile devices, the effects of language learning have been gradually recognized, because it includes all the advantages of e-learning, and eliminates the limitations of learning at any time and space. Mobile assisted language learning has the potential to promote the construction of constructivist, collaborative, environmental sensitive and informal learning [7–9].

4   The Flaws of Mobile Assisted Language Learning

Compared with traditional learning mode, MALL has many advantages, but it also has some shortcomings, and understanding these shortcomings will be better for us to use the new devices for language teaching.

Firstly, some schools may be lack of technical support for MALL. Mobile devices like smart phones and IPADs have the merits of portability compared with desktop computers, but the relatively small size of the screen is the sacrifice of long-time learning input. For information display volume, larger reading training and writing training will undoubtedly cause barriers.

Secondly, learners' may have some negative attitudes toward MALL. The desire of language teachers in the process of using the new technology is always good, but the attitude of the students to the new learning style will also have an important influence on the learning effect. However, learners' attitude is largely determined by teachers' design of learning activities. If teachers can make full use of new equipment and provide appropriate activities to help students achieve the desired learning objectives, then the students will have more positive attitude [10].

Thirdly, teachers have challenges to apply MALL. The contemporary students are called "digital natives", because they are growing in a generation of digital age, and they tend to use new electronic equipment in learning. But foreign language teachers have some obstacles to apply some functions of these devices and combine them with the methods of foreign language teaching, which requires teachers not only learn teaching skills, but also have a good understanding of the mobile learning technology, and the ability to carry out independent research, according to the actual situation of students, so as to achieve the purpose of using mobile devices to enhance the effects of language learning. To many foreign language teachers, this is undoubtedly a huge challenge.

5 Summary

At present, MALL is a new research topic with the development and wide use of mobile technology in teaching. How to combine mobile devices and language teaching, and how to improve foreign language learning effect are very important fields that foreign language teachers need to think about quite carefully. The most prominent application of mobile language learning is that it extends language learning to the outside of the classroom at any time and place, and can also combine class learning and extracurricular learning. Thus, mobile language learning occurs largely outside of classroom, which requires language teachers can fully master the functions of mobile technology, and also help students use these devices rationally to improve learning effects. In addition, foreign language teachers should also encourage students to develop their own MALL strategies, because this is the foundation for their extracurricular independent and autonomous learning. MALL research in language teaching and learning area is still at the initial stage, and there are still many problems to conduct in-depth study. But as a new means of language learning, the potential of MALL is enormous, future research needs to further construct more efficient framework for mobile assisted language learning.

References

1. Stockwell, G.: Using mobile phones for vocabulary activities: examining the effect of the platform. Lang. Learn. Technol. **14**, 95–110 (2010)
2. Reinders, H.: Twenty ideas for using mobile phones in the language classroom. Engl. Teach. Forum **48**, 20–25 (2010)

3. Barrs, K.: Fostering computer-mediated L2 interaction beyond the classroom. Lang. Learn. Technol. **16**, 10–25 (2012)
4. Wong, L.H., Looi, C.-K.: Vocabulary learning by mobile-assisted authentic content creation and social meaning-making: two case studies. J. Comput. Assist. Learn. **26**, 421–433 (2010)
5. Demirbilek, M.: Investigating attitudes of adult educators towards educational mobile media and games in eight European countries. J. Inf. Technol. Educ. **9**, 235–247 (2010)
6. Li, M., et al.: On the connotation and current situation of mobile learning. J. Inf. Commun. **15**, 53–57 (2018). (in Chinese)
7. Zhao, D., et al.: Research on the application of mobile terminal in intelligent classroom teaching practice. J. Huanggang Norm. Univ. **10**, 82–85 (2019). (in Chinese)
8. Wu, S.: Research on the current situation and future trend of mobile learning at home and abroad. Sci. Educ. Guide **9**, 160–162 (2019). (in Chinese)
9. Li, Y., et al.: Research on the development and countermeasures of mobile learning platform in the field of continuing education. Comput. Telecommun. **7**, 19–23 (2019). (in Chinese)
10. Zhang, L., et al.: Research on the mobile learning mode based on smart phones. China Educ. Technol. Equip. **5**, 36–38 (2019). (in Chinese)

Development of Industrial Chain of Internet of Things Based on 5G Communication Technique

Yanping Chen and Minghui Long[✉]

Wuhan Donghu University, Hubei, China
757088198@qq.com

Abstract. The article simply introduces the 5G communication technique, at the same time around its impact on development of the Internet of things industry chain are analyzed, and finally based on 5G communication technique support the development of the Internet of things industry and its application in the Internet of things industry strategy on the discussion, 5G to communication and fusion of Internet of things industry chain development help.

Keywords: 5G communication technique · Internet of things · Industrial chain

1 Introduction

In increasing material and cultural life level today, mobile network and Internet of things has been woven into the every aspect of the real life. And people to the requirement of network communication and Internet are becoming more and higher. So, if you want to meet people's diversified needs. It is important to develop the 5G communication technique. Apply it to the Internet of things industry chain, to the service quality of Internet of things, and further improve operation efficiency, and realize transformation of the Internet of things industry chain as soon as possible. In 5G communications perspective study with development on Internet of things industry chain, also is very realistic.

2 Summarize of 5G Communication Technique

5G communication technique is an innovative mobile communication technique which developed on the basis of 4G (fourth generation) mobile communication technique. Compared with the previous generations of mobile communication technique. 5G communication technique not only has strong stability, high speed of data transmission loses, large system capacity, the advantage of low delay. At the same time, it also can reduce energy consumption. The effect of cost savings, tremendous improvements in performance, in terms of losing data transfer rate.

J. MacIntyre et al. (Eds.): SPIoT 2020, AISC 1283, pp. 680–684, 2021.
https://doi.org/10.1007/978-3-030-62746-1_100

3 Impact of that the Internet of Things Industry Chain on 5G Communication Technique

3.1 Hardware Industry Develops Rapidly

Without hardware support for the development of the Internet of things industry chain, the era of 4G Internet of things industry chain, although has obtained the preliminary improvement of domestic content couplet factory hardware manufacturers also than more. Because it is very wide in the field of Internet of things application. Industry field for Internet of things hardware demand is relatively large. So the supply of all kinds of hardware has been difficult to full market demand [1–3]. There are a lot of restrictions on the development and popularization of the Internet of things. However in the age of 5G communications, Millimeter Wave technique (Millimeter Wave communication technique) as the key part of 5G mobile communication technique obtained the rapid popularization. The Internet of thing industry chain in the hardware industry also obtained the very good development. Compared with the traditional microwave communication system, the components of millimeter wave system on letter much smaller size; millimeter wave communication components were also more likely to implement integration of miniaturization. In such cases, the content couplet factory hardware vendors of hardware equipment production can reach large promotion. Industry developed quickly, and achieves the valuable in the whole industry chain of advacement.

3.2 Communication Standards are Gradually Unified

The industry chain has been formed because there are more and more development of the Internet of things. But there is a different in the field of mobile communication standard is not unified, which makes the Internet of things industry chain presented the sensor/chip manufacturers, telecom, systems integrators and other module differentiation phenomenon. It can't solve this problem as soon as possible. Then under the mode of independent type competition, Internet of things will gradually lose the whole development of the industrial chain advantage, and there will be some enormous obstacles on the long-term development of the Internet of things industry. And under 5G communication eras, because the 5G mobile communication technique research itself need for TDD, FDD, and other mobile communication standard in videos, and a number of standardization research of the project. Thus able to create a good condition for the unity of the mobile communication standard, and with the further development of 5G mobile communication technique. The future of mobile communication technique standard is expected to achieve global unification. The problems of the module points on the Internet of things industry have to be solved. Under the promotion of 5G technique, the development effect of non-independent framework is more ideal [4, 5].

3.3 Extensive Application of Advanced Technique

The construction of 5G mobile communication network needs to be supported by a variety of emerging technologies. After the commercialization of 5G communication technique, its various emerging key technologies will also be widely applied in the area of the Internet of things. Although has only just been developed in recent years, but by its to the realization of the 2 d and 3 d technique for comprehensive optimization model, set up large-scale MIMO system. So the spatial resolution and the degree of deep mining of spatial dimension resources, there are obvious advantages by compared with the existing MIMO system. And in the area of Internet of things, with the help of Massive advantages of MIMO technique in spatial dimension resources mining.

4 Development of Internet of Things Industry Chain Under 5G Communication Technique

4.1 Smart Home

Affected by the high quality of life of modern electrical appliances furniture industry has always been content league one of the most main application fields of the network technique, in the era of 4 g. Because the price of furniture, home appliances products are usually is higher. So consumers while on the Internet of things technique with the support of intelligent furniture. Appliances with strong purchase intention and the demand, but at the time of consumption will still be in commodity prices.

4.2 Construction Industry

With the increasingly serious environmental damage and pollution problem of building construction, many construction projects have begun to try to use the Internet of things technique to accurately control various construction equipment, and hope to achieve the goal of green and energy saving through this way. At the moment, because the Internet of things technique is still imperfections. Therefore in the process of construction and demolition. Construction equipment control accuracy is not ideal, the environmental pollution and destruction of the control effect is not ideal. But in 5G mobile communication technique and application. Internet equipment control accuracy will be under the influence of high data transfer rate and low latency increase greatly. And through construction equipment precise control to realize the green building will become a reality.

4.3 Medical Field

In the past few years, technique of Internet of things has started has been applied in medical equipment, and makes the doctor can remote control medical robots or portable equipment to complete the remote diagnosis and treatment. As a result of 4G mobile communication network signal quality is not stable. It is easy to influence the diagnosis and treatment of sex in a row. At the same time, network delay problems will cause the remote medical instrument feedback timeliness drops greatly. Therefore the practical application of telemedicine will still be subject to many restrictions. 5G technique has brought all-round improvement in the network, which to a considerable degree solves the demands of high efficiency, and stability of medical treatment, under 5G network remote medical treatment.

5   5G Communication Technique is an Effective Strategy to Promote the Industrial Chain Development of the Internet of Things

5.1   Adjust the Deployment of Local Industrial Chain

Although the development of 5G mobile communication technique will provide important technical support for the development on the industry chain of the Internet of things, the deployment of the local industrial chain still needs to be adjusted reasonably if 5G communication technique and the industry chain of the Internet of things are to be effectively integrated. On the one hand, because of 5G mobile communication technique has not been fully commercialised, so during the deployment of iot industry chain, still need to give full consideration to the regional industrial advantages and disadvantages, and make full use of local advantages of industry module layout, to ensure the stability of the Internet of things industry chain, for the application of 5 letter g mobile communication technique to create a good basic condition. On the other hand, as the consumer group of the Internet of things has undergone great changes, the industrial chain layout of the Internet of things needs to be further investigated and analyzed based on the needs of consumers, and a reasonable industrial chain layout strategy should be determined based on the results of big data analysis.

5.2   Optimize the Development Plan of the Internet of Things

With the support on 5G mobile communication technique, there are many new business models have gradually taken shape in the area of Internet of things. The ubiquitous characteristics of 5G will not only increase the number of access devices to the Internet of things. But also produce a business model of the Internet of things show diversified characteristics. Therefore, the development of Internet of things industry in the future must take into full consideration the accelerated iteration and evolution of 5G communication, artificial intelligence, big data and other technologies, and make a new plan for the development of Internet of things industry chain based on diversified business models just to adapt to various new business models. For example in the field of smart home, intelligent washing machine can rely on 5G mobile communication network, the clothes in the washing machine intelligent identification. According to the characteristics of automatic washing mode choice, and even can with clothes and information feedback to the garment enterprises. Such as state of the old and new, guide them to improve technique and design, and in the new business mode, the plans of the development of the Internet of things have to also make some corresponding adjustment.

5.3   Strengthen Information Security Protection

From the Internet of the operating system, the letter of agreement, hardware equipment and other aspects of research and development of the network information safety hidden trouble of testing technique and security protection technique, the power of the unity of the whole industry chain to make good the Internet of things in the whole ecosystem. And body on the other hand, it needs to set up users privacy security defense system of the whole life cycle in based on the Internet of things. For the Internet of things device research and development, the chip production, software design, and many other link to develop safety technical standard, is the information security protection will become the core tasks for the development of the Internet of things, from the source fully avoid the security of user privacy information disclosure problem.

6 Conclusion

To sum up, 5G communication technique in the hardware industry, communications standard unified. Advanced technique application has brought great help to the development of the Internet of things industry chain in areas such as health care. Construction, intelligent furniture also has a very good application prospect, but it wants to make the whole industry chain to achieve long-term stable development. Deployment, also need to pay attention to the local industry chain upfront costs control, information security and other issues.

References

1. Tang, J.: A brief analysis of 5G wireless communication technique and its significance to the industrial chain development of the Internet of things. China New Commun. **21**(18), 1–2 (2019). (in Chinese)
2. Lin, W.: Development of Internet of things industry based on 5G communication technique. Commun. World **26**(07), 108–109 (2019)
3. Ning, X.: Research on the industrial chain of internet of things for 5G communication technique. Digit. Commun. World (02), 125 (2019)
4. Gao, F., Zhang, X.: Research on 5G wireless communication technique and its impact on the industrial chain development of internet of things. Digit. Commun. World (10), 75 (2018)
5. Lu, F.: Research on the industrial chain development of internet of things from the perspective of 5G communication technique development. China's Strateg. Emerg. Ind. (04), 49 (2018). (in Chinese)

Path Choice of Smart City Construction from the Perspective of Economic Growth

Lingyu Chi[✉]

Department of Economics, Shanghai University, Shanghai, China
1105192241@qq.com

Abstract. As an important part of economic development, urban construction can't be ignored in economic development. Technology plays an important role in the construction of smart cities. It is undeniable that under the technological innovation model of China, policies have a limited effect on economic development. How should China's future smart city economic development policy be formulated? By establishing a theoretical model, it is found that when technological innovation models include imitative innovation, policies have a limited effect on economic growth. The theoretical model shows that the horizontal effect of factor accumulation cannot be ignored in the policy formulation process, but the policy direction should be changed from the past "quantity" to "quality", that is, to improve the efficiency of capital and population utilization in cities.

Keywords: Smart city · Technological innovation · Urban construction · Economic growth

1 Introduction

Since reform and opening up, China's economy has developed rapidly in the past 40 years. Many literatures have found through research that factors such as technological innovation, factor accumulation, and financial development level have played an important role in economic growth. As an important part in economic development, the role of urban construction in economic development cannot be ignored. In the past, China's urbanization mainly developed by means of urban area expansion and spatial concentration of population and other factors. With the development of time, problems such as population expansion, traffic congestion, resource shortage, and environmental degradation have gradually emerged. It is of great practical significance to explore a new type of urban development model. In 2012, China began to explore a new type of urban development model of smart cities, and established the first batch of smart city pilots in 90 prefecture-level cities across the country. As a city development model in the new era, smart cities rely on emerging information science and technology such as the Internet of Things, cloud computing, the Internet, and big data to realize the interconnection and high integration of core systems of the city's main functions, thereby forming a highly integrated and intelligent Collaborative city management network and

J. MacIntyre et al. (Eds.): SPIoT 2020, AISC 1283, pp. 685–689, 2021.
https://doi.org/10.1007/978-3-030-62746-1_101

perception system. The shift in this model highlights the important role of technology in cities.

Should the development of cities be promoted by the growth rate of factor accumulation or should they be improved by technological innovation? From a theoretical point of view, the marginal return of factors is diminishing. Whether the accumulation of factors has maintained the trend of increasing marginal returns in the course of economic development in the past 40 years remains to be verified. On the other hand, as is said in the Solow model, the third source of economic growth, productivity growth has a certain driving force for income growth. Technology has played an important role in the development of China's smart cities, but China still has a certain gap compared with the developed country, the United States.

With the intensification of international competition, a country's scientific and technological strength and innovation ability have played an increasingly important role in national competitiveness. Science and technology are of great significance to the construction of smart cities. The development of China's information security technology industry is a necessary condition for building a smart city security system, but the core technologies used by China's smart cities are facing non-autonomous introductions. The support of innovation policy emphasizes the importance of independent innovation, but at the same time, it is undeniable that imitative innovation accounts for a proportion of technology. At present, the research on the construction of smart cities is mostly focused on the development of technology, such as big data construction, blockchain construction, this paper establishes theoretical models and further introduces different ways of technological innovation in China and the United States, however, the theory indicates that we cannot emphasize too much the development of technological innovation and ignore factors accumulation.

2   Literature Review

At present, there are many studies on the construction of smart cities, and some scholars have made suggestions from the perspective of transportation in smart cities. Parallel trend of automotive automation and electrification represents smart travel in smart cities, the most important part of infrastructure construction is how to fill the gap in charging capacity for more and more electric vehicles [1]. Chen and other scholars analyzed traffic data throughout the life cycle, combined with the collection, analysis, discovery, and application of traffic data, using big data technology to guide urban transportation planning, construction, management, operation and decision support [2]. Some scholars started with information construction, the structural composition and functional data flow of the blockchain smart city information resource sharing and exchange model, the role of intelligent big data platforms is very important to the construction of smart cities [3]. Knowledge also played an important role. It is the key component of a smart vision is to include some perspectives while excluding others, further research should focus on political and controversial construction and the use of knowledge in the vision process [4]. Research on the overall framework of smart cities is another perspective, a common, shareable and integrated conceptual framework is proposed, a unified portal platform that can balance multiple stakeholders, including the government, citizens and businesses, as well as for common, custom and other application modes [5].

This article proposes smart city construction from the perspective of technological innovation. Many factors are involved in the process of technological innovation, and different forms of technological innovation will be generated based on different perspectives. This article mainly divides technological innovation into independent innovation and imitation innovation from the perspective of obtaining technology sources. Such a combination of technological innovations can most effectively achieve the goal of technological progress, in order to study the impact of technology sources on its output in the construction of intelligent cities in China.

Imitative innovation is important for the development of late-developing countries, but when the technology gap narrows, a country's technological foundation strengthens, and the cost of imitative innovation abroad increases, and imitative innovation faces diminishing scale benefits—less technological opportunities and increased imitative costs [6, 7]. At this time, we should not continue to adopt imitative innovation to promote technological progress. Although imitative innovation is effective in catching up in the short-term, it is not conducive to long-term economic growth. Only independent innovation is the driving force to support long-term economic growth [8]. Independent innovation first appeared in the endogenous growth model, with continuous improvement by Romer, Grossman and Helpman, and Aghion and Howitt [9–11]. At present, most scholars agree that independent innovation is conducive to economic growth, and R&D plays a role in promoting total factor productivity [12]. The technological innovation methods of independent innovation are very important for the development of China's economy. Imitative innovation can only play a promotion role when it is far from the frontier of technology. Based on this, a theoretical model is established to explain the uses of technology to urban development and construction.

3 Model

3.1 The Technological Progress Model of the US

This article builds a simple technology selection model based on the neoclassical development theory to compare the urban development levels of China and the US. It is assumed that the construction of intelligent cities in the United States is a model of independent innovation to promote technological progress. The only input factor of production is labor.

Suppose L_Y is the number of laborers participating in the urban production hypothesis, and L_A is the number of laborers participating in the technological innovation required for urban development, $L = L_Y + L_A$. Let γ_A be the proportion of labor force engaged in R & D, $\gamma_A = \frac{L_A}{L}$. The city's total output is equal to the number of laborers participating in output production multiply the productivity level, that is, $Y = A1 - \gamma_A L$. The form of its labor average is $y = A(1 - \gamma_A)$.

Assuming that the rate of technological progress is a function of the number of laborers involved in R&D, that is, $\widehat{A} = L_A/\mu$. $\widehat{A}$ represents the growth rate of productivity, L_A represents the number of laborers involved in research and development, μ represents the price of a new invention, expressed in units of labor. This formula shows that the growth rate of productivity is directly proportional to the number of laborers involved in research and development, and inversely proportional to the cost of a new invention.

As long as γ_A is constant, the average labor output level y is proportional to the technological level A. Increasing the proportion of the population participating in research and development will increase the output growth rate. If the cost of a new invention is low, economic growth will be faster. If a city puts more resources into research and development, in the short term, the city's output will decline, but it will be conducive to long-term growth, because the output will eventually recover, and it will exceed the level that γ_A can reach without change. It can be seen that for cities with technological progress driven by independent innovation, technological progress has a growth effect.

3.2 The Technological Progress Model of China

Now consider a model that adds imitative innovation, that is, a model of technological progress in Chinese cities. If the total labor force of the cities in the two countries is the same but the technical level is not the same, it is recorded as A_1 and A_2, respectively. Cities in China and the United States both obtain new technology through two ways, creating new technology or imitating technology. The variable A indicates the level of technology. Therefore, the A value of the United States will be higher than China. We assume that the proportion of the labor force engaged in R & D in the United States is higher than that in China, $\gamma_{A,1} > \gamma_{A,2}$. At the same time, suppose that the two countries have labors of the same size. Suppose μ_i represents the cost of the invention. Now we Focus on the situation in China.

Let μ_c be the cost of China's acquisition of a new technology through imitation. A key assumption is that as the technology gap between China and the US widens, the cost of imitation technology in China will decrease. Because there are more technologies that are easy to replicate and easy to imitate. U.S. and China technology growth rates are as follows:

$$\widehat{A_1} = \frac{L_A}{\mu} = \frac{\gamma_{A,1} * L}{\mu_i} \tag{1}$$

$$\widehat{A_2} = \frac{L_A}{\mu} = \frac{\gamma_{A,2} * L}{\mu_c} \tag{2}$$

In a steady state, both countries will grow at the same rate. If China raises $\gamma_{A,2}$, the immediate effect is to reduce China's output level, because the proportion of labor force participating in general labor is small. However, the increase in R & D investment will lead to a faster growth of A_2, so that y_2 will grow faster. When China's technological level approaches the level of the United States, the high growth of China is temporary. Once the A_1/A_2 ratio reaches a new steady state level, China's growth rate will return to that $\gamma_{A,2}$ level before change, the change in technology at this time has only horizontal effects. Changes in "policy" will result in temporary changes in output growth rates. Although the output level is different from what should be achieved when the policy is not changed, the output growth rate will eventually return to the level before the policy change.

4   Conclusion

How should China's future smart city economic development policy be formulated? The policy of changing the steady state growth rate of per capita income has a growth effect, so the progress of technology and efficiency has a growth effect; and the progress of factor accumulation will only increase the level of per capita income, only has the horizontal effects and has no growth effect. The conclusion of the model proves that technological progress will only bring horizontal effects to China, and eventually the output growth rate will return to the original level. The long-term equilibrium of a growth model is an endless result, but this does not mean that it should be the core of policy analysis. For policy purposes, long-term results are relatively insignificant, as the rate of convergence in long-term equilibrium can be extremely slow. In the policy formulation process, the horizontal effect of factor accumulation cannot be ignored, but the policy direction should be changed from the past "quantity" to "quality", that is, to improve the efficiency of capital and population utilization in cities.

References

1. Zhao, D., Thakur, N., Chen, J.: Optimal design of energy storage system to buffer charging infrastructure in smart cities. J. Manag. Eng. 36(2), 04019048 (2019)
2. Chen, X., Wang, H.H., Tian, B.: Visualization model of big data based on self-organizing feature map neural network and graphic theory for smart cities. Clust. Comput. 22(6), 13293–13305 (2018)
3. Lulu, L.: Research on the application of blockchain technology in the field of bidding. Eng. Econ. 28(11), 68–72 (2018). (in chinese)
4. Hoop, E.D., Oers, L.V., Becker, S., et al.: Smart as a global vision? Exploring smart in local district development projects. Archit. Cult. 7(2), 1–19 (2019)
5. Li, C., Liu, X., Dai, Z., Zhao, Z.: Smart city: a shareable framework and its applications in China. Sustainability 11(16), 4346 (2019)
6. Barro, R.J., Sala-I-Martin, X.: Technological diffusion, convergence, and growth. J. Econ. Growth 2(1), 1–26 (1997)
7. Perez-Sebastian, F.: Public support to innovation and imitation in a non-scale growth model. J. Econ. Dyn. Control 31(12), 3791–3821 (2007)
8. Aghion, P., Harris, C., Vickers, H.J.: Competition, imitation and growth with step-by-step innovation. Rev. Econ. Stud. 68(3), 467–492 (2001)
9. Romer, P.M.: Increasing returns and long-run growth. Journal of Political Economy 94(5), 1002–1037 (1986)
10. Grossman, G.M., Helpman, E.: Quality ladders in the theory of growth. Rev. Econ. Stud. 58(1), 43–61 (1991)
11. Aghion, P., Howitt, P.: A model of growth through creative destruction. Econometrica 60(2), 323 (1992)
12. Yan, O., Lingxiao, T.: Economic analysis of the innovation path of big countries. Econ. Res. 52(09), 11–23 (2017)

Computer Audit Quality Control in Big Data Era

Lingling Jiang[(⊠)]

School of Accounting, Harbin University of Commerce, Harbin, China
jyn7777@sina.com

Abstract. In the process of the rapid development of the Internet, the degree of informatization has continued to deepen, and it is compatible with the arrival of the era of big data. In the era of big data, some problems can be found in time by using computer networks and data collection and processing. At this stage, the use of computer-based audit quality control analysis can quickly find some omissions in the audit process, and at the same time can efficiently and targetedly resolve existing problems, so computer audits are widely used. The use and achieved good results, the importance of computer audit quality control analysis in the era of big data is more prominent.

Keywords: Big data era · Audit · Control analysis

The rapid development of the Internet not only promotes the rapid development of the real economy and the virtual economy, but also deeply affects people's lifestyles and working methods. It has produced huge changes in different industries and fields. An era of big data the real arrival. In the process of adapting to social changes, computer audits have gradually changed and adapted to economic and social changes, so as to better achieve and achieve the purpose of auditing, and also have new requirements for the quality control analysis of computer audits. Through the quality control and analysis of computer audits, we continuously solve problems encountered in the real world, and verify the reliability of computer audits in a timely manner. In addition to solving problems, we continuously improve the quality of computer audits. Continuously improve the quality of computer audits.

1   Impact of the Big Data Era on Computer Audit

1.1   The Impact of Computer Audit Methods

From a macro perspective, auditing is of irreplaceable importance in the operation of the national economy. It is related to the long-term stability and prosperity of the national economy. The purpose is to discover problems in economic development and solve existing problems in a timely manner. In the micro aspect, enterprises and other legal persons Organizations will also set up corresponding audit institutions and departments, whose purpose is also to find problems within organizations such as enterprises, improve the management and operation level of enterprises, and achieve long-term development

J. MacIntyre et al. (Eds.): SPIoT 2020, AISC 1283, pp. 690–694, 2021.
https://doi.org/10.1007/978-3-030-62746-1_102

of enterprises. Therefore, the importance of audit is much higher than that of simple ones. Literally, it can be said that the level of auditing directly or indirectly determines the development status of the country's economy and enterprises [1–4].

With the popularization of computers, the Internet has become a new driving force for economic development, which also directly affects the conversion of auditing methods. Traditional auditing methods are naturally based on manpower. Whether it is the audit department of a country or an enterprise, it mainly depends on Specialized auditors perform daily audit work, but in the era of big data, the rapid development of computers has naturally produced computer audit methods. This method not only greatly liberates manpower, but also significantly improves audit efficiency. Normative and professional aspects have also played a significant role in driving and promoting.

1.2 The Use of Big Data Helps Reduce Auditor Fraud

In the past manpower-based auditing, people undoubtedly played a decisive role. Once human factors dominated, in some cases, they often have some negative impacts. The most direct phenomenon is such as auditor fraud. The traditional audit method is a relatively prominent issue. Because in the traditional audit process, all aspects and aspects are completely completed by people, it is easy to produce some fraud. In the era of big data, computer audit, The fraud of auditors has been greatly reduced, because in the era of big data, it can more accurately reflect the existing problems. At the same time, with the help of computers, there is a special audit system that eliminates a large number of human factors and greatly reduces human intervention. Machine-based objective things are the main body. Just as in traditional auditing methods, if several criminals collude with each other, it is easy to reduce the audit fairness or even unfairness. In the era of big data, data is There is a real objective existence, and simply changing the data will cause a chain reaction, which naturally reduces the possibility of fraud by auditors [5–8].

2 Problems in Computer Audit in the Era of Big Data

2.1 Audit Technical Issues

Computer auditing in the era of big data is an important change to the traditional auditing method. It transforms the manual auditing method into a computer data-based auditing, with technology as the carrier and support, bringing convenience. At the same time, the technical issues of auditing have gradually become more important and prominent.

Behind the rapid development of big data is the support and guarantee of modern science and technology. Therefore, the most noteworthy and preventive problem of computer auditing in this era is of course technical issues. While informationization brings convenience and efficiency, auditing China must ensure reliability, precisely because of informatization, if there is a problem with the reliability of auditing, the impact will be more serious. In the big data period, auditing accounting data is mainly stored on disks and other media, compared to traditional the paper storage method, technology is its support, and the technical requirements are high. It can be said that technical problems are the main sticking point of computer audit in the era of big data. If the technical problems of audit can be completely resolved and guaranteed, there is no doubt that computers there will also be a good guarantee of the quality of the audit.

2.2   Audit Method Issues

In the era of big data, computer audit methods have naturally changed. Although not abandoning manual methods completely, most of the work is done by computers. At the same time, the original manual recording of data is transformed into computer input data, and then the data is then processed. In this process, the audit methods are still diversified, including not only submission audits, roving audits, and joint audits.

At a deep level, the application of big data on the one hand brings convenience and efficiency to auditing, but it also causes some potential risks. With the help of computers, multiple auditing methods are more convenient, especially for roving auditing. Auditors only need to bring a computer to complete most of the audit work. However, at the same time, it will also challenge the standardization and institutionalization of audits, which will have a negative impact on roving audits. Over-reliance on computers and big data will also cause lack of communication and coordination, easily lead to misunderstanding or some problems that should be avoided but not avoided. Although everything is mainly data in the era of big data, effective communication and coordination can only ensure Timeliness and validity of data [9, 10].

2.3   Auditor Issues

Computer auditing in the era of big data places higher requirements on the comprehensive quality of auditors. In traditional auditing methods, there is no requirement for the computer level of auditors, but in the extensive application of computer auditing, they are proficient in basic computer operations and the proficient application of office software is the basic requirement. It can be seen that auditors must keep pace with the times and continue to learn to adapt to changing auditing work.

However, at present, there are large differences in the level of auditors in different regions and different departments. In the era of big data, some auditors can adapt to economic and social changes in a timely manner, and continuously improve their professional level and professional ability. Facing the changes in the environment calmly, but in some areas with relatively low economic levels and older auditors, in the era of big data, the use of computers to complete audit work has shown some difficulties and inadequacy. The existing problems will undoubtedly have a direct impact on the audit work.

3   The Countermeasures of Computer Audit Quality Control in the Era of Big Data

3.1   Improving Auditing Technology

The key to the control of computer audit quality in the era of big data lies in the technical level. Therefore, it is necessary to improve the level of auditing technology. Technology is the support in the era of big data. The improvement of the technical level can be started from the following two aspects: First, we must continue to develop computers. To enhance the level of intelligence in the era of big data, the development and upgrading of the industry will inevitably drive the development of technology. Therefore, in terms of policies, we must vigorously support the development of the computer industry to achieve a higher level of intelligence in big data; secondly, we must To ensure the reliability of the computer, the control of computer audit quality must also be based on reliability, so the current main work for computers is to improve its performance and continuously improve reliability, and at the same time, do a good job of planning for special conditions. The era of big data is still relatively new and many experiences are relatively lacking, so it is necessary to formulate some emergency plans.

3.2 Innovative Audit Methods

In the context of big data, the existence of diversified auditing methods. In this case, it is necessary to inherit the traditional reasonable and efficient auditing methods, and at the same time, to innovate and continuously innovate auditing methods to adapt to the economy. Changes in society and changes in technology.

For the time being, the influence of new media is very significant. In the daily audit work, whether it is the audit work of state agencies or the audit work of enterprises and institutions, it is a very innovative way to use new media. This method can not only effectively carry out audit work, but also increase the public's understanding of audit work, thereby increasing their recognition of audit work, and effectively promote audit work in the future; at the same time, it must be fully utilized in the era of big data. The intelligent and convenient feature can completely replace the tedious work in the past with the computer. It must be daring to try, support and encourage creative thinking. In addition, it can also make full use of the Internet to enhance the interaction of audit work and make the public more Support and recognize audit work.

3.3 Enhance the Training of Audit Talents

We must effectively combine the background of the era of big data, do a good job in training talents, and effectively improve the quality of computer audits. To strengthen the training of talents, we must be problem-oriented, and we must dare to face up to the problems in the current computer audit work. Solve with firm confidence. For auditors with low computer skills, on the one hand, centralized training and learning can be organized by the corresponding institutions and departments, and at the same time, individuals must be encouraged to actively exercise and study in their spare time, and truly improve their computers. Level, dare to practice, practice first; on the other hand, theory guides practice, it is necessary to strengthen the study of theory, the training of talents cannot be separated from the guidance of theory, and a comprehensive study of various theoretical systems in the era of big data. Effective combination of practice and promotion of talent training.

4 Concluding Remarks

The analysis of computer audit quality in the era of big data is naturally very different from the traditional method. Under the new changes, it is necessary to clarify the impact of big data, dialectically look at the positive and negative aspects, and Dare to face and correct the problem, so as to better promote the development of the audit work, covering up the problem does not solve the problem. It is important to promptly and effectively propose coping strategies for the existing problems. The real solution is the key. When solving some problems, we must not only provide theoretical guidance but also insist on practice. Only by comprehensively considering these factors and properly handling them can we promote the improvement of computer audit quality in the era of big data.

References

1. Wang, Y.: Computer audit quality control in the era of big data. Audit Mon. **05** (2016)
2. Huang, Y.: Analysis of the impact of audit informationization on audit quality in the era of big data. Sci. Technol. Econ. Mark. **11** (2016)
3. Yue, Y.: Discussion on computer audit quality control in the era of big data. Taxation (2019)
4. Zhang, X.: Thoughts on using big data to promote computer audit. Financ. Econ. (Acad. Ed.) **09** (2018)
5. Tang, Y.: Research on computer audit based on information system. Popul. Stand. **12** (2019)
6. Xu, Y.: On the quality control of computer audit in the era of big data. Comput. Prod. Distrib. **11** (2018)
7. Song, Y.: Method of using big data to advance computer audit. Electron. Technol. Softw. Eng. **02** (2019)
8. Chen, H.: How to use big data to advance computer audits. Sci. Technol. Innov. Herald **01** (2018)
9. Chang, K.: Research on computer audit risk. Mod. Econ. Inf. **03** (2018)
10. Zhao, S.: Thoughts on computer audit talent training and teaching in the big data era. Knowl. Econ. **10** (2018)

MOOC System in the Era of Big Data Improves the Effectiveness of College Physical Education

Jianfeng Zhang[1] and Xuxiu Gao[2(✉)]

[1] Tianjin University of Technology and Education, Tianjin 300222, China
[2] Hebei University of Technology, Tianjin 300401, China
gaoxuxiu19800201@163.com

Abstract. Objective: To use MOOC system to better serve college physical education classroom teaching, effectively improve the effectiveness of teaching, and explore the application of online open physical education courses in college physical education. Methods: Literature, logic analysis, chart and comparison method were used. Results: The MOOC curriculum reform has promoted the fundamental transformation of teacher education mode and student learning mode, cultivated students' lifelong learning ability, and put school teaching on the track of quality education. Conclusion: The MOOC system is conducive to: (1) enriching on-campus course resources, saving teaching and learning time, and promoting the sharing of teaching resources; (2) reduce the cost of school teaching operation, improve the efficiency of physical education, and alleviate the problem of insufficient physical education teachers; (3) it can provide students with an efficient and convenient online learning environment, presenting a panoramic, cross-border and interactive online learning experience.

Keywords: Big data · MOOC system · College physical education

1 Introduction

At present, a new round of physical education curriculum reform is being gradually in the whole nation. Compared with the previous curriculum reform, the MOOC curriculum reform to promote the education mode of the teachers and students learning mode. Fundamental transformation, cultivate the students' lifelong learning ability, for students to further obtain knowledge, create conditions for development, make the school teaching work on the track of quality education. In the current physical education classroom teaching practice, teachers actively try to participate in the MOOC system teaching of new courses, reform learning methods, but in the actual teaching process, there are new problems. To effectively improve college physical education must promote the current development of students, and at the same time promote the coordinated development of students' knowledge, skills, process, methods, emotions, attitudes and values, so as to have an impact on the long-term development of students.

Since the second half of the 1990s, with the needs of the development of domestic reform, curriculum and teaching in our country theorists introduces constructivism

J. MacIntyre et al. (Eds.): SPIoT 2020, AISC 1283, pp. 695–699, 2021.
https://doi.org/10.1007/978-3-030-62746-1_103

psychology and postmodern curriculum theory. Extracting the effective teaching of new standards, such as "the teachers and students together to participate in the activities of innovation"; "equal dialogue between teachers and students", "the study background, and will be linked to teaching and students' real life, create the significance of learning" [1]. With the further development of quality-oriented education, many schools have realized the importance of learning methods, carried out subject experiments on the guidance of learning methods, and achieved results. For example, the experiment on guiding the study method carried out by LiuYang GuanQiao middle school takes the basic task of research as "to find out the rules of students' learning, teach students to master learning methods, develop learning ability, and improve learning efficiency and quality". In this context, the concept of "effective teaching" is proposed. There are some influential western teaching theory and model, such as Dewey child centered, experience restructuring essence, for the teaching activity and the practice as the basic teaching organization. Teaching of Bruner's structuralism practice teaching idea, the class of teaching optimization and bloom of target classification, we can found that the teaching effect of exploration. Further studied effective teaching, and he believed that effective teaching should ensure students' interest, motivation and autonomy. At the same time, many experts and scholars have also carried out researches on the guidance of students to strengthen learning methods. For example, in his "new teaching system", educator Zankov took "making students understand the learning process" as one of the five teaching principles. In 1976, j. Fraser proposed metacognitive theory in cognitive development, emphasizing the important role of individual self-awareness and self-regulation in learning; Krafky proposed "model teaching" [2], emphasizing teaching for transfer, etc.

By paying attention to the exploration of the guidance of MOOC system teaching methods, improving the teaching effect, and training teachers according to the characteristics of effective teaching behaviors. Teachers can be promoted to return to the state of "research", promote teachers to form reflective consciousness, and achieve the goal of changing teachers' teaching methods.

2   Research Objects and Methods

2.1   Research Objects

MOOC system classroom teaching is the main channel of school education and teaching, the main way to promote the development of students, and also the main object of this paper. MOOC are short for Massive open online course. "M" is Massive, the middle two "O" are Open and Online, and "C" stands for Course [3]. See Fig. 1 for the schematic diagram of MOOC system decomposition.

2.2   Research Methods

This study mainly adopts the methods of literature, logical analysis, chart, comparative classification and so on.

Fig. 1. The schematic diagram of MOOC system decomposition

3 Characteristics of MOOC System

1. Multimedia Enters the Campus and Establishes An Online and Offline Platform for Teacher-Student Interaction, Which Provides Great Convenience for Teacher-Student Interaction [4].
2. The deep integration of information technology and physical education is realized.
3. Create "flipped" teaching mode online and offline.
4. The combination with information industry technology enables all kinds of teaching information to be exchanged and Shared efficiently. Moreover, combining with the practical teaching and operation of physical education classes, the cultivation of learning autonomy can be strengthened in teaching [5].

4 Disadvantages of Traditional College Physical Education

1. Teaching objectives are not clear, and the purpose of each link is not strong. The randomness of teaching is great.
2. Focus on results and ignore the teaching process.
3. Teaching methods blindly advocate innovation, and blind and ineffective activities occupy too much classroom time.
4. The limited time for exercises in class, the wide range of students' interests and the lack of independent exercises after class all seriously affect students' interest and enthusiasm in learning sports [6].
5. There are limited available sports venues, few related equipment, and few sports items for students to choose [7].
6. The teacher did not make adequate preparation for class and failed to solve problems in class.

5 Research Major Ideas and Innovations

In the era of MOOC, the current teaching mode that relies on teachers' teaching and students' passive acceptance should be changed. Teachers should be changed from teachers to mentors and promoters to highlight students' dominant position and truly teach students according to their aptitude [8]. Teachers should strengthen the guidance of learning methods and the effectiveness of teaching.

The development and progress of students is an important index to measure the teaching effect. It has three meanings: First is "effective". It is the evaluation of the consistency between the results of teaching activities and the expected teaching objectives. Second, "efficiency", that is, unit time teaching effect is good; the third is "utility", that is, the effective realization of three-dimensional goals.

6   The Practical Significance of Using MOOC System to Improve the Effectiveness of College Physical Education

Effectiveness of MOOC system classroom teaching refers to whether students have made real progress and development in physical health, motor skills, sports participation, mental health and social adaptation after receiving physical education.

6.1   Optimizing the Classroom Structure is the Key to Improve the Teaching Effect

The structure of physical education teaching should be arranged reasonably according to different tasks, teaching materials, environment, students' actual conditions and the changing law of human activity, so as to be grasped and applied flexibly in practice.

6.2   To Improve the Effectiveness of Physical Education Classroom Teaching from the Perspective of Students' Psychology

Physical education teaching arrangements to consider students, so that students quickly understand the learning content. Teachers should step into students' roles, understand their difficulties, and help them make the transition from difficult to easy.

6.3   Teaching Methods Should be Effective

To ensure that it can stimulate students' interest in learning, broaden their horizons, and enable students to connect with the society, so as to lay a solid foundation for students to enter the society in the future [9].

6.4   Strengthen the use of Effective Practices and Feedback

Effective practice, on the one hand, can better complete the teaching task; On the other hand, it is also an important means for students to consolidate their knowledge and improve their skills. In the classroom teaching must carry on the practice, simultaneously must have the methodology practice.

6.5 The Teaching Process Should be Interactive

Through the communication, feedback, re-communication and re-feedback between students and teachers, the problems existing in the teaching process are actively solved, so that the teaching objectives and teaching contents can form a dynamic development model [10].

6.6 Scientific Classroom Teaching Evaluation is the Guarantee to Improve the Teaching Effect

Classroom teaching evaluation is an operational activity. Physical education classroom teaching evaluation mainly adopts the method of qualitative analysis and quantitative analysis.

References

1. Huang, C.: Research on curriculum design of critical thinking in universities from the perspective of general education curriculum. Shanghai Normal University, May 2019
2. Yang, L.: The integration of information technology and mathematics curriculum in the cultivation of metacognitive ability of middle school students – taking the sixth grade of primary school as an example. China West Normal University, May 2019
3. Tian, Y.: The application of MOOC + flip in college physical education teaching under the background of Internet + Research on classroom teaching mode. Sci. Technol. Inf. **34** (2019)
4. Gong, X.: Discussion on the reform of college physical education curriculum model under the background of "Internet+". J. Chengdu Norm. Univ. **9** (2019)
5. Tan, D.: Discussion on flipped classroom model of college physical education under the background of "Internet+" era. J. Jilin Radio TV Univ. **7** (2019)
6. Ren, J.: Analysis on the application of micro-course teaching mode in college physical education class. Martial Arts Res. **9** (2018)
7. Zhang, J.: An analysis on the application of micro-class teaching mode in college physical education class. Theoret. Innov. **1** (2019)
8. Wang, L.: Discussion on P.E. teaching mode in Universities in MOOC era. Comput. Inf. Technol. **2** (2018)
9. Lv, S.: Analysis of college physical education teaching mode under the background of Internet+. Sports Front. **11** (2018)
10. Dai, P.: Research on flipped classroom teaching model based on MOOC. J. Zhoukou Norm. Univ. **3** (2018)

Risks and Prevention in the Application of AI

Hongzhen Lin[1] and Wei Liu[2($\boxtimes$)]

[1] School of Hengda Management, Wuhan University of Science and Technology, Wuhan, China
[2] Subject Construction Office, Zhongnan University of Economics and Law, Wuhan, China
liuewei@foxmail.com

Abstract. Aim of this study is to explore the risks in the design and application of artificial intelligence and put forward corresponding preventive measures. Using the method of literature research deeply, correctly understand and analyze the problem of the risks in the design and application of artificial intelligence. AI business applications bring convenience to human, but also bring a lot of risk problems to social development. There are risks such as security issues, consumer privacy issues, moral hazard and so on. The causes of risks in AI design and application are incomplete laws and regulations and imperfect the ways of supervision and relief and limitations of artificial intelligence technology and lack of new ethics and morality matching artificial intelligence. The conclusion is that effective risk prevention measures must be taken to prevent risks in time. Safety designs should be strengthened to reduce risk from the source. The supervision and management system of artificial intelligence must be improved to curtail moral hazard.

Keywords: Artificial intelligence · Design and application · Risk · Security

1 Introduction

As a new technology, artificial intelligence technology has been applied in many fields. It not only greatly facilitates people's life, but also brings many risks to people's life, such as security risks, ethical risks and so on. Only by taking effective measures to ensure the safety of AI, can it bring benefits to human beings rather than harm [1]. At present, the research mainly focuses on technology and its application prospect. There are not many researches from the perspective of the application risk of artificial intelligence technology, which is the theme of this paper. On the basis of previous studies, this paper mainly studies from the perspective of risk, aiming at the existing risks, through the dual role of technology and system, puts forward the corresponding preventive measures by identifying the risks in the AI application and taking measures to prevent them, we can use AI technology to better serve the public.

2 Main Risks of AI Design and Application

2.1 Security Risks Caused by Artificial Intelligence

Risk has become the hallmark of our times and the all inclusive background of understanding the world. The fundamental choice of risk society lies in risk governance. We

J. MacIntyre et al. (Eds.): SPIoT 2020, AISC 1283, pp. 700–704, 2021.
https://doi.org/10.1007/978-3-030-62746-1_104

should bring risk governance into the agenda as soon as possible, cultivate and establish the ideas, systems, mechanisms, methods and guarantee conditions of risk governance in the government and the whole society, form the consensus and joint forces of risk governance, and finally achieve good risk governance [2]. With the development of computer network and the popularization of Internet, artificial intelligence develops in many fields. The rapid development of artificial intelligence not only promotes the economic growth, but also profoundly changes the way of life and causes security issues [3]. Study the security of artificial intelligence and how to control the development of artificial intelligence to make it beneficial to human society and avoid harm to human beings [4].

2.2 Risk of Consumer Privacy Disclosure

Artificial intelligence technology has been widely used in various industries. Artificial intelligence has significant personification characteristics, which leads to the identification and undertaking of civil liability of artificial intelligence products, such as intelligent robots and autopilot cars. It is necessary to establish special ethical standards, registration publicity system and recall disposal mechanism, establish the principle of responsibility fixation for the difference among producers, sellers and users, and emphasize the burden of proof of producers, and record and save the electronic data of the cause and process of infringement under the premise of protecting the right of privacy [5]. Businesses have enough big data and deep learning algorithm. They will draw the portraits of acquaintances through deep learning algorithm according to the big data information of acquaintances.

2.3 Moral Hazard in AI Design and Application

In terms of enterprises, capital is profit-making. Interests can be divided into legitimate interests and illegitimate interests. It is legal for market agents to develop artificial intelligence technology in pursuit of legitimate interests, which should be protected, so as to encourage people to continuously develop new technologies and promote social development. However, there are some market players who use artificial intelligence technology to pursue improper economic benefits, which is the source of moral hazard. In fact, the development of artificial intelligence depends on the capital investment of large enterprises and capitalists to a great extent. In the process of the development of artificial intelligence, it may pay more attention to economic interests than social interests.

3 The Causes of Risks in AI Design and Application

3.1 Incomplete Laws and Regulations

The future world is a world of science and technology competition. If you want to be at the forefront of science and technology, the key lies in who first master the AI technology and strengthen relevant legislation. At the legislative level, although China has introduced laws and regulations related to artificial intelligence, it still fails to form a systematic legal provisions. On the one hand, it is not clear which subject should bear the tort liability of AI. Producers, users, owners and even artificial intelligence itself may become the subject of tort liability, and the academic community has been exploring to give artificial intelligence the qualification of legal subject to bear tort liability. The imperfection of the legal system leads to the conflict of liability of AI tort. On the other hand, the definition of privacy right and personal information right is unclear.

3.2 Imperfect the Ways of Supervision and Relief

In addition to legislation, the development of artificial intelligence needs an effective supervision and management system, and administrative supervision is very important. After the infringement, the administrative organ shall supervise, manage and punish the enterprises or institutions that may have the risk of infringing the right of privacy. As far as administrative supervision is concerned, it is easy to lead to ineffective supervision due to the lack of clear legal provisions. Due to the unclear legal provisions of the regulatory authorities, if there is no clear division of responsibilities between different departments, it is easy to shift responsibilities, affect the effectiveness and efficiency of supervision, and damage the interests of enterprises, especially citizens. The absence and dispersion of the risk subject of artificial intelligence require the establishment and improvement of the good governance mechanism with the responsibility in place, joint and several, and bottom-up. At the same time of the development of science and technology, keep up with the pace of science and technology, and establish a reasonable and legal supervision and management system.

3.3 Limitations of Artificial Intelligence Technology Itself

Technology itself risk is the most important reason in the risk society, artificial intelligence technology as a new technology, because of its particularity will bring huge risks to the society. With the development of science and technology, its progress is beyond imagination. Any technology has its limitations and technical loopholes. The technical loopholes of network and artificial intelligence are still common. The risk of AI technology is an important embodiment of modern risk society. The risk of AI technology has its own characteristics, which will bring serious consequences [6]. The uncertainty of AI bring risks. Artificial intelligence technology, for people in modern society, is not only limited, but also uncertain and unknowable. For hundreds of years, people have been striving for science and technology.

4   Countermeasures of Risk Prevention

4.1   Perfecting Relevant Legislation

In the era of artificial intelligence, the protection of ethics by law is still inseparable from the coordinated protection of legislation, law enforcement and justice. From the perspective of human being as a whole, the principle of subjectivity requires human beings to be able to control robots well, which is also the premise of realizing the principle of security. Relevant legislation should embody the principle of subjectivity and constructiveness [7]. In legislation, we should implement the principle of interest balance and give consideration to the interests of different subjects. In legislation, the main responsibility should be clear, and the general responsibility of data controller and processor in data protection should be clear.

4.2 Improving the Supervision and Management System of Artificial Intelligence

Any new science and technology needs a set of matching supervision and management system from its initial stage to its mature development. In recent years, with the rapid development of artificial intelligence, its application field covers every corner of people life. What we can do is to speed up the development of artificial intelligence technology standards and effective supervision and management, reduce or reduce the potential danger of artificial intelligence. Artificial intelligence is in urgent need of good risk governance.

The technical attribute, development attribute, social attribute and governance attribute of AI require the implementation of the good governance policy of intrinsic safety, development priority, fairness and fairness, and overall consideration [8]. The government tends to be dominant in the supervision and management system of artificial intelligence. We should fully understand and utilize the benefits of social development and progress brought by robots, and at the same time comprehensively evaluate the social risks that may be generated by robot technology [9].

4.3 Strengthening Safety Design to Reduce Risk from the Source

In the design and development of artificial intelligence products, it is necessary to require the designers and manufacturers of artificial intelligence to have correct value orientation, so as to avoid developing intelligent products that do not meet the security requirements. The principle of active protection and prevention should be adhered to in the design of technical safety. Enterprises are required to form a set of security protection mechanism in advance, and designers and operators are required to maximize the protection of users; security by putting the interests of users first.

The existence of AI ethics can provide a mechanism to protect Privacy [10]. Privacy protection is embedded in the design. That is to say, privacy protection mechanism is required to be a part of the operation structure of information system and business practice rather than an additional part [11]. In addition, cultivate the moral consciousness and social responsibility consciousness of scientific and technological personnel through moral education. Scientific and technological personnel shall abide by scientific and technological ethics and professional ethics, and fulfill the "notification and prevention obligations" [12].

5   Conclusion

Science and technology as the first productivity of human beings, every major technology will affect and change the way of production and life of human beings. The development of artificial intelligence technology has become the trend and direction of society and history. Artificial intelligence can be used not only to benefit the public, but also to bring many risk problems. In order to make AI better applied in business and improve our quality of life. We can only make science and technology better serve human beings if we can prevent troubles in advance, improve the guidance of new ethics and moral values, build a human environment for the development of artificial intelligence, improve the supervision and management mechanism and legal system, and realize the new ecology of the harmonious development of artificial intelligence in which human and computer coexist.

Acknowledgement. This research was supported by the National Social Science Foundation of China, Name of the project: "Entrepreneurial law education research based on risk control in science and Engineering University" (Grant No. BIA170192.).

References

1. Yong, D.Y.: Artificial intelligence security problems and their solutions. Philos. Trends **9**, 99–104 (2016). (in Chinese)
2. Zhang, C., Chen, Z., Xie, Y.: Risk society and risk governance. Teach. Res. **5**, 5–11 (2009). (in Chinese)
3. Xia, W.: Discussion on the security of artificial intelligence. Inf. Comput. (Theoret. Ed.) **407**, 131–132 (2018). (in Chinese)
4. Wang, M.: Artificial intelligence security problems and solutions. Sci. Technol. Wind **11**, 8 (2018). (in Chinese)
5. Zhang, T.: Research on civil liability for damage caused by artificial intelligence products. Soc. Sci. **4**, 103–112 (2018). (in Chinese)
6. Zhang, W.: The risk of artificial intelligence technology from the perspective of risk society. Manag. Technol. Small Medium-Sized Enterp. **563**, 97–98 (2019). (in Chinese)
7. Du, Y.: Outline of robot ethics research. Philos. Sci. Technol. Res. **35**, 110–115 (2018). (in Chinese)
8. Tang, J.: Research on the good governance of risk in artificial intelligence. China Adm. **406**, 48–54 (2019). (in Chinese)
9. Du, Y.: Research on moral responsibility in robot ethics. Sci. Res. **11**, 10–15 (2017). (in Chinese)
10. Guan, Z.: Reasons for the existence of science and technology ethics. Sci. Technol. Guide **24**, 87 (2014). (in Chinese)
11. Hua, J.: Research on legal practice, technical support and commercial application of the mechanism of "protecting privacy through design". Intell. J. **2**, 116–122 (2019). (in Chinese)
12. Wang, Y.: Moral judgment and ethical suggestions of artificial intelligence. J. Nanjing Norm. Univ. (Soc. Sci. Ed.) **4**, 29–36 (2018). (in Chinese)

Improvement of Microblog Recommendation System Based on Interaction Strategies of Agricultural E-Commerce Enterprise

Xu Zu[✉], Yangyang Long, Renji Duan, and Qingxin Gou

Business School, Sichuan Agricultural University, Chengdu 611830, China
403008983@qq.com

Abstract. In the web 3.0 era, social media and e-commerce are converging. Microblog platform designs recommendation system (RS) based on consumer characteristics, which helps to promote marketing effect. Different from the existing RS research results, this paper improves the RS from a new perspective of interaction strategies of agricultural e-commerce enterprise. The aim is to find the high interaction effect from emotional tendencies in the microblog interaction strategies, so as to enhance the interaction of recommendation information and realize the user participation of enterprise microblog in the platform.

Keywords: Social media · Microblog interaction strategy · Recommendation system · Emotion tendency · Agricultural e-commerce enterprise

1 Introduction

Enterprises can apply recommendation system (RS) in many business applications to attain business profits. By combing the existing research results of RS, most of them focus on the prediction of RS application effect and the construction of RS in different situations [1].

Social media has become a major trend of the Internet and attracted extensive attention from various fields, especially business [2]. With the increasingly mature understanding of social media platforms, a large amount of enterprise generated content (EGC) has entered people's vision [3]. With the integration of e-commerce and social media, social media has accumulated a large amount of user characteristic data, which provides a reference for the social media platform to build a companies' RS. In China, Sina Weibo has also become an important social media platform. Excellent agricultural e-commerce companies such as three squirrels, BESTORE have emerged as the times require, which also caters to the trend of the future online consumer market. However, it is worth noting that the interaction effect of enterprise microblogs needs to be improved.

Traditional recommendation systems focus on improving accuracy of predictions, while existing scholars have understood that accuracy alone is not sufficient to make users happy [4]. In order to ensure the interesting recommendation and user participation, the

J. MacIntyre et al. (Eds.): SPIoT 2020, AISC 1283, pp. 705–709, 2021.
https://doi.org/10.1007/978-3-030-62746-1_105

study focuses on the microblogs of agricultural e-commerce enterprises with the potential of online shopping market based on Sina Weibo. This study classifies the text emotional tendencies of microblogs with high interaction effects, and compares the differences of microblog interaction strategies, so as to build recommendation models under different emotional tendencies and achieve high interaction effects of microblog recommendation content.

2 Literature Review

2.1 RS in Microblogs

One of the research results of RS in microblogs can be focused on behavior orientation. Behavioral orientation research can be further broken down into two specific categories: the study of user attention or friend behavior and the study of user behavior data, which is mainly the user's originality, reposting, commenting, collecting, likes and clicking data, etc. to reflect the user's interest attributes [5]. Taking behavior orientation as the starting point, this study will analyze the different effects of enterprise interaction strategies based on microblog user behavior data, and help the platform build and improve RS to enable better business-customer interaction.

2.2 Microblog Interaction Strategy

The division of microblog interaction strategies can be divided into social interaction and task-oriented interaction, which has been widely recognized [6, 7]. Yan & Chang further divided based on grounded theory. Among that, social interaction content was subdivided into three categories: general knowledge, professional knowledge, and emotional communication; task-oriented interaction was subdivided into product interaction, corporate image interaction, and co-creation activity [8]. Existing scholars also explore the interaction effects of microblogs, using likes, comment, or repost of each post [5]. Based on this, combined with the latest research results, social interaction and task-based interaction are regarded as the interaction strategies of sina Weibo platform in this study.

2.3 Microblog Emotional Research

The microblog information often conveys the author's emotional state or the tendency of emotional communication on a specific topic [8]. Microblog text can be used as an important corpus material, which is more conducive to emotional analysis. Smith and Petty argue that positive and negative content can gain attention and awareness [9]. Huffaker divides emotions into positive, negative, and neutral and confirms that emotional messages get more feedback [10]. This study will divide microblog emotions into three categories: positive, negative and neutral, and analyze the effects of interactive strategies combined with the emotions of the enterprise microblog contents.

3   Research Design

3.1   Sample Sources

This study takes five representative agricultural e-commerce companies in the Sina Weibo platform as the research objects. As shown in Table 1.

Table 1. Relevant information of five representative companies

Enterprise	Tmall fans (thousand)	Microblog fans (thousand)	Number of microblogs
Baicaowei	18,416	650	9,044
Three squirrels	24,863	810	13,147
Laiyifen	3,222	1,150	12,330
BESTORE	21,584	710	13,505
Wuhan Loulan	1,481	220	6,601

Note: The collection deadline of April 8, 2019.

3.2   Data Collection and Processing

The process of collecting and cleaning microblog data are as follows. First, The microblog data were mainly collected by Python 3.7. Next, the collected data were cleaned up. Finally, 47917 effective microblog data were identified. The process of data classification is as follows. First, this study used Python 3.7 to create machine language learning classifier, which classified the text content based on the microblog interaction strategy types. Then, ROSTCM6 was used to perform emotional recognition on the results of the word segmentation, and the results were verified by the results identified by Weka software.

4   Research Process

4.1   Data Descriptive Statistics

There are differences in the usage frequency of microblog interaction strategies for agricultural e-commerce companies. This study finds that professional knowledge rarely appears in agricultural e-commerce microblog interaction strategies. General knowledge (33.8%) in social interaction and co-creation activity (38.0%) in task-oriented interaction are the two main types. The proportion of emotional communication (6.5%) in social interaction and corporate image interaction (3.9%) in task-oriented interaction is low.

This study uses Boxplot of SPSS 24.0 to identify and eliminate singular values from repost, comment, and likes in all microblog interaction strategies. Then, one-way ANOVA is used to explore the mean difference of repost, comment or likes in different interaction strategies. Co-creation activity has the highest mean in repost (92.89), comment (81.60) or likes (69.38), and the mean in repost (55.81), comment (33.20) or likes (43.94) for product interaction ranks second. Characteristics of three variables are similar between corporate image interaction and general knowledge ($P > 0.05$), and the mean in repost (9.34), comment (21.49) or likes (28.07) for emotional communication is the lowest. In general, the interaction effect of task-oriented interaction strategy is more significant.

4.2 Emotion Analysis

This study selects general knowledge and product interaction and co-creation activity as samples of microblog emotion analysis. It can be seen that positive emotions account for the highest proportion of the three strategies, reaching 81.37%, indicating that most contents of enterprise microblogs show the positive side to consumers; in addition, negative and neutral emotions accounted for only 11.63% and 7%, respectively.

Then, this study uses one-way ANOVA to further explore. First, in positive emotion aspect, the mean in repost (105.28), comment (89.13) or likes (70.29) for co-creation activity ranks first. The mean in repost (63.02) or comment (35.77) for product interaction ranks second. The mean in repost (17.65) or comment (22.39) for general knowledge ranks third. To sum up, it can be seen that the interaction effect (repost, comment and likes) of co-creation activity is the best in positive emotion aspect. Second, it can be seen that the interaction effect (repost (29.95), comment (48.88) and likes (68.01)) of co-creation activity is the best in neutral emotion aspect. Third, in negative emotion aspect, the mean in repost (64.97), comment (59.37) or likes (64.29) for co-creation activity ranks first.

5 Conclusions

Enterprise microblog interaction strategies have different interactive effects. Taking agricultural e-commerce enterprises as an example, the specific conclusions are as follows. First, under positive or negative emotional tendency, it can be seen that good interaction effects are mainly focused on task-oriented interaction strategy, among which co-creation activity has the best interaction effect, followed by product interaction. However, under neutral emotional tendency, the co-creation activity in task-oriented interaction strategy still have good interaction effect, but product interaction and general knowledge have similar effects. Therefore, the improvement design of the microblog RS taking agricultural e-commerce enterprise as an example is shown in Fig. 1.

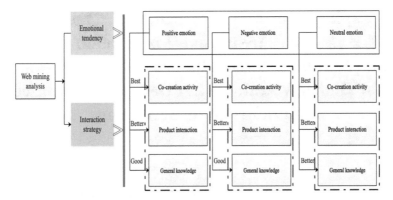

Fig. 1. RS framework - based on agricultural e-commerce enterprise

Acknowledgements. Sichuan Provincial Key Research Base of Philosophy and Social Sciences—Sichuan Center for Rural Development Research General Project (CR1910); Sichuan Provincial Key Research Base of Philosophy and Social Sciences—Research Center for Sichuan Liquor Industry Development General Project (CJY19-03); Sichuan Provincial Key Research Base of Philosophy and Social Sciences—Southwestern Poverty Reduction and Development Research Center Key Project (SCP1802); Sichuan Provincial Key Research Base of Philosophy and Social Sciences—Sichuan Agriculture Featured Brand Development and Communication Research Center General Project (CAB1810).

References

1. Alt, R., Demirkan, H., et al.: Smart services: the move to customer orientation. Electron. Mark. **29**(1), 1–6 (2019)
2. Guan, W., Gao, H., et al.: Analyzing user behavior of the micro-blogging website Sina Weibo during hot social events. Phys. A **395**(1), 340–351 (2014)
3. Zhang, J.: Voluntary information disclosure on social media. Decis. Support Syst. **73**(3), 28–36 (2015)
4. Zhang, Z., Zheng, X., et al.: A framework for diversifying recommendation lists by user interest expansion. Knowl. Based Syst. **105**(4), 83–95 (2016)
5. Kim, D.-H., Spiller, L., Hettche, M.: Analyzing media types and content orientations in Facebook for global brands. J. Res. Interact. Mark. **9**(2), 4–30 (2015)
6. Bass, B.M.: Social behavior and the orientation inventory: a review. Psychol. Bull. **68**(4), 260–292 (1967)
7. Köhler, C.F., et al.: Return on interactivity: the impact of online agents on newcomer adjustment. J. Mark. **75**(2), 93–108 (2011)
8. Yan, X., Chang, Y.: The impact of enterprise microblog interactive strategy on consumer brand relationship-based on the rooted analysis of Sina microblog. J. Mark. Sci. **9**(1), 62–78 (2013). (in Chinese)
9. Smith, S.M., Richard, E.: Petty: message framing and persuasion: a message processing analysis. Pers. Soc. Psychol. Bull. **22**(3), 257–268 (1996)
10. David, H.: Dimensions of leadership and social influence in online communities. Hum. Commun. Res. **36**(4), 593–617 (2010)

Exploration and Construction of "One Ring, Three Deductions" Innovation and Entrepreneurship Talent Cultivation Model for Higher Vocational Art Design Major Based on Information Technology

Bingjie Liu[✉] and Yawen Tang

Jiangxi V&T College of Communications, Nanchang, China
66863378@qq.com, 357219086@qq.com

Abstract. Modern information technology is widely used in art design majors of higher vocational education and plays an important role. This article takes the professional art design major as the research object, and takes the artistic design innovation and entrepreneurship talent cultivation mode as the starting point. The exploration and construction of the "one ring and three deductions" innovation and entrepreneurship talent cultivation model for higher vocational art design majors in information technology. First, through literature review, summarize domestic and foreign research on keywords such as information technology, innovation and entrepreneurship education, and art design talent training models, and then discuss the "one loop and three buckles" theory related to this article. The teaching methods have changed from simple explanations in the past to modern multimedia presentations using information technology. This has increased the interest of the classroom, stimulated students' interest in learning, and improved learning efficiency.

Keywords: Modern information technology · Higher vocational art design major · Innovation and entrepreneurship · Talent training model

1 Introduction

Higher vocational colleges are actively exploring the road of innovation and entrepreneurship education that suits them, and have theoretically and practically pushed the development of innovation and entrepreneurship education in colleges and universities, and have achieved some results and successful experiences. However, there are still many problems in the process of establishing a scientific, rational and distinctive talent training system for innovation and entrepreneurship. Difficulties have been encountered from theoretical exploration to practical operation. The value system deviates, the talent training model is single, the innovation and entrepreneurship education and professional

J. MacIntyre et al. (Eds.): SPIoT 2020, AISC 1283, pp. 710–714, 2021.
https://doi.org/10.1007/978-3-030-62746-1_106

education are separated from each other, and various problems such as the low interest of teachers and students in innovation and entrepreneurship courses have seriously hindered the development of innovation and entrepreneurship education in vocational colleges.

Modern information technology has a huge impact on education. It has changed the traditional teaching mode and improved the learning method [1]. By using information technology in teaching reasonably, it can make the abstract and difficult knowledge in the art design specialty more intuitive and understandable. To a large extent, this has transformed the teaching methods of teachers and improved the quality of professional teaching. The teaching of art design major is closely related to modern information technology [2, 3]. Modern information technology is an important way to learn and create art design. In recent years, it has been exploring the impact of information technology on the reform of professional teaching methods and seeking how to better use information technology. Ways to promote teaching. Modern information technology first had an impact on the reform of teaching concepts [4]. Modern education pays attention to the cultivation of students' comprehensive quality, especially the education of higher vocational colleges focuses more on cultivating the students' comprehensive quality ability to better meet the needs of social and economic development. In the era when information technology is widely used in universities, the traditional indoctrination and one-way teaching concept is no longer suitable for modern teaching [5].

This article takes the art design innovation and entrepreneurship talent training mode as the starting point, and based on the characteristics of the art design industry development, uses questionnaires and factor analysis to analyze the status quo, problems, attribution, and art of the cultivation of innovative entrepreneurship talents in art design majors in universities.

2 Method

2.1 Existing Vocational Education Talent Training Models

"School-enterprise cooperation order style" training refers to the employment contract signed between the company and the school according to job requirements [6, 7]. Training methods for education and educational activities. This training method tends to a kind of hiring training, which directly trains cooperative enterprises to train talents, making the training targeted and practical. The students' training goals and careers are clear, the teaching process and work process are precisely linked, the practice is guided, and the employment is guaranteed. This training method is a higher-standard school-enterprise cooperation method. Enterprises can directly participate in the design of the talent training process, which can not only improve the quality of talent training in schools, but also cultivate talents for themselves, greatly improving the enthusiasm of both sides of the school and enterprises, and the teachers and students of colleges and universities directly participate in corporate technology. In research and development and scientific research projects, it provides intellectual support for enterprise development [8]. At the same time, the "school-enterprise cooperation order-style" training also has higher requirements for both parties to the cooperation. On the one hand, schools must have sufficient education and teaching resources to meet the requirements of enterprises for the cultivation of required talents; on the other hand, enterprises also need to provide sufficient numbers of jobs, diverse job categories, attractive salary and development prospects to meet Student needs. This training method makes the two parties closer to each other and realizes the optimal allocation of resources [9, 10].

2.2 Innovation of Training Mode of Innovative and Entrepreneurial Talents in Higher Vocational Colleges

In vocational education, the cultivation and training of students' practical ability is mainly achieved through the combination of work and study. Therefore, when carrying out innovation and reform of the training mode of innovative entrepreneurial talents, it must also rely on the combination of work and study to achieve. On the one hand, in the practical teaching of students, not only must the development direction of the talent training platform and studio built under the combination of school and enterprise be changed, that is, the projects undertaken by these approaches must change the traditional direction of the past. To undertake more innovative projects. In this way, when teachers and students participate in these projects, they can not only enhance their professional practice abilities, but also understand the cutting-edge innovation development direction and requirements of the major. At the same time, in such a working and teaching environment, students will gradually get used to using creative thinking to solve work problems, opening up new paths in professional practice, and providing conditions for students' innovation and entrepreneurship.

3 Experiment

Step1: Enrich teaching content and innovate teaching methods. In terms of teaching content, corresponding innovation and entrepreneurship courses should be added to maximize the close integration of entrepreneurial theory and entrepreneurial practice. For example, the introduction of case teaching, classroom discussion teaching in theoretical teaching, or on-site simulation teaching, project training teaching, etc., to strengthen students' understanding and digestion of knowledge from multiple aspects. In practice teaching, we can cooperate with "elastic online teaching", "theory + practical project deduction teaching", etc. to realize the construction of online and offline entrepreneurial bases and platforms, so as to improve students' innovative and entrepreneurial practice and adaptability.

Step2: Combining regional development, highlighting founding characteristics. The important sign of the school running level and school running ability of higher vocational colleges is the distinctive school running characteristics. In view of the fact that China's regional economic development levels are not the same, regional economic structure is quite different, and other practical disadvantages, innovation and entrepreneurship education can be oriented to regional economic construction and social development, and truly highlight their own personality and characteristics.

Step3: Strengthen school-enterprise cooperation. As an effective model for cultivating talents in higher vocational education and teaching, the implementation of school-enterprise cooperation focuses on cultivating talents who are "adequate in theory, useful in technology, easy to use, and durable", and this has a lot in common with innovative entrepreneurial talents. This requires higher vocational colleges to integrate education and teaching with corporate training in the process of cultivating innovative and entrepreneurial talents, to achieve the unity of talents and social needs, to serve regional economic development, and to broaden the future of students.

4   Discuss

This article uses questionnaires to investigate the training of innovative and entrepreneurial talents in higher vocational art design majors. When asked "how is the current innovation ability", only 6.6% of students answered very well, 73.8% of students answered generally, 11.5% of students answered poorly, and 8.2% of students answered very poorly, as shown in Table 1. Therefore, in the recognition of innovation ability, most students' performance of innovation ability is average.

Table 1. Questionnaire for cognition of creative ability of higher vocational art and design majors

Evaluation	Well	General	Worse	Very poor
Frequency	12	135	21	15
Percentage (%)	6.6	73.8	11.5	8.2

When asked "what is innovation and entrepreneurship", 41.5% of students think that innovation and entrepreneurship is innovative to start a new business. For new projects, 20.8% of students think that innovation and entrepreneurship means that the nature of their work is innovative, as shown in Fig. 1. It can be seen that many students do not really understand the connotation of innovation and entrepreneurship and the content of innovation and entrepreneurship, and there is narrowness and one-sidedness.

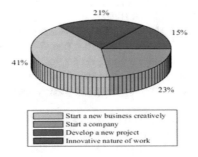

Fig. 1. Survey results of vocational art design major innovation and entrepreneurship cognition

According to the survey results, "insufficient entrepreneurial funds", "inadequate entrepreneurial ability, insufficient experience", and "inadequate entrepreneurial intention" have become the main factors affecting the innovation and entrepreneurship of art design students. Therefore, how to develop innovation and entrepreneurship education, how to cultivate the ability of innovation and entrepreneurship, how to solve the problem of venture capital difficulties, and so on have become the major breakthrough points in the training of innovative and entrepreneurial talents in higher vocational art design majors.

5   Conclusion

Higher vocational colleges are actively exploring the road of innovation and entrepreneurship education that suits them, and have strongly promoted the development of innovation and entrepreneurship education in colleges and universities, and have achieved some results and successful experiences. However, there are still many problems in the process of establishing a scientific, reasonable and distinctive talent training model for innovative entrepreneurship education. Difficulties have been encountered from theoretical exploration to practical operation.

Acknowledgements. Research topic on teaching reform of higher education in Jiangxi Province in 2019: exploration and construction of the "one ring and three deductions" creative and entrepreneurial talent training model (No. JXJG-19-53-11).

References

1. Schmitz, A., Urbano, D., Dandolini, G.A.: Innovation and entrepreneurship in the academic setting: a systematic literature review. Int. Entrep. Manag. J. **13**(2), 369–395 (2017)
2. Hatef, E., Sharfstein, J.M., Labrique, A.B.: Innovation and entrepreneurship: harnessing the public health skill set in a new era of health reforms and investment. J. Public Health Manag. Pract. JPHMP **24**(2), 99–101 (2017)
3. Roig-Tierno, N., Kraus, S., Cruz, S.: The relation between coopetition and innovation/entrepreneurship. RMS **12**(2), 379–383 (2017)
4. de Lurdes Calisto, M., Sarkar, S.: Innovation and corporate entrepreneurship in service businesses. Serv. Bus. **11**(3), 581–600 (2017)
5. Ebrahimi, P., Mirbargkar, S.M.: Green entrepreneurship and green innovation for SME development in market turbulence. Eurasian Bus. Rev. **7**(4), 1–26 (2017)
6. Da Wan, C., Morshidi, S.: International academics in Malaysian public universities: recruitment, integration, and retention. Asia Pac. Educ. Rev. **19**(3/4), 1–12 (2018)
7. Banbury, P.: Review managing information technology in secondary schools. Comput. Bull. **39**(6), 31 (2018)
8. Theis, T.N., Wong, H.-S.P.: The end of Moore's law: a new beginning for information technology. Comput. Sci. Eng. **19**(2), 41–50 (2017)
9. Purwanto, M.I., Wibowo, F.W.: Analysis of XYZ information technology university employees welfare in Islamic perspective. Adv. Sci. Lett. **23**(5), 4981–4985 (2017)
10. Fadel, S., Rajab, H.: Investigating the English language needs of the female students at the faculty of computing and information technology at King Abdulaziz University in Saudi Arabia. Engl. Lang. Teach. **10**(6), 69 (2017)

Research Progress of Neuroimaging Techniques in Organizational Behavior Under the Background of Smart City

Shimin An and Zhen Zhang[✉]

Department of Economics and Management, Lanzhou University of Technology, Lanzhou 730050, Gansu, China
13818702070@163.com

Abstract. The rapid development of information technology will not only affect social life, business operation and other aspects, but also have an important impact on the academic research field of theoretical progress, method innovation and other aspects. This paper discusses the application value of neuroscience technologies in the development of organizational behavior in China, and introduces the commonly used neuroimaging techniques in foreign countries. This paper proposes that in order to further carry out theoretical innovation, knowledge reengineering and enterprise management model innovation, neuroimaging techniques can be used as an important supplementary means to expand the toolbox of traditional organizational behavior measurement methods and play an important role in theoretical progress and innovation as well as management model reform.

Keywords: Smart city · Neuroimaging techniques · Organizational behavior · Organizational Neuroscience

1 The Origin, Unique Role and Development of Organizational Behavior are Studied from the Perspective of Neuroscience

With the development of information technology, such as Internet of things, cloud computing and mobile broadband, technology informatization is undergoing major changes and new breakthroughs. Under the path of information technology oriented urban development, the progress of theory, knowledge and methods in the academic field represents the theoretical and ideological level, which is also moving towards information innovation. In order to discuss the application value of neuroscience in organizational behavior research, we propose to expand the traditional measuring toolbox and make a modest contribution to the progress of academic research.

The field of brain science has been developing and accumulating for more than a century. Intercross with management, economics, cognitive science and other disciplines has also provided an unprecedented opportunity to explore the relationship between behavior and deep brain action. In recent decades, tremendous advances in science and technology

J. MacIntyre et al. (Eds.): SPIoT 2020, AISC 1283, pp. 715–719, 2021.
https://doi.org/10.1007/978-3-030-62746-1_107

have made human exploration of specific brain functions and behavioral related neural activities become an inevitable trend in the development of history. Neuroscience will occupy a seat in the field of management in the future [1], and organizational behavior as an important support for management, drawing on and using the theory and methods of neuroscience will also become the inevitable result of its discipline development.

Historically, the lack of early brain imaging technology and the high cost of hardware facilities limit the application of neuroscience methods in the organizational environment. However, with the rapid development of neuroscience in the past decade, more and more high-precision technologies with cost advantages have been born. This enables researchers to conduct more in-depth research in the organizational environment with relatively less cost and more scientific means [2]. According to the study of Powell (2011), Senior et al. (2011) and Matthews et al. [5], we consider that there are three key advantages in studying organizational behavior from the perspective of neuroscience: [1] studying the neural mechanism of the brain can better understand the ontological structural basis of behavior [2] neuroimaging technology can enhance the robustness and accuracy of traditional measurement methods [3] obtaining neural data can enhance the ability to predict and interfere with important behaviors in an organization.

As a completely new academic concept, Organizational Neuroscience (ON) was first proposed by Becker (2010). The author defined it as "a wise way to cross the great gap between neuroscience and organizational science after careful consideration". It is emphasized that neuroscience methods play a complementary role instead of a substitution [6] in traditional research methods of organizational behavior. Then some scholars advocate combining neuroscience methods with traditional measurement methods, and on the one hand, they can conduct a more in-depth study of the biological mechanism of organizational behavior; on the other hand, they have a better understanding of the biological mechanism of organizational behavior. The two way of measuring symbiosis can improve the robustness of a single method [4]. In view of this view, scholars later made a scientific proof of it. For example, in a study of leadership complexity, the authors used questionnaires and quantitative electroencephalogram (qEEG) to conduct a comparative study of leaders' adaptive decision making. The consistency of the two methods is verified [7]. Some scholars have conducted a neuro level study of the empathy and Machiavelli's empathy. The author has solved the long-standing controversial issue [8] by using traditional self-evaluation methods and functional magnetic resonance imaging (fMRI) research methods. Foreign scholars have appealed to academia to pay close attention to the great potential of neuroscience in providing information for organizational research. Through practical action, the feasibility of applying neural science in organizational research has been proved. Neuroscience technology that can be used to study organizational phenomena and employee behavior is usually not achieved by other means [9].

With the increasing enthusiasm for research in neurohistology in foreign academia, some influential research has gradually emerged in China. The first person to put forward the term "Organizational Neuroscience (ON)" in China is Li Hao, Ma Qingguo and Dong Xin [10]. As early as in 2006, Ma Qingguo and Wang Xiao Yi (2006) made a review of the interdisciplinary fusion of cognitive neuroscience. The concept of "neural management" is prospectively introduced, and the non-invasive brain imaging technology commonly used in Cognitive Neuroscience is introduced. Recently, Liu Yuxin, Chen Chen, Zhu Nan, Ji Zheng (2019) have conducted a more in-depth study of ON. He avoids the problem of confusing the word "Neuro Histology", which is commonly used in the early physiological field of China.

2 The Development Track and Introduction of Commonly Used Neuroimaging Techniques in Organizational Neuroscience

X-rays, developed in the 1895, is the technological origin of studying the morphology and structure and organization characteristics of organisms. It is also the technological origin of measuring the static characteristics of brain structure. Subsequently, computer tomography (CT) and computer axial tomography (CAT) are based on the principle of X-ray technology [11]. X-rays, as a means used by neuroscientists to study brain structure in 1890s, and gradually replaced by other technologies due to its radiation hazards and low sensitivity to human body. In the previous studies of neuroscience, two different brain functional imaging techniques, which are based on the principle of neuron activity leading to biomolecular metabolism and blood oxygen concentration, are introduced. In this paper, the two main techniques and other technologies are briefly introduced (PET, fMRI, EEG/qEEG, MEG, TMS/rTMS, tDCS, fNIRS).

Positron emission tomography (PET) is the first major technology to measure brain metabolic function [12]. Since 1990s, magnetic resonance imaging (MRI) has become the mainstream imaging technology in cognitive neuroscience. It can not only help draw the composition of the cerebral cortex, Data can also be used to elucidate the function of human brain in sensory and cognitive stimulation [13]. fMRI has become an important technical means of studying human brain's advanced activities. At present, fMRI is the highest spatial accuracy in all the neuroscience technologies [14, 15]. Electroencephalography (EEG) uses advanced signal processing techniques to obtain electrochemical signals [18] generated by currents in neurons within the scalp and skull. Quantitative electroencephalography (qEEG) has good time resolution and spatial resolution, and hardware devices are cheaper than EEG. When used, it does not cause any harm to human health [16, 17]. Magnetoencephalography (MEG) is closely related to EEG. The difference is that MEG obtains data by measuring the magnetic field produced by this electrical signal [4]. Compared with traditional neural imaging technology, functional near infrared spectroscopy(FNIRS) provides a relatively non-invasive, safe, portable and low-cost method for detecting brain activity [18]. Transcranial magnetic stimulation (TMS) based on electromagnetic induction, puts the closed loop magnet coil near the head of the subject. The magnet produces weak magnetic fields, inhibits or triggers the activity of the brain region [19]. Transcranial direct current stimulation (tDCS) is another non-invasive brain stimulation technology. It sends a fixed low amplitude direct current to the scalp by placing two or more electrodes above the head, changing the excitability of the cortex and the current flow [20] through the target neurons.

3 The Application Value of Neuroimaging Techniques in Organizational Behavior Research

Over a long period of time, the empirical study of organizational behavior has been heavily dependent on the results of single measurement methods [21], which may hinder the innovation process and quality of management theory and organizational research. Therefore, it is necessary to enrich the measurement methods in organizational research. As Senior mentioned, the multidisciplinary theoretical basis of neuroscience can provide researchers with the ability to explore undetected tissue phenomena. It can help researchers better understand the previous organizational phenomena. The most important thing is that scientific brain imaging technology can solve the current theoretical disputes and contradictions in a sufficiently robust way. In other words, researchers can conduct tests in a broader analytical framework, so that new research hypotheses [22] can be established and tested.

Understanding the neural basis of supporting organizational behavior helps to better develop design training programs. QEEG technology has been applied to leadership development research [23]. For example, Communication with a strong sense of social responsibility will enable team members to invest more in their work and enhance their work motivation [24]. Until now, such training mainly exists in commercial enterprises. But in the future, neuroscience technologies is likely to be widely used in academic research, such as how to use neural feedback training based on neuroscience technology to improve work related skills.

References

1. Dinh, M.L., et al.: Distribution of glial cells in the auditory brainstem: normal development and effects of unilateral lesion. Neuroscience **278**, 237–252 (2014)
2. Waldman, D.A., Wang, D., Fenters, V.: The added value of neuroscience methods in organizational research. Organ. Res. Methods **22**(1), 223–249 (2019)
3. Powell, T.C.: Neurostrategy. Strateg. Manag. J. **32**(13), 1484–1499 (2011)
4. Senior, C., Lee, N., Butler, M.: PERSPECTIVE—organizational cognitive neuroscience. Organ. Sci. **22**(3), 804–815 (2011)
5. Matthews, P.M., et al.: A practical review of the neuropathology and neuroimaging of multiple sclerosis. Pract. Neurol. **16**(4), 279–287 (2016)
6. Becker, W.J., Cropanzano, R.: Organizational neuroscience: the promise and prospects of an emerging discipline. J. Organ. Behav. **31**(7), 1055–1059 (2010)
7. Hannah, S.T., et al.: The psychological and neurological bases of leader self-complexity and effects on adaptive decision-making. J. Appl. Psychol. **98**(3), 393 (2013)
8. Bagozzi, R.P., et al.: Theory of mind and empathic explanations of Machiavellianism: a neuroscience perspective. J. Manag. **39**(7), 1760–1798 (2013)
9. Beugré, C.D.: Brain and human behavior in organizations: a field of neuro-organizational behavior. Neuroeconomics Firm **289** (2010)
10. Li, H., Ma, Q., Dong, X.: Neurohistology: conceptual analysis, theoretical development and research prospect. Manag. World **8**, 164–173 (2016)
11. Balthazard, P.A., Thatcher, R.W.: Neuroimaging modalities and brain technologies in the context of organizational neuroscience. Organ. Neurosci. **7**, 83–113 (2015)
12. Raichle, M.E.: Positron emission tomography. Annu. Rev. Neurosci. **6**(1), 249–267 (1983)

13. Loued-Khenissi, L., Döll, O., Preuschoff, K.: An overview of functional magnetic resonance imaging techniques for organizational research. Organ. Res. Methods **22**(1), 17–45 (2019)
14. Amaro Jr., E., Barker, G.J.: Study design in fMRI: basic principles. Brain Cognit. **60**(3), 220–232 (2006)
15. Huettel, S.A., Song, A.W., McCarthy, G.: Functional Magnetic Resonance Imaging, vol. 1. Sinauer Associates, Sunderland (2004)
16. Silva, F.L., Niedermeyer, E.: Electroencephalography. Basic Principles Clinical. Lippincott Williams & Wilkins Company, Philadelphia (1999)
17. Taylor, J.G., Fragopanagos, N.F.: The interaction of attention and emotion. Neural Netw. **18**(4), 353–369 (2005)
18. Tivadar, R.I., Murray, M.M.: A primer on electroencephalography and event-related potentials for organizational neuroscience. Organ. Res. Methods **22**(1), 69–94 (2019)
19. Irani, F., et al.: Functional near infrared spectroscopy (fNIRS): an emerging neuroimaging technology with important applications for the study of brain disorders. Clin. Neuropsychol. **21**(1), 9–37 (2007)
20. Robertson, E.M., Takacs, A.: Exercising control over memory consolidation. Trends Cognit. Sci. **21**(5), 310–312 (2017)
21. Woods, A.J., et al.: A technical guide to tDCS, and related non-invasive brain stimulation tools. Clin. Neurophysiol. **127**(2), 1031–1048 (2016)
22. Scherbaum, S., Dshemuchadse, M., Goschke, T.: Building a bridge into the future: dynamic connectionist modeling as an integrative tool for research on intertemporal choice. Front. Psychol. **3**, 514 (2012)
23. Senior, C., Lee, N., Braeutigam, S.: Society, organizations and the brain: building toward a unified cognitive neuroscience perspective. Front. Hum. Neurosci. **9**, 289 (2015)
24. Kershaw, C., Wade, B.: Brain change for optimal leadership. Biofeedback **39**(3), 105–108 (2011)

Recommendation Strategies for Smart Tourism Scenic Spots Based on Smart City

Peilin Chen[✉]

China University of Labor Relations, Beijing 100048, China
Peilinchen@126.com

Abstract. With the widespread popularization of smart tourism in smart cities and the increasing requirements of tourists for the quality of tourism services, services such as smart online tourism and smart mobile tourism have also gradually emerged. The ways to build a smart city include: smart public services and smart society management. In order to quantify the recommendation of tourist attractions in smart cities, this paper proposes a personalized tourist attraction recommendation algorithm based on Bayesian network learning. In order to solve the problems of new users and new attractions in smart cities, this algorithm uses the user's demographic information, user-attraction rating information, and attractions attributes. The Bayesian probability model is used to calculate the probability that a user will visit each attraction. Finally, this algorithm is experimentally verified with the traditional algorithm on the Ctrip data set. The results show that the algorithm is more effective in processing new users and new attractions. Good performance, survey results show that user satisfaction has reached more than 80%.

Keywords: Smart city · Scenic spot recommendation · Bayesian network learning · Social network

1 Introduction

"Tourism landscape under the concept of smart city" is a new concept derived from the summary of "smart city", "smart tourism", and "tourist road" [1–3]. Foreign countries have accurately and scientifically defined smart tourism. Smart tourism is based on modern information technology to meet the individual needs of tourists and provide high-quality services to achieve the sharing of tourism and social resources.

The construction of smart cities is linked to smart tourism, both of which serve the same purpose—to build a beautiful tourism environment [4, 5]. It is found through research that the development laws of the two have the essential attributes of coordination and service, construction and optimization. In the development of smart tourism in smart cities, look at issues from a connected perspective, control the whole, create the integration of smart tourism design, give full play to the functions and attributes of the two, and take smart industry information management as a guarantee. Encourage related

© The Editor(s) (if applicable) and The Author(s), under exclusive license
to Springer Nature Switzerland AG 2021
J. MacIntyre et al. (Eds.): SPIoT 2020, AISC 1283, pp. 720–724, 2021.
https://doi.org/10.1007/978-3-030-62746-1_108

industries to innovate to promote the upgrading of the industrial structure. As part of road construction, smart tourism landscape design should improve subjective quality and creativity, develop scientific and technological innovation, and make modern technology more convenient and convenient for tourists, enterprises and governments [6, 7]. The public carrier created by modern technology carries on the design and creation of forms to serve the public and promote the deepening of cultural propaganda. At the same time, the resources on the entire tourism road can be systematically integrated and deeply developed and activated.

This paper proposes a personalized tourist attraction recommendation algorithm based on Bayesian network learning. The algorithm first uses collaborative filtering algorithms and demographic methods to process users; then uses content-based methods to process attraction data to initially build models; subsequently, the Bayesian network is used to build the final model; finally, the model is based on Ctrip data Verification on the set, compared with the traditional method, this method has advantages in solving the problem of new users and new attractions.

2 Smart City Tourism Recommendation Strategy

(1) Location-based social network recommendations

Location-based social network services allow users to check in and share information within a circle of friends [8, 9]. When users are traveling, they usually use check-in data for travel tracks composed of photos or tag information. Information generates many tourism trajectories and plays an important role in the field of tourism recommendation research, such as mobility prediction, urban planning, and traffic management. The current development of digital cameras and mobile phone cameras provides support for users to share on social media websites, increasing the possibility for users to share photos and videos to Weibo, Flickr and YouTube. This rich location information supports tourism research.

(2) Location social network recommendation

By mining user check-in location trajectories and user social activity data, the basic framework of location-based social network recommendation systems, recommendation methods and application types based on different network levels are summarized, compared, and analyzed [10]. The task of a recommendation system in a location-based social network is to mine hidden patterns in the network so that users can fully enjoy the social experience. The location social network recommendation system consists of three modules: an input module, a processing module, and an output module.

(3) Content-based recommendation

Generally, the content of items is displayed in the form of a vector space model, and the items in the model are usually regarded as keyword vectors. If the content of an item is regarded as such an entity, such as a director, an actor, etc., it can usually be regarded as a keyword vector. But if the content of the item is presented in the form of text, it is necessary to introduce natural language processing technology to extract keywords. First, we need to use the word segmentation technology to segment the text to turn the word stream into a word stream, then detect some entities (such as person names, place names, organizations, etc.) from the word stream, and then perform keyword fusion, that is, these entities and some important words Fusion, then calculate the weight ratio of keywords, rank the keywords, and finally generate a keyword vector.

3 Research Method

(1) Personalized attraction recommendation algorithm learned by Bayesian network
Whether the attraction recommendation meets the user's interest will directly affect the user's evaluation and mood of the attraction, so the recommendation of the user's attraction is far more complicated than the recommendation of other tangible products. In the research of personalized attraction recommendation algorithms, it is necessary to strengthen the mining of user interests, and design a personalized recommendation algorithm that can accurately reflect user interest and improve user satisfaction. In order to improve the accuracy of attraction recommendation, this paper uses a personalized tourist attraction recommendation algorithm learned by Bayesian. This algorithm analyzes user log files and uses Bayesian networks to learn the dependencies between attractions to build a personalized attraction recommendation model. To provide users with online real-time information recommendations. Define the formula (the user's implicit rating of the attraction) Ru, a represents the user's rating of the attraction a, and its calculation formula is as follows:

$$R_{u,a} = \gamma \frac{N_{u,a}}{\sum_{j \in A} N_{u,j}} \tag{1}$$

$R_{u,a}$ represents the number of check-ins of the user u at the scenic spot, $\sum_{j \in A} N_{u,j}$ represents the number of check-ins of the user A, A represents the set of user attractions, and γ is an adjustment factor to prevent The value of $R_{u,a}$ is too small.

The data captured by Ctrip Online is used as the data set used in this article. It captures records of 1,291 users and 31,110 tourist attractions. The user attributes include user identification, gender, age, location, sign-in time. The attraction description, the location of the attraction, the attraction label, and the number of check-ins indicate an attraction record. The data set process is as follows:

Perform a pre-filtering process on the data set to filter out users with low travel records, so that the accuracy of recommendations can be improved in the end. The data set is then divided into a training set and a test set according to a ratio of 8:2. The training set is used to establish a personalized attraction recommendation model, and the test set is used to verify the personalized attraction recommendation model.

4 Discussion

4.1 Clustering of Attractions

The data set contains the attribute information related to the attractions. These attribute information need to be used to build a vector space model, and then use DBSCAN to process the attractions. The attraction category is recorded as Pc, and the clustered attraction categories are replaced. For the original attractions, the clustering results will eventually replace the attractions in the personalized rule user-attractions matrix. As shown in Table 1 below.

Table 1. Clustering results of photo attractions with geographic information

Clustering parameters		Attractions			Clustering results	
Eps (km)	MinPts	Great wall	Tiananmen square	The summer palace	The number of categories	Noise rate
0.5	20	341	261	2439	53	71.3%
1	20	428	625	2437	25	58.6%
3	20	672	604	3718	16	37.3%

As shown in Table 1 above, taking Beijing's scenic spots as an example and using DBSCAN algorithm to process their data, the clustering results of scenic spots are generally good. It can be seen from the table that by using the DBSCAN algorithm to cluster the scenic spots, the overall clustering result obtained is still relatively satisfactory to the user, and the satisfaction has reached more than 80%.

4.2 Experimental Analysis of Travel Recommendation Results

After the data required in the experiment is processed, the personalized tourist attraction algorithm proposed in this paper is analyzed experimentally on the data set. Through experimental analysis, it is found that the average absolute error of the recommendation results is related to the number of adjacent users selected in the experiment. Therefore, the number of different adjacent users is selected in the experiment. The performance of the algorithm is measured by the average absolute error.

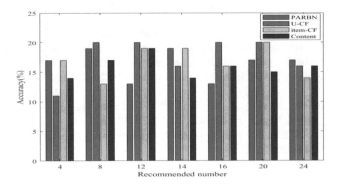

Fig. 1. Impact of number of recommendations on accuracy

As shown in Fig. 1 above, experiments show that when the number of tourist attractions recommended to users is different, the accuracy and recall of their algorithms are also different. Therefore, the algorithm proposed in this paper is compared with the user-based algorithm, the item-based algorithm, and the content-based algorithm, and it is found that the algorithm has higher accuracy and recall rate. As the number of recommendations continues to increase, the accuracy rate and recall rate also gradually increase, but when it increases to 12, the accuracy rate starts to decrease.

5 Conclusions

In order to better predict the next most likely tourist attraction for tourists, this paper proposes a personalized tourist attraction recommendation algorithm based on Bayesian network learning. The algorithm first uses collaborative filtering algorithms and demographic methods to process users; then uses content-based methods to process attraction data to initially build models; subsequently, the Bayesian network is used to build the final model; finally, the model is based on Ctrip data Verification on the set, compared with the traditional method, this method has advantages in solving the problem of new users and new attractions.

References

1. Ruíz, M.A.C., Bohorquez, S.T., Molano, J.I.R.: Colombian tourism: proposal app to foster smart tourism in the country. Adv. Sci. Lett. **23**(11), 10533–10537 (2017)
2. Ariani, Vitria: Socio-cultural impacts of tourism: a case study desa tugu puncak. J. Comput. Theor. Nanosci. **23**(4), 3178–3180 (2017)
3. Hashem, I.A.T., Chang, V., Anuar, N.B.: The role of big data in smart city. Int. J. Inf. Manag. **36**(5), 748–758 (2016)
4. Rogerson, Christian M., Rogerson, Jayne M.: City tourism in South Africa: diversity and change. Tour. Rev. Int. **21**(2), 193–211 (2017)
5. Zhang, P., Cai, Z., Zhang, C.: Potential spatial accessibility to urban scenic spots based on E2SFCA. J. Geomat. **40**(1), 76–79 (2015)
6. Liang, Xuedong., Liu, Canmian, Li, Zhi: Measurement of scenic spots sustainable capacity based on PCA-entropy TOPSIS: a case study from 30 provinces, China. Int. J. Environ. Res. Public Health **15**(1), 10 (2017)
7. Honggang, Xu., Zhu, Dan, Bao, Jigang: Sustainability and nature-based mass tourism: lessons from China's approach to the Huangshan scenic park. J. Sustain. Tour. **24**(2), 1–21 (2015)
8. Yang, Anna., Tang, Dongmei, Jin, Xiulong: The effects of road building on arbuscular mycorrhizal fungal diversity in Huangshan scenic area. World J. Microbiol. Biotechnol. **34**(2), 30 (2018)
9. Lipin, L.M.: Hawaii's scenic roads: paving the way for tourism in the islands by Dawn E. Duensing (review). Hawaii. J. Hist. **50**(50), 172–174 (2016)
10. Lehtonen, J.: Self before self on the scenic model of the early embodied self. J. Conscious. Stud. **23**(1), 214–250 (2016)

Control Strategy of Environmental Control System in Power Transmission

Pin Xia[✉]

Chongqing Vocational College of Transportation, Chongqing 402247, China
Zhichengjingling@163.com

Abstract. The development of electric power technology has provided a lot of convenience for people's lives, and people are becoming more and more dependent on electric power technology. The use of environmental control system control in power transmission is of great significance. An efficient environmental control system can not only detect the temperature and humidity of the power transmission site environment, but also effectively control the temperature and humidity of the site environment, and improve the safety and stability of power transmission. This article gives an overview of the power transmission and its environmental control system structure, and also expounds the role of the environmental control system in power transmission. It analyzes the control strategy of the power transmission environment control system. Use for reference.

Keywords: Power transmission · Environmental control system · Stability · Safety

1 Introduction

Many economic activities in the current society and people's daily life do not leave the support of power equipment, and the widespread use of power equipment requires stable power transmission to ensure. In the power transmission link, many uncontrollable factors will be encountered, such as the influence of external factors such as regional environment, temperature, weather, and air humidity, and the influence of internal factors such as power terminal boxes and instrument boxes. Affects the stability of power transmission. In order to ensure that power can be safely delivered to the actual destination, it is necessary to control the temperature and humidity in the power transmission process. Therefore, in the current power transmission process, an environmental control system is generally used to collect and analyze the environmental temperature and humidity information during the transmission process.

2 Overview of Power Transmission and Environmental Control System

(1) Overview of power transmission

J. MacIntyre et al. (Eds.): SPIoT 2020, AISC 1283, pp. 725–729, 2021.
https://doi.org/10.1007/978-3-030-62746-1_109

Power transmission is essentially the transmission of wireless energy and power. The reason why power transmission can be achieved is because of the mutual conversion between electromagnetic induction and energy during the transmission process. The power transmission environmental control system mainly regulates and controls radio frequency, microwave, electromagnetic induction, electromagnetic resonance, etc. [1], so as to promote the entire environment of power transmission to become more stable, so as to improve the safety factor of power transmission. Power transmission environmental control includes short-range, medium-range and long-range. At present, my country's power transmission has gradually developed into wireless power transmission. The continuous dispersion of radio waves, the actual efficiency of wireless power and the transmission distance are the main obstacles facing the current power transmission process. In the process of power transmission, it is difficult to gather the relevant electromagnetic waves, and the directionality is not good. There is a possibility of transmitting in different directions. In the process of power transmission, the air in the relevant area is a medium for the entire coupling. Changes in the environment such as temperature and humidity in the area may affect the efficiency and stability of power transmission.

(2) Structure of power transmission environmental control system

In the process of power transmission in my country, the development time of control systems including environmental control systems is relatively late. The power transmission environmental control system mainly monitors the environmental conditions such as the transmission environment, the access control system of the equipment room where the transmission equipment is located, water leakage, smoke sensing, temperature and humidity. The power transmission environmental control system uses power transmission resources to monitor the transmission from the power point to the user [2]. Even if there is no professional staff on duty at the user end, they can use the wireless transmission resources they already have to avoid investing maintenance resources in power transmission as much as possible. All-round monitoring is extremely necessary, and the software graphical man-machine interface can be set on the monitoring module side of the power point.

The network structure of the power transmission environment control system depends on its operation, maintenance and management mode. Since operation and maintenance have certain compatibility, in addition, operation and maintenance also need to consider the flexibility required for the development of the environment control system. Including the local monitoring center SC, regional monitoring station SS, end office monitoring unit SU three-level tandem network [3]. Environmental monitoring, information collection, monitoring, stability monitoring, security monitoring and other monitoring modules and front-end mechanisms have become monitoring units, which are the basic units that constitute the power transmission environmental control system. They are usually connected by special data lines. System acquisition is realized by distributed microcomputer. The monitoring module is directly connected with the monitored equipment. The monitoring module is mainly responsible for collecting the monitoring information of the monitoring objects of the environmental control system, and is also responsible for issuing various control commands. The monitoring module sends the alarms and monitoring information of the monitored equipment to the front-end machine of the environmental control system. After receiving the information, the front-end machine analyzes and sends the processed data to the monitoring station. The monitoring unit displays the alarm information and operation status of the monitored device in real time, and the system maintenance personnel sends the corresponding operation instructions to the monitored device based on the monitoring situation.

The communication network connects the monitoring unit and the monitoring station of the power transmission environment. The information of the monitored environment is collected by the monitoring unit. The monitoring station receives these information data, analyzes and processes the information data, and sends them to the monitoring center. The operation status of the monitored equipment environment is monitored and controlled by the monitoring station. The highest authority of the entire power transmission environmental control system is in the monitoring center. The monitored and control system can be processed by it. In addition, it can also statistics, analyze and manage the data of the environmental control system.

3 Application of Power Transmission Environmental Control System

Power transmission began to combine intelligent environment and comprehensive security monitoring technology to design a feasible environmental control system, which can monitor and control the dynamic environment during power transmission. The application of environmental control system provides many guarantees for the safety and stability of power transmission. Its application is mainly reflected in the following two aspects.

First, it promotes the development of power transmission towards interaction, automation, and information. Wireless power transmission requires that the entire distribution network can be digitized, informatized and automated in all facilities within the station, which is the main manifestation of its intelligence. The environment control system has an efficient data integration collection system [4], which can collect data in the power transmission environment, and then send the data to the analysis system of the environment control system. The power transmission environment monitoring staff can make according to the data analysis results Reasonable decision.

Second, discover abnormal situations in power transmission in time to reduce the probability of accidents. The power transmission is controlled by an environmental control system, which can grasp and track the dynamic environment and security data in power transmission in real time. If there is an environmental abnormality problem in power transmission, you can find and make active response in time, which greatly reduces the probability of power outages and other accidents caused by abnormal power transmission environment.

4 Control Strategy of Power Transmission Environmental Control System

There are many control methods for power transmission environmental control systems, and these control methods affect each other. Therefore, in the actual environmental monitoring and control process of power transmission, it is not necessary to formulate a comprehensive control strategy, but need to combine the power transmission area. Climatic conditions and the actual status of power transmission environmental control equipment to formulate highly feasible control strategies, rational and scientific analysis of power transmission site environmental data, and overall planning of control plans. The control strategy of the power transmission environmental control system can start from the following three aspects.

(1) Monitor the temperature of bus and cable

Busbars and cables are the two main power equipment involved in power transmission. The temperature changes of the busbars and cables will affect the stability and safety of power transmission. In order to understand the real-time temperature of the busbars and cables, special staff will usually arrange on-site contact Bus and cable, measure the temperature of the bus and cable. The current power transmission environmental control system generally has a line infrared bus temperature measurement technology and a wireless temperature measurement cable temperature technology. The temperature measurement personnel and the bus and the cable do not need to be in direct contact, they can use wireless communication technology to monitor the problem of the bus and the cable, and track their temperature changes in real time. If the temperature is too high or too low, the environmental control system will perform the corresponding Failure warning.

(2) Monitor the operation of unmanned pumps

In the traditional power transmission process, special staff need to be arranged to inspect the operation of the unmanned pump to understand the start and stop status of the unmanned pump. The power transmission environmental control system has water pump linkage technology [5], which can monitor the operation status of unmanned pumps in real time, and the monitoring results are directly returned to the monitoring site without the need for safety personnel to patrol, reducing the labor cost input in the power transmission environmental control. In addition, the application of underground water leakage monitoring technology in the power transmission environmental control system can timely discover the adverse effects of underground water leakage on the entire distribution network.

(3) Monitor cable corrosion and toxic gas concentration

The cables of the distribution network are generally exposed to the outside. The ozone and other gases contained in the atmosphere are in contact with the cables for a long time, which will corrode the cables. The ozone detection technology in the power transmission environmental control system is used to provide timely protection when the ozone content is too high. Prevent the cable from being corroded greatly and extend the service life of the cable. In addition, after the cable is used for a long time, the cable trench will contain sulfur dioxide. When the sulfur dioxide concentration is too high, it will threaten the life safety of the cable trench construction personnel. The sulfur dioxide detection technology of the environmental control system can understand the concentration of sulfur dioxide in the cable trench and reduce the Safety accidents caused by high sulfur dioxide concentration [6–8].

5 Conclusion

The extensive application of the power transmission environmental control system can effectively manage the operation of the intelligent distribution network, provide a comprehensive safety monitoring and intelligent environment platform for the operation and maintenance management of power transmission, and have a positive significance for the safe and stable transmission of power. In the actual control process of the environmental control system, in addition to monitoring the bus temperature, cable temperature, unmanned pump operation, cable corrosion, toxic gas in the cable trench, etc., we should also pay attention to the application of traditional network technology. The construction method is optimized to prevent abnormal situations by protecting the environment inside the power transmission. In addition, the environmental control system should take the initiative to monitor the internal and external environment of power transmission by video, and timely stop and deal with the invasion and destruction of power transmission.

Acknowledgements. This work is partially supported by the Research Project Foundation of Chongqing Education Committee (Research on the energy consumption control of the BAS based on the subway station, Grant KJQN201805703).

References

1. Fan, X., Gao, L., Mo, X., et al.: Research status and application review of wireless energy transmission technology. J. Electr. Eng. Technol. **034**(007), 1353–1380 (2019)
2. Zhou, W., Wang, X., Li, M.: The technology of wireless energy transmission in power system. Digit. Users **024**(006), 100 (2018)
3. Zhou, J., Zhang, C., Hao, J.: Design of intelligent control system for power equipment operating environment based on C8051F120 single chip microcomputer. Comput. Digit. Eng. **047**(003), 696–699,705 (2019)
4. Wang, Y.: Discussion on control strategy of metro environmental control system. Mod. Urban Rail Transit (9) (2017)
5. Hu, X.: Research on the application of control technology in the environmental control system of rail transit stations. Commod. Qual. **008**, 146 (2018)
6. Chen, X., Zhu, Y., Shen, J.: Input-output dynamic model for optimal environmental pollution control. Appl. Math. Model. **83**, 301–321 (2020)
7. Johnstone, L.: The construction of environmental performance in ISO 14001-certified SMEs. J. Clean. Prod. **263**, 121599 (2020)
8. André, M., De Vecchi, R., Lamberts, R.: User-centered environmental control: a review of current findings on personal conditioning systems and personal comfort models. Energy Build. **222**, 110011 (2020)

MATLAB Software in the Numerical Calculation of Civil Engineering

Xiaowen Hu and Qiang Zhou$^{(\boxtimes)}$

Nantong Institute of Technology, Nantong 226002, Jiangsu, China
huxiaowen19870528@126.com, tb19870629@163.com

Abstract. Matlab provides a variety of matrix calculation and operation, including various numerical calculation toolbox commonly used in the calculation of structural reliability, the constitutive relationship of reinforced concrete and failure criteria, and analyzes the factors affecting the accuracy in the process of nonlinear solution. In this paper, from the two aspects of the basic principle and the analysis of the related calculation examples, the Matlab program for calculating the structural reliability is developed by using MATLAB, which makes the MATLAB language applied in the reliability calculation.

Keywords: Matlab software · Reinforced concrete · Numerical analysis · Civil engineering

1 Introduction

The mechanical problems in engineering structure belong to the category of nonlinear deformation, in essence, that is to say, the internal variation of the studied engineering structure system and the external factors are nonlinear causalities, and the linear assumption is only a simplification of the practical problems [1]. With the higher and higher requirements of engineering structure for model accuracy, the research of nonlinear problems becomes a science. The key content of the research in the field of technology and engineering application.

The reinforced concrete structure is the most widely used engineering structure. Its mechanical properties are affected by many factors, such as water-cement ratio, aggregate strength, etc. it is a kind of building material with complex mechanical properties. Therefore, it is difficult to carry out an accurate internal force analysis of a reinforced concrete structure. For a long time, the research of reinforced concrete structure by scholars from all over the world stays on the level of linear elasticity theory. In recent years, with the development of the computer and finite element method, a numerical calculation has been widely used in practical construction projects with its advantages of fast, low cost, and easy to realize. On this basis, a large number of general nonlinear finite element analysis software developed rapidly, and the finite element method has become an important means to study the concrete structure.

J. MacIntyre et al. (Eds.): SPIoT 2020, AISC 1283, pp. 730–734, 2021.
https://doi.org/10.1007/978-3-030-62746-1_110

As a large-scale and general finite element analysis software, MATLAB has great advantages in nonlinear analysis, and it not only has the advantages of other finite element analysis software, such as fast numerical calculation, high accuracy of results, and low analysis cost, but also has a more user-friendly operation interface and visual results, Especially in the nonlinear analysis of reinforced concrete structure, it can get more accurate and practical results, and it tends to be widely used in the field of structural analysis. Based on the constitutive model of concrete material provided by MATLAB, this paper mainly introduces the fixed dispersion crack model used in the concrete tension area. Its essence is to "disperse" the actual concrete crack into the whole unit model, treat the concrete material as anisotropic material, use the material constitutive model of concrete to simulate the effect of crack, and carry out some reinforced concrete Nonlinear finite element simulation of concrete members.

2 Constitutive Model of Concrete

Due to the complexity of reinforced concrete materials and load effects, the existing various concrete constitutive models, failure criteria, the constitutive relationship of reinforcement, and the interaction model of reinforced concrete are all based on the model test, based on some simplifications and assumptions, and the mathematical and mechanical models which are consistent with the model test results. Based on different assumptions, different finite element software uses different models in the nonlinear analysis of reinforced concrete, each with its characteristics [2].

MATLAB has a strong ability in the reinforced concrete analysis. It provides three concrete constitutive models: concrete SM eared cracking, concrete damaged plasticity and cracking m model for concrete in ABAQUS/explicit. Among them, concrete SM eared cracking is widely used. This paper will discuss the constitutive model.

Concrete SM eared cracking is a constitutive model that describes the compression of concrete with an elastic-plastic model and simulates the tension of concrete with a fixed dispersion crack model. The nonlinear behavior of concrete is simulated by elastic cracking in tension and isotropic hardening in compression. When the material is mainly compressed, the behavior of concrete can be analyzed by isotropic hardening elastic-plastic theory, and the simple yield surface and associated flow rule expressed by equivalent hydrostatic pressure P and equivalent partial stress Q can be used. The model greatly simplifies the actual mechanical behavior of materials. The model is based on classical plastic theory. The yield surface in compression is the Drucker Prager form. It is a straight line in the plane and a circle in the deflection plane. Its function expression is shown in formula (1).

$$f_c = q - \sqrt{3a_0 p} - \sqrt{3}\tau_c \qquad (1)$$

where: $p = 1I_1/3$; $q = 3J_2$;

$$a_0 = \sqrt{3}\frac{1 - r_{bc}^\sigma}{1 - 2r_{bc}^\sigma}; \quad \tau = \left(\frac{1}{\sqrt{3}}, \frac{3}{a_0}\right)\sigma_c$$

Parameter a_0 reflects the contribution of hydrostatic pressure to yield; r_{bc}^σ is the input constant, generally 1.16; τ is the hardening parameter; I_1 is the first invariant of stress tensor; J_2 is the second invariant of stress bias.

3 The Constitutive Relation of Reinforcement and the Bond Between Reinforcement and Concrete

In MATLAB, the reinforcement in the concrete structure can be realized by the reinforcement element (rebar), which can add a separate reinforcement element, or add the reinforcement attribute to the element attribute to define the reinforcement of the composite model, or embed the link element or membrane element into the concrete element by the embedded method to define the reinforcement. The reinforcement element is a one-dimensional strain bar element, so the bilinear ideal elastic-plastic constitutive model is adopted. The characteristic of this model is that the mechanical properties of reinforcement and concrete are independent. The basis of the two different materials working together is that there must be enough bond strength between them. Here, the interaction between reinforcement and concrete and the bond-slip effect is realized by embedding the reinforcement unit into the concrete unit through the embedded technology [3].

4 Stress Analysis and Calculation Example of the Reinforced Concrete Simple Supported Beam

The simply supported beam is 1800 m long, with a section size of 180 m × 100 m. The concrete strength grade is C25. The longitudinal reinforcement and stirrup are made of hpb235 reinforcement. The mechanical parameters of concrete and reinforcement are taken from the standard value of code for the design of concrete structures. See Fig. 1 for details of the simply supported beam.

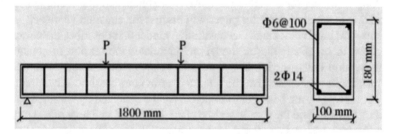

Fig. 1. Reinforcement drawing of the simply supported beam

The c3d8r element and t3d2 element are used to simulate the bond relationship between the reinforced concrete and the embedded concrete element. To prevent the stress concentration on the loading surface and the support during the loading process, the steel gasket is set at the loading position and the support during the modeling to increase the contact area and stiffness. The model diagram is shown in Fig. 2.

Here, the Sanz formula commonly used in the structural calculation is used for constitutive model calculation [4]:

$$\sigma = k_3 f_c \frac{A(\varepsilon/\varepsilon_0) + (D-1)(\varepsilon/\varepsilon_0)^2}{1 + (A-2)(\varepsilon/\varepsilon_0) + D(\varepsilon/\varepsilon_0)^2} \tag{2}$$

where: $k_3 = 1$; $f_c = 11.9$; $A = 1.7388$; $D = 0.5$; $\varepsilon_0 = 0.002$

Fig. 2. Model drawing of the simply supported beam

5   Analysis of Finite Element Results

The load P (P = 16 kn) is applied to the gasket of reinforced concrete simple supported beam model, and the analyzed Mises stress is shown in Fig. 3. It can be seen from the figure that the maximum value of Mises stress is 41.16 mpa, which appears on the bearing base plate. Also, there is large stress around the loaded base plate. The stress value is about 3–14 mp, and changes from large to small from the loading base plate to the bearing direction, thus forming a stress arch between the two bearings. The stress value alternates from small to large, forming a force transmission path, This is similar to the stress model of arch truss with web reinforcement in traditional theory. Because of the small stress on the shoulder of the beam, the value of Mises stress is small, and the value of Mises stress at the bottom of the beam is also small under the action of the tension bar. The relative error between this result and the theoretical value calculated by the mechanics of materials is not too large, which does not exceed the allowable value of the code. Therefore, this simulation analysis has a reliable theoretical basis and is very credible.

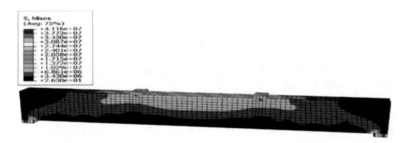

Fig. 3. M ISEs diagram of the simply supported beam

6   Conclusions

In the finite element analysis, the bond between the concrete and the reinforcement is treated by the embedded technology. This method effectively simplifies the modeling, conveniently solves the bond relationship between the reinforcement and the concrete, and makes the modeling more efficient and easy, but it can not reflect the characteristics of the friction between the concrete and the reinforcement changing with the increase of the load, In particular, it can not achieve the slip between the reinforcement and the concrete, which is easy to lead to large error. The error between the analysis result and the theoretical value is not only related to the bond treatment but also related to some other value assumptions, such as the convergence of finite element analysis, the rationality of parameter value in simulation, the division form and quantity of finite element elements, the pouring quality and loading of test specimens, etc. How to use the finite element software to simulate the relationship between reinforcement and concrete more closely needs to be further studied.

According to the comparison between the analysis results of the reinforced concrete simple supported beam and the theoretical value of mechanical calculation, the results of the numerical simulation can reflect the mechanical properties of the model, which also proves the correctness of using MATLAB nonlinear finite element analysis software for structural analysis, and proves that using MATLAB software to analyze the nonlinear problems of the reinforced concrete structure is very effective. However, how to simulate the engineering structure under the actual stress state and get more accurate and closer to the actual results remains to be further studied and learned in the future. However, it is not easy to simulate concrete as accurately as possible in the finite element software. Scholars at home and abroad put forward the concrete constitutive model based on various theories. But up to now, no theory has been accepted to describe the constitutive relation of concrete completely.

Acknowledgments. Construction system science and technology project of Jiangsu housing and urban-rural development department (source) Research and implementation of the information management system of construction engineering laboratory based on MATLAB (name) 2019ZD047 (No.).

References

1. Zhang, G., Su, J.: Nonlinear analysis of reinforced concrete based on ABAQUS. Sci. Technol. Eng. **10**, 5620–5624 (2008)
2. GB50010-2002 code for design of concrete structures
3. Li, W., Sihua, D.: Nonlinear analysis of bending failure experiment of concrete beams based on ABAQUS. Inf. Technol. Civil Eng. **03** (2010)
4. Southeast University, et al.: Principles of concrete structure design. China Building Industry Press (2008)

Low Latency V2X Application of MEC Architecture in Traffic Safety

Zunyi Shang[✉]

Network Information Center, Dalian Jiaotong University, Dalian 116208, Liaoning, China
szy@djtu.edu.cn

Abstract. To address the drawbacks of the traditional "end, tube, cloud" architecture of V2X (Vehicle to Everything) with slow real-time response and vertical transmission of information, this paper has carried out research from architecture adjustment, introduces the edge computing layer to sink the computational power required for latency-sensitive services to the network edge and extend the transmission of information from vertical to horizontal, realizing the transformation from V2N (Vehicle To Network) to V2V (Vehicle To Vehicle), combining the advantages of the exclusive band of C-V2X (Cellular Vehicle to Everything), fusing MEC (Mobile Edge Computing) and C-V2X to form a new low-latency V2X architecture, which can be applied to many mainstream traffic scenarios and effectively improve Traffic safety.

Keywords: Traffic safety · Vehicle networking · Low latency · Edge computing · C-V2X

1 Introduction

With the rapid growth of automobile ownership, traffic safety has become the most important branch of urban public travel safety, among many urban road conditions, intersections are the most complex [1]. Vehicles, non-motorized vehicles and pedestrians in different directions need to cross intersections in a limited time. Therefore, intersections are usually the bottleneck of frequent traffic accidents and traffic efficiency, traffic accidents about 30%–50% occur at intersections [2]. In order to manage traffic congestion and traffic safety problems, on the basis of adjusting the distribution of roads and increasing road networks and other infrastructure construction, it is necessary to further seek new breakthroughs, and V2X centered on vehicles, roads, pedestrian, roadside units, network access, cloud is the main direction to solve the above problems. Traditional V2X using "end, tube, cloud" architecture [3], through the establishment of third-party OBU or manufacturer T-BOX and back-end data collection system to provide a platform for users or service providers, can achieve the vehicle and cloud service side of the information interaction, including information collection, voice services, vehicle rescue, navigation services, etc., Vehicle manufacturers are currently equipped with such V2X systems, with V2N to form a relatively closed-loop service ecology, but for traffic safety,

J. MacIntyre et al. (Eds.): SPIoT 2020, AISC 1283, pp. 735–739, 2021.
https://doi.org/10.1007/978-3-030-62746-1_111

it is more important to require lower delay of information transmission and processing, real-time traffic analysis, accident prediction, information transmission, action response all need to be completed in millisecond delay to ensure the effectiveness and timeliness of traffic safety, this paper proposes to change two key aspects of V2X architecture and network mode research, the goal is to reduce the overall delay, and expand low latency V2X applications.

2 MEC Architecture

However, from the actual use of the communication method, the existing "end, tube, cloud" three-layer model of vehicle networking has certain limitations in terms of real-time data transmission and immediate response [4]. Mainly reflected in the transmission delay caused by long network links, uneven mobile network signal, network jitter caused by intensive access, and reduced response capacity caused by concurrent processing of large amounts of data in the cloud, the traditional end-to-cloud direct interaction method is difficult to meet the needs of traffic safety for real-time communication and data processing.

Through continuous practice and summary, we believe that due to the current network conditions and the computing power of in-vehicle computing units, it is necessary to adjust the vehicle networking architecture to ensure low latency and high reliability of V2X, to this end, MEC is introduced to bring computing and data processing capabilities of cloud transactions in traditional three-tier architecture of vehicle networking down to the edge of the network [5]. Decoupling the real-time interaction between the edge and the central non-latency sensitive backend services to achieve decentralized network services, The V2X is divided into four layers, namely Perception layer, Communication layer, Edge computing layer and Cloud layer, The tasks carried by the layers are shown in Fig. 1. The MEC runs at the edge of the network close to the user's terminal, significantly reducing the transmission delay of C-V2X services, providing powerful computing and storage capabilities and improving the user experience. For example, C-V2X services in the driver safety category impose stringent communication latency requirements, and deploying such services on an MEC can significantly reduce business response times compared to deploying them on a central cloud. In addition, MEC can also provide online auxiliary calculations for vehicle/roadside/pedestrian terminals for fast task processing and feedback.

The edge computing layer connects vehicles, pedestrian, roads and other units within a certain range to form a distributed regional centralized network. The edge server acts as the central processor and communicator in the network, interacting with the OBD, TBOX and other in-vehicle control units carried by vehicles in the region in real time through 5G, LTE-V2X and other communication methods, collecting the location and operation information of each vehicle and transmitting these information data back to other vehicles in the region, so that the vehicles can know the location, speed, operation trajectory and driving trend of all vehicles in the region from each other, realizing inter-vehicle interconnection V2V. At the same time, MEC can expand automotive application services by designing standardized interfaces for third-party applications to provide network information sharing functions, including limited user information, network information, vehicle status information, operational data, location information, etc.

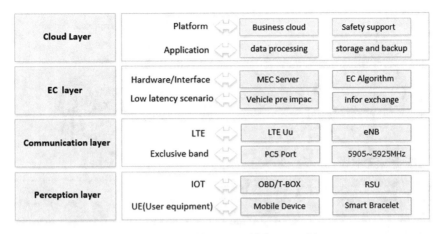

Fig. 1. MEC architecture vehicle networking

3   C-V2X Networking

There are two types of V2X networking technology, one is the IEEE-led DSRC, the other is the 3GPP-led C-V2X, DSRC is based on IEEE802.1P standard, C-V2X is based on cellular communication technology [6]. The former technology is mature but the delay jitter, high speed and high density scenario reliability is poor, the latter has mobility, strong reliability, and at the same time has forward compatible 5G evolution route, the future can support autopilot, C-V2X self-organized network mode is fast and flexible, autonomous and controllable [7]. In November 2018, the Ministry of Industry and Information Technology (MIIT) issued "Regulations on the Management of the Use of 5905 ~ 5925 MHz for Direct Communication in V2X" [8]. which planned dedicated frequency resources with a total bandwidth of 20 MHz in the frequency band for direct communication in intelligent networked vehicles based on the formation of LTE evolution V2X (direct communication between vehicle and vehicle, vehicle and pedestrian, vehicle and road), and at the same time, made provisions for the management of related frequencies, stations, equipment, interference coordination, so that edge computing servers can directly realize self-organized network and direct communication with vehicle OBU and roadside RSU, and the data traffic of communication does not pass through the base station and core network, reducing communication delay, reducing the pressure on base stations, easing the core network load, improving spectrum utilization and system throughput [9]. The MEC and C-V2X converged network topology diagram is shown in Fig. 2. It can further enhance the C-V2X end-to-end communication capability and provide auxiliary computing and data storage support for C-V2X application scenarios. Proprietary frequency band solves the limitations of network dependence on operators, reduces the signal interference of band mixing, and can serve the full function of the network for V2X.

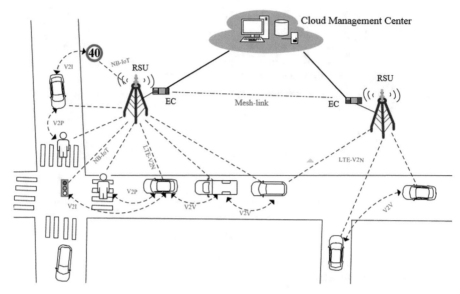

Fig. 2. MEC/C-V2X network topology

4   Application of Low Latency V2X in Traffic Safety

Based on the implementation of V2V, basic road information, such as speed limit, traffic limit, signage, signal status, etc., can be obtained through the third-party system authorization interface to realize V2I (Vehicle-To-Infrastructure) interconnection between vehicles and infrastructure [10]. The edge servers can also be connected to the operator's MES equipment through integrated small 5G base station or Uu interface to obtain pedestrian status and movement trajectory within the road section, thereby enhancing vehicle-to-pedestrian perception, realizing V2P and improving pedestrian safety. Through the edge server and the vehicle, road, pedestrian, etc. to interact with the perception, and then through the computer system to establish a set of holographic mastery of the state of the vehicle, pedestrian and objects within the road section, to achieve fault vehicle reminder, collision prevention warning, road conditions, warning of violations and other functions to enhance the traffic safety management capacity.

The specific application scenarios of low latency V2X include traffic safety, traffic control, emergency evacuation, accident diversion, third-party services, UBI insurance, city management, etc., which can effectively reduce vehicle-vehicle and vehicle-pedestrian collisions, evacuate special vehicles and proximal vehicles at the accident scene, release targeted real-time information, ease traffic congestion, promote points strategy and insurance strategy and other business models to the ground, cultivate good and safe driving habits, and fundamentally solve traffic safety problems.

5   Summary

The design of MEC and C-V2X integration architecture extends the information service-based functions of traditional vehicle networking, solves the demand for low latency for traffic safety, establishes a new vehicle networking ecology with MEC as the underlying technology, and promotes the improvement of traffic management and traffic safety governance level with driving safety and driving habits as the core improvement.

References

1. Li, J., Cheng, H., Guo, H.: Survey on Artificial Intelligence for Vehicles. Autom. Innov. **1**, 2–14 (2017). https://doi.org/10.1007/s42154-018-0009-9
2. Xiaoping, Wu., Deng, Shuai, Xiaohong, Du: Green-wave traffic theory optimization and analysis. World J. Eng. Technol. **2**, 14–19 (2014)
3. Tolba, A.: Content accessibility preference approach for improving service optimality in internet of vehicles. Comput. Netw. **152**, 78–86 (2019)
4. Chang, S.: Key technologies and development trends of 5G optical networks. Appl. Sci. **9**(22), 4835 (2019)
5. Xue, J., Shao, H., Ma, Q.: Resource allocation for system throughput maximization based on mobile edge computing. In: International Conference on Electronics and Electrical Engineering Technology, pp. 177–181, September 2018
6. Tang, L., Zhao, M., Bian, Z., Li, C.: Overview on C-V2X test standard analysis and design of test solutions. Autom. Digest **7**, 46–51 (2019)
7. Gallo, L., Haerri, J.: Unsupervised long-term evolution device-to-device: a case study for safety-critical V2X communications. IEEE Veh. Technol. Mag. **12**(2), 69–77 (2017)
8. China's MIIT publishes regulations for direct communication of internet of vehicles. https://en.imsilkroad.com/p/119878.html.2018,11
9. Lunyuan, C.: Application of D2D communication system based on android and JXTA on the internet of vehicles. J. Phys. Conf. Ser. **1486**, 042014 (2020)
10. Diewald, S., Leinmüller, T., Atanassow, B., Breyer, L.P., Kranz, M.: Mobile device integration and interaction with V2X communication. In: 19th World Congress on Intelligent Transport Systems (ITS) (2012)

Cloud Computing Technology for the Network Resource Allocation on the Research of Application

Gen Zhu and Xiang Fang[✉]

Hubei University of Medicine, Shiyan 442000, Hubei, China
fangxiang0326@qq.com

Abstract. The information revolution has brought about the rapid development of the Internet technology services. Where users obtain required hardware platforms software and services through the network in an on-demand and easily scalable manner are becoming increasingly popular. This article introduces the three architecture models four deployment models five architectures five key technologies of cloud computing and explains how to integrate and share the network through cloud computing, flexibly configure and call software and hardware resources. Under resources it greatly reduces the difficulties encountered in the construction of power costs space costs facility maintenance costs, software and hardware construction costs data maintenance costs, etc., improves network resource utilization, reduces costs and provides users with large-scale massive data storage with handling of services.

Keywords: Cloud computing · Architectural model · Deployment model · Architecture · Key technologies

1 Introduction

"Except God, everyone must speak with data." Not only people but the whole world is becoming more and more digital. The information revolution has developed in depth and surges of data are surging. The number is huge, the types are heterogeneous, and the trend is fast and unprecedented [1]. The 21st century is a century of information explosion, known as the "first year of information" [2]. With the rapid development of Internet technology, services that obtain required hardware, platforms, software, and services and other resources through the network in an on-demand and easily scalable manner are becoming popular. This is called cloud computing. The network that provides resources for users is called "cloud", and distributed large-scale computer clusters and server virtualization software and other technologies provide users with the required computing and analysis capabilities [3]. Cloud computing provides services from IT infrastructure (that is, well-known hardware services) to upper-layer applications (that is, software services), spanning the two fields of IT (information industry) and CT (communication industry) [4]. The cloud computing industry chain is composed of

J. MacIntyre et al. (Eds.): SPIoT 2020, AISC 1283, pp. 740–744, 2021.
https://doi.org/10.1007/978-3-030-62746-1_112

three parts: front-end cloud computing services, back-end cloud computing data center construction, and cloud security [5]. In the cloud computing service field, there are three vendors: IaaS PaaS and SaaS. Software and hardware infrastructure providers, system integrators and data center operators are all involved in the construction of cloud computing data centers.

2 Cloud Computing Architecture Model

There are three cloud architecture models for cloud computing services: IaaS (Infrastructure as a Service) infrastructure as a service, PaaS (Platform as a service) platform as a service, and SaaS (Software as a service) software as a service [6]. IaaS is the lowest-level architecture of cloud services, and the infrastructure includes hardware facilities such as computers, networks, storage, load balancing devices, and virtual machines. Through cloud computing related technologies, it is unified into a virtual resource pool and provided to users and SaaS and PaaS providers. Users obtain corresponding services through the Internet. IaaS allows users and SaaS and PaaS providers to no longer bear the cost of hardware resources required for use. At the same time, they have high scalability, adaptability and flexibility. These services are for end users. Hardware resources can be expanded or contracted according to user needs [7]. IaaS is an important cornerstone for the development of the Internet and the information industry in the future. The development of the Internet and cloud computing has brought about an increase in the demand for IaaS and promoted the development of IaaS. Another cloud service intermediate architecture is PaaS. In this model the managed service provider provides the database Web services development tools, execution runtime, operating system, software and hardware equipment, technical guidance etc. Use the work platform to help customers, customers do not need manually allocate resources but need to pay the rental fee. At the same time users need to follow certain programming principles to customize application software that suits their own characteristics according to their own business data storage, etc., that reduces costs and obtains high-quality and personalized services. PaaS services are hailed as the "operating system" of the future Internet and are also the most active areas of cloud computing technology and application innovation. Compared with IaaS services PaaS services have stronger business stickiness for application developers. The focus is not on direct economic benefits, but more on building and forming a close industrial ecology.

The SaaS model includes software parts such as virtual desktops, various practical applications, content resource management e-mail, software and others. SaaS services use web technology and SOA architecture to provide users with multi-tenant and customizable applications through the Internet. In this model, the supplier is responsible for installing, managing, and operating the various software required by the user of the cloud service. The customers only need log in and use these products through the cloud. The development, maintenance, management and data storage of software used by users are all managed by cloud providers, which greatly shortens the channel chain of the software industry and enables software vendors to change from software production to application service operators. With the continuous increase of software varieties, its application is becoming more and more popular [8]. Cloud computing technology is based on three special cloud computing service models, so it has the characteristics of popularity, efficiency, flexibility, and user-friendliness.

3 Cloud Computing Deployment Model

Cloud computing has four deployment models: public cloud, private cloud community cloud and hybrid cloud, each of them has unique functions to meet different requirements of users [9]. The public cloud has the lowest cost, the private cloud has the highest security, the high flexibility of the hybrid cloud can meet the ever-changing user needs, and the community cloud has a higher purpose. For colleges and universities, which kind of deployment method to choose requires a comprehensive evaluation of the school's own development strategy, business needs and other factor.

4 Cloud Computing Architecture

The architecture of cloud computing consists of 5 parts from bottom to top: resource pool layer, platform layer, application layer, SOA building layer (also called user access layer) and management layer [10]. Its essence is to provide services to users through the network, so services are the core of its architecture. The resource pool layer refers to cloud computing services at the infrastructure level. These services hide the complexity of physical resources and provide virtualized resources for cloud computing [11]. The platform layer provides users with the encapsulation of resource pool layer services, so that users can build their applications. SOA (Service-Oriented Architecture, service-oriented architecture) building layer in cloud services each level of cloud computing service needs to provide corresponding access interface. SOA building layer is also called user access layer for the convenience of users Cloud computing services and various support services required by users. The application layer is to provide users with software services. Common types are divided into organizational applications and personal applications. The management layer provides management functions for all levels of cloud computing services.

5 Key Technologies of Cloud Computing

Five key technologies of cloud computing: virtualization technology, distributed massive data storage, massive data management technology, programming methods, cloud technology platform management technology [12]. Cloud computing breaks the physical division between devices through virtualization technology, deploys tasks on a resource pool composed of physical devices and uses technologies such as the Internet, a central remote server and applications to allow users to dynamically obtain the corresponding information according to their own needs. Services to realize the dynamic structure and services. The business computing model built on virtualization technology. Cloud computing is a business computing model built on virtualization technology. Through virtualization technology, the division between physical devices is broken to achieve dynamic architecture. Tasks are deployed on a resource pool composed of computers and maintained by the Internet and a central remote server. The application and data technology required by users allows users to obtain corresponding services according to their needs [13]. The cloud computing provider provides general network application services. These applications can be accessed through a browser, software, or a program with web service functions. The data generated by these software and applications are stored on the provider's server. Cloud computing allows users on the network to share the hardware and software resources on the network, providing automated, highly detailed and personalized services, which has transformed the resource allocation and distribution pattern and improved the sharing efficiency.

6 Conclusions

Cloud computing is setting off a revolution in data technology, bringing us into the real information age, bringing a lot of reflection to production, life, education and even the way of thinking. It also marks that humanity has taken the road of seeking a quantitative understanding of the world a big step. Cloud computing achieves cost reduction, high efficiency, large-scale mass data storage and processing services through unified sharing and flexible invocation of software and hardware resources on the network, transforming the IT industry from the past self-sufficient "natural economy" era to scale The "commodity economy" era of socialization, service, and sharing has brought far-reaching influences on the reorganization of the industrial chain and the adjustment of the industrial structure [14]. Cloud computing can solve various problems encountered in development and construction, including greatly reducing power costs, space costs, facility maintenance costs, software and hardware construction costs, and data maintenance costs, with limited resources, rational use of resources on the network, and improve Overall computing power and resource utilization.

Although cloud computing has unique advantages in many aspects, cloud computing resources are huge, the number of servers is large and distributed in different locations, and multiple applications are running at the same time. How to effectively manage these servers to ensure that the entire system can provide 24 h of uninterrupted Service is a huge challenge. If there is no high security, high reliability as a guarantee, then everything is empty talk. Cloud computing security is a top priority, and it is necessary to protect the cloud computing from the aspects of access authentication setting mechanism. Data security mechanism network stability mechanism trade secret security mechanism standardization of cloud computing standards and support of laws and regulations.

Acknowledgements. This work was supported by the Scientific and Technological Project of Shiyan City of Hubei Province (sysk202057).

References

1. Guo, W., Gong, J., Jiang, W., et al.: OpenRS-Cloud: a remote sensing image processing platform based on cloud computing environment. Sci. China Tech. Sci. **53**(z1), 221–230 (2010). https://doi.org/10.1007/s11431-010-3234-y

2. Qian, L., Yu, J., et al.: Overview of Cloud Computing. IOP Conf. Ser. Mater. Sci. Eng. **677**(4), 042098 (2019). https://doi.org/10.1088/1757-899x/677/4/042098

3. Sun, D.-W., Chang, G.-R., Gao, S., et al.: Modeling a dynamic data replication strategy to increase system availability in cloud computing environments. J. Comput. Sci. Technol. **27**(2), 256–272 (2012). https://doi.org/10.1007/s11390-012-1221-4

4. Chen, Z., Han, F., Cao, J., et al.: Cloud computing-based forensic analysis for collaborative network security management system. J. Tsinghua Univ. **18**(1), 40–50 (2013). https://doi.org/10.3969/j.issn.1007-0214.2013.01.005

5. Wu, H., Zhang, W., Zhang, J., et al.: A benefit-aware on-demand provisioning approach for multi-tier applications in cloud computing. Front. Comput. Sci. **7**(4), 459–474 (2013). https://doi.org/10.1007/s11704-013-2201-8

6. Jaybhaye, S.M., Attar, V.Z.: Adaptive workflow scheduling using evolutionary approach in cloud computing. Vietnam J. Comput. Sci. **7**(02), 179–196 (2020). https://doi.org/10.1142/s2196888820500104

7. Liang, Q., Wang, Y.-Z., Zhang, Y.-H.: Resource virtualization model using hybrid-graph representation and converging algorithm for cloud computing. Int. J. Autom. Comput. **10**(6), 597–606 (2013). https://doi.org/10.1007/s11633-013-0758-1

8. Mishra, N., Singh, R.K.: DDoS vulnerabilities analysis and mitigation model in cloud computing. J. Discrete Math. Sci. Cryp. **23**(2), 535–545 (2020). https://doi.org/10.1080/09720529.2020.1729503

9. Vijaya, C., Srinivasan, P.: A hybrid technique for server consolidation in cloud computing environment. Cybern. Inform. Technol. **20**(1), 36–52 (2020). https://doi.org/10.2478/cait-2020-0003

10. Qi, H., Shiraz, M., Liu, J.-Y., et al.: Data center network architecture in cloud computing: review, taxonomy, and open research issues. J. Zhejiang Univ. (Engl. Ed.) (Part C: Comput. Electron.) **15**(9), 776–793 (2014). https://doi.org/10.1631/jzus.c1400013

11. Yanli, C.H.E.N., Lingling, S.O.N.G., Geng, Y.A.N.G.: Attribute-based access control for multi-authority systems with constant size ciphertext in cloud computing. China Telecom. **13**(2), 146–162 (2016)

12. Dokuz, A.S., Celik, M.: Cloud computing-based socially important locations discovery on social media big datasets. Int. J. Inform. Technol. Decis. Making **19**(02), 469–497 (2020). https://doi.org/10.1142/s0219622020500091

13. Cheng, C., Li, J., Wang, Y.: An energy-saving task scheduling strategy based on vacation queuing theory in cloud computing. J. Tsinghua Univ. Nat. Sci. Ed. **20**(1), 28–39 (2015). https://doi.org/10.3969/j.issn.1007-0214.2015.01.004

14. Gunawan, W., Setyawan, E.: Applying effective cloud computing maturity model (CCMM). IOP Conf. Ser. Mater. Sci. Eng. **725**(1), 012091 (2020). https://doi.org/10.1088/1757-899x/725/1/012091

Talent Evaluation Model of College Students Based on Big Data Technology

Lei Tan[✉] and Yi Guan

Dalian Neusoft Information Institute, 8 Software Park Rd, Dalian, China
tanlei@neusoft.edu.cn

Abstract. At present, most of the talent assessment methods are expert evaluation, leaderless group discussion, psychological test, structured interview and so on. It is easy to have misunderstanding of social perception such as primacy effect and Hawthorne effect. The weak degree of control leads to more subjective influence on the evaluation results. In addition, visualization is difficult to achieve, such as large amount of data and complex statistical analysis process, which often makes the traditional personnel evaluation more difficult to achieve a high degree of fit between trend prediction and reality. In recent years, the rapid development of big data technology makes it possible to establish a talent evaluation model based on big data technology, which can comprehensively consider the aggregation and integration of multi-source data. In the study and life of college students, a large amount of data will be accumulated, which will be conducive to the prediction of future talent behavior after collection and analysis. This paper attempts to collect college students in and out of the classroom, formal and informal learning environment, learning activities and life performance, etc., to obtain the data related to the students' comprehensive quality, carry out comprehensive data and evaluation on the comprehensive quality and ability of college students, so as to provide the basis for students' development and enterprise classification and selection. The platform will provide reference solutions for solving the employment difficulties of college students and accurate recruitment of enterprises.

Keywords: Big data technology · College students · Talent evaluation

1 Introduction

Since the 21st century, the rapid development of China's higher education has trained a large number of high-quality talents for social construction. However, affected by many factors such as system and mechanism, the supply side and industrial demand side of talent training in schools can not fully adapt to the quality, structure and level. According to statistics, in 2018, the number of college graduates will exceed 8 million. How to realize the reasonable planning of College Students' career and the accurate recruitment of enterprises have become a widespread concern of the society. The college students' talent evaluation model based on big data technology provides a feasible method to solve this problem.

© The Editor(s) (if applicable) and The Author(s), under exclusive license
to Springer Nature Switzerland AG 2021
J. MacIntyre et al. (Eds.): SPIoT 2020, AISC 1283, pp. 745–749, 2021.
https://doi.org/10.1007/978-3-030-62746-1_113

2 Analysis on the Necessity and Feasibility of Big Data in the Field of College Students' Talent Evaluation

Talent evaluation is a kind of personnel management activity which measures and evaluates people's knowledge level, ability and tendency, personal characteristics, work skills and development potential. Talent evaluation is the basic link of human resource management, and scientific evaluation personnel is the starting point of all personnel work [1].

At present, most of the talent assessment methods are expert evaluation, leaderless group discussion, psychological test, structured interview and so on. It is easy to have misunderstanding of social perception such as primacy effect and Hawthorne effect. The weak degree of control leads to more subjective influence on the evaluation results. In addition, visualization is difficult to achieve, such as large amount of data and complex statistical analysis process, which often makes the traditional personnel evaluation more difficult to achieve a high degree of fit between trend prediction and reality [2, 3].

The current and early talent evaluation mainly focuses on the investigation of shallow professional ability (such as skills and knowledge), and pays little attention to the employees' potential values, professional quality and personality. In the big data talent evaluation mode, we can cover shallow and potential ability elements by some means.

2.1 Necessity Analysis

Technology brings revolutionary changes to education. In the future, many fields are inseparable from modern information technology. Information technology makes the collection and analysis of human resource data more efficient and convenient. We predict that big data technology will replace the traditional talent evaluation and become the inevitable choice of human resource management in the future.

2.2 Feasibility Analysis

In recent years, with the development of information technology, more evaluation subjects can participate in education evaluation, so as to collect more multidimensional evaluation data. In addition, it is also possible to collect process data accumulated in the process of education.

With the continuous innovation of data collection means, the continuous improvement of data sharing awareness, the continuous increase of students' learning behavior through the Internet, the use of various teaching management software and platforms (formative assessment system, educational administration system, online course platform, etc.) and the emergence of various social products (especially the popularity and continuous use of mobile app applications), etc., the construction of these digital campus The popularity of design and information technology products makes it possible to collect students' massive information conveniently and quickly [4].

3   Implementation of Big Data Talent Evaluation

Combined with the development of data analysis technology and learning science, this paper intends to build a multi-source comprehensive quality evaluation model based on big data technology. The model comprehensively considers the aggregation and integration of multi-source data, expands the collection range of students' objective information to multiple perspectives such as in class and out of class, formal and informal learning environment, learning activities and life performance, and obtains data reflecting students' comprehensive quality, forming a systematic big data of College Students' growth. After that, by integrating and standardizing the data that can reflect the comprehensive quality of students, a talent evaluation model for college students is established, which will provide reference solutions for solving the employment difficulties of college students and accurate recruitment of enterprises. The evaluation model of College Students' talents can be studied in stages as follows:

3.1   Early Stage of Talent Evaluation: Design Talent Evaluation Scheme Based on Big Data

The difference between big data talent evaluation and traditional talent evaluation lies in the multi-dimensional or full dimension of behavior data collection, that is, in the design of talent evaluation scheme, almost all factors related to talent performance should be covered [5].

Data collection should run through all levels of professional ability. Professional ability generally refers to all the individual factors that can affect an employee's work performance and behavior performance. Both iceberg model and onion model can be used as reference models to understand professional competence. We take onion model as an example. Referring to onion model, the ability can be divided into five levels: professional personality style and tendency, professional values, professional quality, professional knowledge and skills, and professional behavior.

4   The Conclusion and Enlightenment of the Research on the Model of College Students' Talent Evaluation

The results of comprehensive quality evaluation based on big data can be used in individual career planning and personality development, enterprise recruitment and qualitative training, university running service improvement and government macro governance, so as to realize the potential value of big data talent evaluation.

4.1   Providing a Reference for the Career Planning and Personality Development of College Students

Big data talent evaluation is an important means to discover and cultivate college students' good personality by observing, recording and analyzing the overall development of college students. The application of big data technology can conduct digital portraits for individuals and then for college students [6]. The digital portraits of college students are of great value to students themselves, teachers, universities and recruitment enterprises. Students can better understand themselves, understand their abilities, hobbies, strengths and weaknesses, know their strengths, and make adjustments for their own learning and career planning.

4.2 Providing Data Support for More Accurate Recruitment and Targeted Training for Enterprises

According to the digital portrait, the employing enterprise can carry out intelligent and personalized screening on the evaluated personnel, so as to accurately match the positions, and truly achieve the matching of personnel and posts. Through comparative analysis, we can find the talents needed by enterprises.

In addition, the value of employees with different abilities and qualities is different for the enterprise. Combining with the comprehensive quality evaluation of college students, enterprises carry out targeted training and training for different individuals. Make the best use of people. Adopt multi-channel career development path to carry out personalized training for talents [7]. At the same time, enterprises can dynamically predict the professional level of employees through big data technology, design training suitable for their development, and improve the stability of employees [8].

4.3 Providing Services for School Running and Improvement and Government Education Governance

Based on the big data technology, the university student talent evaluation system also has an important service function. On the basis of individual digital portrait, the evaluation system can also draw the group digital portrait of college students by cluster analysis, which mainly serves the macro education governance of the University and the government [9, 10].

Acknowledgements. 2018–2019 Liaoning Science and Technology Innovation Science and Technology Think Tank Project, Liaoning Province Science and Technology Talents Innovation and Entrepreneurship Ecological Environment, project number LNKX2018-2019C37;

The key project of the Social Science Planning Fund Project of Liaoning Province in 2018, the construction of talent ecological environment in Liaoning Province based on digital transformation, project number L18ARK001.

References

1. Ke, L.: Opportunities, challenges and transformation and upgrading of human resource management in the era of big data. J. Jinhua Vocat. Tech. Coll. **15**(4), 35–40 (2015). (in Chinese)
2. Expósito-Langa, M., Molina-Morales, F.X., Capo-Vicedo, J.: New product development and absorptive capacity in industrial districts:a multidimensional approach. Reg. Stud. **45**(3), 319–331 (2011)
3. Hamilton, E.: Entrepreneurial learning in family business: a situated learning perspective. Small Bus. Enterp. Dev. **3**(1), 45–68 (2011)

4. Zhang, X.: Investigation and analysis of network psychology of college students. Int. J. Min. Sci. Technol. **23**(2), 184–191 (2013). (in Chinese)
5. Hoff, K.E., Ervin, R.A.: Extending self- management strategies: the use of a classwide approach. Psychol. Schs. **50**(2), 151–164 (2012)
6. Zhang, X.: Substantial study on the occupation development of gender equality. Int. J. Mining Sci. Technol. **22**(5), 625–628 (2012). (in Chinese)
7. David, W., Jerry, S., Jeffrey, H.: Global self—esteem in relation to structural molds of personality and affectivity. J. Pers. Soc. Psychol. **83**(1), 183–197 (2002)
8. Sun, X., Xue, G.: Construction and validation of self-management scale for undergraduate students. Creative Educ. **2**(02), 142 (2011). (in Chinese)
9. Conger, J.A., Ready, D.A.: Rethinking leadership competencies. Leader Leader **2004**(32), 41–47 (2004)
10. Botwin, M.D., Buss, D.M.: Structure of actreport data: is the five–factor model of personality recaptured. J. Pers. Soc. Psychol. **56**, 988–1001 (1989)

Application of Cloud Class in Comprehensive English Teaching in the Context of Internet Plus

Linhui Wu[✉]

Department of Applied Foreign Languages, Chengdu Neusoft University, Chengdu, Sichuan, China
15413460@qq.com

Abstract. The development of information technology in college education is constantly changing the learning style of college students. The teaching mode based on internet plus background integrates two ways of Online Self-learning and Face-to-face Teaching Approach. It is the central research topic in the reform of professional English teaching. This paper will focus on how the cloud class education platform can be applied in Comprehensive English Teaching under the background of Internet plus. The results show that the model can motivate the learning enthusiasm of students, develop their self-study skills, and reinforce teachers' role of guidance in teaching.

Keywords: Cloud class · Comprehensive english · Internet plus

1 Introduction

As the booming of information technology in college education, information technology is ceaselessly changing the learning methods of college students. The current situation inevitably render educators cogitative on how to create the teaching mode based on information setting, exert teachers' leading function, improve students' learning enthusiasm, develop dynamic learning atmosphere, and upgrade the diversified development of English major basic teaching [1]. These issues are the problems that cannot be denied in the teaching reform of English major and the improvement of English teaching standard. As the essential course of English major, Comprehensive English course occupies a very important position in English teaching. Along with the flourishing of teaching information technology, the traditional single classroom teaching mode can no longer satisfy the teaching requirements of the course [2]. Therefore, the study aims to use cloud class learning platform to study its application in Comprehensive English and further improve the measures in blended learning.

2 Introduction of Cloud Class Teaching Platform

Cloud class is a cost-free interactive teaching software combined with artificial intelligence. On the basis of the mobile Internet settings, it can realize the real-time interaction

J. MacIntyre et al. (Eds.): SPIoT 2020, AISC 1283, pp. 750–754, 2021.
https://doi.org/10.1007/978-3-030-62746-1_114

function between teachers and students, push resources and assign after-class tasks. The delicate motivation and evaluation system can stir up students' interest in self-learning with mobile devices, and elaborate learning activity records can realize the process evaluation of students' learning [3]. It can still offer teachers with superb teaching and research mega data, and realize artificial intelligence on the grounds of technology of personalized intelligent study and teaching assistant function. Cloud class mainly has the following characteristics: 1) It is a free mobile application software; 2) It is completely designed for mobile teaching scenarios, with good mobile application experience, and can also be used on the web version; 3) Teachers can easily manage their own classes on any mobile device, effectively realize resource sharing, and teachers can push various electronic resource packages related to teaching content before class to the cloud platform for students to download and use. 4) Students can get experience value every time they study resources and participate in activities. Teachers can increase their experience value by praising students, which is helpful to stimulate their learning enthusiasm and desire to participate [4]. 5) It has a variety of classroom teaching auxiliary functions, including classroom testing, voting, discussion and brainstorming activities. 6) It can record students' learning behavior in details. When the tasks are completed, teachers can set corresponding experience values to record the students' participation. After the activities, teachers can output detailed learning analysis reports from the platform, summarize and generate process evaluation results, so as to facilitate teachers' teaching reflection.

3   Current Situation of Comprehensive English Teaching

Being the foundation course of English major, Comprehensive English has always been the top priority of English teaching reform [5]. Teachers of English major in our university have been on the way of teaching exploration, looking for a suitable teaching mode for English Majors in application-oriented universities. Although some achievements have been made in the early stage of reform, Comprehensive English course has been dominated by the previous classroom teaching, supplemented by students' after-school learning, and developed into a multi-dimensional teaching mode of classroom teaching, platform interactive learning, subject competition project embedding and students' extracurricular learning, especially the combination of platform interactive learning plate and subject competition with comprehensive English curriculum and platform. At the same time, the evaluation method of Comprehensive English curriculum has been adjusted accordingly, and the evaluation subject, evaluation objective, evaluation content and evaluation method have been promoted in turn. The evaluation subject has been shifted from the previous teacher assessment to the way in which the leader of the study group, student group leaders and students themselves participate in the assessment. The evaluation objectives are diversified, mainly including the language knowledge application, learning attitude and learning skills. The evaluation content typically involves the assessment of students' learning procedure, learning skills and learning effectiveness. The evaluation method mainly consists of process evaluation and terminal evaluation.

However, on the basis of diversified educational information technology and personalized learning methods, there are still some problems to be settled in the teaching of Comprehensive English: 1) students' self-learning ability is insufficient and the curriculum dependence is strong. Many students put their English learning on the limited teaching time in the classroom, and there is not enough autonomous learning after class. 2) Students' English level is uneven, and the traditional teaching mode cannot satisfy the personalized requirements of students. 3) The traditional teaching still takes up most of the class time. 4) Teachers lack of timely and effective supervision of students' learning.

4   Application of Cloud Class in Comprehensive English Teaching

Combined with the characteristics of Comprehensive English course, especially the particularity of the epidemic teaching, this study takes cloud class learning platform as the carrier to explore its role in Comprehensive English course [6]. As a free learning platform, cloud class can be used for mobile platform and PC platform [7]. Teachers can create and manage classes through any mobile device to achieve online and offline teaching interaction and feedback, which provides a very good platform for blended English teaching [8]. According to the teaching practice, this study will explore and study the use of cloud class before, during and after class.

4.1   Guidance Before Class

In the course preparation stage, teachers create cloud classes on the computer, and each teaching class will get a class number accordingly. After the teacher releases the class number to the students, the students join the corresponding class according to the class. On the cloud class platform, students can read the preview tasks published by the teacher according to the unit classification, and upload preview resources (such as unit background knowledge, vocabulary expansion, cultural points, etc.). Students receive preview tasks according to their groups and complete the tasks assigned by the teachers of this unit. At the same time, the group leader collects the difficult problems encountered by the students in the preview, and publishes them in the discussion area of cloud class as a unit, and the teacher answers questions online to solve the problems encountered by the students in the preview. This process realizes online and offline blended learning, students can clearly understand the learning task and the teaching objectives of this unit, and teachers can also master the learning situation of students at this stage.

4.2   Learning in Class

In the learning stage, teachers mainly use inquiry teaching mode, combined with cloud class platform to clear the blind spot of knowledge for students and promote the accumulation of knowledge of students.

According to the students' preview situation mastered by the teachers in the pre class guidance stage, teachers create corresponding learning tasks according to the specific situation of each class in the learning stage of the class.

In this stage, it mainly involves text understanding and appreciation, language knowledge, unit project and summary. The tasks of text comprehension and appreciation include text structure and topic analysis; language knowledge tasks include comprehension and translation of long and difficult sentences, core vocabulary and grammatical structure; unit project tasks include topic report, role play, unit test self-evaluation and explanation, etc. These tasks are assigned to students according to different teaching stages. Students can get the tasks through the task release column of cloud class, and prepare tasks in groups. The group leader organizes the group members to discuss the tasks assigned by the group through WeChat or QQ to share the learning materials and exchange the learning results, so as to prepare for the display of classroom learning achievements, and timely feedback the completion status on the cloud class platform. Teachers can pay attention to students' task completion in software, and answer questions on line that students may encounter. In this context, teaching is no longer a single teacher to transfer knowledge, but to encourage students to connect the original knowledge with the new knowledge. At the same time, it can promote the accumulation of students' knowledge and cultivate their ability to think independently and solve problems. Teachers can give the corresponding experience value according to the task completion status uploaded by students on the platform. In this way, students are the center. Finally, teachers guide students to summarize and reflect on the classroom, and answer questions and offer guidance.

4.3   Consolidation After Class

After class online learning, the teacher will demonstrate the problems students encountered in the cloud class platform to consolidate practice. After the end of each unit, students finish the unit project and unit test. The unit test of cloud class is very flexible, which can complete the arrangement of subjective and objective questions. After completing the test, students can view the test results in real time and get the corresponding experience value. The experience value obtained by students will be included in their usual performance, and the whole link is closely linked. Such a way not only stimulates the sense of competition among students, but also promotes the learning group to continuously improve the outcome of completing tasks, which greatly improves the learning motivation of students. Teachers can also observe students' learning progress at any time in the cloud class platform and reflect on teaching.

5   Conclusion

The blended teaching mode represented by cloud class can take an active role in English teaching [9]. It has adjusted the traditional classroom mode, emphasizing the combination of leading bodies, and has brought the students' learning motivation and teachers' role as guidance. Meanwhile, this teaching style also puts forward higher demands for teachers. Teachers should not be aware of teaching method, but also how to promote students' learning. Teachers should focus on the following issues: 1) how to promote students to form effective knowledge construction [10]. When creating different learning situations through multimedia, teachers should pay attention to trigger students' positive emotional experience, deeply understand and transfer the knowledge to be mastered, and construct high-level thinking and cognition. 2) Give feedback students' problems in learning actively, and provide different communication strategies in different stages of students' learning. Mixed teaching has changed the traditional classroom interaction mode [11]. Teachers should exert their energy on the procedure of communication with students, and integrate classroom communication with online communication. 3) Be aware of the process management of students' learning. The use of blended teaching mode requires teachers to apply the network platform with experienced skills, supervise the students' learning situation through the platform statistical data, record and feedback the students' learning process. Promote the formative evaluation of students. Such regulation can also be counterproductive to students, so that they can better understand the fun of active learning.

References

1. Wenjuan, Q.: Research on blended learning evaluation based on cloud class in public courses of vocational colleges. Vocat. Educ. Forum **6**, 669–670 (2018)
2. Yueguo, G.: Analysis of multimedia and multimodal learning. Foreign Lang. Audio Vis. Teach. **2**, 10–12 (2007)
3. Yan, L., Ning, M.: Process based evaluation and practice of blended teaching reform based on cloud class. Educ. Modernization **15**, 60–61 (2018)
4. Xiaoli, L., Guodong, J.: exploration of college english teaching reform based on mobile learning. China Vocat. Tech. Educ. **29**, 90–91 (2016)
5. Haiying, Y.: Exploration and practice of blended learning mode based on network course platform. Coll. Educ. Forum **11**(11), 86–87 (2012)
6. Fang, Y., Xing, W., Wenxia, Z.: Analysis of college english with blended teaching mode. Foreign Lang. Audio Vis. Teach. **1**, 62 (2017)
7. Yao, Yu.: Research on the mobile english teaching mode under internet plus background. Mod. Commun. **10**, 70 (2018)
8. Huanrun, Q.: The construction of multidimensional teaching mode of thinking ability in college english teaching in internet plus era. Mod. Inform. Technol. Foreign Lang. Teach. **6**, 114–117 (2017)
9. Xiaoying, F., Ruixue, W.: A review of research on blended teaching at home and abroad. J. Distance Educ. **3**, 13 (2018)
10. Jianbo, Z.: Research on the application of cloud class+n in college english mobile teaching mode. Lang. Policy. Lang. Educ. **2**, 90–110 (2018)
11. He, G., Ren, Y.: A preliminary study on the mechanism of college english teaching reform. Heilongjiang High. Educ. Res. **15**(3), 123–135 (2010)

The Application of 3D Printing Technology in Sculpture

Jingxue Yu[✉]

Department of Sculpture, Hubei Institute of Fine Arts, Wuhan 430205, Hubei, China
yujingxue475@sina.com

Abstract. With the development of society, we have entered the era of digital information. The popularization and application of computers is also important for art, and provides inspiration and production methods for artistic creation. Sculpture requires extremely high craftsmanship and complex production processes, and makers are required to be years of training and experience. 3D printing technology use computer data as a basis to convert virtual data into real objects, directly skipping the traditional and complicated process. The rise of 3D printing technology makes sculpture completely bid farewell to the manual era and enter the era of digital design and manufacturing. Sculptors design sculpture models by relying on computers, which is conducive to promoting the development and innovation of sculpture art. 3D printing technology can display the pictures depicted in the hearts of artists and sculptors in the form of real objects completely, which can maximize the artistic expression.

Keywords: Sculpture · 3D printing technology · Computer

The sculpture has a history of thousands of years and belongs to Chinese intangible cultural heritage. It has gone through thousands of years of development, but the process of sculpture production has not changed much [1]. The sculpture is still made by hand. The past technological advances have not brought great changes to the sculpture process. Sculpture is famous for its complicated production process and craftsmanship [2]. Many people want to learn to design sculpture, but they give up due to its complicated process. The 3D printing technology has transformed a purely manual technology into a high-tech work controlled by computer. The promotion and popularization of 3D printing technology enables people to design their own sculptures like photographs, which has promoted sculpture technology, an intangible cultural heritage to a certain extent [3]. 3D printing technology has changed the traditional single creative method, and can quickly form a complete sculpture on the basis of modeling, reducing the time required to get from the plan to the actual object. It has rich printing materials, which can give sculptors and artists more creative inspiration, so that they can produce more diverse sculpture crafts.

J. MacIntyre et al. (Eds.): SPIoT 2020, AISC 1283, pp. 755–759, 2021.
https://doi.org/10.1007/978-3-030-62746-1_115

1   Problems in Application of 3D Printing Technology

1.1   High Cost, Long Working Hours and Low Efficiency

3D printing technology is an expensive technology, which costs a lot. Due to the difficulty of developing materials and the narrow application area, the cost of 3D printing technology is high and the manufacturing efficiency is low [4]. Compared with the traditional processing method, the traditional processing is to reduce the materials to process on a certain basis. Generally, the difference between the raw embryo and the final processed parts is relatively small, and the processing is relatively fast; and 3D printing technology is to process the parts' materials separately, and finally stack them to synthesize the final parts. And the cost is relatively high. 3D printing technology has the advantages of distributed production, but it has no advantages in scale production. In some items that need mass production, 3D printing technology can not play its role.

1.2   Low Quality and Accuracy

In the field of sculpture, the accuracy of sculptures produced by 3D printing technology is still unsatisfactory, so 3D printing technology can not replace the traditional sculpture production for the time being. 3D printing adopts a technique of stacking materials layer by layer. Even if each material is stacked carefully, it can't be compared with sculptures made by traditional craftsmen [5]. As the manufacturing method of 3D printing technology is not perfect, the materials printed by 3D printing technology cannot get rid of the defects of rough surface, imperfect size and shape. We can't make meaningful crafts that can be spread in the world. We can only use them as a means of entertainment to encourage people to get close to sculpture and understand its culture. It can also be a new type of sculpture that sculptors can create and improvise.

1.3   Material Limits

Another defect of 3D printing technology is that it comes from the materials it makes. At present, the materials used for 3D printing are mainly plastics, gypsum, metals and ceramics. Due to the limitation of printing materials, printing technology can not be applied to some materials needed in industrial production. The limitation leads to the fact that 3D printing technology can not be widely used, but can only be used in few special aspects [6]. On the other hand, some materials can only be printed by their corresponding 3D printers, that is, one printer can only print several materials, which leads to the need to make a special printer suitable for different materials when printing other materials. It is undoubtedly a time-consuming and labor-consuming problem. Although 3D printing technology has made great progress, the above problems are still a big obstacle to the development and promotion of 3D printing technology.

2   Solutions to the Problems in the Application of 3D Printing Technology

2.1   Solutions to the Problems of High Cost, Long Working Hours and Low Efficiency

To solve the problem of high cost, a certain amount of materials can be added to the basic materials (hairy embryo), which can reduce the demand for materials to be printed and make up for the disadvantage that 3D printing cannot be mass produced [7]. In view of the problem of long working hours and low efficiency, we can design a machine that can move at high speed to shorten the production process of printing materials, and can also coordinate the design of materials needed for each part for batch production to speed up its production. The production of materials should not be done as a whole, but should be reasonably planned and carried out in batches, and the materials should be divided into different small parts, so that a batch of materials can be used for multiple parts and the final assembled parts can be more precise and meticulous.

2.2   Solutions to the Problems of Low Quality and Accuracy

For problems of quality, we can carefully check the printing process of the printer to create a more delicate texture. By constantly changing the printing strength of the printer, we can find the most intact strength of manufactured materials for mass production, which can prevent the printed materials from being incomplete or damaged. Sampling survey should also be carried out strictly for the manufactured materials to filter out defects and leave good products with quality assurance. And a report should be made for each sampling survey to point out the problems in the process, and improve the printing technology [8]. In terms of accuracy, the resolution of the printer can be appropriately increased to produce more exquisite and detailed materials, so that the assembly of final parts is more complete.

2.3   Solutions to the Problems of Material Limitations

In terms of the problem of material limitation, new materials can be developed, which is also the development direction of the country for material production. Through the research and development of new materials with good printing performance, it can also meet the requirements of traditional materials to produce more artistic sculptures and improve the application value of 3D printing technology [9]. New printed materials can be better preserved to prevent hard-made sculptures from being stored for a long time. New materials also need to be light-weight and not easily destroyed. A perfect sculpture often contains the story and experience of a sculptor. It will be heartbreaking if it is accidentally broken or destroyed in the future. And the new printing materials are best to be green and harmless, which can also protect the environment, which is in line with our current national conditions. In addition, the 3D printer should be updated and optimized, so that one printer can print a variety of materials, which can well avoid the trouble of constantly changing the printer to print different materials.

3 Value and Advantages of Sculpture Technology in the Field of Sculpture

Nowadays 3D technology is involved in many fields, and many sculptors have discovered its application value in sculpture. Applying these modern technologies to traditional artistic creation has a kind of cooperation that breaks through the boundary, and can promote sculpture, an intangible cultural heritage that gradually fades out of people's vision, and make it enter people's daily life again. Many artists have used this new technology, but they all use different methods. Some people continue to use 3D printing technology to improve and rework the works based on the sculpture they have made. Some people build a model after the sculpture is completed and print the complete sculpture model; others use computer modeling and then 3D printing to get a complete model [10]. The interactive fusion of modern technology and traditional art has expanded the language of sculpture to a certain extent, and has made innovation and new development in the advancement of sculpture technology. 3D printing has greatly changed the creation of sculpture. Some classic sculptures made by masters may be difficult to restore in the eyes of ordinary people, but 3D printing technology can restore the original appearance of the sculpture to a great extent, which makes later sculptors understand the master more clearly and learn the technique of sculpture from it [11]. In the past, large-scale sculptures required a lot of manpower and material resources. which are time-consuming and laborious, and the sculptures are likely to fail to achieve proportion coordination. But with the aid of 3D printing technology, large-scale sculptures have become a breeze. As long as each part of the material required for large-scale sculptures is made, these materials are finally combined together to construct a perfect sculpture.

Compared with traditional sculpture technology, 3D printing technology has more convenience. (1) It can print out the large body shape of sculpture in a short time, which is convenient for more complete supplement of sculpture. (2) Print with computer is convenient, which saves manpower and material resources. (3) 3D printing technology is more environmentally friendly and pollution-free. Many printing materials are organic and recyclable materials that are harmless to the environment, and will not cause pollution to the environment [12]. (4) The 3D printing technology adopts the superimposed sculpture mode. The demand for materials is not as much as the traditional sculpture technology, which saves materials and avoids wasting money. (5) The materials printed by 3D printing technology are more accurate than those made by traditional crafts, which can measure materials in smaller counting units. Only more accurate materials can produce more perfect sculpture crafts. (6) The sculpture made by 3D printing technology is easier to operate, and there are not so many complicated steps and processes. As long as the proportion of modeling and planning is well done on the computer, we can make our own sculpture with one click.

4 Conclusion

To sum up, with the development of the times, the traditional sculpture technology is gradually replaced by 3D printing technology. Today, the society has entered a fast way of life. Compared with the long process of traditional sculpture production, the production of 3D printing technology is undoubtedly an inevitable result. In today's fast life style, people pay more attention to efficiency and time, and all their energy is spent on work, not too much energy is wasted in making a sculpture, and the sculpture technology that needs time and energy has gradually faded out of people's view with the development of society. But the application of 3D printing technology in sculpture makes the abandoned traditional technology return to people's daily life. Compared with traditional sculpture which must be completed by experienced craftsmen with complex technology, sculpture now seems to be an entertainment method that everyone can try. As long as people design the sculpture in their mind on the computer, they can quickly print out a complete object. 3D printing technology can promote sculpture technology, an intangible cultural heritage, so that people can feel the charm of sculpture. Only when people establish their interest in sculpture, can they spend time to understand the development history and cultural background of sculpture, so as to prevent the sculpture technology from being abandoned. 3D printing technology is a kind of technology that connects traditional technology and modern technology. It plays a milestone role in the development of sculpture, and establishes a bridge between traditional sculpture and modern sculpture, so that sculpture can adapt to today's society and get better development, which is worthy of promoting and using.

References

1. Leng, K.: Research on Urban Digital Sculpture Design Based on 3D Printing Technology
2. Yin, W.: Application of 3D printing technology in sculpture. Sculpture (3) (2017)
3. Dong, C.: Research on the Application of 3D Digital Modeling in Sculpture
4. Xue, C.: Application and Research of Digital Technology in Sculpture Creation (2015)
5. Ming, Y.: Research on the application of 3D printing technology in the preliminary design of urban landscape sculpture. City Architect. **26**, 326 (2016)
6. Liu, X.: Research on the application of 3D printing technology in public sculpture. Cult. Mon. (6) (2018)
7. Xu, L., Guanrong, C.: An exploration of the in-depth application of 3D printing technology in sculpture creation. New Gener. Theory Ed. (004), 28 (2019)
8. Li, G.: Research on the application of 3D printing technology in modern sculpture creation. Art Works (32) (2018)
9. Zhen, W.: Research on the Potential Characteristics of Concrete Media in Contemporary Urban Sculpture
10. Lei., H.: Application of Computer Reverse Engineering Concepts and Techniques in Sculpture Practice
11. Juanjuan, H.: Thinking about the application of 3D printing technology in mathematics teaching. J. Suzhou Educ. Inst. **17**(4), 183–184 (2014)
12. Yi, W.: The application and thinking of 3D printing technology in art. Chinese Fine Arts (003), 55 (2016)

The Development Strategy of Current Medical Scientific Research Information

Qingxi Huang and Peng Ruan[✉]

The First Affiliated Hospital of Chengdu Medical College, Chengdu, Sichuan, China
1098998556@qq.com

Abstract. Due to the impact of COVID-19 pandemic, there is a high demand for hospitals all over the world, and hospitals also bear huge medical pressure and responsibility. Medical research is the driving force to promote the continuous development of medicine, and how to do a good job in the development of medical research informatization is one of the keys. This study from China's large public hospitals as the research object, pointed out the shortcomings and problems of hospital scientific research management, put forward that the development of information technology is the relatively backward solution of scientific research management in hospitals, and then put forward the development strategy of medical research informatization based on the current situation.

Keywords: Medical research project · Information development · Information technology · Hospital management

1 Introduction

According to the development goals of the Chinese government, it is planned that by 2020, China's digital construction will not be as good as the fast track, the construction of Digital China will achieve remarkable results, and the level of informatization development can enhance the international competitiveness of enterprises [1]. However, due to the influence of COVID-19 pandemic [2], the development of medical informatization has been greatly hindered. The reason is that hospitals are busy dealing with COVID-19 pandemic, treating patients with COVID-19 pandemic and preventing the spread of the epidemic. In the current process of hospital development and upgrading, the level of information development is also one of the core indicators of hospital evaluation [3]. At present, some large public hospitals are still in the business of diagnosis and treatment, and the development level can not meet the overall requirements of the state [4]. The development of medical information is not enough. The scientific research level of a hospital determines its scientific and technological innovation ability [5]. Therefore, information management of medical research management can not only solve the management problems that restrict the development of scientific research [6], but also help to play a great role in fighting against the COVID-19 pandemic epidemic.

J. MacIntyre et al. (Eds.): SPIoT 2020, AISC 1283, pp. 760–764, 2021.
https://doi.org/10.1007/978-3-030-62746-1_116

2 Problems in the Management of Medical Scientific Research in Public Hospitals

2.1 Problems in Management Methods

At present, the common methods of medical research management include: registration and filing with paper materials, paper document notification, telephone notification, etc. For example, the filing of experimental materials, papers, project documents, etc., is often lack of electronic filing, so that some important documents and materials are damaged and lost due to staff changes and office relocation; various paper notices are relatively cumbersome, and need to send people to each department to sign for receipt, which is also not conducive to custody and transmission; there are many disadvantages of telephone notification When we call the relevant personnel, we can't get through, because the clinical medical staff may be in the process of diagnosis and treatment, surgery, etc., and the content of the phone is often remembered, and it is also common to forget or mistake the time and place later. At present, when applying for scientific research projects, the scientific research management department usually holds mobilization meetings, and often because medical staff are busy with diagnosis and treatment and surgery, resulting in many shortages Seats.

2.2 Analysis of Some Abnormal Phenomena

First, the enthusiasm of medical staff in scientific research is not high, the number of scientific research projects declared and the rate of project approval are low. Because of the inconvenience of scientific research declaration and other work, the enthusiasm of some medical staff is not high. Second, the project is approved, but the implementation is less, and the project completion rate is low. Some projects have the situation that only the project is approved but not implemented. After some project leaders have successfully promoted their professional titles with the aid of project projects, they have been neglected in management. The completion of many projects is worrying, and can not meet the completion requirements. It is difficult for scientific research management departments to effectively supervise and assist scientific research projects by traditional means, thus losing the management effect for some scientific research projects, and the scientific research management work is just like a void.

In addition, the output of medical papers is low and the quality of publication is low. Due to the lack of adequate scientific research management, the output of scientific research papers will be less, and the quality of published papers will be low. Many authors are not clear about what is the core journal and how to choose a reasonable contribution. However, the lack of corresponding measures to guide the scientific research management departments and help the authors to submit contributions reasonably leads to the papers published in lower level journals, and even some authors are published in supplements and illegal journals.

2.3 The Dissemination of Scientific Research Information Is Blocked

Scientific research information is blocked, which leads to the failure of sharing, and the different parts act on their own and lack of cooperation. In fact, medical scientific research management also involves the reimbursement of scientific research funds by financial departments, asset registration management of state-owned assets management department, and even related to personnel title work. Because the main information still relies on traditional manual transmission, it is difficult to share information in real time. For example, the management of scientific research project funds may appear that the project funds are not received or the funds have been overspent, but the scientific research management departments do not know. Although the financial management department has its own management system, the system can not be opened to other departments. The biggest problem of information sharing lies in the transmission of scientific research information to medical staff. Sichuan University is not timely, resulting in information failure.

2.4 The Management Personnel of the Department Do not Pay Enough Attention to Informatization

The sources of management personnel are complex and lack of professionalism. Because medical research managers are often selected from clinical medical staff, there are also certain public health management talents. This kind of talents often carry out scientific research management according to policies, personal experience and their own feelings. Although they also encounter management problems and problems and are trying to solve them, they often rely on their own efforts to make a limited improvement. Therefore, many scientific research managers know little about information construction, and never think of using information construction to solve the problems of scientific research management. The professionals who are proficient in computer and information construction are often concentrated in the information department and network office of the hospital. Similarly, due to the lack of work coordination and information sharing, departments often act on their own affairs, so although there are talents for information construction in the hospital, the scope of their play is limited to one department, which can not promote other departments.

3 Conclusions

3.1 Development of Information Management Software

The main tool of informatization is management software. The design of hospital scientific research information management system should be combined with the characteristics and needs of the hospital itself. At present, many major scientific research project applications have realized online declaration and management. The design of the hospital's own scientific research information management system should be combined with the actual situation of the hospital to carry out personalized and humanized system design with reference to the application system of superior competent departments. For example, specialized hospitals and general hospitals should be different, and the design complexity of large hospitals and small hospitals should also be different, so as to avoid the waste of resources or excessive design. The construction, use and improvement of scientific research information management system should have strict rules and regulations to ensure the safe and effective operation of the whole system. For example, the prevention of viruses and hacker attacks. At the same time, the authority of users should be strictly checked. For example, the staff of scientific research management department should have different authority, and the authority of the person in charge of scientific research project and its members should also be different, so as to effectively prevent the interference and serial modification of human factors. Only the safe operation system has enough authority and effectiveness.

3.2   Establishment of Management System for Informatization Development

Management system is to ensure the implementation of information laws and regulations, enhance the implementation of information.

First of all, in the future, all kinds of project declaration, conclusion, thesis publication, reimbursement registration, academic conferences, etc. will be reported and fed back through the information system, so as to fully realize the goal of scientific research management and scientific research office with network, electronic and paperless. Speed up the elimination of traditional and backward scientific research information management means. Secondly, scientific research management departments provide powerful scientific research assistance to scientific and technological personnel through information management platform. For example, the preliminary examination and examination of papers, the guidelines for contributions to core journals, the sample display of excellent papers and outstanding project applications, etc. through the information management platform, scientific and technological personnel can learn by themselves and improve their scientific research ability. In addition, a good reward and punishment system for scientific research should be established. The intensity of punishment and reward must be able to attract enough attention of scientific and technological personnel. For example, the key reward project, which does not finish on schedule, will be given economic punishment.

3.3   Establish Internal Information Sharing Mode

Hospital management departments should strengthen cooperation with relevant departments. For example, we should cooperate with the financial department in the management and use of scientific research funds, strengthen the cooperation with the personnel management department in the identification of scientific and technological achievements, and strengthen the technical cooperation with the information department. At the same time, it is also necessary to strengthen the communication with clinical departments and even with each scientific and technological personnel through the information construction. When the hospital has reached a consensus on information construction, the cooperation between departments and the development of information construction will become smooth.

References

1. Funayama, M., Kurose, S., Kudo, S., Shimizu, Y., Takata, T.: New information technology (IT)-related approaches could facilitate psychiatric treatments in general hospital psychiatry during the COVID-19 pandemic. Asian J. Psychiatry **54** (2020)

2. State Council. Notice on printing and distributing the 13th five year plan for national informatization [EB/OL]. http://www.gov.cn. Accessed 27 Dec 2016
3. XiaoYing, Z., Zhang, P.: Discussion on the role of information construction in public hospital reform. Chin. Hosp. Manage. **36**(9), 76–77 (2016)
4. Lina, W., Zhang, D.: Investigation on the current situation of informatization construction in 37 hospitals in Henan Province. China Health Stat. **34**(3), 492–493 (2017)
5. Cai, X.: Problems and countermeasures in the management of scientific research funds in public hospitals. Friends Accountants **30**(27), 79–81 (2013)
6. Information Technology - Data Warehousing and Mining: Reports summarize data warehousing and mining findings from taichung veterans general hospital (integrating feature and instance selection techniques in opinion mining). Comput. Technol. J. (2020)

On the Development of the Industry Trend of "Ai+Education"

Dongmei Liang[(⊠)]

Marxism College, Jilin Institute of Business and Technology, Changchun, Jilin, China
373020893@qq.com

Abstract. Artificial intelligence+education can redesign and present a new learning space and environment. With the help of Internet of things, big data, cloud computing, vr/ar and other technologies, through learning measurement and online learning platform, learners can acquire more and more effective learning experience and results, and get practical knowledge and skills of problem solving. At present, in the process of promoting "Ai+education" jointly by schools, enterprises and governments, there are many new educational industry trends, and all of them are developing towards the education of the masses, the third party educational services, the support of standard construction, and the formation of the educational industry chain.

Keywords: Artificial intelligence+Education · Industry trend · Development

1 Introduction

Establishing a learner centered educational environment, providing precise push education services, and realizing the customisation of daily education and lifelong education requires the trend of "Ai+education", which is fully co operated by the government, enterprises and schools. The government is the vane of the development of "Ai+education", and formulating guiding policies and standards for the sustainable development of "Ai+education". It provides scientific guidance and service for enterprises to develop "Ai+education" products. Enterprises rely on advanced AI technology in the education industry to provide diversified personalized educational products and services for the society, and open the door to school education [1]. Promoting education is gradually facing the society and the market. Schools provide excellent teaching theories and advanced learning resources for the realization of "Ai+education", providing a systematic educational theoretical guidance for the realization of "Ai+education"; at the same time, the school is also the main practice base of "Ai+education".

At present, under the situation of "Ai+education" promoted by the government, enterprises and schools. A number of new educational trends have emerged, and all of them are developing towards the development of education, the third party education service, the support of standard construction, and the formation of the education industry chain.

J. MacIntyre et al. (Eds.): SPIoT 2020, AISC 1283, pp. 765–768, 2021.
https://doi.org/10.1007/978-3-030-62746-1_117

2   All the Intelligent People in Education Create Lots of People

On July 17, 2019, Premier Li Keqiang chaired a State Council executive meeting. The meeting held that the need to build a support platform for promoting entrepreneurship and innovation for all. We should use the "Internet +" to actively develop new models such as public innovation, crowdsourcing, crowdsourcing and crowdsourcing to promote the docking of production and demand, effectively converge resources and promote economic growth, and form a new pattern of innovation driven development [2]. "Ai+education" is based on "Internet+education". It is the further development of "Internet+education". On the basis of open education ecology, it emphasizes individualized education service. "Internet+education" has greatly increased the type and quantity of educational users. This undoubtedly brings pressure to personalized education services. On the one hand, "Ai+education" adopts advanced AI technology to provide personalized educational services for the majority of users. On the other hand, through the "wisdom of all" mechanism, it integrates the wisdom of the masses and optimizes its services. It provides open and sharing educational services through the development of educational creation space and platform, and gathers talents of various kinds of AI, education and other industries. It supports users to directly participate in educational innovation, realize group intelligence, and promote "artificial intelligence+education" innovation with "all intelligence". Education public funding is also a new industry trend of "Ai+education". It takes the "Ai+education" as the starting point, pays attention to creativity, relies on the strength of the masses, and has the advantages of resources on the line and offline. Let more people who love education learn what they can do. All the chips will make these "Ai+education" ideas more perfect, enabling enough supporters and support funds to verify their value [3]. To promote the development of "Ai+education".

3   Third Party Education Service

The "13th Five-Year plan" of educational informatization in China points out that we should comprehensively carry out the supervision and evaluation of the regional education informatization and the third party evaluation, and implement the policies of all the public funds that can be used to purchase information resources and services in all localities. It can be seen that the third party education service has been included in the national education policy level. The third party education service. The third party evaluation is one of the functions of the third party education service. With the development of "Ai+education", the third party education service is becoming more and more specialized and its functions are constantly improving. Under the mode of "human intelligence+education", the enterprise has the advantage of AI technology. The main purpose is to develop intelligent products for education, and to provide users with the most needed educational products and services. Users mostly want to get the most convenient or effective educational intelligent service. For users, the third party education service mainly provides users with service support in the process of using enterprise education products to solve the problem of users' use. Users can focus on their own development, but also provide intelligent exami nation, education evaluation, credit certification and other services [4]. For enterprises, the third party education service can provide constructive suggestions for the development of enterprise education intelligent products by collecting user's application data and according to user feedback. The third party education service can not only make education intelligent product R & D enterprises more focus on their core business, but also provide fine and humanized educational support services for users. With the development of artificial intelligence education market, it will become a new trend of "Ai+education".

4 Support for Standard Construction

With the promotion of the government, enterprises and schools, many advanced achievements of "Ai+education" have emerged at home and abroad. These achievements have promoted the development of "Ai+education", and have also stimulated the requirement of establishing "Ai education" standards, so as to standardize the products of AI+education and ensure their universality. In the thirty-seventh International Conference on "iso/iec JTC1 sc36 education informatization International Conference", experts put forward the proposal of artificial intelligence education application, and made the relevant standards of "Ai+education". It has become an international consensus. In order to ensure the good development of AI+education, at present, the relevant standards of "Ai+education" have been planned in China. The Education Information Technology Standards Committee has begun to set up national standards such as "smart classroom" and "smart campus" [5]. These standards will provide strong guidance for "Ai+education".

5 Formation of Education Industry Chain

With the formation of the educational industry chain, the Internet education industry has shown the "chessboard pattern". The vertical chessboard is a deep education pattern: The Geek academy, Shanghai online school, shark Park, etc. the horizontal chessboard is a comprehensive platform education pattern: Tencent education, Baidu education, NetEase education, etc. there are still some small and excellent chessboard scattered points: the education pattern of stagnation point, for example, APIs test library, happy college entrance examination, super curriculum and so on [6]. "Ai+education" constantly optimizes the "chessboard education pattern". The "Ai+education" industry chain based on Internet education mainly includes educational resources and content providers, technology developers and platform operators. Educational resources and contents are the material basis of "Ai+education". Educational resources, content providers, including audio-visual book publishers, education and training institutions, online learning platforms, schools, etc., are mainly responsible for providing all kinds of professional education resources and contents. Artificial intelligence technology is the technological basis of "Ai+education", and technology developers are mainly responsible for using AI technology advantages to assist education [7]. Research and development of personalized intelligent education products or services. Platform operators mainly operate in the form of providing third party education services, promote and apply personalized intelligent education products or services, and provide educational support services for users through windows. With the continuous promotion of "Ai+education", its industrial chain is gradually showing a trend of integration. A number of educational comprehensive application enterprises, which include educational resources, platform operation and technology development, have emerged, such as: Science and technology, University of science and technology, and so on. They cooperate with schools, publishers, etc. to build up educational service platform with their own technological advantages, develop teaching resources independently, and provide users with windows of intelligent educational products. Promote the optimization of the "Ai+education" industry chain.

6   Conclusions

With the progress of time and technology, artificial intelligence is developing constantly. Countries and enterprises all over the world are fully aware of the potential contained in artificial intelligence, and make greater efforts to increase their research and application in artificial intelligence. China is advancing rapidly with the times into the ranks of artificial intelligence, and has made great efforts to promote the application research of artificial intelligence in various fields of society. All these provide a good foundation for the development of China's "Ai+education" industry.

References

1. Liu, X., Ying, N.: Reflections on daily crisis management in colleges and universities. China High. Educ. **23**, 56–67 (2003)
2. Sina: The State Council: actively develop new models, such as [eb/ol], 24 August 2017. http://finance.sina.com.cn/china/20150917/010923268468.shtml
3. Sohu: Can public funding blossom in the education industry? [eb/ol], 24 August 2017. http://www.sohu.com/a/74916231_361784
4. 360doc personal library. The potential, bureau, style and way of Internet Education [EB/OL], 24 August 2017. http://www.360doc.com/content/16/0227/00/2369606_537682503.shtml
5. Liu, X., Ying, N.: Reflections on daily crisis management in colleges and universities. China High. Educ. **23**, 89–95 (2003)
6. Bai, T., Xu, Z.: Countermeasures of crisis management in universities. J. S. China Univ. Technol. **2**, 125–134 (2005)
7. Zhao, Y., Rajabov, B., Yang, Q.: Innovation and entrepreneurship education reform of engineering talents in application-oriented universities based on cross-border integration. Int. J. Front. Eng. Technol. **1**(1), 56–61 (2019)

Research on UAV Control Improvement

Moude Yang, Jiezhuo Zhong, and Longjuan Wang[(✉)]

School of Computer Science and Cyberspace Security, Hainan University, Haikou, China
wanglongjuan@hainanu.edu.cn

Abstract. With the development of the Internet, people are more and more inclined to combine the Internet with the actual objects, so the Internet of things is produced. In the Internet of things, we combine the UAV with the Internet to control. In the current mainstream control of UAV, the handle is the most extensive. Generally speaking, the flight control of UAV can be controlled by control handle and computer, both of which generate flight route by control operation. This article is about the innovation of UAV control flight mode. The way we want to improve the control is by giving us our route and letting the UAV fly in our route. And this way is more game oriented and market-oriented.

Keywords: UAV · Horizontal · Vertical control

1 Introduction

At present, the vast majority of small UAVs on the market are mainly controlled by two major control components. One is the use of a joystick, which allows the user to control the flight of the UAV through the left and right up and down direction of the handle, and the other is remote control by computer. This is the route set by the computer before flying, which can be said to be a prescribed dead route. What we want to do is to show real-time flexible control. The initial goal is to create a coordinate map on the plane and fly according to the route we draw.

2 Introduction to the Internet of Things

The Internet of things [1] can be defined as [2] to collect any object or process that needs to be monitored, connected and interacted in real time through various devices and technologies such as information sensors, radio frequency identification technology, global positioning system, infrared sensor, laser scanner, etc., so as to collect the information needed for sound, light, heat, electricity, mechanics, chemistry, biology, position, etc.

Through all kinds of possible network access, the ubiquitous connection between objects and objects, objects and people can be realized, and the intelligent perception, identification and management of objects and processes can be realized. The Internet of things (IOT) is an information carrier based on the Internet and traditional telecommunication networks [3]. It enables all ordinary physical objects that can be independently addressed to form an interconnected network.

J. MacIntyre et al. (Eds.): SPIoT 2020, AISC 1283, pp. 769–772, 2021.
https://doi.org/10.1007/978-3-030-62746-1_118

3 Introduction to UAV Operation

Most of the traditional UAVs [4] operate by the control handle and the route under the control of the computer. However, in our actual use, we are not as good as using the route drawn by our hands, whether in terms of suddenness or viewing. And the route we draw in the iPad is random and has more market development prospects.

4 Analysis of Current Situation at Home and Abroad

Dajiang has accounted for more than half of the domestic and foreign UAV market share. At present, the newly launched "Yu mavic air 2" and "Yu mavic mini" are all controlled by the control handle. However, it is almost impossible for us to draw the simulated flight route on the iPad, and then let the UAV fly through the route we draw through the sensor. This way is almost non-existent in the market, so there is a huge market.

5 Basic Principle

The working principle of UAV is to use radio remote control equipment [5] and its own attached control program to control the UAV. When using the handle to control the UAV to fly, it controls its direction by controlling the front handle's forward and backward, left and right. For our route flying, we only need to divide the route into the most basic routes, i.e. front, back, left, right, up and down. Next, let's describe in detail how to realize the route from drawing to actual operation.

The basic flight routes are: forward, backward, left, right, up and down. When using the joystick to fly, the reason why we can complete a series of seemingly cool flights is actually only through the combination of the front and rear left and right keys of the handle. In our iPad, we use a program that we've designed to divide a route we've drawn into two parts: horizontal and vertical. Before dividing the direction, we have to establish a time axis, which does not specify how to fly at this moment, but records what operations need to be performed in both horizontal and vertical directions at a certain time. For example, at time t, the route we drew needs to fly up, forward and right, so we need to fly in these three directions at the same time, and the combination of the three forms a path of flying forward and up to the right [6].

Any route can be converted into two basic routes: horizontal and vertical [7]. In the horizontal direction, if we use the handle to control flight, we will find that when the vertical direction remains unchanged, we are equivalent to flying in a horizontal plane, and on the horizontal plane, we only need to push the front, back, left and right on the control handle. Therefore, the horizontal direction movement can be carried out by program decomposition to move the route in the horizontal direction with the same time axis as the vertical direction.

Now let's look at the vertical direction. We can see that in flight, what we intuitively feel is moving up and down, so in this direction, we only need to determine the up and down direction at each moment based on the vertical direction of the starting point.

We have divided the routes we have drawn into routes that are easy to control. Now we need to study how to transmit our routes to UAVs and let them obey our command.

First of all, we determine that our flight route is composed of horizontal and vertical directions [8]. The most important problem of synthesis is the problem of time point. We can't combine the horizontal route at t0 and the vertical route at T1! Therefore, in the process of sensing, we must pay attention to the horizontal and vertical routes corresponding to each time point. We can learn from how the handle controls the UAV to fly. After dividing the route, we can simulate the control of the UAV in our iPad.

It can be said that we use the iPad route control flight and use the handle control flight is a pair of reciprocal process. The handle flight is to use the handle movement to control the flight of the UAV, and then produce a real flight route. The innovative control method of our article is to draw the flight route of the UAV first, and then analyze it through the route of the UAV How to control. In fact, the control is very simple. We can reverse the route to how the handle operates through our program, and then transmit the operation control we analyzed to the UAV through the sensor, so that the UAV can fly according to our assigned route.

6   Effectiveness of the Route

In the actual operation process, if we draw a route casually, this may happen. There may be obstacles at some position of the route. However, if the UAV continues to carry out the route we have drawn at this time, the result must be that the UAV is damaged. Therefore, we have to improve the scheme. There are two kinds of improvement schemes at present.

7   Improvement

In the first scheme, the route given by us is calculated by our computer when piloting the UAV. Therefore, in general, if the route fails to work, the program will give an alarm, display the invalid route, and prompt the user to change the route. This effectively reduces the number of invalid routes. The second scheme is self-improvement scheme. The route input by the user is always valid. However, when our UAV encounters obstacles in flight, it will make self correction according to our built-in program. For example, when the UAV is flying according to the route drawn by us, if there is an obstacle ahead, according to our built-in program, whether the up, down, left and right routes ahead can be searched Through, if possible, through, and then return to the original route. The realization of these two schemes only needs the program designed in iPad to reverse the route to the flight controlled by UAV handle.

8   Conclusion

The flight control of UAV is a great innovation and improvement in the Internet of things. For the vast majority of civil UAVs in the current market, the handle occupies the majority of the market. However, the handle may bump into obstacles due to people's mistakes in the operation process, and the machine may be destroyed in the end. But for the route we draw manually in iPad, it not only increases the game of UAV, but also through our built-in program, if the route we draw is wrong, the program will automatically alarm or correct itself during flight, which greatly improves the situation of UAV failure due to obstacles. In the control mode of the UAV, the operation is simple and convenient, more interesting than the handle, so it has a great market prospect.

Acknowledgments. This work was partially supported by Hainan Provincial Natural Science Foundation of China (618QN218), the Science Project of Hainan University (KYQD(ZR)20021).

References

1. Cao, Y., Jie, H., Huang, X., et al.: Application of Internet of things technology in condition monitoring of power transmission and transformation equipment. J. Electr. Power Sci. Technol. **027**(003), 16–27 (2012)
2. Zhihong, Q., Yijun, W.: Overview of wireless sensor networks for Internet of things. J. Electr. Inform. **01**, 215–227 (2013)
3. Ni, G.: Research on Internet of things technology and application. Sci. Technol. Econ. Guide (036), 9 (2017)
4. Jie, S., Zongjian, L., Hongxia, C.: UAV low altitude remote sensing monitoring system. Remote Sens. Inform. **000**(001), 49–50 (2003)
5. Wu, Q., Qingnan, H., Xu, Y.: Research and implementation of improving the stability of UAV attitude sensing system. Ind. Control Comput. **031**(005), 43–45 (2018)
6. Yuan, W., Zheng, X., Ang, H.: Autonomous navigation control of flight trajectory of quadrotor UAV. Jiangsu Aviat. (4), 5–7 (2014)
7. Liu, D., Xi, L., Sun, X.: Research on vertical attitude control of vector pull vertical takeoff and landing UAV. Comput. Eng. Appl. **53**(001), 260–264 (2017)
8. Xiaoming, T., Mingge, X., Bingrong, L., et al.: Review of UAV multi-sensor image fusion. J. Naval Aeronaut. Eng. **20**(5), 505–509 (2005)

Smart Sleep Mattress Based on Internet of Things Technology

Yi Yue, Jiezhuo Zhong, and Longjuan Wang[✉]

School of Computer Science and Cyberspace Security, Hainan University, Haikou, China
wanglongjuan@hainanu.edu.cn

Abstract. Internet of things (IOT) is a kind of network that connects any object with the Internet through radio frequency identification (RFID), infrared sensor, global positioning system, laser scanner and other information sensing equipment according to the agreed protocol, so as to realize the intelligent identification, positioning, tracking, monitoring and management of items. With the high rhythm of modern people's life, the pressure increases, and the quality of sleep is getting lower and lower. However, sleep is closely related to health and should not be ignored. The smart mattresses on the market either have pure detection function and are used in special environment such as medical treatment, or can only achieve the effect of relieving sleep through materials. In the upsurge of smart home, using the technology of Internet of things, we designed a mattress with more intelligent functions, such as detecting sleep quality, having intelligent wake-up service, and realizing intelligent control of app and voice. The smart mattress system mainly has the detection function, wake-up function, and manipulation customization function. The mattress has a detection of human sleep time and state. It can make the user wake up without pressure and reduce the sleepiness through the physical movement of the mattress, improve the efficiency and ensure the health of the biological clock. It can also intelligently control some customizable operations of the mattress through voice and software, such as wake up time, mattress soft and hard, this intelligent mattress has a variety of environment and conditions for sleep needs at the same time, the use of modern mechanical technology, network connection, to achieve many practical functions, pay attention to health management, convenient for human beings, improve the quality of life and sleep quality. In practical application, the technology is more practical and has a certain market prospect.

Keywords: Internet of things · Mattress · Smart home

1 Introduction

With the high rhythm of modern people's life, the pressure increases, and the quality of sleep is getting lower and lower. However, sleep is closely related to health and should not be ignored. The smart mattresses on the market either have pure detection function and are used in special environment such as medical treatment, or can only achieve the

J. MacIntyre et al. (Eds.): SPIoT 2020, AISC 1283, pp. 773–776, 2021.
https://doi.org/10.1007/978-3-030-62746-1_119

effect of relieving sleep through materials. In the upsurge of smart home, we use the technology of Internet of things to design a mattress with more intelligent functions, such as detecting sleep quality, having intelligent wake-up service, and realizing intelligent control of app and voice.

2 Introduction of Key Technologies

2.1 Introduction to the Internet of Things

The Internet of things, referred to as IOT, is the "Internet connected with all things". It is an extended and extended network based on the Internet. It combines various information sensing devices with the Internet to form a huge network, realizing the interconnection of people, machines and things at any time and place [1].

The Internet of things is an important part of the new generation of information technology. IT industry is also called pan Internet, which means the connection of things and everything. Thus, "the Internet of things is the Internet connected with things". The definition of the Internet of things is a kind of network that connects any object with the Internet, carries out information exchange and communication through radio frequency identification, infrared sensor, global positioning system, laser scanner and other information sensing equipment according to the agreement, so as to realize the intelligent identification, positioning, tracking, monitoring and management of goods [2].

2.2 Speech Recognition Technology

ASR (automatic speech recognition) is a speech as the research object, through speech signal processing and pattern recognition, the machine can automatically recognize and understand human spoken language or characters [3].

3 Structure of Smart Sleep Mattress System

3.1 Basic Functions of Ontology

The smart mattress system mainly has the functions of detection, wake-up and manipulation.

The mattress has a detection of people's sleep time and state. The physical movement of the mattress can make the user wake up without pressure and reduce the sleepiness, improve the efficiency and ensure the health of the biological clock. It can also intelligently control some customizable operations of the mattress through voice and software, such as the time of getting up and the soft and hard mattress.

3.2 Smart Mattress Structure

The smart mattress is divided into two parts: the mattress itself and the connected app. The inner part of the mattress is provided with a bed board which can supplement and reduce the spring, and the upper part of the bed can be lifted up. Sound is set inside the bed to broadcast voice or music. The mattress is equipped with heart rate, infrared, gravity sensors and voice signal receiver.

4   Specific Functions of Smart Sleep Mattress

4.1   Sleep Monitoring

The detection of human sleep depth by mattress is generally based on the decrease of physical activity and sensory sensitivity. However, it is difficult to accurately measure the depth of sleep at present, so the monitoring data are only approximate sleep time. Sleep monitoring is to monitor people's movements by body movement recorder, and calculate them in a systematic way. The total value is recorded every 2 min. At the same time, posture data and heart rate are recorded. Sleep state can be judged by calculation. The data can be realized by some existing algorithms, such as sleep detection based on heart rate signal [4], portable sleep monitoring system algorithm [5], and corresponding sleep detection algorithm based on sound and ECG signal [6]. After getting the final result, the best sleep time of users will be calculated scientifically and reasonably to give users a sleep management plan. If there are abnormal conditions such as palpitation and abnormal breathing, the mattress alarm and app connection will be used to alarm in case of accidents.

4.2   Wake up

The way to wake up is mainly through playing music on the mattress, and properly lifting the upper part of the mattress from a suitable angle, so that people gradually wake up from sleepiness. On the one hand, the timing of wake-up lies in people's active arrangement, such as getting up on time when going to work and school; on the other hand, there is no special arrangement, but too much useless sleep will lead to fatigue, dizziness and waste of time. In order to develop a healthy biological clock and improve the work efficiency during the day, the mattress will wake up timely under the right sleep time One step combined with the previous step of sleep monitoring to calculate, and the use of mechanical principles to achieve hardware design and adjustment. In addition, people can take the initiative to use the "get up" command to lift the upper part of the bed, so that the physical recovery of mental consciousness, reduce sleepiness.

4.3   Manipulation Interaction

The mattress has voice and app control functions. With speech recognition technology, people can command the mattress to do something by speaking, such as "wake me up at seven tomorrow morning.". At this point, the mattress will recognize the command and set the wake-up time. The app can also set commands, also can see a detection and calculation of sleep, clearly recognize their sleep quality, and improve sleep quality with the help of mattresses. In the operation, we can also adjust the hardness of mattress, which will be achieved by increasing and decreasing the spring at the bottom of the mattress. In addition, people can order songs in the form of voice or app, and play appropriate music with the bed's own stereo. The process is shown in Fig. 1

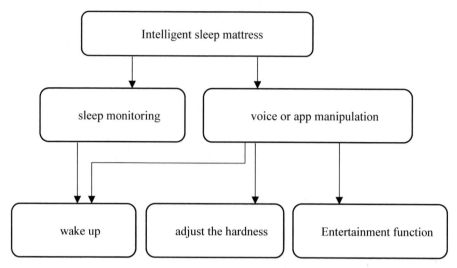

Fig. 1. Process of manipulation interaction

5 Conclusion

Smart home is more and more widely used in our life, to provide us with convenience. This smart mattress has all kinds of environment and conditions needed for sleep at the same time. By using modern mechanical technology and network connection, it has achieved many practical functions. It pays attention to health management, which facilitates human beings and improves the quality of life and sleep. In practical application, the technology is more practical and has a certain market prospect.

Acknowledgments. This work was partially supported by Hainan Provincial Natural Science Foundation of China (618QN218), the Science Project of Hainan University (KYQD(ZR)20021).

References

1. Chen, L., Xinghong, J., Gang, D.: Technical characteristics of Internet of things and its wide application. Sci. Consult. **9**, 86 (2011)
2. Yigang, J.: Research on the application of Internet of things technology in environmental monitoring and early warning. Shanghai Constr. Technol. **6**, 65–67 (2010)
3. Bingqing, Y., Gan, Y., Zhou, X.: Speech recognition technology. Digit. Commun. World **18**(2), 43–44 (2020)
4. Mianshi, L.: Design and implementation of sleep monitor based on heart rate signal (2018)
5. Cao, Z.: Design and implementation of algorithm for portable sleep monitoring system. Harbin Institute of technology (2012)
6. Ren, Y.: Research on sleep monitoring method based on sound and ECG signals (2018)

On the Application and Research of Electronic and Electrical Technology in Power System

Xiaoyun Sun[✉]

Wuhan Railway Vocational College of Technology, Wuhan 430205, China
sunxiaoyun67_2020@163.com

Abstract. Under the background of the rapid development of modern social economy and technology, the application scope of information technology is expanding constantly, which not only improves the real life of modern people, but also makes small changes People's production structure. Especially in the back-ground of more and more significant application characteristics of electrical and electronic technology, the application scope is expanding, so in order to make electrical and electronic technology conform to the whole power system
The development needs to optimize the electrical technology in the production of power system, so that the role of electrical and electronic technology can be fully played. In this paper, through the specific analysis of electronic and electrical. The importance of technology application, and in-depth analysis of the application of electrical and electronic technology in the power system, in order to promote the more stable and healthy development of the whole power system.

Keywords: Keyword: electrical and electronic · Power system · Technology

1 Introduction

Microteaching is translated as "microteaching" in China. Microteaching originated from the American educational reform movement in the 1960s. It was developed on the basis of the role-playing teaching method proposed by two professors of Stanford Univers [1].

With the continuous development of the society, the demand for electric power is gradually increasing. It has become an important power source for the innovation and development of electrical and electronic technology. Electrical and electronic technology. Technology is an important technology to promote the stable and healthy development of the whole power system power. Therefore, by making full use of electrical and electronic technology, the whole electricity can be effectively guaranteed. The power system is more stable and safe to operate and standardize the operation of the power system. In modern times Under the background of the rapid development of society, the operation of the whole power system has great impact on the electronic and electrical technology. The application of electrical and electronic technology in power system is increasing It is of great practical significance to carry out in-depth study. This not only helps To promote the stable development of the whole power system provides important technical support, and To promote the healthy operation of the power system.

J. MacIntyre et al. (Eds.): SPIoT 2020, AISC 1283, pp. 777–781, 2021.
https://doi.org/10.1007/978-3-030-62746-1_120

2   The Importance of the Application of Electronic and Electrical Technology

Due to the continuous operation of the power system, electrical and electronic technology has a very important practical significance. In this way, electronic technology and electrical technology are integrated, so that the advantages of various technologies can be brought into full play, and create good conditions for promoting the development of power system. In general, in the whole power system. In the process of development, the importance of electrical and electronic technology is embodied in the following aspects: first, through the comprehensive integration of resources in the operation of all kinds of power systems, optimize the efficiency of all kinds of resource allocation, so as to enhance the energy consumption [2]. Therefore, when the electric and electronic technology is used in the operation of the power system, the security and stability of the whole system can be guaranteed. At the same time, it can also control all kinds of faults very accurately, ensure the normality of all operation indexes of the power system, reasonably reduce the energy consumed in the operation process of the power system, and truly achieve the purpose of optimizing resource allocation; secondly, it can innovate the application of technology to lay a solid foundation for optimizing the power system. Especially in the process of the continuous development of modern power market, the demand standard for power is constantly improving, so it is necessary to optimize the power system. Only in this way can the function of energy be brought into full play in the process of improving operation index. Electronic and electrical technology, a new technology, can promote the development of power system towards modernization and intelligence in the process of application. Under the background of the rapid development of modern science and technology, the development trend of the whole power industry has also been raised to the best standard, so the reasonable application of electrical and electronic technology in the power system will reduce the volume of electrical components to a certain extent, and provide an important boost for the intelligent development of the whole power system.

3   Application Strategy of Electrical and Electronic Technology in Power System

3.1   Power Generation Link

In the process of power generation, it is necessary to effectively control the static excitation, so that the technical staff can make full use of the variable-frequency power supply to reasonably control the frequency of the current, so as to ensure that the whole power system has the maximum generating power in the process of stable operation, so as to maximize the regulation role of the technical staff in the whole power system development process. At the same time, in the operation process of the whole power generation system, we can make full use of the thermal resources, and according to the actual situation, we can use the frequency control device to deal with the problems of low efficiency and large energy consumption in the operation process of the power generation system, which can provide an important basis for enhancing the power generation rate of the thermal power generation system.

In the operation of the whole power system, there are many factors that affect the power generation frequency. For example, the flow speed or pressure in the region will affect the whole hydropower generation system to a certain extent. At the same time, these factors are difficult to be effectively controlled by people, so in the whole power system operation process, the frequency of hydropower generation. It will be difficult to control in a stable situation for a long time. One of the basic requirements of power transmission is to maintain the stability of frequency, so when affected by these factors, the whole water conservancy and power generation system can not be sustained and healthy development. In this way, in the process of hydropower generation, we can flexibly use electrical and electronic technology to strengthen the control of power generation frequency, and solve the problem that power generation frequency cannot be effectively controlled from the source. At the same time, for other different types of power generation systems, such as solar power generation technology, is an important core part of promoting the development of the whole power system [3].

Comparing solar energy resources with other types of resources, solar energy resources have an overwhelming advantage. Especially in the background of the rapid development of modern society, solar power generation technology continues to optimize, so it plays an important role in alleviating the social energy crisis in the development of modern society.

4 Transmission Link

The development of flexible AC technology is relatively late, but in recent years, it has been recognized by all countries in the world, and has gradually become the backbone of the modern power transmission process. However, due to the complexity of the application of flexible AC technology and its small application, it is difficult to meet the specific requirements of modern power transmission process. By combining the electric power technology closely in the whole flexible AC transmission link, the relevant technical staff can not only simplify the steps of the electric and electronic technology in the whole power system operation process, but also ensure the stability of the flexible AC current transmission in the power system operation process (Fig. 1).

An obvious application mark of electrical and electronic technology in the transmission link is the thyristor converter valve. In this way, under the background of the deepening application of HVDC in the process of power transmission, the efficiency and level of the whole power transmission system are gradually improved [4]. Especially in the context of the expanding application scale of thyristor converter valve, it not only replaces the specific application of direct transformer in the transmission link to a certain extent, effectively controls the cost of power transmission process, but also can maximize the mobility of various current conversion equipment in the process of HVDC transmission link. For example, the current electrical and electronic technology in the application process is mainly. It is applicable to power transistors, and the working frequency of the transistor can reach hundreds or even kilohertz, which shows that its working frequency is huge.

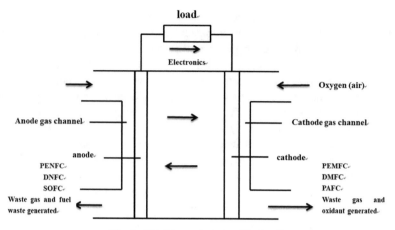

Fig. 1. Working principle of fuel cell

5   Power Distribution Link

To ensure the stability and efficiency of power system operation, the most important key is to ensure the safety and stability of power distribution system. In view of the safety of distribution system is closely related to power control. Therefore, in the actual power production process, only on the premise of ensuring the normal distribution and transportation, can it be delivered to users. Of course, in this process, many factors may lead to the unstable state of voltage and current, especially for the distribution link in the store. If this link is unstable, it will not only lead to serious waste of power resources, but also damage other power facilities, and then make users suffer unnecessary losses. In this regard, as a technician, in the daily maintenance work, it is necessary to strictly control the frequency and voltage in the shop. Once it is found that it presents an unstable state, it should be timely and effectively adjusted to enhance the power safety as much as possible and reduce the loss of users.

6   Conclusions

For the use of electric energy, one part of the electric energy will be used after being distributed to the user by the Distribution Director, while the other part will be lost in the transmission process, and the damage on the way of transportation is undoubtedly a great waste of electric energy. Therefore, in order to ensure the high utilization rate of electric energy, it is necessary to reduce the unnecessary electric energy and enhance its efficiency. At present, China's existing electronic and electrical technology has been able to make an effective wood ratio for this, and can greatly improve the actual utilization rate of electric energy, and then ensure the user's electricity safety.

In a word, electronic and electrical technology is very important in the development of power system in China. Therefore, China's power system should strengthen the effective use of related electronic and electrical technology, so as to ensure the safety of power consumption in China and add power to the sustainable development of China's power enterprises.

References

1. Haiyan, L.: Application and exploration of electrical and electronic technology in power system. Intern. Combust. Engine Accessories **21**, 122–123 (2017)
2. Xu, G.-Y.: Application of electrical and electronic technology in automatic control of reactive power compensation. China Chem. Trade **9**(7) (2017)
3. Shaowei, Q.: Characteristics and application of electrical and electronic technology in power system. Mod. Ind. Econ. Informatization **7**(9), 48–49 (2017)
4. Xu, C.: Application of electronic and electrical technology in power system. Digital Users **23**(2) (2017)
5. Liu, C.: Application of electronic and electrical technology and networking technology in power system. Electron. Technol. Softw. Eng. (5) (2018)
6. On the application of electronic and electrical technology in power system. Electron. World (22), 182 (2017)

Driving the Application of Bioinformatics Under the Development of Cloud Technology

Cui Yingying[✉]

Laiwu Vocational and Technical College, Jinan 271100, Shandong, China
yingying_300@163.com

Abstract. The development of bioinformatics has produced a large number of biological data for the analysis of biological problems.It provides a lot of information, but it also provides a higher storage and computing power for data requirement. Cloud computing can provide unlimited storage and computing power with low cost and high performance.This paper analyzes the characteristics of cloud computing, introduces its application in bioinformatics, and puts forward some suggestions for researchers who need to apply cloud computing.

Keywords: Cloud computing bioinformatics

1 Introduction

The application of next generation sequencing technology has produced a large number of sequencing data, which has brought new challenges to biology, especially bioinformatics in data storage, management and search. For a long time, the capacity of computer storage and processing data has been growing faster than that of biological data. However, after 2003, due to the development of sequencing technology, the cost of sequencing has been greatly reduced, resulting in a large number of biological data, and the capacity of computer storage and computing is gradually unable to meet the needs of big data. This promotes the application and development of cloud computing, which enables users to rent hardware devices and software according to their needs, avoiding a large amount of capital investment and management investment in hardware devices [1].

2 Cloud Computing Definition

"Cloud" is a service network that connects cloud computers or servers through virtual technology. Data storage and analysis are done by servers or computers on the "cloud" side. Liu Peng, an expert in cloud computing in China, gives the following definition: "cloud computing is a business computing model, which distributes computing tasks on a resource pool composed of a large number of computers, enabling users to Computing power, storage space and information services are needed. " According to the level of

J. MacIntyre et al. (Eds.): SPIoT 2020, AISC 1283, pp. 782–786, 2021.
https://doi.org/10.1007/978-3-030-62746-1_121

resource sharing, cloud computing service modes are divided into three types, infrastructure as a service. Platform as a ser V ice and software are as a ser V ice. LAAS (infrastructure as a service1 service: Base.

Infrastructure as a service. It integrates infrastructure such as virtual host, storage device, network device and other resources to become a service platform for users. IAAs is located at the bottom of the network and provides users with computing devices and storage devices that are distributed on demand and paid on demand. PAAS (platform as a ser V ice) provides a service platform for users to control the environment in which applications operate, and can apply, test and develop software on the platform. SaaS (soft FW are as a service) provides software for users to use on the service platform. Users only use software and do not master the network infrastructure such as operating system and hardware. Users do not have to install their own software, just need to connect to the browser to the public service platform. The supplier will install the required software according to the user's requirements, and be responsible for the software upgrade and maintenance [2].

The main advantages of cloud computing:

(1) Free users from the task of installing and testing software. Cloud computing platform can provide software and hardware services according to the needs of users. Users do not need to consider the complex hardware architecture under the network, just need to pay attention to the calculation and analysis.
(2) Renting computing resources on demand allows users to pay less. On the cloud computing platform, users can rent a small number of machines at the beginning, and then increase or decrease the rented machines with the increase or decrease of demand. The cost paid by the user is the actual cost of renting the machine.
(3) Cloud computing facilitates data sharing and analysis among researchers. Different researchers may have different versions of software installed on local servers, so it is difficult to share data and software. Cloud computing enables users who log in to the same platform to share the operating system and all software data, and ensures the version of the software to be updated synchronously.

3 Application of Cloud Computing in Biological Information

We introduce the application of cloud computing in bioinformatics in terms of IaaS, PaaS and SaaS.

3.1 IaaS

Users can rent virtual hosts on cloud computing to control computing, storage and other hardware devices, and establish the required computing environment. And a large number of bioinformatics tools can be packaged as virtual images to be used on the virtual host of rented cloud computing, which can easily carry out a variety of data analysis. For example, a virtual host with pre configured and automatic bioinformatics processes provided by ciov R can run on a local computer or on a cloud computing platform [3]. This virtual machine is based on Ubuntu and biolinux, with grid engine and H adoop as job scheduling, ergaf is as workflow system, and many open-source bioinformatics software, such as blast, 16S rrn a, etc. Users can also develop their own software to run on virtual machines. Bioconductor is an open source bioinformatics library about R language, which provides a series of software packages for microarray data analysis. Users can download the image provided by B ioconductor and install it on the rented cloud computing platform.

3.2 PaaS

Galaxy cloudm an and eoulsan can be regarded as PaaS. Galaxy integrates a series of simple and easy-to-use tools to provide a simple web page for data analysis. Galaxy-cloudman packages the software tools of galaxy into a mirror image, which can be used in aw s (a m AZON W eb ser V ice). Users can install other software installed on galaxy Cloudman to their own cloud computing platform, and even define plug-ins on galaxy Cloudman. By adding additional tools, you can extend the default functions and test and use them. In this sense, G alaxy Cloudman can be regarded as Paas. Eoulsan integrates many next-generation genetic data analysis tools, such as BWA, bwotie, soap2, gsna P, edger, and dedeq in one framework. At the same time, it also supports user-developed plug-ins for data analysis.

3.3 SaaS

Many traditional bioinformatics tools, such as BLA st, u CSC g enom e browser, can log in to the server with only one browser and use the corresponding services. They can also be called SaaS. These services are generally provided by the developers of software tools, with poor scalability. We mainly introduce the scalable bioinformatics tools applied to the cloud computing platform. Short sequence (reading segment) matching refers to matching the short sequence obtained from sequencing to the reference genome, which is the first step of many sequencing data analysis, such as SNP recognition and gene expression profile analysis. C loudb urst, C Louda lign Er, Se Al and C rossbow are all software based on MAPR education, which can match millions of sequences. Cloudburst developed by Schatz with "seed and extend" algorithm can determine the number of mismatches. Cloud brunst imitates the algorithm of RM AP, but its speed is increased by 30 times. But cloudburst does not support fastq files and cannot process data from bisulfite sequencing and (double) end sequencing. C Louda aligner makes up for this and is 35% to 8O% faster than cloudburst. Sea l integrates BWA, which can remove repeated sequences in sequence matching, which is very useful for SNP recognition and later analysis. Crossbow using mapreduee integrates bowtie and soap SNP, and can match billions of sequences in a few h.

Differential expression analysis can be used to find genes with obvious differences in different samples, and RNA sequencing (RNA SEQ) can be used to quantify the level of gene expression in samples. Myrna is a software for computing large-scale RNA sequencing on cloud computing platform. It integrates sequence matching, normalization, cluster analysis and statistical model, and directly outputs gene expression levels of different samples and genes of different expression levels. However, the biggest defect of Myrna is that it can't correctly match the short sequence to the splicing site of exon.

But FX makes up for that. FX uses the improved matching function to analyze RN a data, and outputs different gene expression levels in the format of RPK m or bpk M.

4 Problems Faced by Cloud Computing

Cloud computing provides powerful computing power, but its own characteristics also make its development face some difficulties and constraints. The application of cloud computing in bioinformatics is still in the initial stage. Although there are a certain number of bioinformatics tools, there are still many analysis can not be completed, and many tools need to be upgraded or developed. The privacy and security of data on cloud computing are also aspects that users need to consider. In particular, some biological data involve patients' privacy, but many countries have no laws to protect such data privacy. Cloud computing service providers need to make some rules to protect users' data.

5 Suggestions for Application of Cloud Computing

For users who will use cloud computing, three aspects need to be considered: data scale, security privacy and cost.

Data size and security privacy: first of all, consider whether your data size exceeds the processing capacity of the local computer. Now the local PC can handle gigabytes of data, and the server can handle hundreds of gigabytes of data at a time. If users are familiar with parallel computing technology, they can process terabytes of data. But if your data is larger and you're not proficient in parallel computing, it's hard for local computers and servers to handle, so you can consider cloud computing. If users want to transfer data to the cloud computing platform, they need to consider the security and privacy of the data. For example, whether the privacy of patients will be disclosed, whether cloud computing service providers can ensure the security of data, etc.

Cost: the cost of cloud computing is generally calculated according to the amount of computing resources used and the length of use time. You should evaluate the cost of using cloud computing before using it. Users should consider the cost of all stages, such as data transmission, storage, analysis, etc.

At present, both cloud computing and bioinformatics are in rapid development, and the application of cloud computing in bioinformatics is more and more extensive and in-depth. In particular, with the large-scale increase of biological data, biologists must distinguish useful information from a large number of data. This requires a strong storage capacity and computing analysis capabilities, cloud computing can be a good solution to this problem. The combination of cloud computing and bioinformatics will greatly promote the development of biology.

Acknowledgements. Department of Education, Shandong Province 2019"Qingchuang Science and Technology Program" – research achievements of doctor innovation team for wetland water environment restoration in Shandong Province(Project number: 2019KJE010).

References

1. Liu, P.: Cloud Computing [M], vol. 5, 2nd edn. Electronic Industry Press, Beijing (2011)
2. Luo, J.: Explore how to use information technology to cultivate students' core literacy [D]. New course (2). Issue 03 (2019)
3. Song, H.: Information technology and students' core literacy.
4. Fan, G.: A practical research on the use of information technology to cultivate primary school students' mathematical core literacy.

Computer Network Information Security and Protection Strategy Under the Background of Big Data

Liu Guanxiu[✉]

School of Information Technology of Shangqiu Normal University, Henan Shangqiu 476000,
China
gxliu920@163.com

Abstract. The security protection of computer network information under the background of big data era is more important. This paper summarizes the network security in the era of big data, analyzes the factors affecting computer network security under the background of big data era, and puts forward some suggestions on network information security protection strategy.

Keywords: Big data era · Computer network information security · Firewall technology · Antivirus software

1 Introduction

The era of "big data" has been constantly mentioned in recent years. With the continuous expansion of the global coverage of computer network technology and the deepening of its application, the degree of globalization and informatization is getting higher and higher. Data has penetrated into various industries and fields. The high development of the Internet and information industry reveals the arrival of the era of big data. "Big data" penetrates various fields and industries. Through the comprehensive perception, preservation and sharing of data, a digital world is constructed. Under this condition, people's way of looking at the world and the decisions made by the industry are more based on the reference of factual data. The arrival of the era of big data has caused great social changes [1].

Network security under the background of the times "Big data" is formed under the application of computer network technology. Computer network technology plays an important role in all sectors of society. It plays a key role in social life and production development, so it is very important to protect computer network information security. Under the background of big data era, the computer network information security protection involves different technologies and different fields, and the combination of various technologies can provide certain security guarantee for the computer network. The threats existing in the operation of computer network system require the establishment of a special computer network information protection system. Only by constantly improving the computer network security protection system under different security threats can we meet the needs of security protection strategy [2].

J. MacIntyre et al. (Eds.): SPIoT 2020, AISC 1283, pp. 787–791, 2021.
https://doi.org/10.1007/978-3-030-62746-1_122

2   Factors Affecting Computer Network Security Under the Background of Big Data Era

2.1   Natural Disaster

The computer has its fixed external equipment, and the machine equipment itself does not have the ability to resist the damage brought by the external environment. This leads to the computer in the face of natural disasters or accidents caused by vibration, pollution, water and fire threats, pollution and lightning threats cannot guarantee the safety of machine components, therefore, the vulnerability of the machine itself is one of the factors affecting computer network security.

2.2   The Openness of Network Itself

The computer network has the characteristics of openness under the requirements of wide application, and its own openness determines the vulnerability of the computer network system itself. Under the characteristics of openness, the security of TCP IP protocol adopted by the Internet is relatively low, which results in the weakness of network security foundation, and the service and data functions cannot meet the requirements when running low security protocols. Therefore, the openness of the network itself is one of the factors affecting the computer network security.

2.3   Operation Error

The computer needs to be in the user's specific operation in order to complete the function play. And the user's operation has a certain subjectivity, which may bury a security risk for the network information security. In the user's specific operation, due to the user's own security awareness and operation technology, there will be different situations in the user's password setting and correct operation, so the user's operation error is one of the factors affecting the computer network information security.

2.4   Hacker Attack

Hacker attack is an important factor affecting the security of computer network information. There are two kinds of man-made malicious attacks. One is active and the other is targeted destruction, which destroys the selected target, resulting in the lack of integrity and effectiveness of target information. The other is to passively crack and intercept the target information, which will not affect the normal operation of the computer network. Two kinds of man-made malicious attacks will bring important data loss and bring great harm to computer network security. Man made malicious attacks will lead to poor use of information network, which may lead to system paralysis, thus affecting social life and production.

2.5 Computer Virus Invasion

In the era of big data, the openness of computer network is more obvious, which makes the concealment of computer virus invasion more prominent. Computer virus has the characteristics of concealment, execution and storage. Once the virus is loaded into the program, the infectivity, latent, powerful destructive and trigger ability of the virus itself will also be revealed, which will bring great harm to the network data. The transmission vectors of network virus are mainly floppy disk, CD-ROM and hard disk. Through these transmission ways, the virus will run in the program if it spreads in the data flow. For the highly harmful viruses, the greater the threat to the data, such as panda burning incense and CIH virus, have brought great impact on the network operation. Therefore, computer virus is one of the factors that affect the security of computer network information.

2.6 Spam and Information Theft

Spam information is mainly transmitted by mail and news. Through the dissemination of spam information, it is mandatory to spread information such as business, politics and religion to others. The main reason of information being stolen in computer network is that it is invaded by spyware. The difference between spyware and computer virus is that spyware will not damage computer system, but mainly steal system and user information, which involves the security of computer and user information, and threatens the stable and smooth network operation environment. Therefore, spam information and information theft is one of the factors that affect the security of computer network information.

3 Analysis of Computer Network Security Protection Strategy Under the Background of Big Data Era

Pay attention to account security protection. The consideration and implementation of computer network security protection strategy should start from the factors affecting computer network security. The protection requirements of internal factors of network security should pay attention to account security management. Account security management includes the management of a variety of account types, through the protection of computer system account, online banking account, Tencent account and email account, the effective management of account security is realized. The security protection and management of account requires users to enhance the awareness of computer security use and management. In the case of increased security awareness, the complexity of account setting will make the password more difficult to disclose. Secondly, setting special symbols in the account security management can effectively avoid the situation that the password is similar because of the simple setting of the password. Secondly, we should pay attention to the length of the password and replace it regularly [3].

Network firewall technology, network firewall technology is a kind of internal protection measures for network access control, its role is reflected in preventing external users from using illegal means to enter the network system, so as to protect the internal network environment to a certain extent, and provide a certain guarantee for the stability of the network operation environment. Firewall technology can check the data in the network transmission on the basis of secure network interaction. The implementation of security measures through the operation within the given program determines the transmission blocking or permission of the target network data. Firewall is divided into address translation type, proxy type, detection type and packet filtering type according to different technology differences. In different aspects of technology to form targeted firewall access control, can organize the threat outside the internal network environment, for the normal operation of the network environment to provide a certain guarantee.

4 Ways to Use Network Security

The use of anti-virus software is to cooperate with the firewall to detect the harmful information. The application of anti-virus software in the modern computer network environment has considerable universality and practicality. Anti virus software targeted at the known virus to effectively kill, can detect some hacker attacks, improve the security of network use. In the process of using anti-virus software, we should pay attention to its upgrading. The upgrading of anti-virus software makes it have the latest anti-virus function, and can effectively ensure the network information security in the use of computers.

Network monitoring and monitoring, intrusion detection technology in recent years more widely used at the same time, technology is also constantly developing, the role of intrusion detection technology is to detect whether the monitoring network in use has been abused or will be invaded. The analysis techniques used in intrusion detection include statistical analysis and signature analysis. The performance of signature analysis is to detect the weakness of the system. The statistical analysis rule refers to the use of statistical theory to judge the action mode in the stable operation of computer system, so as to determine whether the operation action is within the safe range. Network monitoring and monitoring technology in the use of computer networks for computer network information security protection provides a certain detection technology foundation.

4.1 Encryption of Data Preservation and Circulation

Data preservation and circulation are universal in the computer network. The security protection of data preservation and data circulation is the requirement of computer network security protection strategy in the era of big data. The way of data storage requires file encryption. File encryption technology is to improve the security of information system and the confidentiality of data needs, and to prevent data from being stolen and destroyed. Encryption protection in data circulation refers to the use of digital signature technology. The purpose of data signature technology is to provide encryption service for the security of data circulation and transmission. There are two kinds of encryption services in data signature technology, one is line encryption, the other is end-to-end encryption. Line encryption pays more attention to the security protection of line transmission. In line transmission, the use of different encryption keys increases the security protection strength of the target information that needs to be kept secret. The end-to-end encryption needs to use encryption software. On the basis of encryption technology, the sender encrypts the target file in real-time through the encryption software, and transfers the security information by converting the plaintext in the file into ciphertext. When the target information reaches the destination, the receiver needs to decrypt the ciphertext with the key, these ciphertexts are converted into plaintext that can be read directly.

5   Conclusions

Under the background of big data era, more attention should be paid to the security of computer network information in the preservation, dissemination and circulation. The openness and coverage of computer network require the establishment of a stable and safe computer network operating environment. Strengthening the computer security information technology is of great significance to establish a healthy and stable computer application environment.

References

1. Gao Nuo, P.: Research on computer network information security and protection strategy [J]. Comput. Digit. Eng. **1**, 121–124178 (2011)
2. Hongmei, W., Huijuan, Z., Aimin, W.: Research on computer network information security and protection strategy [J]. Value Eng. **1**, 209–2.10 (2015)
3. Lei, Wang: Research on computer network information security and protection strategy [J]. Comput. Knowl. Technol. **19**, 4414–4416 (2014)

Design and Implementation of Legal Consultation System Based on Android Platform

Xiaoyu Lu(✉)

Ji Lin Justice Officer Academy, Jilin Chanchun 130062, China
lxy880421@163.com

Abstract. At present, the legal consciousness of the people in our country is becoming stronger and stronger, and the demand for legal related business is also increasing. Therefore, all kinds of lawyer affairs have been established. At present, there is a large amount of legal consulting business in Shenyang, and the traditional legal consulting business processing method is difficult to adapt to the current actual needs. In view of the smart phone application has become the current mainstream, so the research and development of legal consulting business software system using smart phone platform has become a development direction with broad market prospects. This paper analyzes the current situation of the legal consultation business in China, investigates and studies different ways of legal consultation business, A mobile legal consulting system is developed based on Android mobile phone platform. The mobile client software of the legal consulting system is designed and implemented. The user needs analysis, overall design, database design, detailed design, coding and other processes are completed; The core functions of the legal consultation system are implemented based on Android mobile operating system with Java language, including the user's login and registration functions, legal information query functions, lawyer consultation functions, case analysis functions and information management functions.

Keywords: Legal consulting · Android · Mobile internet · Remote consulting

1 Introduction

As a powerful weapon to protect the legitimate rights and interests of the people in modern society, law has been paid more and more attention by the people. With the development of economy, the speed of social development in China has been continuously improved, and the people's material living standard has been constantly improved [1]; High, its demand for social services, entertainment and other spiritual aspects is also rising, and the process of urbanization has led to more and more complex disputes and conflicts between people, between people and organizations, and the business needs for legal consultation are also changing. With the rapid development of the legal system in our country, people's legal consciousness has been enhanced, which has helped the prosperity of the legal consultation market in our country. Information technology plays an increasingly important role in people's daily life. The development of computer

J. MacIntyre et al. (Eds.): SPIoT 2020, AISC 1283, pp. 792–796, 2021.
https://doi.org/10.1007/978-3-030-62746-1_123

hardware technology makes its price more and more low, and it has really entered the common people's home. The huge impact of network technology makes the development speed of modern society exceed people's imagination. Traditional concepts and methods are difficult to meet the needs of the development of the times. Therefore, it is an irreversible trend to adopt new technology and new methods to carry out traditional business reform.

2   Feasibility Analysis of Legal Consultation System

According to the technical characteristics and actual research and development of the system, the feasibility analysis of the legal consulting system is carried out. The feasibility analysis is of great significance to the decision-makers and managers of the project. Generally speaking, the system feasibility analysis is carried out from the aspects of market, cost input, expected income, technology and operation to determine whether the system is necessary for investment and development Sex and feasibility. The system feasibility analysis can effectively reduce the risk of R & D, which has important reference significance for the follow-up development of the system. Technically speaking, the legal consulting system developed in this paper is based on Android smartphone platform in Java language [2].

With the popularity of the Internet, network applications become more and more frequent, so Internet-based application development has become a mainstream mode, which is the B/S mode. Compared with the C/S mode, it has a very flexible feature. As long as it can access the Internet, it can realize remote access through the Internet anytime and anywhere, without being restricted by the region. Compared with C/S mode, browser's access mode enables users to realize program use at any place, which has great flexibility. At the same time, due to its great flexibility, it leads to certain instability in the use process. The main performance is as follows:

(1) Due to the great openness of the Internet, the hidden danger of network security has become a restricting factor. At present, with the various forms and development of Internet hackers and Trojans, network security issues are becoming more and more important. In the process of application development of B/S mode, we need to consider how to effectively avoid being attacked by network, especially some important numbers According to.

(2) The hardware of Internet remote server has a high cost, and the deployment of Web mode is more complex than that of LAN C/S mode, which is one of the important factors in technology and cost. Therefore, in many application environments, B/S mode has certain limitations.

C/S mode, strictly speaking, belongs to a new development situation, that is, application development based on mobile terminal, which has the following characteristics:

1. The terminal equipment has been changed from the traditional PC to the mobile devices such as smart phones and iPad tablets, which have the characteristics of wireless transmission;

2. C/S mode is still adopted, and HTTP protocol is used to realize the connection and data interaction with remote server to complete data query, reading, display and related processing;

3. In view of the size of the smart phone, the design of the interface and the realization of the function should be considered. Figure 1 shows the software development architecture based on mobile terminal.

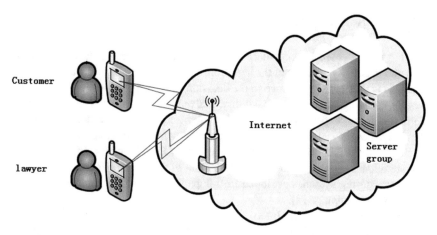

Fig. 1. Shows the software development architecture based on mobile terminal

3 Design of Legal Consultation System

The design of legal consultation system, system design is an important link between the preceding and the following, it carries on the user demand analysis, and provides important guidance for the follow-up system research and development. System design usually includes system outline design, database design and other contents, which is a very critical design process. According to the results of user requirements, the structure, function, database and interface of the system will be designed [3].

System structure design is to design the hardware structure and software structure of the system. For the system, it is clear about its structure and convenient for subsequent development. This system mainly includes the following main elements:

1. Client hardware and software, the client is composed of Android smartphone and application software. The user installs the client software of the legal consulting system on Android smartphone. The user can send relevant data requests through the software platform, and carry out some processing. Send data through HTTP protocol, connect with the server interface, and finally complete complex data interaction and related operations.

2. The server is composed of two parts: the commercial server and the server-side program. It receives the client's request through Tomcat web server and processes it by the server-side program. This process is usually related to the database;

3. Database is an indispensable part of the whole system. Generally speaking, it consists of database management software and related application programs installed on the server hardware. The database interacts with the server through SQL statements, and finally completes the data reading, modification and other related operations. Android smartphone terminal connects with commercial server through wireless network to complete data interaction and related operations, as shown in Fig. 2 is the system structure diagram.

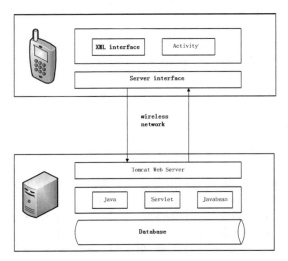

Fig. 2. The system structure diagram

4 System Function Test

System function test is to verify all kinds of operation of application program to determine whether the development of the system meets the expected application requirements of users, and whether all kinds of operation meet the functional design objectives. Function test is the most direct user-oriented process of system test. Generally, the black box test method is used, that is, to use the application program from the perspective of the user, to verify the correctness of various operations, and to modify and debug the program for the parts that do not meet the requirements, repeatedly, until the system test is finally passed.

Combined with the research and development background of the system, the system function test mainly focuses on the system login, registration function, legal information query function, case analysis function, online consulting function and system information management function. The specific test process is as follows:

1. Open the mobile phone and install the legal consulting system. After the installation, click to enter the system;

2. Enter the login page, click the registration button, first test the use of the registration function; after the registration is completed, return to the login interface, enter the user name and password, click login, test the use of the system login function, and enter the main interface after the login is successful;
3. Verify the functions of different types of users, test all the functions and related operations of the system, demonstrate the relevant legal consultation data and self-made test data of the unit in 2013, test whether the legal inquiry, online consultation, case analysis, system management core function modules and related operations of the system are normal, and record the test result;

The system meets the expected functional design objectives, can carry out legal inquiry, case analysis, online consultation and other business processing, and has the functions of login, registration and system information management, so it meets the actual application needs and passes the functional test.

5 Conclusions

The R & D of the system adopts the standard process of software development. Firstly, the feasibility of the system is demonstrated, and the technical route and cost input of the system are analyzed in detail to demonstrate whether the R & D of the system is feasible. After the feasibility demonstration, the detailed user requirements analysis of the system, analysis of the system function, performance, user interface and other design objectives, to make clear guidance for the follow-up development of the system. Then, the architecture design, case analysis and database design of the system are carried out, which provides support for the research and development of the system. On the basis of the system design, build the Android development environment, code the core functions of the system based on the Android smartphone platform and related technologies, complete the detailed design and code implementation of the core functions such as system login, registration, legal information query, case analysis, online consultation and system management, and then carry out the system test analysis, build the test ring Environment, design test cases, test and analyze the function, performance, user interface, database and other aspects of the system. The test results show the correctness of the system development.

References

1. Guozhi, X.: System Science [M], vol. 6, pp. 45–49. Shanghai Science and Technology Education Press, Shanghai (2000)
2. Jining, X.: Systematic law in China [J]. Forum Polit. Law 6, 78–82 (2000)
3. Jining, X.: Conflict between legal regulation technology and social justice [J]. Chin. Law 5, 34–56 (2001)
4. Jining, X.: Prediction and analysis of the development trend of the judicial system's bearing capacity. This research achievement was awarded the "Excellent Thesis Award" at the 1987 University of political science and law Symposium on the anniversary of the University of China. Development Strategy and Systems Engineering, pp. 771–776 (1987)

The Application of Big Data Development in Power Engineering Cost

Wang Linfeng[1]([✉]), Dong Zhen[2], Wang Yanqin[1], Wang Yong[2], and Wang Dongchao[1]

[1] State Grid Hebei Electric Power Company, Economic Research Institute, Shijiazhuang 050000, Hebei, China
wanglinfeng197601@163.com, wangyanqin2020@163.com,
jyy_wangdc@163.com
[2] State Grid Hebei Electric Power Company, Shijiazhuang 050000, Hebei, China
dongzhen197603@163.com, wangyong198104@163.com

Abstract. With the rapid development of modern science and technology, many new and high-end practical technologies have ushered in unprecedented development opportunities, especially computer information technology has made great progress. Because it has its own unique practical advantages, it has been gradually used in the process of domestic power engineering cost, greatly improving the reasonable degree of power engineering cost in China, making enterprises have greater economic benefits. Technical personnel of power engineering cost.

The analysis and grasp of relevant information technology, and finally effectively apply information technology to the process of power engineering cost. Its purpose is to find out the defects in the application process, and to work out reasonable solutions through systematic, comprehensive and accurate research, so as to make it more and more in line with the development requirements in the future application process, and optimize the cost performance ratio of the whole cost. At the same time, it also makes the information technology and power engineering cost technology of our country have a good development, and the gap with the world standardization development is becoming smaller and smaller, which lays a solid foundation for the future development of our society.

Keywords: Information technology · Power engineering cost · Application status · Improvement measures

1 Introduction

In recent years, China's urban industrial construction has made a greater breakthrough with the economic development. Due to the huge demand of the construction market, the role of power engineering construction has become important [1]. However, with the price rise of power facilities, raw materials and other equipment, it has seriously restricted its rapid development. In order to solve this problem, only by making full use of power information technology, can we provide effective reference for power engineering cost by using these power management intelligent information technology, so as to make the

J. MacIntyre et al. (Eds.): SPIoT 2020, AISC 1283, pp. 797–800, 2021.
https://doi.org/10.1007/978-3-030-62746-1_124

design budget of the whole power cost more accurate and reduce the waste of funds in the whole cost process. In addition, computer information technology also provides a very good platform for the scientific exploration and design of power engineering. Many data that are difficult to be calculated by ordinary operators can be utilized use it to help.

2 Application Status of Information Technology in Power Engineering Cost in China

The great decision of reform and opening up has greatly improved the information technology and power engineering technology in China, and also promoted the rapid development of society. In order to meet the urgent needs of power engineering cost, information management system has been established, and it has been fully applied to the process of power engineering cost. Through the establishment of accurate cost data, the use of scientific information analysis technology and advanced information management methods, the power engineering cost has been comprehensively considered to improve the enterprise Economic benefits of the industry.

At present, the application and development of information technology in the cost of electric power engineering is relatively good. Many electric power enterprises are constantly using it to assist the cost research work, which has been widely recognized. However, due to its late development, many application details have not reached the ideal requirements, and a scientific application management system has not been formed, which restricts its rapid development [2].

3 Problems Existing in the Application of Domestic Information Technology in the Cost of Electric Power Engineering

Due to the late reform of science and technology in our country, it still has a lot of shortcomings, which seriously affects the development prospects of power technology. At present, its problems mainly lie in:

3.1 Incomplete Data of Power Cost Information Management

The data of electric power cost information management is not complete, and there are many errors in current cost information system data, so the comprehensive analysis and processing are not achieved. Generally, the investment cost of electric power engineering is determined by the cost information, so the imperfect information will guide the cost work wrongly.

3.2 Information Management Method Is not Reasonable in the Process of Cost

In the process of cost management, the information management method is not reasonable, which makes the whole cost result difficult to meet the ideal requirements. For example, in the process of cost information management in some electric power enterprises, some details are ignored, which eventually causes fatal impact on the whole cost work.

3.3 Dynamic Cost Information Does not Form Monitoring Management, Resulting in Cost Out of Control

Under normal circumstances, cost work is a dynamic project, which will change with the outside world. However, many power enterprises fail to monitor and manage the dynamic cost information and Despise its dynamic impact.

Finally, there is no standardized development research market, and the whole development market is in disorder, which is not conducive to the overall management. For example, at this stage, many cost application technologies are more, forming a vicious competition situation. It is not conducive to its application and development without forming a whole development status quo.

4 Main Measures to Effectively Improve These Application Problems

In order to effectively improve its application effect, it is absolutely necessary to solve these existing problems and phenomena. Specific solutions are needed to solve these problems:

1. Electric power enterprises must establish a complete cost information data management system: for example, organize some professional cost information personnel to sort out and constantly improve all the data related to the cost information foundation of electric power projects, ensure its accuracy and greatly improve its practical application [3].
2. The selection of information management methods in the process of cost management must be based on the actual situation and cannot be separated from the specific cost management requirements: for example, power engineering enterprises must pay attention to some details, not its actual role. In addition, the meaningless information must be removed, and the valuable cost information data should be used to avoid it interfering with the cost work.
3. Power enterprises should have a good grasp of dynamic cost information and monitor and manage it at all times: for example, enterprises should set up a dynamic cost information management and research department to analyze and process all kinds of information in time, and reflect it to the cost research department in the first time, so that they can adjust the cost design scheme in time and improve the whole cost Accuracy.
4. Any technology needs to form an overall research and development system, otherwise it will have a vicious development status quo: therefore, in the future development process, it is necessary to strictly regulate the power cost management market, so that it can develop into an organic integrated whole. For example, we can establish some legal mechanisms to effectively regulate this vicious competitive market, so as to gradually lead them to the standardized development market, and finally make its application scope wider.

5 Application and Development Direction of Information Technology in Power Engineering Cost in the Future

Its development not only promotes the development of the electric power industry, but also meets the interests and needs of the electric power enterprises, which is in line with the requirements of the contemporary social development and construction. Therefore, in the future, in addition to effective solutions and improvements, the application of information technology in the cost of electric power engineering in Feng1'should also clearly determine its development direction, because it is not only related to the quality of the development of electric power engineering cost, but also related to the rapid development of China's modern socialism. In the future development process, information technology should be allowed to form a standardized application system in the cost of electric power engineering. All the application technologies and schemes should be refined to make it more and more standardized and systematic. In addition, it should be internationalized gradually to open the door for the development of China's electric power industry.

6 Conclusions

Since China's accession to the world trade organization, the state attaches great importance to the development of the power industry, which has become the key investment object of the state. However, the design management problems existing in the cost process have not been well solved, and there are generally bad phenomena of cost control, such as bidding and no accounting, construction and other stages did not have a good grasp of the cost work, which brought many additional costs to the enterprise. Although our country has not been fully used at the present stage, as long as we continue to improve and innovate for a long time, and e reasonably introduce some advanced western application technologies, we can improve the application effect of information technology in the cost of power engineering, and also play a role in promoting its development.

References

1. Yan, Z., Feng, Y.: Information technology and its application in power engineering cost [M]. Financial press (2008)
2. Zhengyou, H., Jianwei, Y., Tianlei, Z.: Application of Intelligent Information Processing in Power Grid Fault Diagnosis [M]. Southwest Jiaotong University Press, Chengdu (2013)
3. Electric power engineering cost and quota management department. Price information of equipment and materials for 20 kV and below distribution network engineering [M]. China Electric Power Press (2012)

Design and Research on the Management of the Electricity System of WebGIS

Wu Jianhui and Meng Xiangnan[✉]

State Grid ShangQiu Power Supply Company, Shangqiu 476000, China
zmdxn869@163.com, mengxiangnan1002@163.com

Abstract. Power industry is the pillar industry of national economy. GIS has become an indispensable part of power information construction with its unique spatial analysis ability and visual expression ability. The construction goal of power GIS is to improve the visual management level of power equipment, enhance the real-time supervision ability of equipment and lines, realize the integration of the whole network information, and provide decision support for power production and operation. All kinds of traditional GIS applications are based on the data structure of point, line and surface spatial relationship analysis. The GIS services and web applications provided by traditional GIS cannot meet the needs of power professional applications. Based on the study of traditional GIS, a new power GIS platform is designed and developed. This paper proposes a quad tree data index solution based on tile element association, which can reduce the depth of spatial index search tree and significantly accelerate the speed of element retrieval; Research the real-time data push technology and the integrated display technology of data and graph, realize the real-time on-line monitoring of equipment status based on the theme, improve the shortcomings of traditional real-time system in server query pressure, real-time performance can not be guaranteed; research the best i-path algorithm model for emergency repair of faulty equipment, A parallel search algorithm based on region partition is proposed, which has the advantages of simple preprocessing, high concurrency and fast convergence.

Keywords: Power GIS · OGC specification · Operation and maintenance management · Online monitoring: dispatching path

1 Introduction

China's GIS technology level is relatively backward compared with foreign countries, and the terrain complexity of power equipment distribution determines that foreign advanced power GIS solutions can not well adapt to the needs of China, which reflects the importance of China's independent property rights power GIS solutions. Based on the in-depth study of foreign advanced open-source GIS frameworks and toolkits, Combined with the new characteristics of electric power GIS and the construction goal of "three sets and five major" of State Grid Corporation of home appliances, the sharing platform of electric power WebGIS is developed from the bottom to further share In

J. MacIntyre et al. (Eds.): SPIoT 2020, AISC 1283, pp. 801–805, 2021.
https://doi.org/10.1007/978-3-030-62746-1_125

order to support the construction of operation and maintenance management application platform, this paper discusses the design ideas and implementation methods of key issues in the construction of sharing platform and application platform, which has guiding significance for the construction of power WebGIS sharing platform and business application platform in the new environment.

2 WebGIS and Related Technologies

WebGIS is the product of Internet and WWW technology applied in GIS development. WebGIS not only has the function of most traditional GIS software, but also has the unique function of Internet. From any node of the Internet, users can access the spatial data in WebGIS site; users can browse remote GIS data and applications without installing GIS software on the local computer, and carry out various spatial information retrieval and spatial analysis (Fig. 1).

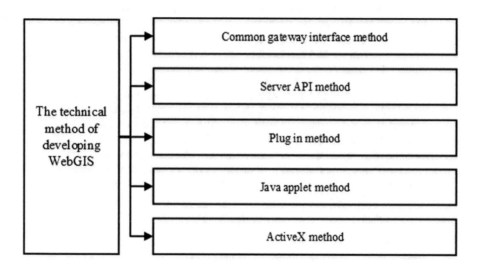

Fig. 1. Technical method of developing WebGIS

At present, there are many different technical methods applied to develop WebGIS, including CGI (Common Gateway Interface) method, server API method, plug-ins method, Java applet method and ActiveX method. The following is a brief description and comparison of these technologies.

CGI is a specific standard used between web server and client browser. It allows web users to start a program (called CGI program) that exists in the host of web server through the command of web page, and receive the output of this program. When the user sends a request to the web server, the web server forwards the request to the GIS service program running at the back end through CGI, and the GIS server generates the result and hands it to the web server. The web server then passes the result to the client for display. However, the disadvantage of this method is to restart a new service process for each customer request. When there are multiple user requests, it will cause the server burden to increase. In fact, due to design reasons, most GIS software cannot be directly connected to the web as CGI program, but it can be solved by the following two technologies:

(1) Start the back-end batch drawing software with CGI. The user can directly input instructions in the computer terminal line by line to draw. However, every user's requirement is to start the corresponding GIS software. If the software is large, the waiting time will be long.
(2) CGI starts the back-end GIS software. The information exchange between CGI and back-end GIS software is completed by IPC inter process communication. Because GIS software is message driven, CGI can only drive GIS software to perform specific operations by sending messages, and does not need to restart every time.

3 The Introduction of WebGIS Implementation Mode

In the implementation of WebGIS, two problems should be considered, that is, controlling the amount of data transmitted by the network and the interaction between the browser and the user. At present, the application development model of WebGIS based on Internet/Intranet is generally the structure of client server and the browser/server structure developed on this basis. The architecture has developed from the first two-tier structure to the multi-tier structure. The existing WebGIS mainly has three kinds of implementation modes, that is, the application system based on server, the application system based on client and the hybrid system based on server/client. Various operation mechanisms, modes and work efficiency are different.

There are many limitations in pure server mode and client mode: pure server mode is limited by Internet bandwidth and network communication ability; while client mode is limited by client computing ability. Therefore, it is a common mode to combine the server mode and the client mode to achieve a balance between the server and the client to maximize efficiency. In the first mock exam, the large database is used to perform the task of complex analysis, and the task is controlled by the client. The server and the client are allocated different tasks according to their performance to give full play to their respective advantages. This hybrid mode consists of three parts, including web server, client and database server. Web server is responsible for web services, database server is responsible for database management and data services, and other tasks are completed by clients. What client users see and operate are vector graphics, and they are connected with database seamlessly. Roaming, zooming, querying and analyzing on the client are all completed on the client without the participation of web server and data server, while database management, data service and responsible spatial processing are completed on the database server.

4 System Analysis and Design

4.1 System Design Objectives

In order to provide high-quality services for enterprises and users at all levels, improve the service level of the power supply department, consolidate and improve the normal operation mechanism of high-quality power services, fully implemented the urban power supply guarantee project, and established a service mechanism to respond quickly to the needs of users, so as to do a good job in power supply guarantee. In order to improve the service level, the electric power bureau has set up a service platform with multiple functions such as information inquiry, accident repair, service complaint and business acceptance, so that it can effectively make full use of various information resources and provide effective and rapid services for users at all levels.

The design goal of this system can realize the spatial data management and attribute data management of the distribution network facilities through the Web way, unify and manage all kinds of graphs, maps and data attribute information of the distribution network facilities, make the distribution network equipment and the geographical graphs of the urban area dynamically publish on the Internet through the web technology, realize the data sharing, and provide the visual management for the working departments of the electric power bureau Methods, provide scientific basis for analysis and decision-making, and realize intelligent business management.

4.2   Use Case Model

The use case model describes the system functions understood by external executors. The use case model is used in the requirement analysis stage. It describes the functional requirements of the system to be developed, understands the system from the perspective of external executors, and drives the development work of each stage after the requirement analysis. It not only ensures the realization of all the functions of the system in the development process, but also is used to verify and test the developed system. Role description, by demand analysis, the user categories that can be determined by the system include public users, customer service department and bureau MIS users, etc.

4.3   Determination of User Object

(1) Public users. The public users can browse the map information published by the distribution network through the browser, and can zoom in, zoom out and roam the map. In the follow-up project, the power supply information and maintenance plan issued by the power supplier can also be inquired.

(2) customer service. Customer service department, as a window for customer service, handles customer's fault repair information in a timely manner. The customer service department can browse the map information of the distribution network, realize the operation of map generation, reduction and roaming, etc.: the customer service department can also query the basic information of the distribution network, find the address through the device name, or find the relevant equipment through the address name, so as to realize the two-way query of attributes and geographical conditions.

(3) Office users. Bureau users can realize all the operation functions of the above customer service department, including browsing the map information of the distribution network through the browser, zooming in, zooming out, roaming and other operations on the map. In the follow-up project, the work instruction can also be released through the network.

5   Conclusions

GIS is an interdisciplinary subject and technology of information science, space science and geography science. WebGIS is an organic combination of GIS and network technology. With the rapid development and popularization of Internet, the application of WebGIS continues to expand and deepen, greatly expanding the field of GIS research and application.

WebGIS has a wide range of potential and advantages in the process of urban distribution network informatization. All kinds of equipment of urban distribution network have obvious spatial distribution characteristics, which is suitable for GIS integrating spatial data and geographical analysis methods Development and application: the main objects such as wires and cables, transformers and supporting equipment are mostly represented by vector elements such as points and lines in GIS. The data modeling is simple, the amount of data is not large, and the calculation of the system is relatively small, which can achieve more complex functions.

References

1. Fuling, B.: Principles and Methods of Geographic Information System [edition], pp. 3–7. Surveying and Mapping Press, Beijing (1996)
2. Nan, L., Renyi, L.: Geographic Information System [edition], pp. 5–8. Higher Education Press, Beijing (2002)
3. Jian, N., lingkui, M., Yuchuan, W., et al.: Power Geographic Information System [edition], pp. 21–23. China Power Press, Beijing (2004)
4. Caixin, S., Bro, Z., et al.: Electric Power GIS and Its Application in Distribution Network, 1st edn., pp. 197–200. Science Press, Beijing (2003)
5. Shuobon, B.: Principles and Methods of GIS Software Engineering [edition], pp. 115–117. Science Press, Beijing (2003)

Application of Improved CMAC Algorithm in International Trade Analysis Module Construction

Zhendong Zhu[✉]

Xiangsihu College of Guangxi University for Nationalities, Nanning 530008, Guangxi, China
frankpep@126.com

Abstract. With the development of science and technology and the deepening of globalization, the algorithm of CMAC neural network is analyzed. Although the basic CMAC neural network has defects, it is greatly constrained by the training samples during offline training. In the self-learning process, the activation weights are modified without difference, which will cause the output jump change, However, compared with other intelligent algorithms, it has the unique advantages of fast convergence, strong generalization ability and similar input to achieve similar output. With the increasingly fierce competition among enterprises, the politics, economy, society and environment are complex and changeable. The dynamic environment will inevitably affect the business content, organizational form and management mode of enterprises, and then affect the post setting, post division and the work content, work characteristics and job requirements of employees. The dynamic organizational environment requires the enterprise to continuously develop the overall flexibility and the ability to adapt to the dynamic environment, and puts forward the requirements for the work analysis to adapt to the dynamic adjustment. The work analysis results obtained by the traditional work analysis method are only static processing of the work information, and the analysis results cannot be adjusted dynamically. A new work analysis is a huge project, which will waste a lot of time, human, material and financial resources, which makes the human resource management of the enterprise encounter a major problem.

Keywords: CMAC neural network · Work analysis · Process analysis · Work module

1 Introduction

In the process of modern international trade, with the continuous improvement of product quality requirements, the modular management of the corresponding trade system is becoming more and more complex. In such a social environment, fuzzy control, neural network control and other intelligent control plays an increasingly important role. However, the control of intelligent international trade management system requires fast, continuous and stable control. Because CMAC neural network has unique advantages in self-learning speed, close input and close output, it has been discussed and studied by

J. MacIntyre et al. (Eds.): SPIoT 2020, AISC 1283, pp. 806–810, 2021.
https://doi.org/10.1007/978-3-030-62746-1_126

scholars and experts [1]. After the research in the last century, the research focuses on the generalization ability, structure and parameter selection, control accuracy, process mapping and convergence of CMAC neural network. The constructed trade analysis module system can give early warning of customer churn, help retain customers to a certain extent, and reduce customer churn rate.

2 Function Demand Analysis of Trade Analysis Module

In this module, we introduce the more mature data mining algorithm CMAC neural network based on information theory to conduct in-depth analysis and Research on customer churn. We can deduce a list of customers that are easy to lose, and then combine the value score of each customer, the company can take some retention measures to improve the company's marketing volume.

Customer segmentation, e-commerce enterprises provide products and services that meet the needs of different customers to meet diverse needs, so that different customers are satisfied with the enterprise, we need to segment customers according to different standards, in order to achieve long-term and stable relationship between customers and enterprises. Although it is impossible for members of each small group to achieve the same consumption behavior, they can also show certain commonness. After mastering the commonness of these small groups, enterprises can formulate targeted marketing strategies to guide the development of the company [2].

Customer value degree, customer value degree, that is, the enterprise gains from the purchase of customers. For enterprises, an occasional customer and a regular customer have different customer value. We can predict customer value according to customer's consumption behavior and specific characteristics. To analyze the profit value of customers to enterprises, we need to obtain a considerable number of samples based on traditional experience, and then use decision tree classification algorithm to conduct in-depth research and analysis. After obtaining the decision tree model, we can predict for general customers.

3 Application of CMAC Neural Network to the Construction of International Trade Module

CMAC neural network is proposed by albus according to the biological model of cerebellum. As a kind of neural network, it is more in line with human's thinking form. In terms of self-learning, CMAC neural network needs less modified weights than BP and RBF neural network, and its convergence speed is faster, and it has strong generalization ability to achieve similar input and achieve similar output. CMAC neural network takes the input variables of the system as a pointer, and stores the relevant information in a group of storage units. It is essentially a look-up table technique for mapping complex nonlinear functions. The enterprise has accumulated a large amount of original data, recorded the valuable information such as customers' browsing behavior, concerned content, trading method, etc., but how to play the utility of these data, mining the mode or law of business activities is an important means for each enterprise to enhance its competitiveness. Therefore, the analysis of trade activities has become an important part of the work of enterprises. In this paper, we make full use of the large amount of original sales data, customer browsing behavior log and other important resources accumulated in the past, and use the intelligent data analysis algorithm of CMAC neural network in data mining to build a trade analysis platform.

3.1 Improved Algorithm Am-CMAC and Convergence Analysis

In CMAC neural network, the concept storage space a is mapped from binary to the associated coordinate system with the input x as the abscissa scale and the membership μ as the ordinate. The input interval $[x_{min}, x_{max}]$ of x is divided into $N + C - 1$ perceptual intervals with width δ. S, the corresponding interval with width φ in the associated coordinate system is defined as the associated domain a, where $(i = 1, 2, , \ldots, N)$.

The calculation formula of δ is:

$$\delta = \frac{x_{min}, x_{max}}{N + C - 1} \tag{1}$$

The calculation formula of Ø is:

$$\emptyset = \frac{x_{min}, x_{max}}{N} \tag{2}$$

Definition 2 in the related coordinate system, establish the fuzzy membership function $\lambda_J = \mu_J(x)$, where $j = 1, 2, \ldots, N + C - 1$.

$$\mu_j(x) = \begin{cases} \frac{2}{(2C-1)\delta}x + \frac{2}{(2C-1)}\left(C - j - \frac{x_{min}}{\delta}\right), x_{min} + (j - C)\delta \leq x \\ \frac{2}{(1-2C)\delta}x + \frac{2}{(2C-1)}\left(C + j - 1 + \frac{x_{min}}{\delta}\right), x_{min} + \left(j - \frac{1}{2}\right)\delta \leq x \end{cases} \tag{3}$$

$\mu_J(x)$ is the fuzzy membership of input x about a_j; if the perceptron is not activated, its fuzzy membership is 0 (Fig. 1).

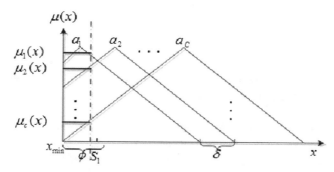

Fig. 1. Schematic diagram of associated fuzzy membership function

The current international trade situation is changing rapidly, and the business volume is also adjusting at any time. Reasonable post setting should fully consider the requirements of dynamic adjustment of business volume, analysis and design of system framework structure, customer churn and user value. The constructed trade analysis module system can give early warning of customer churn, help retain customers to a certain extent, and reduce customer churn rate.

4 Design and Implementation of Customer Churn Analysis Module in International Trade

CMAC algorithm technology is used to establish a prediction model of customer churn in international trade, so as to find out the most important factors to promote customer churn, generate a list of customer churn prediction, and then combine the value of these customers, derive a reasonable retention measures to help the company to make appropriate decision-making strategies, and maintain high profits and customer loyalty. The methods that can be used are data mining classification and clustering. The difference between the two is that the classification needs to know the sample classification in advance, while clustering does not. The decision tree method in this paper is simple and easy to understand. The representative algorithms include ID3 algorithm, C4.5 algorithm and so on. The main steps of building a data mining customer churn prediction model are as follows: data integration, data processing (data de-noising, data extraction, data generalization, etc.) and CMAC algorithm implementation. The relevant steps are as follows [3].

Data integration, customer data and transaction data needed for customer churn analysis are stored in customer index table, order table and other tables. Therefore, first of all, we need to denoise these data, remove the unwanted attributes such as customer phone, age and other information, and extract useful information. In data processing, one of the principles of information extraction is to reflect the information related to customer type, customer background and transaction, and then put the extracted information into a new table after sorting, and put it into data warehouse for the use of mining algorithm. If some options provided by the business platform are not filled in when the customer is registering or in the actual transaction, there may be many blank data items in the data extracted from this kind of customer. The system needs to provide some measures to deal with these blank data items, either remove the record or fill in the default value. The selection of the default value depends on different attributes, and the average value is selected for general numerical data. After generalization, the data type in the customer information data table may not meet the requirements of data mining algorithm, and it still needs to be transformed.

4.1 Design and Implementation of International Trade Segmentation Analysis Module

Customer segmentation can also be achieved by clustering or classification algorithm. In this system, we choose to use CMAC algorithm to segment customers. The required source data is the same as the source data of customer churn analysis, but it also needs to remove noise and information extraction, discrete processing. Then, these data are analyzed and processed by CMAC algorithm, and the results can be displayed by graphics. Users can assist their business activities by viewing the graphical clustering results and the data information in each group. To analyze the profitability of customers, we must first find a standard that can effectively measure the profitability of customers. This problem is relatively difficult because different users and staff have different views and understandings. The method of this system is to accumulate market experience and classify the customer profitability in the sample data. After these classification samples reach a certain number, CMAC algorithm can be used for learning prediction, so as to generate a decision tree model for general users to predict [4].

5 Conclusions

Based on the development and design of CMAC algorithm international trade management system platform, the unified business processing scheme and CMAC algorithm data analysis method are applied in the designed system, and data mining is introduced into the system's trade analysis module to provide intelligent guidance for decision makers. The next step is to further study the algorithm, optimize the algorithm and improve the efficiency of operation.

References

1. Yan, W.: Design of personalized intelligent recommendation system based on data mining technology [J]. Microcomput. Appl. **2**, 119–121 (2019)
2. Lin, L., TU, G., Yang, F.: Application of decision tree based data mining in CRM [J]. Comput. Technol. Autom. **1**, 67–69 (2006)
3. Fan, J., Yang, Y., Wen, P.: Application of C4.5 algorithm in online learning behavior evaluation system [J]. Comput. Eng. Des. **6**, 946–948 (2006)
4. Zhangwei, L., Chen, L., Cao, Z.: Research on collaborative filtering algorithm in intelligent recommendation system under big data environment. Comput. Prog. Skills Maintenance **11**, 78–79 (2016)

Design and Research of Electroless Copper Plating in Microelectronics

Dong Zhenhua[✉] and Li Xiaoming

Department of Information Engineering, Laiwu Vocational and Technical College,
No.1, Shancai Street, Laiwu, Jinan 271100, Shandong, China
31378843@qq.com, lzytjc@163.com

Abstract. The principle of electroless copper plating technology and its application status and production process in through-hole metallization, inner copper foil treatment, electronic packaging technology and electromagnetic wave shielding in printed circuit board (PCB) are briefly introduced. The development of electroless copper plating in microelectronics is prospected.

Keywords: Electroless copper · Printed circuit board · Copper foil · Electromagnetic wave shielding · Packaging

1 Introduction

Electroless copper plating has the advantages of good adhesion, corrosion resistance, electrical reliability, heat resistance, electromagnetic shielding and even coating distribution. At the same time, the electroless copper plating technology is also suitable for the metallization of non-conductor surface, which is widely used in the through-hole metallization of printed circuit boards, inner copper foil treatment, electromagnetic wave shielding and electronic component packaging technology. With the development of high technology, a variety of new technologies and surface treatment technologies are integrated. For example, laser and UV induced electroless copper plating technology will make the electroless copper plating technology more widely used in the manufacturing and repairing process of microelectronic products.

2 Basic Principle of Electroless Copper Plating

The basic principle of electroless copper plating is that the copper ions in the solution are reduced to metallic copper crystals with the help of appropriate reducing agents, thus depositing on the surface of the substrate. At present, formaldehyde is used as reducing agent in most electroless copper plating. Formaldehyde can only reduce in alkaline medium with pH > 11. The reaction mechanism of electroless copper plating is as follows:

$$Cu^2 + HCHO + 3OH^- \rightarrow Cu + HCOO^- + 2H_2 0$$

J. MacIntyre et al. (Eds.): SPIoT 2020, AISC 1283, pp. 811–815, 2021.
https://doi.org/10.1007/978-3-030-62746-1_127

Because formaldehyde is toxic, formaldehyde free electroless copper plating came into being. Since 1990s, hypophosphite has become a new reducing agent for electroless copper plating. The reaction mechanism is as follows.

Acid Electroless Copper Plating:

$$Cu^{2+} + H_2PO_2^- + H_2O \rightarrow Cu + 2H_2PO_3^- + 2H^+$$

Alkaline electroless copper plating:

$$Cu^{2+} + H_2PO_2^- + 2OH^- \rightarrow Cu + H_2PO_3^- + H_2O$$

2.1 Process Principle of Electroless Copper Plating

The rapid development of computer industry in electronic industry and the requirement of high efficiency EMI shielding of electronic components in aviation industry greatly promote the research of electroless copper plating technology. Electroless copper plating was first reported by narcus in 1947. The technology similar to today's electroless copper plating was first reported by Cahill. Tartrate alkaline copper plating bath with formaldehyde as reducing agent was used. Electroless copper plating has the advantages of good corrosion resistance, electrical reliability, good adhesion, heat resistance and electromagnetic shielding. At the same time, electroless copper plating is also suitable for the surface metallization of non-conductor surface. The advantages of uniform coating distribution make it widely used in electronic industry. At present, it is widely used in through-hole metallization, inner copper foil treatment, electromagnetic wave shielding and electronic component packaging technology in electronic industry.

Electroless copper plating is a method that copper ions in the solution are reduced to Copper Crystals and deposited on the surface of the substrate with the help of appropriate reducing agents. At present, formaldehyde is used as reducing agent in most electroless copper plating. Formaldehyde can only be reduced in alkaline medium with pH > 11. As formaldehyde is toxic, formaldehyde free electroless copper plating has become a trend for environmental protection. Since 1990s, hypophosphite has become a new reducing agent for formaldehyde free copper plating. Electroless plating is an autocatalytic process, but copper itself has no catalytic activity, so palladium should be deposited on the surface without catalytic activity to play a catalytic role.

3 Application of Electroless Copper Plating in Electronic Industry

3.1 Conducting Treatment PTH for Through Hole and Inner Wall of Thin Tube

With the rapid development of microelectronics industry and computer industry, the requirements for electronic circuits and devices are not only high performance, but also miniaturization of their structures. The wiring width must be reduced and double-sided or multi-layer printed circuit boards should be used. Metallization of through-hole in PCB by electroless copper plating is the main process in PCB production. Electroless copper plating is used for conducting treatment of through hole and inner wall of thin tube on printed circuit board, which is usually divided into two categories, namely, thin copper layer and thick copper layer. The thickness of the thin copper layer must be more than 0.5 μm, and the copper coating on the inner wall should have no holes and uniform thickness [2]. The electroless copper plating bath without reducing promoter can obtain 0.1 μm thick copper coating after immersion in the copper plating solution for 3 min, and then the copper plating layer of 0.5 μm can be obtained by electrodeposition of copper with micro current of 1A/DM 2 in copper sulfate plating solution. Thick copper layer is electrodeposited on the basis of electroless thin copper layer to reach the required thickness.

3.2 Inner Copper Foil Treatment

When making multilayer printed circuit boards, blackening treatment is usually needed to ensure the bonding force between the inner copper layer and the insulating resin. The so-called blackening treatment is to soak the inner copper foil in a strong alkali solution containing chlorite as the main reagent to obtain black copper oxide. Although black copper oxide has good binding force with epoxy resin, it has poor affinity with polyimide, BT resin and PPE resin with excellent heat resistance and dielectric property. In addition, the copper oxide formed by blackening is easy to dissolve and form holes in the treatment solution containing hydrochloric acid and chelating agent. In order to solve the above problems, a needle like crystal copper plating layer was obtained by using hypophosphite as reducing agent to replace the blackening treatment of the inner copper layer [3].

4 Electromagnetic Wave Shielding

In recent years, with the large-scale use of communication equipment, semiconductor equipment and electronic instruments, electromagnetic wave pollution is causing widespread concern. At the same time, in order to protect some precision electronic instruments from electromagnetic interference, the use of electromagnetic shielding layer plays a key role. Due to the price, weight and other reasons, the shell of electronic components is made of plastic, but the electromagnetic interference shielding of plastic is poor. In order to reduce electromagnetic interference, some electronic components need electromagnetic interference shielding cover inside and outside [4]. From 1966, when Lordi first proposed to use electroless copper plating to shield electromagnetic interference, it was widely used in the early 1980s. At present, this technology is still used for electromagnetic shielding. In order to achieve excellent effect of electromagnetic wave sealing, the shell of portable electronic instruments and semiconductor equipment is electroless plated with 1–2 μm thick copper layer with hypophosphite as reducing agent, so as to form needle like crystal or porous copper coating on the shell, then electroless nickel plating, and finally coating with acrylic urethane. After this treatment, the bonding strength of the electroless coating and the coating is up to 200 MPa, while that of the electroless copper plating plus organic coating is only 100 MPa, but the bonding strength is significantly improved, the durability is greatly enhanced, and the sealing effect on electromagnetic wave is also obviously enhanced. This can achieve the purpose of beautiful appearance, good durability and high strength.

4.1 Electronic Packaging Technology

Aluminum as the preferred material for complex circuits and pad metallization has lasted for more than 30 years. However, with the development of microelectronics manufacturing to the fine direction, the disadvantages of high resistance and poor heat dissipation of aluminum appear, and copper just has this advantage. In this way, electroless copper plating is widely used in electronic packaging technology, and the most prominent one is the metallization of ceramic circuit substrate. There are advantages and disadvantages in metallization of microwave and ceramic hybrid circuit substrates, such as thin film, thick film, CO firing [9] and direct copper coating. In order to meet the requirements of power and heat dissipation in packaging, electroless copper plating metallization on ceramic substrate was developed in the 1980s, and then photolithography was used to produce the required circuit graphics; since the 1990s, this technology has become more mature and has been widely used. Process flow: etch activated electroless copper plating with complex photoresist circuit pattern exposure development pattern etching removal photoresist electroless gold plating using this process to make circuit substrate has good conductivity, good thermal conductivity, bonding performance (nickel/gold plating on copper conductor) and soft soldering performance, stable process, convenient production and low cost, Metallization of microwave and hybrid IC substrates is a practical and effective process. This technique is not only suitable for 96% and 99% Al 2O 3 ceramics, but also suitable for metallization of BeO and AlN. Because of its high thermal conductivity (up to 8–10 times of that of Al 2O 3 ceramics), the thermal expansion coefficient of AlN is close to that of silicon, with good insulation, low dielectric constant, good mechanical properties and non-toxic. Therefore, as the preferred substitute of Al 2O 3 and BeO, it is a new kind of electronic packaging material with great potential. In recent years, electroless copper plating on AlN has also been studied. In addition, semi addition or full addition method can be used to make fine conductor circuit patterns with spacing less than 100 μm in packaging manufacturing to meet the needs of high-density interconnection.

5   Conclusions

Considering the environmental pollution caused by formaldehyde in electroless copper plating technology, the process of electroless copper plating with no formaldehyde reducing agent and hypophosphite, cobalt ethylenediamine and other reducing agents has the characteristics of precipitation rate, coating composition and surface morphology that formaldehyde method does not have. At the same time, in order to improve the work efficiency and simplify the process in the manufacturing process of printed circuit boards, the use of laser, ultrasonic, infrared and ultraviolet induced electroless copper plating, combined with microcomputer control, can obtain various metal wiring.

References

1. Deckert, C.A.: Electroless copper plating: a reiew. Plat. Surf. Finish. **82**(2), 48–55 (1995)
2. Dinella, D.: Printed circuit world convention VI, San Francisco, **7** (1993)

3. Qing, L.: Application of electroless multi-functional copper coating in electronic industry [J]. Spec. Equipment Electron. Ind. **27**(2), 27–30 (1998)
4. Violett, J.L.N., et al.: Electromagnetic Compatibility Handbook. Van, Nostrand Reinhold Co., New York (1987)
5. Lordi, G.A.: Electroless plating for electronic applications. Plating **54**, 382 (1967)

Application of Neural Network Algorithm in Building Energy Consumption Prediction

Mao Yueqiang[⊠]

Southwest Forestry University, Kunming 650224, Yunnan, China
maoyueqiang@126.com

Abstract. Building energy conservation is the forefront and research hotspot of urban construction and social development. Comprehensive analysis and evaluation of building energy consumption status is the premise and foundation for energy-saving transformation or energy-saving design. The establishment of prediction model reflecting energy consumption change is an effective way to analyze and recognize the change and development characteristics of building energy consumption from a macro scale, and provide decision-making basis for public building energy conservation Ways and important means. Aiming at the shortcomings of the conventional BP network algorithm, such as slow convergence speed and easy to fall into the local minimum point, LM algorithm with fast convergence speed and stability is used for prediction, and a building electricity consumption prediction model based on BP neural network is constructed. Taking the statistical data of original power consumption of public buildings in a city as samples, the prediction model is simulated and predicted by MATLAB. The results show that the error is within the allowable range.

Keywords: Building energy consumption · Neural network · LM algorithm · MATLAB · Prediction model

1 Introduction

With the development of China's economy, the problem of high energy consumption in office buildings and large public buildings is becoming increasingly prominent. It is of great significance to do a good job in the energy-saving management of state office buildings and large-scale public buildings to achieve the goal of building energy-saving planning in the 12th five-year plan. Building energy saving is the forefront and research hotspot of urban construction and social development. Public building energy saving is an important part of building energy saving. Comprehensive analysis and evaluation of the current situation of building energy consumption is the premise and basis for energy-saving transformation of public buildings or energy-saving design of new buildings. The establishment of prediction model reflecting energy consumption change is an effective way and important means to analyze and understand the change and development characteristics of building energy consumption from a macro scale, and

J. MacIntyre et al. (Eds.): SPIoT 2020, AISC 1283, pp. 816–820, 2021.
https://doi.org/10.1007/978-3-030-62746-1_128

provide decision-making basis for public building energy conservation work. Artificial neural network is an information processing system designed to imitate the structure and function of human brain. It has the functions of self-learning, self-organization, associative memory and parallel processing. BP (back propagation) neural network is one of the most widely used models in neural network. Because of its high ability of self-learning, self-organization and self-adaptive, as well as its own nonlinearity, neural network can approach the nonlinearity between input and output of the system without knowing the exact mathematical model [1].

Therefore, neural network is widely used in prediction model. At present, most of the public building equipment is driven by electricity, and electric energy consumption is the main body of building energy consumption. In this paper, LM (Levenberg Marquardt) algorithm is used to preliminarily discuss the application of neural network in building energy consumption prediction.

2 Neural Network and Its Algorithm

Neural network is a multi-layer forward network with unidirectional propagation. BP network is a kind of neural network with three or more layers, including input layer, middle layer (hidden layer) and output layer; the upper and lower layers are fully connected, and there is no connection between each layer of neurons. When a pair of learning samples are provided to the network, the activation value of neurons propagates from the input layer through the intermediate layer to the output layer, and each neuron in the output layer obtains the input response of the network. Then, according to the direction of reducing the target output and actual error, the connection weights and thresholds are modified layer by layer from the output layer through the intermediate layer, and finally return to the input layer. This algorithm is called "error back propagation algorithm", namely BP algorithm. It has been proved that when the number of hidden layer neurons of BP neural network is increased enough, it can approach any nonlinear function with finite discontinuities with arbitrary precision. BP neural network training is a nonlinear fitting method. The trained neural network can also give the appropriate output for the input which is not near the sample set. Because of the strong nonlinear fitting ability of BP neural network, the learning rules are simple and easy to be realized by computer, so the BP learning algorithm is often used to train the neural network. The conventional BP neural network model has some problems, such as unstable prediction results, difficult to determine the learning rate and falling into local optimum. It needs to be optimized according to the actual characteristics.

2.1 LM Algorithm

The original BP algorithm is called gradient descent method. The parameters move along the opposite direction of the error gradient, so that the error function decreases until the minimum value is obtained. Its computational complexity is mainly caused by the calculation of partial derivatives. The gradient descent method is only linear convergence and slow. There are two ways to improve the algorithm: one is to use heuristic learning methods, such as gradient algorithm with momentum, which can be regarded as the approximation of conjugate gradient method; the other is to adopt more effective optimization algorithms, such as conjugate gradient method and Newton method. LM algorithm is a fast algorithm using standard numerical optimization technology. It is a combination of gradient descent method and Gauss Newton method. It is an improved form of Gauss Newton method. It not only has the local convergence of Gauss Newton method, but also has the global characteristics of gradient descent method. LM algorithm uses approximate second derivative information, and its convergence rate is better than gradient descent method. The LM algorithm is briefly described below [2].

Newton's Law:

$$\Delta x = -\left[\nabla^2 E(x)\right]^{-1} \nabla E(x) \tag{1}$$

Where $\nabla^2 E(x)$ is the Hessian matrix of the error index function $E(x)$; $\nabla E(x)$ is the gradient.

Set error index function:

$$E(x) = \frac{1}{2} \sum e_i^2(x) \tag{2}$$

Where e (x) is the error, then:

$$\nabla E(X) = J^T(x)e(x) \tag{3}$$

$$\nabla^2 E(X) = J^T(x)e(x) + S(x) \tag{4}$$

When the Gauss Newton method approaches the minimum error, the calculation speed is faster and the accuracy is higher. Because LM algorithm uses approximate second derivative information, its speed is better than gradient descent method. Practice shows that LM algorithm can increase the speed of original gradient descent method by tens or even hundreds of times.

3   BP Network Model of Building Energy Consumption

Building energy consumption system itself can be regarded as a highly nonlinear system about time series, so a multilayer BP network can be used to simulate building energy consumption. In order to determine the structure of the model, it is necessary to consider the variable factors which have great influence on energy consumption, and analyze the feasibility of its application in neural network. There are many factors that affect the energy consumption of public buildings, such as outdoor climate conditions, personnel activities, office equipment changes, equipment operation rules and so on. The change of each factor will affect the building energy consumption. If all factors are considered, the system will be too complex and huge, the network learning time will be too long, and it is easy to fall into local minima, which can not achieve satisfactory results. Therefore, we should consider the influence factors according to the actual situation, train the network according to the historical records, and test the trained network through other known data. This paper analyzes the data obtained from the existing energy consumption monitoring system and establishes a prediction model.

3.1 Data Processing

In order to reduce the influence of singular samples on the performance of neural network, the samples are normalized to make the range between [0,1]. The normalization formula is as follows:

$$y = \frac{x - x_{min}}{x_{max} - x_{min}} \tag{5}$$

Where: y is the normalized output of x, x_{min} is the minimum value of x, and x_{max} is the maximum value of x.

3.2 Establishment of Prediction Model

The application of neural network in power consumption prediction can give full play to the advantages of neural network in nonlinear mapping, comprehensively integrate various factors affecting energy consumption, and make the prediction closer to the actual situation. Due to the limited data available, this paper uses the power consumption statistics of public buildings in a city in the first ten months of 2019 as samples, selects six groups of data as training samples to train the neural network model, and takes three groups of data as test samples to test the accuracy of the established model. Building area and power consumption from January to June are taken as input variables in the sample. The target vector is the electricity consumption from July to October. Therefore, there are 7 input vectors and 4 output vectors. Finally, the BP neural network with 12 inputs and 4 outputs is selected for power consumption prediction.

According to Kolmogorov theorem, the number of neurons in the hidden layer is set to 12. The neuron transfer function in the middle layer of the network adopts the S-type tangent function Tansig, and the output layer adopts the S-type logarithmic function logsig. The output interval of this function lies in the interval [0,1], which meets the requirements of the network.

4 Network Training

The network can only be used in practical application after training. Considering the complexity of the network structure and the number of neurons, it is necessary to increase the training times and learning rate. The training parameters are set in Table 1 [3].

Table 1. Training parameters

Training times	Training objectives	Learning rate
1000	10 ~ 11	0.05

The training results are shown in Fig. 1.

After training the BP neural network model with training samples for nearly 290 times, the standard numerical error reaches 1e-11. It can be seen that the established prediction model has good prediction accuracy.

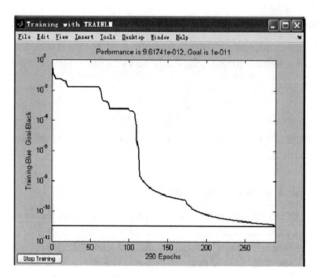

Fig. 1. Training results

5 Conclusions

Through the prediction of building energy consumption, it is helpful to analyze and understand the characteristics of the change and development of building energy consumption from a macro scale, so as to have a good guiding role in building energy conservation work. The building energy consumption can be predicted by a new method of building energy consumption analysis with a small amount of new neural network. Since the building energy consumption monitoring system has just started, the information obtained is limited. This paper only makes a preliminary discussion on this. There is still a long way to go to get accurate power consumption data model, and further practical work is needed.

References

1. Zhe, Z., Zhifei, W., Jianli, Y.: Neural network method for electric power forecasting. Knowl. Jungle **4**, 156 (2008)
2. Zhang Nai, Y., Taiping, Y.: Neural Network and Fuzzy Control, pp. 35–46. Tsinghua Press, Beijing (1998)
3. Lee, K.Y., Cha, Y.T., Park, J.H.: Short term load forecasting using anartificial neural network. IEEE Trans. Power Syst. **7**(1), 124–132 (1992)

Research and Design of Big Data Unified Analysis Platform Based on Cloud Computing

Zhen Zhiming[(⊠)]

Faculty of Science and Technology, Gannan Normal University, Ganzhou 341000, China
qgjxzbt@163.com

Abstract. By introducing the business characteristics of the big data era and the challenges faced by the current big data era, this paper makes a detailed research and design of the big data unified analysis platform based on cloud computing, including the architecture system of the big data analysis platform, the software architecture of the big data analysis platform, the network architecture of the big data analysis platform, and the big data unified analysis platform This paper analyzes the competitive advantages of the unified big data analysis platform based on cloud computing. The big data unified analysis platform based on cloud computing will more effectively support the development of telecom operators in the future.

Keywords: Cloud computing big data master host segment server completely share nothing architecture data protection parallel computing framework (map reduce) high availability

1 He Coming of Big Data Era

With the increasing competition in the communication industry, how to effectively use the huge signaling data to further realize deep operation and accurate marketing has become an urgent task for operators. A data platform with controllable investment can meet the requirements of controllable signaling data storage, and can efficiently analyze and mine the value of signaling data. Big data "big data" is another subversive technological change in IT industry after cloud computing and Internet of things, which will have a huge impact on national governance mode, enterprise decision-making, organization and business process, and personal lifestyle. In the field of research, McKinsey believes that data has become a torrent into every area of the global economy. Big data can become the new assets of enterprises, form an important foundation of competitiveness, and play an important economic role. IDC believes that big data processing will become an essential capability in 2012. Gartner believes that more than 85% of Fortune 500 companies will lose their edge in big data competition in 2015. In March 2012, the Obama administration released the "big data development plan" and defined it as "the new oil of the future". This series of events makes big data another hot word [1].

J. MacIntyre et al. (Eds.): SPIoT 2020, AISC 1283, pp. 821–825, 2021.
https://doi.org/10.1007/978-3-030-62746-1_129

Telecom operators introduce big data technology to realize massive data storage and layering through controllable cost. At the same time, by shortening data processing path and providing super large data processing bandwidth, telecom operators can effectively reduce the response time of data analysis, improve the business value of signaling analysis, and enhance the core competitiveness of operators.

2 Challenges in the Era of Big Data

2.1 Concept of Big Data

(1) large data scale: it is difficult to give an absolute digital standard to determine the size, which may be compared with some fuzzy feelings;
(2) High complexity of data structure: the data of complex data structure can transmit more abundant information;
(3) High degree of data correlation: the level of data association is related to the data Mining degree, if the data association degree is low, no matter how large the amount of data, how complex the structure, can not form big data.

2.2 Problems in the Era of Big Data

(1) Simple script language preprocessing can't analyze the complicated data structure;
(2) Relational database faces embarrassment in the face of big data;
(3) The optimization space of commercial database is limited;
(4) Data quality cannot be effectively monitored;
(5) More and more business requirements compromise to data computing capability [2, 3].

3 Research and Design of Big Data Scheme Based on Cloud Computing

3.1 Design Idea of Big Data Unified Analysis Platform

(1) Build a unified data computing platform in the enterprise;
(2) Enterprise owners can directly control their data instances;
(3) It provides enterprise level data access function directly through entity integration;
(4) Flexible expansion and configuration reduce the average risk of investment.

The big data platform in the cloud era not only supports Pb level, but also massive structured, semi-structured, and even unstructured data storage at Pb level and even ZB level with high performance price ratio and high scalability. At the same time, we also need to be able to mine the value of these data at a high speed, create profits for enterprises, and truly realize that big data equals big value. The storage and unified big data analysis platform based on cloud computing combines database map reduce architecture to build a big data analysis platform for enterprises to efficiently process structured, semi-structured and even unstructured data. Customers can realize the transformation of data assets from cost center to profit center based on this platform and drive business with data (Fig. 1).

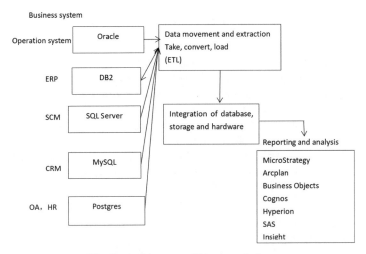

Fig. 1. Architecture of big data platform

3.2 Big Data Unified Analysis Platform Software Architecture

(1) Software architecture

Through the master host and multi node segment host, the database is connected through the Internet. Applications access data through the master host. Each storage node in the network is an independent database, and there is no sharing between them. Data exchange between multi storage nodes and master host.

The segment server of each node is connected through the Internet to complete the same task. From the user's point of view, it is a server system. Its basic feature is that it is connected by segment servers (each segment server is a node) through the Internet. Each node only accesses its own local resources, including memory, storage, etc. it is a completely share nothing architecture. Therefore, it has the best scalability. Theoretically, it has unlimited expansion. The current technology can realize the interconnection of 512 nodes, thousands of nodes CPU. Each node can run its own database and operating system, but each node can't access the memory of other nodes. The information interaction between nodes is realized through the node Internet, which is called data redistribution.

(2) High availability scheme design [4]

The disk sharing architecture allows the system to have multiple servers that are connected to a San or other shared storage device. This architecture requires a narrow data pipeline to filter all I/O information to an expensive shared disk subsystem. In terms of structure, the scalability and performance of "full sharing" or "disk sharing" system are limited. Moreover, the general disk sharing system is complex and fragile, and it is difficult to handle terabyte data.

3.3   Advantage Analysis of Big Data Unified Analysis Platform

(1)   Agility that allows the deployment and redistribution of large amounts of computing resources on demand according to business priorities;
(2)   It can analyze more detailed and diversified low latency data sets (big data), while retaining the subtle differences and relationships within the data, so as to obtain differentiation insights that are conducive to optimizing business performance;
(3)   Organization wide collaboration around key business plans to quickly disseminate best practices and organizational findings.
(4)   Cost advantage: you can use commoditized processing components to analyze big data, so as to take advantage of business opportunities that could not be utilized economically and efficiently before.

According to Gartner's prediction, big data technology is in a period of rapid development in 2012, with continuous technological breakthroughs, and product intensive release or other projects that can produce significant benefits appear rapidly. The big data unified analysis platform based on cloud computing will effectively support the data with high correlation and complex data structure, effectively support Pb level data, effectively reduce the response time of data analysis, and improve the business value of signaling analysis. The big data unified analysis platform based on cloud computing has important strategic and economic significance for the future business and technology development of telecom operators.

4   Empirical Analysis Based on Analytic Hierarchy Process

The analytic hierarchy process (AHP) can effectively transform the complicated and ambiguous relationship into quantitative analysis relationship. In the face of complex choice problems, people often decompose the problem into various constituent factors, and then group these factors into hierarchical structure according to the dominant relationship. Through pairwise comparison, the relative importance of various factors in the hierarchy is determined, and then the overall ranking of the relative importance of decision-making schemes is determined by synthesizing the judgment of decision-makers, so as to make choices and judgments.

The discriminant matrix was constructed by pairwise comparison:

$$
U = \begin{bmatrix}
U_{11}, U_{12}, \cdots, U_{1m} \\
U_{21}, U_{22}, \cdots, U2_{1m} \\
\vdots \quad \vdots \quad \vdots \\
U_{n1}, U_{n2}, \cdots, U_{nm},
\end{bmatrix}
$$

5   Conclusion

According to Gartner's prediction, big data technology is in a period of rapid development in 2012, with continuous technological breakthroughs, and product intensive release or other projects that can produce significant benefits appear rapidly. The big data unified analysis platform based on cloud computing will effectively support the data with high correlation and complex data structure, effectively support Pb level data, effectively reduce the response time of data analysis, and improve the business value of signaling analysis. The big data unified analysis platform based on cloud computing has important strategic and economic significance for the future business and technology development of telecom operators.

Acknowledgements. Science and Technology Research Project of Jiangxi Education Department (161631); Design and implementation of online interactive learning system for teacher qualification in teachers' colleges and universities.

References

1. Ruming, C.: Challenges, values and coping strategies in the era of big data. Mob. Commun. **17**, 14–15 (2012)
2. Chunlei, Z., Nahan, G.: Computer information processing technology in the era of big data. World Sci. **2**, 30–31 (2012)
3. Qilong, Z., Ming, F., Sheng, W.: Parallel scientific computing based on MapReduce model. Microelectron. Comput. **08**, 13–17 (2009)
4. GuiQiang, W., Chaojun, L.: Discussion on statistical analysis of large amount of data based on parallel technology. Comput. Appl. Softw. **3**, 162–165 (2011)

Computer Software Technology Under the Background of Big Data Development

Lv Ying and Zhang Zhifeng[✉]

Laiwu Vocational and Technical College, No.1, Shanghai Street, Laiwu, Jinan 271100, Shandong, China
lvying-1982@163.com, lwvczhang@126.com

Abstract. With the continuous development of science and technology in our country, big data technology is becoming more and more mature. In recent years, the application of computer software in our country is more and more extensive. As an important part of social production, it is of great significance to strengthen the research of computer software technology. Computer technology, as one of the landmark technologies in the 21st century, has a direct relationship with social production and people's quality of life. Therefore, this paper focuses on the application of computer software technology in the era of big data, to improve the data processing capacity and strengthen the economic benefits of social production. Therefore, it is of great significance to strengthen the application of computer software technology in the era of big data.

Keywords: Computer software technology · Big data era · Application · Development

1 Introduction

With the continuous development of science and technology in our country, the application of computer software is more and more extensive, which makes all walks of life out of the traditional management mode and realizes modern management. Nowadays, big data has become the focus of social attention. Through the analysis of big data, it can effectively improve the social production capacity, thus promoting the economic development of our country. For the analysis of current research results of computer software, China's computer software technology is at the top level in the world, and there are some differences compared with some developed countries. To promote the development of computer software technology in our country and integrate computer software technology into big data, we must start with real software technology, and carry out reform and innovation according to existing resources, to realize scientific and reasonable operation mode and promote the healthy development of various industries in our market [1].

J. MacIntyre et al. (Eds.): SPIoT 2020, AISC 1283, pp. 826–830, 2021.
https://doi.org/10.1007/978-3-030-62746-1_130

2   Analysis of the Development of Computer Software Technology in China

With the continuous development of science and technology in China, the role of computer software technology in social production is increasingly prominent. Many enterprises and institutions have realized the positive role of computer soft armor, vigorously developed computer software technology, and in the era of big data, people have also focused on the research work on the storage and database level of big data, which has brought great convenience to the people. Big data has become a major trend in the development of computer software technology. Because big data can provide users with a huge amount of data, that is to integrate network data through big data, to provide users with corresponding convenience services. Big data has strong decision-making power, insight, and process optimization ability, so it will continue to expand the database on the Internet, and the data will gradually accumulate and increase. According to the development of big data in China today, the total amount of big data in China has reached 1 billion T capacity, and it only takes five years to break through the 2 billion T capacity at the current development speed. Therefore, the arrival of big data brings huge development space to computer software technology and has a huge impact on promoting social and economic benefits. In short, big data, as the Internet of things and cloud computing in the IT industry, provides an opportunity for reform. New requirements have been put forward in enterprise management, government public relations, and personnel training, which is also the inevitable trend of development in the era of big data.

3   The Application of Computer Software Technology in the Era of Big Data

In recent years, big data has become a focus topic in society. The main application core in the era of big data is to find rules in the data, to provide customers with more convenient needs. According to the user's use characteristics, to carry out design, production, service, and other content. The key performance of computer information processing technology in the data age is as follows:

3.1   Virtualization Technology

Virtualization technology is mainly realized through virtual resource management, and at the same time, through the optimization of the internal resources of big data, it can improve the efficiency of information processing and provide the flexibility of user operation. In recent years, virtual technology is widely respected in the world, and many research institutions and enterprises vigorously develop virtual technology. Also, virtual technology can not only be displayed in computers but also in people's lives. 15 years and 16 years are called the first year of virtual technology. In these two years, the development of virtual technology in China is particularly rapid, especially in the VR industry, the development of which is more space-intensive. By integrating virtual technology into big data, we can not only improve the research level of virtual technology but also realize the innovation and development of virtual technology. By enriching big data and improving the function of virtual software, we can achieve long-term development [2].

3.2 Cloud Storage Technology

Cloud storage technology is widely used in today's society and can change the disadvantages of traditional storage methods, breaking the limitation of time and space. As long as users can access the network terminal equipment and connect to the network, they can download and check the contents of cloud storage. Cloud storage is a whole composed of multiple storage system units, mainly through a combination of multiple functions, through collaborative work to achieve data storage, that is, the so-called network database. In the era of big data, cloud storage technology can provide users with more convenient information services, and can integrate and classify massive data information. It can be said that cloud storage and big data complement each other, which is the inevitable and central link of data processing in the era of big data.

3.3 Information Security Technology

In the era of big data, more or less of all kinds of data will have a certain relevance, which will affect each other, thus causing a certain threat to information data. It is necessary to adjust the entire data management system network, to improve the security level of the data cluster. Both big data and cloud technology rely on the Internet, but as an open platform, the Internet always contains this certain danger, network viruses, Trojans and other negative software emerge in endlessly. In future development, if people want to ensure the security of big data information, they must constantly promote the development of security technology. Although China's network information technology started late, in the development of decades, China's security technology is also at the top level in the world. In the future development of China's computer software, the application of big data will become more and more extensive, and will gradually become one of the applications recognized by people. Although China's computer software will encounter many challenges, according to the development characteristics of various industries Point and problem content, to put forward a more reasonable development plan, can reduce the security risks of big data information. To give full play to the positive role of computer software technology in the era of big data.

4 Notes on the Development of Computer Software Technology

4.1 Information and Communication Level

By using the prediction and analysis software of IBM SPSS, the loss of customers can be greatly reduced, and the corresponding problems can be found out from the industry operation, to provide a more convenient and convenient platform. For example, in the communication industry, some communication operators need to use computer software technology to integrate and divide customer information and analyze customer habits through big data, to develop relevant services that meet customer needs. Big data plays an important role in information analysis, and it is also an inevitable trend in the development of computer software technology, to promote the development of economic and social benefits of related industries [3].

4.2   Analysis of Enterprise Information Solution

In the application of enterprise operation management software, the data of customers can be obtained effectively, to analyze the risk of customer data and find out the problems. According to the characteristics of the era of big data, to provide better and more professional methods for enterprises, we need to start from the following aspects in the process of data development: first, sampling. Sampling is to sample the production products, select the representative samples, through the positioning of sample capacity, to achieve the development and utilization of product production [4]; second, development. Through the methods of development and exploration, we can analyze the data and strengthen people's cognition of the data. In the process of development, it is likely to involve data import, combination, selection, etc. by using the relevant computer software, we can quickly solve this problem, to improve the development efficiency; third, modify. In the actual operation, people can modify the data information through establishment and selection. In the process of modification, variable transformation and product code transformation can be realized. Then adjust according to the adjusted data to ensure the quality of modification; fourth, the model. Through the use of the model, technology can make the prediction results more reliable and accurate, and this step can effectively solve the corresponding scheme of the enterprise, has an important position in the enterprise decision-making, to improve the economic benefits of the enterprise; fifth, positioning. Positioning is to get the corresponding evaluation data, analyze the data, and make the final evaluation by comparing the evaluation technology and model. In the process of data mining, technicians can develop new operation mode through big data, and guarantee the accuracy of data information through data integration and analysis. The visualization technology can also be used to deal with the data graphically, to realize the dynamic data representation, and deepen people's understanding of the data information [5].

5   Conclusions

With the continuous development of science and technology in China, the arrival of big data brings new opportunities to the development of computer software technology. In the era of big data, it can not only improve the efficiency of information processing but also reduce the operating cost. Now there are a lot of professionals focusing on the development of computer software technology. I believe that with the efforts of scientific researchers in our country, computer software technology in our country can go to a higher level, to move towards a new stage of the information age.

References

1. Zihong, W.: Application of computer software technology in the era of big data. SME Manag. Technol. **09**, 45–46 (2014)
2. Ning, C.: Application of computer software technology in large-scale structure experiments and field test data processing. Ind. Technol. Forum **21**, 111–112 (2013)
3. Tao, L.: Application of computer integration software technology in the integration of groundwater monitoring data. Groundwater **05**, 78–79 (2013)

4. Hongyu, L.: Analysis of the development and application of computer software technology. Enterp. Guide **08**, 34–35 (2016)
5. Guofeng, W.: Research on the transformation of media production mode and communication mechanism in the era of big data. Shandong University (2014)

Application and Development of Digital Image Processing Technology in Jewelry Design

Xinyi Yang[✉]

Yunnan University of Business Management, Kunming 650000, China
wanglei18291826361@126.com

Abstract. The combination of jewelry science and digital image processing technology is more and more close. There will be new, more convenient, and fast solutions to more traditional and unsolved jewelry problems. The application of digital image processing technology can solve problems such as diamond cutting automation, three-dimensional observation of gem contents, artificial intelligence recognition of jewelry, and so on, which has a great impact on the further development of jewelry science.

Keywords: Digital image processing · Automatic recognition · Digital 3D modeling technology · Artificial intelligence recognition technology

1 Introduction

Digital image processing is to use a digital computer or other digital image acquisition equipment to perform some mathematical operations on the electrical signals obtained from image information conversion, to improve the practicability of the image. The research of modern jewelry is characterized by the combination of modern science and technology with traditional jewelry. In the field of jewelry, digital pattern recognition technology is widely used, which makes some classification methods such as diamond decoration degree, shape and size of inclusions, a color grade of diamonds, and so on quantified, which is easier for most people to understand and master [1–3].

2 Application of Digital Image Processing in Jewelry Cutting and Detection

At present, the related technology of foreign jewelry proportion cutting measurement system based on digital image processing has been relatively mature, and the related products have also been put into the market. China University of Geosciences (Wuhan) Institute of jewelry independently developed the jewelry proportion cutting measurement system, whose technology has reached a considerable level abroad [4, 5].

J. MacIntyre et al. (Eds.): SPIoT 2020, AISC 1283, pp. 831–835, 2021.
https://doi.org/10.1007/978-3-030-62746-1_131

2.1 Composition of the System

The jewelry proportion cutting measurement system adopts the methods of image processing and pattern recognition by introducing a digital camera and image digital processing technology to collect and process the image, to achieve the requirements of more accurate and faster jewelry cutting classification. The basic components of the system hardware include the bearing base, CCD camera, lens, computer, auxiliary lighting equipment, stepping motor, stabilized voltage power supply, etc. The high-performance camera system is used to ensure reliable and clear jewelry images in the rotating state. Considering the height and width of the jewelry image and the characteristics of the CCD camera, it is necessary to fix the installation angle and height of the camera on the bearing base. The CCD camera converts the analog signal into a digital signal, which is then transmitted to the computer [6, 7].

2.2 Features and Recognition Difficulties of Jewelry Image

Digital image processing is mainly through point operation, image analysis, and other methods to process the image into a computer acceptable form. Before processing jewelry pictures, it is necessary to understand the characteristics of the pictures. Under natural light, due to the high dispersion and strong fire color of jewelry, clear images of jewelry can not be collected, and the background of the acquired images is very complex (Fig. 1 and Fig. 2). Compared with the natural light, the brightness of the image captured under the auxiliary lighting will be relatively uniform, which is more conducive to image processing. In an ideal condition, the refractive index and reflectivity of jewelry are very high, which can be calculated by the formula $R = (n-1)^2/(n+1)^2$

When the light hits the jewelry Pavilion in parallel, about 1/3 of the light will directly reflect. Even if the light is injected into the diamond, it is mostly emitted from the table or the main facet, not from the pavilion at the other end, so the jewelry image collected from the collection device is completely dark. Not all jewelry cutters can achieve the degree of no light leakage, there will be a certain degree of light leakage, which will form a bright spot in the full dark image, affecting the next analysis.

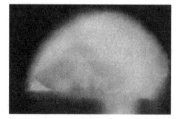   

Fig. 1. Blur contour caused by too strong light   **Fig. 2.** Phenomenon of light leak in the middle

3   Find Special Points

The measurement system of jewelry proportion cutting mainly determines a series of key data of jewelry cutting, such as table width ratio, pavilion depth ratio, full depth ratio, waist thickness ratio, crown angle, etc. Therefore, it is necessary to find the points related to the key data (Fig. 3). After confirming the coordinates of these key points, we can calculate the waist edge diameter, full depth, platform width, crown height, and pavilion depth of jewelry, and then we can get the values of width ratio, crown height ratio, pavilion depth ratio, crown angle, and pavilion angle.

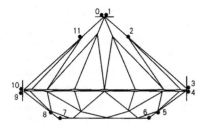

Fig. 3. Key spots of calculation

Advantages and disadvantages of the system

The biggest advantage of the jewelry proportion cutting measuring system is its high speed and accuracy. It takes more than one minute for a diamond to be cut and graded manually. With this system, each jewelry can be measured and printed in less than 10 s, which greatly saves time and improves efficiency.

When we use 10× magnifying glass to classify diamond cutting, we mostly rely on experience judgment, and there is no accurate quantitative data. If a jewelry scale is used, it is relatively complex and time-consuming. With this system, the high pixel CCD camera and accurate calculation will greatly improve accuracy. The cutting and grinding of jewelry require a high degree of symmetry. Therefore, the three-dimensional sketch can be made according to the coordinates of each point collected from different angles, which is helpful to understand the cutting process of jewelry as a whole. With high-precision laser technology, words or marks can be engraved on the waist edge of the finished diamond. jewelry cutting data and 3D mode can be packaged and linked to the Internet, and the buyer can view them remotely [8, 9].

4 The Development of Digital Image Processing Technology in Jewelry Research

4.1 The Combination of Digital Image Processing and Microscopy

With the advent of the microscope, people's vision extends from macro to micro. With the development of computer technology, the idea of constructing three-dimensional spatial information based on two-dimensional image information becomes simple, intuitive, and fast. Digital image processing, combined with microtechnology, can analyze and explain the phenomena that are difficult to be observed by the naked eye, to make jewelry research more in-depth.

Although microtechnology has many advantages, the jewelry microscope also has its defects. Because the depth of the field of the microscope is too small, we can only observe the plane image of each focal length, and can't get a clear stereo image. As the research of jewelry needs to be carried out without damage, only one layer or part of the characteristics of jewelry can be observed. Also, with the increase of magnification, the field of vision of the microscope is smaller and smaller, and only local observation and photography can be done.

At present, the United States and Israel, and other developed countries have combined digital image processing with micro-technology to develop a micro confocal imaging system. The system uses three CCD cameras to observe synchronously. Compared with the traditional single CCD micrograph technology, it can provide more and better color data and higher resolution images, and can observe and analyze research objects from multiple angles to generate corresponding three-dimensional images.

In jewelry research, the micro confocal imaging system can process the contents of different depth of field, realize the stereoscopic reproduction of the contents in the specimen under the microscope, and perfectly splice the images of different field of vision, quantify these data, carry out 2D or 3D measurement, and classify and analyze the contents.

4.2   The Combination of Digital Image Processing and 3D Modeling Technology

When looking at jewelry, whether with the naked eye, $10\times$ magnifier or microscope, you can only see images on the surface or a certain layer. Therefore, we can only describe three-dimensional objects such as jewelry inclusions in a general way. Using digital image processing technology can solve this problem well.

ts working principle is: turn the angle of the jewel 360° and take a picture every time it turns a certain angle, and use digital image processing technology to analyze the collected picture and find out the key points that can describe the outline of the jewel, to draw the three-dimensional stereo image of the jewel in the new coordinate system. The model established in the new coordinate system can perfectly reproduce the characteristics of each part of jewelry, observe jewelry from different angles, and quantitatively and accurately describe the volume, shape, and position of various inclusions. Figure 4 is a 3D model of different angles reconstructed according to the measured points of jewelry.

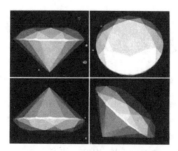

Fig. 4. 3D model recreated on testing spots

The model can be used to divide the scale randomly and simulate the cutting to ensure the maximum utilization of jewelry and its raw stones. With digital image processing, thousands of cutting and grinding schemes can be calculated in a few seconds, and then the impact of different sawing and planing methods on the yield of finished products is analyzed. Finally, a scheme is determined to maximize the use of raw stone. If the 3D model is connected to the CNC cutting machine, the largest and most perfect jewelry can be cut. By using the 3D model and the optical parameters of the corresponding jewelry, the optimal cutting and grinding parameters of the jewelry can be calculated. For example, through the 3D model and the refraction and reflection properties of light, the best cutting and grinding angle of jewelry can be calculated, achieve the best effect of the table size and fire color of jewelry.

5   Conclusions

When testing or studying the finished jewelry, any testing must be carried out on the premise of no damage to the shape or appearance of the gem. Most of the jewelry materials can be identified by well-trained inspectors using conventional gem detection instruments and techniques. However, in some cases, synthetic gemstones which are similar to the corresponding natural gemstones in appearance and gemological characteristics, as well as jewelry materials improved in the laboratory, need to be identified with the help of advanced non-destructive imaging technology and image processing technology.

These NDT technologies mainly include X-ray imaging, cathodoluminescence, UV-IR, and other nonvisible electromagnetic wave detection technologies. These electromagnetic waves interact with gemstones to generate optical signals. These signals can be received by the absorption screen, converted into digital signals, and then processed by digital image processing technology.

References

1. Zhang, H.: Visual c++ Digital Image Pattern Recognition Technology and Engineering practice. People's Post and Telecommunications Press, Beijing (2003)
2. Zhang, H.: Visual c++ Digital Image Processing, 2nd edn. People's Post and Telecommunications Press, Beijing (2003)
3. Yuan, X.: Principles and Methods of Diamond Grading. China University of Geosciences Press, Wuhan (1998)
4. Wang, Y., Li, S., Mao, J.: Computer Image Processing and Recognition Technology. Higher Education Press, Beijing (2001)
5. Fang, R., Xu, C.J., Li, X.: Computer Image Processing Technology and its Application in Agricultural Engineering. Tsinghua University Press, Beijing (1999)
6. Ziegler, J., Lu, T.: Nondestructive testing of jewelry. Jewelry Technol. **13**(42), 45–47 (2001)
7. Liu, Z., Dai, C.: X-ray nondestructive testing of wood defects. Forest. Sci. **20**, 35–37 (1984)
8. Zhang, A.Z., Ye, N.: Newton ring detection based on image pattern recognition technology. Phys. Exp. (23), 33–39 (2003)
9. Hideki, Y.: 3-dimensional internal structure microscope compared with the observation plan of hundreds of frozen biomass samples 0. J. Cryobiol. Eng. **44**(1), 1–9 (1998)

Horizontal Model of Higher Education Management Policy Support System Based on Data Mining

Jin Chen[✉]

Wuhan University of Technology, Wuhan 430070, Hubei, China
1287455332@qq.com

Abstract. This paper expounds the concept of data mining in theory, demonstrates the core application of data mining in education management decision support system, introduces in detail the data collection, preprocessing, the establishment and maintenance of data warehouse, the selection of data mining algorithm, the specific content of decision tree algorithm, VB This paper discusses the theoretical framework of the system model, studies the great role of the combination of data mining and education management, and finally looks forward to the application prospect of data mining technology in education management decision support system.

Keywords: Data mining · Data warehouse · Education management · Algorithm · VB

1 Data Mining Technology

The so-called data mining is a complete process, which is to extract and analyze the hidden, unknown, but intrinsically related and valuable information data from the massive computer application data, which provides the necessary support for decision-making [1].

In the process of data mining, the general contents of each step are as follows:

1. Determine the theme Z, that is, do a good job in demand analysis, clearly define the problem, and have predictability for the problem to be explored.
2. Data extraction: search data information related to the topic, study the quality of data, and determine the type of mining operation to be carried out.
3. Data transformation Z transforms data into an analysis model, which is established for the mining algorithm.
4. Data mining Z is used to mine the transformed data. In addition to improving the selection of appropriate mining algorithms, all other work should be completed automatically.

J. MacIntyre et al. (Eds.): SPIoT 2020, AISC 1283, pp. 836–840, 2021.
https://doi.org/10.1007/978-3-030-62746-1_132

5. Result analysis: to interpret and evaluate the results, the analysis method used generally depends on the data-mining operation, and visualization technology is usually used.
6. Knowledge assimilation Z integrates the knowledge obtained from analysis into the organizational structure of the business information system.

2 Data Mining and Educational Management Decision Support System

We use data mining technology to carry out specific "mining" on the information we have. We can find g generalized knowledge, association knowledge, classification knowledge, predictive knowledge, and deviation knowledge. We can find predictive information in the data, provide the basis for our management to "make decisions", and can strongly support us to make appropriate choices.

The so-called education management is a behavior that follows the objective laws of education and allocates various educational resources reasonably in a specific social environment to achieve educational policies and objectives. Education management is part of social management. In school education management, we introduce data mining technology, which is to effectively mine the massive data of education, to make full use of the advantages of data mining, to find the potential and long-term undiscovered rules in these massive data, and to make data mining for students, which provides a certain degree of reference for the enrollment work of schools.

We can also excavate the problems concerned by the school management. For example, what is the proportion of teachers in all teaching staff: in general, whether a teacher has met the general standards of the school; can analyze the teaching ability of teachers and the optimization of teachers' human resources, In the recruitment of teachers, it provides a general judgment basis for the human resources director, analyzes the potential and Prospect of the applied teachers, to assist the management to make better decisions, optimize education management, and promote the healthy development of education management.

3 Construction of System Architecture

We will design and implement the system model in the following steps:

(1) Analyze the data and establish the data warehouse

The main source of data is the existing archives of the archives, including the staff and students, but also add some data dynamically. It would be better to have the support of educational institutions to enrich data. The requirement of this data is large, true, and accurate [2].

The information of teaching staff mainly includes the following attributes: number, name, gender, nationality, date of birth, political status, working hours, whether it is a headteacher or not, whether it is the school management, the graduate school, the highest education, the highest degree, the highest professional title, the family economic conditions, the teaching evaluation results, the honors obtained The more detailed we carry out data mining, the better. It is shown in the table below (Table 1).

Table 1. Teacher team information

Number	Full name	Date of birth	Political outlook	Graduate School	Title6	Teaching evaluation
T0001	Zhang San	1976-2	Party member	Guangzhou University	Professor	A
T0002	Li Si	1975-2	Party member	Xinjiang University	Associate professor	A
......						

For students' information, there are the following attributes: student number, name, gender, graduation destination, graduation report card, graduation thesis name, graduation thesis score, graduation certificate family, date of birth, political status, major, class, admission time, enrollment score, books, ten years after graduation (optional), twenty years after graduation (optional) 40 years after graduation (optional), family conditions, honors in school, etc.

Each attribute is assigned and initialized. For example, gender Z is defined as a character type variable, M represents male and f represents female. For another example, the teacher's research results are assigned by characters, covering published papers, books, scientific research achievements, etc., which are graded according to their grades and accumulated. Finally, all the results are the research results of the teacher Then the score is weighted and assigned with a, B, C characters according to the evaluation. Assign values to each staff and student, and input them into the data warehouse. We can choose the Microsoft SQL Server as the tool to create a data warehouse. It should be noted that the SQL server group name in the console root directory of SQL Server "enterprise manager" should be consistent with other names. For example, if the analysis server is linlin2006, it should also be linlin2006 [3, 4].

(2) decision tree algorithm

What kind of decision tree method is provided to show the value of the rule? As shown in Fig. 1, a decision tree is formed after data mining for the influencing factors of students' Graduation scores. We can see that the basic components of the decision tree are Z nodes, branches, and leaves.

Whether the condition of "entrance score = good" is the root of the tree, that is, the root node of the decision tree. Different answers to the conditions produce "yes" and "no" branches; whether "hard learning" is the child node of the root node (the number of child nodes is related to the decision tree algorithm); if there is no node under the branch, it is called a leaf to the end of the tree, "graduation score = difference" is one of

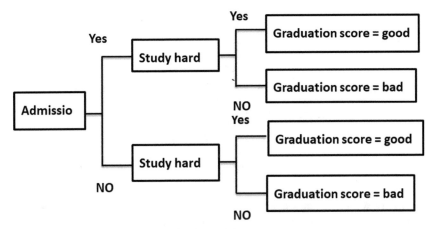

Fig. 1. Decision tree

the leaves. The process from the root to each leaf is the process of classification using a decision tree, which is often called "traversal".

The decision tree algorithm is common in data mining, which is often used in data analysis and prediction. The result of decision tree construction is a binary tree (i.e. each node has two branches) or multiple trees (nodes can contain more than two child nodes). The process of constructing a decision tree, that is, the growth process of the tree is the process of continuously dividing conditional data. Each segment corresponds to a problem (i.e. one node). The internal nodes (nonleaves) of a binary tree are usually expressed as logical judgments. Each subdivision requires to maximize the "difference" between groups. The main difference between various decision tree algorithms lies in the different measurement methods. We just need to think of segmentation as dividing the data into parts. Each data information should not have the same or similar attributes as far as possible, while the data in the same share should have the same attributes as far as possible, that is, they belong to the same category.

(3) The algorithm is programmed with VB to realize the function of each module. The graphic application interface design and core program programming are carried out by VB. The menu design is as follows: import data warehouse, select data warehouse, task establishment, select factors, start running, view results, save results, translation rules, save rules, exit. Mainly through the form and command to achieve each function. Sample space refers to the data space composed of data in the data warehouse of faculty and students.

(4) Testing

According to the requirements of software testing, the system must be tested with data, that is, by comparing the file data with the results of data mining.

First of all, it is assumed that some of the files of teachers in a school are as follows (limited by space, only part of them are included in the file) (Table 2):

Table 2. Comparison of file data and data mining results

Number	Full name	Date of birth	Political outlook	Graduate School	Title6	Teaching evaluation
T0001	Ma Fang	1979-5	Party member	Guangzhou University	Professor	A
T0002	Zeng Yi	1982-2	League member	Yunnan University	Lecturer	B
……						

Secondly, we use the data in the table to check the correctness of each branch and leaf in the tree structure. If most of the match, then the DSS is successful, and the software and algorithm are correct; otherwise, it must be checked and corrected. The workload of this test and the comparison of test results is huge.

4 Looking Forward to the Future

With the continuous development of information technology, the education management decision support system based on data mining will also cover personnel training, human resources, teaching evaluation, education economy, and other aspects, and will gradually develop into the application platform of education management data mining. The decision support system of education management based on data mining will play an increasingly important role.

References

1. Han, J.W., Kamber, M., Ming, F., Xiaofeng, M., et al.: Concepts and techniques of data mining, (trans.). Beijing Machine Press
2. Yaoting, Z., Chang, X.B., Shiwu, Z. (eds.) Introduction and Application of Data Mining: From the Perspective of Statistical Technology. China Statistics Press. [China]
3. Held, D.J., Gang, Q., Lingfeng, Y., Yaojun, S., Tao, Z., Changjian, G.: MS VB6.0 database programming encyclopedia (proofread). Electronic Industry Press, DHN w. Frackowiak [U.S.]
4. Data mining discussion group [China]. www.dmgroup.org.cn

Design and Implementation of Tourism Management System Based on SSH

Baojian Cui[1], Xiaochun Sun[2], and Yuhong Chen[3(✉)]

[1] Inner Mongolia Business & Trade Vocational College, Inner Mongolia, Huhhot 010070, China
cuibaojian_only@126.com
[2] Inner Mongolia University of Finance and Economics, Inner Mongolia, Hohhot 010051, China
sunxiaochun73@sina.com
[3] Inner Mongolia University of Technology, Inner Mongolia, Hohhot 010051, China
chenyh0510@163.com

Abstract. With the improvement of people's living standards, traveling has become a common way of leisure. Tourism has gradually become the mainstay of China's tertiary industry. The amount of daily work information on travel agencies has increased, which brings a certain difficulty to the management of the tourism industry. The combination of tourism and its technology is regarded as a shortcut to success by the tourism industry. However, the similar tourism management software that has been built at this stage is difficult to use, upgrade, and maintain due to its monotonous function and backward software development technology. The application system developed by the SSH integration framework has a clear structure, high development efficiency, and is also convenient for later maintenance and expansion. Therefore, it is a common topic for the tourism and software industry to build a stable and easy to expand tourism management system with an advanced and mature SSH framework.

Keywords: Tourism management system · SSH framework

1 The Technical Foundation of the SSH Integration Framework

SSH integration framework is the technical foundation of building a tourism management system. This chapter focuses on the SSH integration framework. This paper makes a technical preparation for the design of the tourism management system and the realization of the tourism management system [1].

1.1 MVC Mode

MVC model consists of three parts: model, view, and controller. They are data display, input, and output control of the enterprise application system. MVC model makes enterprise application system structure clear, simplifies system R & D cycle, and improves the maintainability and scalability of the system. The structure of the MVC model is shown in Fig. 1

J. MacIntyre et al. (Eds.): SPIoT 2020, AISC 1283, pp. 841–845, 2021.
https://doi.org/10.1007/978-3-030-62746-1_133

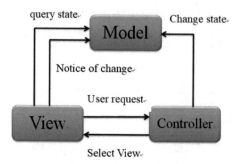

Fig. 1. MVC model structure

1. Model

The model represents the formulation of business rules, business processes, and state processing. The model includes the core functions of enterprise application and encapsulates the state of enterprise application. The process of the model processing business process is a black-box operation for the view layer and control layer, that is to say, the model is also unknown to view or controller. The model layer is the core part of the MVC pattern, which is equivalent to the business layer in the enterprise application system [2].

2. View

The view provides a representation of the model, which represents the interaction interface between the user and the system. For web application system, view. Is the appearance of the application, including HTML pages, JSPS, XML, and so on. The web system can select different views according to their needs. MVC model only deals with the input and output of data, but not the business process. The view can access the read method of the model, but not the write method. Also, it knows nothing about the controller. When changing the model, the view should be notified.

2 Demand Analysis of Tourism Management System

Demand analysis is a key link in the construction of the tourism management system. In this process, system analysts need to explore and master the needs of users. Only when the user needs are clear, the system analyst can analyze and seek the solution of the tourism management system. The task of the demand analysis stage is to determine the function of the tourism management system.

2.1 Requirements for the Construction of a Tourism Management System

1. Construction ideas

The general idea of the construction of the tourism management system is: according to the guidance of the National Tourism Administration on the construction of tourism informatization. It is required to combine the actual operation of some travel agencies and rely on advanced IT technology to achieve tourism management and marketing innovation and create advanced tourism management and technology tourism brands [3].

(1) Based on the technical specifications of the travel agency computer management system and tourism e-commerce issued by the National Tourism Administration. The website construction technical specifications and other relevant standards of tourism information construction are taken as guidance to listen to the opinions and suggestions of relevant experts from tourism authorities on the construction of the tourism management system.

(2) Supported by advanced tourism management concepts and information technology. The advanced nature of the tourism management system is not only. Reflected in the adopted software development technology, more importantly, reflected in the concept of tourism management and tourism services. Based on an in-depth understanding of the management concept of some travel agencies, combined with the advanced software framework [4].

2.2 Feasibility Analysis

1. Large number: the basis of Internet users

(1) According to the content of the statistical report on the development of China's Internet, by the end of 2012, the number of domestic Internet users had reached 564 million, with a total increase of 50.9 million Internet users throughout the year. The Internet penetration rate was 42.1%, an increase of 3.8% points compared with the end of 2011

(2) By the end of 2012, 91.3% of small and medium-sized enterprises surveyed used computers to work online, 78.5% used the Internet, 71.0% used fixed broadband, 25.3% carried out online sales and 26.5% carried out online procurement, and 23.0% used the Internet for marketing and promotion activities.

2. Online tourism transaction and demand environment based on Internet
In recent years, the scale of online tourism transactions based on the Internet in China has reached 200 billion yuan, accounting for the largest proportion in China. 15% of the total tourism revenue, the tourism e-commerce market is in a stable rising period, with great growth potential.

3 Design of Tourism Management System

3.1 System Function Module Structure

In Fig. 2, the tourism management system includes a member operation module, line manager operation module, and schedule management. Operator operation module and system administrator operation module. Among them, the member operation module includes browsing the recommended route, browsing tourist route, querying tourist route, booking tourist route, online payment, unsubscribing tourist schedule, and other sub-modules. Also, due to the limitation of the picture frame, it includes login verification, maintenance of own information, and other sub-modules; The operation module of the route manager includes browsing the recommended route, maintaining the recommended route, browsing tourist route, maintaining tourist route and other sub-modules. The same time, due to the limitation of the map, the operation module of schedule manager does not draw and maintain its information and other sub-modules; the operation module of schedule manager includes maintaining its information and browsing tourist route

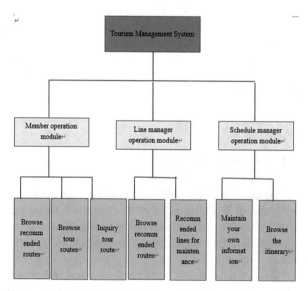

Fig. 2. Shows the function module structure of the tourism management system

3.2   System Database Design

The security assurance of the tourism management system is divided into the following aspects: database access control, system use. Users (members, tour route administrators, tour schedule administrators, system administrators, etc.) have authority control, detailed log mechanism, etc. to realize the supervision of data changes in the system, The tourism management system records the user, operation time, and machine of each operation in detail, and can also record the situation before and after the change of the project (about the tour route and schedule), and can also restore the single-step data according to the log. The tourism management system includes member data table, tour route manager data table, tour route data table, tour schedule data table, etc. (Table 1).

Table 1. tourism management system data

Serial number	Field name	Field meaning	Type and size	Remarks
1	ID	ID number	Int(20)	Primary key
2	Login_No	Member login account	Varchar(12)	
3	Name	Member name	Varchar(15)	
4	PassWord	Member password	Varchar(18)	
5	Sex	Member gender	Varchar(2)	
6	Birthday	Member's birthday	date	
7	FixedPhone	Member phone	Varchar(15)	
8	Maison	Member Mobile	Varchar(15)	

4 Conclusions

The construction of the tourism management system provides a network service window for travel agencies, which can effectively create. Create the brand effect of the travel agency. Due to the acceleration of broadband construction and the popularity of mobile Internet and wireless Internet, the tourism management system is convenient for some time and place of tourism customers to use, and they can enjoy the high-quality services of travel agencies. It can be predicted that the construction of the tourism management system will play a huge role in promoting the image publicity and brand building of the travel agency, greatly improving the service quality of the travel agency, creating and improving the brand advantage of the travel agency, and improving the level of information technology.

References

1. Mao, X., Li, S.: Research on tourism e-commerce platform based on Web Service. J. Hunan Agric. Univ. (Nat. Sci. Ed.) **35**(3), 335–338 (2009)
2. Lei, W., Yan, Z.: Design and implementation of tourism system based on flex and J2EE framework. Softw. Guide **10**, 100–102 (2012)
3. Dong, L., Zhai, Z., Chen, X.: Shenyang tourism information query system. Comput. Syst. Appl. **21**(7), 26–31 (2012)
4. Du, X., Hou, F., Hua, H., Pan, Y.: Research and application of an Android-based tourism information system

Design and Research on Big Data Application of Web Convergence Media

Yingmin Cui$^{(\boxtimes)}$

Department of Computer Science, Private Hualian College, Guangzhou 510665, China
cuiyingmin@163.com

Abstract. As a new data processing solution, big data technology can well complete the processing and value mining of massive data of various types and large scale. The media form formed by the integration of the Internet and traditional radio and television, as an important field of media communication and media data source, big data application will have a profound impact on it. Through the elaboration of the development and characteristics of Internet convergence media in the era of big data, this paper reveals the importance and necessity of the application of big data in Internet convergence media. At the same time, the application technology of big data in Internet convergence media is briefly analyzed.

Keywords: Big data · Internet convergence media · Hadoop

1 Introduction

After cloud computing, the Internet of things, and mobile Internet, "big data" has become a new focus of global attention and is leading a new wave of data technological innovation. Big data has not been defined properly, but the public generally agrees that its four characteristics can be described well, namely volume, velocity, variety, and value. At present, big data technology has spread across many technical fields, from cloud computing, virtualization, and data storage, to database management, data mining, and processing. This has greatly improved the application value of big data and has had a profound impact on various fields including health care, education services, e-commerce, media opinion, etc. In the face of the ever-expanding scale and increasing variety of massive data, the public has deeply felt the arrival of the era of big data [1].

2 Internet Convergence Media in the Era of Big Data

Internet integrated media usually refers to the media form that broadcasting and television institutions use the Internet information dissemination platform to digitize the text, sound, image, video, and other forms of data, and then transmit them on the computer, mobile phone, tablet computer, and other terminals. It is digital, interactive, global, easy to store, easy to retrieve, multimedia, and other characteristics. With the advent of the era of big data, Internet convergence media is undergoing new technological changes, which also shows new features:

1) Data growth is increasingly unstructured

   With the continuous development of Internet technology, network data is developing towards the direction of unstructured. More diverse text formats, rich audio, and video and image display, etc., have shown that unstructured data has become an important data source of Internet convergence media. However, how to effectively analyze unstructured data such as text documents, PDF documents, XML, images, audio, and video has always been a problem in data analysis. As new data storage and processing technology, big data can provide new solutions.

2) Faster information growth

   With the development of interactive technology and mobile Internet technology, Internet convergence media users have realized the operation of information publishing, communication, and interaction by accessing the Internet at any place and at any time [2]. This promotes the increasing speed of network information and the increasing scale of data.

3 Application Technology of Big Data in Internet Convergence Media

The emergence of big data not only leads to the innovation of data technology but also establishes a new ecosystem - big data ecosystem in the application field. The system includes data devices, data collectors, data aggregators, data users, and consumers. Data equipment mainly includes relevant equipment for generating and collecting data; data collectors mainly include entities for data acquisition from relevant equipment and clients; data aggregators are mainly responsible for data analysis and processing, extracting the value and law contained in big data; data users/consumers mainly use and consume valuable data analysis results.

3.1 Big Data Acquisition Technology

The data type structure of big data is mainly divided into structured data, semi-structured data, and unstructured data, as shown in Fig. 1.

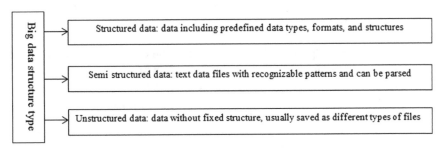

Fig. 1. Big data structure type

1) Structured data

The dynamic structured data in the Internet fusion media has higher research value, can better analyze the user behavior, and summarize the potential value in the data. It uses dynamic data sources of search platform, analytic data source, and regular expression matching data to grab, as shown in Fig. 2.

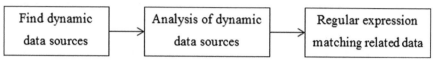

Fig. 2. Data grabbing mode

2) Unstructured data

For the unstructured data in the Internet fusion media, the simple HTML DOM parsing class library provided by PHP5 can be used to parse the HTML of the page, locate the information through the ID, class, tag, etc. of the elements, use the find function to find the elements in the HTML document, and return an array containing the objects to complete the data acquisition.

4 Big Data Analysis Technology

In the face of the rapid growth of data volume brought by Internet convergence media, to better analyze and use these massive data resources, we must use effective data analysis platform and technology. Using the concept and technological innovation brought by big data to deal with data resources in-depth, mining the potential value of internal data, and providing effective guidance for the development of Internet convergence media.

4.1 The Big Data Analysis Method

The main analysis methods based on big data mining are as follows:

(1) Clustering analysis (k-means algorithm), using similarity to group the data to find the commonness in the data;

K-means clustering algorithm (K-means clustering algorithm) is an iterative clustering analysis algorithm. Its steps are: first, divide the data into k groups, then randomly select k objects as the initial clustering center, then calculate the distance between each object and each seed clustering center, and assign each object to the nearest clustering center. Cluster centers and objects assigned to them represent a cluster [3–5]. Every time a sample is allocated, the cluster center will be recalculated according to the existing objects in the cluster. This process will be repeated until a termination condition is met. The termination condition can be that no (or minimum number) objects are reassigned to different clusters, no (or minimum number) cluster centers change again, and the sum of error squares is locally minimum.

The common measure of distance is the square of Euclidean distance:

$$d(x, y)^2 = \sum_{i=1}^{n}(x_i - y_i)^2 = \|x - y\|_2^2$$

Where x and Y represent two different samples, and N represents the dimension (number of features) of the samples. Based on Euclidean distance, the problem that the k-means algorithm needs to be optimized is to make the square sum of errors in the cluster.

(2) Regression analysis, to determine the relationship between input variables and results, mainly linear regression and logical regression;

Linear regression can be solved by the least two multiplication. Find out θ as follows:

$$\theta = \left(X^T X\right)^{-1} X^T \vec{y}$$

Logistic regression is essentially a linear regression model. The logistic regression algorithm mainly uses the sigmoid function to classify data.

Sigmoid function (logistic function):

$$g(z) = \frac{1}{1 + e^{-z}}$$

The most typical construction method of logistic regression is maximum likelihood estimation (Fig. 3).

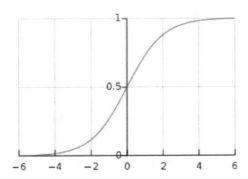

Fig. 3. Sigmoid function

Likelihood function:

$$L(\theta) = \prod_{i=1}^{m} P(y_i|x_i; \theta) = \prod_{i=1}^{m}\left(h_\theta(x_i)^{y_i}(1 - h_\theta x_i)\right)^{1-y_i}$$

(3) Association rule analysis, looking for the relationship between behaviors, mainly apriori, FP growth algorithm;

(4) Classification, label processing objects, mainly including decision tree, naive Bayes, and so on.

4.2 Hadoop Big Data Support Platform

As an open-source distributed computing platform under the Apache foundation, Hadoop provides users with the underlying distributed infrastructure of the system. It has high reliability, high scalability, high efficiency, and high fault tolerance. It uses a computer cluster for data storage and operation. Data can move dynamically between nodes to ensure the dynamic balance of nodes. At the same time, the platform allows node expansion and increases platform capacity. Therefore, Hadoop is a suitable platform for big data storage and processing. Hadoop is a data operation platform based on the Linux system, which has two core systems, namely, distributed file system (HDFS) and MapReduce system.

5 Conclusions

With the continuous innovation in the field of Internet and the improvement of the application ability of new media technology of traditional radio and television institutions, the Internet integrated media, which has both radio and television and Internet genes, is experiencing a stage of rapid development. The structure of big data is more complicated and the scale of data is larger. Big data can well realize the massive storage, efficient processing, no delay search and analysis modeling of the Internet convergence media data, mining the potential value and law behind the big data, and guiding the healthy and orderly development of the Internet convergence media.

References

1. Zhou, B., Liu, W., Fan, C.: Big Data Strategy, Technology, and Practice. Electronic Industry Press, Beijing (2013)
2. Zhong, Y., Zhang, H.: Media transformation driven by big data. News Writ. (12), 11–13 (2013)
3. Yan, C., Song, D.: The era of "big data" may set off a new pattern of media development. China Media Technol. (19), 64–65 (2012)
4. Zhang, Y.: Challenges and opportunities of the domestic media industry in the era of big data. Mod. Media (11), 22–26 (2013)
5. Lu, J.: Hadoop Practice. China Machine Press, Beijing (2011)

Design and Research of Wedding Photography Business System Based on Web

Bing Liu[✉]

College of Art and Design, Shang Qiu Normal University, Shangqiu 476000, China
sheyingliubing@163.com

Abstract. With the improvement of people's living standards, wedding photos have been paid more and more attention to by more and more new people. At the same time, the emerging industry of film studio also began to form. A large number of outsiders began to enter China, bringing new concepts of make-up, photography, and studio management, which made wedding photography develop rapidly in a short period of more than ten years, spread all over the country, and the competition is also extremely fierce. Through the analysis of the composition and function of the wedding photography e-commerce system.

Keywords: Wedding photography · e-commerce · System composition

1 Introduction

Wedding photography has gone through the process of infant market initial construction and has gradually evolved into a special industry of its scale. Since wedding photography was successfully introduced into the mainland by Taiwanese for nearly 20 years, it has been deeply accepted by Chinese people and formed a special and inherent industrial culture. No matter in China's cities and villages, young people's marriage takes a set of wedding photos as a major event before marriage. This has resulted in a huge industrial market of nearly 10 billion yuan per year [1]. As a result of the huge commercial interests of the attraction, from the city to the township, wedding studios have sprung up one after another, another round of reshuffle out, the market competition is very cruel, "natural selection, the survival of the fittest" is so accurate and specific. With the development of time and the continuous improvement of people's aesthetic standards, the requirements for wedding photography are becoming higher and higher, and many small personal studios have emerged as the times require. In the traditional studio, those who can carry out reform according to the situation can be established, and those who can not be flexible are gradually eliminated by the market. Compared with the traditional wedding studio, the style of the personal studio is more personalized, so it is sought after by many fashionable new people [2].

J. MacIntyre et al. (Eds.): SPIoT 2020, AISC 1283, pp. 851–855, 2021.
https://doi.org/10.1007/978-3-030-62746-1_135

2 The Market Concentration of the Wedding Photography Industry is Very High

Due to the closure of the wedding photography market and the shackles of traditional concepts, the enterprises in the wedding photography industry are accepting more and more traditional market tests, which makes us feel the cruelty of the wedding dress market. In this industry, because the scale will significantly affect the size of competitiveness, the more rapid way to improve competitiveness is to adopt standardized low prices or different strategies of the base price. The use of a low price strategy can help open the market and then get more market space. In the scale competition enterprises in the wedding photography industry, the most classic photo album company Jingchen used this successful strategy. After occupying a large part of the market with scale, it launched three international brands in 2003, which expanded the international market rapidly. The reason why Jingchen can achieve such performance is related to the low-cost differentiation strategy and brand differentiation strategy. It can be seen that no matter it is the scale competitive advantage type or the product leading advantage type enterprise, as long as it wants to be bigger and better, it will finally move towards the brand competitive advantage type. At present, for the development of e-commerce applications, it is recognized that there are three main development directions: B2B, B2C, and C2C. B2B refers to a market in which products, information, or services are traded among enterprises. It has two kinds of modes, one is the direct e-commerce between enterprises. The second is all kinds of commercial activities with the help of third-party websites. According to the survey, three-quarters of all e-commerce transactions made on the Internet in the world account for B2C trade for every 10 billion US dollars. Therefore, B2C e-commerce is most worthy of our discussion and attention, which is determined by its strong development potential. With the development of society, time and market, old-fashioned, low-level and original business methods will be gradually replaced by sound and scientific business strategies, because the social demand is constantly updated, market competition is increasingly fierce, and the rights of the buyer and the seller are gradually replaced, which makes the operator have to keep up with the market changes and constantly change their business strategies and roles. With the rapid development of the network, e-commerce based on the network has made great progress, especially B2C e-commerce. There is no doubt that B2C has become the main e-commerce activities of contemporary enterprises. Guide platform, it is a revolution, can successfully lead the development of the wedding photography industry in China? The survey found that the current domestic B2C information level is very low, many enterprises in the operation of information technology lack of consideration of Web factors, thus limiting the development of B2C [3].

3 Business System Overview

The business system is to replace the previous face-to-face cash transactions, using the network electronic tool to achieve the relationship between the sale of physical goods, under the rapid development of the Internet, it has led to the rapid development of e-commerce. There are two main parts of e-commerce, which use the network to conduct e-commerce and use it for various activities in the business circle [2, 4]. E-commerce is a very convenient commodity trading mode. It takes information products such as the network as the trading platform, which provides a quick trading way for the "big users". Using all kinds of network transmission software that we are familiar with, we can trade B.

4   Data Analysis of Web Business System

In today's society, a huge amount of data is involved every day, but few of them are saved every day. Therefore, data mining (DM) is a technology that uses random data to mine with potentially useful information. Data mining can deal with various types of data, including database data, text data, graphics, and image data. Extract useful information from huge data to achieve decision support, etc., and provide help for trend prediction and decision support. One of the data mining functions:

Association analysis: in a transaction set, if there are two or more items that have a certain relationship, then one of the items can be predicted by other items so that the potential association between items can be mined.

Classification association rules and sequence analysis are widely used, and the research is more in-depth with the development and application of database knowledge discovery and data mining technology. However, any work has its shortcomings. In general, when considering the user's shopping, the shopping activities are independent of each other, and there is a lack of a systematic and effective method to study and discuss from the whole. Association rules show the relationship between items in the transaction data transaction set, which reflects the interaction between the items purchased by users and the behavior pattern of consumers. In the system shopping basket, related product recommendation, commodity shelf design, can be optimized. The transaction is each transaction, with a unique transaction ID number corresponding to it. It is called TID. The data item is each commodity. The collection of some commodities is called a shopping basket, which is called itemset. The association rules can be formally defined that the given set I = {i1, I2, ..., im} is the item source and the transaction database D = {T1, T2, ..., Tn}. Each transaction Ti ($1 \leqslant I \leqslant n$) is a set of items, Ti ∈ I. Association rules are logical implication x y in the following form. Where X, Y are two itemsets, X ∈ I and Y ∈ I, and X ∩ Y = φ. Itemsets have the feature of support count, that is, the number of transactions containing a specific itemset. Mathematically, the support count δ (X) = {Ti I XTi, Ti ∈ D} I, where I * I denotes the cardinality of the set. It is not hard to see that its essence reflects the connection between a set of data items. Association rules have two important attributes, support, and confidence. The support degree is S, the support degree of association rule X ⇒ Y is x, and Y is the probability that the two item sets appear at the same time in transaction database D.

$$s(X \Rightarrow Y) = \frac{\sigma (X \cup Y)}{|D|}$$

Confidence degree C, the confidence degree of association rule x y is the probability of occurrence of item set X in transaction databased at the same time. It's a conditional probability:

$$c(X \Rightarrow Y) = \frac{\sigma(X \cup Y)}{\sigma(X)}$$

If the support degree of the association rule (s_ Min) threshold and minimum confidence (C_ If the given value is greater than min, it is called the minimum value. In a transaction set, mining association rules is a process that requires support and confidence to meet strong rules.

5 System Target Requirements

Demand analysis is to analyze the needs of customers, all for the needs of customers, all for the needs of customers, all for the needs of customers is the purpose of demand analysis that we strive for. In short, we try to realize what customers need in the development system, and we don't waste time in the development of what customers don't need. This is the problem that the application system needs to solve. The system requirements analysis stage is a comprehensive investigation of the application of the system management system, which is an important stage to determine the system objectives and user requirements.

5.1 System Structure Design

As a kind of network structure mode, B/S structure is called browser/server mode, which is born with the rise of the Internet. B/S structure has become one of the most important architecture modes of the network application system. The main reason is that it realizes the unity of the client and the core part of the system function on the server, which makes the realization, maintenance, and operation of the system simple and convenient. The simple system expansion operation realizes zero maintenance and zero installation of the client, that is, no need to install professional software, only the computer can access the Internet, no matter when and where it can be operated. Through what mode of e-commerce system development and implementation? We will analyze the following aspects.

(1) Environmental performance. The core part of the B/S structure system function is concentrated on the server, which realizes the low requirement of the client computer and high requirement of server performance.
(2) Maintenance and upgrade are simple and convenient. B/s upgrade and maintenance can be implemented on the network server, or remote maintenance, upgrade, and sharing of the system can be realized through computers in different places, provided that you have a computer connected to the network. Because the software developed by B/S architecture only runs on the management server, the client can avoid downloading any professional software, and the workload of upgrading and maintenance can be avoided to change according to the change of users. After the analysis of the two architecture models, each has its advantages. Considering the actual situation of the e-commerce system, this e-commerce system development

will adopt B/S architecture. Its purpose is to reduce the maintenance and upgrade costs of the system in the later period, and to realize the construction of the system's foreground network by making reasonable use of the enterprise's network advantages. Besides, the application of B/S architecture mode not only enables the data and information sharing among all departments of the enterprise but also enables them to have good communication. The B/S structure of the e-commerce system is shown in Fig. 1:

Fig. 1. B/S structure of an e-commerce system

6   Conclusions

With the continuous development of e-commerce, the normal development and interests of wedding photography enterprises are related to the survival of enterprises. How to develop a set of wedding photography business website system at a lower cost is the research direction of this paper.

References

1. Yu, X., Zhang, X.: Design and implementation of web-based e-commerce system. Comput. Digit. Eng. **08**, 78–80+117 (2010)
2. Liu, S.: Interpretation of the most important strategic resource of Chinese wedding photography industry - professional Manager. Bus. Manag. (5) (2006)
3. Peng, S., Liu, C.: Customer resource management of wedding studio. Manag. Forum (2) (2007)
4. Zhang, W.: Network marketing, sales sword of wedding film studio in the information age – a successful case analysis of network marketing. Mark. Plan. (10) (2006)

Development of a Virtual Experiment Teaching System Based on LabVIEW

Saying Ren[✉]

Ji Lin Justice Officer Academy, Changchun 130062, Jilin, China
siying_06@163.com

Abstract. This paper mainly uses LabView 8.5, a virtual instrument development software, to design and develop a signal analysis experiment system based on the sound card, which has the function of collecting, analyzing, displaying, and storing sound data through the ordinary sound card. To improve the readability and running speed of the program, a modular programming idea is adopted. A large number of "dynamic loading sub VI" programs are used. Firstly, the system is divided into three parts, then the program is further refined, and debugging is a successful one by one. Finally, it is connected with the designed framework program to form the final system program. And use it to build the audio signal acquisition and analysis virtual experiment system, the practice has proved that it can be used in the specific experiment teaching.

Keywords: LabVIEW · Experimental teaching: test system · Virtual instrument · Signal analysis

1 Introduction

With the rapid development of information science, the problems that need to be dealt with are more and more complex, and the requirements for signal analysis are higher and higher. With the rapid development of electronic technology, computer technology, and network technology and its application in the field of electronic measurement technology and instruments, new test theories, new test methods, new test fields, and new instrument structures continue to appear. Traditional instruments are becoming more and more powerless. The design of the signal processing circuit is very complex and difficult to update. When the new calculation method comes out, the traditional instruments can not be upgraded, can not meet the requirements of scientific researchers, and bring extra research costs to the research work. The signal analysis system of the virtual instrument can completely solve this problem. Using a virtual instrument development platform to develop various "virtual instruments" is not only cheap, simple, and easy, but also interactive, operable, and realistic. Software is an instrument that reflects the basic characteristics of the virtual instrument [1].

Test and measurement is the basis of the development of science and technology, and also an important part of teaching experiments in Colleges and universities [2]. With the

J. MacIntyre et al. (Eds.): SPIoT 2020, AISC 1283, pp. 856–860, 2021.
https://doi.org/10.1007/978-3-030-62746-1_136

development of electronic technology and the popularization of computer technology, the rapid improvement of the intelligent degree of the instrument makes the traditional test instrument in function, efficiency and accuracy can not meet the current test requirements [3].

2 Comparison Between Virtual Instrument Experiment System and Traditional Experiment

Traditional instruments have many problems, such as "long technology update period", "instrument function can not be customized", "difficult to connect with other equipment", "high development and maintenance cost", etc. so today, with higher and higher requirements for signal processing, users hope to establish a signal analysis system on the virtual instrument platform to solve the above problems [4]. Also, according to the survey, there are "less experimental information provided by traditional instruments, the high error rate of experimental results due to manual reading", "traditional instruments are unable to carry out remote experiments, to share equipment resources", "it is difficult to update experimental equipment, and most of the equipment lags behind the needs of curriculum construction" The content of the experiment focuses on the verification and imitation training of the theory, the unification of the experimental content of the students, and the lack of the cultivation of the students' innovative consciousness and the improvement of the comprehensive ability. To a great extent, it restricts the development of experimental teaching and the improvement of personnel training quality. It requires educators to develop experimental teaching instruments that can meet the requirements of modern experimental teaching and are of good quality and low price, to improve the level of experimental teaching and cultivate innovative talents with high quality and high skills. At present, in domestic universities, virtual instruments are gradually entering science and engineering classes and laboratories. More and more schools set up high-grade virtual laboratories by purchasing virtual instruments products from Ni company of the United States, but the cost is.

3 The Construction of a Virtual Experiment System

Module Establishment of the Virtual Test System

According to the organizational structure of the system, the typical hardware structure of the data acquisition system based on the virtual instrument is sensor → signal conditioning → data acquisition card → computer. In the acquisition process, the physical quantity to be measured is first converted into electric quantity by the sensor, and preprocessed by the signal conditioning part; then the analog signal is converted into a digital signal by the data acquisition card, in addition to the amplification, sample holding, and multiplexing functions. The structure of the data acquisition system is shown in Fig. 1. Through the analysis, we can know that in the signal analysis system, there are many modules to be called repeatedly. Without modular programming, there will be a lot of repetitive work. This not only wastes a lot of valuable time but also makes the designed

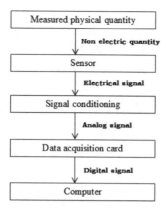

Fig. 1. Structure of the data acquisition system

program disorderly, which makes it difficult to debug and modify the program. It can be seen that modular programming is very necessary for this design.

The data acquisition system is composed of a sensor, signal conditioning, data acquisition card (usually integrated with multi-channel analog switch, program-controlled amplifier, timer, sampling/holding device, A/D converter, and D/A converter), computer, and peripheral equipment.

Sampling Frequency. Sampling frequency refers to the acquisition times of analog signal by data acquisition card in unit time, which is an important technical index of data acquisition card. According to the Nyquist sampling theorem, to reproduce the original input signal without distortion, the sampling frequency f must be at least twice the effective frequency FMX of the highest input signal. When setting the sampling frequency, it is enough to set the sampling frequency twice the highest frequency component of the collected signal in theory. In engineering practice, 7–10 times is selected. Sometimes, to restore the waveform better, the higher sampling frequency can be selected appropriately.

Several Digits. The number of bits refers to the number of bits in the output binary of the A/D converter. When the input voltage is increased from $U = 0$ V to full-scale value $U = U_{fs}$, the digital output of an 8-bit (b = 8) A/D converter changes from 8 "0" to 8 "1", a total of 2^b states are changed, so the A/D converter produces a least significant bit (Least Significant Bit, LSB) digital quantity output change quantity, corresponding input quantity is:

$$U_{min} = 1LSB = q = \frac{U_{fs}}{2^b} \tag{1}$$

Where: q is the quantized value; U_{fs} is the full-range input voltage, usually the A/D converter power supply voltage.

4 Design of Signal Source Selection Module

The front panel of the signal source selection module is shown in Fig. 2. The signal source selection module is mainly composed of three sub-modules: "real-time acquisition signal", "historical heavy load signal" and "simulation signal". Different sub-modules are selected by triggering the event structure of different options of tab control.

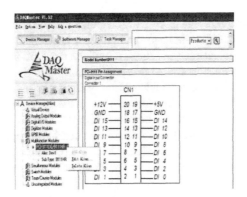

Fig. 2. Signal source selection module front panel

Help module design, help interface is to let users have more understanding of the system, mainly introduces the main functions of the signal analysis system and matters needing attention in operation. Also, in the menu button of the program, help documents are compiled by using dialog function, the system explains the operation steps of the experiment, through the description and tip of each button Dialog box to edit the function description of each button, so that when the mouse is parked on the button, it will automatically prompt its function.

4.1 The Effect of the Virtual Instrument Experiment System on the Reform of Experiment Teaching

The concept of the virtual instrument is different from that of EDA simulation software. It can completely replace the traditional desktop measurement and test instrument. The virtual instrument in EDA simulation software is pure software and simulation. The application of virtual instrument technology in experimental teaching has a wide range of advantages for promoting experimental teaching. Its main role is reflected in the following aspects: Virtual Instrument experimental system can be applied to basic laboratories of different levels of disciplines, as well as professional laboratories, which can play a great role in experimental teaching. The experiment teaching with the virtual instrument test system is highly efficient and targeted. The function and scale of the virtual instrument can be customized by the user according to the experiment requirements. We can create a simple and bright front panel and a flow chart with exact function according to the needs of the experiment content. The same kind of instrument can also generate front panel with different functions according to different experimental requirements. In the experiment, students can quickly get familiar with the operation of virtual instruments and use more energy and time to test data and analyze results. This is of great benefit to improve the efficiency of the experiment.

5 Conclusions

The virtual instrument experiment system integrates the functions of a virtual oscilloscope, signal generator, spectrum analyzer, and other instruments. These virtual instruments are used in the experiment teaching to replace conventional instruments. In the actual teaching, various panels can be designed according to the experimental requirements, the functions of the instruments can be defined, the output test results can be expressed in various forms, real-time simulation analysis can be carried out, and the traditional instruments can be used Compared with the device, the frequency domain analysis function is added. Under the same hardware condition, the new instrument function can be formed by modifying or adding software modules. The hardware is open, and the performance can be improved by upgrading the hardware.

References

1. Chen, X., Zhang, Y.: LabVIEW 8.20 Programming from Entry to Mastery. Tsinghua University Press, Beijing (2007)
2. National Instruments: LabVIEW User Manual. National Instruments, Texas (2003)
3. Li, J., et al.: Typical Application Examples of Virtual Instrument Design and Control. Electronic Industry Press, Beijing (2010)
4. Liu, M.: Research and implementation of physical experiment teaching based on network environment. Master's degree thesis of engineering, Hehai University (2005)

Author Index

Printed in the United States
By Bookmasters